WORLD **ENERGY** OUTLOOK 2012

INTERNATIONAL ENERGY AGENCY

The International Energy Agency (IEA), an autonomous agency, was established in November 1974. Its primary mandate was – and is – two-fold: to promote energy security amongst its member countries through collective response to physical disruptions in oil supply, and provide authoritative research and analysis on ways to ensure reliable, affordable and clean energy for its 28 member countries and beyond. The IEA carries out a comprehensive programme of energy co-operation among its member countries, each of which is obliged to hold oil stocks equivalent to 90 days of its net imports. The Agency's aims include the following objectives:

- Secure member countries' access to reliable and ample supplies of all forms of energy; in particular, through maintaining effective emergency response capabilities in case of oil supply disruptions.
- Promote sustainable energy policies that spur economic growth and environmental protection in a global context – particularly in terms of reducing greenhouse-gas emissions that contribute to climate change.
- Improve transparency of international markets through collection and analysis of energy data.
- Support global collaboration on energy technology to secure future energy supplies and mitigate their environmental impact, including through improved energy efficiency and development and deployment of low-carbon technologies.
- Find solutions to global energy challenges through engagement and dialogue with non-member countries, industry, international organisations and other stakeholders.

IEA member countries:

Australia
Austria
Belgium
Canada
Czech Republic
Denmark
Finland
France
Germany
Greece
Hungary
Ireland
Italy
Japan
Korea (Republic of)
Luxembourg
Netherlands
New Zealand
Norway
Poland
Portugal
Slovak Republic
Spain
Sweden
Switzerland
Turkey
United Kingdom
United States

The European Commission also participates in the work of the IEA.

International Energy Agency

International Energy Agency
9 rue de la Fédération
75739 Paris Cedex 15, France

www.iea.org

In the Foreword to last year's *World Energy Outlook*, I stressed the need for detailed, meticulous, objective and original analysis as a foundation for sound energy policy making. Important political decisions shaping the energy future have since been taken, in many cases drawing on the analysis in the *World Energy Outlook.* They include the specification, in many countries, of principles to guide the development of unconventional gas resources and new commitments, from all parts of society, to boost investment towards the goal of providing universal energy access.

Dr. Fatih Birol and his team in the Agency's Directorate of Global Energy Economics again present us this year with detailed information on what has happened across the energy sector and visions of where we are heading. They do so with the greatly valued participation of many experts from across the world, whose commitment and contribution I acknowledge and commend.

We have been privileged this year to work with representatives of the Iraqi government and other experts on Iraq to produce a special study of that country's energy prospects. Many people do not realise that Iraq is, already, the third-largest oil exporter in the world. The future level of Iraq's oil (and natural gas) production will not only determine the country's economic future but will also shape future world oil markets.

Efficiency in energy use is just as important to the future energy balance as unconstrained energy supply. Despite years of persistent pursuit, the potential of energy efficiency remains largely untapped. This year we demonstrate the full extent of that potential through an *Efficient World Scenario*, showing what can be achieved simply by adopting measures which fully justify themselves in economic terms. By 2035, we can save energy equivalent to 18% of global energy consumption in 2010. Global GDP would be 0.4% higher. The financial savings on reduced fuel consumption would be in excess of $17 trillion. We would have reduced oil consumption by nearly 13 mb/d. And we would have gained precious time in which to act decisively in pursuit of our global climate goal.

Water is needed for energy and energy is needed to produce and supply water. In an increasingly water-constrained world, we analyse this mutual dependence for the first time this year. We quantify energy's call on global water supplies and set it in context. A volume of water which exceeds that carried annually by the Ganges River is needed by the energy sector each year.

Finally, one remarkable feature of the market, to which we draw attention this year, is an energy transformation in the United States. Due to a combination of new technology, free markets and policy action, the United States is well on the way to becoming self-sufficient in energy, in net terms. We show how this is coming about and discuss its far-reaching implications.

I have only touched here on the scope of this *WEO*. All policy makers in energy and related fields will find here material to increase their knowledge and contribute to the quality of their decisions.

Maria van der Hoeven
Executive Director

This publication has been produced under the authority of the Executive Director of the International Energy Agency. The views expressed do not necessarily reflect the views or policies of individual IEA member countries.

This study was prepared by the **Directorate of Global Energy Economics** of the International Energy Agency in co-operation with other offices of the Agency. It was designed and directed by **Fatih Birol**, Chief Economist of the IEA. **Laura Cozzi** co-ordinated the analysis of energy efficiency, climate change and modelling; **Tim Gould** co-ordinated the analysis of oil, natural gas and Iraq; **Amos Bromhead** co-ordinated the subsidy analysis and contributed to the analysis of global trends; **Marco Baroni** co-ordinated the power and renewables analysis; **Paweł Olejarnik** contributed to the hydrocarbon-supply modelling and Iraq analysis; **Dan Dorner** co-ordinated the analysis of Iraq and energy access, **Timur Gül** co-ordinated the transport analysis and contributed to the energy efficiency analysis; **Matthew Frank** co-ordinated the energy-water analysis and contributed to the analysis of Iraq. Other colleagues in the Directorate of Global Energy Economics contributed to multiple aspects of the analysis and were instrumental in delivering the study: **Sabah Al-Khshali** (Iraq); **Mustafa Al-Maliky** (Iraq); **Ali Al-Saffar** (Iraq); **Prasoon Agarwal** (efficiency and buildings); **Christian Besson** (oil, natural gas and Iraq); **Alessandro Blasi** (energy-water and Iraq); **Violet Shupei Chen** (subsidies and energy-water); **Ian Cronshaw** (coal and natural gas); **Dafydd Elis** (power and Iraq); **Capella Festa** (oil, natural gas and Iraq); **Fabian Kęsicki** (climate and efficiency); **Shinichi Kihara** (efficiency); **Jonas Kraeusel** (transport and subsidies); **Jung Woo Lee** (subsidies and efficiency); **Breno Lyro** (bioenergy and transport); **Chiara Marricchi** (power and renewables); **Jules Schers** (energy access); **Timur Topalgoekceli** (oil, natural gas and Iraq); **Brent Wanner** (power and renewables); **David Wilkinson** (power, efficiency and renewables); **Peter Wood** (oil, natural gas and Iraq); **Akira Yanagisawa** (efficiency, industry and Iraq); **Sandra Mooney** and **Magdalena Sanocka** provided essential support. More details about the team are at *www.worldenergyoutlook.org/aboutweo/team*.

Robert Priddle carried editorial responsibility.

The study benefited from input provided by numerous IEA experts. In particular, the Energy Efficiency and Environment Division, including Philippe Benoit, Robert Tromop, Lisa Ryan, Grayson Heffner, Sara Pasquier, Vida Rozite, Aurelien Saussay, Yamina Saheb and Nora Selmet, provided input to the efficiency analysis. Other IEA colleagues who also made significant contributions to different aspects of the report include Carlos Fernandez Alvarez, Manuel Baritaud, Paolo Frankl, Cristina Hood, Alexander Koerner, Christopher Segar, Cecilia Tam, Johannes Truby, Nathalie Trudeau, Laszlo Varro and Hirohisa Yamada. Experts from the OECD also contributed to the report, particularly Jean Chateau, Bertrand Magné, Rob Dellink and Marie-Christine Tremblay. Thanks also go to the IEA's Communication and Information Office for their help in producing the final report, to Bertrand Sadin and Anne Mayne for graphics and to Debra Justus for proofreading the text.

The special focus on Iraq would not have been possible without the close co-operation received from the federal government of Iraq, the regional and provincial governments and officials across many government bodies in Iraq. We are particularly grateful to H.E. Dr. Hussain Al-Shahristani, Iraq's Deputy Prime Minister for Energy, for his strong support from the inception of this study through to its conclusion. H.E. Thamir Ghadhban, Chair of the Prime Minister's Advisory Commission in Iraq, provided indispensable input as keynote speaker and chair of our high-level workshop in Istanbul.

We are indebted to many Iraqi senior officials and experts for their time and assistance. Special thanks are due to the leadership and staff of the Ministry of Oil and, in particular, to Fayadh Hassan Neema, Director of the Technical Directorate at the Ministry of Oil, who was an invaluable interlocutor throughout, and to Falah Alamri, Director General of the State Oil Marketing Organization. H.E. Dr. Ashti Hawrami, Minister of Natural Resources in the Kurdistan Regional Government, provided valuable support and insights to the IEA team during their visit to Erbil. We greatly appreciated the welcome afforded the team in Basrah by Ahmed Al-Hassani, Deputy Governor of Basrah Governorate, and by the Provincial Council.

Ambassador Fareed Yasseen, Iraq's Ambassador to France, and his staff provided unstinting assistance to the completion of this study and the team's visits to Iraq, as did Dr. Usama Karim, Senior Advisor to Iraq's Deputy Prime Minister for Energy. Sincere thanks also to Tariq Shafiq, Managing Director of Petrolog and Associates and one of the founding fathers of Iraq's petroleum industry, for his expert counsel.

Maria Argiri, Janusz Cofala (International Institute for Applied System Analysis), Xiaoli Liu (Energy Research Institute), Michael McNeil (Lawrence Berkeley National Laboratory), Trevor Morgan (Menecon Consulting) and Paul Waide (Waide Strategic Efficiency) provided valuable input to the analysis.

The work could not have been achieved without the substantial support and co-operation provided by many government bodies, organisations and energy companies worldwide, notably: American Iron and Steel Institute; Department of Resources, Energy and Tourism, Australia; Booz & Company; Cement Sustainability Initiative of the World Business Council for Sustainable Development; Daimler; Ministry of Climate, Energy and Building, Denmark; Enel; Eni; Institute for Industrial Productivity; The Global Building Performance Network; Ministry of Foreign Affairs, Italy; Iron and Steel Federation, Japan; The Institute of Energy Economics, Japan; Ministry of Economy, Trade and Industry, Japan; Ministry of Energy, Mexico; Ministry of Economic Affairs, the Netherlands; Ministry of Foreign Affairs, Norway; Ministry of Petroleum and Energy, Norway; Norwegian Agency for Development Cooperation; Parsons Brinckerhoff; Peabody Energy; Petrobras; Ministry of Economy, Poland; Schlumberger; Schneider Electric; Shell; Siemens; Energy Market Authority, Singapore; Statoil; Toyota; Ministry of Energy and Natural Resources, Turkey; Department for International Development, United Kingdom; Foreign and Commonwealth Office, United Kingdom; Executive Office of the Secretary-General, United Nations; Department of Energy, United States; Department of State, United States; and Vattenfall.

Workshops

Many international experts participated in a number of workshops that were held to gather input to this study, resulting in valuable new insights, feedback and data:

- Golden Rules for a Golden Age of Gas: Warsaw, 7 March 2012
- Iraq Energy Outlook: Istanbul, 4 May 2012
- Fuelling the Future with Energy Efficiency: Tokyo, 10 May 2012
- Measuring Progress Towards Universal Energy Access: Paris, 25 May 2012

More information at *www.worldenergyoutlook.org/aboutweo/workshops.*

IEA Energy Business Council

Special thanks go to the companies that participated in meetings of the IEA Energy Business Council (EBC) during 2012, which provided significant insights to this study. The EBC brings together many of the world's largest companies in terms of energy exploration, production and use, ranging from commodities companies, automobile manufacturers, wind and solar producers and financial institutions. Further details may be found at *www.iea.org/energybusinesscouncil.*

Peer reviewers

Many international experts provided input, commented on the underlying analytical work and reviewed early drafts of each chapter. Their comments and suggestions were of great value. They include:

Majeed A. Abdul-Hussain	Parsons Brinckerhoff
Saleh Abdurrahman	National Energy Council of Indonesia
Emmanuel Ackom	UNEP Risø Centre
Ali Aissaoui	APICORP
Natik Al-Bayati	Prime Minister's Advisory Commission, Iraq
Abdulilah Al-Emir	Deputy Prime Minister's Office, Iraq
Jabbar Allibi	Independent consultant
Alicia Altagracia Aponte	GE Energy
Marco Arcelli	Enel
Sefa Sadık Aytekin	Ministry of Energy and Natural Resources, Turkey
Christopher Baker	Department of Climate Change and Energy Efficiency, Australia
Peter Bach	Danish Energy Agency, Ministry of Climate, Energy and Buildings
Amit Bando	International Partnership for Energy Efficiency Cooperation
Paul Baruya	IEA Clean Coal Centre
Georg Bäuml	Volkswagen
Morgan Bazilian	National Renewable Energy Laboratory (NREL)
Carmen Becerril	Acciona Energia
Laurent Bellet	EDF
Kamel Bennaceur	Schlumberger
Somnath Bhattacharjee	Institute for Industrial Productivity

Alexey Biteryakov	GazpromExport
Edgar Blaustein	EU Energy Initiative Partnership Dialogue Facility
Maike Böggemann	Shell
Nils Borg	European Council for an Energy-Efficient Economy
Valentina Bosetti	Fondazione Eni Enrico Mattei (FEEM)
Jean-Paul Bouttes	EDF
Albert Bressand	Columbia School of International and Public Affairs, United States
Abeeku Brew-Hammond	Energy Commission, Ghana
Nicholas Bridge	British Ambassador to the OECD
Nigel Bruce	World Health Organization, Switzerland
Peter Brun	Vestas
Anne Brunila	Fortum Corporation
Mick Buffier	Xstrata Coal
Stephan Buller	Siemens
Ron Cameron	OECD Nuclear Energy Agency
Peter Candler	Foreign and Commonwealth Office, United Kingdom
Kevin Carey	World Bank
Guy Caruso	Center for Strategic and International Studies, United States
Robert Cekuta	Department of State, United States
Sharat Chand	The Energy and Resources Institute, India
Surya Prakash Chandak	United Nations Environment Programme
Suani Teixeira Coelho	University of São Paulo, Brazil
Heather Cooley	Pacific Institute, United States
Alan Copeland	Bureau of Resource and Energy Economics, Australia
John Corben	Schlumberger
Eduardo Luiz Correia	Petrobras
Joppe Cramwinckel	World Business Council for Sustainable Development, Switzerland
Jos Delbeke	Directorate General Climate Action, European Commission
Tim Depledge	BP
Carmine Difiglio	Department of Energy, United States
Jens Drillisch	KfW Development Bank
Richard Duke	Department of Energy, United States
Adeline Duterque	GDF SUEZ
Jonathan Elkind	Department of Energy, United States
David Ensign	Department of State, United States
Peter Evans	GE Energy
Nikki Fisher	Anglo American Thermal Coal
Philippe Fonta	World Business Council for Sustainable Development, Switzerland
Gabriele Franceschini	Eni
Peter Fraser	Ontario Energy Board, Canada
Ajay Gambhir	Grantham Institute for Climate Change, United Kingdom
Amit Garg	India Institute of Management, Ahmedabad

Dario Garofalo	Enel
Francesco Gattei	Eni
Arunabha Ghosh	Council on Energy, Environment and Water, India
Dolf Gielen	International Renewable Energy Agency
Adriano Gomes de Sousa	Ministry of Mining and Energy, Brazil
Rainer Görgen	Federal Ministry of Economics and Technology, Germany
Hugo Leonardo Gosmann	Ministry of Mining and Energy, Brazil
Christopher Guelff	Foreign and Commonwealth Office, United Kingdom
Cecilia Gunnarsson	Volvo
Wenke Han	Energy Research Institute, China
David Hawkins	Natural Resource Defence Council, United States
Hamah-Ameen Hawramany	Ministry of Electricity, Kurdistan Regional Government
James Hewlett	Department of Energy, United States
Nikolas Hill	AEA
Takashi Hongo	Mitsui Global Strategic Studies Institute
Michael Howard	Ministry of Natural Resources, Kurdistan Regional Government
Tom Howes	European Commission
Marco Hüls	GIZ, Germany
Steven Hunt	Department for International Development, United Kingdom
Ruba Husari	Independent consultant
Jana Hybaskova	Ambassador, European Union Delegation to Iraq
Taichi Ito	Mitsubishi UFJ Lease and Finance Corporation, Co. Ltd.
Akihiro Iwata	New Energy Technology Development Organisation, Japan
Anil Jain	Planning Commission, Government of India
James Jensen	Jensen Associates
Jan-Hein Jesse	Clingendael Institute
Ahmed Mousa Jiyad	Independent consultant
Catrin Jung-Draschil	Vattenfall
Marianne Kah	ConocoPhillips
Sami Kamel	General Electric
I.A. Khan	Planning Commission, Government of India
Ali H. Khudhier	South Gas Company, Iraq
David Knapp	Energy Intelligence
Kenji Kobayashi	Japan Automobile Importers Association
Hans Jørgen Koch	Ministry of Climate, Energy and Buildings, Denmark
Beejaye Kokil	African Development Bank
Keiichi Komai	The Energy Conservation Center, Japan
Raëd Kombargi	Booz & Company
Michihisa Kono	Siemens
Annemarije Kooijman-van Dijk	University of Twente, the Netherlands
Robert P. Kool	Chair of IEA Demand-Side Management Programme
Ken Koyama	The Institute of Energy Economics, Japan
Rakesh Kumar	PTC India

Ken Kuroda	Mitsubishi Corporation
Takayuki Kusajima	Toyota Motor Corporation
Nicolas Lambert	European Commission
Jeff Larkin	Parsons Brinckerhoff
Massimo Lombardini	European Commission
Hirono Masazumi	The Japan Gas Association
Steve Lennon	Eskom
Eoin Lees	The Regulatory Assistance Project
Wenge Liu	China Coal Information Institute
Joan MacNaughton	Energy Institute, United Kingdom
Claude Mandil	Former IEA Executive Director
Ritu Mathur	The Energy and Resources Institute, India
Kenichi Matsui	The Institute of Energy Economics, Japan
Takahiro Mazaki	Mitsubishi Corporation
Siddiq McDoom	Natural Resources Canada
Hamish McNinch	Independent consultant
Pedro Antonio Merino	Repsol
Lawrence Metzroth	Arch Coal
Ryo Minami	Ministry of Economy, Trade and Industry, Japan
Vincent Minier	Schneider Electric
Bogdan Mitna	Daimler AG
Arne Mogren	European Climate Foundation
Salah Mohammed	Rumaila Operating Organization, Iraq
Klaus Mohn	Statoil
Simone Mori	Enel
Ed Morse	Citigroup
Richard Morse	Stanford University, United States
Asri Mousa	Independent consultant
Omar Moussa	Schlumberger
Steve Nadel	American Council for an Energy-Efficient Economy
Yugo Nakamura	Bloomberg New Energy Finance
Naokazu Nakano	Sumitomo Metal Industries
Anne Arquit Niederberger	Policy Solutions
Hans Nijkamp	Shell
Shohei Nishimura	Japan Oil, Gas and Metals National Corporation
Petter Nore	Norwegian Agency for Development Cooperation
Patrick Nussbaumer	United Nations Industrial Development Organisation
Per Magnus Nysveen	Rystad Energy
Ulrich Oberndorfer	Federal Ministry of Economics and Technology, Germany
Patrick Oliva	Michelin
Bram Otto	Shell
Shonali Pachauri	International Institute for Applied System Analysis, Austria
Binu Parthan	Sustainable Energy Associates
Andrew Pearce	Parsons Brinckerhoff

Volkmar Pflug	Siemens
Friedbert Pflüger	King's College London
Christian Pichat	Areva
Roberto Potí	Edison
Venkata Ramana Putti	World Bank
William Ramsay	Former IEA Deputy Executive Director
Oliver Rapf	Buildings Performance Institute Europe
Birgitta Resvik	Fortum Corporation
Brian Ricketts	Euracoal
Marc Ringel	European Commission
Nick Robins	HSBC
Hans-Holger Rogner	International Atomic Energy Agency, Austria
Piet Ruijtenberg	ParPetro Consulting
Ralph Samuelson	Asia Pacific Energy Research Centre
George Sarraf	Booz & Company
Koichi Sasaki	The Institute of Energy Economics, Japan
Steve Sawyer	Global Wind Energy Council
Deger Saygin	Utrecht University, the Netherlands
Stefan Scheuer	Coalition for Energy Savings
Hans-Wilhelm Schiffer	RWE
Martin Schöpe	Federal Ministry for the Environment, Nature Conservation and Nuclear Safety, Germany
Kristin Seyboth	Intergovernmental Panel on Climate Change
Baoguo Shan	State Grid Energy Research Institute, China
P.R. Shukla	Indian Institute of Management, Ahmedabad
Ralph Sims	Massey University, New Zealand
Stephan Singer	WWF International
Kelly Smith	Natural Resources Canada
Christopher Snary	Department of Energy and Climate Change, United Kingdom
Pil-Bae Song	Asian Development Bank
Leena Srivastava	The Energy and Resources Institute, India
Simon Stolp	World Bank
Michael Stoppard	IHS CERA
Greg Stringham	Canadian Association of Petroleum Producers
Cartan Sumner	Peabody Energy
Renata Nascimento Szczerbacki	Petrobras
Minoru Takada	United Nations
Kazunori Takahashi	Hitachi Ltd.
Kuniharu Takemata	Electric Power Development Co., Ltd. (J-POWER)
Nobuo Tanaka	Former IEA Executive Director
Josue Tanaka	European Bank for Reconstruction and Development
Chee Hong Tat	Energy Market Authority, Singapore
Tessa Terpstra	Ministry of Foreign Affairs, the Netherlands
Wim Thomas	Shell

Simon Trace	Practical Action, United Kingdom
James Turner	Department of Energy and Climate Change, United Kingdom
Jim Turner	Department of State, United States
Oras Tynkkynen	Member of Parliament, Finland
Thamir Uqaili	Independent consultant
Aad van Bohemen	Ministry of Economic Affairs, the Netherlands
Coby van der Linde	Clingendael Institute, the Netherlands
Noe van Hulst	Energy Academy Europe
Wim J. van Nes	SNV Netherlands Development Organisation
Kristi Varangu	Natural Resources Canada
Livia Vasakova	Directorate-General for Energy, European Commission
Antonio Verde	Ministry of Foreign Affairs, Italy
Pierre Verlinden	Trina Solar
Frank Verrastro	Center for Strategic and International Studies, United States
Louise Vickery	Department of Resources, Energy and Tourism, Australia
Daniele Violetti	United Nations Framework Convention on Climate Change
Faisal Wadi	South Oil Company, Iraq
Henry Wang	Saudi Basic Industries Corporation, Saudi Arabia
Peter Wells	Cardiff Business School
Jacob Williams	Peabody Energy
Stephen Wilson	Rio Tinto
Steven Winberg	CONSOL Energy
Peter Wooders	International Institute for Sustainable Development, Switzerland
Xiaojie Xu	Academy of Social Sciences, China
Zhang Yuzhuo	Shenhua Group
Craig Zamuda	Department of Energy, United States
Alex Zapantis	Rio Tinto
Krzysztof Żmijewski	Public Board for Development of Low-emission Economy, Poland

The individuals and organisations that contributed to this study are not responsible for any opinions or judgements contained in this study. All errors and omissions are solely the responsibility of the IEA.

Comments and questions are welcome and should be addressed to:

Dr. Fatih Birol
Chief Economist
Director, Directorate of Global Energy Economics
International Energy Agency
9, rue de la Fédération
75739 Paris Cedex 15
France

Telephone: (33-1) 4057 6670
Email: weo@iea.org

More information about the *World Energy Outlook* is available at *www.worldenergyoutlook.org*.

TABLE OF CONTENTS

Part C: IRAQ ENERGY OUTLOOK 385

A new global energy landscape is emerging

The global energy map is changing, with potentially far-reaching consequences for energy markets and trade. It is being redrawn by the resurgence in oil and gas production in the United States and could be further reshaped by a retreat from nuclear power in some countries, continued rapid growth in the use of wind and solar technologies and by the global spread of unconventional gas production. Perspectives for international oil markets hinge on Iraq's success in revitalising its oil sector. If new policy initiatives are broadened and implemented in a concerted effort to improve global energy efficiency, this could likewise be a game-changer. On the basis of global scenarios and multiple case studies, this *World Energy Outlook* assesses how these new developments might affect global energy and climate trends over the coming decades. It examines their impact on the critical challenges facing the energy system: to meet the world's ever-growing energy needs, led by rising incomes and populations in emerging economies; to provide energy access to the world's poorest; and to bring the world towards meeting its climate change objectives.

Taking all new developments and policies into account, the world is still failing to put the global energy system onto a more sustainable path. Global energy demand grows by more than one-third over the period to 2035 in the New Policies Scenario (our central scenario), with China, India and the Middle East accounting for 60% of the increase. Energy demand barely rises in OECD countries, although there is a pronounced shift away from oil and coal (and, in some countries, nuclear) towards natural gas and renewables. Despite the growth in low-carbon sources of energy, fossil fuels remain dominant in the global energy mix, supported by subsidies that amounted to $523 billion in 2011, up almost 30% on 2010 and six times more than subsidies to renewables. The cost of fossil-fuel subsidies has been driven up by higher oil prices; they remain most prevalent in the Middle East and North Africa, where momentum towards their reform appears to have been lost. Emissions in the New Policies Scenario correspond to a long-term average global temperature increase of 3.6 °C.

The tide turns for US energy flows

Energy developments in the United States are profound and their effect will be felt well beyond North America – and the energy sector. The recent rebound in US oil and gas production, driven by upstream technologies that are unlocking light tight oil and shale gas resources, is spurring economic activity – with less expensive gas and electricity prices giving industry a competitive edge – and steadily changing the role of North America in global energy trade. By around 2020, the United States is projected to become the largest global oil producer (overtaking Saudi Arabia until the mid-2020s) and starts to see the impact of new fuel-efficiency measures in transport. The result is a continued fall in US oil imports, to the extent that North America becomes a net oil exporter around 2030. This accelerates the switch in direction of international oil trade towards Asia, putting a focus on the security of the strategic routes that bring Middle East oil to Asian markets. The

United States, which currently imports around 20% of its total energy needs, becomes all but self-sufficient in net terms – a dramatic reversal of the trend seen in most other energy-importing countries.

But there is no immunity from global markets

No country is an energy "island" and the interactions between different fuels, markets and prices are intensifying. Most oil consumers are used to the effects of worldwide fluctuations in price (reducing its oil imports will not insulate the United States from developments in international markets), but consumers can expect to see growing linkages in other areas. A current example is how low-priced natural gas is reducing coal use in the United States, freeing up coal for export to Europe (where, in turn, it has displaced higher-priced gas). At its lowest level in 2012, natural gas in the United States traded at around one-fifth of import prices in Europe and one-eighth of those in Japan. Going forward, price relationships between regional gas markets are set to strengthen as liquefied natural gas trade becomes more flexible and contract terms evolve, meaning that changes in one part of the world are more quickly felt elsewhere. Within individual countries and regions, competitive power markets are creating stronger links between gas and coal markets, while these markets also need to adapt to the increasing role of renewables and, in some cases, to the reduced role of nuclear power. Policy makers looking for simultaneous progress towards energy security, economic and environmental objectives are facing increasingly complex – and sometimes contradictory – choices.

A blueprint for an energy-efficient world

Energy efficiency is widely recognised as a key option in the hands of policy makers but current efforts fall well short of tapping its full economic potential. In the last year, major energy-consuming countries have announced new measures: China is targeting a 16% reduction in energy intensity by 2015; the United States has adopted new fuel-economy standards; the European Union has committed to a cut of 20% in its 2020 energy demand; and Japan aims to cut 10% from electricity consumption by 2030. In the New Policies Scenario, these help to speed up the disappointingly slow progress in global energy efficiency seen over the last decade. But even with these and other new policies in place, a significant share of the potential to improve energy efficiency – four-fifths of the potential in the buildings sector and more than half in industry – still remains untapped.

Our Efficient World Scenario shows how tackling the barriers to energy efficiency investment can unleash this potential and realise huge gains for energy security, economic growth and the environment. These gains are not based on achieving any major or unexpected technological breakthroughs, but just on taking actions to remove the barriers obstructing the implementation of energy efficiency measures that are economically viable. Successful action to this effect would have a major impact on global energy and climate trends, compared with the New Policies Scenario. The growth in global primary energy demand to 2035 would be halved. Oil demand would peak just before 2020 and would be almost 13 mb/d lower by 2035, a reduction equal to the current production of Russia and

Norway combined, easing the pressure for new discoveries and development. Additional investment of $11.8 trillion (in year-2011 dollars) in more energy-efficient technologies would be more than offset by reduced fuel expenditures. The accrued resources would facilitate a gradual reorientation of the global economy, boosting cumulative economic output to 2035 by $18 trillion, with the biggest gross domestic product (GDP) gains in India, China, the United States and Europe. Universal access to modern energy would be easier to achieve and air quality improved, as emissions of local pollutants fall sharply. Energy-related carbon-dioxide (CO_2) emissions would peak before 2020, with a decline thereafter consistent with a long-term temperature increase of 3 °C.

We propose policy principles that can turn the Efficient World Scenario into reality. Although the specific steps will vary by country and by sector, there are six broad areas that need to be addressed. Energy efficiency needs to be made clearly visible, by strengthening the measurement and disclosure of its economic gains. The profile of energy efficiency needs to be raised, so that efficiency concerns are integrated into decision making throughout government, industry and society. Policy makers need to improve the affordability of energy efficiency, by creating and supporting business models, financing vehicles and incentives to ensure that investors reap an appropriate share of the rewards. By deploying a mix of regulations to discourage the least-efficient approaches and incentives to deploy the most efficient, governments can help push energy-efficient technologies into the mainstream. Monitoring, verification and enforcement activities are essential to realise expected energy savings. These steps would need to be underpinned by greater investment in energy efficiency governance and administrative capacity at all levels.

Energy efficiency can keep the door to 2 °C open for just a bit longer

Successive editions of this report have shown that the climate goal of limiting warming to 2 °C is becoming more difficult and more costly with each year that passes. Our 450 Scenario examines the actions necessary to achieve this goal and finds that almost four-fifths of the CO_2 emissions allowable by 2035 are already locked-in by existing power plants, factories, buildings, etc. If action to reduce CO_2 emissions is not taken before 2017, all the allowable CO_2 emissions would be locked-in by energy infrastructure existing at that time. Rapid deployment of energy-efficient technologies – as in our Efficient World Scenario – would postpone this complete lock-in to 2022, buying time to secure a much-needed global agreement to cut greenhouse-gas emissions.

No more than one-third of proven reserves of fossil fuels can be consumed prior to 2050 if the world is to achieve the 2 °C goal, unless carbon capture and storage (CCS) technology is widely deployed. This finding is based on our assessment of global "carbon reserves", measured as the potential CO_2 emissions from proven fossil-fuel reserves. Almost two-thirds of these carbon reserves are related to coal, 22% to oil and 15% to gas. Geographically, two-thirds are held by North America, the Middle East, China and Russia. These findings underline the importance of CCS as a key option to mitigate CO_2 emissions, but its pace of deployment remains highly uncertain, with only a handful of commercial-scale projects currently in operation.

Trucks deliver a large share of oil demand growth

Growth in oil consumption in emerging economies, particularly for transport in China, India and the Middle East, more than outweighs reduced demand in the OECD, pushing oil use steadily higher in the New Policies Scenario. Oil demand reaches 99.7 mb/d in 2035, up from 87.4 mb/d in 2011, and the average IEA crude oil import price rises to $125/barrel (in year-2011 dollars) in 2035 (over $215/barrel in nominal terms). The transport sector already accounts for over half of global oil consumption, and this share increases as the number of passenger cars doubles to 1.7 billion and demand for road freight rises quickly. The latter is responsible for almost 40% of the increase in global oil demand: oil use for trucks – predominantly diesel – increases much faster than that for passenger vehicles, in part because fuel-economy standards for trucks are much less widely adopted.

Non-OPEC oil output steps up over the current decade, but supply after 2020 depends increasingly on OPEC. A surge in unconventional supplies, mainly from light tight oil in the United States and oil sands in Canada, natural gas liquids, and a jump in deepwater production in Brazil, push non-OPEC production up after 2015 to a plateau above 53 mb/d, from under 49 mb/d in 2011. This is maintained until the mid-2020s, before falling back to 50 mb/d in 2035. Output from OPEC countries rises, particularly after 2020, bringing the OPEC share in global production from its current 42% up towards 50% by 2035. The net increase in global oil production is driven entirely by unconventional oil, including a contribution from light tight oil that exceeds 4 mb/d for much of the 2020s, and by natural gas liquids. Of the $15 trillion in upstream oil and gas investment that is required over the period to 2035, almost 30% is in North America.

Much is riding on Iraq's success

Iraq makes the largest contribution by far to global oil supply growth. Iraq's ambition to expand output after decades of conflict and instability is not limited by the size of its resources or by the costs of producing them, but will require co-ordinated progress all along the energy supply chain, clarity on how Iraq plans to derive long-term value from its hydrocarbon wealth and successful consolidation of a domestic consensus on oil policy. In our projections, oil output in Iraq exceeds 6 mb/d in 2020 and rises to more than 8 mb/d in 2035. Iraq becomes a key supplier to fast-growing Asian markets, mainly China, and the second-largest global exporter by the 2030s, overtaking Russia. Without this supply growth from Iraq, oil markets would be set for difficult times, characterised by prices that are almost $15/barrel higher than the level in the New Policies Scenario by 2035.

Iraq stands to gain almost $5 trillion in revenue from oil exports over the period to 2035, an annual average of $200 billion, and an opportunity to transform the country's prospects. The energy sector competes with a host of other spending needs in Iraq, but one urgent priority is to catch up and keep pace with rising electricity demand: if planned new capacity is delivered on time, grid-based electricity generation will be sufficient to meet peak demand by around 2015. Gathering and processing associated gas – much of which is currently flared – and developing non-associated gas offers the promise of a more efficient gas-fuelled power sector and, once domestic demand is satisfied, of gas exports. Translating oil export receipts into greater prosperity will require strengthened institutions, both to ensure efficient, transparent management of revenues and spending, and to set the course necessary to encourage more diverse economic activity.

Different shades of gold for natural gas

Natural gas is the only fossil fuel for which global demand grows in all scenarios, showing that it fares well under different policy conditions; but the outlook varies by region. Demand growth in China, India and the Middle East is strong: active policy support and regulatory reforms push China's consumption up from around 130 billion cubic metres (bcm) in 2011 to 545 bcm in 2035. In the United States, low prices and abundant supply see gas overtake oil around 2030 to become the largest fuel in the energy mix. Europe takes almost a decade to get back to 2010 levels of gas demand: the growth in Japan is similarly limited by higher gas prices and a policy emphasis on renewables and energy efficiency.

Unconventional gas accounts for nearly half of the increase in global gas production to 2035, with most of the increase coming from China, the United States and Australia. But the unconventional gas business is still in its formative years, with uncertainty in many countries about the extent and quality of the resource base. As analysed in a *World Energy Outlook Special Report* released in May 2012, there are also concerns about the environmental impact of producing unconventional gas that, if not properly addressed, could halt the unconventional gas revolution in its tracks. Public confidence can be underpinned by robust regulatory frameworks and exemplary industry performance. By bolstering and diversifying sources of supply, tempering demand for imports (as in China) and fostering the emergence of new exporting countries (as in the United States), unconventional gas can accelerate movement towards more diversified trade flows, putting pressure on conventional gas suppliers and on traditional oil-linked pricing mechanisms for gas.

Will coal remain a fuel of choice?

Coal has met nearly half of the rise in global energy demand over the last decade, growing faster even than total renewables. Whether coal demand carries on rising strongly or changes course will depend on the strength of policy measures that favour lower-emissions energy sources, the deployment of more efficient coal-burning technologies and, especially important in the longer term, CCS. The policy decisions carrying the most weight for the global coal balance will be taken in Beijing and New Delhi – China and India account for almost three-quarters of projected non-OECD coal demand growth (OECD coal use declines). China's demand peaks around 2020 and is then steady to 2035; coal use in India continues to rise and, by 2025, it overtakes the United States as the world's second-largest user of coal. Coal trade continues to grow to 2020, at which point India becomes the largest net importer of coal, but then levels off as China's imports decline. The sensitivity of these trajectories to changes in policy, the development of alternative fuels (*e.g.* unconventional gas in China) and the timely availability of infrastructure, create much uncertainty for international steam coal markets and prices.

If nuclear falls back, what takes its place?

The world's demand for electricity grows almost twice as fast as its total energy consumption, and the challenge to meet this demand is heightened by the investment needed to replace ageing power sector infrastructure. Of the new generation capacity that is built to 2035, around one-third is needed to replace plants that are retired. Half of all new capacity is based on renewable sources of energy, although coal remains the leading global fuel for power generation. The growth in China's electricity demand over the period to 2035 is greater than total current electricity demand in the United States and Japan. China's coal-fired output increases almost as much as its generation from nuclear, wind and hydropower combined. Average global electricity prices increase by 15% to 2035 in real terms, driven higher by increased fuel input costs, a shift to more capital-intensive generating capacity, subsidies to renewables and CO_2 pricing in some countries. There are significant regional price variations, with the highest prices persisting in the European Union and Japan, well above those in the United States and China.

The anticipated role of nuclear power has been scaled back as countries have reviewed policies in the wake of the 2011 accident at the Fukushima Daiichi nuclear power station. Japan and France have recently joined the countries with intentions to reduce their use of nuclear power, while its competitiveness in the United States and Canada is being challenged by relatively cheap natural gas. Our projections for growth in installed nuclear capacity are lower than in last year's *Outlook* and, while nuclear output still grows in absolute terms (driven by expanded generation in China, Korea, India and Russia), its share in the global electricity mix falls slightly over time. Shifting away from nuclear power can have significant implications for a country's spending on imports of fossil fuels, for electricity prices and for the level of effort needed to meet climate targets.

Renewables take their place in the sun

A steady increase in hydropower and the rapid expansion of wind and solar power has cemented the position of renewables as an indispensable part of the global energy mix; by 2035, renewables account for almost one-third of total electricity output. Solar grows more rapidly than any other renewable technology. Renewables become the world's second-largest source of power generation by 2015 (roughly half that of coal) and, by 2035, they approach coal as the primary source of global electricity. Consumption of biomass (for power generation) and biofuels grows four-fold, with increasing volumes being traded internationally. Global bioenergy resources are more than sufficient to meet our projected biofuels and biomass supply without competing with food production, although the land-use implications have to be managed carefully. The rapid increase in renewable energy is underpinned by falling technology costs, rising fossil-fuel prices and carbon pricing, but mainly by continued subsidies: from $88 billion globally in 2011, they rise to nearly $240 billion in 2035. Subsidy measures to support new renewable energy projects need to be adjusted over time as capacity increases and as the costs of renewable technologies fall, to avoid excessive burdens on governments and consumers.

A continuing focus on the goal of universal energy access

Despite progress in the past year, nearly 1.3 billion people remain without access to electricity and 2.6 billion do not have access to clean cooking facilities. Ten countries – four in developing Asia and six in sub-Saharan Africa – account for two-thirds of those people without electricity and just three countries – India, China and Bangladesh – account for more than half of those without clean cooking facilities. While the Rio+20 Summit did not result in a binding commitment towards universal modern energy access by 2030, the UN Year of Sustainable Energy for All has generated welcome new commitments towards this goal. But much more is required. In the absence of further action, we project that nearly one billion people will be without electricity and 2.6 billion people will still be without clean cooking facilities in 2030. We estimate that nearly $1 trillion in cumulative investment is needed to achieve universal energy access by 2030.

We present an Energy Development Index (EDI) for 80 countries, to aid policy makers in tracking progress towards providing modern energy access. The EDI is a composite index that measures a country's energy development at the household and community level. It reveals a broad improvement in recent years, with China, Thailand, El Salvador, Argentina, Uruguay, Vietnam and Algeria showing the greatest progress. There are also a number of countries whose EDI scores remain low, such as Ethiopia, Liberia, Rwanda, Guinea, Uganda and Burkina Faso. The sub-Saharan Africa region scores least well, dominating the lower half of the rankings.

Energy is becoming a thirstier resource

Water needs for energy production are set to grow at twice the rate of energy demand. Water is essential to energy production: in power generation; in the extraction, transport and processing of oil, gas and coal; and, increasingly, in irrigation for crops used to produce biofuels. We estimate that water withdrawals for energy production in 2010 were 583 billion cubic metres (bcm). Of that, water consumption – the volume withdrawn but not returned to its source – was 66 bcm. The projected rise in water consumption of 85% over the period to 2035 reflects a move towards more water-intensive power generation and expanding output of biofuels.

Water is growing in importance as a criterion for assessing the viability of energy projects, as population and economic growth intensify competition for water resources. In some regions, water constraints are already affecting the reliability of existing operations and they will increasingly impose additional costs. In some cases, they could threaten the viability of projects. The vulnerability of the energy sector to water constraints is widely spread geographically, affecting, among others, shale gas development and power generation in parts of China and the United States, the operation of India's highly water-intensive fleet of power plants, Canadian oil sands production and the maintenance of oil-field pressures in Iraq. Managing the energy sector's water vulnerabilities will require deployment of better technology and greater integration of energy and water policies.

PART A
GLOBAL ENERGY TRENDS

PREFACE

Part A of this *WEO* (Chapters 1-8) presents energy projections to 2035. Global energy demand, production, trade, investment and carbon-dioxide emissions are broken down by region or country, by fuel and by sector. There are three core scenarios presented in Part A: the New Policies Scenario, the Current Policies Scenario and the 450 Scenario.

Chapter 1 defines the scenarios and sets out their underlying assumptions.

Chapter 2 defines the results of the projections for global energy in aggregate and discusses their implications for energy security, the environment and economic development. While results for each of the scenarios are shown, more focus is given to the results of the New Policies Scenario, to provide a clear picture of where planned policies, if implemented in a relatively cautious manner, would take us. The chapter ends with a special feature on the resurgence of oil and gas production in the United States and what it means for the world's energy landscape.

Chapters 3-7 analyse the outlook for each of the main energy sources and carriers – oil, gas, coal, electricity and renewables.

Chapter 8 presents our analysis of climate change mitigation. This includes a review of recent developments in climate policy and trends in carbon-dioxide emissions, followed by a focus on the 450 Scenario, which sets out a global energy pathway compatible with limiting warming to 2 °C. Comparisons to other *WEO-2012* scenarios are given in order to illustrate the extent of the additional efforts that would be necessary to achieve the climate goal.

Understanding the scenarios

What are the drivers of our energy future?

Highlights

- Though there are some underlying trends which are difficult to shift, government policy action has a strong influence on energy markets. Our scenarios accordingly reflect different levels of government action. Our central scenario, the New Policies Scenario, takes into account existing policy commitments and assumes that those recently announced are implemented, albeit in a cautious manner. The Current Policies Scenario assumes no implementation of policies beyond those adopted by mid-2012. The 450 Scenario assumes policy action consistent with limiting the long-term global temperature increase to 2 °C. As a special feature of *WEO-2012*, we also present an Efficient World Scenario, reflecting adoption of all economically viable steps to improve energy efficiency.

- The level of economic activity plays a key role in global energy trends, introducing important uncertainties to our projections, particularly in the near to medium term. We assume a modest rebound in global GDP in the near term and an average rate of growth in real terms of 3.5% per year through to 2035, with growth slowing gradually as the emerging economies mature.

- Demographic change affects the size and composition of energy demand, directly and through its impact on economic growth and development. World population is assumed to rise from 6.8 billion in 2010 to 8.6 billion in 2035 – another 1.7 billion energy consumers, mainly in Asia and Africa. India's population exceeds 1.5 billion in 2035, after overtaking China soon after 2025.

- Energy prices are another key determinant of energy trends. In the New Policies Scenario, the average IEA crude oil import price approaches $125/barrel (in year-2011 dollars) in 2035. The ratio of natural gas to oil prices (on an energy-equivalent basis) remains lower than the historical average, due in large part to brighter prospects for unconventional gas, which also contributes to a degree of price convergence between regional gas markets. Coal prices rise less rapidly than oil and gas prices, as a result of strong competition from gas in power generation and abundant coal resources. In the New Policies Scenario, CO_2 prices in parts of the OECD rise to $45/tonne in 2035 and CO_2 pricing (or shadow pricing influencing investment decisions) is gradually introduced in some other countries. In the 450 Scenario, more regions adopt CO_2 pricing and prices are markedly higher.

- The rate at which the efficiency of current energy technologies improves and new technologies are adopted will be a crucial determinant of future energy trends. Key uncertainties relate to the prospects for carbon capture and storage, solar power, advanced biofuels, advanced vehicle technologies and nuclear power.

Defining the scenarios

Drawing on the latest data and policy developments, this year's *World Energy Outlook* presents projections of energy trends through to 2035 and insights into what they mean for energy security, the environment and economic development. Over the *Outlook* period, the interaction of many different factors will drive the evolution of energy markets. As outcomes are hard to predict with accuracy, the report presents several different scenarios, which are differentiated primarily by their underlying assumptions about government policies. The starting year of the scenarios is 2010, the latest year for which comprehensive historical energy data for all countries were available at the time of writing, though preliminary data for 2011 have been incorporated in many cases.

The ***New Policies Scenario*** – our central scenario – takes into account broad policy commitments and plans that have already been implemented to address energy-related challenges as well as those that have been announced, even where the specific measures to implement these commitments have yet to be introduced (Table 1.1). New commitments include renewable energy and energy efficiency targets, programmes relating to nuclear phase-out or additions, national targets to reduce greenhouse-gas emissions communicated under the 2010 Cancun Agreements and the initiatives taken by G-20 and Asia-Pacific Economic Cooperation (APEC) economies to phase out inefficient fossil-fuel subsidies. The New Policies Scenario assumes only cautious implementation of current commitments and plans: for example, countries that have set a range for a particular target are assumed to adopt policies that are consistent with reaching the less ambitious end of the range. In countries where climate policy is particularly uncertain, it is assumed that the policies adopted are insufficient to reach the declared target. This caution stems from the widely recognised distinction between political commitments and implementing action or, in some cases, the limited details available about how new initiatives might be designed. Some targets involve a range, while others are conditional on funding or on comparable emissions reductions in other countries. And some targets relate to energy or carbon intensity, rather than absolute energy savings or emissions reductions.

To illustrate the outcome of our current course, if unchanged, the ***Current Policies Scenario*** embodies the effects of only those government policies and measures that had been enacted or adopted by mid-2012. Without implying that total inaction is probable, it does not take into account any possible, potential or even likely future policy actions. Several of the policy commitments and plans that were included in the New Policies Scenario in the *World Energy Outlook 2011 (WEO-2011)* have since been enacted, so are now included in the Current Policies Scenario. These include, for example, energy and climate targets in China's 12th Five-Year Plan for the period 2011-2015 and new feed-in tariffs for renewable energy technologies in Japan.

The basis of the ***450 Scenario*** is different. Rather than being a projection based on past trends, modified by known policy actions, it deliberately selects a plausible energy pathway. The pathway chosen is consistent with actions having around a 50% chance of meeting the goal of limiting the global increase in average temperature to two degrees Celsius (2 °C) in the long term, compared with pre-industrial levels. According to climate experts, to meet this goal the long-term concentration of greenhouse gases in the atmosphere needs to be limited to around 450 parts per million of carbon-dioxide equivalent (ppm CO_2-eq) – hence the scenario name. For the period to 2020, we assume policy action to implement fully the commitments under the Cancun Agreements. After 2020, OECD countries and other major economies set emissions targets for 2035 and beyond that collectively ensure an emissions trajectory consistent with ultimate stabilisation of the greenhouse-gas concentration at 450 ppm. This is in line with the agreement reached at the United Nations Framework Convention on Climate Change (UNFCCC) 17th Conference of the Parties (COP-17) in December 2011 to establish the "Durban Platform on Enhanced Action", which is intended to lead to a new climate regime. We also assume from 2020 that $100 billion in annual financing is provided by OECD countries to non-OECD countries for abatement measures.

Table 1.1 ▷ Definitions and objectives of the *WEO-2012* scenarios

	Current Policies Scenario	New Policies Scenario	450 Scenario	Efficient World Scenario
Definitions	Government policies that had been enacted or adopted by mid-2012 continue unchanged.	Existing policies are maintained and recently announced commitments and plans, including those yet to be formally adopted, are implemented in a cautious manner.	Policies are adopted that put the world on a pathway that is consistent with having around a 50% chance of limiting the global increase in average temperature to 2 °C in the long term, compared with pre-industrial levels.	All energy efficiency investments that are economically viable are made and all necessary policies to eliminate market barriers to energy efficiency are adopted.
Objectives	To provide a baseline that shows how energy markets would evolve if underlying trends in energy demand and supply are not changed.	To provide a benchmark to assess the potential achievements (and limitations) of recent developments in energy and climate policy.	To demonstrate a plausible path to achieve the climate target.	To explore the results of improving energy efficiency in every way that makes economic sense.

The ***Efficient World Scenario***, the results of which are presented in Chapters 10-12, has been developed especially for the *World Energy Outlook 2012 (WEO-2012)*. It enables us to quantify the implications for the economy, the environment and energy security of a major step change in energy efficiency. It is based on the core assumption that all investments capable of improving energy efficiency are made so long as they are economically viable and any market barriers obstructing their realisation are removed. The scale of the opportunity is determined, by sector and region, on the basis of a thorough review of the technical potential to raise energy efficiency, and our judgement of the payback periods that investors will require in order to commit funds to energy efficiency projects.

The projections for each of the scenarios are derived from the IEA's World Energy Model (WEM) – a large-scale partial equilibrium model that is designed to replicate how energy markets function over the medium to long term.[1] The WEM comprises six modules, covering final demand (by sector and sub-sector), power and heat generation, refining and other transformation activities, supply (by primary energy source), carbon-dioxide (CO_2) emissions and investment in energy supply infrastructure. Improvements to the WEM for *WEO-2012* include more detailed representation of energy efficiency policies, and fuel and technology choices in the industry and buildings sectors; development of a specific module for road-freight transport, which endogenously determines fuel-economy improvements; more detailed modelling of wholesale electricity pricing for baseload, mid-load and peak load and of end-user prices, based on wholesale prices, network costs, renewable subsidies and retailing costs; more detailed modules for biomass supply and biofuels production; and the inclusion of a separate module for Iraq (see Chapter 13). Some 3 000 policies have been reviewed in developing the assumptions for each of the scenarios presented. The most important ones are summarised in Annex B, categorised by scenario.

Non-policy assumptions

Economic growth

Demand for energy is strongly correlated to economic activity, so our projections are highly sensitive to the underlying assumptions about the rate of growth of gross domestic product (GDP). Rarely has the near-term outlook for the global economy been so uncertain. The world economy bounced back strongly in 2010 from the recession of 2008-2009, but experienced a slowdown in 2011, with some economies dipping back into recession. According to the International Monetary Fund (IMF), only a gradual strengthening of economic activity is expected during the course of 2012 (IMF, 2012a). Furthermore, downside risks to the economic outlook remain acute. An escalation of the euro-zone crisis or failure of the United States to avoid the looming "fiscal cliff" (which would lead to automatic tax increases and spending cuts) could have harmful spill-over effects on the rest of the global economy. There are also clear signs of slowing in the key emerging economies, including China, India and Brazil, which have served as engines of global growth over the past decade. The threat of higher oil prices that could result from heightened geopolitical tensions is another threat to the global economy.

This *Outlook* bases its medium-term GDP growth assumptions primarily on IMF projections, with some adjustments to reflect information available from regional, national and other sources (Table 1.2). Longer-term GDP assumptions are based on projections made by various economic forecasting bodies, as well as our assessment of prospects for the growth in labour supply and improvements in productivity. We assume world GDP (in purchasing power parity [PPP] terms)[2] will grow by an average of 3.5% per year over the period 2010-

1. A full description of the WEM is available at *www.worldenergyoutlook.org/weomodel*.

2. Purchasing power parities measure the amount of a given currency needed to buy the same basket of goods and services, traded and non-traded, as one unit of the reference currency – in this report, the US dollar. By adjusting for differences in price levels, PPPs, in principle, can provide a more reliable indicator than market exchange rates of the true level of economic activity globally or regionally.

2035. This is marginally lower than in last year's *WEO*, although growth in several countries and regions – including Africa and Latin America, Korea and Australia – has been revised upwards. Our assumed rate of growth in world GDP slows gradually as the emerging economies mature.

Table 1.2 ▷ Real GDP growth assumptions by region

	Compound average annual growth rate			
	1990-2010	2010-15	2010-20	2010-35
OECD	**2.2%**	**2.1%**	**2.2%**	**2.1%**
Americas	2.5%	2.6%	2.7%	2.4%
United States	2.5%	2.5%	2.6%	2.4%
Europe	2.0%	1.5%	1.8%	1.8%
Asia Oceania	1.9%	2.0%	2.0%	1.8%
Japan	0.9%	1.2%	1.2%	1.2%
Non-OECD	**4.9%**	**6.1%**	**5.9%**	**4.8%**
E. Europe/Eurasia	0.5%	3.9%	3.8%	3.4%
Russia	0.4%	4.0%	3.9%	3.5%
Asia	7.5%	7.5%	7.0%	5.5%
China	10.1%	8.6%	7.9%	5.7%
India	6.5%	7.3%	7.1%	6.3%
Middle East	4.3%	3.7%	3.9%	3.8%
Iraq	3.1%	10.0%	10.6%	6.9%
Africa	3.8%	4.4%	4.6%	3.8%
Latin America	3.4%	4.2%	4.1%	3.4%
Brazil	3.1%	3.6%	3.8%	3.6%
World	**3.2%**	**4.0%**	**4.0%**	**3.5%**
European Union	1.8%	1.3%	1.7%	1.8%

Note: Calculated based on GDP expressed in year-2011 dollars in PPP terms.

Sources: IMF (2012b); OECD (2012); Economist Intelligence Unit and World Bank databases; IEA databases and analysis.

The non-OECD countries as a group continue to grow much more rapidly than the OECD countries through to 2035, driving up their share of world GDP. China continues to set the pace during the first half of the projection period, although its growth rate is likely to tail off as the country gets richer and its working-age population starts to shrink. By the mid-2020s, India overtakes China to become the fastest-growing *WEO* region, thanks to its rapidly growing population, rising labour participation rates and its earlier stage of economic development. Nonetheless, India's growth rate slows, too, as its economy matures. As explained in Chapter 13, variations in the assumptions about national GDP are an integral part of the analysis presented in our special feature on Iraq, reflecting the dominance of oil revenues in the Iraqi economy; but in each of the scenarios modelled, the assumed GDP growth rate is well above the rate for the Middle East as a whole.

SPOTLIGHT

What is the biggest source of uncertainty for energy prospects?

As Niels Bohr said, "it is tough to make predictions, especially about the future". In common with all attempts to describe future market trends, the projections in *WEO-2012* are subject to a wide range of uncertainties. Indeed, it is unlikely that the future will follow any of the precise paths described in our scenarios. But that is not the aim of the *WEO*: none of the scenarios is a forecast. Each is intended to demonstrate how markets could evolve under certain conditions. How close those scenarios are to reality not only hinges on how well the model represents the way energy markets work and the validity of the assumptions that underpin that model, but also on the occurrence of "game changer" events in the economy at large. Many past forecasts and outlooks have been confounded by events such as the 1997-1998 Asian financial crisis, the US subprime mortgage crisis and subsequent recession and the more recent "Arab Spring".

Key drivers of energy markets are hard to predict, in part because they interact with each other. Yet some are inherently easier to predict than others, especially in the near term. Population, even by region, is unlikely to deviate much from the assumptions used for this *Outlook*. And we can be reasonably confident about how technology is likely to evolve in the short to medium term, even if there are surprises, such as the improved technologies that have recently unlocked huge unconventional gas and oil resources in the United States and elsewhere.

Economic prospects in the near term are much more uncertain and, given the strong link between economic growth and energy demand in most countries, are perhaps the biggest source of uncertainty in the medium term – even more so than usual, given the fragile state of the global economy. But most forecasts of economic growth at the world and regional levels over the long term fall within a relatively narrow range, even if there may be significant divergence between countries.

History has shown that energy prices are notoriously difficult to predict. But even if prices follow a very different path than what we have assumed, the overall impact on demand may not be very pronounced: demand for some forms of energy is less sensitive to price changes than demand for many other goods. Nevertheless, the dramatic five-fold increase in the oil price that we have witnessed over the last decade could still have an unexpectedly large impact on long-term demand, especially if (as our assumptions suggest) these price levels persist.

In the very near term, even government policy is reasonably predictable, as changes tend to be incremental: short of a breakthrough agreement on climate change or a major geopolitical event, it is unlikely that any policy moves in the next twelve months would have a large impact on key global energy trends in the medium term. But in the longer term, policy making is the area in which the greatest uncertainty exists, particularly when it comes to issues such as the extent to which action is taken to mitigate climate change, developments in energy subsidies, decisions on nuclear power, and the pricing and production strategies of the major oil and gas exporters.

Population

Population growth is an important driver of energy use, directly through its impact on the size and composition of energy demand and indirectly through its effect on economic growth and development. The rates of population growth assumed for each region and in all the scenarios are based on the most recent projections by the United Nations (UNPD, 2011). World population is projected to grow from an estimated 6.8 billion in 2010 to 8.6 billion in 2035, or by some 1.7 billion new energy consumers. In line with the long-term historical trend, population growth slows over the projection period, from 1.1% per year in 2010-2020 to 0.8% per year in 2020-2035 (Table 1.3). Yet population still increases substantially in absolute terms each year: the projected annual increase in 2035 is 56 million, compared with 78 million in 2010.

Table 1.3 ▷ Population assumptions by region

	Population growth*			Population (million)		Urbanisation rate	
	2010-20	2020-35	2010-35	2010	2035	2010	2035
OECD	**0.5%**	**0.3%**	**0.4%**	**1 237**	**1 373**	**77%**	**84%**
Americas	0.9%	0.7%	0.7%	474	571	82%	87%
United States	0.8%	0.7%	0.7%	314	377	82%	88%
Europe	0.4%	0.2%	0.3%	560	599	74%	82%
Asia Oceania	0.2%	-0.1%	0.0%	203	203	74%	81%
Japan	-0.1%	-0.4%	-0.3%	127	118	67%	75%
Non-OECD	**1.2%**	**0.9%**	**1.0%**	**5 606**	**7 183**	**45%**	**57%**
E. Europe/Eurasia	0.1%	-0.1%	0.0%	335	331	63%	70%
Russia	-0.1%	-0.4%	-0.3%	142	133	73%	78%
Asia	0.9%	0.6%	0.7%	3 583	4 271	39%	53%
China	0.3%	0.0%	0.1%	1 345	1 387	47%	65%
India	1.3%	0.9%	1.0%	1 171	1 511	30%	43%
Middle East	1.8%	1.4%	1.6%	199	293	67%	74%
Iraq	3.0%	2.5%	2.7%	30	58	66%	71%
Africa	2.3%	2.0%	2.1%	1 032	1 730	40%	53%
Latin America	1.0%	0.7%	0.8%	456	558	80%	86%
Brazil	0.8%	0.4%	0.5%	195	224	87%	92%
World	**1.1%**	**0.8%**	**0.9%**	**6 843**	**8 556**	**51%**	**61%**
European Union	0.2%	0.1%	0.1%	502	518	74%	81%

* The assumed compound average annual growth rates are the same for all scenarios presented in this *Outlook*.

Sources: UNPD and World Bank databases; IEA databases.

Almost all of the increase in global population is expected to occur in non-OECD countries, mainly in Asia and Africa. India overtakes China soon after 2025 to become the most populous country and its population exceeds 1.5 billion in 2035. Africa sees the fastest rate of growth, its population increasing over the *Outlook* period by more than the current population of Europe. Russia is the only major non-OECD country that experiences a decline in its population. The population of the OECD increases by 0.4% per year on average over 2010-2035. The share of the world's population living in urban areas continues to increase, from 51% in 2010 to 61% in 2035, with implications for the amount and type of energy demanded. People living in urban areas in the developing world typically have higher incomes and better access to energy services, and therefore use more energy. This income effect usually outweighs energy efficiency gains that come from higher density settlements in urban areas. By contrast, city and rural residents in developed countries tend to enjoy similar levels of energy service.

Energy prices

Price remains an important determinant of energy trends. Actual prices paid by energy consumers affect the amount of each fuel they choose to consume and their choice of technology and equipment used to provide a particular energy service, while the price received by producers affects their production and investment decisions. Our assumptions differ across the scenarios, reflecting the impact of government policies on the demand and supply of each fuel.[3] The assumed paths for each form of energy reflect our judgment of the prices that would be needed to encourage sufficient investment in supply to meet projected demand over the *Outlook* period. Although the price paths follow smooth trends, prices in reality may be expected to deviate from these assumed trends – widely at times – in response to economic, energy market or geopolitical developments. For example, if the exceptionally challenging economic environment at present leads to a further deterioration in global growth, this could prompt a slump in energy demand that might temporarily drive prices well below the levels we assume.

Our assumptions about international fossil fuel prices (summarised in Table 1.4) are used to derive average retail prices in end-uses, power generation and other transformation sectors in each region. These end-use prices take into account local market conditions, including taxes, excise duties, CO_2 emissions penalties and pricing, as well as any subsidies for fossil fuels and/or renewables. In all scenarios, the rates of value-added taxes and excise duties on fuels are assumed to remain unchanged throughout the projection period, except where future tax changes have already been adopted or are planned. In the 450 Scenario, administrative arrangements (price controls or higher taxes) are assumed to be put in place to keep end-user prices for oil-based transport fuels at a level similar to those in the Current Policies Scenario. This assumption ensures that the lower international prices that result from policy action over and above what is assumed in the Current Policies Scenario do not lead to a rebound in transport demand via lower end-user prices.

3. Energy prices are an exogenous (*i.e.* external) determinant of energy demand and supply in the World Energy Model. However, the model is run in an iterative manner, adjusting prices so as to ensure that demand and supply are in balance in each year of the projection period.

Table 1.4 ▷ Fossil-fuel import price assumptions by scenario (dollars per unit)

			New Policies Scenario					Current Policies Scenario					450 Scenario				
	Unit	2011	2015	2020	2025	2030	2035	2015	2020	2025	2030	2035	2015	2020	2025	2030	2035
Real terms (2011 prices)																	
IEA crude oil imports	barrel	107.6	116.0	119.5	121.9	123.6	125.0	118.4	128.3	135.7	141.1	145.0	115.3	113.3	109.1	104.7	100.0
Natural gas																	
United States	MBtu	4.1	4.6	5.4	6.3	7.1	8.0	4.6	5.5	6.4	7.2	8.0	4.4	5.5	6.9	7.6	7.6
Europe imports	MBtu	9.6	11.0	11.5	11.9	12.2	12.5	11.2	12.1	12.9	13.4	13.7	10.9	10.8	10.4	10.0	9.6
Japan imports	MBtu	14.8	15.0	14.3	14.5	14.7	14.8	15.3	14.7	15.2	15.6	16.0	14.9	13.5	12.9	12.5	12.2
OECD steam coal imports	tonne	123.4	108.5	112.0	113.0	114.0	115.0	110.0	115.0	119.2	122.5	125.0	105.3	97.5	89.0	78.0	70.0
Nominal terms																	
IEA crude oil imports	barrel	107.6	127.0	146.7	167.6	190.4	215.7	129.7	157.4	186.6	217.4	250.3	126.3	139.0	150.0	161.2	172.6
Natural gas																	
United States	MBtu	4.1	5.0	6.7	8.7	11.0	13.8	5.0	6.7	8.8	11.1	13.8	4.8	6.7	9.5	11.7	13.2
Europe imports	MBtu	9.6	12.1	14.1	16.4	18.8	21.6	12.3	14.9	17.7	20.6	23.6	11.9	13.2	14.3	15.4	16.6
Japan imports	MBtu	14.8	16.4	17.5	19.9	22.6	25.5	16.8	18.1	20.9	24.0	27.6	16.3	16.6	17.7	19.3	21.1
OECD steam coal imports	tonne	123.4	118.8	137.4	155.4	175.6	198.5	120.5	141.1	163.8	188.7	215.7	115.3	119.6	122.4	120.2	120.8

Notes: Gas prices are weighted averages expressed on a gross calorific-value basis. All prices are for bulk supplies exclusive of tax. The US price reflects the wholesale price prevailing on the domestic market. Nominal prices assume inflation of 2.3% per year from 2011.

In recent years, momentum has been building to reform fossil-fuel subsidies. In September 2009, G-20 leaders gathered at the Pittsburgh Summit committed to "rationalize and phase out over the medium term inefficient fossil-fuel subsidies that encourage wasteful consumption". In November 2009, APEC leaders meeting in Singapore made a similar pledge, thereby broadening the international commitment to reform. Our assumptions concerning the phase-out of fossil-fuel subsidies vary by scenario. In the Current Policies Scenario, these subsidies are assumed to be phased out only where explicit policies are already in place. In the New Policies Scenario, they are assumed to be phased out by 2020 (at the latest) in all net energy-importing countries and more gradually in those exporting countries that have announced plans to do so. In the 450 Scenario, fossil-fuel subsidies are phased out by 2035 (at the latest) in all regions except the Middle East, where they are reduced to a maximum subsidisation rate of 20% by 2035.

Oil prices

Since trading at lows of around $30/barrel during the height of the financial crisis in late 2008, oil prices have followed a general upward trend with some periods of sharp fluctuation. By early October 2012, prices for benchmark Brent and West Texas Intermediate futures were trading at around $115/barrel and $93/barrel, respectively. In the New Policies Scenario, the average IEA crude oil import price – a proxy for international oil prices – rises to $120/barrel (in year-2011 dollars) in 2020 and $125/barrel in 2035 (Figure 1.1). This rising trend reflects the mounting cost of producing oil from new sources, as existing fields are depleted, in order to satisfy increasing demand. In the Current Policies Scenario, substantially higher prices are needed to balance supply with the faster growth in demand, reaching $145/barrel in 2035. In the 450 Scenario, lower oil demand means there is less need to develop oil from costly fields higher up the supply curve in non-OPEC countries. As a result, the oil price is assumed to level off at about $115/barrel by 2015 and then decline gradually to about $100/barrel by 2035.

Figure 1.1 ▷ Average IEA crude oil price

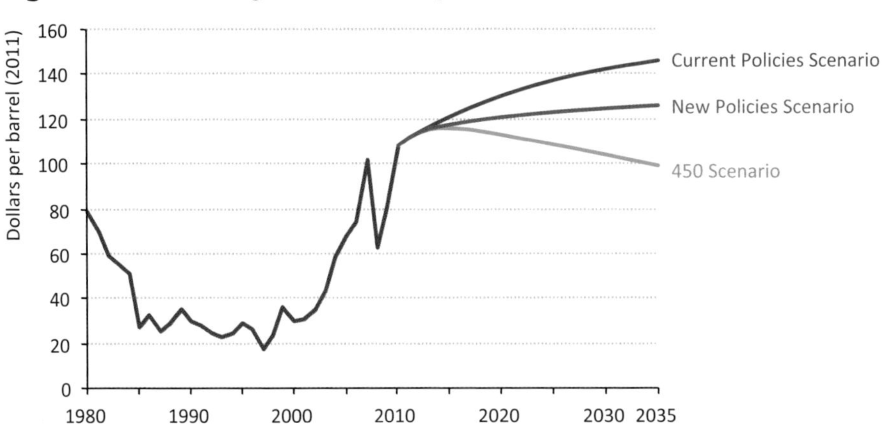

Note: In the 450 Scenario, administrative arrangements are assumed to be put in place to keep end-user prices for oil-based transport fuels at a level similar to the Current Policies Scenario.

Natural gas prices

Natural gas prices in North America, the United Kingdom and, to a somewhat lesser extent, Australia, are established through hub-based pricing, meaning they move in line with local supply and demand. Traditionally, most gas in continental Europe has been traded under long-term contracts with oil-price indexation, but that is changing with a growing share of prices now being set by gas-to-gas competition. Oil-price indexation remains the predominant pricing mechanism in Japan and Korea. Outside the OECD, a wide range of different natural gas pricing mechanisms is in use and in some cases prices are subsidised, for example in the Middle East and Russia.

Differences in pricing mechanisms, limited arbitrage options, the cost of transport between regions and local gas market conditions mean that prices across different regional markets can diverge markedly. At their lowest levels in 2012, natural gas prices in the United States briefly dipped below $2 per million British thermal units (MBtu), which at the time were around one-fifth of import prices in Europe and one-eighth of import prices in Japan. Significant price differentials across these main markets have reinvigorated debate about how quickly gas prices will move away from oil indexation and whether this will result in lower gas prices (relative to oil) and convergence of gas prices globally.

Figure 1.2 ▷ Ratio of average natural gas and coal prices to crude oil prices in the New Policies Scenario

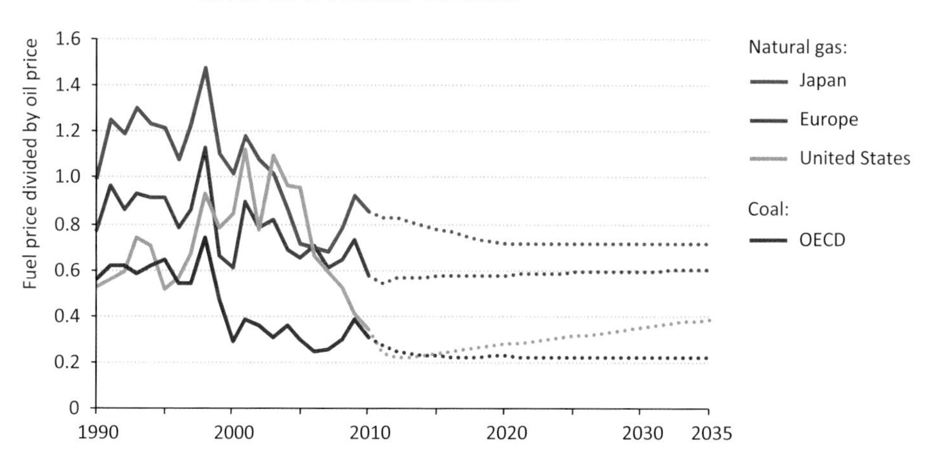

Note: Calculated on an energy-equivalent basis.

The outlook for natural gas prices depends both on demand-side factors, notably competition between natural gas and coal in the power sector, and supply-side factors. The gas prices assumed in this year's *Outlook* point to a narrowing of price differentials between the main regional markets (Figure 1.2). Prices in North America are assumed to remain the lowest, thanks to abundant supplies of relatively low-cost unconventional gas. Real prices there are expected to rise in absolute terms and relative to the other two regions as demand grows in response to lower prices, costs rise and export capacity is built. In Europe, the move towards hub-based pricing helps to moderate gas price increases, but this effect is offset by growing reliance on gas imports from more distant sources, which are likely to cost more to transport. In Asia, increased reliance on local supplies of unconventional gas, notably in China, and on spot purchases of LNG exert some downward pressure on the import prices of LNG and piped gas under long-term contracts. Globally, the gas prices assumed broadly follow the trend in oil prices, though the ratio of the gas price to the average crude oil price remains well below historical averages. Natural gas prices vary across the scenarios, reflecting the different assumptions about policy intervention to curb growth in energy demand and CO_2 emissions. In the New Policies Scenario, prices reach $12.5/MBtu (in year-2011 dollars) in Europe, $14.8/MBtu in Japan and $8.0/MBtu in the United States by 2035.

Steam coal prices

Only a relatively small fraction of coal consumption is traded internationally. While in certain regions international prices play an important role in setting domestic prices, in others, domestic prices are more closely related to indigenous production costs (such as in the United States) or are controlled by the government. The prices of internationally traded coal have become much more volatile in the last few years in response to rapid changes in coal production, use and trade. Following a collapse in prices from record highs in 2008, prices rose strongly from early-2009 through to mid-2011, falling back thereafter on weak demand, due to the recession, and increased availability on international markets.

The outlook for coal prices depends heavily on demand-side factors, the most important being competition between natural gas and coal in the power sector. The average OECD steam coal import price use is used as a general proxy for international prices and also drives our assumptions about coking coal and other coal qualities. In the New Policies Scenario, the average OECD steam coal import price is assumed to remain around $110 per tonne (in year-2011 dollars) to 2015 and then rise slowly to about $115 per tonne in 2035. Coal prices increase less in percentage terms than oil or gas prices, partly because coal-production costs are expected to rise less rapidly and because coal demand levels off by around 2025. Coal prices rise more quickly in the Current Policies Scenario, on stronger demand growth, but fall sharply in the 450 Scenario, reflecting the impact of much stronger policy action to reduce CO_2 emissions (which lowers coal demand).

CO_2 prices

The pricing of CO_2 emissions (either through cap-and-trade programmes or carbon taxes) affects demand for energy by altering the relative costs of using different fuels. Several countries have implemented emissions trading schemes to set CO_2 prices, while many others have schemes under development, some at an advanced stage of design. Other countries have introduced carbon taxes – taxes on fuels linked to related emissions – or are considering doing so. The EU Emissions Trading System (ETS) is currently the world's largest, covering all 27 member states, plus Norway, Iceland and Liechtenstein. CO_2 prices under the programme had been driven to record lows by mid-2012, primarily due to the dampening effect of the economic recession on energy demand. A debate followed on reform options to ensure that prices rise sufficiently to encourage investment in low-carbon technologies. Programmes that put a price on CO_2 emissions are also currently operating in New Zealand and in Australia (having started in July 2012, it will progressively be linked to the EU ETS between 2015-2018). Obligations in California's emissions trading scheme will take effect in 2013, phasing in to cover 85% of state-wide emissions in 2015. Korea has adopted a law (May 2012) to establish a scheme, with trading due to begin in 2015. In its 12th Five-Year Plan, China included plans to put a price on carbon and separately has announced plans to introduce city and provincial level pilot carbon emissions trading schemes in the near future.

The prevalence of carbon pricing and the level of CO_2 prices vary across the scenarios (Table 1.5):

- In the Current Policies Scenario, only the existing and the planned programmes described above are taken into consideration. The price of CO_2 is assumed to rise under each programme over the projection period; in Europe it increases from an average of \$19/tonne in 2011 to \$30/tonne (in year-2011 dollars) in 2020 and \$45/tonne in 2035 (the same price levels in 2020 and 2035 are also reached in Australia and New Zealand).

- In the New Policies Scenario, we assume a carbon price (starting at a low level) is introduced in China from 2020 covering all sectors, in line with the current Five-Year Plan.[4] In addition, we assume that from 2015 onwards all investment decisions in the power sector in the United States, Canada and Japan factor in an implicit or "shadow" price for carbon, to take account of the expectation that some form of action will be taken to penalise CO_2 emissions, although we do not assume that an explicit trading programme is introduced. In these countries, power projects get the go-ahead only if they remain profitable under the assumption that a carbon price is introduced. Given the uncertainty that surrounds future climate policies, many companies around the world already use such an approach as a means of ensuring that they are prepared for the possible introduction of a carbon tax or cap-and-trade programme.

- In the 450 Scenario, we assume that pricing of CO_2 emissions is eventually established in all OECD countries and that CO_2 prices in these markets begin to converge from 2025, reaching \$120/tonne in most OECD countries in 2035. Several major non-OECD countries are also assumed to put a price on CO_2 emissions. Although we assume no direct link between these systems before the end of the projection period, they all have access to offsets, which is expected to lead to some convergence of prices.

4. Recent IEA analysis explores the conditions needed for effective functioning of a CO_2 emissions trading system in China's electricity sector (Baron, *et al.*, 2012).

Table 1.5 ▷ **CO_2 price assumptions in selected regions and countries by scenario** ($2011 per tonne)

	Region	Sectors	2020	2030	2035
Current Policies Scenario	European Union	Power, industry and aviation	30	40	45
	Australia and New Zealand	All	30	40	45
	Korea	Power and industry	23	38	45
New Policies Scenario	European Union	Power, industry and aviation	30	40	45
	Australia and New Zealand	All	30	40	45
	Korea	Power and industry	23	38	45
	China	All	10	24	30
450 Scenario	United States and Canada	Power and industry	20	90	120
	European Union	Power, industry and aviation	45	95	120
	Japan	Power and industry	25	90	120
	Korea	Power and industry	35	90	120
	Australia and New Zealand	All	45	95	120
	China, Russia, Brazil and South Africa	Power and industry*	10	65	95

* All sectors in China.

Note: In the New Policies Scenario, a shadow price for CO_2 in the power sector is assumed to be adopted as of 2015 in the United States, Canada and Japan (starting at $15/tonne and rising to $35/tonne in 2035).

Technology

The types of energy technology that are developed and deployed, for application to energy supply and energy use, will affect investment decisions, the cost of supply of different forms of energy, and the level and composition of future energy demand. Our projections are, therefore, sensitive to assumptions about how quickly existing and new technologies will be deployed and fully used. These assumptions vary by fuel, end-use sector, location and scenario, and are based on our assessment of the current stage of technological development, how far the optimal scope for deployment will be realised and the potential for further gains, as well as our analysis of how effectively different policy assumptions will drive technological advances. While no breakthrough technologies are deployed in any of the scenarios, we assume that technologies that are in use today or are approaching the commercialisation phase will achieve further cost reductions as a result of increased learning and deployment. On the supply side, exploration and production techniques are also expected to improve, which could lower unit production costs and open up new opportunities for developing resources.

Some of the key uncertainties in our scenarios relate to prospects for carbon capture and storage (CCS), solar power, advanced biofuels, advanced vehicle technologies and nuclear power. At present, in liberalised markets that have relatively low gas and/or coal prices, new nuclear reactors are generally not an economically attractive option without some form of government support. Moreover, the accident at the Fukushima Daiichi nuclear power plant in March 2011 has thrown into doubt plans to expand capacity, particularly in OECD countries. Public resistance to nuclear power has grown and several countries have decided, based on their own assessments of nuclear energy's benefits and risks, to phase out their nuclear programmes. Although there is little evidence that events at Fukushima have changed policies in the countries that were expected to drive its expansion, there remains the possibility that tighter safety regulations could push up the cost of new reactors and thereby slow their deployment.

The pace of development of CCS technology remains highly uncertain. It could prove to be critical to the prospects for coal use in many regions, while in the longer term it is also likely to be critical to the prospects for natural gas and energy-intensive industries globally. The technology exists to capture CO_2 emissions from power stations and industrial plants, and to transport and permanently store the gas in geological formations, but only a handful of commercial-scale CCS projects are currently operating. Experience yet to be gained from the operation of further large demonstration projects, particularly in the power sector, will be critical to public acceptance, driving down costs and, hence, to the prospects for its widespread deployment. CCS technology is deployed only on a very limited scale in the power sector in the New Policies Scenario, but much more widely in the 450 Scenario thanks to stronger CO_2 price signals and faster cost reductions.

The resource potential for solar photovoltaics (PV) and concentrating solar power (CSP) is enormous. Their economic exploitation depends on further advances in technology, especially related to energy storage and cost reductions. Prospects for solar PV are particularly encouraging, as costs have declined dramatically in recent years, as experience and economies of scale have shifted costs down the learning curve. Details about cost assumptions for renewables-based generating technologies can be found in Chapter 7.

Electric vehicles and plug-in hybrids are commercially available, but their deployment has been limited. Around 40 000 electric vehicles and plug-in hybrids were sold worldwide in 2011, which represents less than 0.05% of the entire vehicle market. There is still a need for further progress on a number of technical issues, on overcoming drivers' reservations (such as driving range) and on reducing the costs of batteries. While a few plants producing advanced biofuels at demonstration scale already exist, widespread commercialisation hinges on further technological progress and additional policies to encourage investment. In the 450 Scenario, advanced biofuels, including those from lignocellulosic feedstocks, are assumed to reach commercialisation by around 2015 and, in the Current Policies Scenario, by 2025.

Energy trends to 2035

How will global energy markets evolve?

Highlights

- Despite their very different policy assumptions, a number of fundamental trends characterise each of the scenarios presented in this *Outlook*: rising incomes and population push energy needs higher; energy-market dynamics are increasingly determined by the emerging economies; fossil fuels continue to meet the bulk of the world's energy needs, from an ample resource base; and providing universal energy access to the world's poor remains an elusive goal.

- In the New Policies Scenario, our central scenario, global primary energy demand rises by over one-third in the period to 2035. Oil demand reaches 99.7 mb/d in 2035, up from 87.4 mb/d in 2011. Coal demand rises by 21% and natural gas by a remarkable 50%. Renewables are deployed rapidly, particularly in the power sector, where their share of generation increases from around 20% today to 31%. Growth in nuclear power is revised down relative to our previous projections, in large part due to policy moves following Fukushima Daiichi. These trends call for $37 trillion of investment in the world's energy supply infrastructure to 2035.

- China accounts for the largest share of the projected growth in global energy use, its demand rising by 60% by 2035, followed by India (where demand more than doubles) and the Middle East. OECD energy demand in 2035 is just 3% higher than in 2010, but there are dramatic shifts in its energy mix as fuel substitution sees the collective share of oil and coal drop by fifteen percentage points to 42%.

- A renaissance of the US energy sector is reshaping the world's energy landscape, with far-reaching implications. The United States currently relies on imports for around 20% of its primary energy demand, but rising production of oil, shale gas and bioenergy means that it becomes all but self-sufficient in net-terms by 2035. By contrast, most other energy importers become more dependent on imports.

- Growing water constraints are set to impose additional costs on the energy sector and in some cases threaten the viability of projects. In the New Policies Scenario, the volume of water consumed to produce energy increases by 85%, from 66 billion cubic metres (bcm) in 2010 to 120 bcm in 2035. This is more than twice the rate of growth of energy demand, driven by more water-intensive power generation and expanding output of water-thirsty biofuels.

- In the New Policies Scenario, energy-related CO_2 emissions rise from an estimated 31.2 Gt in 2011 to 37.0 Gt in 2035, pointing to a long-term average temperature increase of 3.6 °C. Phasing out fossil-fuel subsidies, which totalled $523 billion in 2011 (outweighing subsidies for renewables by a factor of almost six), would sharply curb growth in emissions.

Global energy trends by scenario

Short-term changes in energy demand and the composition of the fuel mix are largely a function of economic conditions, energy prices and the weather. But longer-term trends, as is shown by the stark contrasts across the main *WEO-2012* scenarios, can be significantly changed by the manner in which governments intervene in markets to tackle energy-related challenges. Nonetheless, across the scenarios several fundamental energy trends persist: rising incomes and population push energy needs higher; energy-market dynamics are increasingly determined by the emerging economies; fossil fuels meet most of the world's energy needs, from an ample resource base; and providing universal energy access to the world's poor remains an elusive goal.

The first of these fundamental trends is that the world's energy needs are set to rise. With the assumed expansion of the global economy of almost 140% (equivalent to some seven times the current economic output of the United States) and an increase of 1.7 billion in the world's population (more than the current population of China and the United States combined), more energy will be needed to satisfy growing demand for energy services, even though new policies and programmes are put in place to encourage energy savings. World primary energy demand increases by 35% between 2010 and 2035 in the New Policies Scenario, or 1.2% per year on average (Figure 2.1 and Table 2.1). This represents a sharp slowdown in the energy demand growth experienced over the past two decades, testament to the anticipated effect that already implemented and planned policies would have on energy markets. Not surprisingly, growth in energy demand is higher in the Current Policies Scenario, in which no change in policies is assumed, at 1.5% per year over 2010-2035. But even in the 450 Scenario, which involves a fundamental transformation of the energy sector, energy demand still increases between 2010 and 2035, albeit by a modest 0.6% per year.

Figure 2.1 ▷ World primary energy demand by scenario

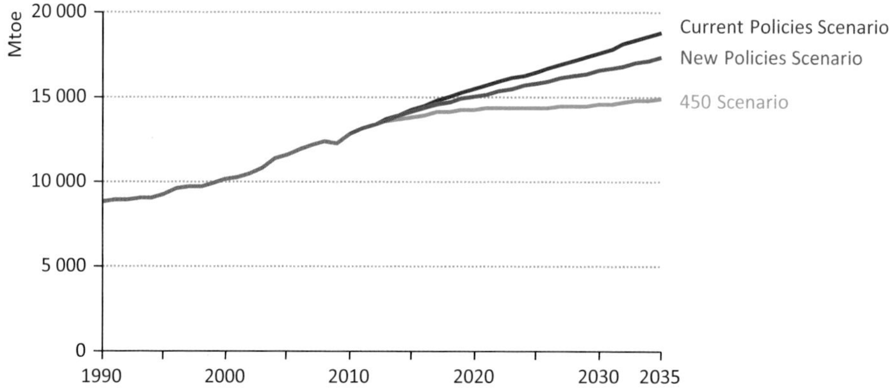

Mtoe
20 000
15 000
10 000
5 000
0
1990
2000
2010
2020
2030
2035
Current Policies Scenario
New Policies Scenario
450 Scenario

The second fundamental trend is that the dynamics of energy markets will be determined by emerging economies. The non-OECD share of global primary energy demand, which has already increased from 36% in 1973 to 55% in 2010, continues to rise in each of the scenarios. This reflects their faster rates of growth of population, economic activity, urbanisation and industrial production, as well as saturation effects that curb increases in demand for energy in the mature economies. The share of global energy demand in non-OECD countries in 2035 averages 64% across the scenarios.

Table 2.1 ▷ World primary energy demand and energy-related CO_2 emissions by scenario (Mtoe)

			New Policies		Current Policies		450 Scenario	
	2000	2010	2020	2035	2020	2035	2020	2035
Total	**10 097**	**12 730**	**14 922**	**17 197**	**15 332**	**18 676**	**14 176**	**14 793**
Coal	2 378	3 474	4 082	4 218	4 417	5 523	3 569	2 337
Oil	3 659	4 113	4 457	4 656	4 542	5 053	4 282	3 682
Gas	2 073	2 740	3 266	4 106	3 341	4 380	3 078	3 293
Nuclear	676	719	898	1 138	886	1 019	939	1 556
Hydro	226	295	388	488	377	460	401	539
Bioenergy*	1 027	1 277	1 532	1 881	1 504	1 741	1 568	2 235
Other renewables	60	112	299	710	265	501	340	1 151
Fossil fuel share in TPED	*80%*	*81%*	*79%*	*75%*	*80%*	*80%*	*77%*	*63%*
*Non-OECD share of TPED***	*45%*	*55%*	*60%*	*65%*	*61%*	*66%*	*60%*	*63%*
CO_2 emissions (Gt)	**23.7**	**30.2**	**34.6**	**37.0**	**36.3**	**44.1**	**31.4**	**22.1**

* Includes traditional and modern biomass uses. ** Excludes international bunkers.

Note: TPED = total primary energy demand; Mtoe = million tonnes of oil equivalent; Gt = gigatonnes.

The third trend is that fossil fuels – oil, coal and natural gas – will continue to meet most of the world's energy needs. Fossil fuels, which represented 81% of the primary fuel mix in 2010, remain the dominant sources of energy through 2035 in all scenarios, although their share of the mix in 2035 varies markedly. It is highest in the Current Policies Scenario (80%) and lowest in the 450 Scenario (63%). Among the fossil fuels, the biggest uncertainty with respect to future use relates to coal: demand increases by 59% between 2010 and 2035 in the Current Policies Scenario, while it declines sharply over the same period, after peaking before 2015, in the 450 Scenario. Despite the persistent demand for fossil fuels in all scenarios, demand for renewable sources of energy rises at a faster rate, giving them a considerably higher share of the energy mix in each case. Their share is highest in the 450 Scenario, reaching 27% in 2035 from 13% in 2010.

The fourth trend is the persistent failure to provide universal energy access to the world's poor. Our latest estimate is that currently almost 1.3 billion people lack access to electricity (around 20% of the global population), while 2.6 billion people rely on traditional biomass for cooking (see Chapter 18). More than 95% of these people are in either sub-Saharan Africa or developing Asia. Neither a business-as-usual approach nor the new commitments so far announced, even if implemented in full, will do nearly enough to achieve universal access to modern energy services within the timeframe of the *Outlook.* We project that

990 million people will still lack access to electricity in 2030 (the date of the proposed goal of universal access to modern energy services under the United Nations Sustainable Energy for All initiative) and 2.6 billion people will still rely on traditional biomass for cooking.

One trend that does vary dramatically across the three scenarios is the trajectory of carbon-dioxide (CO_2) emissions. Based on preliminary estimates, energy-related CO_2 emissions reached a record 31.2 gigatonnes (Gt) in 2011, representing by far the largest source (around 60%) of global greenhouse-gas emissions (measured on a CO_2-equivalent basis). Emissions continue to rise in the New Policies Scenario, putting the world on a path that is consistent with a long-term average global temperature increase of 3.6 °C above levels that prevailed at the start of the industrial era (see Chapter 8 for analysis of the linkages between energy and climate change) (Figure 2.2). As shown by the Current Policies Scenario, without these new policies, we are on track for a temperature increase of 5.3 °C. By design, the 450 Scenario is compatible with around a 50% chance of limiting the temperature rise to 2 °C.

Figure 2.2 ▷ **Global energy-related CO_2 emissions by scenario**

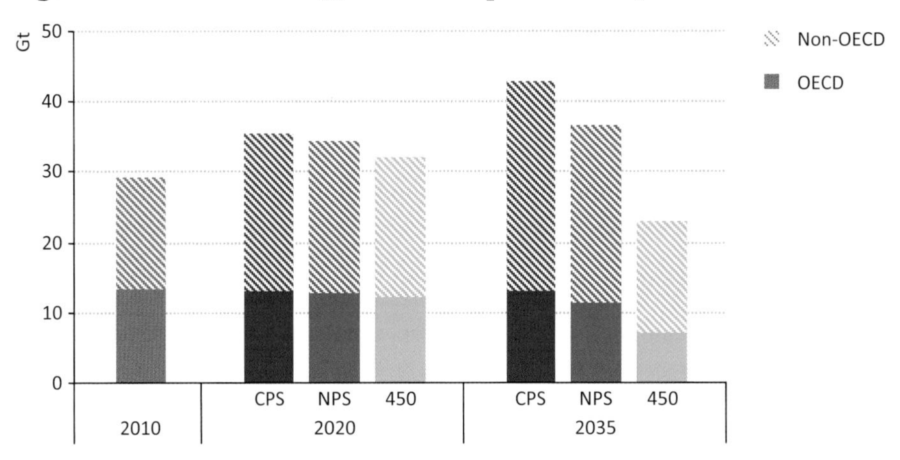

Note: NPS = New Policies Scenario; CPS = Current Policies Scenario; 450 = 450 Scenario.

Energy trends in the New Policies Scenario

Primary energy demand

The remainder of this chapter focuses on the results of the New Policies Scenario, to provide a clear picture of where currently planned policies, if implemented in a relatively cautious way, would take us.[1] Global energy demand in the New Policies Scenario is projected to increase by 35% between 2010 and 2035, rising from nearly 13 000 million tonnes of oil equivalent (Mtoe) to around 17 000 Mtoe (Table 2.2). Growth slows from an average of 1.6% per year in the period 2010-2020 to 1.0% per year in 2020-2035, as measures

1. Annex A provides detailed projections of energy demand by fuel, sector and region for each of the main *WEO-2012* scenarios.

introduced to meet energy security and climate objectives take effect and as economic and population growth rates drop in the key emerging economies.

Table 2.2 ▷ World primary energy demand by fuel in the New Policies Scenario (Mtoe)

	1990	2010	2015	2020	2030	2035	2010-35*
Coal	2 231	3 474	3 945	4 082	4 180	4 218	0.8%
Oil	3 230	4 113	4 352	4 457	4 578	4 656	0.5%
Gas	1 668	2 740	2 993	3 266	3 820	4 106	1.6%
Nuclear	526	719	751	898	1 073	1 138	1.9%
Hydro	184	295	340	388	458	488	2.0%
Bioenergy**	903	1 277	1 408	1 532	1 755	1 881	1.6%
Other renewables	36	112	200	299	554	710	7.7%
Total	8 779	12 730	13 989	14 922	16 417	17 197	1.2%

*Compound average annual growth rate. ** Includes traditional and modern biomass uses.

Fossil fuels account for 59% of the overall increase in demand, remaining the principal sources of energy worldwide (Figure 2.3). Global oil demand reaches 94.2 million barrels per day (mb/d) in 2020 and 99.7 mb/d in 2035 – up from 87.4 mb/d in 2011. Demand growth is slowed by planned policies, including efficiency measures, and by higher prices: the crude oil price rises to $125/barrel (in year-2011 dollars) in 2035. Oil remains the most important fuel in the primary energy mix, but its share drops to 27% in 2035, from around 32% today. Coal met 45% of the rise in global energy demand between 2001 and 2011, growing faster even than total renewables. We project coal demand to continue to rise in the medium term, with growth slowing from around 2020, reaching around 6 000 million tonnes of coal equivalent (Mtce) in 2035, 21% higher than in 2010. Natural gas demand bounced back by a remarkable 7.8% in 2010, after declining sharply in 2009, as a result of the global recession. We project demand to rise from 3.3 trillion cubic metres (tcm) in 2010 to 5.0 tcm in 2035, an increase of 50%. Its share of the global energy mix rises from 22% in 2010 to 24% in 2035, all but catching up with coal.

In 2035, the share of renewables (including traditional biomass) in world primary energy demand reaches 18%, from 13% in 2010. This rapid increase is underpinned by incentives to overcome market barriers, falling technology costs, rising fossil fuel prices and in some cases carbon pricing (see Chapter 1). Most of the growth occurs in the power sector, where their share in total generation grows from 20% to 31%, a near tripling in actual generation. Biofuel use increases from 1.3 mb/d in 2010 to 4.5 mb/d in 2035.

We project nuclear power to supply 12% of the world's electricity in 2035, which is fairly close to today's level. Installed nuclear capacity rises to just over 580 gigawatts (GW) in 2035, with non-OECD countries accounting for 94% of the almost 200 GW net increase. China sees the biggest increase in capacity, with a rise from almost 12 GW in 2011 to 128 GW in 2035, followed by Korea, India and Russia. The future of nuclear power became more uncertain in the wake of the accident at the Fukushima Daiichi nuclear power station in Japan in March 2011. The projections for nuclear power in last year's *Outlook* already took account of immediate reactions, including accelerated nuclear phase-outs in Germany and Switzerland and an abandonment of steps towards building new plants in Italy; but more recent developments have also been taken into account this year, most notably Japan's new Innovative Energy and Environmental Strategy, which includes a goal of reducing reliance on nuclear energy, and developments in inter-fuel competition. These factors contributed to a downward revision in global nuclear generation in 2035 of nearly 300 terawatt-hours (TWh), or about 6% compared to *WEO-2011*, despite the maintained commitment to nuclear power in parts of the world, especially outside of the OECD.

Figure 2.3 ▷ World primary energy demand by fuel in the New Policies Scenario

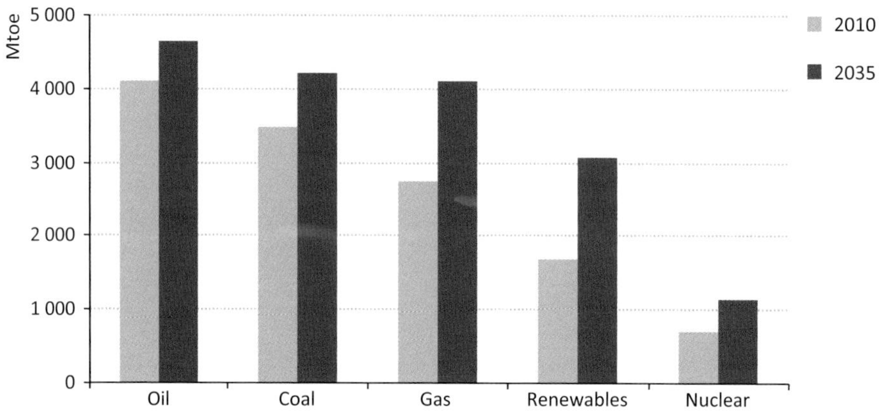

Energy intensity and per-capita consumption

In the two years that followed the economic crisis of 2008, global energy demand grew at a faster rate than the global economy. This disrupted the broad trend of declining global energy intensity[2], measured as energy demand per dollar of GDP, over the last several decades. This discontinuity can be attributed to a number of factors: the financial crisis delayed investment in more efficient buildings, vehicles and appliances; emerging economies, where energy intensity is typically higher, were less affected by the global recession; energy-intensive infrastructure projects received a boost from economic stimulus programmes; some regions experienced unusually cold winters, which pushed up demand for heating; with the downturn, many industrial facilities operated at less than optimal capacity; and lower energy prices in the immediate wake of the downturn reduced incentives to conserve energy. Preliminary data point to a 0.6% improvement in energy intensity in 2011, indicating that the long-running trend may have been restored.

2. See Chapters 9-12 for detailed coverage of future prospects for energy intensity, an aggregate but imperfect measure of energy efficiency.

In the New Policies Scenario, global energy intensity (based on GDP at market exchange rates [MER]) falls by 1.8% per year between 2010 and 2035, underpinned by improvements in energy efficiency, structural shifts within the economy in favour of less energy-intensive activities and the assumed rise in energy prices. Reductions are seen in all regions, but are more pronounced in the emerging economies, where there remains greater scope for gains (Figure 2.4). Between 2010 and 2035, energy intensity declines by an average of 37% and 49% in OECD and non-OECD countries respectively. Yet average energy intensity in non-OCED countries in 2035 of 0.16 tonnes of oil equivalent (toe) per thousand dollars of GDP is still more than twice the OECD level. The most dramatic reduction is in China, with the energy intensity of its economy dropping by 3.6% annually on average, meaning China requires barely two-fifths as much energy in 2035, compared with 2010, to generate a unit of GDP. China's 12th Five-Year Plan includes a target to cut energy consumption per unit of GDP by 16% between 2010 and 2015: we project that target to be met.

Figure 2.4 ▷ World primary energy demand per unit of GDP and per capita in the New Policies Scenario in selected regions and countries

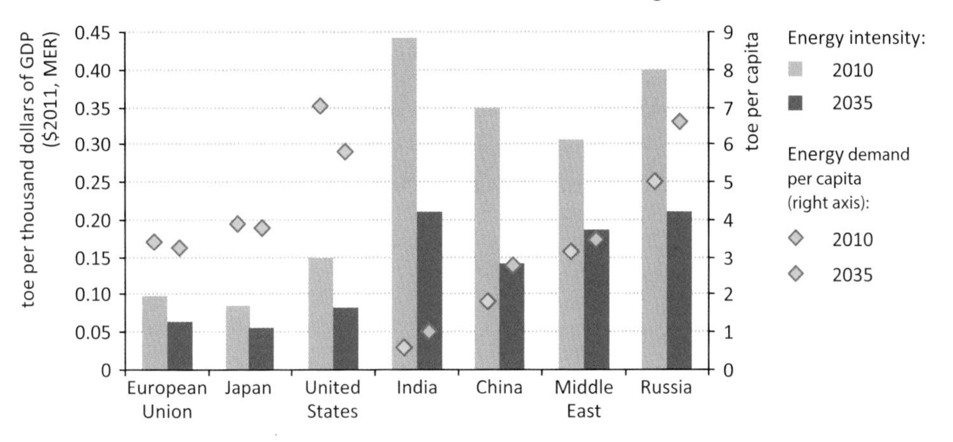

In contrast to energy intensity, per-capita energy demand tends to be much higher in developed countries, as it is closely linked to a country's stage of economic development, even though other factors, such as lifestyle and climate, are also important. For the world as a whole, it was 1.9 toe per person in 2010, 4.4 toe in the OECD and 1.2 toe elsewhere. In the New Policies Scenario, a significant gap remains between per-capita usage in OECD and non-OECD regions in 2035, although there is a convergent trend. Average per-capita energy demand in the OECD drops by 7% between 2010 and 2035, while it increases by 25% in the rest of the world. It declines most rapidly in the United States, from 7.0 toe per person in 2010 to 5.8 toe per person in 2035. Among emerging economies, the fastest increase occurs in India (even though the rate of growth is moderated by switching from traditional biomass to more efficient modern fuels), reaching 1.0 toe per person in 2035, still less than one-quarter of the current OECD average. China also sees a large increase, rising to 2.8 toe per person in 2035.

Regional demand

All but 4% of the projected rise in world primary energy demand between 2010 and 2035 comes from countries outside of the OECD, reflecting faster rates of growth in economic activity, industrial output and population. Nonetheless, the average annual rate of growth in non-OECD energy demand slows through the *Outlook* period, from 2.6% in 2010-2020 to 1.4% in 2020-2035. The 60% increase in non-OECD energy demand over the period results in its share of world demand rising from 55% to 65% (Figure 2.5).

Figure 2.5 ▷ Non-OECD primary energy demand by region in the New Policies Scenario

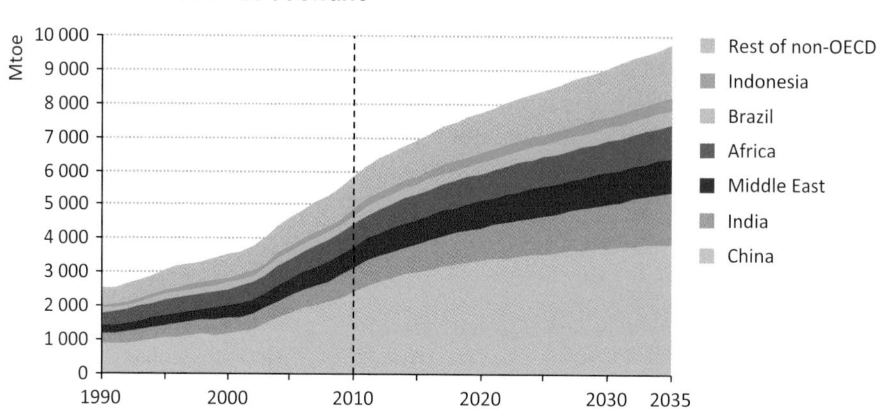

China is set to play a decisive role in global energy markets over the *Outlook* period. In the New Policies Scenario, it makes the biggest contribution (33%) to the growth in global energy use, its demand rising by 60% between 2010 and 2035, with growth decelerating over time (Table 2.3). China consolidates its position as the world's largest energy consumer and, by 2035, its use is 77% higher than the second-placed United States (though China's use is 52% lower on a per-capita basis). China makes a major contribution to the increase in primary demand for all fuels: oil (54%), coal (49%), natural gas (27%), nuclear power (57%) and renewables (14%). Its reliance on coal declines from 66% of the country's primary energy use in 2010 to 51% in 2035, driven by the assumed continuation of provisions in the 12th Five-Year Plan to increase the proportion of natural gas and non-fossil energy in the mix.

Energy use in India, which recently overtook Russia to become the world's third-largest energy consumer, more than doubles over the *Outlook* period. India makes the second-largest contribution to the increase in global demand after China, although, to put the two in perspective, the increase in India is just 57% of that in China. Despite the difficulties India is currently facing in opening new coal mines, coal remains the dominant fuel in its primary energy mix, with a share that remains at around 43%. Efforts to diversify the fuel mix achieve growing shares for nuclear power and natural gas. In line with ongoing programmes to improve rural electrification and encourage the use of efficient cook stoves, the share of the population with access to electricity rises from 75% to 90% (see Chapter 18). Given that so much of the projected growth in global energy demand comes from China and India, our analysis is highly sensitive to the assumptions made for the parameters that will shape those countries' energy consumption patterns, particularly the level and composition of economic growth, and developments in policies related to energy security, energy diversity and the environment (Figure 2.6).

Figure 2.6 ▷ Share of China and India in net increase in global primary energy demand by fuel and CO_2 emissions in the New Policies Scenario, 2010-2035

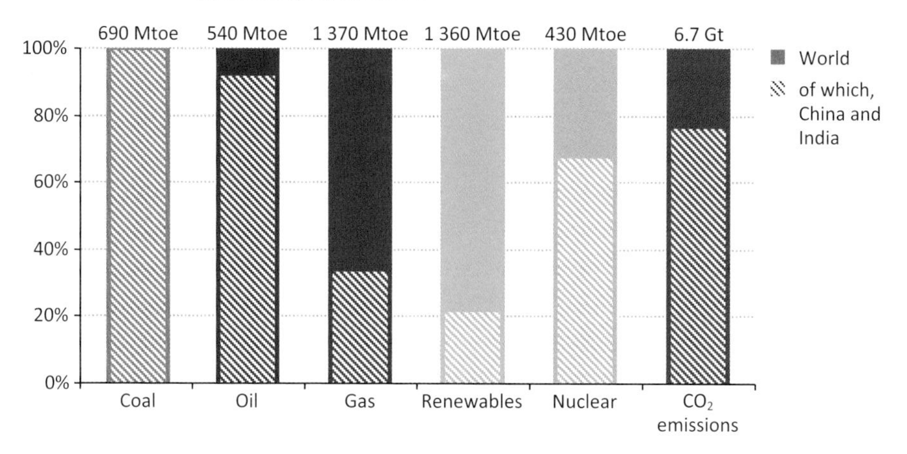

Russia remains the world's fourth-largest energy user throughout the *Outlook* period. In the New Policies Scenario, its total primary energy demand expands by 23%, at an average pace of 0.8% per year. Russia remains among the world's highest per-capita energy consumers, at 6.6 toe in 2035. This is a function of the cold climate, heavy reliance on energy-intensive activities and relatively inefficient (although improving) energy production and consumption practices. Fossil fuels, particularly natural gas, remain dominant in Russia's domestic energy mix, with a share of 86% in 2035, down from 91% in 2010 as nuclear and renewables make inroads.

Energy demand growth in the Middle East remains strong, underpinned by an expanding population, rising incomes, rapid development and, in many cases, heavy subsidies for fossil fuels. The region's energy consumption rises by almost two-thirds in the New Policies Scenario, growing at 1.9% per year on average between 2010 and 2035. Iraq sees particularly rapid growth, with its energy demand increasing more than four-fold over the period (see Chapter 15). Per-capita consumption in the Middle East is almost 85% of the OECD average in 2035, up from 72% in 2010. The domestic energy mix in the region remains heavily dependent on oil and natural gas.

Africa currently accommodates 15% of the world's population, but accounts for just 5% of global primary energy demand. In the New Policies Scenario, these statistics are even more unbalanced by 2035: Africa's share of the global population rises to one-fifth but its share of global energy demand is essentially unchanged. The continent's total primary energy demand increases at 1.4% per year over the period, with the energy mix remaining heavily dependent on traditional biomass, although natural gas grows in importance. We project that more than 650 million people in Africa, mainly in sub-Saharan Africa, will still lack reliable access to electricity in 2030 and around 880 million will be without access to clean cooking facilities.

Table 2.3 ▷ World primary energy demand by region in the New Policies Scenario (Mtoe)

	1990	2000	2010	2015	2020	2030	2035	2010-35*
OECD	**4 521**	**5 292**	**5 404**	**5 465**	**5 530**	**5 553**	**5 579**	**0.1%**
Americas	2 260	2 695	2 677	2 751	2 792	2 795	2 806	0.2%
United States	1 915	2 270	2 214	2 246	2 260	2 206	2 187	0.0%
Europe	1 630	1 765	1 837	1 817	1 829	1 835	1 847	0.0%
Asia Oceania	631	832	890	897	909	923	927	0.2%
Japan	439	519	497	472	465	450	447	-0.4%
Non-OECD	**4 058**	**4 536**	**6 972**	**8 158**	**9 001**	**10 424**	**11 147**	**1.9%**
E. Europe/Eurasia	2 617	999	1 137	1 209	1 250	1 349	1 407	0.9%
Russia	880	620	710	750	774	837	875	0.8%
Asia	1 589	2 248	3 936	4 808	5 400	6 351	6 839	2.2%
China	881	1 196	2 416	3 020	3 359	3 742	3 872	1.9%
India	317	457	691	837	974	1 300	1 516	3.2%
Middle East	210	365	624	715	792	935	1 012	1.9%
Iraq	21	28	38	77	113	145	160	5.9%
Africa	388	496	690	750	819	932	984	1.4%
Latin America	331	429	586	675	740	856	905	1.8%
Brazil	138	184	262	309	346	413	444	2.1%
World**	**8 779**	**10 097**	**12 730**	**13 989**	**14 922**	**16 417**	**17 197**	**1.2%**
European Union	1 633	1 683	1 713	1 681	1 678	1 667	1 670	-0.1%

*Compound average annual growth rate. ** Includes bunkers.

OECD energy demand in 2035 is just 3% higher than in 2010, but fuel substitution leads to marked shifts in the primary energy mix (Figure 2.7). The share of fossil fuels declines from 81% in 2010 to 70% in 2035. The shift is even more pronounced for oil and coal, their collective share dropping by fifteen percentage points to 42%. OECD oil demand is 33 mb/d in 2035, down by around 21% from 42 mb/d in 2011, while coal demand falls by 24%. By contrast, natural gas and renewables both experience rising demand and their shares of the OECD's primary energy mix increase through to 2035. The biggest change is for renewables, which make up one-third of OECD power generation in 2035, compared with 18% in 2010. Following recent moves in Japan and parts of Europe to reduce reliance on nuclear power, the share of nuclear power in OECD power generation falls from 21% in 2010 to 19% in 2035. This represents a slight increase in absolute terms, as growth in North America and Korea offsets reductions elsewhere.

SPOTLIGHT

Are we on track to achieve the targets in the UN Sustainable Energy for All initiative?

Despite rising energy use across the world, a significant share of the world's population still remains without access to electricity or clean cooking facilities, representing a serious barrier to social and economic development (see Chapter 18). Fortunately, international concern about the issue is growing. The UN General Assembly declared 2012 to be the "International Year of Sustainable Energy for All" and the UN Secretary General launched his Sustainable Energy for All (SE4All) initiative, which includes three targets to be met by 2030: (i) double the rate of improvement in energy efficiency; (ii) double the share of renewable energy in the global energy mix; and (iii) ensure universal access to modern energy services.

The IEA and World Bank are co-leading a project to develop metrics against which progress towards the SE4All targets will be measured. The results of this work are not yet available, but do the projections in this *Outlook* provide optimism that we are on track to meet these targets? The answer on all counts is no – greater policy efforts are needed.

Based on recent years, during which energy intensity – a crude proxy for energy efficiency – has increased rather than decreased, it appears that much needs to be done to meet the energy efficiency target. However, it is not impossible: our Efficient World Scenario (presented in Chapter 10) shows that by applying known and economically viable technical solutions, energy efficiency could be improved at 2.6 times the average annual rate of the last 25 years.

Renewables (excluding traditional biomass) made up 13% of global primary energy demand in 2010. This share increases sharply in each of our scenarios, but it is only in the 450 Scenario – in which the share of renewables reaches 23% by 2030 – that the second SE4All target is somewhat close to being achieved. These projections rely on falling technology costs, rising fossil fuel prices, carbon pricing and continued subsidies. However, if traditional biomass is included in the definition of renewables, it becomes even more challenging.

Our latest estimate is that nearly 1.3 billion people currently lack access to electricity and 2.6 billion people rely on traditional biomass for cooking. We project that these numbers will have changed little by 2030, even though modern energy services will have been extended to many more people. To meet the third of the SE4All targets, we estimate that cumulative investment of almost $1 trillion is required – an average of almost $50 billion per year, more than five times the $9 billion of investments in 2009 (IEA, 2011). The target of universal modern energy access by 2030 is achievable and would have only a minor impact on global energy demand and carbon dioxide emissions, but much more funding will be needed.

Figure 2.7 ▷ OECD primary energy demand by fuel in the New Policies Scenario

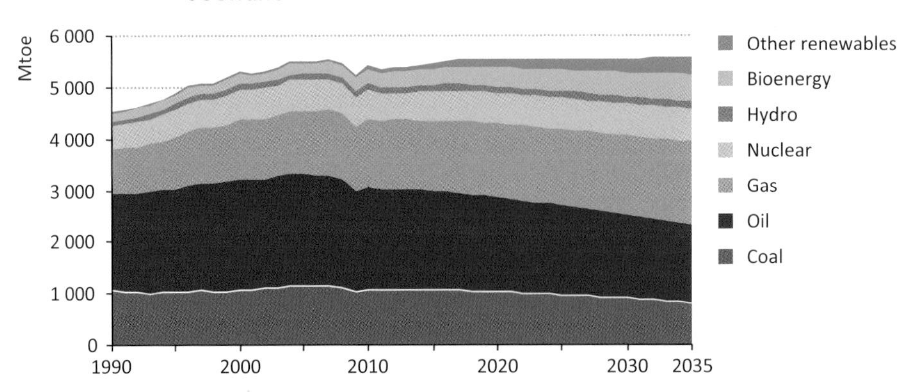

Primary energy consumption in the United States is slightly lower in 2035 than in 2010. Oil demand ends the period 5.0 mb/d lower than 2011, at 12.6 mb/d, with the bulk of the savings arising in the transport sector, driven by improvements in fuel economy and increased use of ethanol and biodiesel. These projections are underpinned by an assumed extension of the US Corporate Average Fuel Economy (CAFE) standards. Coal use, which is constrained by relatively cheap gas and concerns over local air quality and greenhouse-gas emissions, is 17% lower in 2035 than in 2010. Natural gas demand increases by just 0.5% per year on average, largely because the favourable price differentials that currently exist between gas and other fuels are assumed to diminish over time. However, natural gas could play a much bigger role if these differentials persist, triggering more use in power generation, industry and also in transport. The share of renewables in total primary energy demand rises by 9.5 percentage points to 15%, thanks largely to the assumed extension of federal renewable electricity production tax credits (which are at present scheduled to expire at the end of 2012) and state renewable portfolio standards.

Primary energy demand in the European Union ends the *Outlook* period marginally lower than 2010. The shift in the EU energy mix is much more marked. Coal demand plunges by almost half, with most of the decline occurring in the power sector, where the economics of coal deteriorate as the price of CO_2 rises to $45/tonne in 2035 (compared with an average of around $19/tonne in 2011) and renewables are bolstered by sustained government support. Oil demand falls from 11.6 mb/d in 2011 to 8.7 mb/d in 2035 on improvements in fuel economy and growing biofuels use. Demand for natural gas rises, ending the projection period 15% higher than in 2010, with most of the growth in the power sector. Nuclear's share of total generation falls by six percentage points to 22%. The share of renewables in power generation rises from 21% in 2010 to 43% in 2035. Wind generates two times as much electricity as coal in 2035. The European Union has set targets to cut emissions of greenhouse gases by 20% (compared with 1990 levels), increase the share of renewables in the energy mix to 20% and cut energy consumption by 20%, all by 2020. In the New Policies Scenario, all but the third of these goals are met (see Chapter 9 for our analysis of prospects for EU energy efficiency).

Following the accident at the Fukushima Daiichi nuclear power plant, Japan undertook a full review of its national energy policy which led to the release in September 2012 of the Innovative Strategy for Energy and the Environment, which includes the goal of reducing reliance on nuclear energy. As not all of the details of the new strategy were available as this analysis was completed, we have assumed that most of the existing nuclear reactors will be brought back into service progressively and that their lifetimes will be limited compared with our previous estimates. Based on these and other assumptions, Japan's total primary energy demand in the New Policies Scenario declines by 10% between 2010 and 2035. Oil demand falls from 4.3 mb/d in 2011 to 3.1 mb/d in 2035, driven by efficiency gains in transport and the uptake of alternative vehicle technologies. Demand for coal drops sharply, while demand for natural gas increases. Renewables make up 14% of total primary energy supply in 2035, up from just 4% in 2010, under the impetus of a renewed policy push to diversify the energy mix. Nuclear power accounts for 20% of total generation in 2020 and 15% in 2035. Total nuclear generation in Japan falls from 288 TWh in 2010 to 174 TWh in 2035, which is 57% lower than projected in *WEO-2011*.

Sectoral demand

Inputs to the power sector to generate electricity accounted for 38% of global primary energy use in 2010, the single largest element of primary demand (Figure 2.8). In the New Policies Scenario, this share rises to 42% in 2035. Demand for electricity is pushed higher by population and economic growth, and by households and industries switching from traditional biomass, coal, oil and natural gas to electricity, for reasons of convenience, efficiency and practicality. The fuel mix within the power sector changes considerably, with low- and zero-carbon technologies becoming increasingly important. Fossil fuels account for 63% of total inputs to power generation in 2035, down from 75% in 2010. Coal's share declines the most although it still remains the dominant fuel of the power sector globally. The share of oil, already low at 6%, falls to just 2% in 2035. However, it still remains an important fuel in the power sector in some regions, notably the Middle East. By contrast to the other fossil fuels, the share of natural gas in power generation rises marginally on greater availability, lower prices and its relatively favourable operational and environmental attributes.

Total final consumption[3] (the sum of energy consumption in the various end-use sectors) grows at an average annual rate of 1.2% through to 2035. The buildings sector, which uses energy for heating, cooling, lighting, refrigeration and for powering electrical appliances, is currently the single largest final end-use consumer. The demand in this sector is projected to grow at an average annual rate of 1.0% through to 2035, an overall increase of 29%. The bulk of the growth is in non-OECD regions, in line with faster population growth rates, rapidly increasing markets for electrical appliances and less stringent building standards than in the OECD. Global growth in energy use in buildings is underpinned by a 52%

3. Total final consumption is the sum of consumption by industry, transport, buildings (including residential and services) and other (including agriculture and non-energy use). It excludes international marine and aviation bunkers, except at the world level where these are included in the transport sector.

Figure 2.8 ▷ **The global energy system, 2010** (Mtoe)

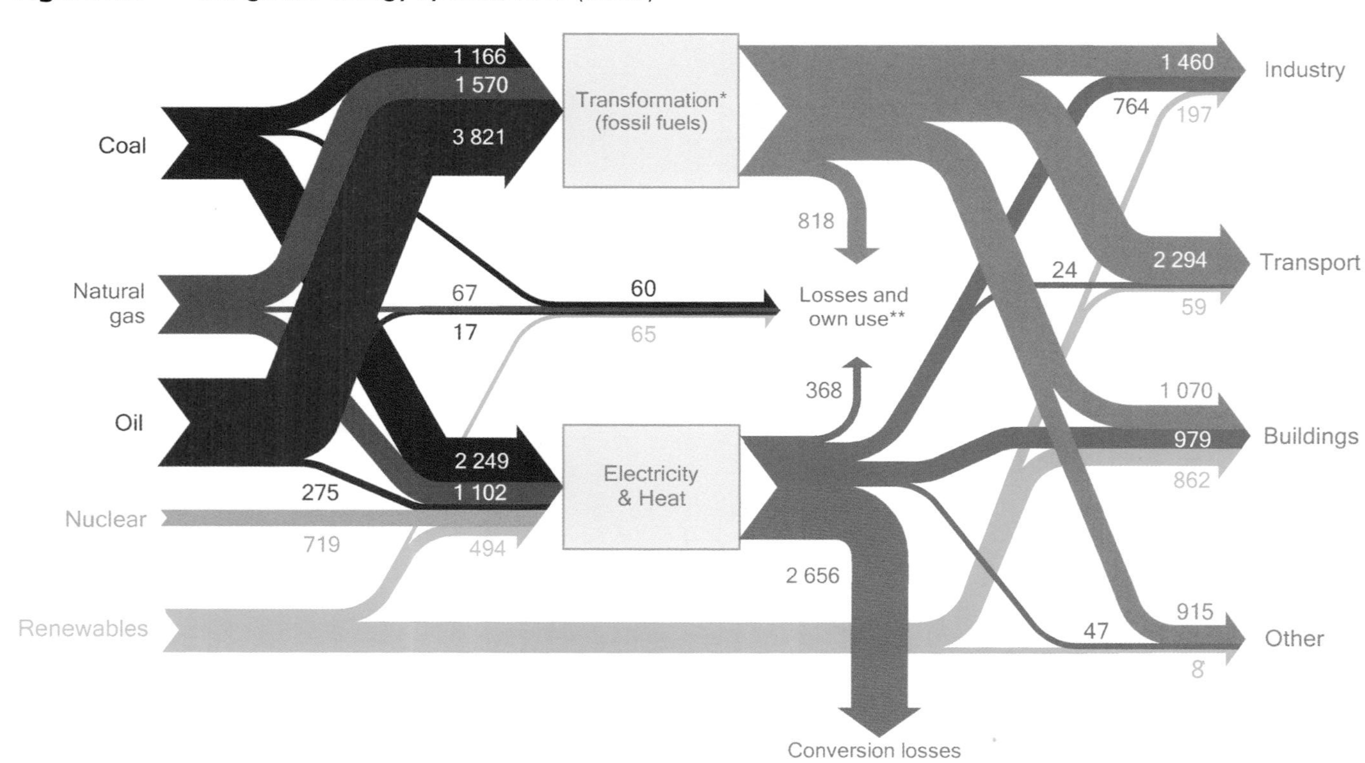

* Transformation of fossil fuels from primary energy into a form that can be used in the final consuming sectors. ** Includes losses and fuel consumed in oil and gas production, transformation losses and own use, generation lost or consumed in the process of electricity production, and transmission and distribution losses.

increase in residential floor space and a 116% increase in the value of services. Electricity's dominance of energy use in buildings grows, mainly at the expense of traditional biomass, which becomes a less important energy source for households in developing countries.

Energy demand in the industry sector grows faster than in any other sector, despite continuing efficiency gains and an eventual slowdown in the growth of industrial output. Within the sector, iron and steel remains the largest consumer globally until the end of the *Outlook* period when it is overtaken by chemicals. The share of electricity in the industrial fuel mix rises from 26% in 2010 to 32%, at the expense of coal and oil. Non-OECD countries account for 93% of the increase in industrial energy demand, with the fastest rates of growth occurring in India and Indonesia. Industrial energy demand in China overtakes that of the entire OECD before 2015, but growth slows towards the end of the *Outlook* period.

Global transport energy demand increases by 38% from 2010-2035, growth of 1.3% per year on average. The growth rate slows over the *Outlook* period, due to higher oil prices, efficiency improvements, vehicle saturation in mature markets and the reduction of subsidies in some emerging economies. A more-than-doubling of transport energy demand in non-OECD regions is only slightly offset by a fall of 13% in the OECD. Of all transport modes, road transport remains the largest consumer, accounting for around three-quarters of the sector's total energy demand in 2035, underpinned by a two-fold increase in the passenger vehicle fleet to 1.7 billion in 2035. Passenger vehicle ownership per 1 000 people in China climbs from around 40 in 2010 (compared with just four in 2000 and with close to 660 at present in the United States) to 310 in 2035 – a critical source of global oil demand.

Energy supply and trade

Energy resources

Taking into account our energy price assumptions and our expectations for advances in technology and extraction methods, we judge the world's endowment of energy resources to be sufficient to satisfy projected energy demand to 2035 and well beyond. Investors in energy projects are exposed to a wide array of risks, including geological, technical, regulatory, fiscal, market and geopolitical risks. As a result, harnessing the necessary investment, technology and skilled workforce is expected to be an ongoing challenge. At certain times, sectors and places, investment will undoubtedly fall short of what is needed (though there will also be occasions when the reverse occurs).

Fossil fuel resources remain plentiful (Table 2.4). Coal, in particular, is extremely abundant. Proven reserves of coal, essentially an inventory of what is currently economic to produce, are much greater than those of oil and gas combined, on an energy basis. They are sufficient to supply around 132 years of production at 2011 levels (BGR, 2011). Proven reserves of natural gas are also more than enough to meet our projected demand, totalling 232 tcm. Almost half of the world's proven natural gas reserves are located in just three countries – Russia, Iran and Qatar. Proven reserves of oil amount to 55 years at 2011 rates of production, with OPEC countries' reserves representing 71% of the total.

Table 2.4 ▷ Fossil-fuel reserves and resources by region and type, end-2011

	Coal* (billion tonnes)		Natural gas (tcm)		Oil (billion barrels)	
	Proven reserves	Recoverable resources	Proven reserves	Recoverable resources	Proven reserves	Recoverable resources
OECD	427	10 657	28	193	244	2 345
Non-OECD	576	10 551	205	597	1 450	3 526
World	1 004	21 208	232	790	1 694	5 871
Share of non-OECD	*57%*	*50%*	*88%*	*76%*	*86%*	*60%*
R/P ratio (years)	132	2 780	71	241	55	189

* For coal, the data are for 2010.

Notes: R/P ratio = Reserves-to-production ratio based on 2011 levels of production. Resources are remaining technically recoverable resources.

Sources: BGR (2011); O&GJ (2011); USGS (2000); USGS (2012a); USGS (2012b); IEA databases and analysis.

Ultimately recoverable resources, the measure of long-term fossil fuel production potential used in this *Outlook*, are considerably higher than proven reserves. As market conditions change and technology advances, some of these resources are set to move into the proven category, providing further reassurance that the resource base will not constrain production for many decades to come. In particular, large volumes of unconventional oil and gas are expected to be proven in many parts of the world, diversifying the geographical distribution of reserves. The costs of supply will undoubtedly be higher than in the past, as existing sources are depleted and companies are forced to turn to more difficult sources to replace lost capacity.

Renewable energy sources are abundant and capable of meeting a large proportion of the growth in projected energy demand. Likewise, known resources of uranium, the raw material for nuclear fuel, are more than adequate to meet the projected expansion of the world's nuclear reactors throughout the *Outlook* period (NEA and IAEA, 2010).

Production prospects

In the New Policies Scenario, non-OECD regions account for all of the net increase in aggregate fossil fuel production between 2010 and 2035, even though the United States shows a marked increase (Figure 2.9). This reflects the geographical distribution of fossil fuel resources and the broad trends projected for the level and composition of energy demand. The increase in production of or output from non-fossil fuels – nuclear and renewables – is also heavily skewed towards non-OECD regions.

Oil production (net of processing gains) rises from 84 mb/d in 2011 to 92 mb/d in 2020 and then 97 mb/d in 2035. There is a significant change in the composition of production, both in terms of the types of oil that are produced and the geographic location of production. Crude oil production – the largest single component of oil production – falls slightly between 2011 and 2035. But this is more than outweighed by sharp increases in the output of natural gas liquids (NGLs), thanks to rising gas production, and unconventional oil, chiefly oil sands in Canada. Several members of the Organization of the Petroleum

Exporting Countries (OPEC) and a handful of non-OPEC countries, notably Brazil, Canada, Kazakhstan and the United States, are responsible for the bulk of the increase in global oil production. OPEC's share in total production rises from 42% in 2011 to 48% in 2035. Iraq registers the largest growth, with output rising from 2.7 mb/d in 2011 to 8.3 mb/d in 2035 (see Chapter 14 for analysis of Iraq's supply potential).

Figure 2.9 ▷ Fossil fuel production in selected regions in the New Policies Scenario, 2010 and 2035

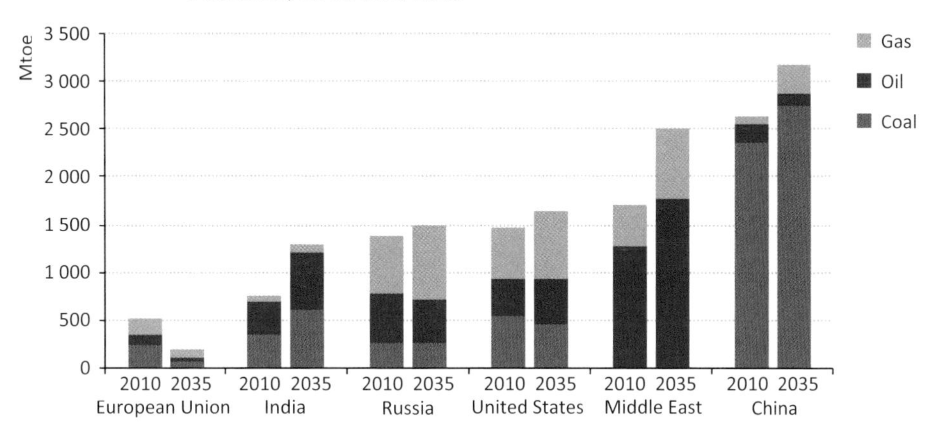

Among the fossil fuels, production of natural gas increases at the fastest rate, rising from 3.3 tcm in 2010 to 5.0 tcm in 2035, an increase of 51%. Natural gas production in the OECD rises by 23% between 2010 and 2035, with sharp increases in Australia and the United States more than offsetting declining output in the European Union. Non-OECD production rises by 67%, with the Middle East making the largest contribution, followed by China, Africa and Russia. Unconventional sources represent 48% of incremental output, pushing their share of total supply from 14% in 2010 to 26% in 2035. Unconventional gas could play an even more important role if concerns about the environmental and safety aspects of hydraulic fracturing can be overcome (IEA, 2012).

Coal production increases from around 5 100 Mtce in 2010 to 6 000 Mtce in 2035, with most of the growth occurring prior to 2020. Production continues to be dominated by non-OECD countries, with their combined share of output rising from 73% in 2010 to around 80% in 2035. China, which accounted for almost half of global production in 2010 in absolute tonnage terms, remains the largest producer in 2035 and makes the biggest contribution to the growth in supply, followed by India and Indonesia. By contrast, production falls by some 10% in the OECD, as increased supply in Australia is more than offset by sharply lower output in Europe and a modest decline in the United States.

Renewables are the fastest growing energy source over the *Outlook* period. Electricity generation from renewables almost triples between 2010 and 2035, approaching that of coal. The increase comes primarily from wind and hydropower. In 2035, renewables supply 43% of total electricity in the European Union, 27% in China and 23% in the United States.

Biofuels production increases from 1.3 mb/d in 2011 to 4.5 mb/d in 2035, but at the end of the period is still equal to less than 5% of oil demand, on an energy-equivalent basis. The United States and Brazil remain the largest producers, but output also rises sharply in China. Advanced biofuels, including those made from lignocellulosic feedstocks, grow in importance after 2020, when they are assumed to become commercially viable. Despite cost reductions achieved through greater deployment and increases in fossil-fuel prices, these projections are contingent on ongoing subsidies. This is because for many regions and technologies, energy derived from renewable sources is projected to remain more costly than energy from fossil fuels for decades to come. We estimate subsidies to renewables to rise to almost $240 billion in 2035, from $88 billion in 2011 (see Chapter 7).

Water is essential to many forms of energy production: in power generation; in the extraction, transport and processing of fossil fuels; and, increasingly, in irrigation for crops used to produce biofuels. We estimate that water withdrawals for energy production in 2010 were 580 billion cubic metres (bcm), with the power sector accounting for over 90% of the total. Of that, water consumption – the volume withdrawn but not returned to its source – was about 70 bcm. In the New Policies Scenario, withdrawals increase by 20% between 2010 and 2035, but consumption rises by a more dramatic 85%. This is more than twice the rate of growth of energy demand, reflecting a move towards more water-intensive power generation and expanding output of biofuels. As described in Chapter 17, energy production in some regions is already constrained or suffering as a result of limited water availability.

Inter-regional trade

In the New Policies Scenario, the geographical mismatch between the location of oil, natural gas and coal resources and the location of the main demand centres drives an expansion of international trade.[4] In addition to the rise in absolute terms, there are important changes in the geographical patterns of trade, implying a geographical and political shift in concerns about the cost of imports and supply security. Inter-regional oil trade increases by almost 20%, from 42 mb/d in 2011 to 50 mb/d in 2035. Developing Asia sees the biggest increase: China's imports rise from 4.9 mb/d to 12.3 mb/d and India's from 2.5 mb/d to 6.9 mb/d. By contrast, net imports into the OECD are cut almost in half, to 12.4 mb/d, mainly on account of the situation in the United States, where rising domestic output and improved transport efficiency slash net-import requirements to just 3.4 mb/d in 2035, from 9.5 mb/d in 2011. The United States moves from being the world's second-largest oil importer today to just the fourth-largest soon after 2030, behind China, the European Union and India. Net exports from the Middle East, which remains the biggest exporting region throughout the *Outlook* period, rise from 21 mb/d in 2011 to 26 mb/d in 2035, with Iraq making the largest contribution to the growth.

4. All trade projections refer to net trade between major *WEO* regions.

Box 2.1 ▷ How have the New Policies Scenario projections changed since *WEO-2011*?

Our energy projections change each year due to changes in key assumptions, including GDP, energy prices, population and technological developments, as well as changes in base year data and developments in policy making over the previous twelve months. In aggregate, our *WEO-2012* projections for world energy demand are not substantially different compared with last year's edition. Primary energy demand in 2035 in the New Policies Scenario is about 1% higher (Figure 2.10). But there are important differences in the relative contributions the different fuels make to meet demand growth. Demand for coal and natural gas have been revised upwards, while nuclear prospects have been revised down by 6% as a result of changes in inter-fuel competition in power generation and policy changes after Fukushima. Renewables in power generation achieve a slightly lower annual growth rate, although not significantly altering their share of total generation in 2035.

There are also some important changes at the regional and country level. Projected energy demand growth in the OECD has been revised down, due to the combined effects of slower GDP growth and more rapid gains in energy efficiency. Demand for oil falls faster than before, while natural gas and renewables both play bigger parts in meeting the region's energy needs. Projections for oil and natural gas production in the OECD have been revised slightly upwards, due to brighter prospects in the United States and Canada. By contrast to the OECD, non-OECD energy demand is revised upwards compared with *WEO-2011*, in part due to upward revisions of the statistics for the base year. There are small differences in the non-OECD energy mix, with natural gas in particular having a bigger role in 2035.

Figure 2.10 ▷ Change in key *WEO-2012* projections for 2035 compared with *WEO-2011* in the New Policies Scenario

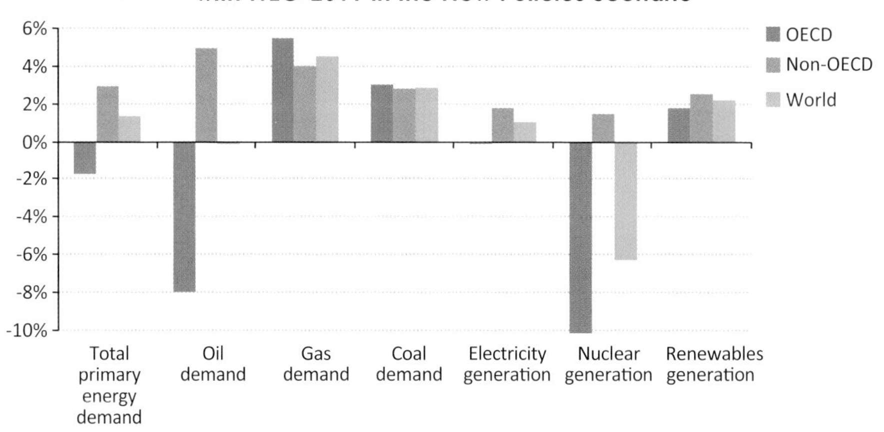

Natural gas trade grows from 675 bcm in 2010 to about 1 200 bcm in 2035 – an increase of 77%, compared with growth in production of 51%. The European Union remains the largest importer: imports make up 85% of its consumption in 2035, up from 62% at present, with the largest growth in supply coming from Russia. China becomes the world's second-largest importer, with imports growing from just 15 bcm in 2010 to 226 bcm in 2035. Global trade in liquefied natural gas (LNG) doubles, to almost half of gas trade in 2035. The United States emerges as an LNG exporter before 2020 and Canada also starts to export LNG in the same time frame. Although exports from North America are projected to remain fairly limited, they are important in boosting the competitiveness of traded gas markets, providing buyers with additional supply options.

Trade in hard coal (coking and steam coal) rises from 833 Mtce in 2010 to a plateau of around 1 100 Mtce by 2025, reflecting the projected levelling off of global demand for coal. Coal trade as a share of global demand is 19% in 2035, barely higher than in 2010. This is lower than the levels for oil and natural gas because the largest coal consumers hold their own significant resources and because transporting coal is relatively costly. Based on preliminary data, China – already the largest coal consumer and producer – is set to become the world's largest coal net-importer, ahead of Japan. In our projections, its imports peak at about 190 Mtce in 2015 and then decline progressively through to 2035, as domestic demand stagnates and indigenous production edges up. India then becomes the biggest coal importer by 2020, as transport bottlenecks and delays in opening new mines continue to constrain indigenous production. India's imports reach almost 315 Mtce in 2035 – about 34% of its hard coal use and 28% of global trade. On the export side, Australia and Indonesia supply 65% of inter-regional hard coal trade in 2035, up from 60% today.

Energy-related CO_2 emissions

The trends projected in the New Policies Scenario have direct implications for climate change, because of the large contribution of fossil fuels to total anthropogenic greenhouse-gas emissions (around 60% in 2010). Preliminary IEA estimates indicate that global CO_2 emissions from fossil-fuel combustion reached a record high of 31.2 gigatonnes (Gt) in 2011, an increase of 1 Gt, or 3.2%, on 2010. China made the largest contribution to the increase, with its emissions rising by 9%, to 7.9 Gt. However, China's efforts to reduce its carbon intensity — the amount of CO_2 emitted per unit of GDP — during the period covered by its 11th Five-Year Plan (2006-2010) substantially slowed growth in its emissions.

In the New Policies Scenario, energy-related CO_2 emissions remain on an upward path, increasing by 23% compared with 2010 to 37.0 Gt in 2035, a trajectory that is likely to be consistent with a long-term average global temperature increase of 3.6 °C. Emissions in the OECD decline by 16% between 2010 and 2035, reaching 10.4 Gt; the OECD's share of global emissions falls from 41% in 2010 to 28% in 2035. The largest reduction in absolute terms within the OECD occurs in the United States: it emits 19% less CO_2 in 2035 than in 2010, primarily due to the higher share of renewables in the power sector and the implementation of fuel-efficiency standards (Figure 2.11). By contrast, emissions in non-OECD regions rise by just over half. China is poised to emit more CO_2 than the United States and the European Union combined in cumulative terms over the period, while its per-capita emissions increase to the average level of the OECD by 2035. India's emissions more than double.

Figure 2.11 ▷ **Energy-related CO_2 emissions in selected countries and regions in the New Policies Scenario, 2010 and 2035**

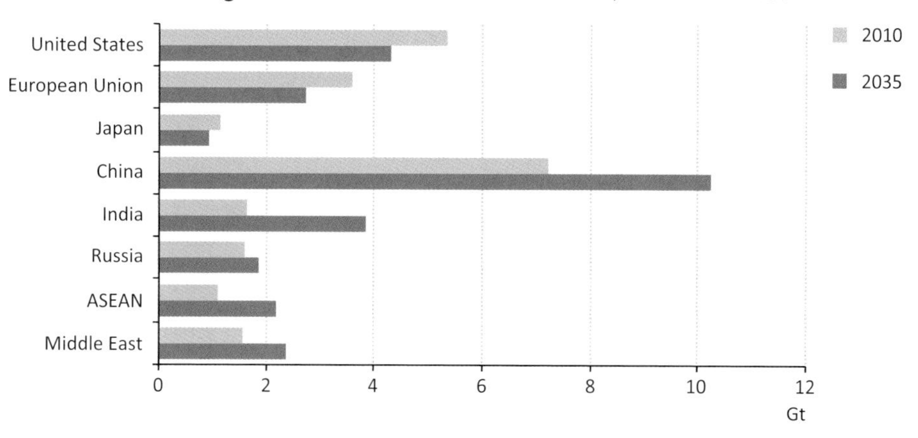

Economic implications

Fossil-fuel subsidies

The evolution of energy prices will be an important determinant of future energy trends, as prices affect consumer demand for fuels and financial incentives to invest in energy efficiency. A key issue in this respect is how quickly fossil-fuel subsidies, which still remain commonplace and large in many countries, will be phased out. Subsidies can result in an economically inefficient allocation of resources and market distortions, while often failing to meet their intended objectives. By protecting parts of the market, they can also make the rest of the market more volatile. Moreover, the prospect of higher international prices means that fossil-fuel subsidies could represent a growing burden on state budgets. And for net exporting countries, subsidies could act to restrict export availability by continuing to inflate domestic demand, leading to lower export earnings in the longer term.

In 2011, fossil-fuel consumption subsidies worldwide are estimated to have totalled \$523 billion, \$111 billion higher than in 2010 (Figure 2.12).[5] By comparison, financial support to renewable energy amounted to \$88 billion in 2011 (see Chapter 7). The increase in the cost of fossil-fuel subsidies between 2010 and 2011 primarily reflects higher international energy prices and rising consumption of subsidised fuels. The estimated subsidy bill would have been even higher had it not been for policy interventions to reform

5. Based on an IEA global survey that identified and analysed economies that subsidise fossil-fuel consumption.

subsidy programmes in a number of countries, most notably in Iran. Oil products attracted the largest subsidies, totalling $285 billion (or 54% of the total), followed by fossil-fuel subsidies reflected in the under-pricing of electricity at $131 billion. Natural gas subsidies were also significant, reaching $104 billion. Comparatively, subsidies to coal end-use consumption were small, at $3.2 billion.

Figure 2.12 ▷ **Economic value of fossil-fuel consumption subsidies by fuel**

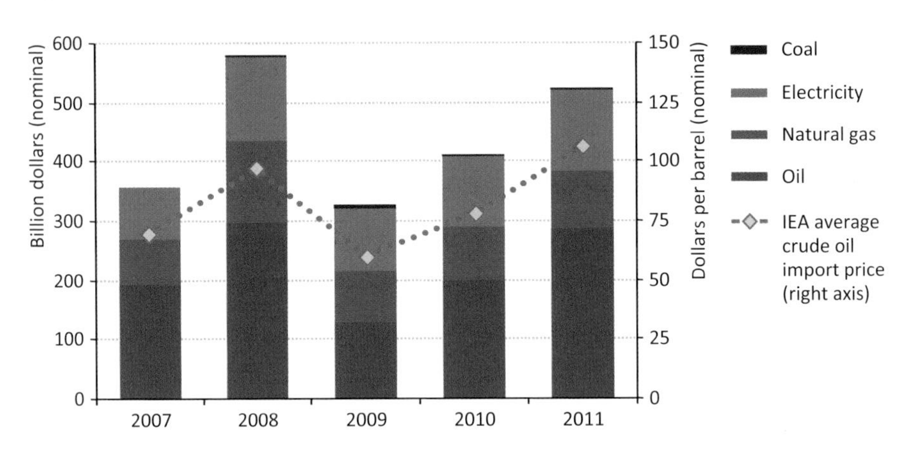

Note: Electricity subsidies include only those resulting from under-pricing of fossil fuels consumed in power generation.

Fossil fuels were subsidised at a weighted-average rate of 24% in 2011 in the economies identified, meaning that consumers paid only 76% of the reference or unsubsidised price, based on international prices. Subsidies were most prevalent in the Middle East, amounting to 34% of the global total (Figure 2.13). At $82 billion, Iran's subsidies were the highest of any country despite the introduction in late-2010 of major energy-price reforms. The world's two largest energy exporters – Saudi Arabia and Russia – had the next-highest subsidies, at $61 billion and $40 billion respectively.

These estimates capture the value of subsidies that reduce end-user prices below those that would prevail in an open and competitive market.[6] Such subsidisation occurs whether energy is imported at world prices and then sold domestically at lower, regulated prices, or, in the case of countries that are net exporters of a given product, where domestic energy is priced below international market levels. In the latter case, our estimates capture the corresponding opportunity cost, *i.e.* the rent that could be recovered if consumers paid world prices. Separate to the subsidies captured by the IEA estimates, the OECD is tracking budgetary transfers and tax expenditures that provide support to fossil fuels (OECD, 2011).

6. Full details of the IEA's methodology for estimating subsidies are available at *www.worldenergyoutlook.org/resources/energysubsidies.*

Figure 2.13 ▷ Economic value of fossil-fuel consumption subsidies by fuel for top 25 countries, 2011

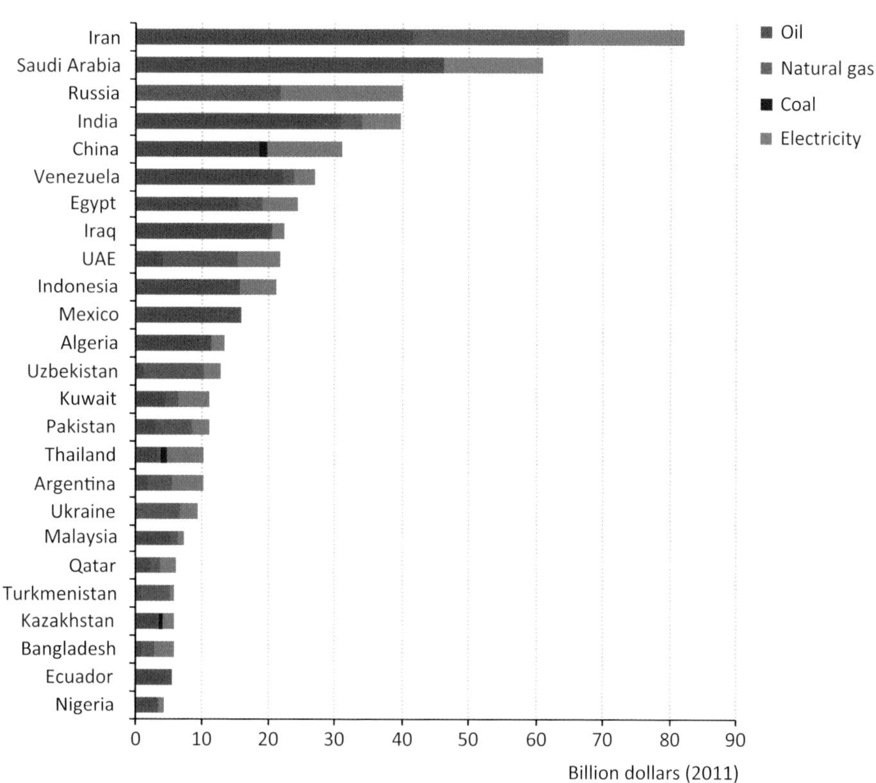

G-20 and APEC (Asia-Pacific Economic Cooperation) member economies have made commitments in recent years to phase out inefficient fossil-fuel subsidies. Many countries outside these groupings have also committed to subsidy reform, in most cases because high energy prices have made subsidies an unsustainable fiscal burden on government budgets. While an encouraging start has been made, much remains to be done to fulfil the commitments, both in terms of defining the fossil-fuel subsidies to be phased out and following through with durable and well-designed reform efforts. Some notable examples of recent reform efforts include:

- Iran initiated the first step of a major shake-up of energy pricing in December 2010, although it still remains the world's largest subsidiser of fossil fuels. The five-year programme aims to increase oil product prices to at least 90% of the level of Persian Gulf FOB (free on board) prices, natural gas prices to between 65% and 75% of export prices and electricity prices to their full cost of supply. After the first stage of reforms, gasoline prices increased by 300% (and by 600% for volumes above a certain quota), while natural gas prices rose by some 900%. Despite initially promising results, the second phase of the programme has been postponed due to concerns about further exacerbating already high levels of inflation.

- Indonesia currently regulates the prices of kerosene, liquefied petroleum gas (LPG) in small containers, some grades of diesel and gasoline, and electricity. Efforts to reform these subsides are being stepped up, in part because the country is becoming increasingly dependent on oil imports. In April 2012, an attempt to increase gasoline and diesel prices by 33% was unsuccessful, due to public protests. Legislative provision was later made to allow fuel prices to be raised if the average crude price over a six-month period is at least 15% above $105 per barrel, the base price in the 2012 state budget. Indonesia also plans to keep track of subsidised fuel use by vehicles, although the necessary system has yet to be put in place. Indonesia has, meanwhile, banned state-owned and (certain) company vehicles from using subsidised fuels, is promoting gas as a substitute for kerosene and diesel, and is reducing electricity use in state-owned buildings and in street lighting.
- Nigeria, after reducing diesel and kerosene subsidies in 2011, completely removed gasoline subsidies in January 2012. This resulted in several weeks of nation-wide protests, which the government sought to appease by cutting prices by one-third (which left them still 50% higher than pre-removal levels). Although the subsidies were reduced, the problems that arose highlighted the requirements of best practice approaches to pricing reform, including public consultation prior to the price increases and measures to ease the burden on poorest segments of the population.
- Sudan cut subsidies to gasoline and fuel oil in June 2012, as part of austerity measures following a sharp drop in oil-export revenues. This occurred after South Sudan assumed control, following independence in July 2011, of over three-quarters of the country's oil production. Under the new arrangements, gasoline prices doubled and fares on public transport increased by some 35%. Again, there were widespread public protests. The government has, nonetheless, since announced additional measures, which will lead to further increases in prices for gasoline, diesel and LPG.

Spending on fossil-fuel imports

For many countries, persistently high energy prices and high or increasing reliance on imports mean that spending on imports of fossil fuels has become a significant economic burden, particularly at a time of economic weakness. Local factors have also increased spending in certain cases. In Japan, for example, spending reached record highs as power generators turned to expensive, oil-indexed and spot-priced natural gas and, to a lesser extent, heavy fuel oil, to make up for the loss of nuclear capacity after Fukushima.

In the New Policies Scenario, annual spending on imports of fossil fuels increases from $2.0 trillion in 2011 to $3.0 trillion in 2035, a rise of 47%. Currently, oil accounts for 81% of the total fossil-fuel import bill, followed by natural gas (12%) and coal (7%). Through the *Outlook* period, the share of natural gas increases progressively, but oil costs remain bigger. In the United States, the projected decline in oil imports and the country's emergence as a gas exporter result in a fall in the overall energy-import bill, despite rising international prices, from $364 billion in 2011 to $135 billion in 2035 (Figure 2.14). By contrast, all of the other major importers experience persistently high or increasing import costs. China's fossil-fuel import bill amounts to around $700 billion in 2035, up from $234 billion in 2011. India's spending exceeds $400 billion in 2035, compared with $120 billion in 2011. When viewed as a proportion of GDP, spending in 2035 is highest in India (5.6%) and China (2.5%).

Figure 2.14 ▷ Spending on net imports of fossil fuels in the New Policies Scenario

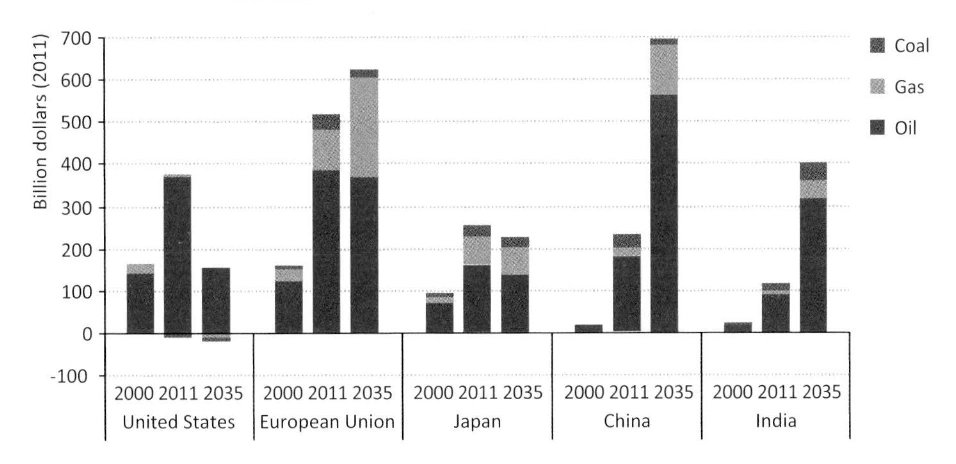

The rise in fossil-fuel import bills in the New Policies Scenario is matched by a commensurate increase in export earnings by fossil-fuel producing countries. The impact on the global economy will largely depend on how these earnings are spent, including the extent to which these "petrodollars" are recycled through purchases of goods and services from other countries or by investing in foreign assets. OPEC oil-export revenues, which reached a record of $1.1 trillion in 2011, increase to $1.2 trillion in 2020 and $1.6 trillion in 2035. Russia, which remains the largest individual energy exporter throughout the *Outlook* period, sees its revenues from oil, natural gas and coal exports rise from $380 billion in 2011 to $410 billion in 2035; an increasing share of this comes from markets in the east, but Europe remains its major market.

Investment in energy-supply infrastructure

In the New Policies Scenario, cumulative investment of $37 trillion (in year-2011 dollars) is needed in the world's energy supply system over 2012-2035, or $1.6 trillion per year on average (Table 2.5). This includes investment to expand supply capacity and to replace existing and future supply facilities that will be exhausted or become obsolete; it does not include demand-side spending, such as on purchasing cars, air conditioners and refrigerators. This large sum, in absolute terms, amounts to 1.5% of global GDP on average to 2035 and financing it in a timely manner will depend on attractive investment conditions, notably in terms of the return available on investment.

Table 2.5 ▷ Cumulative investment in energy-supply infrastructure in the New Policies Scenario, 2012-2035 (billion in year-2011 dollars)

	Coal	Oil	Gas	Power	Biofuels	Total	Share of GDP
OECD	204	3 341	3 720	6 787	206	14 258	1.0%
Americas	79	2 666	2 337	2 852	131	8 065	1.3%
Europe	6	551	924	2 797	73	4 351	0.8%
Pacific	119	124	460	1 138	2	1 842	0.7%
Non-OECD	963	6 641	4 854	10 080	149	22 687	2.1%
E. Europe/Eurasia	36	1 239	1 455	1 182	4	3 917	3.5%
Russia	23	745	987	717	-	2 472	3.5%
Developing Asia	844	1 036	1 425	6 768	74	10 147	1.6%
China	634	576	577	3 712	43	5 541	1.3%
India	93	202	199	1 620	19	2 133	2.2%
Middle East	0	1 074	498	577	-	2 149	2.5%
Africa	56	1 604	936	745	1	3 342	4.3%
Latin America	27	1 688	540	808	69	3 132	1.9%
Inter-regional transport	57	259	103	-	22	422	n.a.
World	1 224	10 242	8 677	16 867	357	37 366	1.5%

Non-OECD countries require 61% of investment in the energy sector as a whole. The share in the OECD, where production and demand increase much less rapidly, is disproportionally high, reflecting higher unit costs for capacity additions and the need to replace significant amounts of ageing infrastructure. The United States requires the most investment, at 17% of the world total, followed by China at 15%. Investment in the power sector (generation, transmission and distribution) absorbs $17 trillion, or 45% of the total. Renewables account for around 60% of the investment in power generation, far more than their share of incremental generation, because of their high capital intensity and relatively low output. Investments in the oil and gas sectors combined total some $19 trillion, or around 51% of global energy investment. While far from negligible, coal investment is small relative to other sectors.

US developments redefining the global energy map

A dominant narrative over recent years has been the growing influence of emerging economies in the global energy system, a development which is poised to continue over the coming decades. But a striking new trend now emerging is the resurgence of oil and gas production in the United States, where output had been widely assumed, even as recently as a few years ago, to be in inevitable decline. Together with efficiency measures that are set to curb oil consumption, this energy renaissance has far-reaching consequences for energy markets, trade and, potentially, even for energy security, geopolitics and the global economy.

Diverging trends in import dependency

The United States became a net importer of oil in the mid-1940s and has, since the mid-1970s, accounted for around one-quarter of the world's oil trade, as demand consistently outstripped domestic supply. The United States has also been an importer of natural gas (from Canada) and, although more than covering its coal needs, just over 80% of its primary energy demand is currently met by energy produced domestically. This picture is changing. Advances in technology have unlocked production of unconventional gas, which has grown at a phenomenal rate: the increase in US unconventional gas production over the last five years is comparable to the current annual gas exports of Russia. Around the same time, oil demand in the United States started to fall back from a historic high of 20 mb/d. While oil production has been on an upward trend since 2008, in large part because of deployment in the production of light tight oil of the same technologies used to produce shale gas. Use of renewables is also increasing, boosted by policies to support electricity at the state level and by a nation-wide policy that mandates significant increases in the use of biofuels. In our New Policies Scenario, the result is that, by 2035 the United States is 97% energy self sufficient in net terms, as exports of coal, gas and bioenergy (biofuels and wood pellets) help offset (in energy equivalent terms) the declining net imports of oil (Figure 2.15).

Figure 2.15 ▷ Net energy self-sufficiency in selected countries and regions in the New Policies Scenario

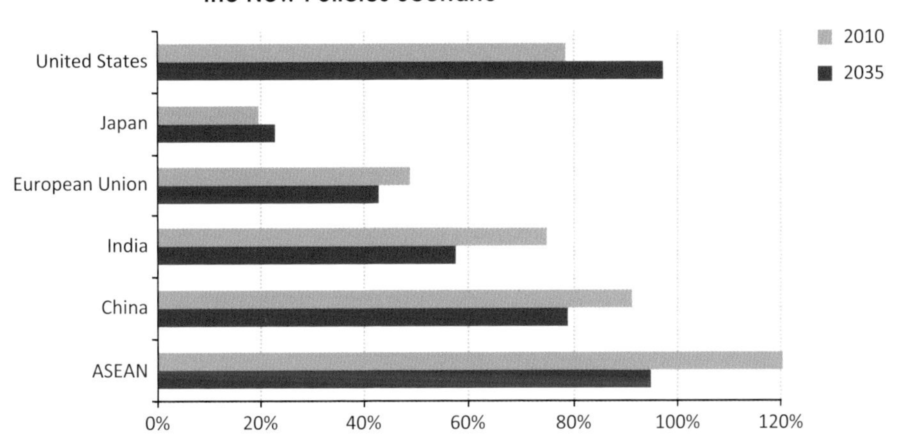

Note: Self-sufficiency is calculated as indigenous energy production (including nuclear power) divided by total primary energy demand.

By reversing the trend towards greater dependence on imported energy, the United States stands out from most other major energy consuming regions and countries, which become less self-sufficient, *i.e.* having an even greater dependence on imports. The change, relative to 2010, is largest in India, the ASEAN region and China, where a growing share of incremental energy will need to come from imports. Japan, in particular, is set to remain highly dependent on imported energy, though, over time, a combination of greater indigenous production from renewable energy sources, greater energy efficiency and declining domestic energy demand outweigh the assumed effect of reduced nuclear generation post-Fukushima (in the short to medium term, Japan is importing more fossil fuels to offset at least in part the loss of nuclear capacity).

The contrast is particularly striking if the analysis is limited to oil and gas. While the United States reverses the trend towards greater import dependence, China, India, ASEAN and the European Union all see a steady move towards greater reliance on imports (or, in the case of ASEAN gas supply alone, a reduced surplus of gas for export) (Figure 2.16). Japan and Korea are already almost entirely reliant on oil and gas imports. By contrast, the United States is projected to reduce its reliance on imported oil from more than 50% of consumption today to less than 30% in 2035, while becoming a net exporter of gas. North America, as a whole, even becomes a net exporter of oil from around 2030 in the New Policies Scenario.

Figure 2.16 ▷ Net oil and gas import dependency in selected countries in the New Policies Scenario

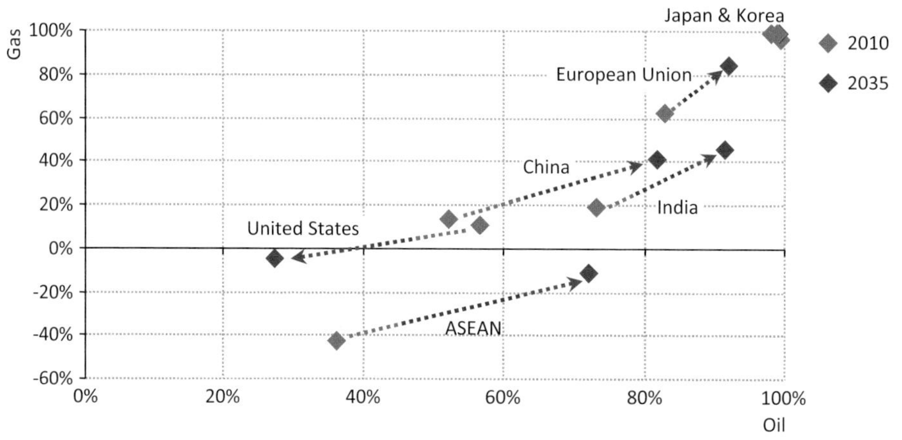

Note: Import dependency is calculated as net imports divided by primary demand for each fuel.

The projected reduction in US net oil imports, from 9.5 mb/d in 2011 to 3.4 mb/d in 2035, is a product of changes both on the supply side and demand sides (Figure 2.17). The impact of increased supply, mainly due to expanded production of light tight oil, predominates early in the projection period. After 2020, though, the bulk of the reduction is attributable to developments on the demand side – a big improvement in the fuel-efficiency of vehicles – as well as the use of biofuels and natural gas in the transport sector. Falling oil imports (and rising natural gas exports) will have important implications for the US trade deficit – net oil imports were equivalent in value in 2011 to around two-thirds of the 2011 deficit in goods – as well as broader benefits from the economic stimulus provided by higher production. But reduced dependence on imported oil does not mean, of course, that the United States will be immune to developments in international markets. Oil prices are set globally, so consumers in the United States will continue to feel the effects of worldwide price fluctuations. Moreover, there is not necessarily a direct correlation between levels of import dependency and the security or insecurity of energy supply.

Figure 2.17 ▷ **Reductions in net oil imports in the United States by source in the New Policies Scenario**

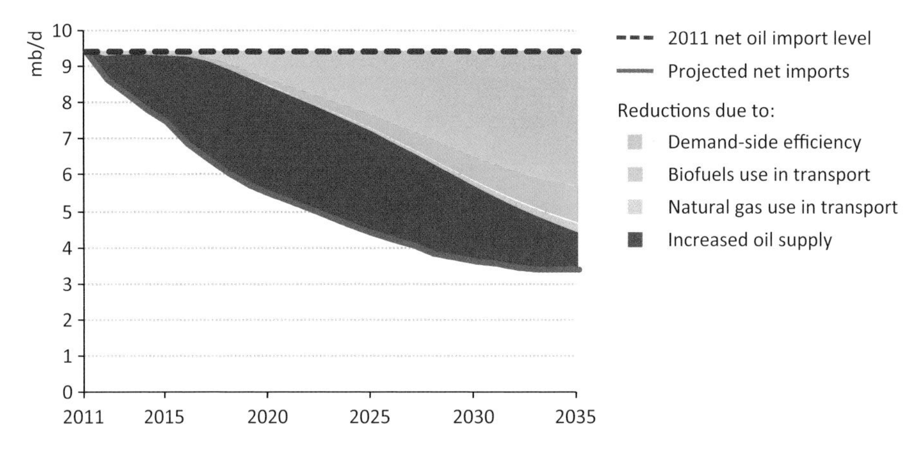

The impact of unconventional gas

The surge in unconventional gas production has been a game-changing development in North American natural gas markets, bringing prices to historic lows and improving dramatically the competitiveness of gas against other fuels. Natural gas is now making a substantial additional contribution to economic activity and employment, both within the energy industry and by lowering energy costs for other industrial sectors. Where it replaces coal, as in power generation, natural gas has also contributed to a decline in US emissions of CO_2. The impact on energy costs is felt both directly and via electricity prices, which are lower than in many other IEA countries. Although energy represents only a small share of costs for most US manufacturers, a competitive source of gas can be a crucial consideration in attracting investment in energy-intensive manufacturing. This can be the case for aluminium, paper or iron and steel, or in industries requiring gas as feedstock, such as the petrochemicals sector, where feedstock costs can represent over 80% of total operating expenses.

Developments in the United States have had a significant impact on global energy markets. Just a few years ago, the United States was expected to become a major importer of LNG. Instead, LNG imports have shrunk to a tiny proportion of demand and the United States and Canada are set to become LNG exporters once the infrastructure is in place in the middle of this decade. Projects that were being undertaken in exporting countries with a view to the North American market have had to seek alternative destinations, allowing other importers to benefit from relatively plentiful LNG supplies. This has presented a challenge to gas exporters selling gas on an oil-indexed price basis under long-term contracts. They have found that buyers were not only exercising any flexibility under the contracts to reduce deliveries in order to switch to cheaper alternatives, but also pushing for changes to the way in which gas is priced to more closely reflect market realities.

The impact of unconventional gas in North America has not been unambiguously good news for gas in other markets. Natural gas has been sufficiently cheap to displace coal in power generation in many parts of the United States. This coal became available for export and some capacity existed to ship it to other markets (the same is not yet true for gas, where exports are currently constrained by the time required to gain the necessary regulatory approvals and to construct LNG terminals). Based on preliminary data, net coal exports from the United States could approach 100 million tonnes (Mt) in 2012, compared with less than 10 Mt in 2006. The result has been to put downward pressure on coal prices in other regions. In Europe, low coal prices (and low CO_2 prices) in 2011, which drove a dramatic increase in coal use of some 7%, contributed to a dismal year for the European gas industry, with gas consumption falling by an estimated 10%.

The longer-term impact of unconventional gas will depend on how opportunities for supply develop, in particular the extent to which the North American experience is replicated elsewhere. The potential is considerable and there are strong incentives for countries currently relying on imported gas to develop indigenous resources. However, outside North America and Australia, the unconventional gas business is still in its formative years, with questions still to be answered about the extent and quality of the resource base and unsatisfied concerns about the environmental impact of producing unconventional gas. If these concerns are not addressed properly, there is a very real possibility that public opposition will halt the unconventional gas revolution in its tracks. In the New Policies Scenario, unconventional gas supply is projected to expand significantly, accounting for almost half of the increase in global gas output. By bolstering and diversifying sources of supply, tempering demand for imports (as in China) and fostering the emergence of new exporting countries (as in the United States), unconventional gas can be expected to lead to more diversified trade flows, putting pressure on conventional producers and traditional oil-linked pricing mechanisms for gas.

The direction of oil trade and supply security

The reduction in US oil imports – in combination with surging oil consumption in emerging economies – is set to be a major driver for changing patterns of global oil trade. The way that oil trade evolves will have implications for market dynamics and for trade flows along some key strategic maritime and pipeline transportation routes. The main shift is already visible, as oil is drawn increasingly towards Asia-Pacific markets and away from the Atlantic basin. Successive *Outlooks* have highlighted the accelerating pace of oil imports to China, India and the rest of developing Asia, matched by rising exports from the Middle East. The evolution of Iraq's exports in recent years illustrates the trend (see Part C). As recently as 2008, only one-third of Iraq's oil went to Asian markets, with the rest divided between Europe and North America. By 2011, the share going to Asian markets had risen above 50%. Over the projection period, Middle Eastern suppliers to Asia are joined in increasing volumes by producers in Eurasia, notably Russia and Kazakhstan, Brazil and also Canada (if it builds infrastructure to allow exports from its west coast).

Growing trade has the virtue of consolidating global interdependence; but it brings the risk of short-term supply interruptions, particularly if geographic supply diversity is reduced and reliance on a few strategic supply routes is increased. In the New Policies Scenario, an increasing share of global oil trade is set to transit through the Straits of Hormuz, the world's most important maritime oil-shipping route, where oil transportation rises from close to 18 mb/d in 2010 (or 42% of global trade in oil) to almost 25 mb/d in 2035 (or 50% of projected trade) (Figure 2.18).[7] As more oil flows eastwards from the Middle East to meet rising demand in Asia, there is also growing reliance on the Straits of Malacca, where oil transit volumes as a share of global trade rise from 32% in 2010 to 45% in 2035. By contrast, US dependence on supplies that transit chokepoints diminishes as US net imports drop and new opportunities become available to source oil from closer to home, notably from Brazil. In the New Policies Scenario, US imports from the Middle East, which have already declined from around 2.8 m/d in 2000 to 2 mb/d in 2011, fall to just 0.3 mb/d in 2025.

Figure 2.18 ▷ Share of inter-regional oil and gas trade through key choke points in the New Policies Scenario

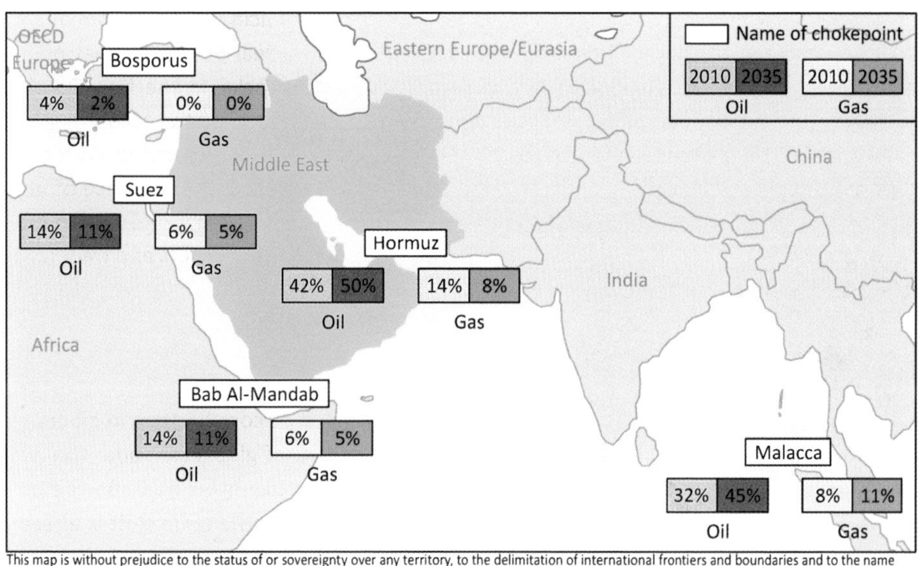

This map is without prejudice to the status of or sovereignty over any territory, to the delimitation of international frontiers and boundaries and to the name of any territory, city or area.

Note: Data is based on the volumes for which going through the chokepoints would be the shortest route; in reality not all trade takes (nor will take) the shortest route as other factors may result in flows in directions that are not consistent with relative transport costs. The above data include flows going through pipelines by-passing the chokepoints and not just those going through the shipping lanes.

7. Volumes are based on inter-regional trade. Actual volumes would be higher due to intra-regional trade.

The prospect of increased reliance on trade through these chokepoints raises the issue of the possibility of finding alternative routes to market and increases the importance of the policies of net importing countries to reduce the likelihood of disruptions in supply or provide insurance if they do occur. In the case of the Straits of Hormuz, two new pipelines have recently been commissioned that reduce reliance on the Straits: a converted gas pipeline traversing Saudi Arabia to Yanbu on the Red Sea, with a capacity of up to 2 mb/d, and a 1.5 mb/d line in the United Arab Emirates to the port of Fujairah on the Indian Ocean. But these and other planned or actual pipelines in the region could not alone sustain projected oil export levels from the Gulf countries, were this maritime route to be closed. Countries that are increasingly reliant on imports will be wise to engage actively in efforts to ensure the security of international shipping routes. In the case of China, addressing this concern will form a natural part of the strategy underlying the growing presence of its national oil companies in the upstream sector in Africa and the Middle East and its strengthening of political and economic ties with these regions. Building oil stocks and contributing to plans for their co-ordinated use are other features of well-based plans to secure long-term energy needs. All oil-consuming countries stand to gain from such efforts, as any disruption to supplies, whether flowing eastwards to Asia or elsewhere, would affect the prices paid on international markets by all consumers.

Oil market outlook

What will drive growth?

Highlights

- Global oil demand increases steadily in the New Policies Scenario to 99.7 mb/d in 2035, up from 87.4 mb/d in 2011. The rate of demand growth is slowed by government policies, including efficiency measures, and higher prices: the crude oil price rises to $125/barrel (in year-2011 dollars) in 2035. Demand grows briskly in non-OECD countries, with China alone accounting for around 50% of the net increase worldwide, more than offsetting a steady decline in OECD regions brought about by efficiency gains, inter-fuel substitution and saturation effects.

- All of the net growth in global oil demand comes from the transport sector in emerging economies. Although passenger light-duty vehicles (PLDVs) remain the biggest component of transport demand, road freight demand increases more quickly, approaching the level of demand from PLDVs today by 2035. Fuel-economy standards for trucks are much less widely adopted than for PLDVs.

- Oil production, net of processing gains, is projected to rise from 84 mb/d in 2011 to 97 mb/d in 2035, the increase coming entirely from natural gas liquids and unconventional sources. Output of crude oil (excluding light tight oil) fluctuates between 65 mb/d and 69 mb/d, never quite reaching the historic peak of 70 mb/d in 2008 and falling by 3 mb/d between 2011 and 2035. Light tight oil production grows above 4 mb/d in the 2020s, mainly from the United States and Canada.

- Non-OPEC output rises from under 49 mb/d in 2011 to above 53 mb/d after 2015, a level maintained until the mid-2020s. The increase is due to rising unconventional supplies, mainly from light tight oil and Canadian oil sands, natural gas liquids, and a jump in deepwater production in Brazil. After 2025, non-OPEC output falls back to 50 mb/d in 2035. The United States overtakes Russia and Saudi Arabia before 2020 to become, until the mid-2020s, the world's largest oil producer. Output by OPEC countries collectively accelerates through the projection period, particularly after 2020, from 36 mb/d in 2011 to 46 mb/d in 2035, their share of world oil production rising from 42% to 48%. Iraq sees the biggest absolute increase in output.

- Worldwide upstream oil and gas investment is expected to rise by around 8% in 2012, relative to 2011, reaching a new record of $619 billion – more than 20% up on 2008 and five times the level of 2000. Higher costs, which have risen 12% since 2009 and more than doubled since 2000, explain part of the increase. Our projections call for upstream oil and gas investment to remain at similarly high levels, with an average investment requirement of $615 billion per year for 2012-2035.

Demand

Primary oil demand trends

After rebounding sharply in 2010, primary oil demand grew by a more modest 0.7% in 2011 to reach 87.4 million barrels per day (mb/d).[1] The trajectory that oil use follows over the coming decades differs considerably by scenario, reflecting the different assumptions about government policies to curb rising demand and emissions. In the Current Policies and New Policies Scenarios, oil use increases in absolute terms to 2035, driven mainly by population and economic growth in the emerging economies, but it falls in the 450 Scenario, in response to strong policy action to curb fossil-energy use (Figure 3.1). The share of oil in total world primary energy demand falls in each scenario, most sharply in the 450 Scenario, where it reaches 25% in 2035 – down from 32% in 2011. The share falls to 27% in the New Policies Scenario and in the Current Policies Scenario.

Figure 3.1 ▷ World oil demand and oil price* by scenario

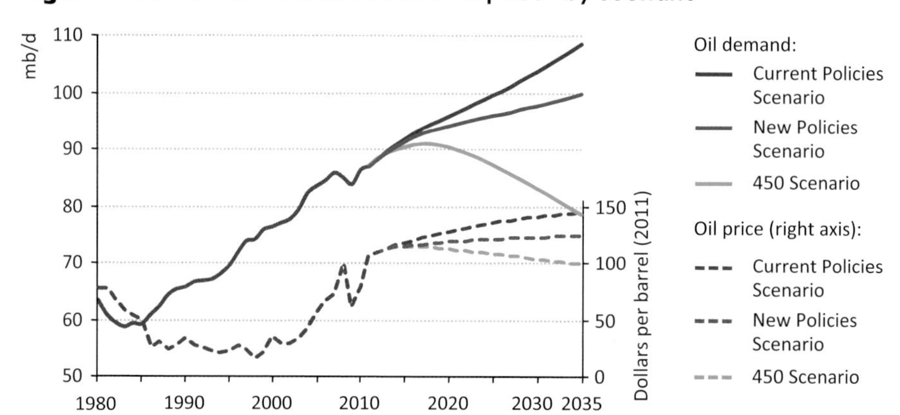

* Average IEA crude oil import price.

In the New Policies Scenario, demand rises progressively to 99.7 mb/d by 2035, 12.3 mb/d up on the 2011 level (Table 3.1). Measures to promote more efficient oil use and switching to other fuels, together with higher prices that result from price rises on international markets, reduced subsidies in some major consuming countries and increased taxes on oil products, help to offset much of the underlying growth in demand for mobility, especially in non-OECD countries. Several policy measures to curb oil demand, in particular in the transportation sector, have been announced over the past year and are taken into consideration. They include the extension of fuel-economy standards for passenger light-duty vehicles (PLDVs) to 2025 in the United States and to 2020 in Japan.

1. Preliminary data on total oil demand only are available for 2011 by region; the sectoral breakdown of demand is available up to 2010. All sectoral oil demand data presented for 2011 is therefore estimated.

Table 3.1 ▷ Oil and total liquids demand by scenario (mb/d)

			New Policies		Current Policies		450 Scenario	
	1990	2011	2020	2035	2020	2035	2020	2035
OECD	39.5	42.1	39.4	33.3	40.2	37.6	38.0	26.0
Non-OECD	22.9	38.4	47.1	57.1	48.1	61.3	45.1	45.0
Bunkers*	3.9	6.9	7.7	9.3	7.7	9.6	7.5	8.0
World oil	66.3	87.4	94.2	99.7	96.0	108.5	90.5	79.0
Share of non-OECD	*34%*	*43%*	*49%*	*55%*	*49%*	*55%*	*48%*	*52%*
World biofuels**	0.1	1.3	2.4	4.5	2.1	3.7	2.8	8.2
World total liquids	66.4	88.8	96.6	104.2	98.2	112.2	93.3	87.2

* Includes international marine and aviation fuel. ** Expressed in energy-equivalent volumes of gasoline and diesel.

Oil demand trends across the scenarios diverge most after 2020. While the majority of policies and measures that have been enacted or announced to curb oil demand result in action in the period to 2020 or, at the latest, 2025, most of the impact of these policies is felt during the last decade of the projection period. This is because of inertia and technology lead times: it can take many years for new and more efficient technologies that affect the use of oil to be developed and commercialised, while consumers buy new equipment or vehicles only when the existing ones are retired. For example, the full effect on oil demand of an improvement in the average fuel economy of new cars is not felt for at least fifteen years – the average lifetime of cars. In other sectors like industry, buildings or power generation, technology lifetimes can be much longer, such that the fuel substitution or fuel savings that result from a change in policy are even more protracted.

While government policy is an important factor behind long-term trends in oil demand, other factors play a key role as well, notably economic activity, population, prices and technology (see Chapter 1). All of these factors are inter-linked. Economic and population growth continue to push up demand for personal mobility and freight, but technology and fuel prices influence how transport services are provided. Oil-based fuels dominate transport energy use today, but this could change as new technologies are deployed, while more efficient vehicle technologies can help to decouple rising demand for mobility from fuel use. The use of oil in stationary applications – in power generation, industry and buildings – is similarly affected by economic and population growth, but here competition from other fuels plays a larger role. Generally, oil products continue to struggle to compete outside the transport sector in all three main scenarios, as a result of high oil prices relative to other fuels, with most power generators and end users switching away from oil where possible; by 2035, only 2% of global power generation is provided by oil. Whatever long-term path oil demand takes, it will surely continue to fluctuate from one year to the next in response to volatile oil prices, economic swings, weather variations and one-off events, such as natural disasters.

The average IEA crude oil import price – a proxy for international prices – required to balance demand and supply differs markedly across the scenarios, according to differences in how policy intervention affects underlying market conditions. Starting at $108 per barrel (in year-2011 dollars) on average in 2011, the price, which is derived through iterations of the World Energy Model (WEM), reaches $120/barrel in 2020 and $125/barrel in 2035 in the New Policies Scenario (in real terms). The price rises more quickly in the Current Policies Scenario, reaching $145/barrel in 2035: higher prices are needed for supply to keep pace with higher demand, as existing reserves are depleted faster and oil companies are forced to turn to more costly new sources of oil sooner. In the 450 Scenario, the oil price peaks at $115/barrel by around 2015 and then slides to $100/barrel as radical policy action causes oil demand to fall steeply.

Regional trends

Following the broad pattern of recent years, the outlook for oil demand varies markedly by region. While overall global oil use rose during the course of 2011, demand continued to fall in most OECD countries. Demand dropped by 1.9% in the United States, due to the economic downturn as well as a mild winter, and fell by 2.2% in Europe, as a result of the economic impacts of the Euro zone crisis. By contrast, Japan saw strong growth in the direct burning of crude oil and residual fuel oil (as well as natural gas) to compensate for the suspension of output from nuclear power stations. Overall, OECD oil demand fell by 1.3%, in contrast to the 3% overall increase seen in the non-OECD regions.

In all *WEO* scenarios, OECD oil demand continues to fall over the *Outlook* period, most rapidly in the 450 Scenario. In the New Policies Scenario, OECD demand falls from 42 mb/d in 2011 to 33 mb/d in 2035, mainly as a result of substantial gains in vehicle fuel efficiency and fuel switching in the transport sector, resulting from a combination of standards and other government measures, as well as higher pump prices (Table 3.2). Demand in non-transport sectors falls, mainly due to oil's declining competitiveness compared with other fuels. Oil demand falls in all OECD regions (Figure 3.2). The United States sees the largest absolute decline, with oil use falling from 17.6 mb/d in 2011 to 12.6 mb/d in 2035 (this fall of 5 mb/d is partially offset by an increase of 1 million barrels of oil equivalent per day in the use of biofuels). Demand falls as quickly in percentage terms in Japan, where oil use for power generation – boosted in the short term by the effects of the accident at the Fukushima Daiichi nuclear plant – falls away by 2035.

In contrast to the projected 8.8 mb/d decline in the OECD, the non-OECD countries in aggregate see their demand rise by 18.8 mb/d between 2011 and 2035 in the New Policies Scenario. As a result of their faster demand growth, total oil use in the non-OECD countries overtakes that of the OECD before 2015. It is about 70% higher by 2035, with the non-OECD countries accounting for 63% of world oil use (excluding international bunkers), compared with just under half today. Strong economic and population growth in the non-OECD countries, coupled with the huge latent demand for personal mobility, more than offsets important efficiency gains in transport.

Table 3.2 ▷ Oil demand by region in the New Policies Scenario (mb/d)

	1990	2011	2015	2020	2025	2030	2035	2011-2035 Delta	2011-2035 CAAGR*
OECD	**39.5**	**42.1**	**41.2**	**39.4**	**37.4**	**35.2**	**33.3**	**-8.8**	**-1.0%**
Americas	19.8	22.2	22.0	21.2	20.0	18.6	17.5	-4.7	-1.0%
United States	16.4	17.6	17.5	16.6	15.4	13.9	12.6	-5.0	-1.4%
Europe	12.8	12.6	12.0	11.4	10.9	10.4	10.0	-2.6	-1.0%
Asia Oceania	6.9	7.3	7.2	6.7	6.4	6.1	5.9	-1.4	-0.9%
Japan	5.1	4.3	4.1	3.7	3.5	3.2	3.1	-1.2	-1.4%
Non-OECD	**22.9**	**38.4**	**43.2**	**47.1**	**50.5**	**53.9**	**57.1**	**18.8**	**1.7%**
E. Europe/Eurasia	8.8	4.8	5.0	5.2	5.3	5.5	5.6	0.8	0.6%
Russia	4.9	3.1	3.2	3.2	3.3	3.4	3.5	0.4	0.5%
Asia	6.2	18.3	21.3	23.8	26.2	28.6	30.9	12.6	2.2%
China	2.3	9.0	11.0	12.7	13.9	14.7	15.1	6.1	2.2%
India	1.2	3.4	3.8	4.3	5.0	6.2	7.5	4.1	3.3%
Middle East	2.9	6.8	7.5	8.1	8.6	8.9	9.4	2.7	1.4%
Africa	1.8	3.1	3.4	3.8	4.0	4.2	4.5	1.4	1.6%
Latin America	3.2	5.5	6.0	6.3	6.5	6.6	6.8	1.3	0.9%
Brazil	1.4	2.4	2.6	2.7	2.8	2.9	3.1	0.7	1.1%
Bunkers**	3.9	6.9	7.2	7.7	8.2	8.7	9.3	2.4	1.2%
World oil	**66.3**	**87.4**	**91.6**	**94.2**	**96.1**	**97.7**	**99.7**	**12.3**	**0.6%**
European Union	n.a.	11.6	11.0	10.3	9.8	9.2	8.7	-2.9	-1.2%
World biofuels***	0.1	1.3	1.8	2.4	3.0	3.7	4.5	3.1	5.1%
World total liquids	66.4	88.8	93.4	96.6	99.1	101.4	104.2	15.5	0.7%

*Compound average annual growth rate. **Includes international marine and aviation fuel. *** Expressed in energy-equivalent volumes of gasoline and diesel.

Figure 3.2 ▷ Oil demand growth by region in the New Policies Scenario

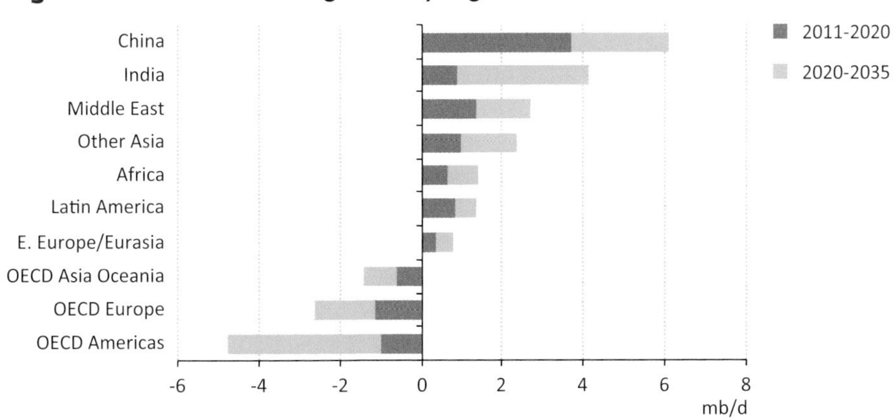

Asia is set to remain the main driver of non-OECD demand growth. China contributes the biggest increase in demand in absolute terms in the New Policies Scenario, with its consumption surging from 9.0 mb/d in 2011 to 15.1 mb/d in 2035 – an average annual increase of 2.2%. This increase is equal to just about half of the net increase in oil demand worldwide. Chinese oil demand exceeds that of the United States towards the end of the 2020s and its growth offsets large parts of the oil savings achieved in OECD countries: the increase to 2020 alone more than offsets demand reductions in the European Union over the whole projection period. Demand grows even faster in other emerging economies, notably India, Africa, Southeast Asia and the Middle East, though much less rapidly than over the last two decades, mainly because of slower economic and population growth. Oil demand growth in India is particularly large, though, and offsets around 80% of the savings realised in the United States until 2035.

Box 3.1 ▷ The economic implications of runaway oil-demand growth in Saudi Arabia

Saudi Arabia, in common with many countries in the Middle East, faces a growing need to rein in rampant oil demand. Between 2000 and 2011, Saudi oil consumption almost doubled, to 2.8 mb/d, partly because of surging use of oil for power generation. Demand for electricity, much of it for air-conditioning, has been growing by more than 7% per year since 2000, outstripping capacity to generate power from natural gas. This has forced generators to burn residual fuel and crude oil directly. In addition, extremely low pump prices have encouraged wasteful use of transport fuels; at 14 US cents per litre for gasoline and 7 US cents per litre for diesel, Saudi Arabia currently has the lowest gasoline prices in the region and among the lowest in the world. Annual per capita consumption of oil was 4.7 tonnes (35 barrels) in 2011 – the highest in the world.

In the medium term, the growth in oil use for power is expected to slow as new gas supplies become available. The recent completion of the second phase of a large-scale project to develop the Karan gas field is expected to yield enough gas to displace at least 200 kb/d of oil once it reaches full capacity. Other projects to expand non-associated gas production are planned, though low gas prices may discourage investment in gas exploration and development. The New Policies Scenario assumes that the Saudi government reduces oil and gas subsidies to a limited degree, which helps to encourage more efficient energy use and spurs more gas production. But cutting subsidies and raising prices is not an easy task. Although a signatory to the 2009 G-20 call for inefficient fossil fuel subsidies to be phased out, the Saudi government has not yet announced any plans to raise oil or gas prices, though electricity price rises are planned.

The under-pricing of oil products can carry a high economic price in the longer term. If no new policies are implemented, then Saudi-Arabia's oil demand would be around 400 kb/d higher by 2035, with a corresponding impact on export volumes. Given the large share of government spending that is financed by oil-export revenues, any reduction in export volumes could have an impact on the economy and the government's ability to spend more on domestic welfare and services.

The pace of demand growth in the Middle East is critical to the outlook for the region's oil exports and, therefore, to its economic prospects. As a result of strong economic growth – partly thanks to high international prices – and widespread subsidies that have kept the price of oil products in domestic markets low, the Middle East, already a big oil producing region, has become a major oil consumer too. Its oil demand grew by around 60% between 2000 and 2011, an average rate of 4.2%, or 225 thousand barrels per day (kb/d), per year, outstripping every *WEO* region, except China. In the New Policies Scenario, the rate of demand growth in the Middle East slows to 1.4% over the *Outlook* period, but these projections hinge on governments following through on plans to reduce fuel subsidies, expand gas supplies to allow power generators and industrial consumers to switch away from burning oil and introduce other measures to promote energy efficiency. Failure to do so would have far-reaching implications for the key producers' capacity to export oil and fuel regional development and social programmes – not least in Saudi Arabia (Box 3.1). There is enormous potential across the region for oil to be used more efficiently (see Part B).

Sectoral trends

More than 50% of global primary oil demand today is concentrated in the transport sector, compared with around 8% each in the buildings and industry sectors and 7% in power generation (Figure 3.3). As noted, for the non-transport sectors, high oil prices provide a strong economic signal to reduce oil use, either by using it more efficiently or using other fuels. Oil consumption in industry is only 8% higher in 2035 than in 2011, largely as a result of switching to natural gas and, to a lesser extent, electricity. While the use of natural gas for heat and steam generation in industry is already widespread in the United States and Europe today, the projected gap between oil and gas prices creates considerable scope for gas to capture most of the increase in demand for these purposes in Asia and other non-OECD regions. Up to now, this shift has been limited by relatively high gas prices and lack of supply. In some countries, including China, coal may also be preferred to oil products, due to local availability. There are, nonetheless, limits to the substitution of oil in industry, particularly in the petrochemicals industry and in other non-energy applications.

Oil currently meets about 11% of global energy needs in buildings, amounting to an estimated 7.7 mb/d, split about equally between OECD and non-OECD countries. Patterns of oil use in buildings vary across regions, reflecting to a large extent their different stages of economic development: while more than 80% is used in the services sector and for space heating in OECD countries, much of the oil used in buildings in non-OECD countries is in the form of kerosene and liquefied petroleum gas (LPG) for cooking, in particular in Africa. Over time, the use of oil in buildings worldwide is expected to decline, due to the use of more efficient appliances and increasing substitution with electricity and gas; the main exception is in Africa, where switching from traditional fuels continues to boost demand for oil products (see Chapter 18). Oil demand in buildings is projected to reach a peak in the coming few years and then dwindle gradually, reaching 7 mb/d in 2035, or about 8% of total energy use for buildings.

Figure 3.3 ▷ **World oil demand by sector in the New Policies Scenario**

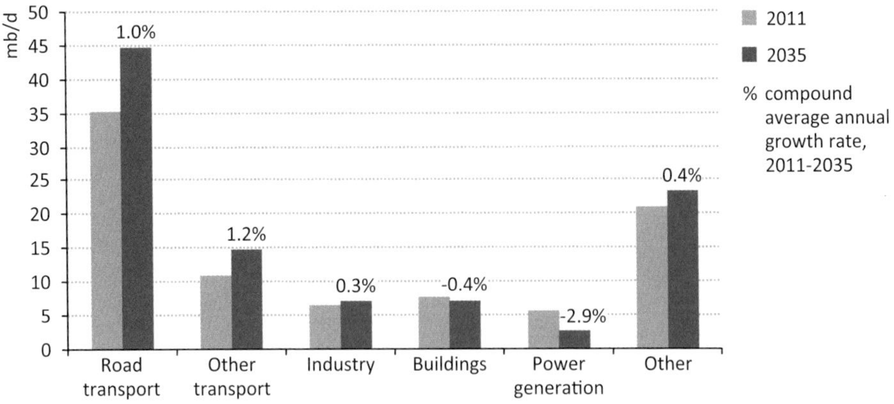

Note: Other includes non-energy use, including feedstocks for industry.

The use of oil in other sectors, including power generation, is likely to stagnate due to the fuel's declining competitiveness. The share of oil in power generation has been falling steadily for many years, reaching 6% in 2011, and is projected to fall further, to just 2% by 2035 in the New Policies Scenario. In absolute terms, oil-fired generation drops by about half, with oil consumption falling from above 6 mb/d to around 3 mb/d. Oil use for power all but disappears in the OECD, but remains significant in some non-OECD regions, including the Middle East.

The transport sector accounts for well over half of global oil consumption today and this share is expected to rise further in the coming decades. In the New Policies Scenario, total transport oil demand rises from an estimated 46 mb/d in 2011 to about 60 mb/d in 2035, its share of total oil demand reaching 60%. Growth in demand for transport services – especially in the non-OECD countries, where the level of car ownership is still much lower than in OECD countries – is expected to outweigh the effect of large improvements in fuel economy and the growing penetration of alternative fuels, such as biofuels, natural gas and electricity (in battery electric vehicles and trains).

Within the transport sector, the fleet of passenger light-duty vehicles accounts for the bulk of road transport oil use today and these vehicles are projected to remain the leading consumer of oil in the transport sector through to 2035 (Figure 3.4). Nonetheless, oil demand for road freight grows more quickly, almost reaching the current level of PLDV demand by 2035 (see the focus on freight below). After road transport, aviation (domestic and international) and maritime transport are the other major consumers of oil in the transport sector. There have been only a few cases of governments intervening to curb demand growth in these sectors, mainly because applying regulations, such as standards or taxes, would require high levels of international co-operation. There has been strong resistance from some countries, notably China, India and the United States, to recent moves to include airline companies in the EU Emissions Trading System.

Figure 3.4 ▷ **World transport oil demand by sub-sector in the New Policies Scenario**

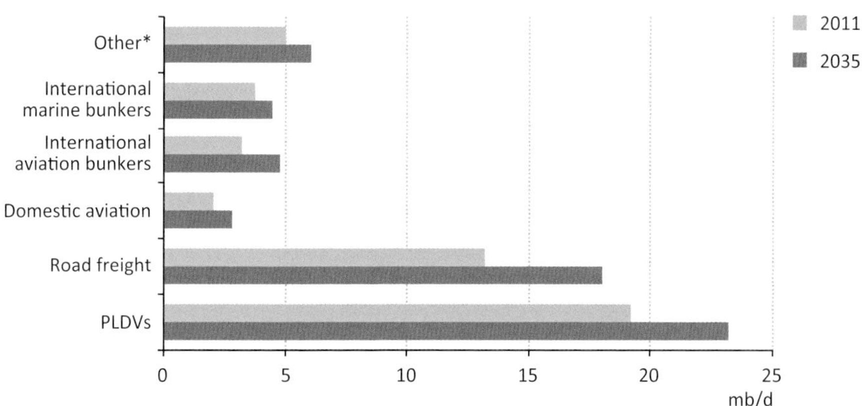

* Other includes other road, domestic navigation, rail, pipeline and non-specified transport.

Fuel economy regulation has been the main policy response in recent years to rising oil prices and to concerns about energy security and climate change. Some of the regulations in place today have a long pedigree. In the United States, for example, corporate average fuel-economy standards go back to the 1970s; the first standard was introduced in Japan in 1979, while the European Union introduced mandatory standards in the early 2000s having previously negotiated voluntary agreements with the car industry to improve vehicle efficiency. These standards have become more stringent in recent years. Outside the OECD, only China has taken steps to adopt standards (which will take effect in two stages in 2015 and 2020), though India plans to do so. Over time, there is likely to be more focus on standards for heavy-duty vehicles, given the prospects of rapid growth in oil demand from the road freight sector. Today, such standards exist only in Japan (since 2006) and in the United States (since 2011).

The savings in oil consumption that can be achieved by the adoption of fuel-economy standards are substantial. We estimate that the standards that have already been adopted together with those currently planned for PLDVs in the United States, Japan, the European Union, China and India alone, are set to save a cumulative total of 17 billion barrels of oil over the *Outlook* period, compared with a baseline assuming that fuel economy of new PLDV sales remains unchanged at 2010 levels (Figure 3.5).[2] Those savings are roughly equivalent to the current proven crude oil reserves of Qatar and Oman.

The global PLDV fleet continues to expand as incomes and population rise, from an estimated 870 million in 2011 to 1.7 billion in 2035. Most of the growth arises in China, where the fleet jumps from around 60 million to more than 400 million; the fleet in the leading PLDV markets of the European Union and the United States (where the recession

2. As fuel economy policies were enacted in different years in all regions considered, 2010 was chosen as a common starting point. The analysis thereby neglects savings that have been achieved prior to 2010 as a result of enacted policies. These savings are not insignificant.

led to a decline in the PLDV fleet, due to increased scrappage and lower sales) grows only modestly, reaching a combined 560 million in 2035 compared with 450 million in 2011. India's fleet is projected to reach around 160 million vehicles in 2035, up from only around 14 million in 2011. Most of the increase comes after 2020 (Figure 3.6). Despite this growth, vehicle ownership in China and other non-OECD countries remains well below the level of OECD countries, constrained in part by difficulties in building sufficient new roads (Box 3.2).

Figure 3.5 ▷ Cumulative oil savings from vehicle fuel-economy standards in selected regions, 2010-2035

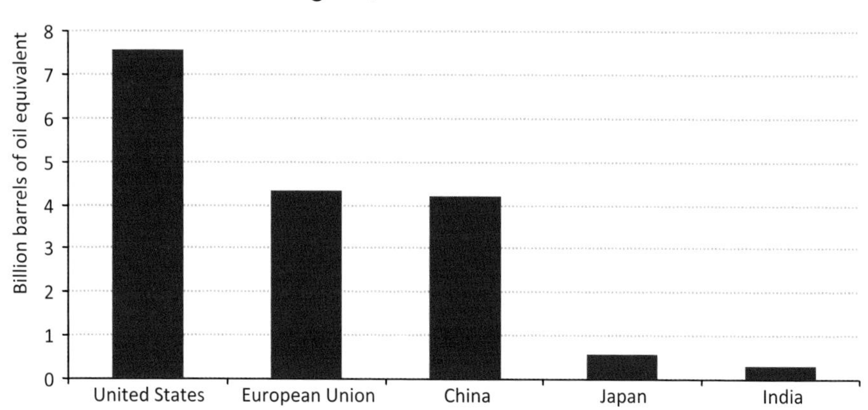

Notes: Savings are relative to the average fuel consumption level of PLDVs in 2010 in each region. They are largest in regions with strict standards and high vehicle sales volumes.

Box 3.2 ▷ Cars everywhere, but where will they go?

Today, there are around 45 million paved lane-kilometres of roads available worldwide (IEA, 2012a), about 30% more than a decade ago. Urban areas in most countries in the OECD and elsewhere have traditionally suffered from congestion, but in many cases the problem has worsened considerably in recent years, as car ownership rates and driving have risen. City authorities have adopted various measures to try to combat congestion, such as parking restrictions and charging for the use of inner-city roads, with varying degrees of success.

Although the number of PLDVs is projected to double between now and 2035 in the New Policies Scenario, the size of the road network is projected to expand by only 40%, to reach 62 million paved lane-kilometres. Some 80% of new roads are in non-OECD countries, more than half of which are in China and India alone. The total investment required amounts to over $20 trillion over the projection period. Despite this projected impressive build up of road infrastructure, congestion problems and pollution are expected to worsen. China's road occupancy increases by more than 70% on average, overtaking the level of the United States; that of India more than triples. Increasing congestion would tend to offset part of the effect of improvements in vehicle fuel economy, as stop-and-go driving in heavy traffic is more energy-intensive than driving on open, uncongested roads.

The PLDV market remains dominated by internal combustion engine (ICE) vehicles in the New Policies Scenario, broadly consistent with the projections in last year's *WEO*. By 2035, around three-quarters of all vehicles sold are conventional ICE vehicles, followed by hybrids at more than 20%, and natural gas vehicles, at 3%. Electric vehicles (EVs) and plug-in hybrids reach a combined share of 4% in 2035. Even though there have been recent policy discussions in India about adopting a target of 6-7 million EVs by 2020 under a National Electric Mobility Mission, policy support generally is not sufficient yet to generate expectations of higher levels of EV market penetration (IEA, 2011).

Figure 3.6 ▷ PLDV fleet in selected regions in the New Policies Scenario

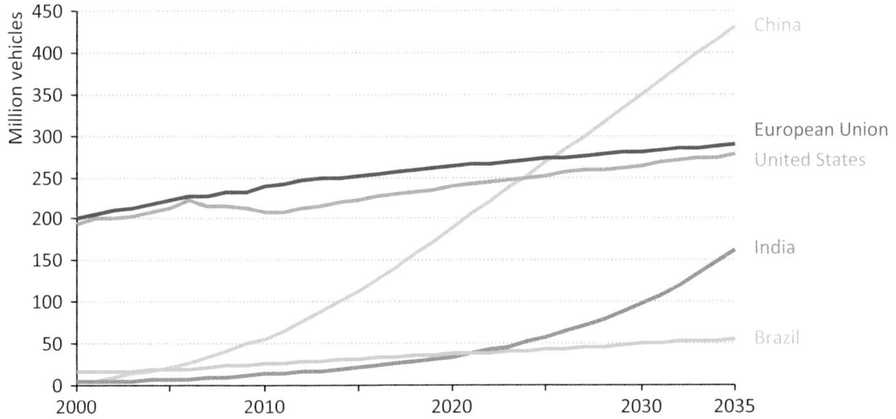

Special topic: heavy freight road transport

Road freight plays a key role in global transport demand for oil in the New Policies Scenario, accounting for almost 40% of the overall increase in demand between 2011 and 2035. Total road freight oil consumption (including trucks and light commercial vehicles[3]) rises from 13 mb/d in 2011 to 18 mb/d in 2035, approaching the level of PLDVs today. In recognition of the growing importance of road freight, we summarise below the results of a detailed analysis of the prospects for the sector undertaken for this year's *Outlook*.

In principle, goods can be carried by all main modes of transport – road, rail, shipping or aviation. In practice, there are a number of factors which determine the most appropriate mode for freight, including the required speed of delivery, the size and weight of goods and the availability and capacity of infrastructure for point-to-point delivery. While rail is the predominant mode of freight transport in large countries, like the United States, China or Russia, goods are more often transported by vans and trucks in densely populated countries, such as in Europe, or in regions with a low density of rail networks, like Africa (IEA, 2009). In all cases, however, final delivery is made mostly by road.

3. Freight trucks are a type of heavy-duty vehicle, the other type being buses. Light-duty vehicles (LDVs) are split between commercial (LCVs) and passenger vehicles (PLDVs). For the purpose of this chapter, we define a truck as a commercial freight vehicle with a gross vehicle weight of more than 3.5 tonnes. We refer to heavy freight trucks as trucks with a gross vehicle weight of more than 16 tonnes.

The most commonly used overall indicator of freight activity in economic planning is the tonne-kilometres, *i.e.* the weight of goods transported, multiplied by the distance transported. In these terms, the growth in the amount of goods transported by road in non-OECD countries has outpaced that of OECD countries over the past decade, and the gap is set to widen – in the New Policies Scenario, the growth of China alone, up to 2035, is as large as that of the OECD countries as a whole (Figure 3.7). Historically, tonne-kilometres have correlated strongly with income. Although this correlation has begun to weaken in some markets in recent years, because of saturation effects, it is expected to remain fairly constant at the global level over the *Outlook* period. The load carried per vehicle is expected to fall in non-OECD countries as a whole, as efforts are made to reduce overloading, even though the use of progressively larger trucks to optimise logistics will offset much of this reduction. In OECD countries, the load carried per vehicle is set to rise through optimisation of logistics.

Figure 3.7 ▷ Incremental road freight growth by region since 2000 in the New Policies Scenario

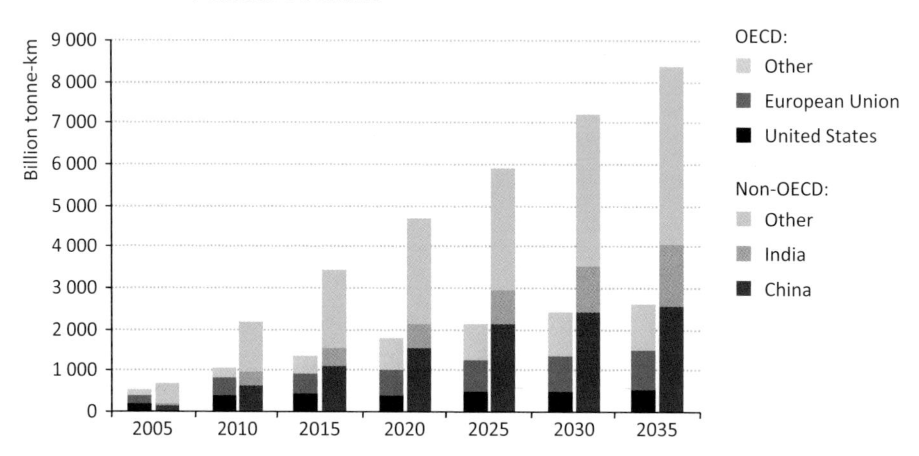

There are two main types of trucks weighing more than 3.5 tonnes: long-haul and delivery/distribution. Long-haul trucks typically weigh more than 16 tonnes (heavy trucks), have two or three axles and have a horsepower rating of between 250 and 800. The heaviest of them are operated essentially all year round, usually covering more than 100 000 km per year and sometimes twice as much. Average fuel consumption is between 25-50 litres of diesel per 100 km (l/100 km), depending on the vehicle technology, the geography of the operating area, average load and driving behaviour. Delivery trucks vary in weight between 7.5-25 tonnes and generally last much longer than long-haul trucks, as they operate for fewer hours each day and cover shorter distances (typically about 40 000-60 000 km per year). Other types of trucks vary considerably, according to their purposes, as does their average annual level of operation, from about 5 000 km for special vehicles and fire-fighter trucks, up to about 60 000 km for construction trucks. Fuel consumption also varies, from about 30 l/100 km in normal operations, up to more than 80 l/100 km.

The truck market is more diversified than that for PLDVs, partly because of the large variety of different purposes served, and partly because requirements are much more stringent for heavy-duty vehicles with regard to operation and endurance, which often requires tailored solutions.

The largest markets for trucks today are in the United States and China, each having sales of around 1.6 million in 2011 (Figure 3.8). Sales in the European Union, Japan and India are much smaller, at around 500 000 or less, though India's market is growing rapidly (sales almost doubled over the two years to 2011). There are big differences in the size of trucks sold across countries: in the United States, for example, only about 7% of the trucks sold in 2011 were heavy trucks weighing more than 16 tonnes. In China, these heavy trucks accounted for around two-thirds of the market, reflecting their use in the rapidly expanding manufacturing and construction sectors.

Figure 3.8 ▷ Freight truck sales and stock by region in the New Policies Scenario

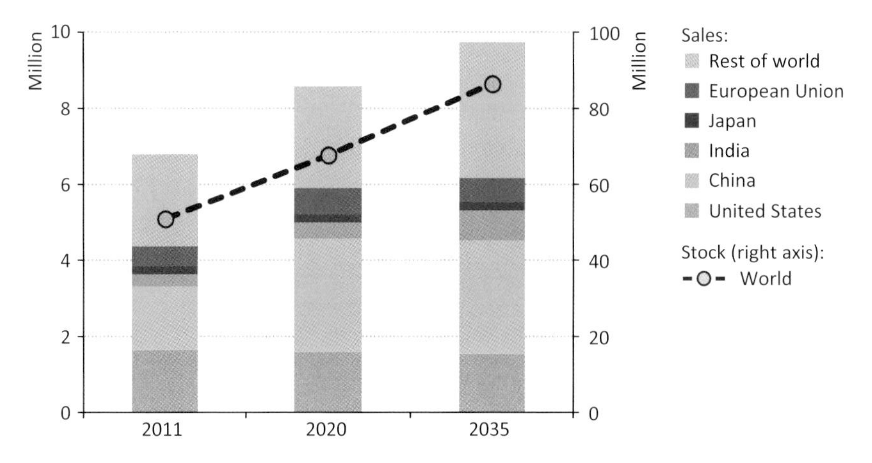

The global market for trucks is much smaller than for other types of road vehicle: of around 80 million vehicles sold worldwide in 2011, trucks accounted for only around 7 million. We estimate that there are some 50 million trucks on the road today, compared with more than 850 million PLDVs. The truck stock is projected to reach 90 million in 2035 in the New Policies Scenario. The average lifetime of trucks varies widely according to the type of vehicle and how it is used, resulting in big differences across regions. Long-haul trucks, for example, are usually retired after around eight to ten years in OECD countries, *i.e.* by the end of their economic lifetime. But in most non-OECD countries, they are generally used for much longer, sometimes more than 20 years. The notable exception is China, where long-haul trucks currently operate on average for only six to seven years because of heavy overloading and safety reasons.

The truck manufacturing industry is much less globalised than that for PLDVs. Manufacturers include a few companies that operate globally, including Daimler, Volvo, Iveco, MAN/Scania, Navistar and Paccar, as well as companies that focus on single, national markets, such as

Dongfeng, Foton, FAW and CNHTC in China, Kamaz in Russia and Ashok Leyland, Tata and Eicher in India. Together, the companies cited held a total world market share of about 70% in 2010.[4] The industry is likely to consolidate in the coming years as more stringent environmental regulations lead to higher development and production costs, particularly in BRICS countries.[5]

Despite the small size of the truck fleet, fuel consumption is very large because of the sheer size and weight of the vehicles and their long operating times. More than 60% of all the diesel consumed globally in road transport is by trucks, and it is by far the most important fuel used by trucks, accounting for more than 90% of total fuel use, or 9 mb/d; gasoline accounts for 6% and biofuels for most of the remainder (both as additives and as a stand-alone fuel). The dominance of diesel in truck fuel consumption increases with vehicle size; long-haul heavy trucks (weighing more than 16 tonnes) use diesel almost exclusively.

Over the *Outlook* period, diesel maintains its role as the dominant fuel for trucks in the New Policies Scenario. Diesel consumption by trucks rises from 9 mb/d in 2011 to more than 13 mb/d in 2035, its share of total road freight energy consumption remaining at a level of around 90%. Alternative fuels, including biofuels and natural gas, make inroads and reach a share of more than 5% of truck fuels combined, up from around 2% today. The increasing use of alternative fuels comes mainly at the expense of gasoline, the share of which is reduced from around 6% of all oil-based fuels in road freight today to about 2% by 2035, as trends towards the use of larger trucks increasingly favour the use of diesel over gasoline. But overall, switching to alternative fuels makes little impact on oil demand for road freight, with fuel economy improvements making the biggest contribution to dampening the pace of growth in demand through to 2035 (Figure 3.9).

Figure 3.9 ▷ World freight truck oil demand in the New Policies Scenario

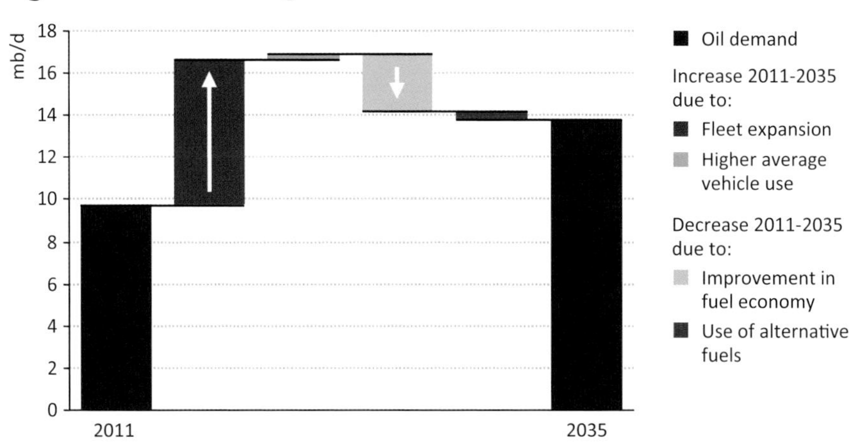

Note: Vehicle use refers to changes in tonne-kilometres driven by vehicle.

4. For trucks of more than 6 tonnes.
5. Brazil, Russia, India, China and South Africa.

Reducing the dependence of road freight on oil by increasing the use of alternative fuels is constrained by the lack of viable fuel options for trucks compared with light-duty vehicles. In the near term, the two main options are biofuels and natural gas (Figure 3.10), as the electrification of trucks is likely to be hampered by the size and weight of the battery that would be required to propel the heavy weight of the vehicle and its load. Increasing the use of biofuels would require an expansion of biodiesel supply. At present, biodiesel production worldwide is only one-fourth that of ethanol, almost half of it in Europe. Advanced biodiesel technologies, which convert biomass to diesel in a more cost-efficient way and offer bigger carbon dioxide (CO_2) savings than conventional biodiesel, are under development but have not yet passed the demonstration stage. Natural gas can be used as a fuel by all types of road freight vehicles. Liquefied natural gas (LNG) is more practical for heavy freight, particularly long-haul trucks, as the size of the fuel tank required for compressed natural gas (CNG) in order to offer an appropriate driving range would be too big. CNG is a viable option for light commercial vehicles and can be an option for trucks in densely populated countries, where range is less of an issue. The use of LNG in road freight has so far been constrained by its high cost, compared with diesel, though this might change if there is sustained fall in the price of natural gas, relative to that of oil (which we do not assume). In the New Policies Scenario, the share of natural gas in road freight worldwide reaches just over 2% in 2035, but some markets see higher shares (for example 10% in heavy trucks in the United States). Lower natural gas prices or lower vehicle conversion costs would be needed to reduce payback periods and stimulate faster growth in gas use in road freight. In addition, developing refuelling infrastructure along main trucking routes (such as "America's Natural Gas Highway" in the United States) requires large upfront investments. To achieve higher shares of LNG in road-freight transport, firm policies are required to address these barriers and create a self-sustaining market.

Figure 3.10 ▷ Alternative fuel use by freight trucks in the New Policies Scenario

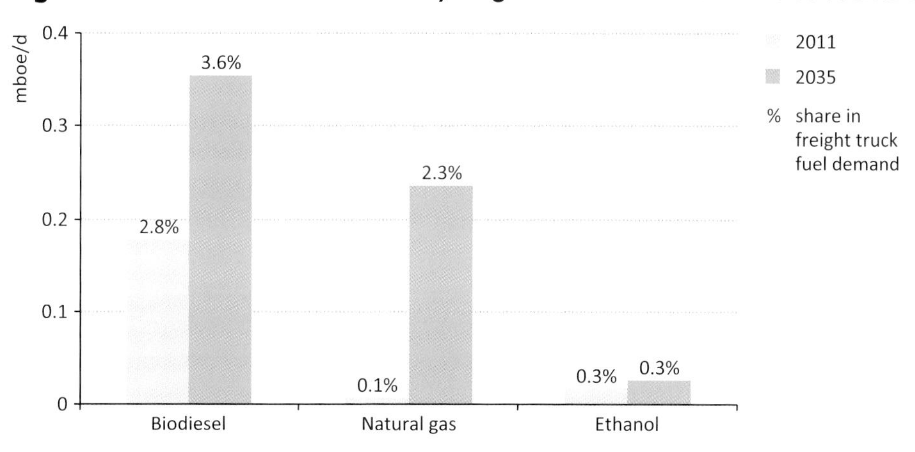

Increasing efficiency is the main option for reducing oil demand from road freight transport.[6] Fuel costs are an important criterion for many truck operators in choosing

6. See Part B for a more detailed discussion about the scope for efficiency gains in transport and other sectors.

their vehicles (in particular in regions with high fuel taxes like Europe), which has driven manufacturers to improve the fuel economy of their models, but there is still considerable scope for further efficiency gains. Recent studies point to a technical potential to improve the fuel economy of trucks by up to 50% until 2020 in the United States and Europe (IEA, 2012b), though the vehicle manufacturers doubt that the potential is that big, with some companies suggesting it may be closer to 30% for long-haul trucks, and then only by 2030. Whatever the potential, what matters – in the absence of regulation – is the payback period for truck operators, who often look to recover their investment within eighteen months (IEA, 2012c). But payback periods vary considerably by vehicle segment and average annual use: while there are a number of options for long-haul trucks, with payback periods of less than two years, there are fewer cost-effective options for urban and service delivery trucks and the largest improvements, such as from hybridisation, mostly come at prohibitively high cost (Hill, *et al.*, 2011).

For this reason, efficiency gains will need to be driven by regulation, in particular in regions with low fuel tax levels. At present, only the United States and Japan have introduced fuel-economy standards for trucks. The European Union and Canada are currently discussing the introduction of such standards. China is in the process of establishing the actual fuel economy of trucks on the road as a basis for developing possible standards. The process is complicated by controversy over the efficiency potential and the pursuit of multiple policy objectives: air pollution policy can reduce the potential to improve fuel efficiency (*e.g.* catalytic converters come at a small efficiency cost), while limits on the maximum size of heavy freight trucks for safety reasons reduce the scope for realising further economies of scale. The lack of fuel-economy standards for trucks in most countries is compounded by a lack of good certifiable information: while fuel economy labelling is common for passenger cars in many OECD countries, it is rare for trucks, complicating the choice in particular for small fleet operators. In countries where fuel economy policy does not exist, transparency on fuel consumption through improved measurement procedures and labelling will be essential to identify the efficiency potential.

In the New Policies Scenario, the average on-road fuel consumption of heavy trucks (exceeding 16 tonnes) is projected to drop by an average of 22% between 2011 and 2035, reaching 28 l/100 km. Efficiency remains highest in the OECD countries, with strong improvements occurring in the United States (29%) and Japan (24%) – again the only countries, as yet, with fuel-economy standards (Figure 3.11). But regions with comparatively high levels of fuel taxes (like Europe) or currently high levels of average fuel consumption (like China or India) also see significant reductions. The bulk of the efficiency improvements in these regions occur in the period to 2020, when the payback periods for the required investments are still short.[7]

7. For regions without fuel-economy standards, the World Energy Model endogenises the investment decision for improving truck efficiency through calculating their economic competitiveness under the oil price assumptions of the New Policies Scenario. For details, see *www.worldenergyoutlook.org*.

Figure 3.11 ▷ **Average on-road fuel consumption of heavy freight trucks by region in the New Policies Scenario**

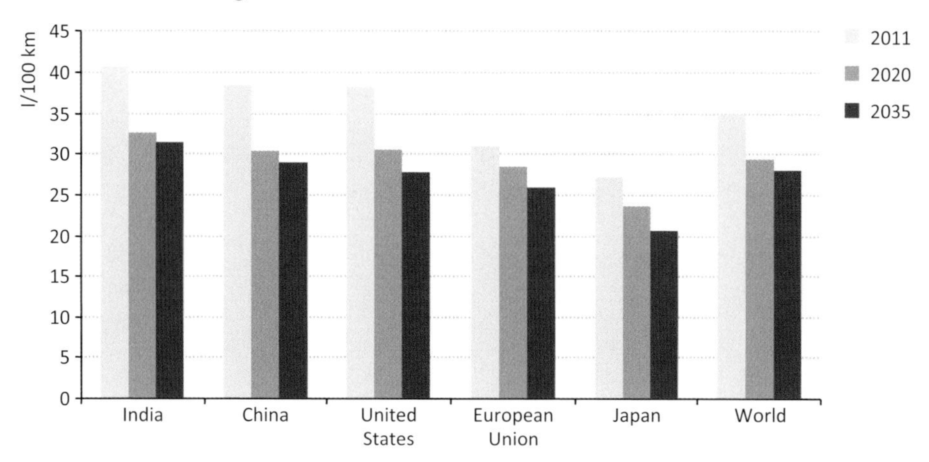

Note: Heavy freight trucks are defined as larger than 16 tonnes. Data for 2011 is estimated.

Supply

Reserves and resources

The world's remaining resources of oil are not a limiting factor in meeting projected demand to 2035, even in the Current Policies Scenario, though the cost of exploiting new sources is set to rise, as the most accessible and, therefore, least costly resources are depleted. According to the *Oil and Gas Journal* (O&GJ), proven reserves of oil worldwide, which provide an indication of the near- to medium-term potential for new developments, amounted to 1 523 billion barrels at the end of 2011 – an increase of 3.6% on a year earlier and a new record (O&GJ, 2011); while according to the *BP Statistical Review of World Energy*, proven reserves increased by 1.9% to 1 653 billion (BP, 2012).

The main difference between the two sets of estimates relates to extra heavy oil in Venezuela's Orinoco belt, as the *Oil and Gas Journal* had not yet taken account of the latest announcement about these reserves, with smaller differences in Russia, China and the United States. With the latest assessment of the Orinoco by Petróleos de Venezuela (PDVSA), the national oil company, Venezuela is now the leading country for proven reserves, having overtaken Saudi Arabia in 2010 (Figure 3.12).[8]

Proven reserves have increased by close to one-third since 2000, with more than half of the increase coming from upgrades of Orinoco reserves and most of the rest from revisions in other OPEC countries (though there are question marks over the estimates for some

8. The new Venezuelan number was included by OPEC in their 2011 Annual Report as the end-2010 proven reserves. Due to timing of publication, it was not included in the IEA's *World Energy Outlook 2011*.

OPEC countries and their comparability with those for other countries).[9] Outside OPEC, Russia and Kazakhstan have increased reserves the most since 2000, while the rest of the non-OPEC countries have broadly maintained their reserves level. OPEC countries account for about 70% of world total reserves. The global reserves-to-production ratio, which is sometimes used as an indicator of future production potential, has increased steadily in recent years, to around 55 years at end-2011.

Figure 3.12 ▷ Proven oil reserves in the top 15 countries, end-2011

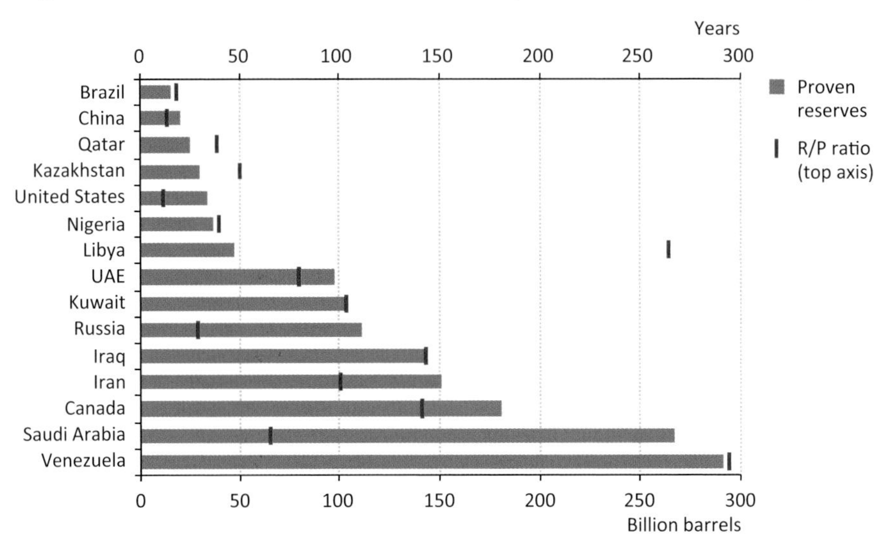

Notes: R/P ratio = reserves-to-production ratio. The R/P ratio does not imply continuous output in the future, nor that production will stop at the end of the period. The ratio fluctuates over time as new discoveries are made, existing reserves are revised and technology and production rates change.

Source: IEA analysis.

More than 70% of the increase in proven reserves since 2000 has come from revisions to reserves in discovered fields, known as reserves growth, with the rest coming from discoveries. Discoveries have picked up in the last few years with increased exploration, driven by higher oil prices, but they still lag production by a large margin. In 2011, 12 billion barrels were discovered, equal to 40% of the oil produced during the year. The average size of discoveries has tended to increase since the end of the 1990s, reversing the historic trend, as exploration has focused on new deepwater locations.

Estimates of remaining recoverable resources (RRR), which include proven reserves plus oil in existing fields that could be "proven up" in the future, as well as technically producible oil that is yet to be found, provide a much better indication of long-term production

9. While some progress has been made in establishing a harmonised system of defining and classifying reserves, the way they are measured in practice still differs by country and jurisdiction. Many oil companies use external auditors of their reserve estimates based on the Petroleum Resource Management System (PRMS) methodology (SPE, 2007) and publish the results, but some national companies do not, making it difficult to assess if PRMS definitions are used consistently.

potential. There is, inevitably, much more doubt about their size and, therefore, about just how much oil remains available for future production. Based on data from a number of sources, including the US Geological Survey (USGS) and the German Federal Institute for Geosciences and Natural Resources (BGR), we estimate that remaining technically recoverable resources (ultimately recoverable resources – less cumulative production to date) at the end of 2011 amounted to close to 5 900 billion barrels – a 9% increase on last year's estimate (Figure 3.13). This estimate takes into account the latest USGS assessments of undiscovered conventional resources worldwide (USGS, 2012a) and of reserve growth potential (USGS, 2012b)[10]; it also takes into account an IEA estimate for potential light tight oil resources (oil produced from shale, or other very low permeability rocks, with technologies similar to those used to produce shale gas), which were not previously included.

Figure 3.13 ▷ Ultimately technically recoverable resources and cumulative production by region in the New Policies Scenario

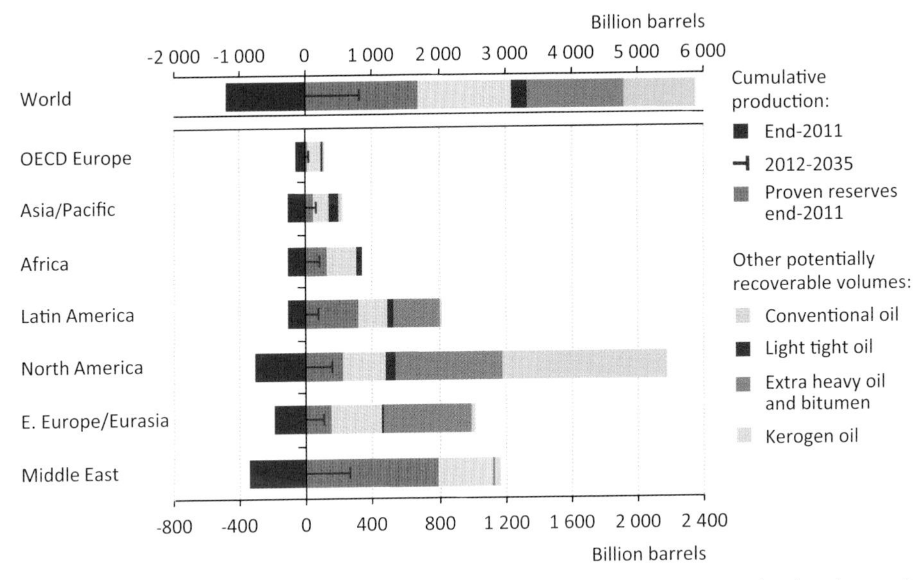

Notes: Proven reserves are usually defined as discovered volumes having a 90% probability that they can be extracted profitably. Ultimately recoverable resources comprises cumulative production, proven reserves, reserves growth (the projected increase in reserves in known fields) and as yet undiscovered resources that are judged likely to be ultimately producible using current technology. Remaining recoverable resources are equal to ultimately recoverable resources less cumulative production. The latter is shown in the chart as a negative number, such that the total of the bars to the right indicates remaining recoverable resources.

Sources: BGR (2011); O&GJ (2011); USGS (2000, 2012a); IEA databases and analysis.

A large proportion of the world's remaining recoverable resources of oil are classified as unconventional, mainly light tight oil, extra-heavy oil, natural bitumen (oil sands) and

10. See Chapter 4, Box 4.2, for more detail about the USGS assessments.

kerogen oil.[11] In total, these unconventional resources amount to an estimated 3 200 billion barrels – equal to more than half of all remaining resources (Table 3.3). Unconventional resource estimates are less reliable than those of conventional resources, as they have generally been less thoroughly explored and studied, and there is less experience of exploiting them. In many cases, considerable technical, environmental, political and cost challenges will need to be overcome for them to be produced commercially. Appraisal and production is most advanced in North America, which largely explains why that region is currently assessed to hold the largest estimated unconventional resources. Other regions that have received less attention because of their large conventional resources, including the Middle East and Africa, may also hold large volumes of unconventional oil.

Figure 3.14 ▷ Liquid fuels schematic

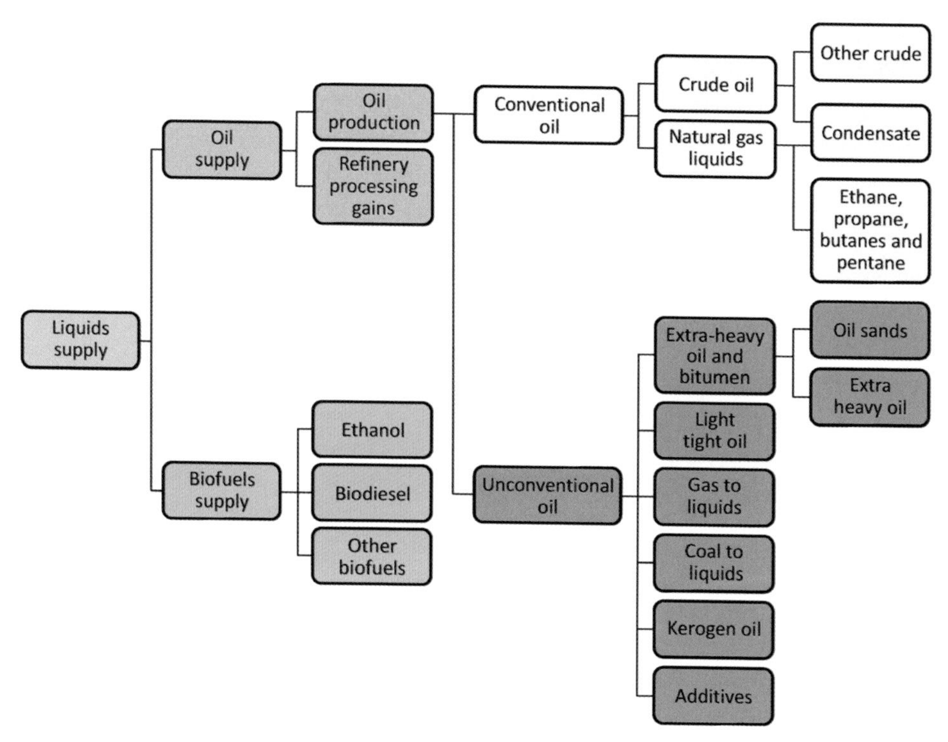

Notes: For more discussion of unconventional oil, refer to *WEO-2010*, Chapter 4 (IEA, 2010). Field condensate (natural gas liquids [NGLs] separated from a gas flow at the well site) is reported as part of crude oil in some countries (OECD in particular) and part of NGLs in others (OPEC in particular).

The volume of remaining recoverable resources is obviously crucial to potential future production. But it is important to allow for the trend that estimates of that volume generally rise over time, as advances in technology open up new areas of production. Also,

11. We use in this chapter the same definitions of liquid fuels as in previous *Outlooks* (Figure 3.14), except that this year we present separate projections for light tight oil and include them into the unconventional oil category; in the past, these have been included in the figures for crude oil. This new classification for light tight oil is coherent with our classification of shale gas as unconventional gas.

technological advances can lower costs, thereby increasing the proportion of technically recoverable resources that can be recovered commercially. Higher prices, which are assumed in the Current and New Policies Scenarios, can also increase the recovery factor. Thus, it is likely that the estimated size of ultimately recoverable resources of all types of oil will increase further in the future.

Table 3.3 ▷ Remaining technically recoverable oil resources by type and region, end-2011 (billion barrels)

	Conventional			Unconventional				Total
	Crude oil	NGLs	Total	EHOB	Kerogen oil	Light tight oil	Total	
OECD	318	99	417	812	1 016	101	1 929	2 345
Americas	253	57	310	809	1 000	70	1 878	2 188
Europe	59	31	91	3	4	18	25	116
Asia Oceania	5	11	16	0	12	13	25	41
Non-OECD	1 928	334	2 261	1 069	57	139	1 264	3 526
E. Europe/Eurasia	352	81	433	552	20	14	586	1 019
Asia	95	26	121	3	4	50	57	178
Middle East	982	142	1 124	14	30	4	48	1 172
Africa	255	52	306	2	0	33	35	341
Latin America	245	32	277	498	3	37	538	815
World	2 245	433	2 678	1 880	1 073	240	3 193	5 871

Note: EHOB = extra-heavy oil and bitumen.

Sources: BGR (2011); O&GJ (2011); USGS (2000, 2012a, 2012b); IEA databases and analysis.

Production prospects

Oil supply – the production of crude oil, NGLs and unconventional oil, plus processing gains – follows the same trajectory as demand in each of the three scenarios (Table 3.4); capacity in excess of production is not modelled explicitly in our World Energy Model and therefore is implicitly assumed to remain unchanged. In the New Policies Scenario, oil production (net of processing gains) rises from 84 mb/d to 97 mb/d in 2035, the increase coming entirely from natural gas liquids (NGLs) and unconventional sources. Output of crude oil (excluding light tight oil) fluctuates between 65-69 mb/d over 2011-2035, never quite reaching the historic peak of 70 mb/d in 2008, and falling by 3 mb/d in that period.

Non-OPEC production in total reaches a plateau around 53 mb/d after 2015 and begins to decline soon after 2025, reaching 50 mb/d by the end of the projection period. Strong growth in unconventional production – mostly from light tight oil and natural gas liquids in the United States and from oil sands in Canada – is insufficient to offset the fall in non-OPEC crude oil production. In contrast, OPEC oil production accelerates through the projection period, particularly after 2020, rising from 36 mb/d in 2011 to 38.5 mb/d in 2020 and then to 46 mb/d in 2035. OPEC's share of world oil production increases from 42% in 2011 to 48% in 2035 – a level still below the historical highpoint of 53% reached just before the first oil shock in 1973. This increase is nonetheless less marked than in last year's *Outlook*, mainly due to the expansion of light tight oil production in non-OPEC countries, particularly in North America.

Table 3.4 ▷ Oil production and oil and liquids supply by type and scenario (mb/d)

			New Policies		Current Policies		450 Scenario	
	1990	2011	2020	2035	2020	2035	2020	2035
OPEC	23.9	35.7	38.5	46.5	39.2	50.5	36.6	35.6
Crude oil	21.9	29.3	29.8	33.8	30.2	36.4	28.4	25.9
Natural gas liquids	2.0	5.7	7.0	9.8	7.3	10.9	6.6	7.5
Unconventional	0.0	0.7	1.8	2.8	1.8	3.2	1.7	2.2
Non-OPEC	41.8	48.8	53.2	50.4	54.3	54.9	51.5	41.1
Crude oil	37.6	39.2	37.1	31.6	37.8	34.5	36.1	25.6
Natural gas liquids	3.7	6.4	8.2	8.3	8.3	8.6	7.6	6.9
Unconventional	0.4	3.2	8.0	10.4	8.2	11.8	7.7	8.6
World oil production	65.7	84.5	91.8	96.8	93.5	105.4	88.2	76.6
Crude oil	59.6	68.5	66.9	65.4	68.0	70.8	64.5	51.5
Natural gas liquids	5.7	12.0	15.2	18.2	15.5	19.5	14.2	14.4
Unconventional	0.4	3.9	9.7	13.2	10.0	15.0	9.5	10.8
Processing gains	*1.3*	*2.1*	*2.5*	*2.9*	*2.5*	*3.2*	*2.4*	*2.3*
World oil supply*	67.0	86.6	94.2	99.7	96.0	108.5	90.5	79.0
World biofuels supply**	0.1	1.3	2.4	4.5	2.1	3.7	2.8	8.2
World total liquids supply	67.1	87.9	96.6	104.2	98.2	112.2	93.3	87.2

*Differences between historical supply and demand volumes are due to changes in stocks. **Expressed in energy-equivalent volumes of gasoline and diesel.

The need to add new capacity is much bigger than the projected increase in output, because of the need to compensate for the decline in production at existing oilfields as they pass their peak and flow-rates begin to drop. Crude oil output from those fields that were in production in 2011 falls by close to two-thirds, to only 26 mb/d by 2035 (Figure 3.15). Thus, the projected production of 65 mb/d in 2035 requires almost 40 mb/d of new capacity to be added over the projection period. Of this capacity, 26 mb/d, or 66%, comes from discovered fields yet to be developed, most of which are in OPEC countries, and the remaining 13 mb/d from fields that have yet to be found, mainly in non-OPEC countries. The need for new capacity by 2020 is 10 mb/d. This analysis is based on production profiles of oilfields according to their type and location. Observed decline rates – the average annual rate of change in production for fields that have passed their peak – are set to rise over the projection period, as output shifts to smaller fields and deepwater developments, where the rate of extraction relative to the size of the field tends to be highest (IEA, 2008). NGLs from gas fields that are currently producing also decline substantially, but this is more than outweighed by a big increase in new NGL capacity associated with higher gas production.

Figure 3.15 ▷ World oil supply by type in the New Policies Scenario

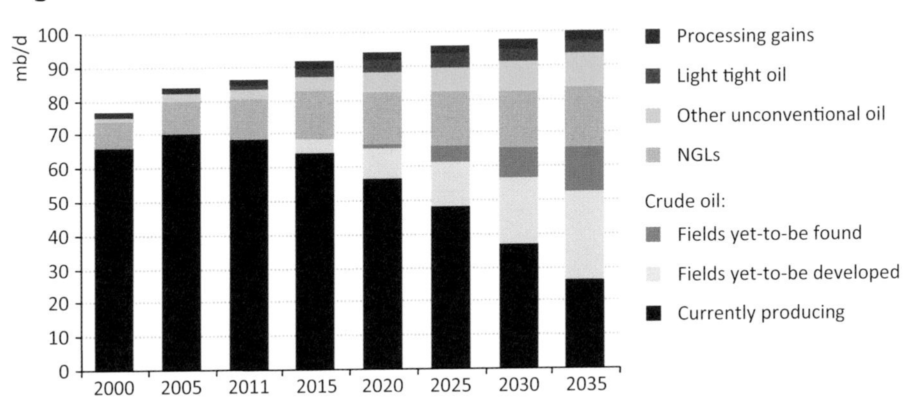

Overall, the total cumulative volume of oil produced in the New Policies Scenario between 2011 and 2035 is close to 800 billion barrels. Adding this to the total historical volume of oil produced by the end-2011 means that, by 2035, just over 2 trillion barrels will have been extracted. This is equal to about 28% of the world's ultimately recoverable resources (or 33% if kerogen oil is excluded from the calculation, as resource estimates outside the United States are uncertain and prospects for commercial production in the United States are likewise uncertain). Of the total amount of oil that is produced between 2011 and 2035, about 5% comes from crude oil fields that have yet to be found (Spotlight).

The output of NGLs is projected to grow strongly in the New Policies Scenario, rising from 12 mb/d in 2011 to just over 18 mb/d in 2035 and accounting for 50% of the net increase in global oil production over that period (Figure 3.16).[12] The main reason for this increase is the growth in natural gas production, particularly in the Middle East, where gas generally has a higher NGL content than in most other regions. North America is a strong contributor to the growth in NGLs supply, as is the anticipated reduction in flaring of associated gas (which tends to be relatively rich in NGLs) in Russia, Nigeria and Iraq, among others. Light tight oil production also grows rapidly, climbing from 1.0 mb/d in 2011 to a peak of 4.3 mb/d in the mid-2020s.

The strong growth in NGLs supply will add to the supply of light products, either for final use or as feedstock to upgrading units in refineries, while increased output of light tight oil will help to lighten feedstock to distillation units. However, these effects are expected

12. NGL production is reported in volume terms, but the balance between demand and supply is calculated on an energy-equivalent basis, which takes into account the lower energy per unit volume of NGLs compared with crude oil.

to be at least partially offset by a rise in the share of extra-heavy oil and natural bitumen in overall oil production (Figure 3.17). This changing production mix will require more investment in upgraders for the heavier crudes and raw bitumen and in condensate and NGL processing facilities for the lighter fluids.

Figure 3.16 ▷ World oil production by type in the New Policies Scenario

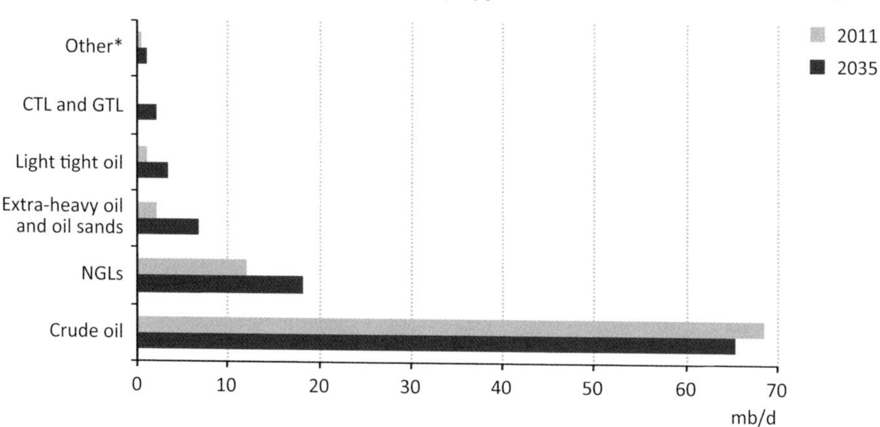

* Additives (including methanol derived from coal) and kerogen oil.

Output of oil sands in Canada grows rapidly, from 1.6 mb/d in 2011 to 4.3 mb/d in 2035, on the assumption that public concerns about the environmental impact of their development can be addressed. Venezuela is the other contributor to increased production of extra-heavy oil, with production from the Orinoco Belt expected to reach 2.1 mb/d in 2035 – a rise of 1.5 mb/d. This represents a conservative assessment of the announced projects, which aim for an increase of 2.4 mb/d by 2017. Modest amounts of bitumen or extra-heavy oil are also expected to come from Russia and China.

Figure 3.17 ▷ World oil production by quality in the New Policies Scenario

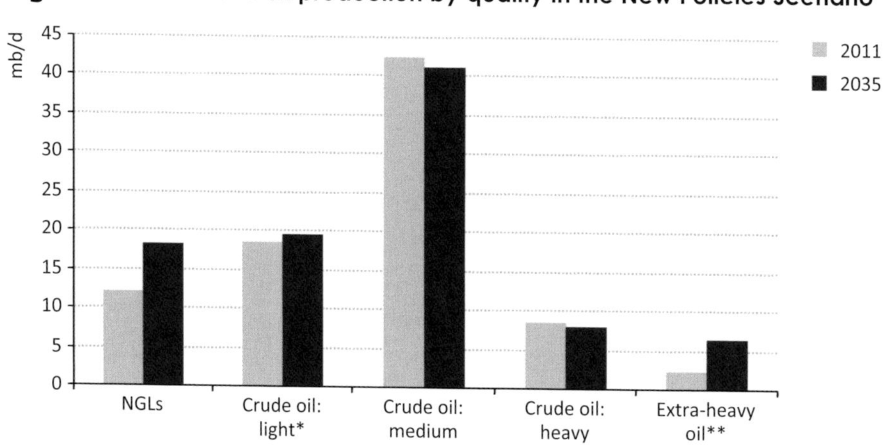

* Including light tight oil. ** Including Canadian oil sands (natural bitumen).

SPOTLIGHT

Are we finding enough crude oil to sustain production?

The number of oilfields that have been found and their average size has been in decline for several decades (with a moderate turn around in recent years). Most of the world's super-giant fields (holding more than 5 billion barrels) were discovered before the 1970s and only one – Lula in Brazil – has been found in the last ten years (if, as generally expected, its reserves are confirmed), and the rate of oil production has exceeded that of discoveries by a wide margin for many years. Does this mean that resources will simply not be big enough to support a continuing rise in output, such as that we project in the New and Current Policies Scenarios and that a rapid decline is imminent?

The simple answer is: no. In the oil-supply module of the World Energy Model, which is used to generate the projections, discovery rates are derived from country-by-country estimates of ultimately recoverable resources. Production in each country is then projected, based on investment in different categories of resource: existing fields (including those in production and those awaiting development) and new fields yet to be found, as well as unconventional resources. Investment is determined by the profitability of each type of project, based on assumptions about the capital and operating costs of different types of projects. Standard production profiles are applied to derive the production trend for each type of field.

In the Current Policies Scenario, in which crude oil production expands most over the projection period, the volume of resources that are developed in fields that are yet to be found is actually well below the average rate of discovery of new fields over the past decade or so. In 2000-2011, an average of 14 billion barrels was found each year; the amount that needs to be discovered (and developed) to meet the Current Policies Scenario projections for 2012-2035 is only 7 billion barrels per year. In other words, our projections allow for the fact that discoveries are projected to continue to decline over the longer term. The volume of oil that needs to be discovered to support the lower projected production in the New Policies Scenario is correspondingly smaller.

In practice, it is far from certain that actual discoveries will fall that much: as mentioned above, the fall in the rate of discovery has levelled off since the 1990s, mainly as a result of higher oil prices, which have stimulated increased exploration activity. And advances in exploration technology, including 3D seismic and sub-salt imaging, have boosted exploration drilling success rates, reducing the number of dry wells, helping to offset the effect of the diminishing average size of fields that are found in mature areas and encouraging exploration in higher-risk, less-explored, areas that may lead to larger discoveries. Indeed, with oil prices assumed to rise over the projection period and further technical advances in prospect, it is not implausible that discovery rates could actually increase over the medium term. Our analysis suggests that, for the period to 2035, it is not the size of the resource base that is at issue for future oil supply, but rather how demand for oil will be influenced by new policy measures, how much resources will cost to develop and whether the conditions for investment – including the oil price – will be conducive to producing them.

The supply of oil from the conversion of natural gas and coal also expands considerably, though volumes remain relatively modest. Gas-to-liquids (GTL) production increases from 76 kb/d in 2011 to around 900 kb/d in 2035; in the near term, output is given a big boost by the ramping up of Shell's 140 kb/d Pearl plant, which started up in 2011 and which will triple global GTL production capacity. Several other plants have been proposed in the Middle East and in North America, where the low price of gas relative to oil has enhanced the economics of such plants. Output from coal-to-liquids (CTL) plants also increases, with most of this coming from China (which, outside South Africa, is the only commercial CTL producer today), Australia and the United States. Production of kerogen oil is projected to take off only towards the end of the projection period, as the technology is still immature and expensive, reaching 200 kb/d in 2035, with half of this coming from the Utah-Colorado area of the United States.

Non-OPEC production

In the New Policies Scenario, non-OPEC production rises in the near term to more than 53 mb/d and then declines gradually to around 50 mb/d by 2035. However, there is a marked shift in both the balance and the type of output (Table 3.5). The biggest increases in non-OPEC supply occur in Brazil, Canada, Kazakhstan and the United States; output falls in most others, notably China, the United Kingdom, Norway and also Russia. Growth in production of unconventional oil, including light tight oil, and NGLs compensates for most of the decline in crude oil output.

Figure 3.18 ▷ United States oil production by type in the New Policies Scenario

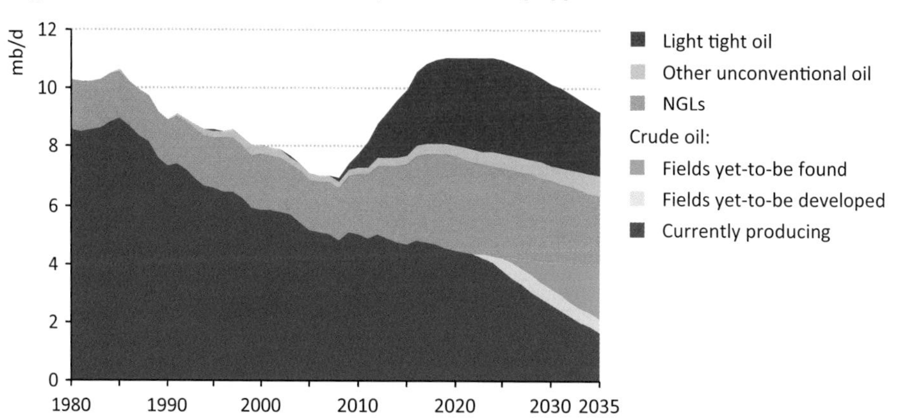

Note: The World Energy Model supply model starts producing yet-to-find oil after it has put all yet-to-develop fields into production. In reality, some yet-to-find fields would start production earlier than shown in the figure.

The resurgence of oil production in North America in the last few years is projected to continue. Production in the United States hit a low of 6.9 mb/d in 2008 and has since been rising strongly, reaching 8.1 mb/d in 2011. We project this upward trend to continue, with production climbing to a plateau of 11 mb/d before 2020 and then declining gradually from the late 2020s to around 9 mb/d by 2035 (Figure 3.18). This is a significant upward revision from last year's *Outlook*. Light tight oil continues to rise into the 2020s (Box 3.3), supplemented by deepwater output and by supplies of unconventional CTL, GTL and kerogen oil, particularly towards the end of the projection period.

Table 3.5 ▷ Non-OPEC oil production in the New Policies Scenario (mb/d)

								2011-2035	
	1990	2011	2015	2020	2025	2030	2035	Delta	CAAGR*
OECD	19.0	18.9	21.0	22.1	22.2	21.7	20.9	1.9	0.4%
Americas	13.9	14.6	17.0	18.6	18.9	18.7	18.1	3.5	0.9%
Canada	2.0	3.5	4.3	4.9	5.4	5.9	6.3	2.7	2.4%
Mexico	3.0	2.9	2.7	2.6	2.6	2.6	2.6	-0.3	-0.5%
United States	8.9	8.1	10.0	11.1	10.9	10.2	9.2	1.1	0.5%
Europe	4.3	3.8	3.4	2.9	2.6	2.3	2.1	-1.7	-2.5%
Asia Oceania	0.7	0.6	0.6	0.7	0.7	0.7	0.7	0.2	1.1%
Non-OECD	22.8	29.9	30.9	31.1	30.8	30.3	29.5	-0.4	-0.1%
E. Europe/Eurasia	11.8	13.7	13.8	13.4	13.4	13.8	13.9	0.2	0.1%
Kazakhstan	0.5	1.6	1.8	2.0	2.7	3.4	3.7	2.1	3.5%
Russia	10.4	10.6	10.5	10.1	9.5	9.3	9.2	-1.4	-0.6%
Asia	6.0	7.7	7.9	7.5	6.9	6.1	5.3	-2.4	-1.5%
China	2.8	4.1	4.3	4.3	4.0	3.3	2.7	-1.4	-1.7%
India	0.7	0.9	0.8	0.7	0.7	0.7	0.6	-0.3	-1.4%
Middle East	1.3	1.6	1.5	1.2	1.0	0.8	0.7	-0.9	-3.3%
Africa	1.7	2.6	2.8	3.0	2.8	2.6	2.4	-0.2	-0.3%
Latin America	2.0	4.2	5.0	6.0	6.8	7.0	7.1	2.8	2.2%
Brazil	0.7	2.2	2.8	4.0	5.0	5.5	5.7	3.5	4.1%
Total non-OPEC	41.8	48.8	52.0	53.2	53.0	51.9	50.4	1.5	0.1%
Non-OPEC market share	*64%*	*58%*	*58%*	*58%*	*57%*	*55%*	*52%*	*n.a.*	*n.a.*
Conventional	41.4	45.6	46.4	45.3	43.7	41.9	40.0	-5.6	-0.5%
Crude oil	37.6	39.2	39.0	37.1	35.4	33.5	31.6	-7.6	-0.9%
Natural gas liquids	3.7	6.4	7.4	8.2	8.3	8.4	8.3	2.0	1.1%
Unconventional	0.4	3.2	5.6	8.0	9.4	10.1	10.4	7.2	5.0%
Canada oil sands	0.2	1.6	2.3	2.9	3.4	3.8	4.3	2.7	4.1%
Light tight oil	-	1.0	2.6	3.8	4.3	4.0	3.4	2.3	5.1%
Coal-to-liquids	0.1	0.2	0.2	0.4	0.7	1.0	1.3	1.1	8.8%
Gas-to-liquids	0.0	0.0	0.0	0.1	0.2	0.3	0.4	0.4	10.2%
Share of total non-OPEC	*1%*	*7%*	*11%*	*15%*	*18%*	*19%*	*21%*	*n.a.*	*n.a.*

* Compound average annual growth rate.

Box 3.3 ▷ The rise and rise of light tight oil

The rise in light tight oil production in the United States in the last few years has been nothing short of spectacular. Production from the Bakken formation had grown to more than 600 kb/d by mid-2012 in North Dakota alone (the formation also extends into the state of Montana in the United States and provinces of Manitoba and Saskatchewan in Canada). Production from the Eagle Ford shale in south Texas, adjacent to the Mexican border, is also expanding rapidly, from almost nothing three years ago to more than 300 kb/d by mid-2012. Combined US production from the Bakken, the Eagle Ford and other emerging plays, such as the Niobrara (which straddles Colorado, Kansas, Nebraska and Wyoming), the Texas Permian basin and the Californian plays, is expected to reach over 3.2 mb/d by 2025 in the New Policies Scenario.

Resources are believed to be large enough to sustain production at that level until about 2030, before a decline in production sets in. The US Energy Information Administration estimates that unproven recoverable resources of light tight oil at end-2009 stood at 33 billion barrels; adding about 2 billion barrels of proven reserves yields a figure for total remaining resources of 35 billion barrels. Some industry sources claim that recoverable resources will end up being much larger, capable of sustaining a higher level of production for longer than we project here. In addition to light tight oil, shale gas plays contribute to the rise in liquid hydrocarbons production in the United States, with the output of NGLs projected to grow by 1 mb/d between now and 2020, reaching 3.2 mb/d, before slowly decreasing to 2.3 mb/d in 2035, as the known liquids-rich gas plays are exhausted and gas production refocuses on dry gas.

Canada has joined the light tight oil boom with production reaching 190 kb/d in 2011 from the Canadian part of the Bakken and from other emerging plays. We project Canadian light tight oil production will reach more than 500 kb/d by 2035. NGLs from shale gas plays also increase significantly, offsetting falling production from conventional gas plays. Outside North America, production of light tight oil and of NGLs from shale is unlikely to make a large contribution to global oil supply before 2020. Even beyond 2020, our projections for light tight oil production outside North America remain small, with only China passing the 200 kb/d mark, Argentina reaching 150 kb/d and all other countries remaining below 100 kb/d, as we have yet to see sufficient progress in confirming resources. But there is clearly some upside potential: the Neuquén basin in Argentina shows promise, while the extension of the Eagle Ford shale into Mexico is also a focus of attention; China could grow a light tight oil industry, in parallel with its efforts on shale gas; and Russia is thought to have significant resources, in particular in the Bazhenov shale in Western Siberia, which are under study. As it uses the same production technologies as shale gas, light tight oil production will require similar attention to the reduction of environmental impacts as unconventional gas developments (see Chapter 4, Focus on Prospects for Unconventional Gas); and development of gas gathering infrastructure to prevent flaring the large quantities of associated gas (Chapter 4, Box 4.4).

Our estimate of ultimate recoverable resources worldwide now includes about 250 billion barrels of light tight oil – around 10% of remaining conventional resources – but this could increase markedly as more is learned about the resources and the associated recovery factors (although some of these resources are in countries that are well endowed with conventional oil). Our current understanding of the resource base inclines us to the view that light tight oil might not turn out to be as important as shale gas, resources of which are about half the size of conventional gas globally.

Production in Canada is projected to rise steadily as oil-sands production grows by two-and-a-half times to 4.3 mb/d by 2035, more than offsetting a decline in conventional output. This projection assumes that environmental concerns can be assuaged and that infrastructure to bring the oil to markets in the United States or to Asia can be built. In January 2012, the US administration rejected a request for authorisation to build the Keystone XL pipeline from Alberta, Canada to the Gulf Coast in Texas to ship oil derived from oil sands and light tight oil from the Bakken Shale. TransCanada, the project sponsor, has since resubmitted a revised application, with a somewhat modified route. The construction of the southern section (entirely within the United States, from Cushing, Oklahoma, to the Gulf Coast) is now underway. Other options for additional Canadian and Bakken oil to reach the Gulf Coast refineries include expansion of the Enbridge Flanagan South pipeline. There are also proposals to pipe the oil to the Canadian west coast for shipping to Asian markets, but they also face opposition on environmental grounds. Without new export capacity, western Canadian oil production would exceed regional consumption and current export capacity before 2016.

In contrast, Mexico's oil production continues to decline in the near term, mainly due to further declines at the super-giant Cantarell field and other mature fields. Cantarell, once the world's second-largest producing field, has been declining extremely rapidly in recent years; it is currently producing only about 400 kb/d, down from a peak above 2.1 mb/d in 2003-2004. Pemex, the national oil company, is now focusing its efforts on further development of the Ku/Maloob/Zaap complex in an area adjacent to Cantarell, where production has risen to over 850 kb/d. In the longer term, Mexican production in total is projected to bottom-out at around 2.5 mb/d by the mid-2020s and then recover slowly, on the assumption that sufficient investment is made in developing new fields, for example in the Mexican sector of the deepwater Gulf of Mexico, and in arresting the decline at existing fields. Production could rebound faster; indeed the Energy Strategy 2012-2026 published this year by the Mexican Ministry of Energy calls for crude oil production alone to grow to 3.3 mb/d by 2026, from 2.6 mb/d in 2011 (Secretaria de Energia, 2012).

A large fall in non-OPEC oil production is projected to occur in Europe, where output almost halves from 3.8 mb/d in 2011 to 2.1 mb/d in 2035 – almost entirely due to dwindling output from North Sea fields. Output falls most sharply in the United Kingdom, where the industry is the most mature. Most UK producing fields are already in long-term decline and the fields that have been found in recent years are generally very small. By 2035,

Box 3.4 ▷ Arctic waters: the final frontier?

Interest in developing offshore oil and gas resources in the Arctic has grown in recent years. An assessment carried out by the USGS in 2008 estimated total undiscovered technically recoverable crude oil resources north of the Arctic circle at 90 billion barrels (14% of the world total) and gas resources at 47 trillion cubic metres (more than a quarter of the global total), containing an estimated 44 billion barrels of natural gas liquids (USGS, 2008). About 84% of the estimated resources are thought to lie offshore. Exploration in Arctic waters has been underway for decades in Russia, and more recently in Norway, Canada and Greenland. No commercial discoveries have been made yet in Canada or Greenland, but several fields have been found in the Norwegian Barents Sea and offshore Russia – notably the Shtokman gasfield, discovered in 1988, which is still awaiting development (see Chapter 4). The first Russian offshore Arctic oilfield slated for production is Gazprom's Prirazlomnoe field in the Pechora Sea, although further drilling has been postponed until 2013. Gazprom is considering other Arctic developments and Rosneft, the state-owned Russian oil company, has recently signed agreements with Eni, ExxonMobil and Statoil to explore Arctic waters.

Arctic oil production is also due to begin in Norway in 2013, when construction work on the Goliat oilfield in the Barents Sea is completed. Goliat, located about 70 km offshore, was discovered in 2000 and is expected to produce 100 kb/d when it reaches plateau, soon after production starts. In April 2011, Statoil and Eni found the Skrugard field, also in the Barents Sea, about 190 km offshore; the field is thought to contain 250 million barrels of recoverable oil, making it marginally bigger than Goliat. Statoil hopes to bring Skrugard into production within five to ten years. In 2010 Russia and Norway reached an historic agreement defining their maritime boundaries in the Barents and Arctic Seas, resolving a 40-year dispute and boosting long-term prospects for exploration in both countries. Elsewhere, Shell's plans to drill in the Chukchi and Beaufort seas off the coast of Alaska have been held back by the complexity of operations and by environmental concerns, with the completion of the first wells likely only in 2013. In July 2012, BP announced that it would not go ahead with a ground-breaking $1.5 billion project to develop the Liberty oilfield in the Beaufort Sea in Alaska, using extended-reach drilling, after concluding it would be too expensive. Cairn Energy carried out an exploration campaign in offshore eastern Greenland in 2010 and 2011 but, after drilling eight wells at a cost of almost $1 billion, has yet to identify commercial quantities of oil.

In view of the technical and environmental challenges and high cost of operating in extreme weather conditions, including the problems of dealing with ice floes and shipping in water that remains frozen for much of the year, we do not expect the Arctic offshore to make a large contribution to global oil supply during the *Outlook* period. However, subject to the trajectory of oil demand, Arctic resources could play a much more important role in the longer term. Technological advances and/or higher oil prices could also result in production taking off before 2035.

UK output is projected to reach just 340 kb/d, compared with 1.1 mb/d in 2011 and its peak of 2.9 mb/d in 1999. Norway's production is also declining, having already dropped from 3.4 mb/d in 2001 to 2 mb/d in 2011, and is projected to fall by another 0.7 mb/d by 2035. However, the pace of decline is expected to be offset to some degree by rising production from the Norwegian and Barents Seas, where some sizeable new fields have been found recently (Box 3.4), as well as from the giant Johan Sverdrup field recently discovered in the, otherwise mature, central North Sea.

Oil production in Russia, which reached 10.6 mb/d in 2011 (its highest levels since the collapse of the Soviet Union), is expected to level off at less than 11 mb/d in the coming years, before beginning a gradual decline to 9.2 mb/d in 2035. Despite the record output, Russia relinquished its position as the largest oil producer in the world to Saudi Arabia in 2011 and, in our projections, is overtaken also by the United States before 2020. Russia has large conventional crude oil resources but they are often located in remote, northerly areas, where operating conditions are very difficult, and so will be costly to develop. New fields in eastern Siberia, together with NGLs from rising gas production (see Chapter 4), are expected to contribute most to new supply, offsetting much of a projected decline from mature producing basins in the Volga-Urals and Western Siberia (see *WEO-2011* for a detailed account of Russian production prospects). The latter regions currently account for well over four-fifths of Russian oil production. These projections are underpinned by an assumption that the licensing and tax regime will be adjusted as necessary to encourage investment, both in existing producing areas and new regions.

Kazakhstan is the driver of increased Caspian production over the *Outlook* period. Production growth, especially in the medium term, comes primarily from the Kashagan and Tengiz fields, pushing total supply up from 1.6 mb/d in 2011 to 3.7 mb/d in 2035. We have reduced our medium-term projections, compared with *WEO-2011,* to reflect project delays. The first phase development of the super-giant Kashagan field is expected to start production in mid-2013, but there are significant uncertainties over the prospects and timing for the larger second phase that could take overall production from the field beyond the 1 mb/d mark. Other Caspian countries see their production decline, except Turkmenistan, where the large increase in gas production brings forth enough additional NGL supply to offset the drop in crude oil production.

Oil production in non-OPEC Latin America increases steadily through the projection period, primarily as a result of strong growth in Brazil. Output falls in the other main producing countries in the region, except Peru; Colombia sees an initial rise, before starting to decline from the late 2010s. Thanks to a series of major deepwater discoveries, Brazil is set to become the fastest-growing oil producer outside the Middle East (Box 3.5). In the New Policies Scenario, total Brazilian oil production is projected to climb briskly, especially during the latter half of the current decade (though not as quickly as targeted by the national oil company, Petrobras, and not as quickly as projected in last year's *Outlook,* due to project delays). Total output reaches 4.0 mb/d in 2020 and continues to climb thereafter, reaching 5.7 mb/d by 2035. Brazil also sees a further big increase in biofuels

Box 3.5 ▷ Brazil's oil boom gathers pace

Brazil offers the brightest prospects for oil production outside OPEC. Output has been growing in recent years, thanks to the development of fields in the Campos and Santos Basins, located off the country's southeast coast. These basins hold the vast majority of Brazil's proven reserves. In 2007, a consortium of Petrobras (the state-controlled oil company), BG Group and Petrogal discovered the Tupi field (since renamed Lula) in the Santos basin, which contains substantial reserves, under a thick layer of salt, some 5 000 metres below the seabed, in 2 000 metres of water. This deepwater "pre-salt" discovery was quickly followed by several others in the same basin, including Iracema (renamed Cernambi), Jupiter, Carioca, Iara, Libra, Franco and Guara. Just how much oil can be recovered from these fields is still not clear. Petrobras, which has important stakes in all the pre-salt finds, estimates that Lula contains 6.5 billion barrels of oil equivalent (bboe) of recoverable resources, Cernambi 1.8 bboe and Guara 1.1 bboe, most of which are oil. If the resources in these three fields alone are proved up, they would increase Brazil's total proven reserves by up to two-thirds.

In 2010, a set of laws was adopted to regulate the development of Brazil's pre-salt reserves. A new public national company, Pré-Sal Petróleo S.A. (PPSA), and a national development fund, into which tax revenues from pre-salt production are to be directed, were established. A production-sharing agreement system was introduced for pre-salt fields, whereby Petrobras will hold a minimum 30% stake and take on the role of operator in every field development (existing fields and future non-pre-salt discoveries are covered by a concession-based system).

The pre-salt fields are expected to drive most of the growth in Brazil's oil production. Pilot projects at Lula and Guara, which were started in 2010 and 2011 respectively, are together currently producing over 100 kb/d; smaller volumes are also being produced from pre-salt fields in the Campos and Espírito Santo basins. Petrobras plans to raise pre-salt production progressively through a large-scale development programme. This will be costly and technically challenging, in view of the water depth, the complexity of drilling through the salt layer, the distance from shore (more than 250 kilometres) and the logistical difficulties of managing such a large-scale programme. The production sharing agreements for the pre-salt fields (and the concession contracts for other fields) are based, in part, on a commitment to local content (up to 65%) and employment. Although the Ministry of Energy carefully monitors that such commitments are reasonable, the level of activity could stretch local capabilities and contribute to delays.

Petrobras currently plans to invest close to $60 billion in developing the pre-salt fields over the period 2012-2016, with additional investment from partners. This spending will cover the installation of 19 offshore platforms and the drilling of up to 500 wells. In its latest business plan, released in June 2012, the company is targeting total oil production of 4.2 mb/d by 2020, almost half from the pre-salt fields. Our projections are more conservative.

production (see Chapter 7), helping to push up total liquids supply from 2.5 mb/d in 2011 to 6.8 mb/d in 2035. Oil output in the other non-OPEC Latin American countries collectively increases in the near term, thanks mainly to Colombia, but starts to decline from the end of the current decade as fields in mature areas, notably in Argentina, are depleted. Argentina's prospects are brightened to a degree by the assumed emergence of a shale gas industry, which brings higher production of NGLs and the start of light tight oil production. The pace of development will depend on supportive conditions for upstream investment and the recent renationalisation of the YPF company has heightened concerns on this front.

3

The prospects for oil production in non-OPEC African countries have improved significantly in the last few years with some important commercial discoveries in several western and eastern Africa countries, notably Cameroon, Ghana, Equatorial Guinea, Republic of Congo, Uganda, Tanzania and, most recently, Kenya, which announced its first onshore find in early 2012 and has licensed deepwater acreage for exploration. Nonetheless, the outlook for production remains very uncertain, as recently discovered fields have yet to be evaluated thoroughly. Reflecting this, non-OPEC African production in total is projected to grow from around 2.6 mb/d in 2011 to about 3.0 mb/d by 2020 and then drop back to 2.4 mb/d, because of declining output in the mature producing countries such as Egypt, Sudan, Chad and Gabon. In the Middle East, non-OPEC production is projected to continue to dwindle (despite the prospect of some kerogen oil production in Jordan and possibly Morocco towards the end of the *Outlook period*) as the main current producers – Oman, Syria and Yemen – all experience declines. China's production holds more-or-less steady near 4 mb/d through to 2025 before commencing a steady decline due to conventional oil resource limitations, despite an increase in offshore production, CTL and light tight oil production. In most other regions, production declines through to 2035.

OPEC production

Oil production in OPEC countries as a group is poised to grow considerably over the longer term, reflecting their large resource base and relatively low finding and development costs, though short-term market management policies will probably continue to constrain the rate of expansion of supply. In the New Policies Scenario, OPEC output rises slowly in the medium term, from 35.7 mb/d in 2011 to 38.5 mb/d in 2020, and then more briskly, to 46.5 mb/d in 2035. The bulk of the increase in OPEC production comes from the Middle East, predominately Iraq,[13] though all OPEC countries in this region see an increase in output over the projection period (Figure 3.19 and Table 3.6). Outside the Middle East, production rises in all member states, except Angola and Ecuador. Crude oil accounts for 40% of the increase in OPEC oil production over 2011-2035, with NGLs contributing another 40% and unconventional oil – mainly Venezuelan extra-heavy oil – the remaining one-fifth.

Saudi Arabia, with the world's largest conventional oil reserves and substantial undeveloped resources, remains well-placed to augment its oil production capacity over the coming decades; but long-term resource depletion considerations and short-term

13. See Chapter 14 in Part C for a detailed discussion of the outlook for oil production in Iraq.

market management policies are assumed to continue to constrain the pace of upstream investment and production growth. Output, which averaged just over 11 mb/d in 2011, is projected to drop back slightly in the medium term, as the call on the country's capacity falls with the strong rise in production from Iraq and other producers, but then rises to 12.3 mb/d in 2035. Almost half of the increase in Saudi oil supply comes from NGLs, resulting from strong projected growth in natural gas production.

Figure 3.19 ▷ Change in oil production in selected countries in the New Policies Scenario, 2011-2035

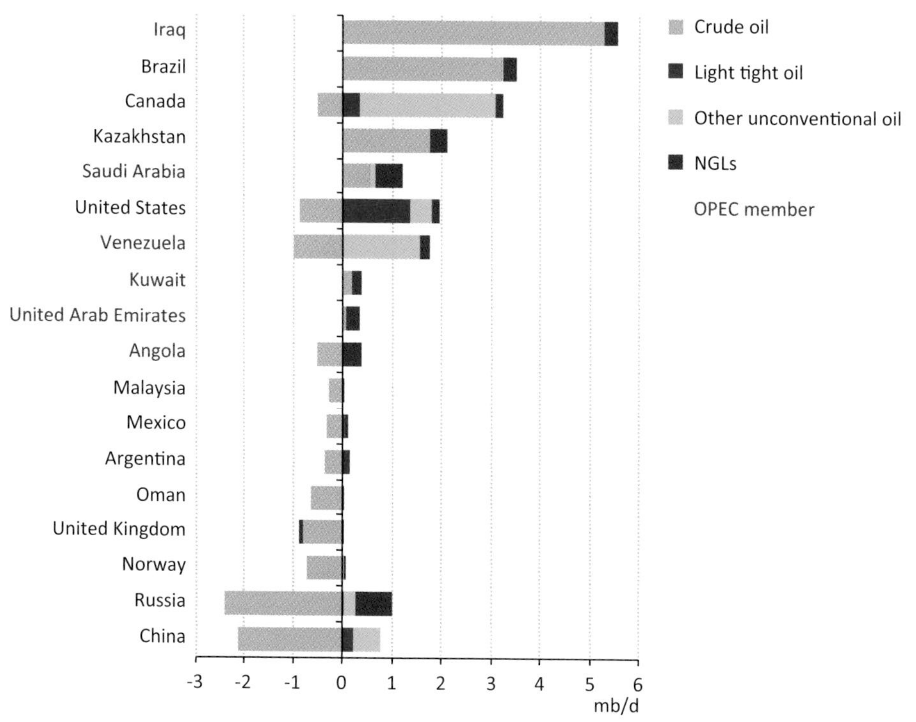

Note: Libya also has a large increase in oil production between 2011 and 2035, as 2011 production was exceptionally low due to the conflict.

These projections take account of Saudi Arabia's stated policy of maintaining spare crude oil production capacity of around 1.5-2 mb/d over the longer term.[14] In 2010, the national company, Saudi Aramco, completed a $100 billion investment programme to expand sustainable crude oil production capacity to about 12 mb/d, of which the most important component was the development of the 1.2 mb/d Khurais field and the 500 kb/d Khursaniyah field. The next major project, the development of the 900 kb/d Manifa offshore field, has been accelerated to supply the heavy crude needed to feed three refineries

14. In August 2012, Saudi Arabia had spare capacity of around 2 mb/d with production (excluding NGLs) running at just under 10 mb/d, according to the IEA's monthly *Oil Market Report* (IEA, 2012e).

that are under construction; it is now due to come on stream in 2014, instead of 2016 as previously planned. Aramco is also upgrading facilities at the heavy oil Safaniyah field, the world's largest offshore oil field. In addition, Chevron is testing steam injection at the Wafra oilfield in the Neutral Zone, which Saudi Arabia shares with Kuwait, ahead of a final decision on whether to implement the technology on a large scale to boost production of heavy oil from the field, where pressure and flow rates are dropping. If approved, this would be the largest steam-injection project in the world. These projects are intended mainly to compensate for declines in production capacity at existing fields, some of which have been producing for more than half a century, rather than to boost overall capacity.

Table 3.6 ▷ OPEC oil production in the New Policies Scenario (mb/d)

								2011-2035	
	1990	2011	2015	2020	2025	2030	2035	Delta	CAAGR*
Middle East	16.4	25.8	26.3	27.8	29.4	31.4	34.4	8.5	1.2%
Iran	3.1	4.2	3.2	3.3	3.6	4.0	4.5	0.3	0.3%
Iraq	2.0	2.7	4.2	6.1	6.9	7.5	8.3	5.6	4.7%
Kuwait	1.3	2.7	2.8	2.7	2.7	2.8	3.1	0.4	0.5%
Qatar	0.4	1.8	1.9	1.8	2.0	2.2	2.5	0.7	1.4%
Saudi Arabia	7.1	11.1	10.9	10.6	10.8	11.4	12.3	1.2	0.4%
United Arab Emirates	2.4	3.3	3.4	3.3	3.3	3.4	3.7	0.4	0.4%
Non-Middle East	7.5	9.8	11.0	10.7	11.0	11.6	12.1	2.3	0.9%
Algeria	1.3	1.8	1.9	1.9	1.9	2.0	2.0	0.2	0.4%
Angola	0.5	1.7	1.8	1.7	1.7	1.7	1.6	-0.1	-0.3%
Ecuador	0.3	0.5	0.5	0.4	0.3	0.3	0.3	-0.2	-2.3%
Libya	1.4	0.5	1.6	1.6	1.8	1.9	2.0	1.5	6.1%
Nigeria	1.8	2.6	2.6	2.4	2.5	2.7	2.7	0.2	0.2%
Venezuela	2.3	2.7	2.6	2.7	2.9	3.2	3.5	0.8	1.0%
Total OPEC	23.9	35.7	37.3	38.5	40.4	43.0	46.5	10.8	1.1%
OPEC market share	*36%*	*42%*	*42%*	*42%*	*43%*	*45%*	*48%*	*n.a.*	*n.a.*
Conventional	23.9	35.0	36.0	36.8	38.3	40.6	43.6	8.7	0.9%
Crude oil	21.9	29.3	29.6	29.8	30.6	31.7	33.8	4.5	0.6%
Natural gas liquids	2.0	5.7	6.4	7.0	7.7	8.8	9.8	4.2	2.3%
Unconventional	0.0	0.7	1.3	1.8	2.1	2.5	2.8	2.1	6.1%
Venezuela extra-heavy	0.0	0.6	0.9	1.4	1.6	1.8	2.1	1.5	5.6%
Gas-to-liquids	-	0.0	0.2	0.2	0.3	0.4	0.5	0.5	11.3%
Share of total OPEC	*0%*	*2%*	*3%*	*5%*	*5%*	*6%*	*6%*	*n.a.*	*n.a.*

* Compound average annual growth rate.

The prospects for production in Iran remain clouded by uncertainty over the impact of international sanctions and the country's upstream investment policies. The country has the resources to support a big increase in production, but a lack of capital and technical expertise have held back investment and led to stagnating output. Rising demand for natural gas for power generation and industrial uses has also limited the availability of gas for injection into oilfields to bolster flow-rates. In 2012, the United States and the

European Union implemented the most comprehensive sanctions yet on the country's oil and financial sectors. This has resulted in a steep drop in exports, now expected to hover near 1 mb/d in the second half of 2012, compared to 2.5 mb/d a year ago. The loss of revenue and even more constrained access to technology and capital is expected to quickly turn this reduced production into a reduced capacity as well, from which the country's production will take several years to recover. In the New Policies Scenario, we assume that Iranian oil production declines significantly in the next few years, but recovers after 2020 – in part thanks to higher NGLs output with rising gas production – on the assumption that the current stand-off eventually gets resolved.

Iraq (discussed in detail in part C) is poised to be by far the largest contributor to growth in world supply. In the Central Scenario for Iraq, which forms part of the overall New Policies Scenario, production grows from 2.7 mb/d in 2011 to 6.1 mb/d in 2020 and 8.3 mb/d in 2035. A High Case anticipates faster growth to more than 9 mb/d in 2020, closer to what is being envisaged in the signed contracts with international oil companies in Iraq. A Delayed Case investigates the consequences of a slower pace of investment, with production growing only to 5.3 mb/d in 2035.

Oil production in the United Arab Emirates (UAE) – the fifth-largest OPEC producer – has been rising steadily in recent years, driven by capacity additions in Abu Dhabi; current UAE crude oil capacity is estimated at 2.8 mb/d while NGL production is running at about 800 kb/d. As a result of weaker global oil-demand growth than expected and OPEC quotas, the government has delayed a plan to raise crude oil capacity to 3.5 mb/d to 2018. Development of a new field – Qusahwira – and the extension of some old giants – Bab and Upper Zakum– are currently being undertaken by the Abu Dhabi National Oil Company and its partners, which between them are expected to add about 250 kb/d by 2014 and another 250 kb/d by 2016. Other companies are also planning capacity increases, though this is likely to take place only after decisions are taken on the renewal of the concessions, due in 2014. National production is projected to reach 3.7 mb/d in 2035, up from 3.3 mb/d in 2011, with most of the increase coming from NGLs.

The outlook for oil production in Kuwait remains uncertain, given continuing political opposition to allowing foreign companies with the requisite expertise to develop the country's heavy oil resources. The country aims to boost overall crude oil capacity to 4 mb/d by 2020, from its current level of 2.7 mb/d, and to maintain it through to 2030, but achieving that will be difficult without the extensive involvement of international companies. For now, the national oil company, Kuwait Petroleum Corporation, is pursuing enhanced technical service agreements as a way of raising production at old producing fields and developing new ones, including the Sabriyah and Umm Niqa fields, which contain heavy sour crude oil. We project production to remain flat over the next few years and then edge up towards 3.1 mb/d by 2035.

In Qatar, projected growth in oil production of 700 kb/d to 2.5 mb/d between 2011 and 2035 comes from NGLs and GTL, with crude oil and condensate output decreasing slightly. The current moratorium on new gas projects is assumed to be lifted after 2015, paving the way for new supplies of NGLs in the 2020's; NGL output has soared in recent years, as a wave of LNG projects has come on stream. GTL capacity will soon reach 174 kb/d, when the 140 kb/d Pearl plant is fully operational, adding to the existing 34 kb/d Oryx plant, commissioned in 2007. GTL capacity is projected to rise to close to 400 kb/d by 2035 (though no firm projects are planned as yet), as this would help Qatar to hedge against diverging price trajectories for LNG and oil.

The outlook for production in Nigeria, which holds large resources, hinges on resolving the persistent problems caused by civil conflict in the key producing regions and improving the transportation infrastructure, the state of which has led to sporadic disruptions to supply. Some 200 kb/d remains shut-in at the time of writing. These problems, the inability of the Nigerian National Petroleum Company to fund its share of investment in joint ventures with international companies, and delays in agreeing a new Petroleum Industry Bill have paralysed investment in recent years. If adopted, new legislation could stimulate growth in output, though this would happen only slowly, given the long lead times for new projects. Output is projected to dip from 2.6 mb/d in 2011 to around 2.4 mb/d by 2020, recovering to 2.7 mb/d by 2035, on the assumption that the required investment is forthcoming. Production in Angola – the other western African OPEC member state – is projected to remain flat through to 2035 at around 1.6 mb/d, on the expectation that new deepwater discoveries yield enough oil to offset declines at existing fields.

Production prospects in Libya are a little clearer than a year ago, with the end of the civil war that began in spring 2011 and rapid progress in restoring damaged capacity. Output is expected to reach close to 1.6 mb/d in 2012, just shy of the 1.7 mb/d produced in 2010, and to stay near that level for a few years before resuming a slow upward path. As existing fields are entering their decline phase, boosting output will depend on new fields and on the success of exploration efforts, which had been stepped up between 2007 and 2011, but have thus far yielded disappointing results. Algeria is facing similar problems, which, together with a fall in discoveries, have led to a slow decline in production since 2007 to around 1.8 mb/d in 2011. A new Hydrocarbons Law adopted in 2005 has failed to stimulate investment in exploration: only two out of ten oil and gas permits on offer were awarded in the last licensing round in March 2011. As a result, the energy minister announced in December 2011 that changes would be proposed to reduce delays in sanctioning projects and improve financial terms. Output is projected to pick up slowly through the projection period, on the assumption that new upstream rules bear fruit.

Oil production in Venezuela has been hindered in recent years by under-investment by the national company, PDVSA, and the government's policy of seeking greater control over natural resources, which has discouraged foreign investment. On the assumption that there is no major change in the political climate, we project output of conventional oil (crude oil and NGLs) to continue to decline from its 2011 level of 2.1 mb/d through to 2020, before stabilising at around 1.3-1.4 mb/d. However, this decline is more than outweighed by a sharp increase in production of extra heavy oil in the Orinoco Belt, which jumps from 570 kb/d in 2011 to 2.1 mb/d in 2035. A number of production blocks have been established, which are being developed in parallel by PDVSA, in partnership with national and international oil companies (PDVSA holds a 60% stake in each case).

Output in Ecuador, the smallest OPEC producer, is projected to decline gradually through the projection period, though development of the Ishpingo-Tambococha-Tiputini complex of heavy oilfields in the Yasuni National Park in the Amazon – if allowed to proceed – could help stabilise production, at least in the medium term. The government has proposed refraining indefinitely from developing the fields in exchange for half of the value of the reserves, which it puts at $3.6 billion, in the form of donations from the international community. By early 2012, it had raised less than $120 million.

Prospects for offshore and deepwater production

The importance of offshore oil fields to future supply derives from the large resources estimated to be located under the sea. Out of our estimate of close to 2 700 billion barrels of remaining recoverable conventional oil (excluding light tight oil), about 45% is in offshore fields. Of those offshore resources, about a quarter are expected to be under deepwater, defined as water with a depth in excess of 400 metres. If we look at non-OPEC countries, the fractions are even higher: 55% offshore, of which one-third is in deepwater. The rapid growth of deepwater developments since 2000 was interrupted in 2010, after the Macondo disaster and oil spill in the Gulf of Mexico. This led to a prolonged moratorium on further deepwater developments in the Gulf of Mexico and a slowdown of development elsewhere, as regulators and companies reviewed safety procedures.

The Macondo disaster has led oil companies, service companies and regulators to review all aspects of offshore operations and procedures in an effort to avoid, or handle effectively, any further such incidents in all producing regions. In the United States, the Department of the Interior has introduced new safeguards, including a requirement to demonstrate the capacity to contain blowouts, and has reformed its management and oversight of offshore oil and gas development, separating safety regulation from developmental decision making. It has also established a new safety and environmental management system. The oil industry, in addition to developing the needed containment capacity, has taken action to improve its safety processes. These changes are expected to entail tougher scrutiny by regulators and higher costs in some instances, which may hinder, to some degree, new developments in the United States and elsewhere. This is compounded by the fact that several other incidents have taken place in the past two years, even though they involved much smaller spills and did not receive the same media coverage as Macondo. For example, in 2011, the Chevron-operated Frade field, offshore Brazil, had a 3 000 barrel spill; ConocoPhilips had two incidents (700 barrels of crude oil and 2 600 barrels of mineral oil based drilling fluid, which was largely recovered) in the Bohai Bay offshore China; Shell had a spill offshore Nigeria, while transferring oil to a tanker; while, in 2009, an extensive spill occurred in the Timor Sea between Indonesia and Australia. These incidents certainly illustrate that offshore drilling continues to test the limits of the industry's safety procedures. Regulators have adopted a tougher stance on incidents, proposing heavy fines, which are still in litigation for the incidents listed above.

Nonetheless, the projected contribution of offshore fields to global crude oil production is projected to be relatively stable over the projection period (Figure 3.20). Deepwater production expands from 4.8 mb/d in 2011 to about 8.7 mb/d by 2035, offsetting a decline in shallow-water production (mainly in the North Sea and the Gulf of Mexico). The expansion of output from deepwater fields will be driven mainly by new developments in Brazil, West Africa and the US part of the Gulf of Mexico. Deepwater production in the United States, which, compared with what production would have been, fell by an estimated 300 kb/d in 2011 as a result of the five-month moratorium on all deepwater drilling, is expected to resume growth in the coming years. This growth will be slow at first as new developments gradually replace decline at existing fields, then accelerate at the end of the decade. In Brazil, deepwater production, mainly from pre-salt fields (Box 3.5), accounts for the bulk of the projected growth in the country's production to 2035, contributing 2.9 mb/d by 2020 and 4.4 mb/d by 2035.

Figure 3.20 ▷ World offshore crude oil production by physiographical location and region in the New Policies Scenario

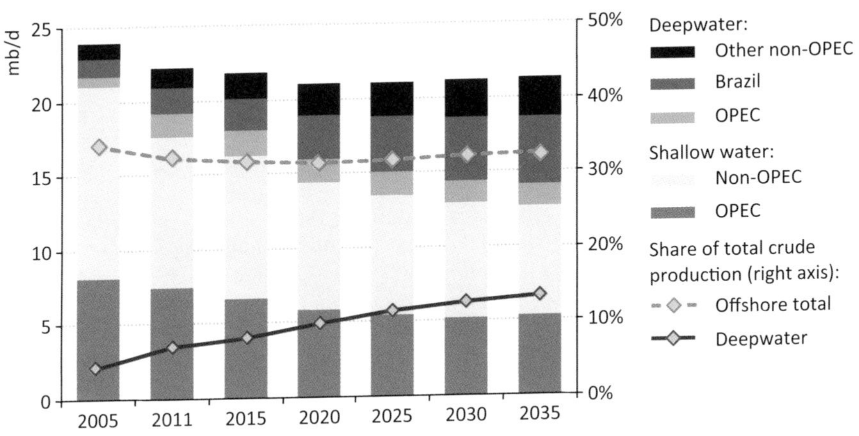

Sources: Rystad Energy AS; IEA analysis.

Trade

The projected trends in demand and supply in the New Policies Scenario result in a significant expansion in inter-regional trade in oil, from 42.1 mb/d in 2011 to 50.2 mb/d in 2035 – an increase of almost 20%. This is almost as big as the projected 12.3 mb/d rise in global oil demand. As a result, the share of oil supply that is traded inter-regionally rises from 48% to 50%. The Middle East, already the biggest exporting region, contributes the largest increase to net exports; they rise from 20.7 mb/d to 25.7 mb/d between 2011 and 2035, the region's share of global inter-regional trade jumping from 49% to 51%. Net exports from Eastern Europe/Eurasia (Russia and the Caspian region) decline slowly, as surging domestic demand outstrips the expansion of production, while net exports from Africa and particularly Latin America increase.

Trends in the net oil-importing regions diverge: imports rise strongly in non-OECD Asia, mainly due to a continued surge in imports into China and India (Figure 3.21). China's imports jump from 4.9 mb/d in 2011 to 12.3 mb/d in 2035, so that it overtakes the United States as the largest oil importer in the world after 2015. India's imports grow less in absolute terms – from 2.5 mb/d to 6.9 mb/d – but at a faster rate. Net imports into Europe decline over the projection period, but only marginally, and remain at 8 mb/d in 2035. By contrast, in the United States, imports fall sharply as a result of falling demand and increasing indigenous production. By the end of the projection period, imports into the United States are only 3.4 mb/d and North America as a whole becomes a net exporting region. By 2035, the bulk of Middle East and African exports are likely to be shipped to Asian markets, though contractual terms and geopolitical factors may, in some instances, result in oil flowing in directions that are not consistent with relative transport costs (which drive trade in our modelling of trade flows).

Figure 3.21 ▷ Net oil imports in selected countries and regions in the New Policies Scenario

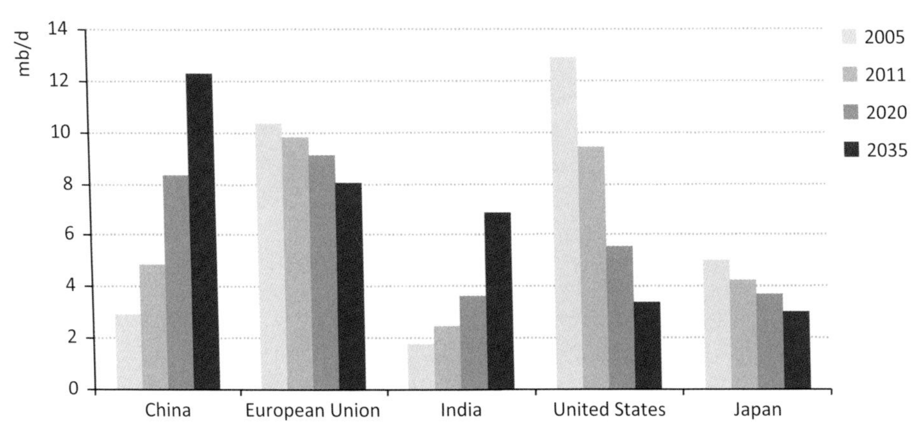

Import dependence, as measured by the share of net imports in demand, increases in all regions except the Americas. In the case of China, imports make up 82% of the country's total oil needs in 2035, compared with only 54% today. In Europe and OECD Asia Oceania, import dependence remains very high. For all importing regions, including the Americas, concerns about oil supply security will remain at the fore in view of the world's increasing reliance on supplies from a small number of producers, the increasing role of the Middle East and the growing reliance on shipping routes that are potentially vulnerable to disruption (see Chapter 2).

Investment in oil and gas

Worldwide upstream oil and gas investment[15] is budgeted to rise by around 8% in 2012 over the level in 2011, reaching a new record of $619 billion – more than 20% higher than in 2008 (Table 3.7).[16] Capital spending rose by 13% in 2011 over 2010. Increased

15. This section covers both oil and gas as companies often do not report such investments separately. Some additional, gas specific, information can be found at the end of Chapter 4.

16. These investment trends are based on the announced plans of 70 oil and gas companies. Total upstream investment is calculated by adjusting upwards their spending according to their share of world oil and gas production for each year.

spending since 2009 reflects a combination of higher oil prices, which increased potential investment returns, and the rising costs of current and planned projects. Our survey of the spending plans of 25 leading companies points to a faster increase in upstream spending in 2012 than downstream. This reflects weak refining margins and the expectation that the downstream overcapacity that has developed in some regions in the last few years may persist for several years (Box 3.6).

Table 3.7 ▷ Oil and gas industry investment by company (nominal dollars)

	Upstream			Total		
	2011 ($ billion)	2012 ($ billion)	Change 2011/12	2011 ($ billion)	2012 ($ billion)	Change 2011-12
Petrochina	27.5	29.3	6%	45.1	48.0	6%
Petrobras	23.0	28.4	23%	43.2	47.3	10%
ExxonMobil	33.1	33.3	1%	36.8	37.0	1%
Chevron	23.9	28.5	19%	26.5	32.7	23%
Royal Dutch Shell	19.1	24.4	28%	23.5	30.0	28%
Gazprom	39.3	24.5	-38%	43.5	27.5	-37%
Sinopec	9.3	12.4	33%	20.6	27.4	33%
Total	16.8	19.2	14%	19.0	24.0	26%
Pemex	17.8	19.0	7%	20.2	22.7	13%
BP	16.4	17.4	6%	20.2	22.0	9%
Eni	12.2	12.8	5%	17.4	17.0	-2%
Statoil	14.7	15.3	4%	16.3	17.0	4%
ConocoPhillips	12.0	14.0	17%	13.3	15.5	17%
BG Group	7.4	8.1	9%	10.6	11.0	4%
CNOOC	6.3	10.2	61%	6.4	10.2	58%
Apache	6.3	7.4	19%	8.0	9.5	19%
Rosneft	6.6	8.0	21%	8.0	9.4	18%
Lukoil	6.6	8.0	21%	8.0	9.4	18%
Occidental	7.5	8.3	10%	7.5	8.3	10%
Chesapeake	5.1	6.1	18%	6.3	7.5	18%
Suncor Energy Inc.	5.9	7.5	26%	6.8	7.5	9%
Anadarko	5.0	6.1	22%	6.6	6.8	4%
Devon Energy Corp	6.7	5.7	-15%	7.5	6.5	-14%
Repsol YPF	2.2	3.6	61%	7.5	4.7	-38%
EnCana	4.3	3.3	-23%	4.6	3.5	-24%
Sub-total 25	335.2	360.6	8%	433.3	462.3	7%
Total 70 companies	462.3	500.4	8%	n.a.	n.a.	n.a.
World	571.9	619.0	8%	n.a.	n.a.	n.a.

Notes: Only publically available data have been included (IEA databases include both public and non-public estimates for all major oil and gas producing companies). The world total for upstream investment is derived by prorating upwards the spending of the 70 leading companies, according to their estimated share of oil and gas production in each year. Pipeline investment by Gazprom is classified as upstream as it is required for the viability of projects. The "Total" column includes both upstream and downstream, as well as other investments (such as petrochemicals, power generation and distribution) for a few companies for which a breakdown is not publicly available. The 2012 figures are based on mid-year budgeted spending plans.

Sources: Company reports and announcements; IEA analysis.

Box 3.6 ▷ Global oil refining faces a major shake-out

A combination of a decline in OECD oil demand and a wave of investment in new plants in non-OECD countries has led to the emergence of a large structural overhang of refining capacity, especially in Europe, driving down utilisation rates and margins. Global crude oil distillation capacity rose by 7 mb/d between 2005 and 2011, to reach 93 mb/d and is set to rise to 101 mb/d by 2017, once all of the planned new capacity comes on stream (IEA, 2012d). Upgrading and desulphurisation capacity will also rise significantly, as refiners strive to meet more stringent fuel quality standards and increase more valuable light product yields. In 2011, global refinery utilisation rates averaged just 81%, down from 86% in 2005, and they are set to fall further, unless refinery closures accelerate: restoring the rate to the average of 83% for 2007-2011 would require nearly 4 mb/d of capacity to be closed or deferred by 2015 in the New Policies Scenario. The fact that a rising share of liquids supply is coming from NGLs, biofuels and GTLs, which can by-pass parts of the refining system altogether, is adding to the downward pressure on utilisation rates.

As has been the case in recent years, the largest additions to refinery capacity will occur in China and the rest of Asia; but around 1.5 mb/d of new distillation capacity, aimed both at meeting surging regional product demand and also export markets, is also due to come on stream in the Middle East by 2015. These capacity additions will keep the pressure on older and less sophisticated refineries in mature OECD markets. Dwindling demand, particularly for gasoline and residual fuel oil, is adding to the sector's woes in these regions. Of Europe's 15.9 mb/d of distillation capacity at the end of 2009, 16% has been sold in the past few years and another 12% is for sale. The major oil companies have been the main sellers, with trading companies and national oil companies, including PetroChina and Rosneft, among the buyers. More than 1.5 mb/d of capacity has been shut (or is now scheduled to shut) in Europe since the economic downturn of 2008, including the 220 kb/d Coryton refinery in the United Kingdom and the 260 kb/d Wilhelmshaven refinery (in 2011), and the 110 kb/d Hamburg (Harburg) refinery (in mid-2013) both in Germany. For the OECD as a whole, 3.5 mb/d of crude distillation capacity has been shut or is scheduled to shut over the same period. In Europe and the United States, Asian companies have been looking into buying existing and mothballed refineries so as to ship facilities overseas.

In the longer term, the refining industry worldwide will have to adapt to further geographical shifts in demand, as well as changes in the mix of demand for refined products and changing crude oil qualities. In particular, there is likely to be a need for more hydrotreating and upgrading capacity to meet the increasing share of low sulphur diesel in refined product demand and to accommodate more heavy oil feedstock in the longer term. Polymerisation and alkylation capacity will also be needed to take more NGLs as feedstock, given the large projected growth in NGL production.

In nominal terms, upstream investment in 2012 is set to reach a level five times higher than in 2000. Although there are signs that cost inflation is slowing down, as global pressures on commodity prices ease, part of this increase is due to higher unit costs of exploration and development, *e.g.* for cement, steel and other construction materials and equipment, as well as skilled personnel, drilling rigs and contracting oilfield services. The IEA's Upstream Investment Cost Index, which measures annual changes in the underlying capital costs of exploration and development, increased by 6% in 2011, taking it to just under 200 – *i.e.* just under twice the level of 2000 and above the previous peak of 2008 (Figure 3.22). Adjusted for costs, annual global upstream investment more than doubled between 2000 and 2011 and is poised to rise by an additional 6% in 2012. In the World Energy Model supply module, real upstream costs vary with the price of oil, technology learning and depletion in the main producing basins; in the New Policies Scenario, real costs rise on average by around 16% to 2035.

Figure 3.22 ▷ Worldwide upstream oil and gas investment and the IEA Upstream Investment Cost Index

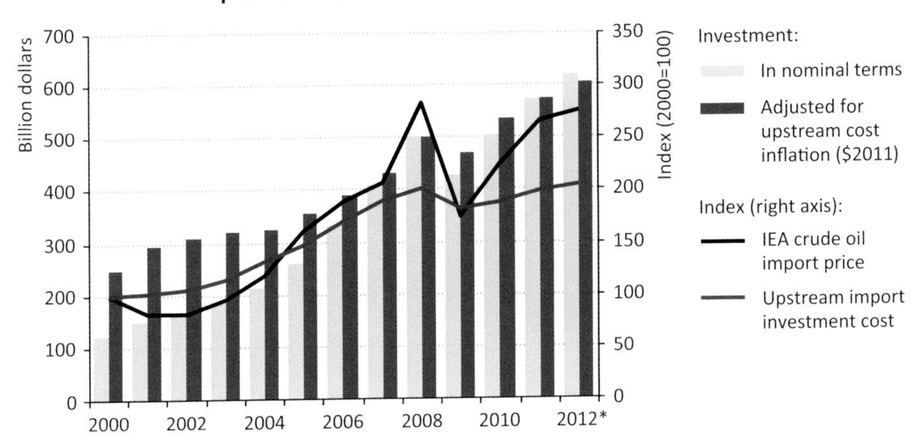

* Budgeted spending.

Notes: The IEA Upstream Investment Cost Index (UICI), set at 100 in 2000, measures the change in underlying capital costs for exploration and production. It uses weighted averages to remove the effects of spending on different types and locations of upstream projects.

Source: IEA databases and analysis based on industry sources.

The projected trends in the New Policies Scenario call for cumulative investment in infrastructure along the oil supply chain of around $10.2 trillion (in year-2011 dollars) over the period 2012-2035, or $430 billion per year (Table 3.8). The upstream sector accounts for almost 90% of this investment. Including gas, total annual upstream capital spending averages around $614 billion per year – about the same as the industry is planning to spend in 2012. Investment will need to remain at that high level for the whole period, with the high requirement underpinned by the need for more capacity to meet higher demand, as well as offset the decline at existing fields; the higher capital cost of new sources of supply, particularly deepwater and unconventional projects in non-OPEC countries; and increased unit upstream costs. These factors more than offset the shift in the balance of investment towards the Middle East and other regions where finding and development costs are relatively low. OPEC countries absorb only about one-fifth of global cumulative upstream oil and gas investment to 2035, despite their large share of incremental capacity.

Table 3.8 ▷ Cumulative investment in oil and gas supply infrastructure by region in the New Policies Scenario, 2012-2035 ($2011 billion)

	Oil			Gas			Total
	Upstream	Refining	Total	Upstream	T&D*	Total	Annual avg upstream
OECD	**3 070**	**271**	**3 341**	**2 547**	**920**	**3 467**	**234**
Americas	2 540	127	2 666	1 768	510	2 278	179
United States	1 822	97	1 919	1 384	386	1 770	134
Europe	456	94	551	561	323	883	42
Asia Oceania	74	50	124	218	87	306	12
Non-OECD	**5 838**	**803**	**6 641**	**3 282**	**1 131**	**4 412**	**380**
E. Europe/Eurasia	1 137	101	1 239	989	370	1 358	89
Russia	682	63	745	661	243	904	56
Asia	588	449	1 036	955	382	1 337	64
China	365	210	576	346	189	535	30
India	59	142	202	116	58	174	7
Middle East	937	137	1 074	240	229	469	49
Africa	1 554	50	1 604	660	61	721	92
Latin America	1 622	66	1 688	438	89	527	86
Brazil	1 083	31	1 113	99	25	124	49
World**	**8 908**	**1 074**	**10 242**	**5 829**	**2 051**	**8 677**	**614**

* T&D is transmission and distribution. ** World total oil includes an additional $259 billion of investment in inter-regional transport infrastructure (pipelines and tankers). World total gas includes an additional $103 billion for inter-regional pipelines and $695 billion for LNG liquefaction, re-gasification and carriers.

According to data we have compiled on the five-year spending plans of the handful of leading companies that publish such data, annual upstream capital spending is set to be stable over the period 2012-2016. Projects now underway, if completed to schedule, point to an expansion of total oil production capacity between 2011 and 2017 of about 9 mb/d (IEA, 2012d). The biggest increases in OPEC capacity are set to occur in Iraq, with smaller contributions from Libya (as capacity is restored to pre-war levels), the United Arab Emirates and Angola. The United States, Canada, Brazil, China and Kazakhstan are set to contribute the most to non-OPEC capacity. In the New Policies Scenario, demand rises by 5.7 mb/d by 2017, implying a substantial increase in the amount of effective spare capacity, all of which is in OPEC countries, from 2.6 mb/d in 2011 to 5.9 mb/d in 2017. Although this suggests easing of oil prices, we assume (as apparently does the oil futures curve) that geopolitical risks will counterbalance the better supplied market to sustain oil prices in this period. In the longer term, beyond 2025, dependence on Middle East will strengthen again, providing support for our oil price projections (Chapter 1).

Natural gas market outlook
Different shades of gold?

Highlights

- Demand for natural gas expands steadily through to 2035 in all three scenarios – the only fossil fuel for which this is the case. In the New Policies Scenario, gas consumption approaches 5 tcm by 2035, from 3.4 tcm today, growing at an annual average rate of 1.6%. The rate of growth, though, varies widely by region: it is three times faster in non-OECD countries than in the more mature OECD markets.

- Active policy support pushes gas demand in China up strongly, from around 130 bcm in 2011 to 545 bcm in 2035. The Middle East and India also see substantial growth in gas consumption, reaching 640 bcm and 180 bcm, respectively, in 2035. In the United States, low prices and abundant supply mean that gas overtakes oil around 2030 to become the largest fuel in the energy mix. New energy efficiency policies and higher prices constrain gas use in Europe and in Japan, with demand in the European Union rising back above 2010 levels only from 2020. Worldwide, the power sector accounts for 40% of incremental demand to 2035.

- The world's resources of natural gas are easily sufficient to accommodate the projected increase in demand, with recent substantial upward revisions to estimates of how much conventional and unconventional gas may be ultimately recoverable. In total, remaining resources amount to 790 tcm, or 230 years of output at current rates. Russia is the main conventional gas resource-holder and remains the largest gas producer and exporter in 2035; conventional output in Iraq, Brazil and new producers in East Africa also grows strongly.

- Unconventional gas accounts in our projections for nearly half of the increase in global gas production to 2035. Most of this increase comes from China, the United States and Australia. Yet the prospects for unconventional gas production worldwide remain uncertain and depend, particularly, on whether governments and industry can develop and apply rules that effectively earn the industry a "social licence to operate" within each jurisdiction, so satisfying already clamorous public concerns about the related environmental and social impacts.

- Rising LNG supplies, increased short-term trading and greater operational flexibility are likely to lead to increasing price connectivity between regions and a degree of price convergence. Opportunities to take advantage of regional price differentials will probably spur more trade between the Atlantic basin and Asia-Pacific markets, which traditionally have been largely distinct from each other. The construction of LNG liquefaction plants in North America opens up the possibility of exports to Asia, helping to reconnect prices between these two regions and accelerating the process of globalisation of natural gas markets.

Demand

Gas demand trends

Consumption of natural gas is set to continue to expand through to 2035, regardless of the direction that government policies take. It is the fastest-growing fossil fuel and the only one for which demand increases in the three main scenarios presented in this *Outlook* (Figure 4.1). This reflects the likelihood that ample supplies will keep gas prices competitive and the fact that it is the least carbon-intensive fossil fuel, so that its use is affected less by policies to curb greenhouse-gas emissions. Nonetheless, there are marked differences in the rates of demand growth across the scenarios, notably after 2020, and between the 450 Scenario on the one hand and the New and Current Policies Scenarios on the other.

Figure 4.1 ▷ World natural gas demand by scenario

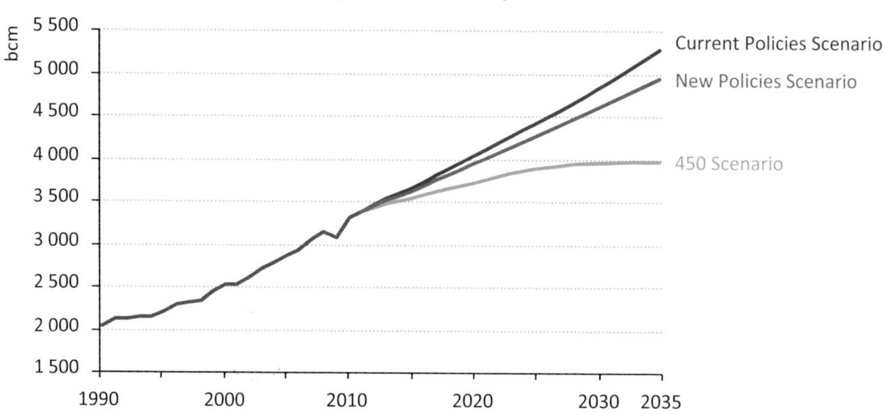

In the New Policies Scenario, world gas demand rises from almost 3.4 trillion cubic metres (tcm) in 2011[1] to just under 5 tcm in 2035 – an annual rate of growth of 1.6% (Table 4.1). Projected demand in 2035 is slightly higher than that of last year's *Outlook*, reflecting in part a higher starting point as global demand growth proved to be even stronger in 2010 than expected. In the Current Policies Scenario (in which government policies remain unchanged), demand grows slightly faster – at 1.9% per year – as no new action is taken to curb either gas or electricity demand (which leads to higher demand for gas in power generation). In the 450 Scenario, radical policies to meet the goal of limiting the global temperature increase to two degrees Celsius (2 °C) trim the average growth rate to only 0.7% per year, with demand levelling off by around 2030.[2]

1. Preliminary data on aggregate gas demand by country are available for 2011. The sectoral breakdown of demand is available for all countries up to 2010.

2. Comparison is also instructive with the 1.8% annual average growth in a Golden Rules Case, examined in a special *WEO* report (IEA, 2012a), in which conditions are put in place, including measures addressing social and environmental impacts, to allow rapid development of global unconventional gas resources.

Table 4.1 ▷ **Natural gas demand by region and scenario** (bcm)

	1990	2010	New Policies 2020	New Policies 2035	Current Policies 2020	Current Policies 2035	450 Scenario 2020	450 Scenario 2035
OECD	1 036	1 597	1 731	1 937	1 759	2 049	1 620	1 505
Non-OECD	1 003	1 710	2 213	3 018	2 275	3 237	2 095	2 466
World	2 039	3 307	3 943	4 955	4 034	5 286	3 716	3 971
Share of non-OECD	*49%*	*52%*	*56%*	*61%*	*56%*	*61%*	*56%*	*62%*

Sources: IEA databases and analysis. Different sources use different definitions of gas volumes; see, for example, Box 8.3 of *World Energy Outlook 2011* (*WEO-2011*) (IEA, 2011).

Global gas demand underwent an unprecedented decline of 2% in 2009 as a result of the worldwide recession, but it bounced back by a remarkable 7.5% in 2010 – the fastest year-on-year increase recorded since IEA statistics began in 1971. This was the result of economic recovery, as well as unusually cold winter weather in the northern hemisphere and high summer temperatures in Asia (which boosted air-conditioning demand).

Preliminary data points to a more modest increase in demand in 2011 of around 2%, below the average annual trend rate of 2.8% seen in 2000-2008. This figure masks a wide divergence in trends across countries and major regions: gas use continued to grow briskly in developing Asia and most other non-OECD regions, but collapsed in Europe, by a record 9%, to levels even lower than in 2009 – the result of the economic slowdown, higher gas (and low coal and carbon) prices and milder weather. Demand was higher in North America, where the boom in shale gas production boosted overall supplies, drove down prices and stimulated demand. In Japan, gas-fired generation surged following the nuclear accident at Fukushima Daiichi, which led to all of the country's nuclear reactors being progressively shut down for scheduled maintenance and "stress tests".

Regional trends

Non-OECD countries will continue to drive changes in future global gas demand, reflecting their more rapid rates of economic growth and the relative immaturity of their gas markets. In the New Policies Scenario, gas consumption expands in every *WEO* region between 2010 and 2035 (Figure 4.2), but growth is nearly three times faster in non-OECD countries, (on average, 2.3% per year) than in the OECD (0.8%) (Table 4.2). Consequently, non-OECD countries account for 80% of the overall increase in world natural gas demand.

Demand in the United States has been growing rapidly since 2009 and it continues to increase in our projections, albeit at a slower annual average pace than at present. Natural gas prices in the United States are assumed to increase, compared with the historic lows seen in 2011-2012, but they remain at levels that promote a continued expansion of consumption (despite the envisaged start of liquefied natural gas [LNG] exports [Spotlight]). A boost to gas-intensive industrial activity, growth in power generation and, from a low base, rising use in the transport sector, mean that gas is the only fossil fuel in the United

States that sees an absolute increase in use over the period to 2035, with coal and oil both on a declining trend. These trends are sufficiently strong that, by 2035, the share of gas overtakes that of oil to become the most important fuel in the United States' energy mix.

Figure 4.2 ▷ Natural gas demand in selected regions in the New Policies Scenario

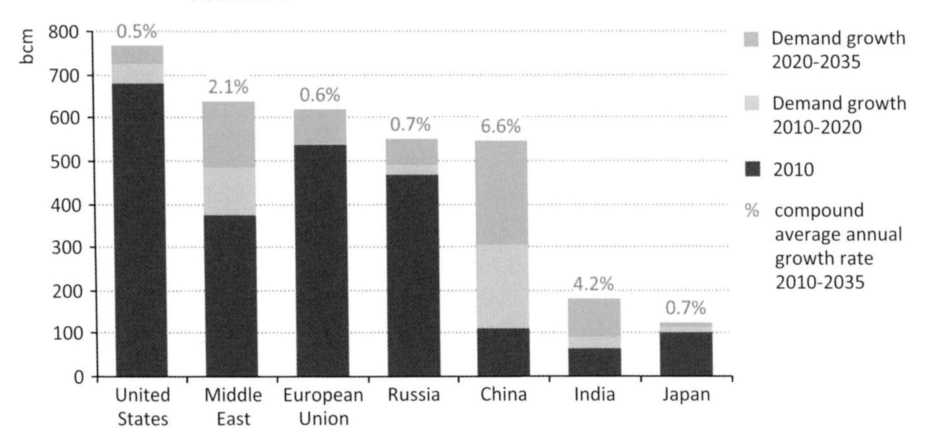

Table 4.2 ▷ Natural gas demand by region in the New Policies Scenario (bcm)

	1990	2010	2015	2020	2025	2030	2035	2010-2035	
								Delta	CAAGR*
OECD	**1 036**	**1 597**	**1 652**	**1 731**	**1 796**	**1 864**	**1 937**	**341**	**0.8%**
Americas	628	845	898	940	962	995	1 032	187	0.8%
United States	533	680	712	728	736	749	766	86	0.5%
Europe	325	569	550	585	619	643	669	100	0.7%
Asia Oceania	82	182	204	206	214	225	236	53	1.0%
Japan	57	104	120	115	118	122	123	20	0.7%
Non-OECD	**1 003**	**1 710**	**1 963**	**2 213**	**2 472**	**2 746**	**3 018**	**1 307**	**2.3%**
E. Europe/Eurasia	737	692	731	747	777	810	842	150	0.8%
Caspian	100	106	117	127	136	142	149	42	1.4%
Russia	447	466	488	492	508	530	549	83	0.7%
Asia	84	393	514	660	801	949	1 111	717	4.2%
China	15	110	195	304	390	469	544	434	6.6%
India	13	64	75	92	116	144	178	115	4.2%
Middle East	87	376	437	485	538	594	640	264	2.1%
Africa	35	103	118	139	153	166	176	73	2.2%
Latin America	60	146	163	182	203	227	249	103	2.2%
Brazil	4	27	30	38	49	63	78	50	4.3%
World	**2 039**	**3 307**	**3 616**	**3 943**	**4 268**	**4 610**	**4 955**	**1 648**	**1.6%**
European Union	368	536	509	540	570	592	618	82	0.6%

* Compound average annual growth rate.

SPOTLIGHT

Where will cheap American gas end up?

The boom in shale gas production in the United States since the mid-2000s has boosted overall gas supplies substantially, removing the anticipated need to import LNG and driving down prices. As a result, the North American gas market is moving to a different rhythm from the rest of the world and this disconnect is reflected in large price differentials with other markets[3]: in June 2012, spot gas was trading at as little as $2.10 per million British thermal units (MBtu) at Henry Hub – the leading US trading hub – compared with $9.90/MBtu in the United Kingdom, $12/MBtu for spot LNG in the Mediterranean and $17.40/MBtu for spot LNG in northeast Asia.

In these market conditions, interest in exporting gas has, unsurprisingly, grown and a number of gas liquefaction projects are planned – though only one has yet gained full regulatory approval, Cheniere's Sabine Pass project in the Gulf of Mexico. Just how much LNG export capacity is built and how much gas is eventually exported hinges on whether regulatory approvals for new export terminals are granted in the United States and whether these regional price differentials are maintained. At both current prices and those assumed for 2020, LNG exports priced off Henry Hub would be profitable, especially to Asian markets, where they would undercut the prevailing oil-indexed prices for alternative supplies. For example, the 2011 notional net-back margin for LNG exported from the US to Japan (the gap between the prevailing market price and notional supply costs) was more than $6/MBtu, though it falls to $4.3/MBtu on the prices assumed for 2020 (Figure 4.3). The margin is much smaller on sales to Europe, at less than $1/MBtu in 2011 and around $1.4/MBtu in 2020.

In the New Policies Scenario, we project that North American exports of LNG (including exports from proposed projects in western Canada) reach 35 billion cubic metres (bcm) by 2020 and more than 40 bcm by 2035, two-thirds of which is expected to go to markets in Asia. The prospect of LNG exports has set off a debate in the United States about the extent to which they will drive up domestic prices, by taking gas off the local market, and the US Department of Energy is waiting to review the results of a price impact study before dealing with the pending export applications.[4] In our projections, 93% of the natural gas produced in the United States remains available to meet domestic demand. We assume that gas prices increase from their current levels to reach $5.5/MBtu in 2020, largely driven by the domestic dynamics of supply and demand (Box 4.4): exports on the scale that we project would not play a large role in domestic price setting.

3. Market fluctuations mean that the spot price differentials vary week-to-week; in early October 2012, the range had narrowed to between $3.20/MBtu (Henry Hub) and $12.80/MBtu (Northeast Asia).

4. Applications for export to countries with which the United States has a free trade agreement are granted without delay, but approval for export to other countries can be limited or blocked if the US Department of Energy finds that they are not in the public interest.

Figure 4.3 ▷ Indicative economics of LNG exports from the United States

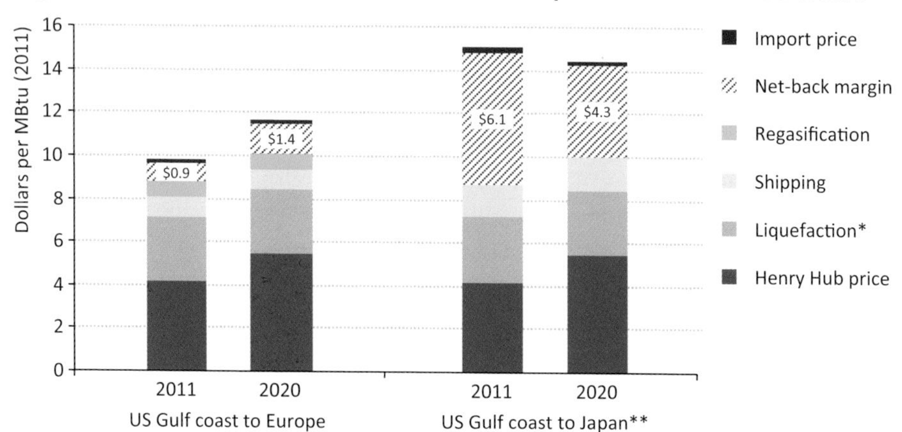

* Includes cost of pipeline transport to export terminal. ** Widening of the Panama Canal, due to be completed in 2014, will allow for more LNG tanker traffic.

Notes: LNG costs are levelised assuming asset life of 30 years and a 10% discount rate. The Japanese import price is for liquefied gas, so it does not include regasification.

In Europe, gas demand recovers slowly in the medium term, returning to the level of 2010 only towards the end of the current decade in the face of relatively high gas prices, strong growth in renewables-based power and low carbon prices (which favour coal in power generation). Thereafter, gas demand grows more strongly, mainly on the back of increasing demand from gas-fired power stations, as policies to reduce CO_2 emissions boost demand for gas relative to other fossil fuels.

The new energy and environment strategy adopted in September 2012 shapes the outlook for gas demand in Japan. In line with this strategy, we have assumed that greater policy emphasis is placed on energy efficiency and renewables, which reduces the role of nuclear power in the energy mix (see the section on nuclear power in Chapter 6). As a result, after the jump in gas demand in 2011 and 2012, gas consumption is projected to remain relatively stable, at around 120 bcm, for much of the projection period. The short-term potential increase in gas demand is limited, in any case, as the country's gas-fired generating plants are already running close to capacity.

Among the non-OECD countries, China is expected to see the strongest growth in gas demand in absolute terms. In the New Policies Scenario, gas use surges from around 130 bcm in 2011 to just over 300 bcm in 2020 and 545 bcm in 2035. The 12th Five-Year Plan, which runs from 2011 to 2015, targets a doubling of the share of gas in the country's primary energy mix, to 8.6%, implying consumption of more than 300 bcm in 2015. We project a lower level of demand, of just under 200 bcm, in view of a number of constraints, including logistical bottlenecks in building transportation infrastructure and a regulatory framework that is subject to review which might lead to profound change (Box 4.1). Other Asian countries also see strong growth in gas demand for power generation and industry.

Box 4.1 ▷ Gas pricing reform in China

At the end of 2011, the Chinese government announced plans to liberalise wholesale gas pricing in order to encourage investment in the upstream sector and in pipeline and LNG import facilities. Under the existing regulatory arrangements, wellhead gas prices and pipeline transportation tariffs have been set by the central government on a cost-plus basis. The regulated wellhead price has effectively been a baseline, above which producers and buyers could negotiate by up to 10%. Transport tariffs have been set according to construction and operating costs, plus a margin, for each individual pipeline. The price of gas delivered to the city gate and to large consumers, such as power generators, thus varies across the country and by supplier according to the actual cost of production, the distance from the source of production and the capacity of the pipeline. End-user prices are also set by the government and vary by type of customer (*e.g.* fertiliser producer, industrial user or local distributer).

The pricing reform announced by the state planning body, the National Development and Reform Commission, in December 2011 aims to establish a single maximum price at the city gate for each location. That price will be linked to the prices of imported fuels: 60% to fuel oil and 40% to liquefied petroleum gas, with a 10% discount to encourage gas use (taking account of distribution and marketing costs). Wellhead prices will be liberalised, allowing producers of all types of gas to negotiate prices with buyers on a net-back basis, taking account of the cost of transportation. The price reform has already been implemented on a trial basis in Guangdong Province and Guangxi Zhuang Autonomous Region, resulting in a city-gate price of 2 740 yuan per thousand cubic metres in Guangdong (around $12/MBtu) and 2 570 yuan in Guangxi ($11/MBtu). The government plans to extend the reform to Shanghai by 2013 and complete the process of implanting it nation-wide by 2015.

Gas demand in Russia and the rest of Eastern Europe/Eurasia grows modestly, compared with other non-OECD countries. These are mature gas markets with slow-growing (or, in some cases, declining) populations and there is a declared political priority, in several countries, to diminish the heavy dependence on gas for power generation. Efficiency gains, as old Soviet-era capital stock is retired, will also limit the extent of increasing gas use. Demand continues to increase strongly in the Middle East, driven by power generation, water desalination and petrochemical projects, as well as own use in LNG and gas-to-liquids production. In some cases, low regulated gas prices have resulted in physical shortages of gas, as demand has outstripped local supply capacity.

Gas use in most African countries – particularly sub-Saharan Africa – is limited by the pace of upstream and transport infrastructure developments, domestically and in neighbouring countries. Low incomes, country risk and large capital needs mean that projects to import gas are hard to justify. In Latin America, as in most other regions, power generation is the main driver of demand: several countries plan to increase the share of gas in the generating fuel mix to balance their reliance on hydropower, which is vulnerable to drought, and oil-fired power stations, which have become very expensive to run as a result of high oil prices.

Sectoral trends

Regardless of how government policies evolve, the power sector will remain the principal driver of gas demand in most regions. In the New Policies Scenario, worldwide use of gas for power (and heat) increases by half between 2010 and 2035 (an average rate of 1.6% per year), accounting for 40% of the total increase in gas demand (Figure 4.4). Gas is well-suited as a fuel for power generation, but the extent of its use will depend critically on the price of the fuel (taking account of any carbon penalties), both in absolute terms and relative to the price of coal and, to a much lesser extent, residual fuel oil and distillate. In the New Policies Scenario, natural gas used mainly in combined-cycle gas turbines (CCGTs) is expected to remain the preferred option for new power stations in many cases, for a combination of economic, operational and environmental reasons: CCGTs have relatively high thermal efficiency, are relatively quick and cheap to build, flexible to operate and emit less carbon dioxide (CO_2) and local air pollutants than other fossil fuel-based technologies (see Chapter 6). CO_2 pricing not only pushes up the relative cost of coal-based power, improving the competitive position of gas, but also encourages renewables, which may be most easily backed up by gas-fired capacity.

Figure 4.4 ▷ World natural gas demand by sector in the New Policies Scenario

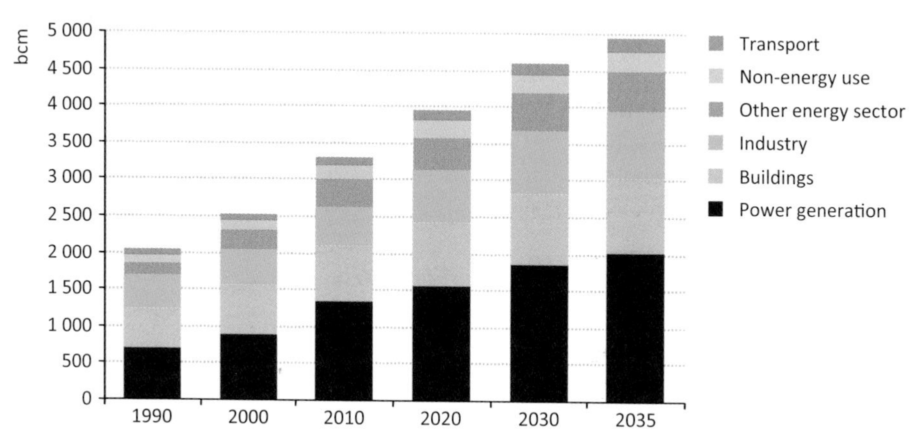

Gas consumption in buildings for space and water heating grows at a more sedate pace of 1.3% per year on average over the projection period, but this remains the largest end-use sector (43% in 2035), even though, in many OECD countries, most of the scope for switching from heating oil and other fuels to gas has been exhausted. The gradual adoption of the most efficient gas-condensing boilers also crimps demand. In many non-OECD countries, the high cost of building local distribution networks acts as a disincentive to expand gas use in buildings, particularly in regions where high average temperatures mean that gas would be used primarily for water heating, rather than space heating. China sees the biggest increase in residential and commercial gas demand, accounting for more than one-third of the global increase in gas use in buildings to 2035, driven by rapid urbanisation, especially in the coastal regions.

Demand for gas in industry is set to grow faster than in any other end-use sector, other than transport (which is growing from a low base). Industrial demand is projected to expand by 1.9% per year over 2010-2035, with most of the increase coming from non-OECD countries, mainly in Asia, Latin America and the Middle East; demand in OECD countries levels off in the 2020s and then remains flat, because of slower growth in industrial output and the adoption of more efficient technologies.

High oil prices and worries about pollution have stimulated interest in using natural gas as a road transport fuel (see Chapter 3), particularly in North America, where gas prices are expected to remain relatively low. The use of gas in compressed or liquefied form worldwide is small, but it has been growing rapidly in recent years in some countries. There were an estimated 15 million natural gas vehicles (NGVs) on the roads in 2011, mainly in Iran, Pakistan, Argentina, Brazil and India. The majority are cars, though buses account for the bulk of the consumption. In the New Policies Scenario, global NGV gas use continues to expand quickly, by around 4.7% per year on average. Nonetheless, the share of gas in world road-transport energy use reaches only 4% in 2035, with consumption remaining concentrated in a small number of countries. North America and India see the biggest increases in volume terms, both adding 13 bcm to their gas demand by 2035. NGV demand also increases markedly in China, where the government is keen to promote gas use to limit local pollution and diversify the energy mix. The potential exists for much faster growth in this sector worldwide, but it would require stronger policy action than we have assumed to promote the necessary investment in distribution infrastructure and persuade customers to switch to NGVs. Despite the environmental advantages of NGVs over conventional vehicle and fuel technologies, expanding the NGV fleet faces several barriers, including the cost of on-board fuel storage and building the infrastructure for delivery and distribution at existing refuelling stations.

Production

Reserves and resources

The world's resources of natural gas are large enough to accommodate vigorous expansion of demand for several decades. Indeed, substantial upward revisions to estimates of proven reserves and ultimately recoverable resources of both conventional and unconventional over the last few years have produced a far more comfortable picture of potential supply, though some uncertainty remains about how much it will cost to develop those resources and whether sufficient and timely investment in productive capacity will be made. At the end of 2011, proven reserves of gas amounted to 208 tcm, according to BP's annual Statistical Review of World Energy (BP, 2012) – an increase of 12 tcm over 2010, thanks mainly to a near-doubling of reserves in Turkmenistan (in the huge Galkynysh field, formerly known as South Yolotan). The slightly smaller figure for global proven reserves given by the Oil and Gas Journal, 191 tcm, does not include a revised figure for Turkmenistan (O&GJ, 2011). In almost all countries, these reserves are entirely conventional; the main exceptions are the United States and Canada, where unconventional gas – shale and tight gas, and coalbed methane – makes up a growing share.

Discoveries of gas fields have continued at a brisk pace, with 3.3 tcm of recoverable resources discovered in 2010 and 2.3 tcm in 2011, including a total of more than 600 bcm in Iran and 500 bcm in Mozambique. The latter continues the trend of large discoveries made in the last two years in offshore East Africa, in which interest has been boosted by its convenient geographical position for LNG exports to Asian markets. Another hot spot of gas exploration is the eastern Mediterranean, where some major finds have been made recently that are all well-placed to serve European markets. The Black Sea is also a theatre of active exploration in its Ukrainian, Turkish, Romanian, Bulgarian and Russian parts.

Our estimate of remaining recoverable resources of gas worldwide has been revised upwards substantially this year, reflecting mainly the results of the updated assessments of undiscovered conventional gas resources and reserves growth carried out by the US Geological Survey, which were recently released (Box 4.2). Drawing on these estimates and those from other sources, we now estimate remaining technically recoverable resources of conventional natural gas worldwide, including proven reserves, reserves growth and undiscovered resources, at just over 460 tcm – an increase of around 60 tcm on last year's figure (Table 4.3). Remaining technically recoverable resources are now estimated at 200 tcm for shale gas – this includes the latest estimates from the US Energy Information Administration (EIA) for the United States (US DOE/EIA, 2012), 81 tcm for tight gas and 47 tcm for coalbed methane. In total, natural gas resources amount to 790 tcm, or more than 230 years of production at current rates (see Box 1.1 in IEA [2012b] for definitions of these types of resources).

Table 4.3 ▷ Remaining technically recoverable natural gas resources by type and region, end-2011 (tcm)

	Conventional	Unconventional				Total
		Tight gas	Shale gas	Coalbed methane	Sub-total	
E. Europe/Eurasia	144	11	12	20	44	187
Middle East	125	9	4	-	12	137
Asia-Pacific	43	21	57	16	94	137
OECD Americas	47	11	47	9	67	114
Africa	49	10	30	0	40	88
Latin America	32	15	33	-	48	80
OECD Europe	24	4	16	2	22	46
World	**462**	**81**	**200**	**47**	**328**	**790**

Notes: Remaining resources comprise proven reserves, reserves growth and undiscovered resources. The resource estimate for coalbed methane in Eastern Europe and Eurasia replaces a figure given in the *WEO-2011*, which included a "gas-in-place" estimate for Russia instead of the estimate for technically recoverable resources. Unconventional gas resources in regions that are richly endowed with conventional gas, such as Eurasia or the Middle East, are often poorly known and could be much larger.

Sources: BGR (2011); US DOE/EIA (2011); USGS (2000); USGS (2012b); USGS (2012c); IEA databases and analysis.

Box 4.2 ▷ USGS conventional oil and gas resource assessment updated

In March 2012, the US Geological Survey (USGS) released a summary of the results of a set of assessments of the undiscovered conventional oil and gas resources of 171 provinces outside the United States, their first comprehensive update since 2000 (USGS, 2012b).[5] The work was carried out between 2009 and 2011. Excluding the United States, the world is now estimated to hold undiscovered, technically recoverable resources of 565 billion barrels of oil, 158 tcm of gas and 167 billion barrels of natural gas liquids (NGLs), based on mean values. These estimates include both onshore and offshore resources. Many new areas were included in the current study, including East Africa, extending the total number of basins (assessment units) from 246 in 2000 to 313. The estimate for undiscovered gas is one-fifth higher than in 2000, but that for oil is 13% lower (largely because of discoveries made in the intervening period).

About 75% of the undiscovered technically recoverable oil, outside the United States, is in four regions: South America and the Caribbean, sub-Saharan Africa, the Middle East and North Africa, and the Arctic provinces of North America. Large undiscovered gas resources remain in all regions, with the largest volume – 46 tcm, or 29% of the world total – in the former Soviet Union.

To supplement these figures, in June 2012 the USGS released a new global estimate for potential additions to oil and gas reserves in discovered fields due to reserve growth (USGS, 2012c). Reserve growth is the progressive increase in the estimated volumes of oil and natural gas that can be recovered from existing fields and reservoirs as a result of better assessment methodologies, new technologies, greater understanding of current reservoirs and other advances. Most reserve growth results from delineation of new reservoirs, field extensions, or improved recovery techniques and recalculations of reserves, due to changing economic and operating conditions. The USGS put total mean unidentified, conventional reserve growth additions in the world outside the United States at 665 billion barrels of oil, 40 tcm of gas and 16 billion barrels of NGLs. Updated estimates of reserve growth for the United States are due to be released later in 2012.

Eastern Europe/Eurasia (mainly Russia) and the Middle East are the largest holders of conventional gas resources, while the biggest unconventional resources are in Asia-Pacific and North America. Overall, Russia has by far the largest gas resources and Europe the smallest. A large part of the world's remaining recoverable unconventional gas lies in countries or regions that are currently net gas importers, such as China and the United States.

5. Detailed reports on the individual assessments by major region can be found on the USGS Energy Resources Program website: *energy.usgs.gov*.

Gas production prospects

Global gas production rises at different rates across the scenarios, in line with demand, rising from 3.3 tcm in 2010 to between 4.0 tcm and 5.3 tcm in 2035, depending on the scenario (Table 4.4). Gas prices vary to balance supply with demand in each case. Non-OECD countries continue to dominate gas production in all three scenarios, their share rising from around 65% in 2010 to between 69% and 72% in 2035. Unconventional gas accounts for a growing share of global production (see next section).

Table 4.4 ▷ Natural gas production by major region and scenario (bcm)

			New Policies		Current Policies		450 Scenario	
	1990	2010	2020	2035	2020	2035	2020	2035
OECD	881	1 178	1 328	1 446	1 334	1 481	1 259	1 230
Non-OECD	1 178	2 106	2 616	3 509	2 700	3 805	2 457	2 741
World	2 059	3 284	3 943	4 955	4 034	5 286	3 716	3 971
Share of unconventional	*3%*	*14%*	*20%*	*26%*	*20%*	*25%*	*20%*	*27%*

Notes: Definitions and reporting of tight gas vary across countries so the split between conventional and unconventional is approximate. Differences between historical supply and demand are due to changes in stocks.

Among the OECD regions, output in North America is projected to continue to expand, thanks mainly to shale gas in the United States. Total US gas production grows from an estimated 650 bcm in 2011 to 800 bcm in 2035 in the New Policies Scenario (Table 4.5), putting the United States ahead of Russia as the largest gas producer in the world between 2015 and the end of the 2020s. Shale gas accounts for almost all of the increase in US output, with conventional gas and coalbed methane output remaining close to current levels in 2035 and tight gas showing a gradual decline. In Canada, gas production is projected to climb gradually through the *Outlook* period, reaching almost 190 bcm in 2035, with higher shale gas and coalbed methane output offsetting a decline in conventional supply and, after 2020, a drop in tight gas production. Mexico's production is projected to rise to around 75 bcm on the back of the development of the country's large unconventional gas resources, with most of the increase coming in the second half of the projection period.

Elsewhere in the OECD, production prospects are brightest in Australia, where output triples over the projection period to over 160 bcm, driven mainly by LNG export projects. Coalbed methane accounts for a large share of the increase in production. Conventional production in the European Union is projected to continue its long-term downward path, only partially offset by output of unconventional gas that starts to pick up around 2020, led by Poland, and approaches 20 bcm by 2035. Gas production in Norway remains at 110-120 bcm per year, with the development of new resources in the Norwegian and Barents Seas more or less compensating for declines at North Sea fields. The eastern Mediterranean is emerging as a promising area for new gas exploration and development, although complex regional politics are likely to make this a slow process. Production in

Israel is set to grow to nearly 10 bcm in 2020 and 19 bcm in 2035, based on several new fields – including the Tamar and Leviathan fields, with combined recoverable resources estimated at 740 bcm – that have been found in recent years. Initially, gas will be supplied to the domestic market, but export projects could be developed later.

Eastern Europe/Eurasia sees the biggest volume increase in overall gas production (just ahead of Asia) over the projection period in the New Policies Scenario. Most of the increase occurs in Russia, but Turkmenistan and Azerbaijan also make a growing contribution. In Russia, where output rises from 677 bcm in 2011 to more than 850 bcm in 2035, production continues to move gradually away from the traditional production areas in Western Siberia towards more challenging and expensive frontiers in the Arctic (mainly the Yamal peninsula, with its huge resources) and in Eastern Siberia (for export to China, which we project to begin in the early 2020s). The bulk of the increase in Russian production goes to export, but uncertainty over the pace of demand growth in Europe, Gazprom's main export market, combined with the emergence of the United States as an LNG exporter, create dilemmas for westward-oriented Russian gas export projects. This is particularly true for the major Arctic LNG projects, the Yamal LNG project proposed by Novatek and the Shtokman project proposed by Gazprom (which has been shelved, according to an August 2012 announcement by Gazprom). These projects can reach Asia-Pacific markets for part of the year via the Arctic northern route, but their reliance on European and Atlantic basin markets at other times would risk displacing a part of Russia's existing exports by pipeline (even with the anticipated extensive use of swaps). We assume that both of these projects will eventually go ahead, with Yamal LNG the first to start operation, towards 2020, on the assumption that the fiscal terms offered by the government are sufficiently attractive to underpin the project economics, and Shtokman only much later in the projection period. Ukraine has the potential to boost its domestic production, both through offshore conventional gas exploration and through development of its onshore shale gas and coalbed methane resources.

Turkmenistan's gas production is projected to continue to rise strongly, driven in large part by development of the Galkynysh field (formerly known as South Yolotan), which now ranks second in the world for reserves after the North Field/South Pars, shared by Qatar and Iran. The phased development of this field, made possible by the commissioning in 2010 of the Central Asia Gas Pipeline to China, means projected growth in Turkmenistan's gas output from 55 bcm in 2011 to 100 bcm soon after 2020 and around 140 bcm by 2035. Output in Azerbaijan is projected to rise to nearly 50 bcm by 2035, reflecting the progress that is being made in defining pipeline routes for the long-discussed Shah Deniz Phase II project and the potential for continued production growth into the 2020s from the Absheron field, deep layers at the Azeri-Chirag-Guneshli complex and continued development of Shah Deniz. Uzbekistan maintains production around 60-70 bcm per year, meeting a moderate increase in domestic demand and maintaining steady exports to China, until about 2030, when resource limitations induce a rapid decline.

Table 4.5 ▷ Natural gas production by region in the New Policies Scenario (bcm)

	1990	2010	2015	2020	2025	2030	2035	2010-2035	
								Delta	CAAGR*
OECD	**881**	**1 178**	**1 239**	**1 328**	**1 360**	**1 395**	**1 446**	**268**	**0.8%**
Americas	643	816	893	970	993	1 026	1 067	251	1.1%
Canada	109	160	165	171	169	174	188	28	0.7%
Mexico	26	50	47	51	57	66	75	25	1.6%
United States	507	604	679	747	765	784	800	196	1.1%
Europe	211	304	267	250	238	226	215	- 89	-1.4%
Norway	28	110	114	118	116	115	113	3	0.1%
Asia Oceania	28	58	80	107	129	143	164	106	4.3%
Australia	20	49	73	102	125	139	161	112	4.9%
Non-OECD	**1 178**	**2 106**	**2 377**	**2 616**	**2 908**	**3 215**	**3 509**	**1 403**	**2.1%**
E. Europe/Eurasia	831	842	893	968	1 057	1 136	1 204	362	1.4%
Azerbaijan	10	17	20	30	43	44	48	32	4.4%
Russia	629	657	675	704	737	808	856	199	1.1%
Turkmenistan	85	46	66	84	110	120	138	92	4.5%
Asia	130	420	502	548	607	684	775	356	2.5%
China	15	95	134	175	217	264	318	223	5.0%
India	13	51	54	62	72	84	97	46	2.6%
Indonesia	48	86	109	109	115	128	143	57	2.1%
Middle East	92	472	565	609	660	722	809	336	2.2%
Iran	23	143	143	150	159	180	219	76	1.7%
Iraq	4	7	13	41	73	82	89	82	10.7%
Qatar	6	121	170	177	187	204	223	102	2.5%
Saudi Arabia	26	81	104	107	108	117	128	47	1.8%
UAE	20	51	57	57	56	58	62	11	0.8%
Africa	64	209	221	277	346	402	428	220	2.9%
Algeria	43	80	83	105	123	140	147	67	2.5%
Libya	6	17	16	20	26	32	37	20	3.2%
Nigeria	4	33	43	58	71	87	94	61	4.3%
Latin America	60	163	195	213	238	271	292	129	2.4%
Argentina	20	42	46	49	58	64	66	23	1.8%
Brazil	4	15	21	32	49	69	87	72	7.3%
Venezuela	22	24	30	37	44	63	73	48	4.5%
World	**2 059**	**3 284**	**3 616**	**3 943**	**4 268**	**4 610**	**4 955**	**1 671**	**1.7%**
European Union	211	201	158	133	116	104	94	- 107	-3.0%

* Compound average annual growth rate.

The Middle East holds enormous resources, but production growth will be constrained in the medium term by a dearth of new LNG export projects, and technical and economic difficulties in developing some fields. Regional production is projected to rise from an estimated 525 bcm in 2011 to 610 bcm in 2020 and then expand more rapidly to 810 bcm by 2035. Iraq (see Chapter 14) and, in the longer term, Qatar and Iran are the biggest contributors to higher production in the region. In Saudi Arabia, four new fields are being developed, including the Karan field, which will together add over 30 bcm/year of marketable gas production by 2014, boosting total capacity by about 40%. But further production increases, which would be needed to keep pace with soaring demand, hinge on finding new fields; recent exploration activity has yielded disappointing results, though Saudi Aramco's recent focus on assessing the kingdom's shale gas resources could change the situation. Low domestic gas prices are also a barrier to the development of resources in many cases.

Prospects for gas production in Iran have been clouded by international sanctions, which have largely closed access to foreign investment and technology. Some progress has been made in bringing new phases of South Pars online in order to meet growing demand from the domestic market, for reinjection into oil fields and for export to Turkey, with production now close to 150 bcm per year. But with the difficulties in completing further expansion phases, and plans to export LNG now effectively on hold because of the sanctions, production is projected to grow only after 2020, responding to large demand increases in India and Pakistan.

There is also considerable potential for production growth in Africa, both in the established producing countries in North and West Africa and in East Africa, where a number of major offshore finds over the past three years look set to provide the basis for production to take off in the coming years (Box 4.3). The main producing African countries today are expected to expand their production in the longer term, despite dwindling output from mature fields and, in some cases, difficulties in developing new reserves.

Algeria's gas production has stagnated over the last few years, largely because of a major accident at the Skikda LNG-export plant in 2004 and, more recently, lower demand in Europe. The government is now fast-tracking new field developments in order to offset output declines at the workhorse Hassi R'Mel field, which still accounts for about half of total marketed gas supply, to meet rising domestic demand and to free up more gas for export. Egypt is also looking to boost its production to keep pace with soaring domestic demand: up to 10 bcm per year of additional supply is due to come from the West Nile Delta over the current decade. Production in Libya is recovering progressively from the effects of the 2011 war, but boosting production beyond pre-war levels is projected to take until after 2015: recent exploration has yielded disappointing results.

Production prospects in West Africa hinge to a large degree on LNG export projects. In Nigeria, which has large reserves, final investment decisions are anticipated in 2012 on Brass LNG and in 2013 on NLNG train 7, but they may be delayed again. Security concerns and uncertainty about a new Petroleum Industry Bill continue to cloud the outlook for the Nigerian gas and oil sector. Equatorial Guinea, already an LNG exporter, and Angola, set to become one, are also considering new projects, though there are doubts about whether their reserves are big enough to support them. Overall, African gas production is projected to expand from around 195 bcm in 2011 to 280 bcm in 2020 and 430 bcm in 2035.

Gas production in Asia grows very strongly, thanks mainly to unconventional gas in China. The Chinese government is keen to boost production of all types of unconventional gas, with the 12th Five-Year Plan setting ambitious targets for coalbed methane and shale gas. Twenty blocks have been included in China's second upstream shale gas bid round, announced in September 2012. We project unconventional gas output, led by shale gas, to increase from 12 bcm in 2011 to more than 230 bcm in 2035, compensating for declining conventional gas production after 2020 and boosting overall production from 103 bcm now to about 320 bcm in 2035. China is also unique in pursuing production of gas from coal (producing synthetic gas, in a process not unlike the "town gas" used in Europe in the early 20th century), with about 30 bcm worth of plants under construction (as "syngas" has a lower calorific content than natural gas, this corresponds to about 10 bcm of natural gas equivalent). We have not yet included this type of "unconventional gas" in our projections, as data availability is limited. Further expansion of this production capacity might be restricted by water availability in the coal producing regions and/or by government policies, for example concerning the large CO_2 emissions. India's production also expands rapidly, doubling by 2035 as shale gas and coalbed methane make a growing contribution.

Box 4.3 ▷ New finds boost hopes for gas production in East Africa

Italy's Eni and the US independent Anadarko have found several large gas fields in the deepwater Rovuma basin of offshore Mozambique since 2010. Eni's finds include the giant Mamba North and South fields, among the biggest gas finds ever made by the company, with total gas in place estimated at around 2 tcm. Anadarko's discoveries, which may be part of a single accumulation, are thought to hold between 1 and 2 tcm. Given the limited size of the domestic market and the long distance from neighbouring South Africa, developing these resources will depend on LNG exports. Anadarko has started front-end engineering work for an initial two-train, 10 million tonne per year plant and expects to take a final investment decision in 2013. Exports, which would most likely go to Asian markets, could start as soon as 2018, stimulated by the presence of Asian companies in the consortium. Eni plans to drill more wells to assess the full potential of the Mamba complex before taking a decision on LNG exports.

BG, Ophir Energy and Norway's Statoil have also found some gas fields just over the maritime border in Tanzania. Based on tentative initial estimates, these fields could hold a total of 340 bcm. A new licensing round, covering 9 deepwater blocks, is anticipated before the end of 2012. Further up the coast, Kenya has already licensed 30 onshore and offshore blocks, with other deepwater blocks soon to be tendered. Apache, a US independent company, will soon begin exploration drilling in the Lamu Basin. The rest of Kenya's deepwater is licensed to Anadarko, France's Total and the UK companies, BG and Ophir Energy. We project that production from Mozambique, Tanzania and Kenya will reach about 45 bcm in 2035, from less than 5 bcm today.

In Latin America, gas output will be driven mainly by domestic needs, as the collapse in US prices has killed off hopes of large-scale export to North America. In Argentina, still the region's biggest gas producer despite recent declines, output rises steadily to 2035, with shale gas making a significant contribution to supply post-2020. However, a worsening of the investment climate in the wake of the government's move to nationalise Repsol's controlling stake in YPF, the country's biggest oil and gas producer, could scupper hopes of a shale gas boom. Prospects appear brighter in Brazil, thanks to a series of major offshore discoveries, including large volumes of associated gas in pre-salt oilfields (see Chapter 3, Box 3.5). The national company, Petrobras, has started production from pilot wells at the pre-salt Lula field, which is now connected to shore by a 3.7 bcm/year pipeline. Total gas production is projected to double between 2011 and 2020, to 32 bcm, and then rise to almost 90 bcm by 2035. Part of this output could be liquefied on a floating LNG plant for shipment to domestic markets in the north of the country or for export, probably after 2020; a decision on this project has been put off to 2013. The outlook for gas production in Venezuela depends critically on the investment climate. Although the country has large reserves of gas, production has been declining in recent years due to a lack of investment. In the longer term, production is projected to rebound, on the assumption that sufficient capital can be mobilised.

Focus on prospects for unconventional gas

Unconventional gas is set to play a central role in meeting rising natural gas demand. In the New Policies Scenario, unconventional gas accounts for close to half of the increase in global gas production between 2011 and 2035, its share of production rising from 16% to about 26%. Shale gas and coalbed methane contribute most of this increase, with the former making up about 14% of total gas production in 2035 and the latter about 7%. The bulk of the increase comes from just three countries – China (30%), the United States (20%) and Australia (12%) – with the United States' share of global unconventional gas production dropping from 78% in 2011 to 44% in 2035 (Figure 4.5). US shale gas production has soared since the mid-2000s, despite a fall in prices to historic lows, boosting the country's overall gas production and making it almost self-sufficient in gas (Box 4.4). It is also spurring interest in applying US technologies and experience to other parts of the world, with potential benefits in the form of greater energy diversity and more secure supply in those countries that rely on imports to meet their gas needs, as well as global benefits in the form of reduced energy costs.

Yet the prospects for unconventional gas production worldwide remain uncertain, for several reasons. In particular, vocal public concerns have been expressed about the environmental and social impacts, and failure to satisfy these concerns could hinder the development of unconventional gas in many parts of the world. Other challenges, including an adverse fiscal and regulatory framework, limited access and proximity to pipelines and markets and shortages of expertise, technology and water (large volumes of which are needed for hydraulic fracturing, a key technique for shale and tight gas) could hold back production. And it remains very uncertain in most cases how costly it will be to produce unconventional gas given that our understanding of the resource base is still relatively limited.

Figure 4.5 ▷ Unconventional gas production in leading countries in the New Policies Scenario, 2035

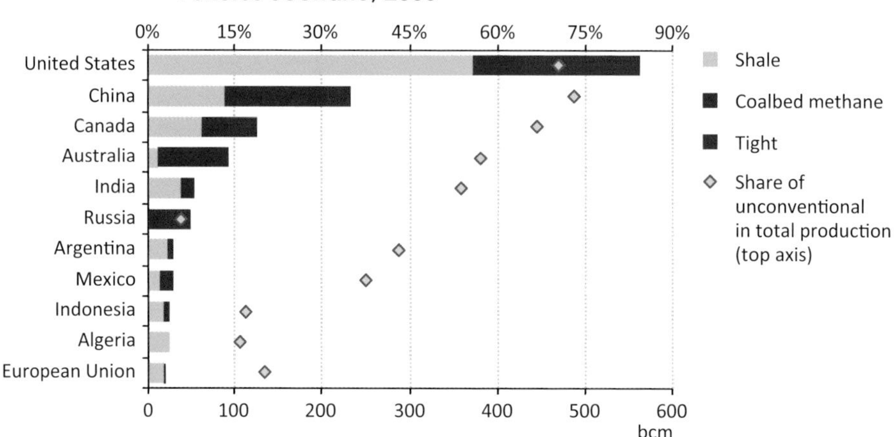

Producing unconventional gas is an intensive industrial process, generally imposing a larger environmental footprint than conventional gas development. More wells are often needed and techniques such as hydraulic fracturing are usually required to boost the flow of gas from the well. The scale of development can have major implications for local communities, land use and water resources. And there are serious public concerns about the risk of air pollution, contamination of surface and groundwater and releases of methane – a potent greenhouse gas – at the point of production (and throughout the natural gas supply chain).

The technologies and know-how exist for unconventional gas to be produced in a way that satisfactorily meets these challenges, but a continuous drive from governments and industry to improve performance is required if public confidence is to be maintained or earned. In light of these uncertainties, in May 2012 the IEA published a *WEO* special report, *Golden Rules for a Golden Age of Gas*, which analysed the prospects for unconventional gas production and proposed a set principles that could allow policy makers, regulators, operators and other stakeholders satisfactorily to address the environmental and social impacts of unconventional gas developments and, thereby, allow investment in them to proceed (IEA, 2012a). They are designed to achieve a level of environmental performance and public acceptance that can maintain or earn the industry a "social licence to operate" within a given jurisdiction, paving the way for the widespread development of unconventional gas resources on a large scale (beyond the levels hitherto indicated in this chapter), boosting overall gas supply and making the golden age of gas a reality.

Box 4.4 ▷ The economics of producing shale gas at \$2/MBtu

The growth of shale gas production in the United States has created a glut of gas supply in the North American market (which currently has almost no LNG export capacity), leading to a plunge in gas prices from above \$12/MBtu in mid-2008 to below \$2/MBtu in early 2012. At these levels, most dry shale gas projects are uneconomic. This should normally trigger a drop in drilling, followed by a fall in gas production, with the market then finding a new price equilibrium at a level that makes shale gas drilling a valid economic proposition. Some of this is happening, as shown by a shift of rigs from gas to oil (Figure 4.6), but, as of mid-2012, there were no signs of a fall in gas output.

Figure 4.6 ▷ Active drilling rigs in the United States, 2002-2012

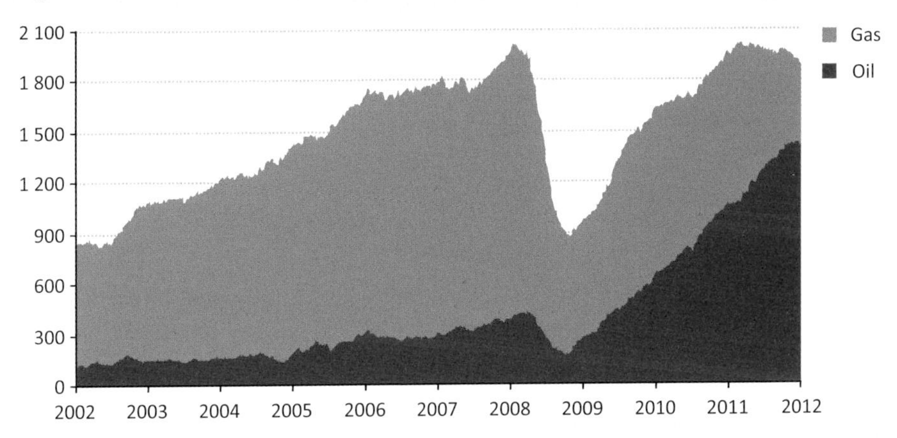

Note: For simplicity, rigs are generally classified as oil or gas based on their main target, though the resulting wells often produce both.

Source: Baker Hughes Rig Count.

Several factors are creating this unexpected market response. Some gas producers still have their production hedged at higher gas prices. Intensive practice of batch drilling in 2010 and 2011 has also left operators with a stock of drilled wells; with most of the cost already sunk, these continued to be connected to production lines. The surge in production of light tight oil (see Chapter 3) brings significant production of associated gas. Finally, plays containing wet gas remain profitable, thanks to robust oil prices.

A typical shale gas well in the United States costs between \$4 million to \$10 million to drill and complete, depending on the depth, pressure and length of the horizontal section. Recovery per well varies between 0.3 billion cubic feet (bcf) and 10 bcf of gas equivalent, with 1 bcf being a typical average in many of the medium-depth plays, such as the Barnett, the Fayetteville, the Woodford, and slightly higher in the Marcellus or the Haynesville (where depth and pressure are higher, with correspondingly higher well costs) (US DOE/EIA, 2012; USGS, 2012a). As output decline is extremely rapid, most of the gas is produced in the first couple of years, so discount rates do not matter greatly to the economics of a project: the commercial decision is essentially determined by whether the value of the gas recovered per well exceeds the well construction costs. If we take a well cost of $5 million and 1 bcf recovery per well, then a dry gas well clearly requires prices in excess of $5/MBtu.[6]

But if the gas contains NGLs, whose value is largely indexed to oil prices[7] (with, for example, a value of $40 per barrel for a mix of ethane, butane, propane and condensate), for the same well the required gas price goes down to $1/MBtu for gas with a 40% liquid content (calculated on an energy-equivalent basis) (Figure 4.7). So wet gas plays with liquid content in excess of 25% are profitable even at current Henry Hub gas prices. The resulting bias towards exploitation of liquid-rich shale gas plays boosts NGLs production in the United States, with our projection peaking at 3 mb/d in 2020. In the latter part of the projection period, however, we expect the richest gas to begin to be depleted, with a resulting move back towards drier gas production (a consideration that underpins our assumption of rising natural gas prices in the United States).

Figure 4.7 ▷ Relationship between break-even price (gas price needed to recover well costs) and the liquid content of the gas produced

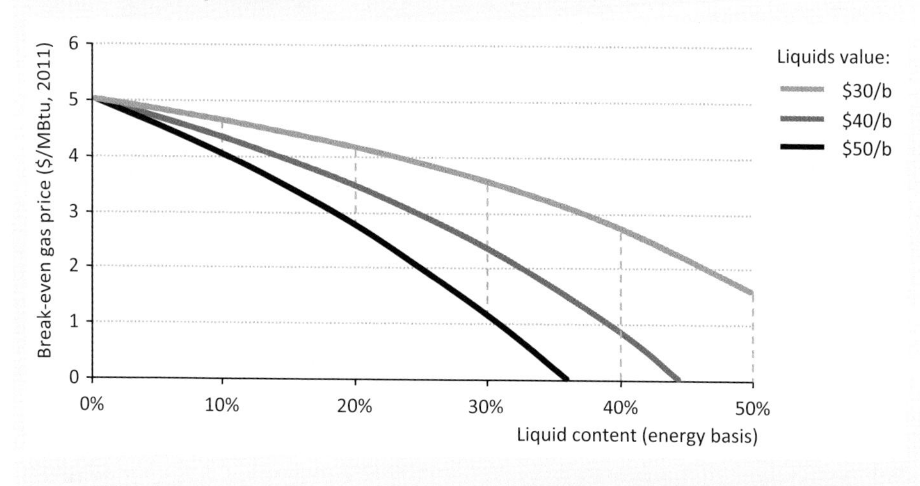

The extreme case is that of associated gas produced from oil wells, such as light tight oil (if more than 50% of the energy content is in the liquid, a well is considered to be an oil well). In these cases, gas is essentially a free by-product (apart from separation and processing costs). Assuming an average gas content of 25% for light tight oil wells, our light tight oil projections account for an additional 25 bcm of gas to the US market in 2015, rising to 35 bcm in 2025, on the assumption that the infrastructure is developed to market the associated gas rather than just flare it.

6. 1 MBtu corresponds to about 1 000 cubic feet for dry gas.

7. Rapid growth in NGLs in North America has de-linked NGLs and oil prices for the moment, but these may reconnect when investment in petrochemical plants and refineries to use more NGL feedstock materialises.

The Golden Rules underline that full transparency, measuring and monitoring of environmental impacts and engagement with local communities are critical to addressing public concerns. Careful choice of drilling sites can reduce the above-ground impacts while also most effectively targeting the productive areas, minimising any risk of earthquakes or of fluids passing between geological strata. Leaks from wells into aquifers can be prevented by high standards of well design, construction and integrity testing. Rigorous assessment and monitoring of water requirements (for shale and tight gas), of the quality of produced water (for coalbed methane [CBM]) and of waste water for all types of unconventional gas can ensure informed and stringent decisions are taken about water handling and disposal. Production-related emissions of local pollutants and greenhouse-gas emissions can be reduced by investments to eliminate venting and flaring during the well-completion phase. We estimate that applying the Golden Rules could increase the overall financial cost of a typical shale-gas well by 7%. However, for a larger development project, with multiple wells, additional investment in measures to mitigate environmental impacts may be offset by lower operating costs.

In the *WEO* special report, we developed a Golden Rules Case in which it is assumed that the necessary conditions are put in place, including the application of the Golden Rules, to allow for an accelerated global expansion of gas supply from unconventional resources beyond the levels announced in the New Policies Scenario. In the Golden Rules Case, the greater availability of gas supply has a strong moderating impact on gas prices and, as a result, demand for gas grows by more than 55% to 2035 (compared with just under 50% in this year's New Policies Scenario), boosting the share of gas in the global energy mix from 21% in 2010 to 25% in 2035. Production of unconventional gas, primarily shale gas, more than triples to 1.6 tcm in 2035, its share in total gas output climbing from 14% in 2010 to 32% in 2035, compared with 26% in this year's New Policies Scenario, which assumes only a partial and slower take up of the Golden Rules (Figure 4.8).

Figure 4.8 ▷ World unconventional gas production by type and scenario/case

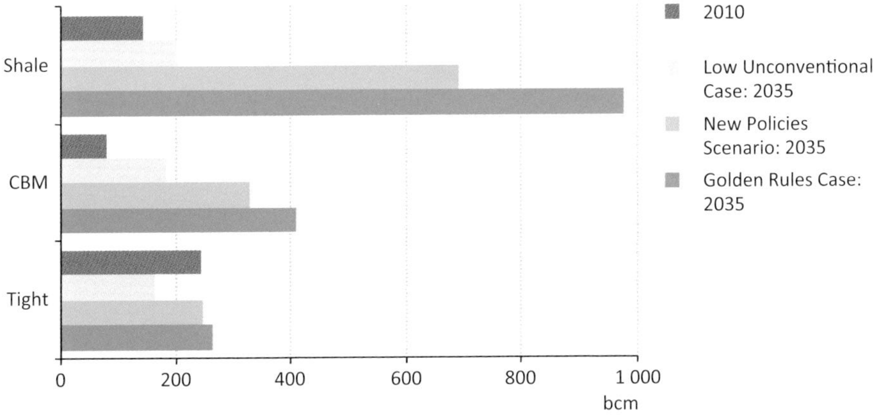

Sources: IEA analysis; IEA (2012a).

The Golden Rules report also presents the results of a Low Unconventional Case, which assumes that – primarily because of a lack of public acceptance – only a small share of unconventional gas resources is accessible for development. In this case, the competitive position of gas deteriorates as a result of lower availability and higher prices. The requirement for imported gas rises and some patterns of trade are reversed, with North America needing significant quantities of imported LNG, while the preeminent position of the main conventional gas resource-holders is reinforced. Although the forces driving the Low Unconventional Case are led by environmental concerns, reduced gas output has the perverse effect of pushing up energy-related CO_2 emissions slightly, compared with the Golden Rules Case, because gas is largely replaced in the global energy mix by coal.

In none of the scenarios and cases considered here – the New Policies Scenario, the Golden Rules Case or the Low Unconventional Case – does the energy mix realise the international goal of limiting the long-term increase in the global mean temperature to 2 °C. However unconventional gas can help to deliver some reductions in CO_2 emissions. During the period 2006-2011, for example, unconventional gas output in the United States increased by 70% and the competitive boost to the position of gas meant significant fuel switching towards gas and away from coal. Together with efficiency gains in power generation and increased output from renewables, this helped to bring CO_2 emissions in the United States down by 7% over this period (Figure 4.9).

Figure 4.9 ▷ Percentage change in selected economic and energy indicators in the United States, 2006-2011

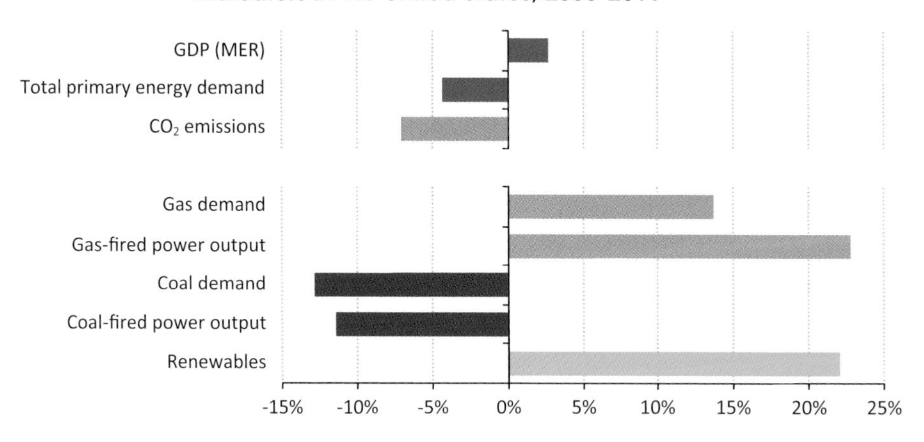

Note: MER = market exchange rate.

Trade

Inter-regional trade

International trade in natural gas is set to continue to expand through to 2035. In the New Policies Scenario, inter-regional gas trade (between major *WEO* regions) increases by nearly 80% between 2010 and 2035, from 675 bcm to nearly 1 200 bcm (Table 4.6). This is a faster increase than that of global production, which grows by 50%. As a result, the share of gas consumption that is traded between regions grows from 20% in 2010 to 24% in 2035.

Table 4.6 ▷ Inter-regional natural gas net trade in the New Policies Scenario

	2010		2020		2035		2010-35
	bcm	Share of demand*	bcm	Share of demand*	bcm	Share of demand*	Delta bcm
OECD	**-419**	**26%**	**-403**	**23%**	**-492**	**25%**	**-73**
Americas**	-29	3%	30	3%	34	3%	64
United States	-76	11%	19	2%	34	4%	110
Europe	-265	47%	-335	57%	-454	68%	-189
Asia Oceania	-125	68%	-99	48%	-72	30%	53
Australia	19	35%	62	59%	108	66%	89
Japan	-100	97%	-114	99%	-123	99%	-22
Non-OECD	**396**	**19%**	**403**	**15%**	**492**	**14%**	**96**
E. Europe/Eurasia	150	18%	221	23%	362	30%	212
Caspian	44	29%	95	43%	144	49%	100
Russia	192	29%	212	30%	307	36%	116
Asia	26	6%	-112	17%	-335	30%	-361
China	-15	14%	-129	42%	-226	41%	-211
India	-12	19%	-30	32%	-81	46%	-69
Middle East	96	20%	125	20%	169	21%	72
Africa	106	51%	138	50%	252	59%	146
Latin America	17	10%	31	15%	44	15%	27
Brazil	-13	47%	-6	15%	9	11%	22
World* **	**675**	**20%**	**860**	**22%**	**1 197**	**24%**	**522**
European Union	-335	62%	-406	75%	-524	85%	-189

* Share of production for net exporting regions. ** OECD Americas includes Chile, unlike the figures for North American exports given in the Spotlight. *** Total net exports for all *WEO* regions, not including trade within *WEO* regions.

Notes: Positive numbers denote net exports; negative numbers net imports. The difference between OECD and non-OECD in 2010 is due to stock change.

Among the importers, we project a big increase in Europe's needs: net gas imports into the European Union rise from 302 bcm in 2011 to 525 bcm in 2035, with the share of imports in total consumption jumping from 63% to 85%. But a larger share of global trade is set to be drawn towards Asia. In our projections, imports to China are set to rise from 30 bcm in 2011 to about 130 bcm in 2020 and 225 bcm in 2035. This increase is equal to nearly 40% of the total expansion in world inter-regional trade. Over the period to 2020, increased imports to China are expected to come by pipeline from Central Asia and Myanmar, and from LNG (seven import terminals are under construction and many others are planned). After 2020, China is expected to begin importing gas from Russia, via pipeline from Eastern Siberia; it already imports Russian LNG from Sakhalin. Elsewhere in Asia, Japan and Korea remain large gas importers and India's gas imports also expand significantly, mainly as LNG but also potentially from the planned TAPI pipeline connection to Turkmenistan (although this will depend on stabilisation of the security situation in Afghanistan).

Among the exporters, Russia continues to be the world's largest gas exporting country, with net export volumes rising to almost 310 bcm in 2035. Among Caspian exporters, exports from Azerbaijan (to Europe) and from Turkmenistan (to Asian markets) increase substantially. Exports from the Middle East grow, led by Qatar and – later in the projection period – by Iran. There is some expansion of existing pipeline exports from North Africa to Europe. But the rise of global trade also sees some new exporting countries and regions emerge, notably in East Africa and in North America. Overall, our projections imply a major re-orientation of global gas trade towards markets in the Asia-Pacific region (Figure 4.10).

In North America, several LNG export terminals are planned in the United States and Canada, in addition to the Sabine Pass LNG project proposed by Cheniere Energy that has already received full regulatory approval. Our projections for LNG exports from North America are considerably lower than the sum of the projects under consideration (which amount to about 200 bcm per year of new export capacity from the United States and almost another 50 bcm per year from Canada). In the New Policies Scenario, competition between suppliers of gas means that North America does not take a large international market share. Nonetheless, the impact on global markets of the increase in North American export capacity on global markets – particularly in the United States where pricing may be linked to US domestic prices – is set to be significant (see next section).

The largest share of inter-regional gas trade (trade between *WEO* regions) is carried by pipeline, with 58% of the total in 2011. Taking intra-regional trade into account (which, taking place over shorter distances, is dominated by pipelines), the share of pipelines in total gas trade is higher, at 68%. Over the projection period, pipeline trade is set to expand as new inter-regional pipeline projects are put into operation. Projects under consideration include lines from Russia to China and Europe, and from the Caspian to Europe and India. In 2035, pipeline trade still represents about half of global inter-regional gas trade.

A larger share of the growth in inter-regional gas trade over the coming decades is set to be taken by LNG. Global LNG trade has been growing rapidly in recent years, with a wave of capacity additions led by Qatar, which has the largest LNG export capacity in the world, at 105 bcm per year. After slower expansion over the next couple of years, commissioning of new plants is set to accelerate again from around 2015 (Table 4.7). There are at present a dozen LNG liquefaction projects under construction – seven of them in Australia – with a combined capacity of 108 bcm per year; when all of them are completed, probably by 2018, they boost global LNG export capacity to around 480 bcm per year.[8]

8. Capacity at end-2011 was 373 bcm per year. The 5.9 bcm per year Pluto LNG plant in Australia was brought on stream earlier this year, boosting capacity in mid-2012 to 379 bcm per year. Around 10 bcm of capacity is expected to be retired in the next few years.

Figure 4.10 ▷ **Net inter-regional natural gas trade flows between major regions in the New Policies Scenario** (bcm)

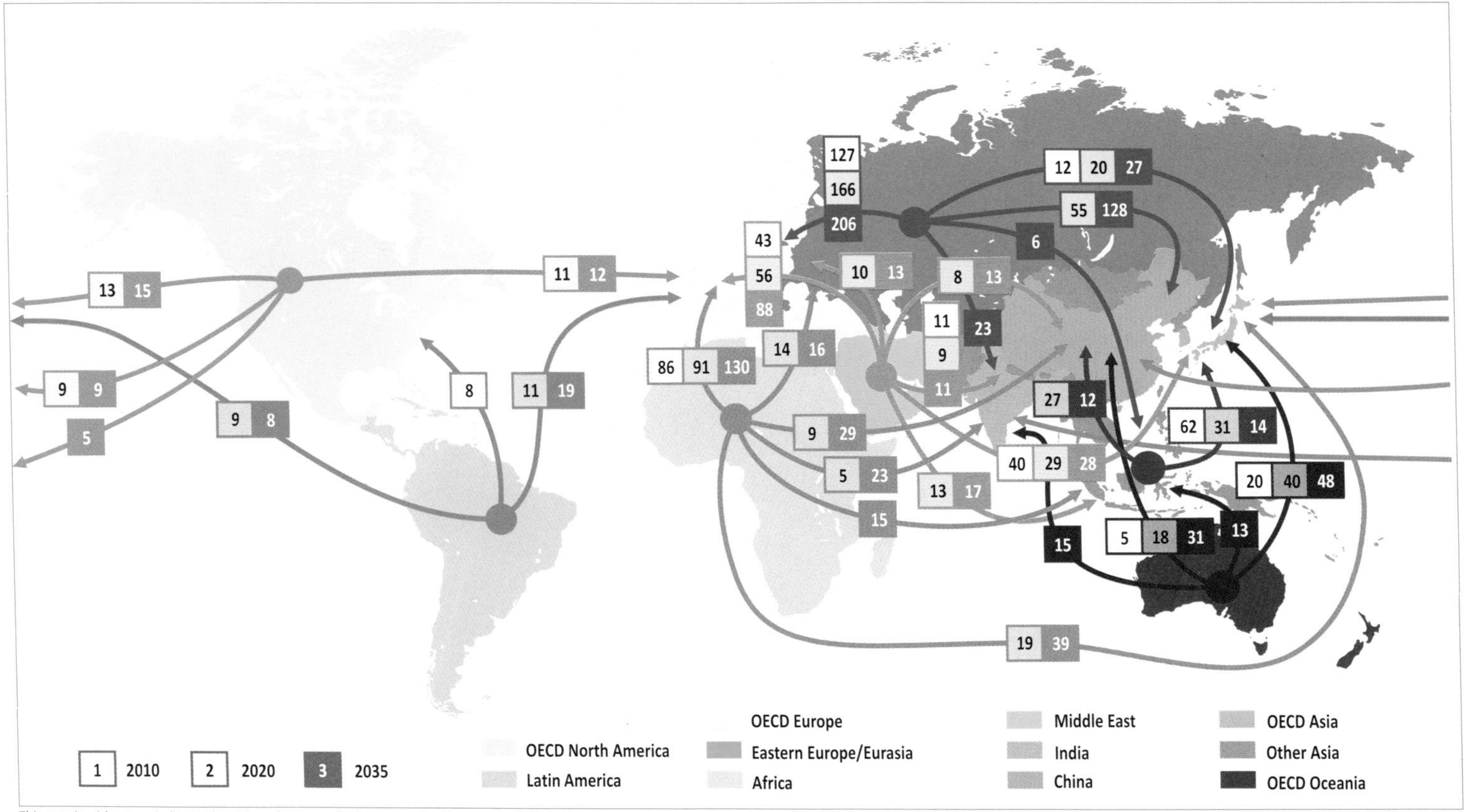

This map is without prejudice to the status of or sovereignty over any territory, to the delimitation of international frontiers and boundaries and to the name of any territory, city or area.

Note: Trade volumes less than 5bcm are not shown.

Australia is set to add close to 80 bcm per year of capacity between 2014 and 2018, including three large plants, with a combined capacity of 28 bcm per year, to be sited at the port of Gladstone in Queensland in eastern Australia, which will process coalbed methane. The Shell-led Prelude plant that is due on stream in 2017 uses floating LNG production and storage technology and is the first project of its kind. Seven other Australian LNG projects with a total capacity of 55 bcm are close to a final investment decision (all of which are conventional gas developments, bar one). Looking further ahead, there are additional LNG export projects planned across the world, and expansion is expected in Africa, Russia, North America, Brazil and parts of the Middle East. Over the projection period, trade in LNG doubles to more than 575 bcm, its share of total inter-regional trade rising from 42% to close to 50%.

Table 4.7 ▷ LNG export projects under construction worldwide (July 2012)

	Project	Operator	Capacity		Start-up date
			Mt/year	bcm/year	
Algeria	Skikda new train	Sonatrach	4.5	6.1	end-2012
	Gassi Touil LNG	Sonatrach	4.7	6.4	2013
Angola	Angola LNG	Chevron	5.2	7.1	Q3 2012
Australia	Gorgon LNG	Chevron	15.0	20.4	2014-15
	Queensland Curtis*	BG	8.5	11.6	2014-15
	Gladstone*	Santos	7.8	10.6	2015-16
	Australia Pacific*	ConocoPhillips	4.5	6.1	2015
	Wheatstone	Chevron	8.9	12.1	2016-17
	Prelude**	Shell	3.6	4.9	2017
	Ichthys	Inpex	8.4	11.4	2017-18
Indonesia	Donggi Senoro	Mitsubishi	2.0	2.7	2014
Papua New Guinea	PNG LNG	ExxonMobil	6.6	9.0	2014-15
Total			**79.7**	**108.4**	

* Coalbed methane based. ** Floating LNG project.

Source: Company announcements.

The pricing of internationally traded gas

Developments in the way internationally traded gas is priced could have an important impact on the actual prices that are paid, investment in new sources of supply and long-term gas supply security. At present, most of the gas that is traded across international borders, whether the gas is physically shipped by pipeline or as LNG, is sold under long-term contracts, usually covering a period of between 10-25 years. Traditionally, gas prices in these contracts have been indexed to the prices of competing fuels, normally oil products or crude oil. But that is changing, at least for gas sold to buyers in Europe (which accounts for one-third of gas traded between *WEO* regions). Increasingly, gas prices are being indexed, at least in part, to published spot prices of gas, usually at hubs in the country or region where the gas is delivered. In addition, a growing share of traded gas worldwide is sold on a spot basis, *i.e.* cash sales of a specified volume of gas at a fixed price for immediate delivery. The average length of contracts is also tending to fall, in part because, as old contracts come to an end, they are rolled over for shorter periods, and because of growing use of medium-term (2-4 year) contracts.

Long-term contracts and oil indexation are instruments used by buyers and sellers to mitigate the volume and price risks they are taking when investing in large projects. The prospect of a large-scale move away from oil indexation and towards spot gas-price indexation (known as hub-based pricing) and increased direct spot or short-term sales by external producers will be driven by market conditions and by the evolution of other approaches to risk management. The latter include, for example, financial and futures markets, cross-participation of buyers in upstream projects and sellers in downstream projects, or the appearance of large "portfolio" players, owning a mix of long-term "baseload" and short-term "swing" gas.

In Europe, falling demand after 2009 opened up a large differential between lower spot gas prices and what was, effectively, an oil-linked gas price under long-term import contracts. This resulted in pressure on suppliers to accept changes to the contractual terms. Norwegian producers were the first to offer more pricing flexibility. Gazprom also granted some important concessions on pricing in early 2012, accepting the partial use of indexation to spot gas prices for a period of three years. In July 2012, Germany's E.On – Europe's largest gas buyer and Gazprom's biggest export customer – announced that it had reached a settlement with the Russian company covering all its long-term contracts, in a deal thought to involve a lower base price and a further increase in the share of hub-based prices in the overall pricing formula. Italy's Eni reportedly renegotiated the price in its contracts with Gazprom earlier in the year. Algeria's national oil and gas company, Sonatrach, has so far resisted pressure to change the pricing terms in its contracts, but it has reportedly granted some concessions on take-or-pay clauses and has indicated that it may be willing to adjust prices, while retaining the principle of oil indexation. Interestingly, the evolution of coal prices can play a role in the evolution of gas contract terms: the low gas prices in the United States have reduced coal use there, freeing up relatively low cost coal for export to Europe, so contributing to reducing gas demand and putting pressure on oil-indexed gas prices – in a sense, a coal-mediated partial convergence between American and European gas prices.

In principle, the transition towards hub-based pricing in Europe would be expected to result in lower import prices than would otherwise be the case. This is almost certainly true in the near to medium term, as supply capacity remains ample. But there may be times when supply shortfalls or sudden surges in demand lead to spikes in prices. And, even if the role of oil-indexed long-term contracts is reduced, a degree of correlation between gas and oil prices could persist in Europe as a result of indirect gas price linkages with the Asia-Pacific markets – at least for as long as gas prices there remain linked to oil prices.

Most of the gas delivered to Japan, Korea, China and India as LNG, as well as China's imports of gas by pipeline from Turkmenistan, is covered by long-term contracts, with indexation based on crude oil prices and destination clauses (limiting the ability of buyers to divert cargoes to other markets). Spot and short-term supplies have been growing (in part because of the unanticipated surge in Japanese demand following Fukushima), but buyers in the region have traditionally placed strong emphasis on long-term security of supply. In some cases, large buyers – including Japanese, Korean and Chinese companies – have taken stakes in the upstream projects in order to share in the rent that might come about as a result of higher oil prices (as well as, in some cases, to learn about the technologies involved). However, there are arguments, similar to those used in Europe, that the rationale for oil-indexed pricing has largely disappeared, and the persistence of much higher LNG prices in those markets, relative to Europe and North America, would surely lead to pressure to seek more favourable pricing arrangements, boosting competition between different sources of gas and improving the competitive position of gas relative to other fuels. So, while oil indexation looks set to remain the dominant pricing mechanism in Asia-Pacific markets for the moment, indexation clauses could evolve to provide more flexibility.

How quickly the move away from oil indexation happens may depend on the availability of LNG for purchase on a spot basis. The share of spot and short-term trade in global LNG supply has been rising rapidly, by 40% in 2010 and a further 50% in 2011, total volumes reaching 83 bcm, or just over one-quarter of total LNG trade (Figure 4.11). This figure could fall back in the next couple of years, as the market tightens with rising demand and fewer LNG export projects coming online, which might mean that buyers prefer to lock in supplies through long-term deals. However, there are indications that spot trade is likely to continue its rise in the longer term.[9] All of the output of the 7 bcm Angola LNG plant, which is about to come online, will be sold onto the spot market, cargo by cargo. And a significant share of the gas that has been contracted from the proposed Sabine Pass plant in the United States could end up on the spot market, as prices are indexed to Henry Hub spot prices and there are no destination clauses. In the longer term, an expansion of spot trade in Asia would help to address one barrier to the adoption of hub-based pricing in the region, as it would provide a sounder basis for a reliable benchmark spot price.

An objection from gas producing and exporting countries to a larger role for hub-based pricing and shorter-term sales is that these mechanisms do not provide adequate security to underpin the very large investments required in new production or transportation capacity. This argument becomes less valid as gas trading at hubs becomes more liquid; several large projects to supply the UK market, including Norway's Ormen Lange field and several LNG receiving terminals, were built on the basis of prices at the main UK trading hub. Oil indexation may also be undermined in the future by the emergence of a gas futures market, which can provide an alternative mechanism to lower the risk associated

9. This prospective increase is due only in part to higher volumes exported at spot prices by producers. There are also (larger) volumes that are acquired on long-term contracts and then resold on a spot basis to take advantage of price peaks.

with large-scale, long-term gas investments. An example of this comes from the US company Chesapeake, which has sold forward $6 billion worth of its future production on a ten-year futures contract.[10] The CME Group (owner, among others, of the New York Mercantile Exchange – NYMEX) has recently introduced a futures contract based on Asian LNG prices, in addition to its Henry Hub contract. The futures market could evolve as a means to underpin investment in the gas industry, accompanying and accelerating a shift in gas pricing mechanisms.

Figure 4.11 ▷ **World short-term LNG trade, 2000-2011**

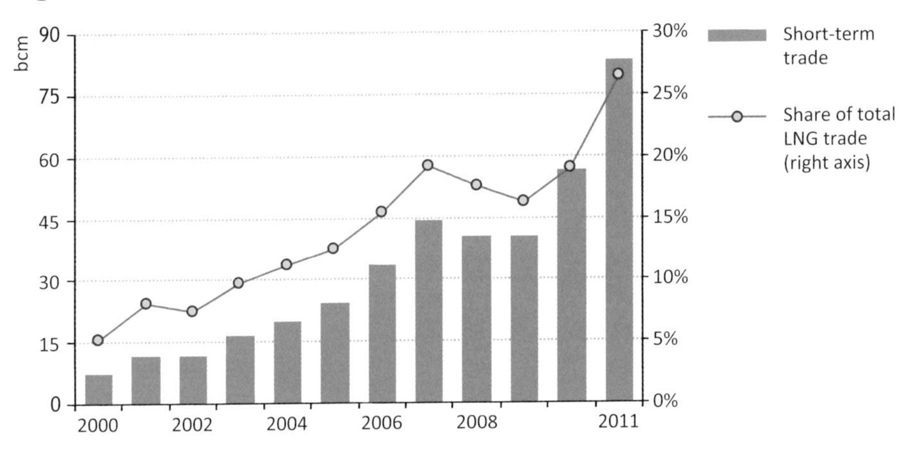

Note: Short-term trade denotes trades under contracts of four-year duration or less, including spot transactions.

Sources: GIIGNL (2012); IEA databases.

Rising LNG supplies, increased short-term trading and greater operational flexibility (resulting in part from increased standardisation of liquefaction, shipping and regasification technologies) are likely to lead to increased price linkages between regions and, probably, a degree of price convergence. Opportunities to take advantage of regional price differentials will probably spur more trade between the Atlantic basin and Asia-Pacific markets, which traditionally have been largely distinct from each other. The construction of LNG liquefaction plants in North America, which would open up the possibility of exports to Asia, would help to connect prices between the two regions and accelerate the process of globalisation of natural gas markets, mirroring the way oil markets have developed historically.

10. This was actually done through so-called "Volumetric Production Payments", in which the payment is made upfront in order to raise cash with the gas delivered over ten years, rather than by means of a futures contract; but it illustrates the role of banks and commodity traders in providing cash for future production.

Investment

Unit capital costs in the upstream and downstream sectors have risen strongly in recent years, because of increased demand for construction services and underlying increases in the prices of basic materials and of labour. LNG construction costs have risen particularly steeply: the capital costs of liquefaction plants under construction, as well as the Pluto plant in Australia that was commissioned in early 2012, range from about $2 800 to $4 000 per tonne of annual production capacity, compared with $1 000 to $2 000 per tonne for plants that came on stream in 2010-2011 (IEA, 2012b). A detailed discussion of upstream oil and gas investment and costs can found in Chapter 3.

The projected trends in gas supply in the New Policies Scenario imply a need for around $8.7 trillion (in year-2011 dollars) of cumulative investment, or $360 billion per year, along the gas supply chain over the period to 2035 (see Table 3.8 and related discussion in Chapter 3). Around two-thirds of the capital, or $240 billion per year, is needed in the upstream to develop new gas fields and to combat decline at existing fields. LNG facilities account for about 9% of total investment (most of it for liquefaction plants) and transmission and distribution networks for the rest. More than half of the investment is needed in non-OECD countries, where local demand and production grows most, even though construction costs there tend to be somewhat lower than in the OECD. The United States and Canada attract nearly 60% of the OECD investment, 25% of the world total.

As ever, it is far from certain that all this investment will be forthcoming, given potential barriers, such as the pace at which producing countries choose to develop their resources, the capacity of national companies fully to develop those resources, constraints on the opportunities and incentives for international companies to invest and geopolitical factors. Investment needs could be greater than projected here, as a result of capital costs increasing faster than assumed (if, for example, upstream costs for gas are affected by tightness in the upstream oil market, driving gas production costs higher). Over time, however, the oil and gas sector, in co-operation with the financial sector, has displayed constant ingenuity in mobilising finance to meet reasonable financing requirements.

Coal market outlook

Asia: the litmus test of global coal markets?

Highlights

- Coal met 45% of the growth in global energy demand over the past decade. Policy decisions will determine whether demand carries on rising strongly or changes course radically. The range of outcomes for coal across the three main scenarios in this year's *WEO* reflects the strength of possible measures to cut coal-related greenhouse-gas emissions and to develop and deploy carbon capture and storage.

- In the New Policies Scenario, global coal demand grows by 0.8% per year to 2035, with growth slowing sharply after 2020 as recently introduced and planned policies to curb use take effect. While coal's share of global primary energy demand falls by nearly three percentage points to less than 25% in 2035, coal remains the second most important fuel behind oil and the backbone of electricity generation.

- Coal use in the New Policies Scenario grows in all major non-OECD regions, with Asian countries at the helm: China and India alone account for nearly 75% of non-OECD growth. China's demand peaks around 2020 and plateaus at that level through 2035, yet still accounts for half of global incremental coal demand over the *Outlook* period. By 2025 India overtakes the United States to become the second-largest coal user. By contrast, almost all major OECD regions see their coal use decline, especially Europe, where demand in 2035 is 60% of the 2010 level.

- Most major coal producers, including China and the United States, see their production slow or even decline over the *Outlook* period. OECD coal output starts to fall around 2020 and is 10% lower in 2035 than in 2010 with declines in Europe and North America offsetting growth in Australia. Non-OECD production carries on rising through 2035. In China, the world's biggest producer, rapid output growth slows around 2020, reaching nearly 20% above the 2010 level by 2035.

- International coal markets and prices remain very sensitive to energy policy and market developments in China and increasingly India. Inter-regional hard coal trade in the New Policies Scenario continues to grow, mainly to meet Asian demand, but levels off after 2020 as global demand growth slows and China's imports fall back. India becomes the world's biggest net importer after 2020, with a correspondingly greater influence in world markets. OECD countries collectively become a coal net exporter, reversing history, as demand declines while exports from Australia and North America remain robust. Coking coal trade, unlike steam coal, is far less affected by policy as there is less scope for its replacement by other less carbon-intensive fuels.

Demand

Overview of global demand trends

Among the three fossil fuels (oil, coal and gas), the outlook for coal diverges the most across the three scenarios. The wide range of outcomes stems from different assumptions underlying the scenarios about energy and environmental policies, reflecting attitudes towards coal and the environmental impact of its use, depending on if it is used efficiently and with effective pollution control technologies. In the key power sector, inter-fuel competition with renewables and gas also affects coal demand in important ways.

In the New Policies Scenario, our central scenario, coal maintains its status as the second most important primary fuel behind oil, despite a nearly three percentage point loss in global market share over the projection period to less than 25%, with natural gas at its heels (Figure 5.1). Coal continues as the backbone of the power sector, especially in non-OECD countries. Global coal demand grows by 0.8% per year over 2010-2035. This is about one-third of the average pace over the past 25 years and much slower than in the last decade. More than four-fifths of the incremental global growth takes place before 2020, as recently introduced and planned policies constraining the use of coal take full effect.

Figure 5.1 ▷ Share of coal in world primary energy demand by scenario

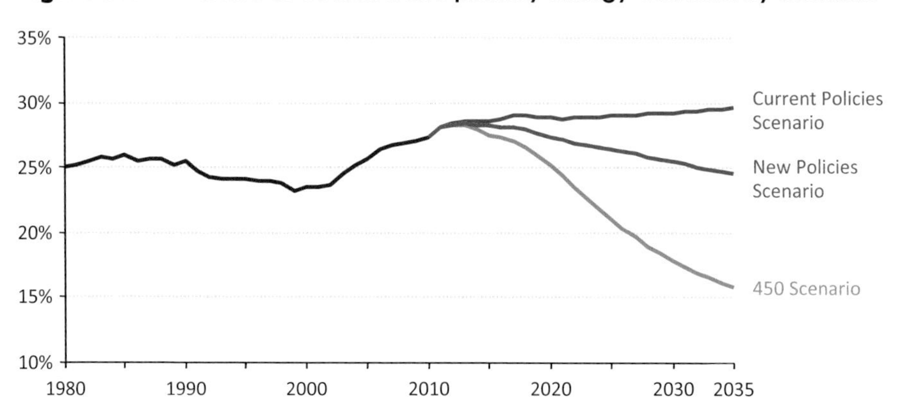

In the Current Policies Scenario, which assumes no change in current policies, coal demand grows by 1.9% per year over the projection period and coal dethrones oil as the leading primary fuel around 2025, settling in at a share of just under 30% of the global energy market by 2035. This prospect is the basis on which some industry players are considering their investment plans. In the 450 Scenario, which involves a fundamental decarbonisation of the energy sector, coal's share in the global primary energy mix collapses to 16% by the end of the *Outlook* period, as it is overtaken by renewables and gas.

In all three scenarios, the coal demand growth trend represents a marked change relative to what has happened over the past decade (Box 5.1). Between 1999-2011, mainly as a result of monumental coal demand growth in China, the share of coal in the global primary energy mix increased five percentage points to 28% – the highest level since IEA statistics began in 1971. Because of the rapid and continuing expansion of coal-fired power generation in emerging economies, particularly in China and India, determined policy action will be needed to reverse this trend.

Well-designed government policies can reduce emissions from coal use, essentially in one of four ways, provided they bring forth the necessary investments:

- Final energy can be consumed in a more efficient manner requiring less primary energy, including coal.
- Existing technologies that use coal much more efficiently can be deployed, particularly in the power sector (which accounts for two-thirds of global coal use).
- Coal's share in the energy mix can be reduced through policies that encourage its replacement by lower emission energy sources, such as gas, renewables and nuclear. This policy can be applied most effectively in the power sector.
- New technologies, such as underground coal gasification and especially carbon capture and storage (CCS) technology – neither of which has been proven on a large commercial scale – can, if given substantial and successful financial support, reduce emissions substantially from coal use in power plants and industrial facilities.

The difference in the projected level of global coal use in 2035 between the Current Policies and 450 Scenarios amounts to 4 550 million tonnes of coal equivalent (Mtce)[1], which is more than 90% of 2010 global coal demand. China, India and the United States remain the dominant coal consumers across the three main scenarios, accounting for around 70% of global coal demand throughout the projection period, making these countries the main shapers of global coal markets and pricing, and giving them strong influence over coal and energy-related carbon dioxide (CO_2) emissions (Table 5.1). This underlines the significance of government policies and consequent national technological uptake, particularly in non-OECD countries, as policy makers grapple with simultaneously achieving affordability, security and environmental goals.

Table 5.1 ▷ Coal demand by region and scenario (Mtce)

			New Policies		Current Policies		450 Scenario	
	1990	2010	2020	2035	2020	2035	2020	2035
OECD	1 544	1 552	1 482	1 181	1 581	1 578	1 312	649
United States	657	718	683	596	720	769	619	308
Europe	645	439	396	266	434	399	326	159
Japan	109	164	147	131	161	152	132	68
Non-OECD	1 644	3 411	4 349	4 845	4 728	6 311	3 787	2 690
China	763	2 288	2 812	2 811	3 068	3 659	2 455	1 505
India	148	405	631	938	682	1 231	531	486
Russia	273	164	172	182	183	225	151	103
World	3 187	4 963	5 831	6 026	6 309	7 889	5 098	3 339
Non-OECD share	*52%*	*69%*	*75%*	*80%*	*75%*	*80%*	*74%*	*81%*
China's share	*24%*	*46%*	*48%*	*47%*	*49%*	*46%*	*48%*	*45%*
India's share	*5%*	*8%*	*11%*	*16%*	*11%*	*16%*	*10%*	*15%*

1. A tonne of coal equivalent represents 7 million kilocalories or is equivalent to 0.7 tonnes of oil equivalent.

Box 5.1 ▷ Boom in global coal use continues apace

World coal demand growth, at 5.6%, remained strong in 2011,[2] following similar growth in 2010; demand was almost flat in 2009 (growth of 0.3%) as a result of the worldwide economic crisis. Global coal use and growth remain mainly a non-OECD affair, with the OECD accounting for just under 30% of global coal demand in 2011 (and demand there actually fell marginally in 2011). China and India, the world's largest and third-largest consumers of coal respectively, together accounted for 80% of non-OECD coal demand in 2011; China alone accounted for more than two-thirds. The United States – the world's second-largest coal user – accounts for 45% of OECD coal demand. Coal demand in the United States fell 4.5% in 2011 to its 1995 level, and has continued to fall further in 2012, as abundant and cheap gas, thanks to the boom in shale gas production, has displaced coal amid generally weak energy demand. OECD Europe, by contrast, has seen a surprising rise in coal use in 2011, as cheaper coal has displaced oil-indexed gas in countries such as Germany, Spain, Turkey and Greece.

Since the start of the 21st century, coal has dominated the global energy demand picture, alone accounting for 45% of primary energy demand growth over 2001-2011 (Figure 5.2). The near 55% increase in coal demand over the past decade was, on an energy basis, equivalent to roughly three times the current coal consumption of the United States. The driving force behind rising global coal use was the power sector in China, India and other non-OECD countries, where, over the past decade, total power output nearly doubled, with 60% of that growth being coal-fired. Despite energy efficiency improvements, electricity demand in China over 2001-2011 continued to grow faster than gross domestic product (GDP). In 2011, China surpassed the United States as the world's biggest power producer. By contrast, coal use has stagnated in the OECD, with the share of coal in power generation declining from 38% to 34% over 2001-2011, while still remaining the leading source of power.

Figure 5.2 ▷ Incremental world primary energy demand by fuel, 2001-2011

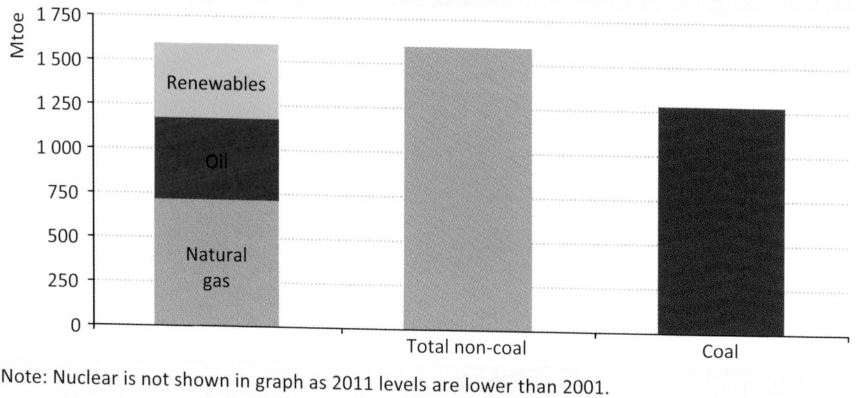

Mtoe
1 750
1 500
1 250
1 000
750
500
250
0
Renewables
Oil
Natural gas
Total non-coal
Coal

Note: Nuclear is not shown in graph as 2011 levels are lower than 2001.

2. For 2011, preliminary data on aggregate coal demand by country are available, while the sectoral breakdown for coal demand is estimated.

Sectoral trends

Today, the use of coal is confined predominantly to generating electricity and fuelling manufacturing – 65% of global coal demand in 2010 was consumed in the power sector and 27% in industry (including blast furnaces/coke ovens). In the New Policies Scenario, the share of power remains constant mainly because at the level of policy intervention assumed, coal often remains the cheapest generating option in the fast-growing Asian countries – especially in the first decade of the projection period (Table 5.2). In some sub-sectors of industry, notably iron and steel, the scope for switching away from coal is much more limited by technical considerations.

5

Table 5.2 ▷ World coal demand by sector in the New Policies Scenario (Mtce)

								2010-2035	
	1990	2010	2015	2020	2025	2030	2035	Delta	CAAGR*
Power generation	1 751	3 213	3 665	3 813	3 856	3 904	3 940	727	0.8%
Industry	680	966	1 120	1 142	1 156	1 168	1 174	208	0.8%
Blast furnace/coke oven**	239	354	398	407	404	399	393	39	0.4%
Coal-to-liquids	20	26	28	60	100	138	185	159	8.2%
Buildings	343	177	186	177	165	151	137	-41	-1.0%
Other	154	227	239	233	222	211	198	-29	-0.5%
Total	3 187	4 963	5 636	5 831	5 901	5 971	6 026	1 063	0.8%
Power generation share	*55%*	*65%*	*65%*	*65%*	*65%*	*65%*	*65%*	*1%*	
*Industry share****	*29%*	*27%*	*27%*	*27%*	*26%*	*26%*	*26%*	*-1%*	

* Compound average annual growth rate. ** Blast furnace and coke oven transformation, and own use. *** The share is calculated based on the sum of industry and blast furnace/coke oven use.

In view of the long lifetimes of coal-using capital stock in industry and even more so in power generation, current large-scale investments in capacity additions will "lock-in" CO_2 emissions for decades, unless CCS technology can be retrofitted to at least some of these plants, and suitable sinks are available (see Chapter 8). For this reason, a rapid shift in investment, at least to the most efficient coal-fired technologies available, is needed to minimise the future environmental and local air pollution impacts of coal. Furthermore, if CCS were not to be widely adopted in the 2020s, to achieve the goal of limiting the long-term global temperature increase to two degrees Celsius would require a heavy reliance on other low-carbon technologies for all new generating plants and acceptance of very high cost measures, such as premature plant closures (IEA, 2011a).

Regional trends

In the New Policies Scenario, all major OECD regions see their coal use decline over the projection period, especially Europe, where coal demand in 2035 is 60% of the 2010 level (Table 5.3). Coal use in the United States sees a slow decline of 0.5% per year early in the projection period, with a faster decline after 2020, as newer gas-fired and other plants successfully compete against coal in the power sector (which currently accounts for more than 90% of United States coal demand), and older coal plants are retired. By contrast, demand grows in all major non-OECD regions, with Asian countries at the helm. China's coal demand peaks around 2020 and plateaus at that level through 2035. Coal demand more than doubles in India and ASEAN countries, with their combined demand by 2035 exceeding that of the OECD as a whole. With this robust growth, India overtakes the United States as the second-largest coal user before 2025.

Table 5.3 ▷ Coal demand by region in the New Policies Scenario (Mtce)

	1990	2010	2015	2020	2025	2030	2035	2010-2035	
								Delta	CAAGR*
OECD	**1 544**	**1 552**	**1 537**	**1 482**	**1 407**	**1 301**	**1 181**	**-371**	**-1.1%**
Americas	701	769	747	740	722	680	638	-131	-0.7%
United States	657	718	691	683	666	630	596	-122	-0.7%
Europe	645	439	428	396	352	310	266	-173	-2.0%
Asia Oceania	198	344	361	346	333	311	277	-67	-0.9%
Japan	109	164	162	147	142	136	131	-33	-0.9%
Non-OECD	**1 644**	**3 411**	**4 099**	**4 349**	**4 494**	**4 670**	**4 845**	**1 434**	**1.4%**
E. Europe/Eurasia	524	309	318	316	313	315	324	15	0.2%
Russia	273	164	171	172	174	176	182	18	0.4%
Asia	992	2 906	3 566	3 797	3 935	4 106	4 269	1 363	1.6%
China	763	2 288	2 758	2 812	2 808	2 824	2 811	523	0.8%
India	148	405	529	631	717	817	938	533	3.4%
ASEAN	18	120	174	230	272	324	374	254	4.6%
Middle East	1	3	4	4	5	5	6	3	2.4%
Africa	106	160	166	183	191	193	196	36	0.8%
Latin America	21	32	45	49	51	51	50	18	1.8%
Brazil	14	21	31	31	29	26	24	4	0.7%
World	**3 187**	**4 963**	**5 636**	**5 831**	**5 901**	**5 971**	**6 026**	**1 063**	**0.8%**
European Union	650	401	394	355	306	259	209	-192	-2.6%

* Compound average annual growth rate.

China

Commanding nearly half of total and incremental global coal demand in the New Policies Scenario, developments in China's coal use, production, transport, trade, costs and the policies that affect them continue to have a profound impact on global coal markets. Coal has played a dominant role in China's energy mix and electricity generation for decades; in 2010, nearly 80% of China's power output was coal-fired. Abundant and relatively cheap domestic coal resources contribute importantly to China's economy, noting that recently imports are making a small but growing contribution. China's share in international coal markets increased by a colossal eighteen percentage points over 2001-2011, accounting for 80% of coal demand growth (Figure 5.3). Over the projection period, China's share in incremental demand is projected to fall to only half, but the country still remains the dominant player in international coal markets. China's coal demand reaches a peak of about 2 850 Mtce around 2020 and then plateaus over the rest of the projection period, reaching a level 23% higher than in 2010.

Figure 5.3 ▷ Incremental coal demand in China and the rest of the world by major sector in the New Policies Scenario

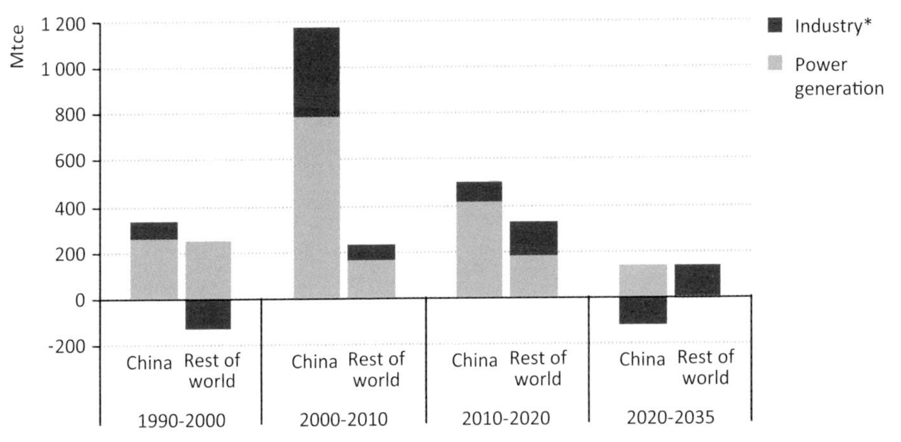

* Includes blast furnace, coke oven transformation, and own use.

On the assumption of continuing diversification of the energy sector envisaged in China's 12th Five-Year Plan, to be brought about by increased investments in nuclear power, renewables and gas, the share of coal in China's electricity generation falls from nearly 80% in 2010 to 55% by 2035 in the New Policies Scenario. Nuclear power plays a major role; 26 of the 62 reactors under construction around the world today are in China. One-third of the global capacity additions to 2020 in non-hydro renewables, such as wind and solar, occurs in China (see Chapter 6). With China holding large unconventional gas resources, there is potential for gas to play a much bigger role in meeting the country's rising power needs if the success of the United States with developing such resources can be replicated. In the New Policies Scenario, China's gas production more than triples, underpinning a rise in the share of gas in power generation from 2% to 8% over 2010-2035 (see Chapter 4). In industry, gas and especially electricity are often preferred to coal, pushing down the share of coal in the sector's energy use from 55% in 2010 to 35% in 2035. Also the shift away from heavy industry (China's crude steel and cement production are projected to peak around 2020) and the slowing pace of economic development in China over the projection period contribute to the lower growth of coal use after 2020.

India

Over the past decade, coal demand in India nearly doubled, making the country the second-largest contributor to global coal use growth (with a share of 12%) and consolidating its position as the third-largest consumer of coal, having overtaken the European Union in 2009 (Figure 5.4). In the New Policies Scenario, demand more than doubles between 2010 and 2035 to 938 Mtce, with India's share of global coal demand rising from 8% to 16%. Before 2025, India displaces the United States as the second-largest consumer of coal.

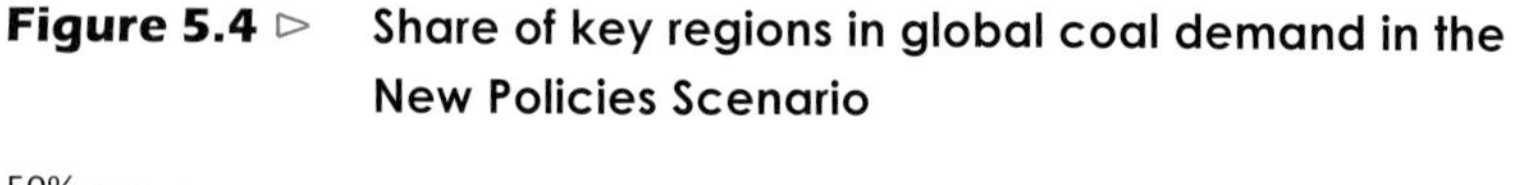

Figure 5.4 ▷ Share of key regions in global coal demand in the New Policies Scenario

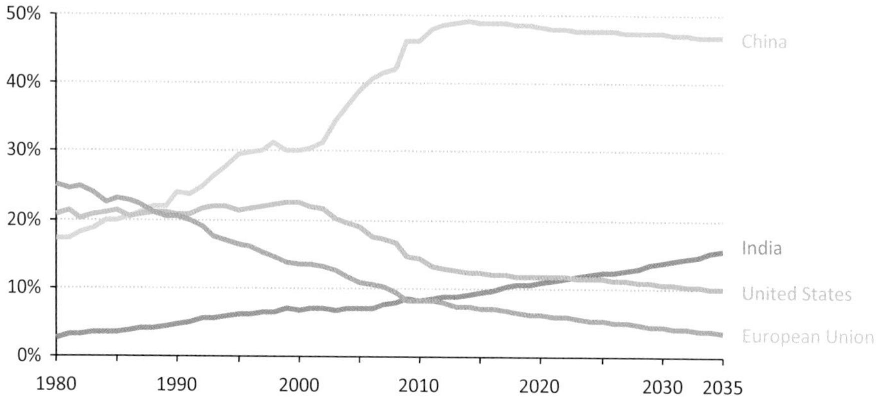

Increasing coal use in India is strongly linked to booming economic activity that drives power demand, as well as crude steel and cement production. Across the country, an estimated 295 million people today still lack access to electricity (see Chapter 18). This number is expected to halve by 2030, and per-capita electricity consumption nearly triples by 2035. Although India's planning authorities support diversification of the energy mix, the focus of new power plant builds is still predominantly on coal. India's 12th Five-Year Plan envisages more than 60 gigawatts (GW) of additional coal-fired capacity over 2012-2017, taking total capacity to around 175 GW. Although these are ambitious plans, it seems certain that India will soon become the world's third-largest coal-fired power producer behind China and the United States. By 2035, coal use in power doubles compared to 2010; in industry, coal use almost triples as a result of massive increases in crude steel and cement production.

United States

In the United States, use of coal in industry is already overshadowed by gas and electricity, such that over 90% of the nation's coal consumption is in the power sector. In the New Policies Scenario, coal-fired generation remains broadly flat at around 1 900 terawatt-hours (TWh) until 2025 before declining to 1 700 TWh in 2035. Although the share of coal in US electricity output declines by nearly fourteen percentage points, to below one-third of total generation by 2035, coal remains the leading source of electricity generation (Figure 5.5). The projected 35% increase in gas-fired generation over the *Outlook* period is supported by continued unconventional gas output growth, which by 2035 matches total US coal production in energy equivalence.

Figure 5.5 ▷ Share of coal and natural gas in US electricity generation in the New Policies Scenario

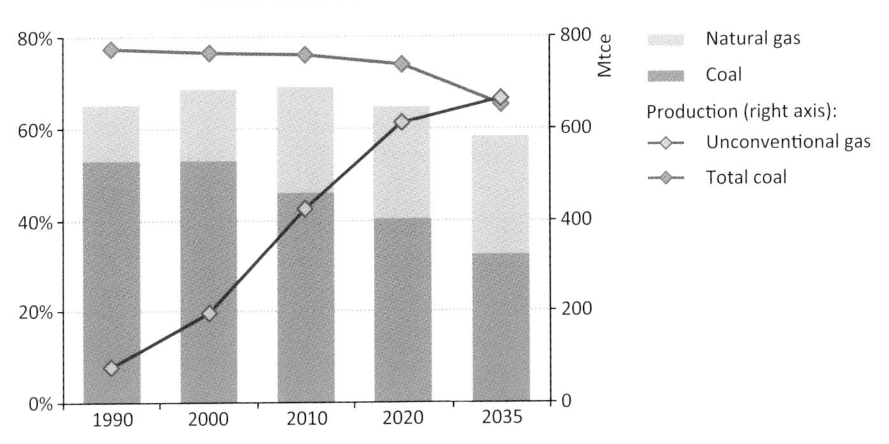

The projected decline in the relative importance of coal in power generation is in line with the long-term historical trend. At its peak in 1988, coal provided nearly 60% of US electricity output; since then, although growing in absolute terms by 30% to its peak in 2005, its share has fallen slowly to about 45% in 2010. In 2011 and 2012, this trend has sharply accelerated; for example, in April 2012, net generation from gas-fired plants was virtually equal to that from coal-fired plants at around one-third of total generation each (US DOE/EIA, 2012). Coal continues to be the prime source of baseload power over the projection period, with gas providing a significant portion of mid- and peak-load power. The assumed increase in gas prices from current very low levels means gas is unlikely to maintain its early 2012 market share. Nonetheless, around one-third of the 340 GW of coal-fired capacity is more than 30 years old and relatively inefficient, and coal-fired generation faces more stringent pollution standards, for example on mercury, which may well lead to accelerated retirements of older plants. There has been hardly any net increase in coal-fired capacity since 2000. By contrast, the last decade saw a near doubling in gas-fired capacity to 425 GW in 2010.

Supply

Reserves and resources[3]

World coal reserves at the end of 2010 amounted to just over 1 trillion tonnes (equivalent to around 140 years of current global coal production), of which hard coal (coking and steam coal) made up over 70% and brown coal (lignite) the rest (BGR, 2011). Total coal resources, including deposits that are yet to be "proven", are about 20 times larger. While hard coal reserves exist in over 70 countries around the world, three-quarters of total reserves are located in just four countries: the United States (31%), China (25%) and around

3. A detailed assessment of coal resources and production technologies was made in *WEO-2011* (IEA, 2011a).

10% each in India and Russia (Figure 5.6). These four countries also account for over 70% of global production, with most of their output going to their domestic markets. Australia and Indonesia, the world's largest hard coal exporters by far, together hold only 7% of global hard coal reserves. While global coal resources will not be a limiting factor for production growth over the coming decades, the costs of production are expected to face further upward pressure as mines currently in production are depleted and new investments shift to less attractive deposits or are located further from existing demand centres or transport infrastructure (IEA, 2011b). Strengthening coal prices and productivity gains from technological improvements may help offset, to some degree, higher production costs.

Figure 5.6 ▷ World hard coal reserves by country, end-2010

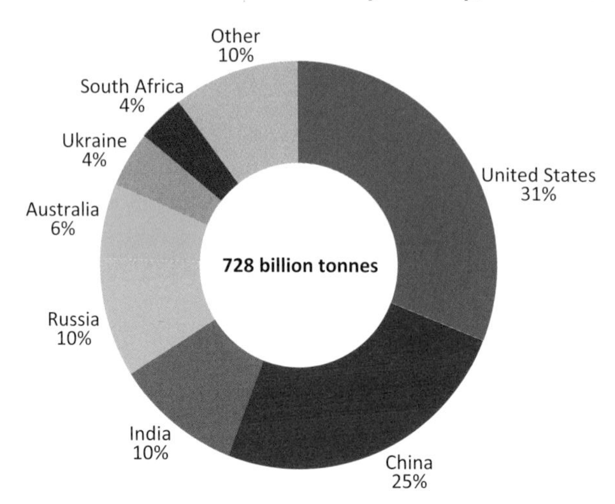

Note: Classification and definitions of hard and brown coal can differ between BGR and IEA due to different methodologies.

Source: BGR (2011).

Overview of global supply trends

Coal production varies markedly across the three main scenarios presented in this *Outlook,* in line with demand. Government policies aimed at curbing greenhouse-gas emissions depress coal demand the most in the 450 Scenario and the least in the Current Policies Scenario. World production of all types of coal (coking, steam, lignite and peat) rises from 5 125 Mtce in 2010 to just over 6 000 Mtce in 2035 in the New Policies Scenario and to nearly 7 900 Mtce in the Current Policies Scenario, falling to 3 340 Mtce in the 450 Scenario (Table 5.4). Different coal price trajectories are assumed in each scenario in order to balance demand with supply. Prices are the highest in the Current Policies Scenario, as stronger demand growth requires more investment in mining and transportation infrastructure, and the lowest in the 450 Scenario.

Table 5.4 ▷ Coal production by type and scenario (Mtce)

	1990	2010	New Policies 2020	New Policies 2035	Current Policies 2020	Current Policies 2035	450 Scenario 2020	450 Scenario 2035
OECD	1 533	1 406	1 403	1 259	1 516	1 689	1 203	690
Steam coal	986	930	962	855	1 048	1 202	804	423
Coking coal	282	278	270	289	279	321	248	226
Lignite	265	198	171	115	188	166	151	41
Non-OECD	1 668	3 718	4 428	4 767	4 794	6 200	3 895	2 649
Steam coal	1 250	3 101	3 762	4 128	4 114	5 493	3 273	2 129
Coking coal	290	517	554	539	565	582	528	470
Lignite	129	100	112	100	115	125	95	51
World	3 201	5 124	5 831	6 026	6 309	7 889	5 098	3 339
Steam coal share	*70%*	*79%*	*81%*	*83%*	*82%*	*85%*	*80%*	*76%*
Coking coal share	*18%*	*16%*	*14%*	*14%*	*13%*	*11%*	*15%*	*21%*
Lignite share	*12%*	*6%*	*5%*	*4%*	*5%*	*4%*	*5%*	*3%*

Note: Lignite, predominantly used for power generation, also includes peat.

The projected breakdown of production by type varies among the scenarios, with steam coal and lignite output constrained more by the stronger policy actions assumed in the New Policies Scenario and especially the 450 Scenario, and their greater impact on coal demand for power generation (Figure 5.7). These trends underline the fact that coking coal is less affected by policy, as there is less scope for its replacement by other less carbon-intensive fuels in the steel industry.

Figure 5.7 ▷ Change in world coal production by type and scenario, 2010-2035

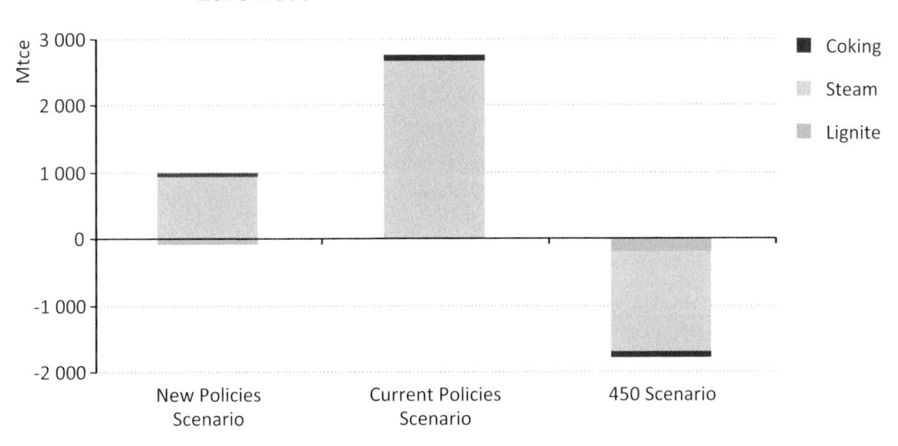

Long-term trends in coal trade are also highly sensitive to government energy and environmental policies, largely through their effect on demand. Because only a relatively small portion of hard coal is traded internationally (17% in 2010), small changes in demand can have much larger impacts on trade. In the New Policies Scenario, hard coal trade between *WEO* regions grows strongly to 2020, but begins to flatten out after that; by 2035, trade volumes are 35% higher than in 2010 (Table 5.5). In the Current Policies Scenario, trade expands much faster, doubling over the projection period, while, in the 450 Scenario, trade ends up 30% lower in 2035 than in 2010. Coking coal trade is less affected by differences in policy. Despite a doubling in hard coal trade between 1996 and 2010, and continuing growth over the *Outlook* period in both the New Policies Scenario and Current Policies Scenario, only roughly one out of five tonnes of coal produced globally is traded inter-regionally. As a result, small but sudden changes in production (caused, for example, by floods or strikes) in countries exporting coal could impose disproportionately large strains on international markets.

Table 5.5 ▷ World inter-regional hard coal trade* by type and scenario (Mtce)

			New Policies		Current Policies		450 Scenario	
	1990	2010	2020	2035	2020	2035	2020	2035
Steam coal	162	576	829	819	952	1 327	632	372
Coking coal	185	260	281	321	299	377	248	246
Total	309	833	1 095	1 122	1 246	1 702	850	585
Share of production								
Hard coal trade	*11%*	*17%*	*20%*	*19%*	*21%*	*22%*	*18%*	*18%*
Steam coal trade	*7%*	*14%*	*18%*	*16%*	*18%*	*20%*	*16%*	*15%*
Coking coal trade	*32%*	*33%*	*34%*	*39%*	*35%*	*42%*	*32%*	*35%*

* Total net exports for all *WEO* regions, not including trade within *WEO* regions.

Regional trends

Almost all major producers, including China and the United States, see coal output slow or even decline in the New Policies Scenario over the *Outlook* period. Production does increase substantially in some regions; for example, output increases by around 80% in both Indonesia and India (Table 5.6). In aggregate, OECD production starts to fall around 2020 and ends up 10% lower by 2035, compared to 2010, with declines in Europe and North America offsetting an increase in Australia. Non-OECD production carries on rising between 2010 and 2035, by a total of 1 050 Mtce, with nearly 70% of the increase occurring by 2020. China, by far the biggest producer in the world, sees output flatten after 2020 at levels 15% to 18% above that of 2010. India, which in 2010 produced less than half the coal of the United States (on an energy basis), by 2035, with output increasing rapidly, approaches the latter as the second-largest coal producer. Indonesian output overtakes that of Australia in the second half of the projection period, as it contributes one-quarter of incremental global output. Apart from additional growth in established producers like South Africa and Colombia, new output comes from Mongolia and southern Africa, notably Mozambique (Box 5.2).

Box 5.2 ▷ South Africa or southern Africa?

Today, South Africa dominates Africa's coal industry, accounting for virtually all of the continent's output. While it will not happen overnight, the picture could change in the medium term as other southern African nations, including Mozambique, Zimbabwe, Botswana, Tanzania, Zambia, Swaziland and Malawi, are endowed with significant coal reserves. In the New Policies Scenario, production from these other southern African countries, most of which is then exported, rises from 5 Mtce in 2010 to nearly 20 Mtce in 2020 and 36 Mtce in 2035.

Prospects appear brightest in Mozambique, as it is one of the largest undeveloped coal regions remaining in the world and is poised to become a significant exporter of coking coal in the future. International companies, including Rio Tinto and Vale, have already started shipping coal from Mozambique and many others have announced projects to build export facilities, which – if they all come to fruition – would raise capacity from 10 million tonnes (Mt) per year at present to more than 50 Mt/year by early in the next decade. In addition, many others are carrying out exploration. While the profitability of exports is expected to be high, due to abundant and easily extractable coal deposits, developing infrastructure will be a challenge. The ports of Beira and Nacala are far from the coal basins and the Zambezi River is environmentally sensitive, so it is unlikely to carry large volumes of coal barge traffic. Plans to expand rail and port capacity are in advanced stages and big investments have been announced, but it will take several years to build them.

Success in Mozambique could accelerate export development in Zimbabwe, which holds large hard coal reserves, totalling 500 Mt, and resources of 25 billion tonnes, many of which can be mined using low-cost open-cast methods. The country's oldest producer, Hwange Colliery Company, has been mining coal for more than a century, but current production is only 2 Mtce, half its peak in 1991. For the time being, lack of transportation facilities (Hwange is 1 420 kilometres from the port at Maputo) and an adverse investment climate are likely to prove substantial barriers to development, though domestic demand will revive if announced plans to build new coal-fired power plants are realised.

Another country with limited production but plenty of potential is Botswana, with its official estimates showing 21 billion tonnes of resources. Several projects to develop various coalfields have been proposed, but again it will not be easy to build the infrastructure required for exports as Botswana, like Zimbabwe, is a landlocked country. The Trans-Kalahari railway to the west and, hence, to Namibia, offers one export route, but with China and India as the main targets for exports, moving the coal through Mozambique is gaining preference. Financing either project will be a challenge. Coal exports from Botswana to South Africa could prove to be the least costly solution in the short term.

Table 5.6 ▷ Coal production by region in the New Policies Scenario (Mtce)

	1990	2010	2015	2020	2025	2030	2035	2010-2035	
								Delta	CAAGR*
OECD	**1 533**	**1 406**	**1 410**	**1 403**	**1 367**	**1 318**	**1 259**	**-147**	**-0.4%**
Americas	836	815	807	797	780	741	699	-116	-0.6%
United States	775	760	748	737	724	690	652	-108	-0.6%
Europe	526	246	209	182	152	129	104	-142	-3.4%
Asia Oceania	171	345	395	424	436	448	456	111	1.1%
Australia	152	339	389	419	431	443	452	113	1.2%
Non-OECD	**1 668**	**3 718**	**4 226**	**4 428**	**4 534**	**4 653**	**4 767**	**1 049**	**1.0%**
E. Europe/Eurasia	532	406	425	425	410	396	389	-17	-0.2%
Russia	275	257	272	276	269	257	252	-5	-0.1%
Asia	960	3 026	3 473	3 650	3 757	3 883	3 997	972	1.1%
China	748	2 309	2 570	2 645	2 673	2 716	2 734	425	0.7%
India	150	349	400	441	489	548	623	273	2.3%
Indonesia	8	266	378	425	450	469	490	223	2.5%
Middle East	1	1	1	1	1	1	1	0	0.8%
Africa	150	210	229	249	262	269	275	65	1.1%
South Africa	143	206	218	229	238	240	240	34	0.6%
Latin America	25	75	98	103	104	104	104	29	1.3%
Colombia	20	69	91	96	97	96	96	27	1.3%
World	**3 201**	**5 124**	**5 636**	**5 831**	**5 901**	**5 971**	**6 026**	**902**	**0.7%**
European Union	528	234	199	168	135	107	78	-156	-4.3%

* Compound average annual growth rate.

In the New Policies Scenario, cumulative investment in coal-supply infrastructure worldwide over the period 2012-2035 amounts to $1.2 trillion (in year-2011 dollars), or about $50 billion per year on average. Of this investment, almost 80% is in non-OECD countries and over 50% in China alone. Mining (including both surface and underground mine development, machinery and other infrastructure) absorbs around 95% of global coal investment, with ports and shipping accounting for the rest. Although global coal investment needs are significant, at only 3% of total energy-supply investment, they are quite small relative to all other sectors.

OECD countries collectively become a net coal exporter towards the end of the projection period, mirroring the position of non-OECD countries as a collective net importer (Table 5.7). This reversal in the OECD's aggregate role stems from declining imports in Japan, Korea and Europe, while at the same time net exports from Australia grow and those in the United States and Canada remain fairly stable. Among the importing countries, China, which only became a significant net importer in 2009, sees its import needs peak around 2015, before falling to about 40% of peak levels by 2035. India, which currently accounts for around 7% of global inter-regional trade (steam coal imports having more than quadrupled since 2004), overtakes Korea, Japan, China and the European Union by 2020 to become the world's largest net importer of coal (Figure 5.8). India's net coal imports grow five-fold by 2035 compared to 2010. Among the exporters, those where coking coal represents a larger proportion of total sales, such as Australia, United States and Canada, fare better as coking coal trade, unlike global steam coal trade, continues to grow throughout the *Outlook* period.

Table 5.7 ▷ Inter-regional hard coal trade in the New Policies Scenario

	2010		2020		2035		2010-35
	Mtce	Share of demand*	Mtce	Share of demand*	Mtce	Share of demand*	Delta Mtce
OECD	**-119**	**9%**	**-79**	**6%**	**78**	**7%**	**197**
Americas	59	8%	58	8%	61	9%	2
United States	53	7%	54	8%	56	9%	3
Europe	-184	61%	-214	77%	-162	84%	22
Asia Oceania	6	2%	78	19%	179	41%	173
Australia	272	86%	340	85%	382	88%	111
Japan	-164	100%	-147	100%	-131	100%	33
Non-OECD	**140**	**4%**	**79**	**2%**	**-78**	**2%**	**-218**
E. Europe/Eurasia	107	32%	109	32%	65	20%	-42
Russia	101	46%	104	46%	70	34%	-31
Asia	-51	2%	-147	4%	-272	6%	-221
China	-145	6%	-167	6%	-77	3%	68
India	-60	15%	-190	31%	-315	34%	-255
Indonesia	223	84%	337	79%	347	71%	125
Middle East	-2	70%	-3	73%	-5	80%	-2
Africa	43	20%	66	27%	80	29%	36
South Africa	50	24%	66	29%	67	28%	17
Latin America	44	59%	54	53%	55	53%	11
Colombia	64	93%	90	94%	89	92%	24
World**	**833**	**17%**	**1 095**	**20%**	**1 122**	**19%**	**289**
European Union	-158	58%	-187	76%	-130	84%	28

* Production for net exporting regions. ** Total net exports for all *WEO* regions, not including trade within *WEO* regions.

Notes: Positive numbers denote net exports; negative numbers net imports. The difference between OECD and non-OECD in 2010 is due to stock change.

The international coal market remains very sensitive to developments in China. Although China currently represents around 17% of inter-regional hard coal trade, net imports represent only about 6% of its hard coal use. Because of the sheer size of China's coal demand and production, relatively small changes in either its consumption or production have major impacts on the global market. For example, a drop in demand or a rise in production of just 3% could halve China's coal import needs based on current levels. Therefore, the success of China's efforts to curb coal-demand growth, for example by improving the thermal efficiency of its coal-fired power plants or more rapid diversification in the power sector, would have sharp and immediate effects on global international coal trade and prices.

Figure 5.8 ▷ Share of major hard coal importers in global inter-regional trade in the New Policies Scenario

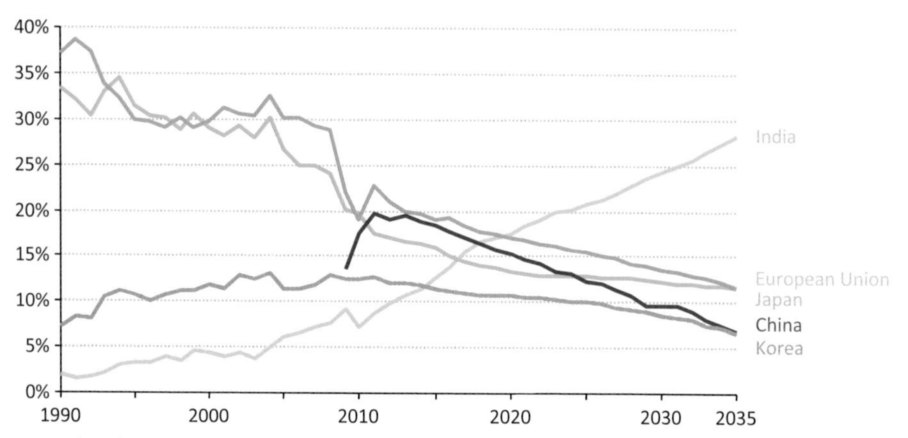

Note: China became a major coal net importer in 2009.

China

China's coal resources are large enough to support rising production levels for many decades. However, with the depletion of reserves in the traditional producing areas in the east, production will need to move further north and west - inevitably placing greater strains on China's coal transportation system, particularly railways. Rising costs of mining and transportation, along with the need to consolidate mining operations, increase efficiency and improve health and safety, have prompted a massive restructuring of China's coal industry over recent years. Under development are fourteen "coal-power bases" that will use long high-voltage transmission lines to ship power generated at the mine mouth to the demand centres in coastal cities. Coal-derived synthetic gas is also poised to enter the energy mix. Recently announced changes to China's 12th Five-Year Plan for 2011-2015 call for an expansion in coal mining capacity to 4.1 billion tonnes, but production will be limited to a plateau of 3.9 billion tonnes (compared with 2011 output estimated at 3.5 billion tonnes) to ease infrastructure constraints, slow resource depletion, address local pollution issues and hold down CO_2 emissions growth. Broadly in line with government plans, three-fifths of China's incremental production in the New Policies Scenario will be added by 2015. Output continues to expand thereafter, albeit at a slower pace, throughout the forecast period to more than 2 700 Mtce by 2035.

China's shift from being a large coal net exporter as recently as 2003 to a sizeable coal net importer was foreseen in *WEO-2007: China and India Insights.* Nonetheless, the turnaround has surprised many by its speed and magnitude. As a net importer, China is already larger than Korea and in 2012 or soon after is expected to overtake Japan, to reach volumes comparable to those of the entire European Union. Even small fluctuations in China's coal trade volumes will continue to shape global markets and prices in the near to medium term. In the longer term, the projected increase in production in the New Policies Scenario outstrips demand growth and drives net imports down from a peak of 190 Mtce around 2015 to only 77 Mtce by 2035 (Figure 5.9). By 2035, China's coal imports represent about 7% of global trade volumes (compared with 17% in 2010), cover less than 3% of national coal demand (6% in 2010) and are equivalent to less than 0.05% of China's GDP (0.27% in 2010).

Figure 5.9 ▷ China's hard coal net trade in the New Policies Scenario

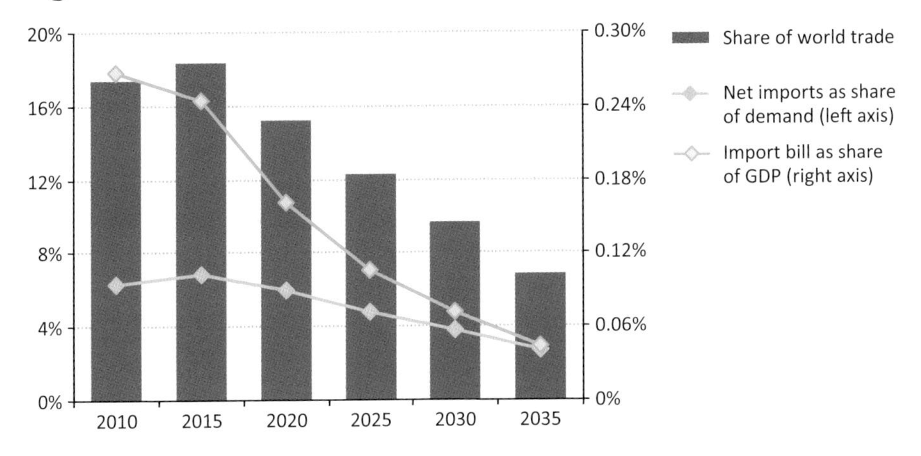

United States

The last two decades have seen major structural changes in the US coal industry, as coal output has shifted west of the Mississippi River. Some 60% of coal is now produced in western states and their production costs are generally lower; for example, as little as $10 per tonne or less in the Powder River basin in Wyoming compared with $50-$70 per tonne or more in the traditional producing areas in the Appalachian Mountains. Part of this mining-cost advantage of the western states is offset by high transport costs to the main demand centres, adding typically $20-$30 per tonne to the delivered cost of coal. Total US coal output remained fairly stable around 800 Mtce between 1995 and 2008, when it peaked at around 830 Mtce. While output has not recovered post 2008, it has not fallen in line with declining demand, as exports have risen. In the New Policies Scenario, production is projected to continue to decline slowly, with the decline accelerating in the second half of the forecast period. The decrease amounts to about 110 Mtce between 2010 and 2035. Steam coal is expected to be most affected. However, continued high levels of coking coal output may soften the impact in the Appalachian area.

The outlook for exports hinges on domestic production costs and global demand trends. Hard coal net exports are projected to remain relatively stable at about 50-60 Mtce over the projection period (Figure 5.10). As US domestic demand weakens over the period, maintaining or growing exports can slow the decline in US coal production, despite being relatively expensive. US exports, mainly through ports along the Atlantic coast or in the Gulf of Mexico, have picked up sharply recently as a result of weak domestic demand and higher prices overseas, which have made it profitable to export to Europe and even to Asia. Net coal exports in 2011 approached 80 Mtce and were one-third higher in the first half of

2012, relative to the same period in 2011. Steam coal represents an increasing share of coal exports. While transport and infrastructure constraints may restrain total export growth, the industry is moving rapidly to reduce these bottlenecks. Shipments have included some western coal exported through Pacific coast ports, including Canada. The substantial price differential between western region mining costs and prevailing Asian prices is currently a powerful incentive for the United States to expand its export infrastructure, especially if supported by long-term export contracts. Such an expansion, were it to occur quickly, could be a powerful influence on international coal trade. However, obstacles to rapidly increasing rail and port capacity appear substantial.

Figure 5.10 ▷ United States hard coal net trade in the New Policies Scenario

Mtce: 0, 20, 40, 60, 80
Right axis: 0%, 5%, 10%, 15%, 20%
1995, 2000, 2005, 2010, 2020, 2035
Hard coal net exports
Share of world hard coal trade (right axis)
Share of net exports in US production (right axis)

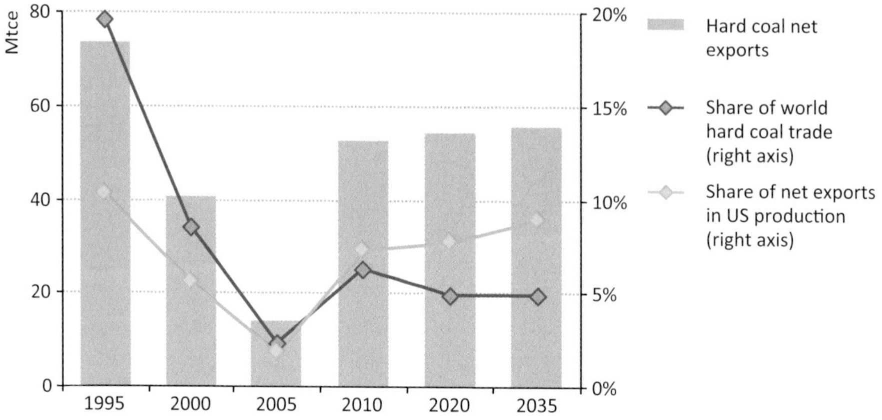

India

Between 2000 and 2011, India's coal output increased by nearly 80% to make it the world's third-largest producer after overtaking Australia. However, towards the end of the past decade, mining output growth slowed sharply, as the nation's coal industry faced several challenges, which will continue to affect output in coming years. Nearly 90% of the country's coal is produced in open-cast operations, but mines are located far from major demand centres and the quality of coal produced is low, with high ash content. Productivity is around a third of South African levels and only a tenth that of the United States. Furthermore, Coal India Limited, the dominant state-owned coal company, has faced multiple difficulties related to: land acquisition, forest clearing, lack of rail capacity and mechanised equipment, as well as labour disputes, mismanagement and bad weather conditions. As a result, an almost four-fold increase in imports occurred between 2000 and 2011 (with most growth occurring in recent years). These supply-side constraints continue to prevent India's coal production from growing in line with rampant demand. In the New Policies Scenario, output reaches nearly 623 Mtce by 2035 or almost 80% higher than in 2010, with growth accelerating somewhat in the second half of the projection period as productivity improves in the nation's coal industry.

The widening gap between production and demand in India in the New Policies Scenario results in hard coal net imports surging more than five-fold by 2035 compared to 2010, to around 315 Mtce, making India the world's largest coal importer around 2020 in the process. As imported steam coal is significantly more expensive for Indian power generators than domestic coal, and given that power prices are regulated, generators that are obliged to import coal have recently incurred substantial financial losses. To offer some immediate help, the Indian authorities have cut the 5% duty on coal imports, at least until 2014, and there are active discussions about embarking on price pooling for imported coal. Another consequence of the shortfall in domestic coal supply, and the resulting exposure to international prices, is that Indian power generators and steel mills have started to invest in the development of their own coal supply sources abroad (Table 5.8).

Table 5.8 ▷ Selected Indian foreign direct investment in coal supply projects

	Project	Investor (share)	Coal type	First production	Peak capacity (Mtpa)
Australia	NRE No.1 Colliery	Gujarat NRE Coking Coal (100%)	Coking	Operational	1
	NRE No.1 Colliery	Gujarat NRE Coking Coal (100%)	Coking	2014	2.2
	Wongawilli Colliery	Gujarat NRE Coking Coal (100%)	Coking	2016	2.5
	Carmichael Coal	Adani (100%)	Thermal	2014	60
	Kevin's Corner	GVK (100%)	Thermal	2015	50
	Alpha Coal	GVK (79%)	Thermal	2015	30
Indonesia	Aries	Essar Group (100%)	Thermal	Operational	4
Mozambique	Benga	Tata Steel (35%)	Both	Operational	20
	Tete Coal	Jindal Steel & Power (90%)	Both	2017	10

Note: Mtpa = million tonnes per annum.
Source: IEA analysis.

Australia

The coal output of Australia, the world's number four coal producer, is closely linked to the fortunes of international markets. The major factors affecting projected export levels, especially in the medium term, are the evolution of demand from importing countries, particularly China and India, the speed of development of emerging suppliers, such as Mongolia and Mozambique, and the development of potential new export outlets from existing exporters, including US supplies through Pacific ports. As noted, the global coal market remains heavily exposed to even small variations in demand, production and policy in major coal-consuming countries. The competitiveness of Australian coal is another factor: costs along the coal-supply chain are likely to continue to rise in the coming years. In the New Policies Scenario, Australia's coal production is projected to rise by a quarter, to reach 419 Mtce in 2020, but then growth slows, so that coal output reaches 452 Mtce by 2035, as global export demand cools (Figure 5.11). Australia's global leadership in coking coal exports proves an asset in a global context of robust coking coal trade growth to 2035.

Figure 5.11 ▷ Australian hard coal exports by type in the New Policies Scenario

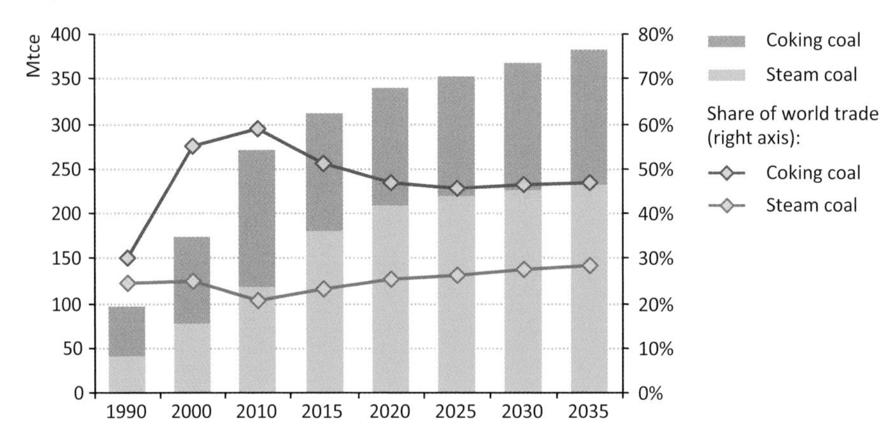

To support coal exports, which have risen by almost 60% over the last decade, significant investments in mining and transportation infrastructure have been made in the main existing mining areas in the key coal-producing states of New South Wales and Queensland. The largely unexploited Galillee and Surat Basins in Queensland will play a large part in expanding exports in the longer term, but challenges remain, including the financing of the mines and supporting infrastructure. Perhaps the biggest challenge is containing cost pressures resulting from the required large influx of labour (including availability of housing, other social infrastructure and water) associated with the massive simultaneous investments in other minerals and energy projects, including liquefied natural gas and iron ore. Some of these projects are in the same relatively remote areas as the coal projects. Additionally, much of the existing rail, road and port infrastructure is being utilised at close to maximum capacity as a result of the rapid growth in exports in recent years. In total, committed port and rail projects to expand coal capacity involve A$9.8 billion (US$9.4 billion) of investment, in addition to A$16.7 billion (US$16 billion) of investment in mining projects (BREE, 2012).

Coal markets and industry trends

Developments in international pricing

International coal prices have become much more volatile in the last few years in response to rapid changes in coal production, use and trade. This situation is likely to persist well into the future. What had been a stable market for nearly a quarter of a century up to around 2004 has seen sharp price swings in both directions, not always correlated to other energy prices. Prices reached record highs in 2008 before collapsing. They then increased steadily from early 2009 through to mid-2011, falling back thereafter in Atlantic basin markets (Figure 5.12). In some cases, short-term price swings have been caused by weather-related disruptions to supply in key exporting countries, such as the flooding in Queensland, Australia in 2011. The inelasticity of parts of coal demand, notably steel mills and power plants, puts further strain on markets.

International prices are normally set in annual negotiations between producers and major customers, such as Japan and Korea, although some large buyers prefer spot deals at prices which can be quite volatile. With domestic production dominating some internal markets, only a relatively small fraction of global coal consumption involves international trade (around 17% in 2011). In these circumstances, any unexpected shortfall in domestic supply in one of the large producing countries can put a sudden and disproportionate additional demand on the international market, as was seen in 2009, when China entered the import market quite suddenly. This has an immediate and sharp effect on the international price.

Prices in the domestic markets of coal-producing countries normally reflect indigenous production costs plus transport, or the price of available competing fuels (such as gas in power plants). This is true in parts of North America, where access to international coal markets is constrained. Where there are facilities to import coal into producing countries, however, such as in some coastal regions of China and India, the international price can be a strong influence on domestic prices, even if imports are quite limited. In countries in which the coal price is controlled by the government, production may respond poorly to changes in demand, since the normal market stimulus of price may be suppressed. Also, care must be taken in comparing prices across different regions, since coal quality can differ significantly in terms of energy content, ash, sulphur and other impurities.

Figure 5.12 ▷ **Average fortnightly steam coal prices in selected ports**

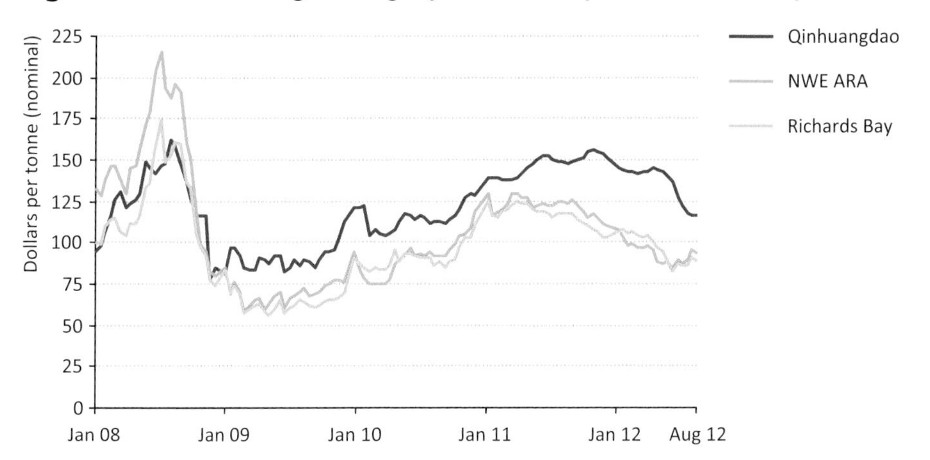

Note: Qinhuangdao is a major coal port in northeast China; NWE ARA is the northwest Europe marker price for Amsterdam-Rotterdam-Antwerp region; Richards Bay is the major South African export port.

Source: McCloskey (2008-2012).

Divergences in regional market conditions since the 2008 global economic crisis have led to a shift in pricing patterns across the world, not entirely related to shipping costs to different markets. While demand in Asia has kept on growing strongly, necessitating increased imports, notably to China, European demand has remained relatively weak, leading to relatively higher prices in Pacific basin markets, as shown by the Qinhuangdao price (Figure 5.12). Increased supply from the United States – where cheap gas has depressed domestic coal demand and pushed coal into the export market – and Colombia have also helped to keep coal prices down in Europe. Prices in Asia can be negotiated on a spot basis, which is preferred in China, or for an annual supply, which is the case for most of the coal bought by Japan and Korea. China now plays a pivotal role in setting import prices in the rest of the world: the price of China's domestic coal delivered in southeast coastal China determine the price received by exporters, such as Indonesia and Australia, selling into that market, which establishes a price marker for other international sales. The strengthening yuan will, of course, also have a bearing on this. Before the global economic crisis, delivered prices of steam coal in northwest Europe (NWE ARA) were set by the free on board (FOB) price of South African coal at Richards Bay plus freight. Since then, South African exporters have been able to gain higher margins by exporting to Asia, reflecting the stronger demand growth in Pacific basin markets.

The outlook for coal prices depends largely on demand-side factors, including growth prospects, but also of some importance is competition between gas and coal in the power sector, notably in North America and Europe. Increasing deployment of renewables and uncertainties around nuclear power (including early retirements, reluctance to grant life extensions or greater public resistance) are likely to affect this competition. A rise in North American gas prices from current very low levels would make coal more competitive in the power sector, and likely reduce coal availability to export markets. This would cause Europe to replace some imports from the United States with other, high cost imports, making coal less competitive in that region. Carbon prices in Europe (currently at very low levels) are also an important factor in European coal demand, and indeed anywhere a carbon price exists. Additional gas supply to Europe tends to be oil-indexed, undermining the many advantages of gas as a power generation fuel in that region, and favouring coal as the marginal power source. Energy policy and market developments in China and increasingly India will remain critical, given the potential for large swings in their coal imports according to the balance of domestic demand and supply. In the case of India, policy reforms will be essential if the pent up demand for electricity is to be met, as shown by the large-scale blackouts in July 2012. In the New Policies Scenario, the average OECD steam coal price – a proxy for international prices – is assumed to fall back to under $110/tonne (in year-2011 dollars) by 2015 and then recover slowly to about $115 in 2035. Coal prices increase less proportionately than gas prices, because coal demand levels off later in the forecast period, which moderates continued upward pressure on coal-production costs. Coal prices are significantly higher in the Current Policies Scenario, driven by higher demand, but lower in the 450 Scenario, as rigorous climate policies lead to a collapse in demand.

Cost and investment developments

As coal production increases, mining companies will need to exploit poorer quality or less accessible deposits, often in areas located further from demand centres and the necessary infrastructure. In the longer term, cost will play an important role in determining prices, both on international and domestic markets. Increasing mining and rail costs, as well as higher sea-freight costs (current costs are low), are expected to put some upward pressure on coal prices. However, the sheer abundance of coal worldwide means that large quantities are available at similar cost levels, reflected in a relatively flat long-run supply cost curve. Although it is possible that technological advances could help to drive the cost curve down to some degree, this effect is likely to be largely, if not entirely, offset by underlying inflation in the cost of materials, equipment and labour.

The cost curve for 2011 is flatter than that shown for 2010 in last year's *Outlook*, as costs in Indonesia at the bottom of the curve have risen faster than elsewhere (Figure 5.13). The average cost is now $12/tonne higher than in 2010, with costs in Russia and the United States having risen by roughly half this amount. Russia has very high transport costs; policy on these costs is a major factor in Russian coal export competitiveness. Some of the cost increase in Australia is due to the appreciation of the Australian dollar against the US dollar, though, as discussed above, the resources investment boom has put upward pressure on costs in remote areas. In most countries, the costs of basic materials and equipment, especially tyres, have risen significantly, driven by the demand from the wider mining and commodities sectors, such as metal ores and precious metals.

Figure 5.13 ▷ Average FOB supply cash costs and prices for internationally traded steam coal, 2011

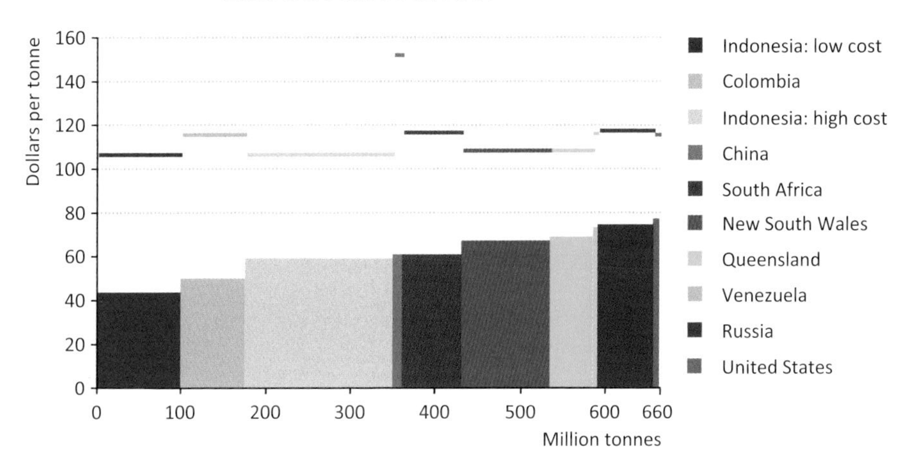

Notes: Prices, costs and volumes are adjusted to 6 000 kilocalories per kilogramme. Boxes represent FOB costs and bars show FOB prices.

Sources: IEA Clean Coal Centre analysis partly based on Marston and Wood Mackenzie.

Despite rising costs, investment in coal mining, excluding mergers and acquisitions, resumed its strong upward path in 2010, following a temporary slowdown in growth in 2009 as a result of the global economic crisis. The 30 leading coal companies worldwide, accounting for around 40% of world coal production, invested a total of $16 billion in 2010 – an increase of $1.5 billion, or 10%, on the 2009 level. Based on initial data and prevailing market conditions, continuing strong investment levels are expected to be seen in reported data for 2011, although a number of companies have lowered investment levels.

Power sector outlook

Winds of change?

Highlights

- Globally, demand for electricity is set to continue to grow faster than for any other final form of energy. In the New Policies Scenario, demand expands by over 70% between 2010 and 2035, or 2.2% per year on average. Over 80% of the growth arises in non-OECD countries, over half in China (38%) and India (13%) alone. In terms of electricity use, industry remains the largest end-use sector through 2035.

- Over the *Outlook* period, average electricity prices increase by 15%, driven by higher fuel prices, a shift to more capital-intensive generating capacity, CO_2 pricing in some countries and growing subsidies for renewables. There are significant regional price variations, with the highest prices persisting in the European Union and Japan, well above those in the United States and China.

- In net terms, global generating capacity expands by almost three-quarters, from 5 429 GW in 2011 to 9 340 GW by 2035. Gas and wind together account for almost 50% of the increase, followed by coal and hydro at about 15% each. Solar PV capacity also expands rapidly, at a rate more than two-and-a-half times that of nuclear, though generation from solar PV increases by half that of nuclear, reflecting the much lower average availability of these plants and its variable nature.

- The total investment cost of the gross generating capacity built worldwide during the *Outlook* period, including replacement of some 1 980 GW of capacity that is retired, is $9.7 trillion (in year-2011 dollars). A further $7.2 trillion is needed for transmission and distribution grids, of which over 40% is to replace ageing infrastructure.

- Coal remains the backbone fuel for electricity generation globally, and its use for this purpose continues to rise in absolute terms. However, its share of total generation falls from 41% in 2010 to 33% in 2035, while the share of gas increases slightly. Renewables' generation almost triples, their share expanding from 20% in 2010 to 31% in 2035 and accounting for 47% of incremental generation. Nuclear power output rises by 58%, though its share falls from 13% to 12%. Global CO_2 emissions from the power sector are about 20% higher in 2035 than in 2010, but average CO_2 intensity falls by 30% as more renewables, gas and efficient plants are deployed.

- Substantial differences in regional trends persist. In the United States, coal demand suffers from the shale gas boom and regulations to promote renewables. In the European Union, the share of renewables in generation more than doubles in 2010-2035 to 43%. In China and India, coal remains the dominant fuel in the generation mix and the leading source of incremental generation. Coal-fired generation in China grows almost as much as generation from nuclear, wind and hydropower combined.

Electricity demand

Global electricity demand increased by 40% between 2000 and 2010, despite a small downturn in 2009 caused by the global economic crisis. Global demand for electricity is set to continue to grow faster than demand for any other final form of energy over the projection period, although the rate of growth differs in each of the three main scenarios according to the nature of government policies related to carbon-dioxide (CO_2) emissions, energy efficiency and energy security. In the New Policies Scenario, world electricity demand expands by over 70% between 2010 and 2035, at an average rate of growth of 2.2% per year. In the Current Policies Scenario, demand rises more quickly, by 2.6% per year, while demand rises annually by an average of only 1.7% in the 450 Scenario (Table 6.1). Growth in demand due to increased electrification of transport in the New Policies Scenario and 450 Scenario is far outweighed by savings achieved in other sectors. Government policies affect electricity demand in complex ways: directly through measures to enhance end-use efficiency of electric appliances and equipment and to promote switching to electricity (for example, in the transport sector) and indirectly through their impact on final prices.

Table 6.1 ▷ Electricity demand* by region and scenario (TWh)

			New Policies		Current Policies		450 Scenario	
	1990	2010	2035	CAAGR 2010-35	2035	CAAGR 2010-35	2035	CAAGR 2010-35
OECD	**6 592**	**9 618**	**11 956**	**0.9%**	**12 635**	**1.1%**	**11 013**	**0.5%**
Americas	3 255	4 659	5 939	1.0%	6 133	1.1%	5 442	0.6%
United States	2 713	3 893	4 769	0.8%	4 892	0.9%	4 374	0.5%
Europe	2 321	3 232	3 938	0.8%	4 247	1.1%	3 676	0.5%
Asia Oceania	1 016	1 727	2 078	0.7%	2 255	1.1%	1 895	0.4%
Japan	758	1 017	1 095	0.3%	1 201	0.7%	976	-0.2%
Non-OECD	**3 494**	**8 825**	**19 903**	**3.3%**	**22 254**	**3.8%**	**16 931**	**2.6%**
E. Europe/Eurasia	1 585	1 350	1 978	1.5%	2 214	2.0%	1 754	1.1%
Russia	909	834	1 234	1.6%	1 405	2.1%	1 092	1.1%
Asia	1 049	5 352	13 705	3.8%	15 431	4.3%	11 438	3.1%
China	558	3 668	8 810	3.6%	10 149	4.2%	7 167	2.7%
India	212	693	2 463	5.2%	2 617	5.5%	2 096	4.5%
Middle East	190	680	1 466	3.1%	1 609	3.5%	1 260	2.5%
Africa	262	569	1 195	3.0%	1 289	3.3%	1 077	2.6%
Latin America	407	875	1 559	2.3%	1 711	2.7%	1 401	1.9%
Brazil	214	451	824	2.4%	904	2.8%	744	2.0%
World	**10 086**	**18 443**	**31 859**	**2.2%**	**34 889**	**2.6%**	**27 944**	**1.7%**
European Union	2 227	2 907	3 415	0.6%	3 694	1.0%	3 220	0.4%

* Electricity demand is calculated as the total gross electricity generated less own use in the production of electricity, less transmission and distribution losses.

Note: TWh = terawatt-hours; CAAGR = compound average annual growth rate.

The bulk of the growth in electricity use to 2035 will arise in non-OECD countries. In the New Policies Scenario, their projected share of incremental global demand is over four-fifths, with China (38%) and India (13%) alone contributing over half of the global growth. In China, which overtook the United States to become the world's largest electricity consumer in 2011, the annual pace of demand growth, which averaged 11.1% between 2005 and 2010, slows to 6.0% in the period to 2020 and to 2.0% in 2020-2035. At an average of 5.2% per year, electricity demand in India grows faster than in any other *WEO* region over the *Outlook* period, due mainly to strong population and economic growth. Average annual electricity demand per capita increases by three-quarters on average in non-OECD countries, from almost 1 600 kWh (kilowatt-hours) in 2010 to 2 800 kWh in 2035. Yet per-capita demand in the OECD remains far higher, though it increases much more slowly, from 7 800 kWh in 2010 to 8 700 kWh in 2035. In sub-Saharan Africa, consumption remains extremely low, at only 500 kWh per capita in 2035. Although the number of people with access to electricity worldwide increases significantly, 12% of the population still lacks access to electricity in 2030, compared with 19% in 2010 (see Chapter 18).

Among end-use sectors, industry remains the largest consumer of electricity in the New Policies Scenario, its demand increasing by 2.3% per year and accounting for over two-fifths of total electricity demand in 2035 (Figure 6.1). Residential demand grows at the same annual rate, while demand for services increases at the slightly lower rate of 1.9% per year. The transport sector grows at the fastest pace, 3.5% per year, thanks to increasing deployment of electric vehicles (trains account for most electricity use in the transport sector today). Even so, the share of transport in total electricity demand reaches only 2.1% in 2035 (compared with about 1.5% in 2010).

Figure 6.1 ▷ World electricity supply* and demand by sector in the New Policies Scenario

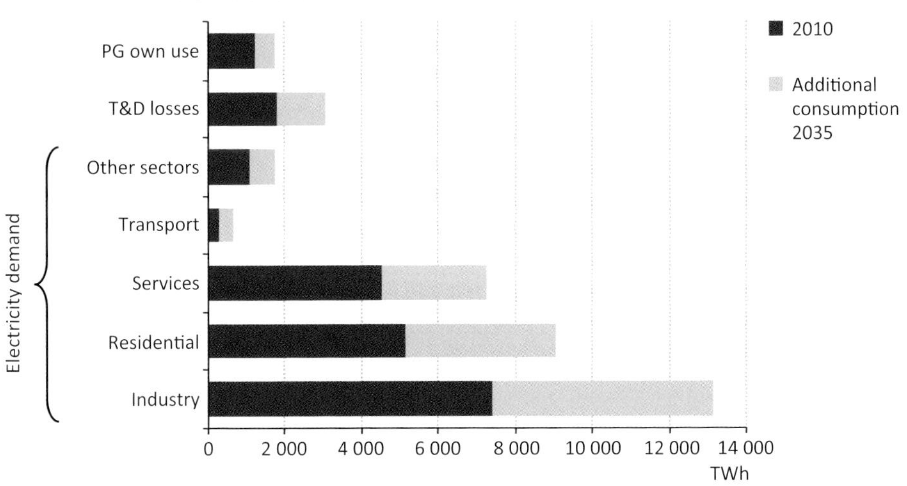

* Electricity supply is defined here as gross generation, including own use by power generators (PG), sufficient to cover demand in final uses (industry, residential, services, transport and other) and losses through transmission and distribution (T&D) grids.

Electricity supply

There are significant differences in the fuel mix between the three main scenarios, with low-carbon technologies playing a much larger role in the 450 Scenario than in the Current Policies Scenario, although even there they increase significantly by 2035 (Table 6.2). This reflects the differences in the policies adopted to drive both more efficient use of electricity and investment in less carbon-intensive power generation capacity.

Table 6.2 ▷ Electricity generation by source and scenario (TWh)

			New Policies		Current Policies		450 Scenario	
	1990	2010	2020	2035	2020	2035	2020	2035
OECD	7 629	10 848	11 910	13 297	12 153	14 110	11 470	12 153
Fossil fuels*	4 561	6 600	6 629	6 401	6 981	7 948	5 931	3 328
Nuclear	1 729	2 288	2 318	2 460	2 299	2 240	2 392	2 982
Hydro	1 182	1 351	1 486	1 622	1 474	1 578	1 521	1 730
Other renewables	157	609	1 477	2 813	1 400	2 343	1 627	4 112
Non-OECD	4 190	10 560	16 325	23 340	17 040	26 255	15 026	19 595
Fossil fuels*	2 929	7 847	11 163	14 528	12 167	18 882	9 522	7 159
Nuclear	283	468	1 125	1 906	1 099	1 668	1 209	2 986
Hydro	962	2 079	3 027	4 054	2 916	3 771	3 137	4 532
Other renewables	15	166	1 010	2 851	858	1 934	1 159	4 918
World	11 819	21 408	28 235	36 637	29 194	40 364	26 497	31 748
Fossil fuels*	7 490	14 446	17 793	20 929	19 148	26 829	15 453	10 487
Nuclear	2 013	2 756	3 443	4 366	3 397	3 908	3 601	5 968
Hydro	2 144	3 431	4 513	5 677	4 390	5 350	4 658	6 263
Other renewables	173	775	2 486	5 665	2 259	4 277	2 785	9 031

* Includes coal-, gas- and oil-fired generation.

In the New Policies Scenario, gross electricity generation increases by over 70% worldwide, or 2.2% per year, from 21 408 terawatt-hours (TWh) in 2010 to almost 36 640 TWh in 2035. Fossil fuels continue to dominate the generation fuel mix. Despite a significant decline in the share of coal in total generation, it nonetheless remains the single biggest source of generation worldwide. Over the projection period, the shares of natural gas and non-hydro renewables increase, resulting in a marked decline in the average CO_2 intensity of generation. There are significant differences in the fuel mix between OECD and non-OECD countries, though the broad trend towards greater diversity is common to both (Figure 6.2). The declining importance of coal and the growing contribution from renewable energy sources are particularly pronounced in OECD countries, where strong support policies are in place for renewables.

Relative costs, which are also influenced by government policies, are the primary driver of the projected changes in the types of fuels and technologies used to generate power. The cost of generation from fossil fuels, especially natural gas, is very sensitive to the price of the fuels, while for nuclear power and renewables the capital cost of the plant is by far the most important driver. Generating costs for each technology vary significantly across regions and countries, according to local cost factors, regulations and fuel prices (IEA, 2009). Carbon prices affect the relative competitiveness of technologies and encourage the use of more efficient technologies. Several countries have already introduced some form of a carbon price and others plan to do so (see Chapter 1). Water scarcity, which can pose reliability risks for coal-fired and nuclear plants that use large amounts of water for cooling, can also influence the generation mix and generating costs (see Chapter 17).

Figure 6.2 ▷ **Share of electricity generation by source and region in the New Policies Scenario**

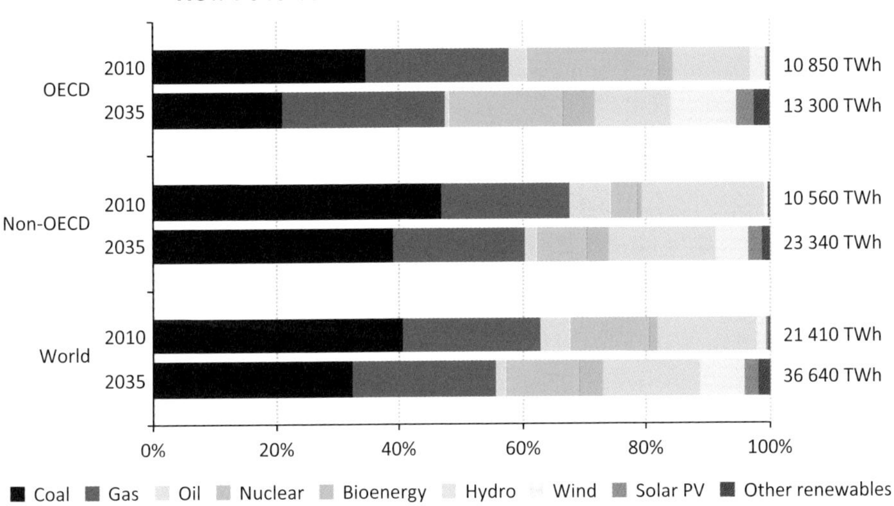

Many forms of government intervention can affect investment in new generating capacity and how existing plants are operated, most notably related to nuclear power and renewables. Policies on nuclear vary considerably across countries: some continue to encourage public and private investment in new capacity, while others ban the use of nuclear energy or have introduced programmes to phase it out. Many countries have adopted specific measures to support renewables-based electricity generation technologies to help them to become more competitive (see Chapter 7).

Capacity retirements and additions

Generating capacity needs to be added to meet rising demand and to replace capacity that will be retired over the projection period. The additional capacity needed in total for a given level of demand depends on the type of plant, as their availability to generate power varies. In the New Policies Scenario, global installed capacity is projected to increase from

5 429 GW in 2011 to about 9 340 GW by 2035 – a net increase of about 3 900 GW, or almost three-quarters. New gross capacity additions are significantly greater – about 5 890 GW – because 1 980 GW is retired over the projection period. In other words, about one-third of new capacity simply replaces that which is retired. Almost two-thirds of the capacity in operation today is still in operation in 2035. Emissions from these plants are effectively "locked-in", unless future policy actions lead to their early retirement or retrofitting with carbon capture and storage (CCS) equipment (see Chapter 8), or changes in fuel prices that affect operational decisions (*e.g.* idling more expensive power plants).

The bulk of the gross capacity additions projected to 2035 comes from coal- and gas-fired power plants and wind power (Figure 6.3). The share of capacity additions from wind and other types of renewables-based technologies is expected to be larger over the *Outlook* period than was the case over the past decade, when coal- and gas-fired power plants dominated gross capacity additions. Of the projected 1 250 GW of gross capacity additions in wind capacity by 2035, half is installed in OECD countries. However, the net increase in capacity is actually slightly bigger outside of the OECD due to greater retirements of existing wind installations in OECD countries. Most of the increase in gas- and (especially) coal-fired capacity, on both a gross and net basis, occurs in non-OECD countries, where electricity demand continues to grow much faster. The global increase in solar photovoltaic (PV) capacity is almost as big as that of hydropower and is more than two-and-a-half times as large as the net increase in nuclear capacity. However, the electricity generated from this new solar capacity is considerably less than the increase in nuclear power generation, at about one-half, reflecting the much lower average availability (capacity factor) of these plants and the variable nature of their output (Figure 6.4).

Figure 6.3 ▷ Power generation gross capacity additions and retirements in the New Policies Scenario, 2012-2035

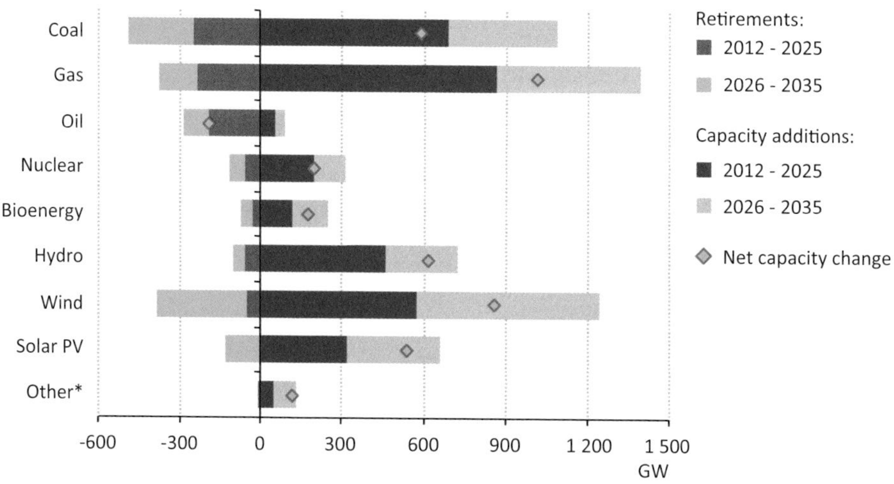

*Other includes geothermal, concentrating solar power and marine.

Around 18% of the gross additions of thermal (fossil-fuelled and nuclear) capacity projected through 2035, or about 520 gigawatts (GW), are already under construction. Almost all of this capacity will be operational by 2017, though some nuclear reactors are due to come online later. Of the thermal capacity being built today, 54% is coal-fired and 30% gas-fired. These figures may understate the role that is likely to be played in the medium term by gas, as the construction lead-times for gas-fired power plants are much less than for coal-fired plants: combined-cycle gas turbines (CCGTs) can usually be built within two to three years, open-cycle gas turbines (OCGTs) in about one to two years, while coal-fired power plants often take more than four years to complete.

Figure 6.4 ▷ World net incremental generation and capacity by type in the New Policies Scenario, 2010-2035

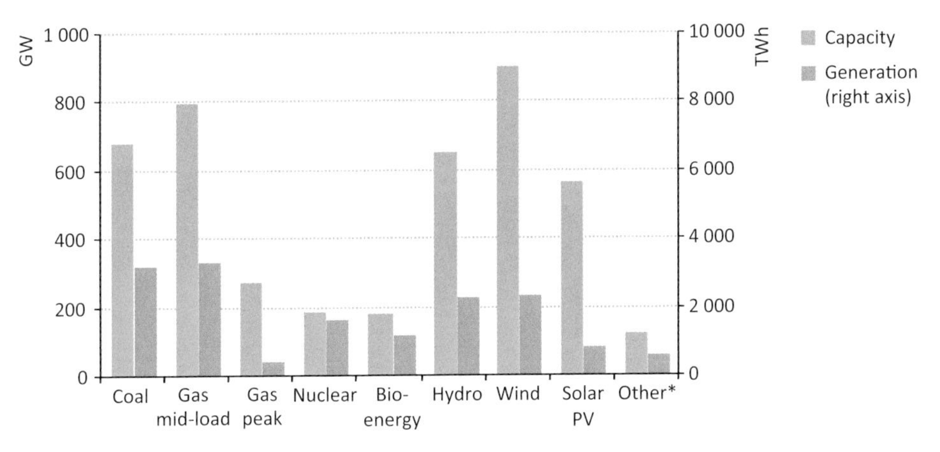

* Other includes geothermal, concentrating solar power and marine.

The bulk of the 1 980 GW of capacity that will be retired between 2012 and 2035 is in OECD countries, where the average age of plants is significantly greater. In non-OECD countries, there are fewer retirements as much of the capacity expansion has occurred in recent years (Table 6.3). Projected retirements are based on assumptions about power plant lifetimes, which vary according to the type.[1] Coal and nuclear plants have the longest technical lifetimes (typically 40-50 years), followed by gas plants (around 40 years) and wind turbines (about 20 years). However, these lifetimes can often be extended by replacing certain equipment. Economic factors and regulations governing power plant operations influence decisions about whether to extend the life of a plant by refurbishing it or simply shutting it down. More coal-fired power plants are retired before 2035 in the 450 Scenario than in the New Policies or Current Policies Scenarios, mainly because of higher CO_2 prices, which worsen their economics, including the economics of refurbishment.

1. Power plant lifetimes are expressed in both technical and economic terms. The technical lifetime corresponds to the operational life of the plant before retirement. The economic lifetime is the time taken to recover the investment in the plant and is usually shorter than the technical lifetime (Table 6.4).

Table 6.3 ▷ Cumulative capacity retirements by region and source in the New Policies Scenario, 2012-2035 (GW)

	Coal	Gas	Oil	Nuclear	Bioenergy	Hydro	Wind	Geo-thermal	Solar PV	CSP*	Marine	Total
OECD	**300**	**187**	**173**	**78**	**49**	**84**	**223**	**7**	**107**	**2**	**0**	**1 208**
Americas	119	109	73	7	17	37	80	4	8	1	0	453
United States	108	100	57	6	12	23	68	3	7	1	-	384
Europe	144	36	51	44	27	37	132	2	73	1	0	548
Asia Oceania	37	42	48	26	5	10	11	1	25	0	0	207
Japan	12	35	44	25	4	7	5	1	22	-	-	155
Non-OECD	**194**	**192**	**114**	**36**	**23**	**20**	**164**	**4**	**20**	**-**	**-**	**768**
E. Europe/Eurasia	91	119	23	32	1	2	4	0	1	-	-	274
Russia	42	84	5	20	1	-	1	0	0	-	-	153
Asia	77	17	27	2	14	9	152	3	17	-	-	319
China	42	1	3	-	5	3	123	0	12	-	-	190
India	27	3	2	1	4	3	27	-	3	-	-	70
Middle East	0	30	39	-	0	1	1	-	0	-	-	70
Africa	21	14	10	-	1	2	2	0	1	-	-	53
Latin America	4	11	15	1	7	7	5	1	1	-	-	52
Brazil	3	1	2	1	5	4	4	-	0	-	-	20
World	**494**	**379**	**287**	**114**	**72**	**105**	**387**	**11**	**127**	**2**	**0**	**1 976**
European Union	151	38	52	42	26	28	130	1	73	1	0	543

*CSP = concentrating solar power.

Table 6.4 ▷ **Cumulative gross capacity additions by region and source in the New Policies Scenario, 2012-2035** (GW)

	Coal	Gas	Oil	Nuclear	Bioenergy	Hydro	Wind	Geo-thermal	Solar PV	CSP*	Marine	Total
OECD	**132**	**543**	**25**	**91**	**107**	**158**	**623**	**23**	**344**	**27**	**14**	**2 087**
Americas	32	267	7	26	50	66	229	11	84	13	2	787
United States	28	205	5	19	40	35	182	8	71	11	1	606
Europe	60	176	2	33	45	71	341	4	171	12	9	923
Asia Oceania	40	100	16	31	12	21	54	8	89	2	3	377
Japan	15	74	16	3	8	15	28	5	72	-	1	236
Non-OECD	**953**	**849**	**68**	**221**	**138**	**564**	**623**	**22**	**318**	**45**	**1**	**3 804**
E. Europe/Eurasia	66	206	1	51	10	30	20	3	6	-	0	393
Russia	33	143	0	34	7	18	6	2	2	-	-	245
Asia	820	346	14	148	99	370	537	15	243	19	1	2 610
China	428	165	2	116	56	193	387	2	122	14	0	1 487
India	251	91	2	25	19	76	108	0	88	4	0	666
Middle East	1	150	35	8	4	13	23	-	23	14	-	271
Africa	59	69	8	6	9	54	17	2	25	10	-	261
Latin America	8	78	11	7	16	97	25	2	21	3	-	269
Brazil	3	46	3	5	11	46	17	-	11	1	-	144
World	**1 085**	**1 392**	**93**	**312**	**245**	**722**	**1 247**	**45**	**662**	**72**	**15**	**5 891**
European Union	54	162	2	33	42	50	324	3	168	12	9	859
Average economic lifetime (years)	30	25	25	35	25	50	20	25	20	20	20	

*CSP = concentrating solar power.

Fossil-fuelled generation

The prospects for generation from fossil fuels (coal, gas and oil) depend critically on the future policy landscape, with their share of total global generation in 2035 ranging from 33% in the 450 Scenario to 66% in the Current Policies Scenario (almost exactly the same share as in 2010). In the New Policies Scenario, it declines to 57%, mainly because of a faster increase in renewables-based generation, even though the absolute level of fossil-fuelled generated power expands by almost half compared with 2010.

Globally, coal-fired generation increases from 8 687 TWh in 2010 to around 11 900 TWh in 2035 in the New Policies Scenario, its share of total generation dropping from 41% to 33%. Coal-fired output increases by over four-fifths in non-OECD countries, outweighing the fall in OECD countries by several times. China continues to contribute the largest increase in coal-fired generation, accounting for two-thirds of the global increase over the *Outlook* period. By 2035, China accounts for 46% of global coal-fired generation, up from 38% in 2010. India overtakes the United States as the second-largest generator of coal-fired power by the end of the projection period (Figure 6.5). In non-OECD countries, coal-fired generation grows most rapidly before 2020, when economic growth is strongest. The decline in coal use for power in the OECD accelerates over the *Outlook* period, in part due to the growing deployment of renewables and gas, rising CO_2 prices and increasingly stringent environmental restrictions.

Figure 6.5 ▷ Coal-fired power generation by selected region in the New Policies Scenario

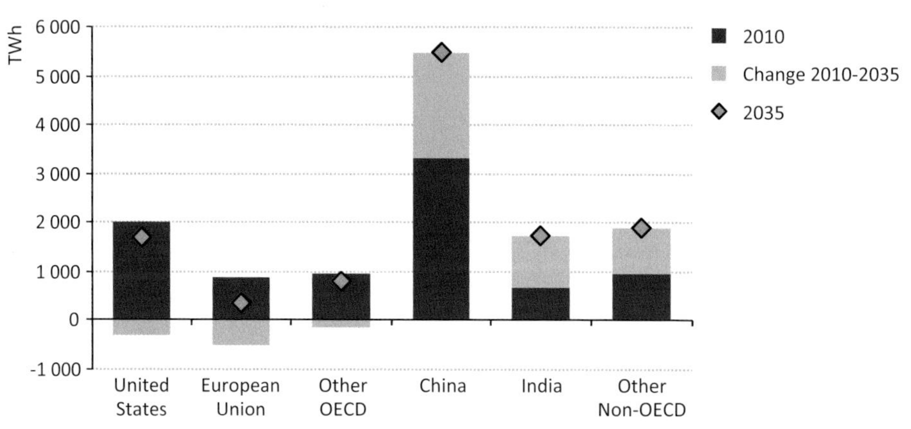

The average thermal efficiency of coal-fired plants increases from 39% in 2010 to 42% in 2035 in the New Policies Scenario, as old plants using subcritical technology are retired and advanced plants, including ultra-supercritical and integrated combined-cycle (IGCC) designs, are built. By 2035, almost 20% of all plants in operation are advanced, compared with only a few percentage points today. There is a small decrease in electricity generation from coal-fired combined heat and power (CHP) plants, and the CHP contribution to coal-fired generation, already small, falls from 6% of total coal-fired generation in 2010 to 4% in 2035. A few gigawatts of CCS-fitted coal plants are installed before 2020 in the New Policies

Scenario, in line with existing plans and funding for demonstration projects. After 2020, support policies and CO_2 pricing in some regions lead to a small additional number of CCS-fitted coal plants being built. In total, these plants have a capacity of 67 GW and generate about 440 TWh by 2035 – around 4% of total coal-fired output.

World gas-fired generation expands from 4 760 TWh in 2010 to about 8 470 TWh in 2035 in the New Policies Scenario, its share of total power generation increasing by one percentage point to 23%. About three-quarters of the projected increase in gas-fired power generation occurs in non-OECD countries, with China alone accounting for 20% of the worldwide growth and the Middle East for 18%. High natural gas prices are expected to constrain demand for new gas-fired generation in some regions, notably in Europe, but gas still has advantages that make it attractive to investors, notably lower capital costs, shorter construction times than coal and nuclear plants and operational flexibility. Subsidised prices boost the competitiveness of gas in the Middle East, where production is projected to continue to expand rapidly (see Chapter 4), while abundant supplies of shale gas keep gas prices low and make gas an attractive generating option in the United States. CCGTs and OCGTs are likely to be used increasingly for mid-merit order and peak load, respectively, as well as providing a significant portion of the flexible capacity that is needed to integrate higher shares of variable wind and solar power into the system, although new coal-fired plant designs also allow for considerable operational flexibility (see Chapter 7).

Figure 6.6 ▷ **Gas-fired electricity generation by region in the New Policies Scenario**

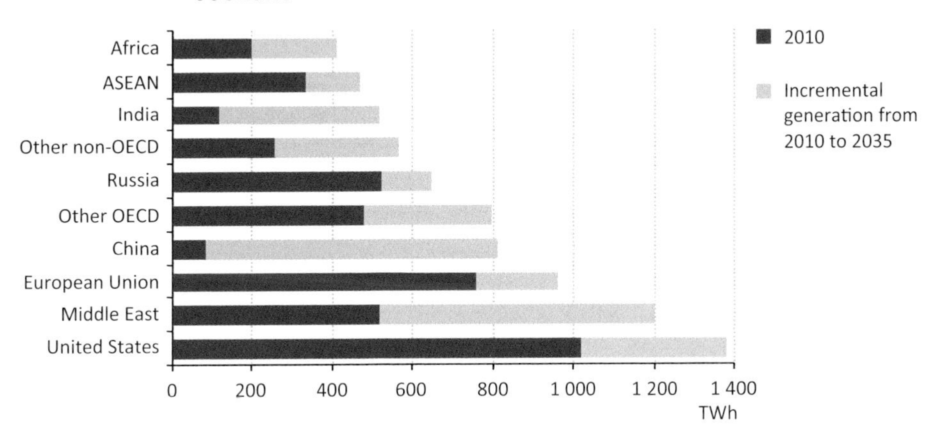

The use of oil for power generation is projected to continue to decline in line with the established long-term trend, its share of total generation dropping from about 5% in 2010 to 1.5% in 2035. Increasing world oil prices, declining subsidies in several countries and the falling relative cost of other generating options render the economics of oil-fired generation increasingly unattractive. Oil use for power generation falls in all regions, most rapidly in the OECD, and at the slowest pace in the Middle East, where subsidies persist and electricity demand grows strongly. In 2035, the Middle East accounts for almost half of world oil-fired generation, up from 29% in 2010.

The growing importance of renewables in the power generation fuel mix has important implications for the way other plants are operated and for power system flexibility. Due to their very low running costs, wind and solar power always come at the top of the merit order – the preferred ranking of capacity for dispatch. As the capacity of these types of plant increases, they will, when available for use, tend to displace fossil fuel-based plants, forcing those plants to ramp up and down more frequently, according to the level of demand, and thereby lowering their capacity factors (the ratio of the output of the plant over a period of time relative to its potential output if it had operated at full nameplate capacity over the entire period).

Nuclear power

The prospects for nuclear power worldwide have been clouded by the uncertainty surrounding nuclear policies after the Fukushima Daiichi accident in March 2011. Several countries have altered their policies in the face of public concerns about the safety of nuclear reactors. Most of these changes were already taken into account in the New Policies Scenario last year and include an accelerated schedule for retiring plants early in Germany, no construction of replacement reactors in Switzerland and a temporary delay in issuing approvals for new plants in China.

In Japan, following the events at Fukushima Daiichi, all reactors were shut down for scheduled maintenance and "stress tests" to review their safety. To date, only two units have been granted permission to restart, with these coming back online in July 2012. In September 2012, Japan released the Innovative Strategy for Energy and the Environment, which includes the goal of reducing reliance on nuclear energy. At the time of writing not all of the details of the new strategy were available. We have assumed in the New Policies Scenario that all existing reactors (except those at Fukushima Daini and Daiichi plants) are re-commissioned progressively over the next few years. Their lifetimes are limited to 40 years in the case of reactors built before 1990 and 50 years for those built more recently (instead of 60 years for all of them, as we assumed previously). In addition, we assume that no new plants are built by 2035, beyond the two that are already under construction (due to be commissioned around 2015). Consequently, total nuclear capacity in Japan falls from 46 GW in 2011 to 24 GW in 2035 – 41 GW, or 63%, lower than in *WEO-2011*. As a result, the share of nuclear in total electricity generation in Japan falls from 26% in 2010 to 20% in 2020 and to 15% in 2035, compared with 33% in 2035 in *WEO-2011*.

Nuclear power output grows at broadly the same pace as projected last year in the rest of the world in the New Policies Scenario, with the exception of the United States and Europe. Production grows slightly less rapidly in the United States, due to the retirement of an additional 5 GW of capacity and fewer plants being built later in the *Outlook* period, as a result of declining competitiveness of nuclear relative to gas (because of lower gas prices), partially offset by the addition of four units that have just been approved and the completion of another unit under construction. As a result, US nuclear capacity reaches

120 GW in 2035 – 5 GW lower than last year's projection and 13 GW above the level in 2011. Installed capacity in 2035 in the European Union is also revised down, from 129 GW to 120 GW, as a result of the reduced competitiveness of nuclear power and slightly higher retirements. In China, it is expected that the moratorium on new approvals will be lifted and the nuclear programme will proceed as planned. At present, there are 64 reactors under construction in the world, totalling 66 GW of capacity.

Figure 6.7 ▷ **Nuclear power capacity by region in the New Policies Scenario**

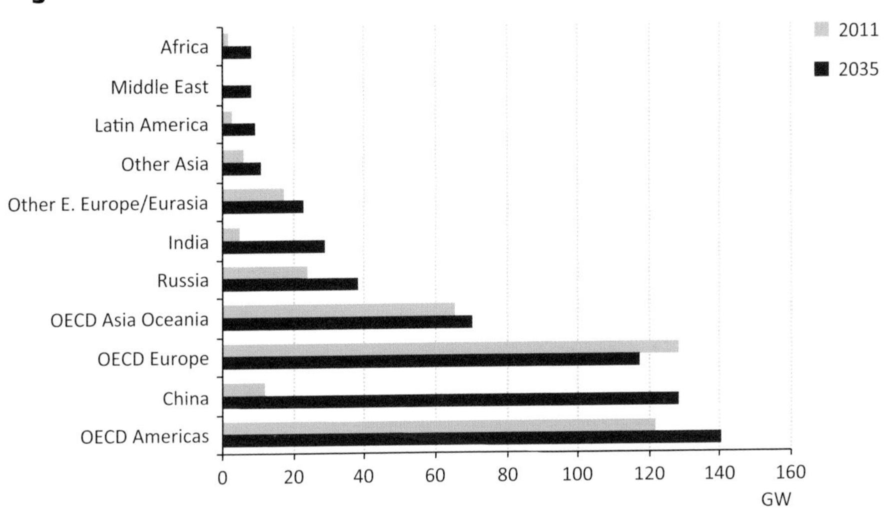

In aggregate, world nuclear capacity in the New Policies Scenario reaches some 580 GW in 2035 – about 50 GW lower than last year's projection. Production rises from 2 756 TWh in 2010 to about 4 370 TWh in 2035, an increase of almost 60%. Correspondingly, the share of nuclear in total generation falls from 13% to 12%.

Renewables[2]

Electricity generation from renewable energy sources almost triples over the *Outlook* period in the New Policies Scenario, driving up their share of total world power generation expanding substantially, from 20% in 2010 to 31% in 2035. Renewables account for 47% of the total increase in generation over that period. Hydropower remains the single biggest renewable source, though its overall share of generation falls slightly, from 16.0% to 15.5% over the *Outlook* period. Wind power grows most in terms of generation.

The rate of expansion of renewables and their importance in overall power generation vary markedly across regions (Figure 6.8). In the European Union, their contribution more than doubles over the projection period to 43%. The United States also experiences a doubling

2. A more detailed analysis of the prospects for renewables for heat and power generation can be found in Chapter 7.

of the share of renewables by 2035, but it still remains below that of most other regions, including both China and India. In most regions, hydropower remains the leading source of renewables-based power.

Figure 6.8 ▷ Share of renewables in electricity generation by region in the New Policies Scenario

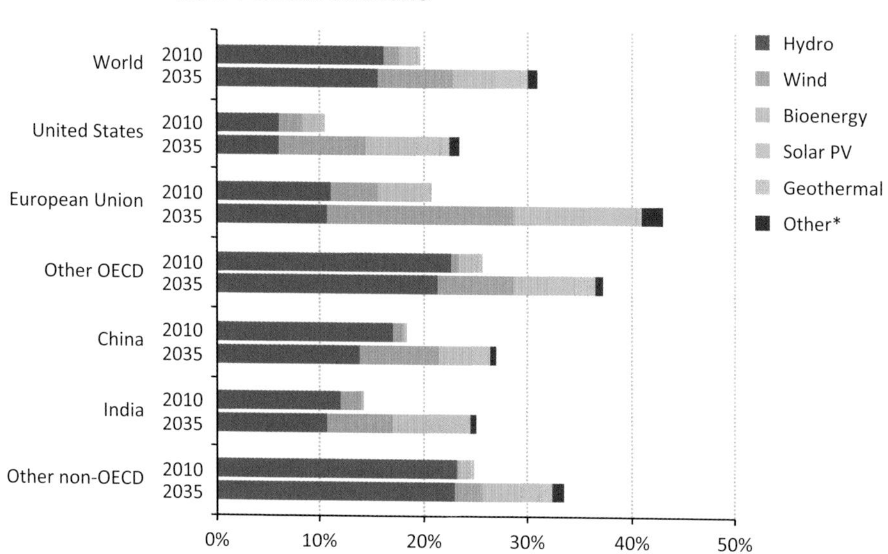

*Other includes concentrating solar power and marine.

Transmission and distribution

Transmission and distribution (T&D) networks continue to evolve and expand in the New Policies Scenario to improve the quality of power delivery to existing customers, to provide access for new end-users and connect new power plants. In the New Policies Scenario, the total length of the T&D system worldwide increases by 25 million kilometres (km) to around 93 million km in 2035. Distribution networks, delivering power over short distances from substations to households, businesses and small industrial facilities, account for about 88% of this increase; transmission grids, transporting power over long distances from generators to local substations near the final customers, account for the remainder.

Robust T&D grids are critical to system flexibility. They are particularly important for accommodating the increasing contribution of variable renewables (see Chapter 7). Future T&D grids are expected to increasingly include smart grids, which use digital communication and control technologies to optimise system operation, lower losses and accommodate new types of load, such as from electric vehicles (Box 6.1). System flexibility can also be improved through strengthened interconnections between regional grids.

Box 6.1 ▷ **Electric vehicles and smart grids**

Although electricity needs for electric vehicles (EVs) are likely to remain small relative to overall load in most regions for many years to come, they could make a big impact on peak load, as motorists seek to recharge their batteries during the evening. Electricity suppliers will need to anticipate the network investments involved.

Recent technological advances in electricity distribution and load management that make use of information and communications technologies, referred to as "smart grids", promise to help make the necessary provision at reasonable cost. Smart-grid technology can enable EV-charging (grid-to-vehicle, or G2V) load to be shifted to off-peak periods, thereby flattening the daily load curve and significantly reducing both generation and network investment needs. Advanced metering will provide customers with the necessary real-time data to support the switch.

In the longer term, smart-grid technology may enable EVs to be used as distributed storage devices, feeding electricity stored in their batteries back into the system when needed (vehicle-to-grid, or V2G), to help provide peak-shaving capability. Technical and practical barriers first need to be overcome, including low battery-discharge rates and storage capacity, and market and regulatory frameworks will also need to adapt. Tariffs need to provide incentives for electricity transmission and distribution companies to invest in appropriate smart-grid technologies, for system operators to take decisions that ensure economically efficient operation of the system taken as a whole and for EV owners to optimise G2V and V2G load.

Investment

Very large investments in electricity-supply infrastructure will be needed over the *Outlook* period to meet rising electricity demand and to replace or refurbish obsolete generating assets and network facilities. In the New Policies Scenario, cumulative investment in the power sector is $16.9 trillion (in year-2011 dollars) from 2012-2035, roughly equivalent to the gross domestic product (GDP) of the entire European Union in 2011 (Table 6.5). This sum is 45% of all energy sector investment. Investment in power plants accounts for 57% of the power sector total, over 60% of it for renewables. Even though they make up just over half of capacity additions, this large share of investment in renewables reflects their higher capital cost, compared with coal- and gas- fired power plants (Figure 6.9).

Unsurprisingly, in view of their faster increase in electricity demand, non-OECD countries account for the greater share, 60% of the cumulative investment in the power sector worldwide. Investment is greatest in China, where total power sector spending is $3.7 trillion, followed by the European Union, United States, and India.

Table 6.5 ▷ Investment in electricity-supply infrastructure by region and source in the New Policies Scenario, 2012-2035 ($2011 billion)

	Coal	Gas	Oil	Nuclear	Bioenergy	Hydro	Wind	Solar PV	Other*	Total Plant	Trans-mission	Distrib-ution	Total T&D	Total
OECD	**451**	**436**	**16**	**360**	**369**	**419**	**1 146**	**717**	**226**	**4 139**	**662**	**1 986**	**2 648**	**6 787**
Americas	207	211	5	115	175	174	411	184	85	1 569	437	846	1 283	2 852
United States	201	170	4	87	156	92	330	158	68	1 266	350	679	1 029	2 295
Europe	145	138	1	133	159	190	630	346	101	1 844	175	778	953	2 797
Asia Oceania	99	87	10	112	34	54	104	187	40	726	50	362	412	1 138
Japan	41	65	9	12	22	38	56	151	15	409	24	192	216	626
Non-OECD	**1 158**	**604**	**58**	**583**	**281**	**1 130**	**983**	**542**	**208**	**5 547**	**1 187**	**3 347**	**4 533**	**10 080**
E. Europe/Eurasia	143	179	1	182	33	63	32	14	6	651	134	397	531	1 182
Russia	74	123	0	119	24	38	10	4	5	397	96	224	320	717
Asia	889	201	9	326	183	701	854	391	99	3 653	802	2 313	3 115	6 768
China	341	82	1	233	94	306	634	193	56	1 939	572	1 200	1 772	3 712
India	347	58	2	71	36	163	160	140	15	992	111	517	629	1 620
Middle East	1	129	36	27	9	27	34	43	47	353	57	166	224	577
Africa	114	42	7	23	20	108	25	51	41	431	89	225	314	745
Latin America	10	54	6	25	35	231	39	43	15	458	104	246	350	808
Brazil	5	33	2	17	24	117	27	22	6	252	69	139	209	461
World	**1 608**	**1 040**	**74**	**942**	**650**	**1 549**	**2 129**	**1 259**	**434**	**9 686**	**1 849**	**5 332**	**7 181**	**16 867**
European Union	133	128	1	134	151	137	603	341	99	1 728	155	688	843	2 571

*Includes geothermal, concentrating solar power and marine.

Figure 6.9 ▷ **Power sector cumulative investment by type in the New Policies Scenario, 2012-2035**

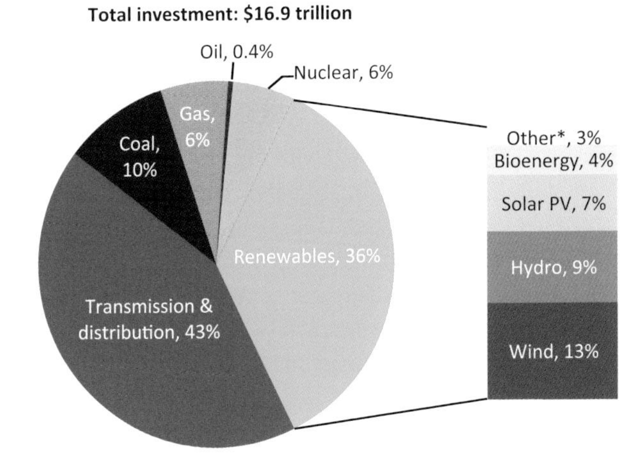

* Other includes geothermal, concentrating solar power and marine.

Cumulative investment in T&D grids from 2012 to 2035 in the New Policies Scenario is $7.2 trillion, representing 43% of the total investment in the power sector. Investment in non-OECD countries makes up more than three-fifths of T&D investment throughout the *Outlook* period due to higher demand growth. China alone represents 40% of the non-OECD T&D investment. Most T&D investment in the OECD goes to replacing and refurbishing T&D assets rather than building new ones, as power markets there are relatively mature. A growing share of T&D investment in all regions is needed to integrate renewables into the system, about 3% of the total over the *Outlook* period. This includes the additional cost of connecting new, often remote and fragmented, sources of supply to the transmission network and of reinforcing other parts of the system (Figure 6.10).

Figure 6.10 ▷ **Annual average investment in T&D infrastructure in the New Policies Scenario**

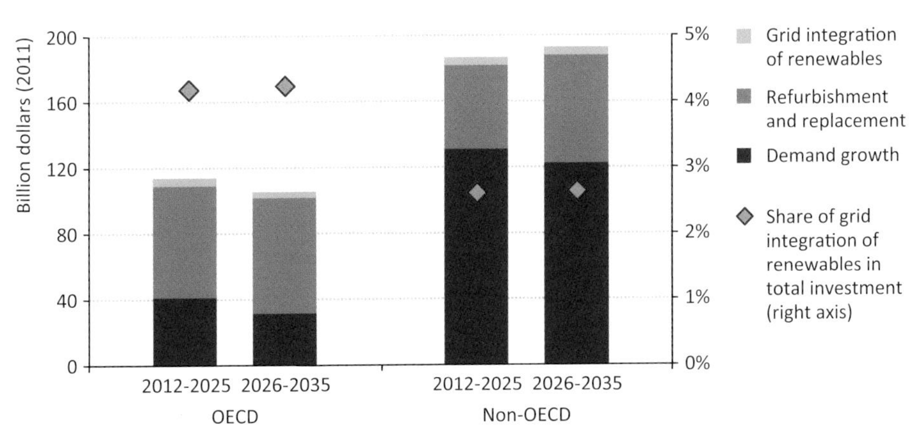

CO_2 emissions

In the three main scenarios in this year's *Outlook*, the increasing penetration of low-carbon technologies in the generating fuel mix and improvements in the thermal efficiency of fossil-fuel-based plants help to slow the growth in CO_2 emissions from the power sector. In the New Policies Scenario, world CO_2 power sector emissions (associated with the production of electricity and heat) climb slowly from 12.5 gigatonnes (Gt) in 2010 to 15.0 Gt in 2035, an increase of 20% compared with an increase in total generation of over 70%. In the OECD, emissions peak before 2015 and then fall steadily, thanks to the declining use of coal, more efficient new plants and the rapid expansion of renewables. Outside the OECD, emissions continue to grow, albeit more slowly as demand growth decelerates after 2020 and nuclear and renewables make a greater contribution to total power generation. China's emissions increase accounts for more than two-thirds of the net increase in global emissions in the power sector (Figure 6.11).

Figure 6.11 ▷ CO_2 emissions from fossil-fuel combustion in the power sector and emission intensity per unit of electricity generation by selected region in the New Policies Scenario

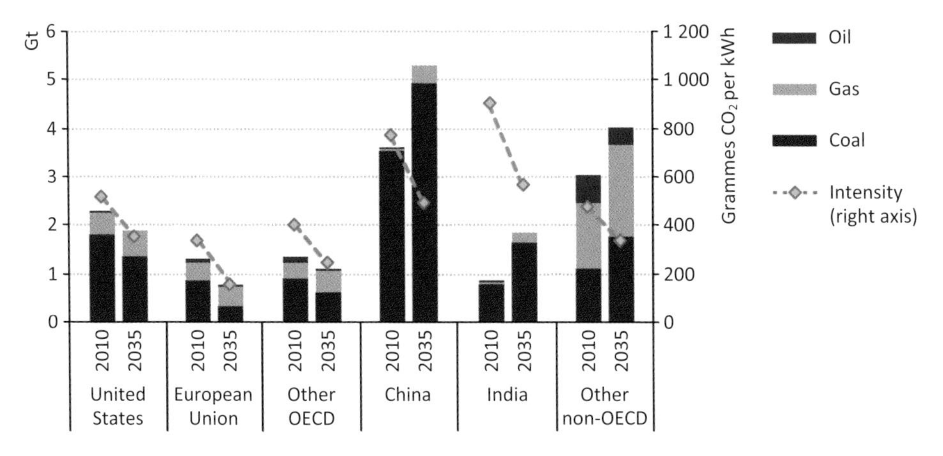

The worldwide shift to lower carbon fuels and more efficient generating technologies in the New Policies Scenario results in a fall in the average CO_2 intensity of electricity generation of about 30%, from 530 grammes of CO_2 per kilowatt-hour (g CO_2/kWh) in 2010 to around 375 g CO_2/kWh in 2035. Were the fuel mix and the efficiency of each type of generating technology to remain unchanged, *i.e.* were CO_2 intensity to remain constant, emissions for electricity generation would rise by 8.1 Gt from 2010-2035, compared with 2.3 Gt in the New Policies Scenario for the same trajectory of electricity demand.

Regional trends

United States

In the United States, the shale gas boom and government regulations are reshaping the power sector. Rapidly expanding domestic gas production has driven down gas prices to a fraction of previous levels, making gas highly competitive with other fuels in the power sector. In the New Policies Scenario, gas-fired generation increases by over 360 TWh from 2010 to 2035, displacing mainly coal- and oil-fired generation, which fall by around 300 TWh and 30 TWh, respectively (Figure 6.12). The other main drivers of the changing fuel mix are support policies for renewables, and existing and proposed regulations aimed at environmental protection. These include the finalised Mercury and Air Toxics Standards, established by the US Environmental Protection Agency, and the Cross-State Air Pollution Rule (currently delayed), regulating emissions of sulphur dioxides and nitrogen oxides. Many of these are therefore refurbished over the projection period. Coal capacity in the United States drops by 80 GW during the *Outlook* period, or almost one-quarter, due to the retirement of almost 110 GW and additions of less than 30 GW. Despite this, coal remains the largest source of power in the United States in 2035.

Figure 6.12 ▷ **United States electricity generation by source in the New Policies Scenario**

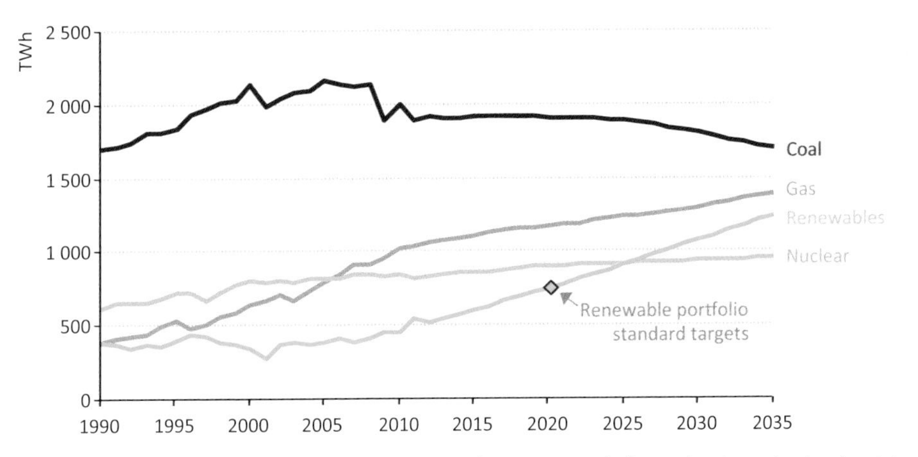

Note: Renewable portfolio standard targets represent the aggregate of all state-level targets plus, for states without targets, holding hydropower output constant and maintaining the share of non-hydro renewable generation over time.

Nuclear power generation is projected to increase slowly through the *Outlook* period, though its share of total generation falls from 19% in 2010 to 18% in 2035. The prospects for the nuclear industry have stagnated in past years, due to its very high upfront capital costs and associated financing, and declining competitiveness with respect to gas-fired plants. However, for the first time since the accident at the Three Mile Island power plant in 1979, the government approved licenses in early 2012 for four new nuclear reactors, located at the Vogtle and VC Summer power plants.

While the prospect of establishing a carbon price in the United States has faded in recent times, the outlook for renewables remains strong. In the New Policies Scenario, generation from renewables roughly triples, providing 23% of total generation in 2035, compared to 10% in 2010. Renewable portfolio standards (RPS) in twenty-nine states and production tax credits for renewables are the main drivers of this growth until 2020, helping push renewable energy technologies to account for almost 16% of total generation by then at the national level. Each RPS includes specific goals for the contribution of renewables to the power sector either in terms of generation or capacity.

The New Policies Scenario requires roughly $2.3 trillion of cumulative investment in the US power sector, including over $1.0 trillion for transmission and distribution. The CO_2 intensity of the power sector declines from nearly 520 g CO_2/kWh in 2010 to around 350 g CO_2/kWh in 2035, mainly due to higher shares of gas and renewables.

Japan

The accident at the Fukushima Daiichi power plant in 2011 significantly altered the outlook of the power sector in Japan by bringing the future of nuclear power into question. In 2010, nuclear provided one-quarter of all the electricity generated in Japan. Post-Fukushima, all of the country's reactors were shut down progressively for scheduled maintenance and "stress tests" to review their safety. This accentuated the shortfall in capacity caused by the permanent loss of the four reactors at Fukushima Daiichi, requiring electricity conservation measures to be stepped up to avoid power shortages. By mid-May 2012, all 50 nuclear power plants in Japan were offline, resulting in zero nuclear power supplied to the grid for the first time in more than 40 years. However, in July 2012, units 3 and 4 of the Ohi nuclear power facility were re-started. Japan released the Innovative Strategy for Energy and the Environment in September 2012, which includes the goal of reducing reliance on nuclear power. In the New Policies Scenario, in the absence of all the details of the new strategy, we have assumed that no nuclear plants are constructed through 2035, beyond the two reactors at Shimane-3 and Ohma that are already at an advanced stage of construction, and shorter lifetimes are applied to existing plants. Over the *Outlook* period, the share of nuclear in total generation recovers to 20% by 2020, but then falls back once again to 15% in 2035 (Figure 6.13).

Continuing and expanding some of the conservation measures helps to limit the need for additional generation to less than 80 TWh between 2010 and 2035. Along with nuclear, fossil fuels contribute a declining share of power generation; combined with nuclear, their share of total generation, drops from 90% in 2010 to around 75% by 2035. Renewables increase to meet the projected increase in demand and the 130 TWh fall in generation from nuclear and fossil fuels. While hydro capacity increases by only 7 GW during the *Outlook* period, non-hydro renewables capacity increases by about 80 GW, led by solar PV and wind. High feed-in tariffs drive solar PV to increase by 51 GW from 2011 to 2035 – equal to over three-quarters of world installed capacity in 2011. Wind capacity increases by some 23 GW – equal to almost half of all the wind capacity in the United States in 2011.

Figure 6.13 ▷ **Japan electricity generation by source in the New Policies Scenario**

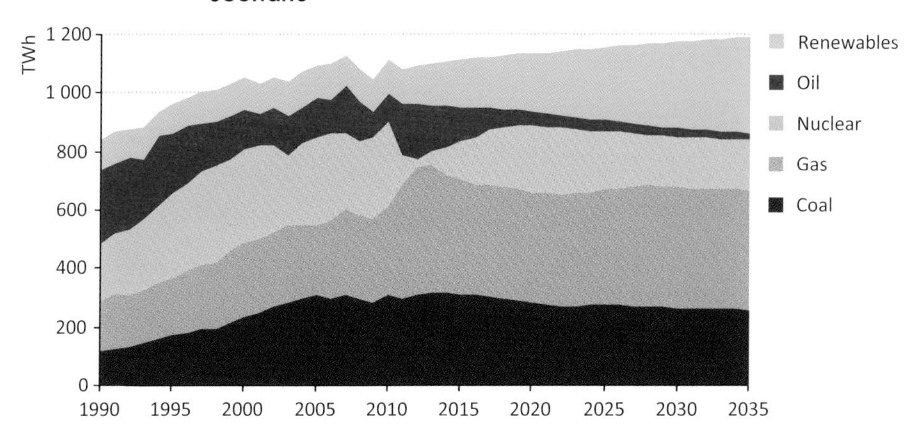

6

European Union

The European Union has been at the forefront of the deployment of renewable energy technologies in the power sector, especially during the last decade, thanks to strong government programmes and incentives – primarily for solar and wind (see Chapter 7). Support for solar PV in Germany and Italy has been particularly robust, with the two countries combined accounting for about 60% of world PV capacity additions in 2011. In the case of wind power, European Union countries were responsible for 23% of capacity additions worldwide in 2011.

The pace of renewables capacity expansion slows in the New Policies Scenario, but still more than doubles over the *Outlook* period, increasing from 325 GW in 2011 to over 670 GW by 2035. In fact, renewables make up the bulk of generating capacity additions during the *Outlook* period, with wind accounting for 38% and PV for 20%. Renewables in total capture three-quarters of the total investment in power generation, and boost their share of total generation from 21% in 2010 to 43% in 2035 (Figure 6.14). Wind power is the dominant source of new generation, providing 60% of the increase from renewables and more than the total net incremental generation to 2035.

In 2011, power generation in the European Union shifted away from gas and towards coal, as a result of higher gas prices and low carbon prices in the EU Emissions Trading System (EU ETS). Gas-fired generation fell by 17%, while coal-fired generation increased by 11%, despite a decline in overall electricity demand caused by the economic downturn. This trend is expected to continue in the short term, even if carbon prices increase during the third phase of the EU ETS (covering the period 2013-2020) when auctions for allowances and tighter limits on the use of offsets and banking of allowances from the current phase are introduced. However, in the longer term, coal-fired generation is projected to drop dramatically, from 26% of total generation in 2010 down to just 9% of generation in 2035. This is due to higher carbon prices, which make coal-fired power less competitive, and greater penetration of renewables. Gas-fired generation regains market share in the longer term, increasing its share of total generation by two-and-a-half percentage points between 2010 and 2035. The share of nuclear in the power generation mix declines with increased retirements, declining from 28% in 2010 to 22% in 2035.

Figure 6.14 ▷ Renewables-based electricity generation by source and share of total generation in the European Union in the New Policies Scenario

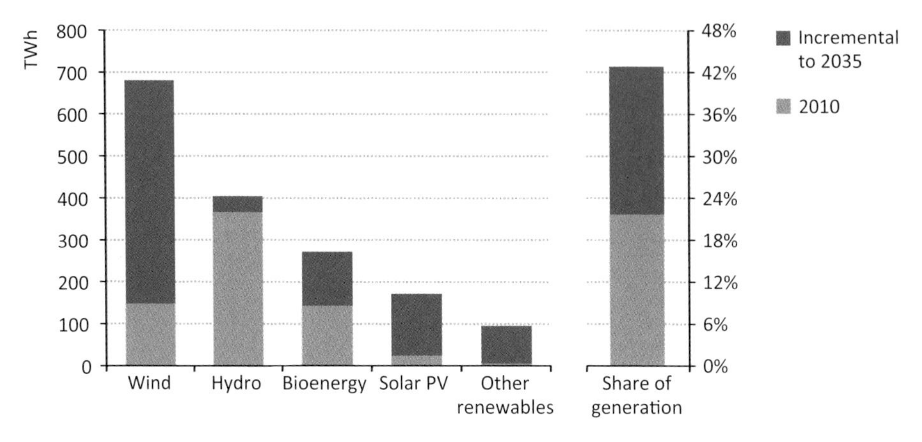

China

The enormous scale of the expansion in power generation in China in the New Policies Scenario is difficult to overstate. Electricity demand increases by almost 5 700 TWh, or 134%, over the *Outlook* period, an increase equivalent to more than the current electricity demand in the United States and Japan combined. This expansion results from continued strong economic growth (albeit slower than over the last two decades), averaging 5.7% per year from 2010-2035. China's 12th Five-Year Plan establishes a series of targets for power generation by technology that guide the development of the power sector to 2015 (see Annex B).

In order to keep pace with the projected demand growth, generation from all fuel sources and technologies increases, except oil, over the projection period. Coal-fired generation increases by nearly 2 160 TWh, or 65%, equal to more than the current coal-fired generation in OECD Americas. Nonetheless, the share of coal in China's total generation falls from 78% to 55%. The shares of gas and nuclear power grow markedly, from only 4% combined in 2010 to 18% by 2035. The share of hydropower in total generation drops from 17% in 2010 to 14% in 2035, despite the additions of almost 200 GW of new capacity, equivalent to the construction of nearly nine projects equal in capacity to the recently completed Three Gorges Dam spanning the Yangtze River, or around one-hundred projects the size of the Hoover Dam in the United States. Overall, China accounts for 25% of the total cumulative capacity additions worldwide over the *Outlook* period; the share is 39% for coal-fired capacity and 37% for nuclear power (Figure 6.15).

Figure 6.15 ▷ Power generating capacity additions by source in China and share of global additions in New Policies Scenario, 2012-2035

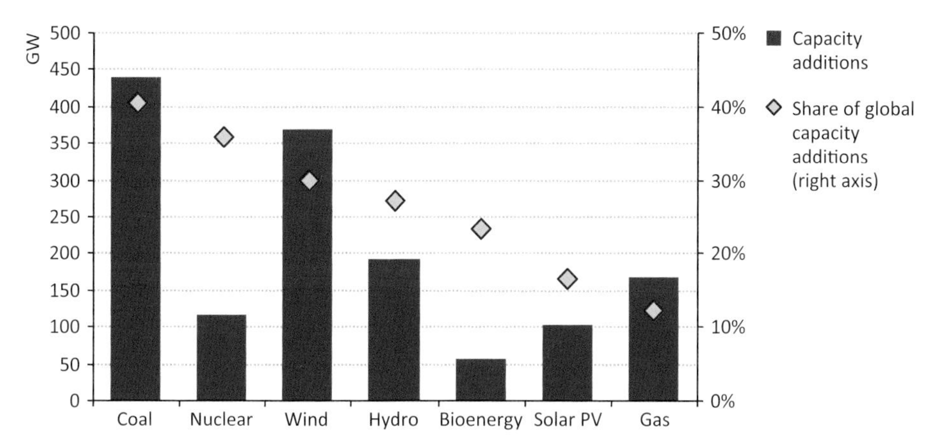

6

China also sees a large increase in non-hydro renewables-based generation in the New Policies Scenario, their combined share in total generation increasing from just over 1% in 2010 to 13% in 2035. Wind power contributes 60% of non-hydro renewables generation in 2035, with capacity reaching almost 330 GW – nearly 40% greater than world capacity in 2011. Bioenergy, solar PV and concentrating solar power (CSP) also play significant roles in the expanding power sector in China, generating over 500 TWh combined in 2035. Generation from non-hydro renewables in China surpasses that in the United States after 2015 and the European Union around 2030.

India

While not as large as the increase in China, electricity demand in India in the New Policies Scenario increases at a faster rate, by 5.0% per year on average, driven by rising population and per-capita incomes. Total demand rises from 693 TWh in 2010 to about 2 450 TWh in 2035. As in China, the rapid expansion in demand requires that all fuels, aside from oil, be part of the supply picture. Coal, already the main source of power, continues to provide more than half of total generation through 2035. India is the world's second-fastest growing market for nuclear power, with annual output increasing by nearly eight times between 2010 and 2035. Gas plays an increasingly important role in meeting demand, its share of total generation growing by three percentage points over the *Outlook* period. Hydropower sees a slight fall in its share of total generation, though capacity nearly triples over the same period. Non-hydro renewables emerge as a significant source of power, their share of total generation jumping from 2.3% in 2010 to 14% in 2035. Wind, solar PV and bioenergy make up about one-third of cumulative capacity additions and investment in power plants through 2035; including hydropower, the share of investment is over half.

Middle East

Gas and oil remain the cornerstones of power generation in the Middle East in the New Policies Scenario, due to plentiful domestic supplies and low (subsidised) prices, always accounting for more than four-fifths of power generation over the *Outlook* period. This share is much higher than in other regions, especially for oil (Figure 6.16). Gas-fired generation more than doubles; its share of total generation increases from 62% in 2010 to close to 70% in 2035. Oil-fired generation declines slightly over time, sharply reducing its share of total generation from 35% in 2010 to 16% in 2035. Combined water and power plants, which produce both freshwater and electricity, see an increasing role in the region, accounting for some 30% of the total capacity additions over the *Outlook* period.

Figure 6.16 ▷ Share of oil and gas electricity generation in selected regions in the New Policies Scenario

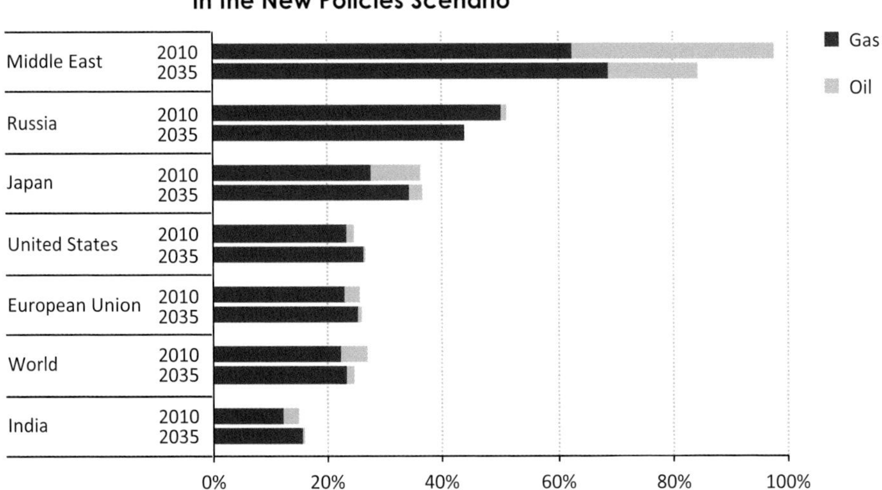

Focus on electricity prices

End-user electricity prices are determined by the underlying costs of supplying electricity and by any taxes or subsidies applied by governments to electricity sales. The costs of electricity supply comprise several elements: the cost of generating the electricity, transmitting it through the grid, distributing it through the network, and retailing it to the final customer (Figure 6.17). Also, in many countries, costs associated with subsidies for renewable energy are passed on to consumers through the electricity price.

Electricity prices differ widely between regions, but several countries have experienced strong increases in electricity prices in recent years. Retail electricity prices grew more quickly than inflation in most OECD countries over the period 2005-2010, including the United States, the United Kingdom and Australia. The main factors responsible for this increase were the rising prices of fuels, materials and equipment, and, in some countries, the introduction of CO_2 pricing.

Figure 6.17 ▷ **Components of the end-user price of electricity**

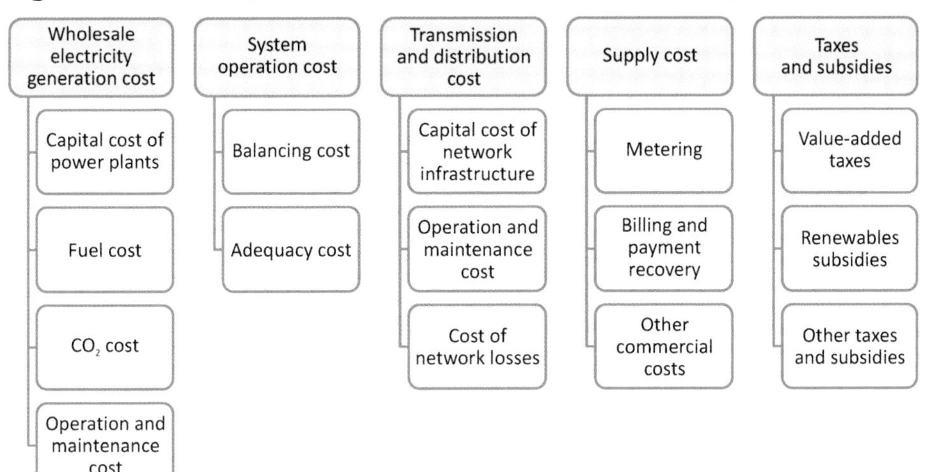

6

The pass-through of growing renewables subsidy costs to consumers has also started to have a material impact on end-user electricity prices in some countries. In Germany, Italy and the United Kingdom for example, the impact on end-user prices has been particularly marked because of the rapid pace of development of renewables in recent years and the big contribution from the relatively-high cost solar PV technology. About one-fifth of the pre-tax price paid by German consumers for electricity in 2011 was due to renewables subsidies; the share was 12% in Italy, 11% in the United Kingdom and 4% in France, according to national data (Figure 6.18).

Figure 6.18 ▷ **Average electricity price to households in selected European countries by cost component, excluding taxes, 2011**

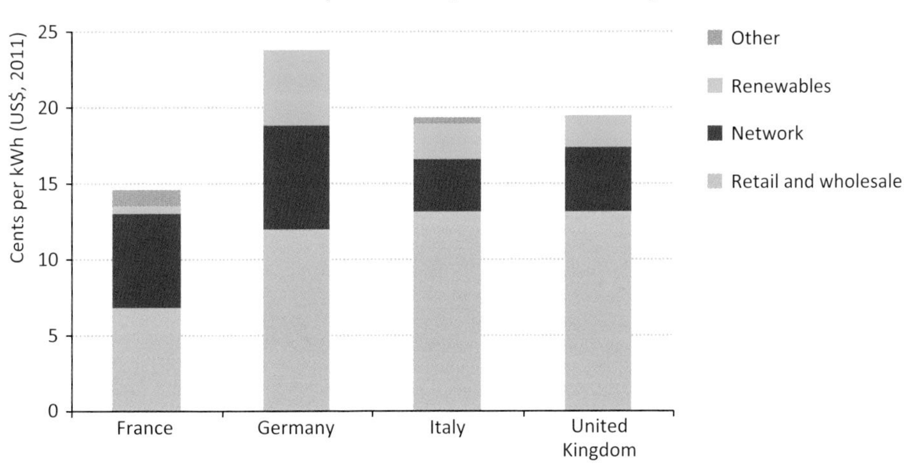

Sources: Ofgem (2012); BDEW (2011); CRE (2011); Autorita'per l'Energia Elettrica e il Gas (2012); IEA (2012).

In our projections, the average cost of electricity is set to continue to rise over the *Outlook* period – regardless of the policy landscape. In the New Policies Scenario, the average end-user price of electricity worldwide (in real terms weighted across all regions and final sectors) increases by 15% between 2011 and 2035, or 0.6% per year (Figure 6.19).[3] Within this average, there are marked variations across countries and regions, reflecting differences in local market conditions and government policies. Three main factors explain the price increases: higher fossil-fuel input costs to generation, higher materials and equipment costs for new projects, and – in some regions – CO_2 prices and costs associated with renewable subsidies. Other factors, such as the costs associated with the abatement of other types of pollution, can also affect prices. In addition to their impact on end-user prices, the growing volumes of variable renewables in the generation mix are introducing new challenges for the design of electricity markets (Box 6.2).

Figure 6.19 ▷ Change in end-user prices by type of consumer and selected region in the New Policies Scenario, 2011-2035

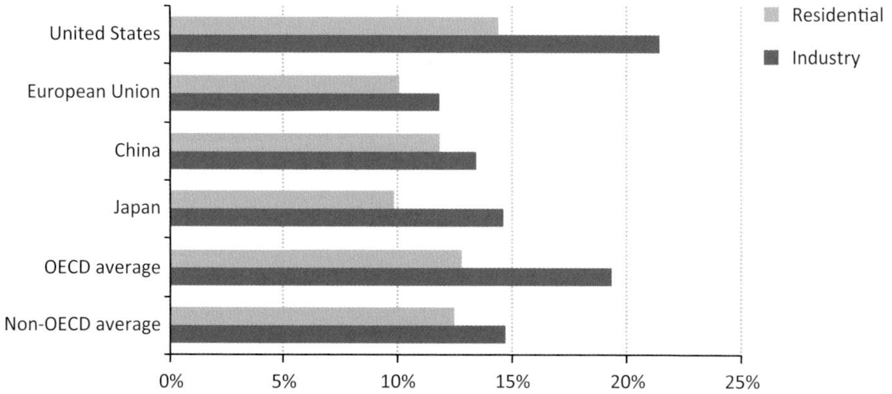

Note: For Japan, change shown is for 2010-2035, due to the extraordinarily high prices experienced in 2011.[4]

Prices differ between consuming sectors as well as between regions. Industrial consumers take far larger quantities of electricity than residential consumers and tend to take them directly from the high-voltage network, rather than from a distribution network, so the price they pay for electricity tends to be lower than that for residential consumers. Because the price paid to generators (the wholesale price) represents a larger proportion of the total to industrial consumers, and the wholesale price rises faster than the other elements of the cost of providing electricity over the projection period, the percentage increase in end-user prices tends to be higher for industrial consumers.

3. This increase excludes the effect of electricity end-user price subsidy removal in some countries. This is in order to allow for a focus on the fundamental price drivers for end-user prices rather than end-use subsidies that are directly determined by government policy. For a discussion of end-use price subsidies, see Chapter 2.

4. Following the Fukushima Daiichi accident in March 2011, Japan shut down the majority of its nuclear reactors. This resulted in a shortage of electrical generation capacity and much heavier reliance on imported gas. This increased wholesale electricity prices considerably, compared to previous years. Given the focus on fundamental cost drivers in this section, we use 2010 as a more typical base year for Japan.

Box 6.2 ▷ Implications of growing renewables shares for power market design

Since the 1980s, a number of countries have taken steps to liberalise their electricity markets. When executed well, market reform can lead to more economically efficient generation and provide investors with the right signals and framework to invest in new capacity.

Electricity demand changes considerably over the course of the day, and the resulting wholesale market prices can also vary significantly over a period of a few hours. At times of low demand, the wholesale price falls to a low level and many generators choose to cease producing. But for variable renewables, the cost of generating an additional unit of electricity is close to zero, as no fuel is required, so there is an incentive for them to produce even if the electricity price at the time is close to zero. Conversely, at times of peak demand all generators have a strong incentive to produce electricity, but variable renewables-based generators cannot respond to high prices by producing more, as they are driven by weather conditions outside their control.

Because generation from renewables cannot be counted upon at all times – a characteristic of particular significance at times of peak demand – other types of quickly dispatchable power plants (*e.g.* open-cycle gas turbines) need to be available to meet supply needs. But some of these have the opportunity to generate electricity only for a few hours of the year and must recover as much as they can of their operating costs and fixed investment costs during those few hours. This means that prices can reach very high levels in these periods (IEA, 2011). If they do not, the incentive to offer supply at those times may be insufficient to warrant new investment and the future adequacy of the power system may be at risk (Mount, *et al.*, 2012).

The increased price volatility associated with the growing element of renewables-based generation has prompted calls for the design of competitive electricity markets to be re-examined to ensure that the risks of investing in other capacity – such as flexible peaking plants, storage, interconnection or demand response – are correctly priced.[5]

One of the potential reforms currently being considered in several markets is the introduction of capacity mechanisms, which explicitly remunerate capacity held available, in addition to the energy generated; and arrangements on these lines have been introduced in some markets. Indeed such a capacity charge was introduced in some systems well before the emergence of variable renewables on a large scale. Whatever their form, these mechanisms have the effect of smoothing generators' revenues over time.

Interconnections between neighbouring markets offer another means of contributing to system adequacy, and electricity markets are, in practice, becoming increasingly integrated. Available generation capacity in one zone can contribute to meet peak

5. This aspect of electricity market design is being addressed by the IEA in the context of its ongoing work on preparing an Electricity Security Action Plan.

demand in another zone and these exchanges are an essential component of integrated electricity markets. Co-ordination is necessary between jurisdictions with and without capacity mechanisms to improve the efficiency of both capacity mechanisms and integrated markets.

While capacity mechanisms are perceived to be effective in maintaining security of electricity supply, it has also been argued that they can be costly to implement and involve heavy-handed regulatory intervention that can favour specific technologies – in particular because the definition of adequate capacity is set administratively by governments, regulators or system operators. The way decisions on required capacity levels are reached can lack transparency. While some countries may adopt capacity mechanisms, other innovations in market design may offer ways to maintain adequate capacity in the presence of a high share of renewables.

The extent of the change in the electricity wholesale price also accounts for the main difference in end-user prices between regions (Figure 6.20). In the European Union, the wholesale price grows by one-third over the projection period, largely due to the rising CO_2 price, which directly accounts for just under half the increase. A shift towards a more capital-intensive generation mix, the need to replace old plants and higher fuel costs are the other main contributory factors. The absolute price increase in the United States is lower than in the European Union, though the percentage increase in the wholesale price is higher, at 50% by 2035. This is partly due to an increase in natural gas in the capacity mix, particularly because of unusually low prices in the baseline year, and other factors, including high volumes of hydroelectric generation in some regions. The growth of the wholesale electricity price in China is more moderate than in the United States or the European Union and starts from a relatively low base. In the period 2011 to 2035 the wholesale electricity price increases by $12 per megawatt-hour (MWh), due to rising fuel and CO_2 prices. Japan had a higher wholesale price in 2010 than any of these three regions, and as the absolute increase to 2035 – $7/MWh – is also smaller, Japan experiences a more modest relative increase than the United States, the European Union and China.

There can be significant variation in wholesale prices within each region. For example, there are several different electricity markets across the United States and Europe, each with different characteristics and sometimes with very different wholesale prices. They can also fluctuate considerably from one year to the next, according to the weather, economic conditions and changes in fuel prices. Other factors can also have an impact on the level of wholesale prices, including market design. The wholesale prices in our projections ensure that new entrants recover their full fixed costs and that all plants recover their variable costs of generation. These are based on smooth fuel price trajectories, resulting in a smooth evolution of electricity prices over time. In reality, prices will inevitably fluctuate around these long-term trends, due to fuel price volatility and cycles of excess capacity followed by relative capacity shortages.

Figure 6.20 ▷ Wholesale electricity prices by region and cost component in the New Policies Scenario

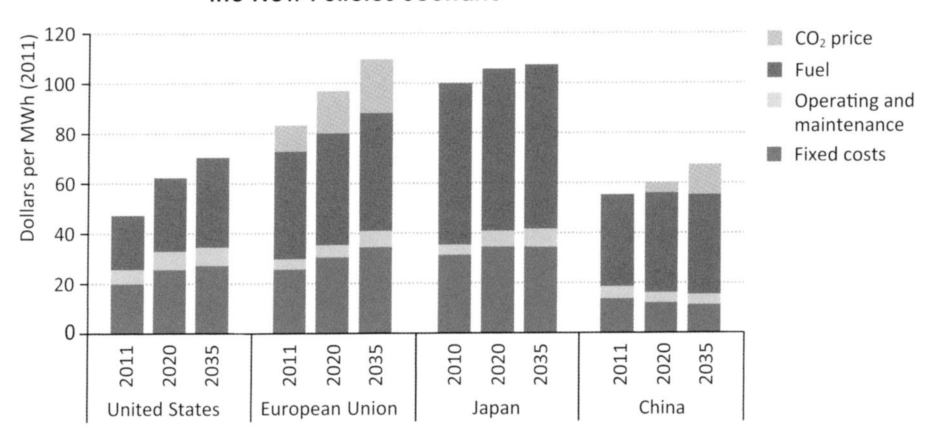

Notes: The shares of each of the four cost components in 2011 are estimated from model results. For Japan, values are for 2010, due to the extraordinarily high prices experienced in 2011.

The presence of variable renewables also has an effect on wholesale prices. The increased deployment of variable renewables in many regions can lower electricity prices at times when generation from variable renewables is high. On the other hand, they can also change the operating patterns of fossil-fuel plants and reduce their capacity factors (see fossil-fuelled generation in this chapter). Lower annual output and increased costs from rapidly adjusting output will increase the costs of production of fossil-fuel plants, which in turn will increase the wholesale price in the long run. As with fuel prices, these two effects vary in size from year to year, depending on weather conditions. The average annual impact of both is taken into account in our projections.

In the New Policies Scenario, subsidies to renewables underpin the growth in electricity generation from non-hydro renewables (see Chapter 7). These subsidies have an effect on the end-user price in a number of regions, depending on the scale of renewables deployment over the projection period and how much of the subsidy is carried by the taxpayer and how much is passed on to the consumer. On average, the impact on household electricity bills of the strong increase in renewables is estimated to be several percentage points on average by 2035. In the United States, the cost of support for renewables grows, from the equivalent of 3% of end-use prices for households in 2011 to 5% in 2035, although under existing policies, the element of the subsidy coming from taxpayers in the form of tax credits is not passed through to the end-user price. In the European Union, rapid growth in renewables means that by 2020, renewables subsidies add 15% on average to the residential end-user price, up from 12% in 2011 (Figure 6.21). Over the course of the 2020s, the additional contribution starts to decline as some technologies, such as onshore wind, become fully cost-competitive and the support for some of the more costly technologies installed around 2010 lapses around 2030. As a result, the additional contribution of renewables subsidies to end-user prices in the European Union falls to 6% by 2035. In Japan, the increase in residential electricity prices due to renewables subsidies is about 4% by the end of the projection period. In China, it is assumed that renewables will continue to be funded by the state, rather than by consumers. Therefore, although there is an economic cost associated with renewable energy, there is no impact on end-user prices. If the cost were to be passed on to consumers it would add the equivalent of 7% to electricity prices in 2035.

Figure 6.21 ▷ Average household electricity prices by region and cost component in the New Policies Scenario

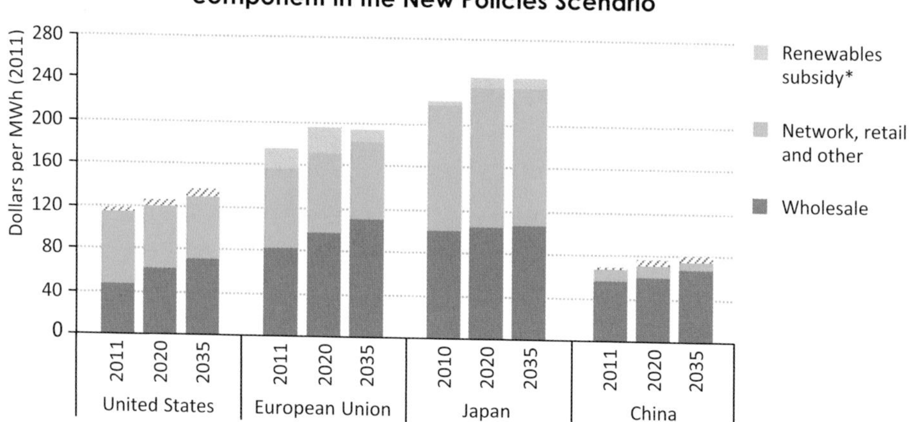

*Hatched areas represent subsidies that are partly or fully borne by taxpayers rather than consumers.
Note: For Japan, base values are for 2010, due to the extraordinarily high prices experienced in 2011.

Similar to the effects on wholesale prices, the impact of renewables subsidies on end-user prices is expected to vary significantly within regions, as well as between regions. This is due to differences in the types of renewable technologies used in different countries or states, differing shares of renewables in the energy mix and differences in policy design.

In addition to the direct costs of subsidies, renewables also give rise to other costs that may be passed on to end-users, including additional generation capacity needed to maintain system adequacy, additional system operation costs and additional network costs (see Chapter 7). Although these system integration costs are not negligible, they are generally smaller than the cost of renewables subsidies. They are included in our wholesale price projections and network tariff projections.

The relationship between renewables subsidies and wholesale prices is complex. Lower wholesale prices negatively impact renewables by delaying their competitiveness and by increasing the level of subsidy required – the difference between the cost of supply from the renewables that are being subsidised and the price achieved for their output in the wholesale market. However, lower wholesale prices can also positively impact renewables, by making subsidies more affordable, as the higher subsidies required are more than offset by lower overall prices. This can be illustrated by an example from the United States. In 2020, with the relatively low assumed gas price of $5.4 per million British thermal units (MBtu) in the New Policies Scenario, the United States wholesale price is $62/MWh. This leads to an average level of the subsidy to renewables equivalent to an additional $6/MWh on residential electricity tariffs (if fully carried by residential consumers), which are about $128/MWh. If the gas price were $2/MBtu higher, the wholesale price would be higher (by $11/MWh) and the cost of supporting renewables would be lower (equivalent to $1/MWh on residential electricity tariffs), with the average household electricity price at $138/MWh.

Figure 6.22 ▷ Total cost per capita of residential sector electricity in selected countries, and share of household income spent on electricity

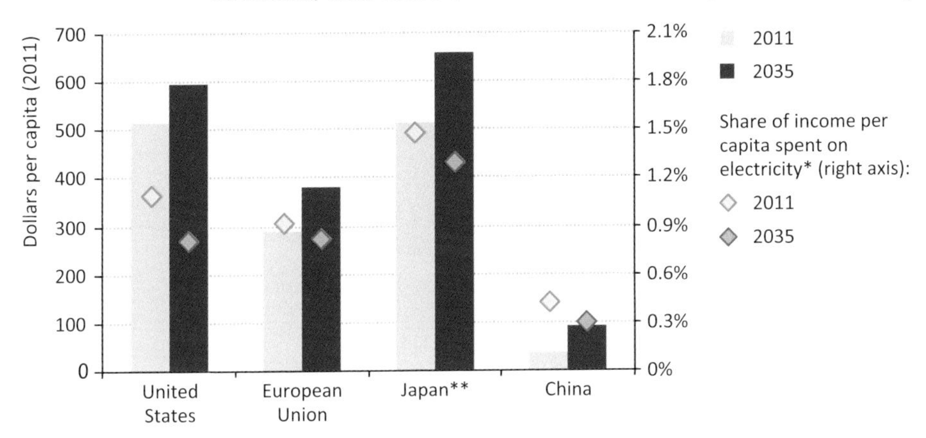

* Share calculated as residential electricity price per capita divided by GDP per capita ($2011, purchasing power parity). ** For Japan, base values are for 2010, due to the extraordinarily high prices experienced in 2011.

The combination of higher prices and, in most regions, higher per-capita consumption is set to push up household electricity bills. In the New Policies Scenario, per-capita household spending on electricity rises by 31% in the European Union, 27% in Japan, and 15% in the United States between 2011 and 2035. In all three cases, per-capita incomes increase more quickly, resulting in the share of income spent on electricity falling over the period (Figure 6.22). In China, household electricity use per capita is more than twice as big in 2035 as in 2011. With the electricity price also increasing, the amount of spending per capita on electricity also more than doubles, though the strong growth of China's economy over the same period means that, even with such a large growth in electricity consumption, the share of national income spent on electricity falls here as well.

Renewable energy outlook

A shining future?

Highlights

- Renewables make up an increasing share of primary energy use in all scenarios in *WEO-2012*, thanks to government support, falling costs, CO_2 pricing in some regions, and rising fossil fuel prices in the longer term. In the New Policies Scenario, electricity generation from renewables nearly triples from 2010 to 2035, reaching 31% of total generation. In 2035, hydropower provides half of renewables-based generation, wind almost one-quarter and solar photovoltaics (PV) 7.5% (even though solar PV generation increases 26-fold from 2010-2035).

- Biofuels use more than triples in the New Policies Scenario, from 1.3 million barrels of oil equivalent per day (mboe/d) in 2010 to 4.5 mboe/d in 2035, driven primarily by blending mandates. Ethanol remains the dominant biofuel, with supply rising from 1 mboe/d in 2010 to 3.4 mboe/d in 2035. Biofuels meet 37% of road transport demand in 2035 in Brazil, 19% in the United States and 16% in the European Union.

- Our assessment indicates that global bioenergy resources are more than sufficient to meet projected demand without competing with food production, although the land-use implications will have to be managed in a sustainable manner. As policy goals exceed the production capacity in some regions, international trade of solid biomass for power generation and biofuels for transport increases about six-fold, driven by imports to the European Union, Japan and India.

- Investment in renewables of $6.4 trillion is required over the period 2012-2035. The power sector accounts for 94% of the total – including wind ($2.1 trillion), hydro ($1.5 trillion) and solar PV ($1.3 trillion) – with the remainder in biofuels. Investment in OECD countries accounts for 48% of the total, focusing mainly on wind and solar PV, while in non-OECD countries most investment is in hydro and wind.

- Renewable energy subsidies jumped to $88 billion in 2011, 24% higher than in 2010, and need to rise to almost $240 billion in 2035 to achieve the trends projected in the New Policies Scenario. Cumulative support to renewables for power generation amounts to $3.5 trillion, of which over one-quarter is already locked-in by commitments to existing capacity (and about 70% is set to be locked-in by 2020). While vital to growth of the industry, subsidies for new renewables capacity need to be reduced as costs fall to avoid them becoming an excessive burden on governments and end-users.

- The deployment of renewables in the New Policies Scenario reduces CO_2 emissions by over 4.1 Gt in 2035, contributes to the diversity of the energy mix, lowers oil and gas import bills, cuts local air pollution and, in most cases, reduces stress on water resources.

Recent developments

The use of renewable energy, including traditional biomass,[1] was 1 684 million tonnes of oil equivalent (Mtoe) in 2010, accounting for 13% of global primary energy demand. This share has remained steady since 2000, but with changing contributions of the different renewable sources. The share of traditional biomass out of total renewable energy fell from 50% in 2000 to 45% in 2010, while biofuels (transport fuels produced from biomass feedstocks) met a growing share of transportation fuel needs. Hydropower, the largest source of renewables-based electricity, remained stable. Electricity generation from wind grew by 27% and solar photovoltaics (PV) by 42% per year on average during this period. The renewables sector has not been immune to the recent global economic crisis, but weaker performance in some regions, for example, in parts of Europe and the United States, was largely offset by strong growth in the rest of the world, notably Asia.

Government policies have been essential to recent growth in renewable energy, especially in the power sector. Environmental concerns have been a key policy driver, targeting emissions reductions of carbon dioxide (CO_2) and local pollutants. Renewables have also been supported to stimulate economies, enhance energy security and diversify energy supply. The main focus has been on the electricity sector, followed by biofuels. In most cases, subsidies have been required as renewables are still more expensive than conventional energy sources. Driven by policies, more than 70 countries are expected to deploy renewable energy technologies in the power sector by 2017 (IEA, 2012).

The United Nations recently launched its Sustainable Energy for All initiative, which calls for a global target of doubling the share of renewable energy by 2030 (along with targets to ensure universal access to modern energy and to double the rate of energy efficiency improvements); the IEA is working with the United Nations to define the baseline for this target and monitor progress. The most significant shift in renewables policy in the last year occurred in Japan, where support has been sharply increased to promote additional renewables capacity and generation to compensate for lower nuclear power output. Recent developments in Japan and other major countries are summarised below.[2]

In 2009, the European Union released the Renewable Energy Directive, which set legally binding targets for the share of renewable energy (covering electricity, heat and biofuels) in gross final energy consumption of each member state by 2020, equating to 20% in total. To ensure that their targets are met, each country is required to prepare an action plan and provide regular progress reports. Renewable energy is expected to continue to be central to EU energy policy beyond 2020. A recent European Commission report indicated that renewable energy could meet 55-75% of final energy consumption by 2050, compared with less than 10% in 2010 (EC, 2011; EU, 2011).

1. Traditional biomass comprises wood, charcoal, crop residues and animal dung mainly used mainly for heating and cooking.

2. Information about other countries can be found in IEA (2012) and in the IEA's Policies and Measures Database: *www.iea.org/policiesandmeasures/renewableenergy/*.

In the United States, both federal and state-level policies push the continued deployment of renewables. Renewable portfolio standards – regulations requiring a specified share of electricity sales from renewables or a minimum amount of renewables capacity – now exist in 29 states and the District of Columbia.[3] The Clean Energy Standard Act of 2012, currently being considered by the US Congress, would set the first nation-wide targets for clean electricity, defined as that produced from renewables, nuclear power and gas-fired generation. The Renewable Fuel Standard, adopted in 2005 and extended in 2007, mandates 36 billion gallons (136 billion litres) of biofuels to be blended into transportation fuel by 2022. The US Environmental Protection Agency (EPA) gave approval in June 2012 for retailers to sell fuel blends containing 15% ethanol (E15), compared with 10% (E10) commonly sold today, creating an opportunity for greater biofuels use. The United States also provides tax incentives (credits, rebates and exemptions), grants and loans to support the growth of renewables. Production tax credits for wind power, biodiesel and advanced biofuels are set to expire at the end of 2012, though extensions are under debate. Solar tax credits are set to expire in 2016. The future of these tax credits remains uncertain. Ethanol import tariffs were removed in late 2011, increasing competition with domestic biofuels.

Japan's renewable energy policy was reviewed and extended through legislation passed in 2009 and a revised Basic Energy Plan in 2010. Following Fukushima Daiichi, Japan released the Innovative Strategy for Energy and the Environment in September 2012, which includes the goal of reducing the role of nuclear power. This would be compensated in part by increasing the deployment of renewable energy. By 2030, the strategy calls for power generation from renewables to triple compared to 2010, reaching about 30% of total generation. In July 2012, Japan launched a new feed-in tariffs system for wind and solar power and other renewables, creating incentives which are among the most generous in the world. Other subsidy mechanisms include investment grants, loans and tax reductions.

In Australia, renewable energy policy focuses mainly on electricity and heat. In 2010, the government set the target of adding 45 terawatt-hours (TWh) of renewables-based electricity and heat by 2020. The target is expected to be met through the issue of renewable energy certificates (tradable certificates of proof that electricity has been generated from an eligible source). In addition, a number of states offer feed-in tariffs for solar PV. Biofuels blending mandates exist in New South Wales and Queensland.

China's renewables policy is laid down in the Renewable Energy Law, passed in 2005 and subsequent amendments. In 2009, China set a target to increase the share of non-fossil energy (nuclear and renewables) in the power sector to 15% by 2020. The 12th Five-Year Plan, covering the period 2011-2015, calls for 70 gigawatts (GW) of additional wind capacity, 120 GW of additional hydropower and 5 GW of additional solar capacity by 2015. An update to the 12th Five-Year Plan, released in July 2012, calls for wind and solar capacity to reach 200 GW and 50 GW respectively by 2020. Targets have been set for the first time for geothermal and marine power. Nine provinces have ethanol-blending mandates of 10%,

3. As of July 2012. Details about incentives at state levels are available at: *www.dsireusa.org*.

supported through production-tax incentives, though incentives for grain-based ethanol were substantially reduced in 2012. A nation-wide diesel fuel blend standard, mandating 5% biodiesel, has been in place since 2011, but on a voluntary basis. It is applied only in Hainan province (USDA, 2011).

In India, the Jawaharlal Nehru National Solar Mission, launched in 2010, is a major policy initiative targeting 20 GW of grid-connected solar power by 2022. The plan also covers off-grid solar power, with special attention on rural electrification, solar lighting and heat (solar water heaters). The main subsidy mechanisms for solar and wind power are feed-in tariffs. The successful expansion of large hydropower, provided for in the country's five-year plan, is uncertain due in part to re-settlement issues. Provisions for small hydro indicate that it will play an increasing role. Non-binding biofuel-blending targets were introduced in 2009, starting at 5% for ethanol and reaching 20% for ethanol and biodiesel by 2017. Current blending targets of 5% are not being met due to high ethanol prices arising from competition for supply from the chemicals and distillery industry.

Brazil relies on capacity tenders to increase renewables-based electricity generation. The ten-year plan for energy expansion through 2020 aims for renewables to account almost 80% of total installed capacity in 2020. This target is expected to be met mainly by hydropower, but also with wind power and biomass. The development of large hydropower projects continues through different programmes. Biofuels is the other major focus of Brazilian renewable energy policy. Mandatory minimum blending levels for ethanol in gasoline were revised down in 2011, from 25% to 20%, because of a reduced sugarcane harvest and record sugar prices in that year. Brazil has had a 5% biodiesel blending mandate since 2010.

Outlook for renewable energy by scenario

The use of renewable energy increases considerably from the 2010 level (1 684 Mtoe) in all scenarios over the *Outlook* period. By 2035, it reaches 3 079 Mtoe in the New Policies Scenario, 2 702 Mtoe in the Current Policies Scenario, and 3 925 Mtoe in the 450 Scenario (Table 7.1). This growth is entirely due to additional supply of modern renewables (all renewables including hydro, except traditional biomass).

Traditional biomass at 751 Mtoe in 2010 falls to 687 Mtoe in the New Policies Scenario, about 650 Mtoe in the 450 Scenario and just under 700 Mtoe in the Current Policies Scenario over the projection period. Despite greater access to modern fuels, many people in non-OECD countries, particularly in sub-Saharan Africa, continue to rely heavily on traditional biomass, essentially for cooking (see Chapter 18). Traditional biomass represents 42% of total primary energy demand in that region in 2035. Its use declines in India and China, as both countries shift towards modern fuels.

In the New Policies Scenario, our central scenario in this *Outlook*, primary energy demand for modern renewables increases from 933 Mtoe in 2010 to 1 459 Mtoe in 2020 and 2 392 Mtoe in 2035, with significant increases across all regions and sectors. Demand increases substantially in the European Union and China, and by 2035, both regions account for 16% of the world's modern renewable energy use. The United States makes up 14% of the total in 2035, boosted by policies to support electricity at the state level and by nation-wide policies mandating large increases in the use of biofuels. In India, demand for modern renewables more than triples between 2010 and 2035.

Table 7.1 ▷ Total primary demand for renewable energy by region and scenario (Mtoe)

			New Policies		Current Policies		450 Scenario	
	1990	2010	2035	2010-35*	2035	2010-35*	2035	2010-35*
OECD	277	443	1 005	3.3%	861	2.7%	1 393	4.7%
Americas	153	199	461	3.4%	402	2.9%	686	5.1%
United States	100	131	338	3.9%	298	3.3%	522	5.7%
Europe	98	208	423	2.9%	373	2.4%	533	3.8%
Asia Oceania	26	36	121	5.0%	86	3.6%	173	6.5%
Japan	15	18	63	5.2%	39	3.2%	89	6.7%
Non-OECD	847	1 241	2 073	2.1%	1 840	1.6%	2 500	2.8%
E. Europe/Eurasia	40	47	103	3.2%	84	2.3%	165	5.2%
Russia	26	22	53	3.6%	41	2.6%	101	6.3%
Asia	497	676	1 133	2.1%	955	1.4%	1 412	3.0%
China	211	284	483	2.1%	401	1.4%	629	3.2%
India	140	182	287	1.8%	247	1.2%	335	2.5%
Middle East	2	2	33	11.5%	19	9.1%	68	14.8%
Africa	196	339	483	1.4%	478	1.4%	500	1.6%
Latin America	112	177	322	2.4%	305	2.2%	355	2.8%
Brazil	66	117	210	2.4%	200	2.2%	230	2.8%
World	1 124	1 684	3 079	2.4%	2 702	1.9%	3 925	3.4%
European Union	74	184	384	3.0%	338	2.5%	481	3.9%

* Compound average annual growth rate.

Note: Includes traditional biomass.

In the New Policies Scenario, global electricity generation from renewable energy sources grows 2.7 times between 2010 and 2035 (Table 7.2). Consumption of biofuels more than triples over the same period to reach 4.5 million barrels of oil equivalent per day (mboe/d) (expressed in energy-equivalent volumes of gasoline and diesel), up from 1.3 mboe/d in 2010. Almost all biofuels are used in road transport, but the consumption of aviation biofuels make inroads towards 2035. The use of modern renewables to produce heat almost doubles, from 337 Mtoe in 2010 to 604 Mtoe in 2035. This heat is used mainly by industry (where biomass is used to produce steam, in co-generation and in steel production) but also by households (where biomass, solar and geothermal energy are used primarily for space and water heating). These overall trends are much more pronounced in the 450 Scenario: renewables-based electricity generation supplies almost half the world's electricity in 2035, while biofuel use grows to 8.2 mboe/d – equivalent to 14% of total transport fuel demand.

Table 7.2 ▷ World renewable energy use by type and scenario

		New Policies		Current Policies		450 Scenario	
	2010	2020	2035	2020	2035	2020	2035
Traditional biomass (Mtoe)	**751**	**761**	**687**	**764**	**697**	**748**	**653**
Share of total biomass	*59%*	*50%*	*37%*	*51%*	*40%*	*48%*	*29%*
Electricity generation (TWh)	**4 206**	**6 999**	**11 342**	**6 648**	**9 627**	**7 443**	**15 293**
Bioenergy	331	696	1 487	668	1 212	750	2 033
Hydro	3 431	4 513	5 677	4 390	5 350	4 658	6 263
Wind	342	1 272	2 681	1 148	2 151	1 442	4 281
Geothermal	68	131	315	118	217	150	449
Solar PV	32	332	846	282	524	376	1 371
Concentrating solar power	2	50	278	39	141	61	815
Marine	1	5	57	3	32	6	82
Share of total generation	*20%*	*25%*	*31%*	*23%*	*24%*	*28%*	*48%*
Heat demand (Mtoe)	**337**	**447**	**604**	**429**	**537**	**461**	**715**
Industry	207	263	324	258	308	263	345
Buildings* and agriculture	131	184	280	170	229	198	370
Share of total production	*10%*	*12%*	*14%*	*11%*	*12%*	*13%*	*19%*
Biofuels** (mboe/d)	**1.3**	**2.4**	**4.5**	**2.1**	**3.7**	**2.8**	**8.2**
Road transport	1.3	2.4	4.4	2.1	3.6	2.8	6.8
Aviation	-	-	0.1	-	0.1	-	0.8
Other***	-	-	0.0	-	0.0	-	0.6
Share of total transport	*2%*	*4%*	*6%*	*4%*	*5%*	*5%*	*14%*

* Excludes traditional biomass. ** Expressed in energy-equivalent volumes of gasoline and diesel. *** Other includes international bunkers.

In all scenarios, the share of renewables in electricity generation is higher than in heat production or road transport throughout the *Outlook* period (Figure 7.1). In the New Policies Scenario, renewables collectively become the world's second-largest source of electricity generation by 2015 (roughly half that of coal) and by 2035 they approach coal as the primary source of electricity. Between 2010 and 2020, generation from renewables grows 5.2% per year, compared with 3.9% per year between 2000 and 2010.

Figure 7.1 ▷ **Share of renewables by category and scenario**

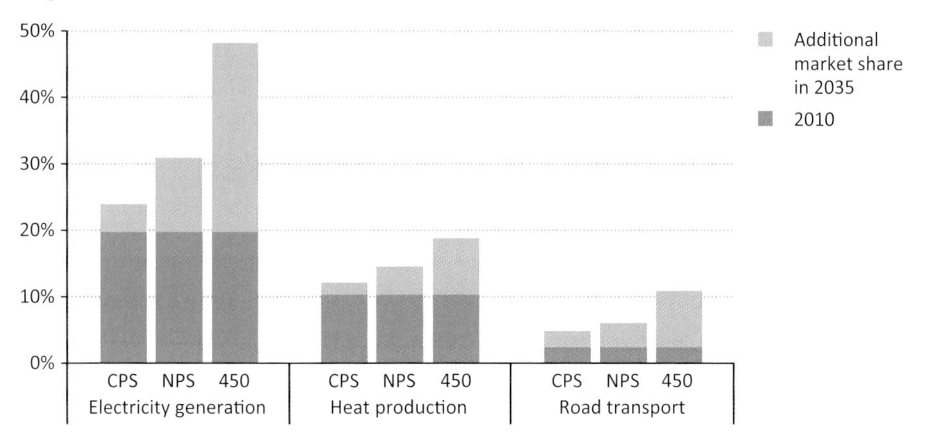

Note: CPS = Current Policies Scenario; NPS = New Policies Scenario; 450 = 450 Scenario.

The global installed capacity of renewable energy sources for electricity production increases from 1 465 GW in 2011 to 3 770 GW in 2035. A total of just over 3 000 GW of renewables capacity, including replacement of older installations, is built from 2012 to 2035 – more than half of total gross capacity additions in the power sector. Replacement for retiring assets amount to 700 GW of capacity over the *Outlook* period, of which 55% is wind, 18% is solar PV, and 15% is hydro. By the end of the *Outlook* period, renewable energy capacity additions exceed 170 GW per year (Figure 7.2).

Figure 7.2 ▷ **World average annual renewables-based capacity additions by type in the New Policies Scenario**

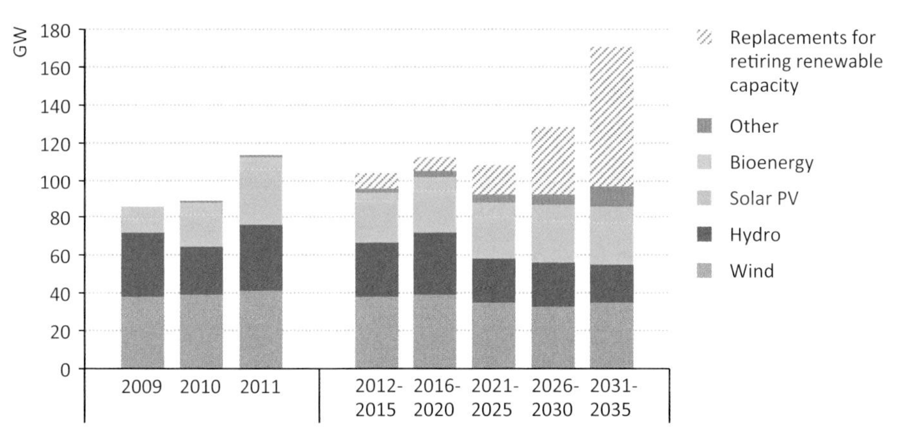

Notes: Net additions plus replacement of retired capacity equals gross capacity additions. Other includes geothermal, concentrating solar power and marine energy.

In 2010, total renewables-based electricity generation was higher in non-OECD countries than in the OECD (Table 7.3). In the New Policies Scenario, total output grows most rapidly in non-OECD countries, though the share of renewables in total generation reaches a higher level in the OECD by the end of the *Outlook* period. In 2035, half of the world's renewables-based electricity comes from hydropower, almost a quarter from wind, 13% from bioenergy and 7.5% from solar PV. Renewables account for 47% of global incremental generation between 2010 and 2035. In the OECD, wind power accounts for about half of the total incremental generation, with the remainder from other renewables, as generation from fossil fuels declines. This leads to a significant change in the electricity mix in the OECD: by 2035, renewables provide one-third of total generation. In non-OECD countries, just over a third of incremental electricity generation is from renewables, taking the total share of renewables generation to 30%. In these countries, hydropower maintains a dominant position, though its share of total generation falls from 20% in 2010 to 17% in 2035, as solar PV and wind power see strong growth.

Table 7.3 ▷ Renewables-based electricity generation by region in the New Policies Scenario (TWh)

	Renewable electricity generation							Share of total generation	
	1990	2010	2015	2020	2025	2030	2035	2010	2035
OECD	1 339	1 960	2 493	2 963	3 444	3 936	4 436	18%	33%
Americas	718	896	1 105	1 297	1 504	1 724	1 953	17%	29%
United States	379	454	600	750	909	1 074	1 238	10%	23%
Europe	472	887	1 138	1 351	1 545	1 734	1 937	24%	44%
Asia Oceania	149	177	250	315	396	477	546	9%	24%
Japan	102	116	161	199	247	292	325	10%	27%
Non-OECD	977	2 245	3 038	4 037	4 904	5 851	6 906	21%	30%
E. Europe/Eurasia	266	309	315	347	391	446	516	18%	22%
Russia	166	170	176	195	224	260	305	16%	21%
Asia	281	1 090	1 688	2 445	3 039	3 663	4 320	17%	27%
China	127	779	1 223	1 789	2 112	2 400	2 689	18%	27%
India	72	136	213	318	466	644	826	14%	25%
Middle East	12	18	28	46	72	119	208	2%	12%
Africa	57	110	141	198	275	374	495	17%	36%
Latin America	361	718	866	1 000	1 127	1 248	1 367	67%	73%
Brazil	211	437	514	585	646	701	754	85%	79%
World	2 316	4 206	5 531	6 999	8 348	9 786	11 342	20%	31%
European Union	310	687	922	1 113	1 285	1 450	1 626	21%	43%

Globally, the production of heat from modern renewables continues to be dominated by bioenergy throughout the projection period. Incentives and obligations for renewables in electricity generation increase the use of bioenergy in combined heat and power production, particularly in OECD countries. Global bioenergy use, excluding traditional biomass, for heat production grows from 294 Mtoe in 2010 to 480 Mtoe in 2035. Solar heat, mainly used in buildings, grows at 5.5% per year from 19 Mtoe to 73 Mtoe over 2010-2035. The largest share of the growth is in China, followed by the European Union and the United States. Geothermal heat, also used mainly in buildings, grows at 7.8% per year from 3 Mtoe in 2010 to 19 Mtoe in 2035.

Outlook by type in the New Policies Scenario

Focus on bioenergy

Demand

Global primary energy demand for bioenergy,[4] excluding traditional biomass, more than doubles from 526 Mtoe in 2010 to nearly 1 200 Mtoe by 2035, growing at an average rate of 3.3% per year. The industrial sector is the largest consumer of bioenergy in 2010 at 196 Mtoe, increasing to over 300 Mtoe in 2035. However, the power sector accounts for a larger share of bioenergy consumption in 2035 (Figure 7.3). Together, these two sectors account for about two-thirds of the additional consumption of bioenergy. Bioenergy consumption to produce biofuels increases by 250% from 2010-2035, reaching almost 210 Mtoe by 2035. The use of traditional biomass declines over time as access to modern fuels increases around the world (see Chapter 18).

Figure 7.3 ▷ World bioenergy use by sector and use of traditional biomass in the New Policies Scenario, 2010 and 2035

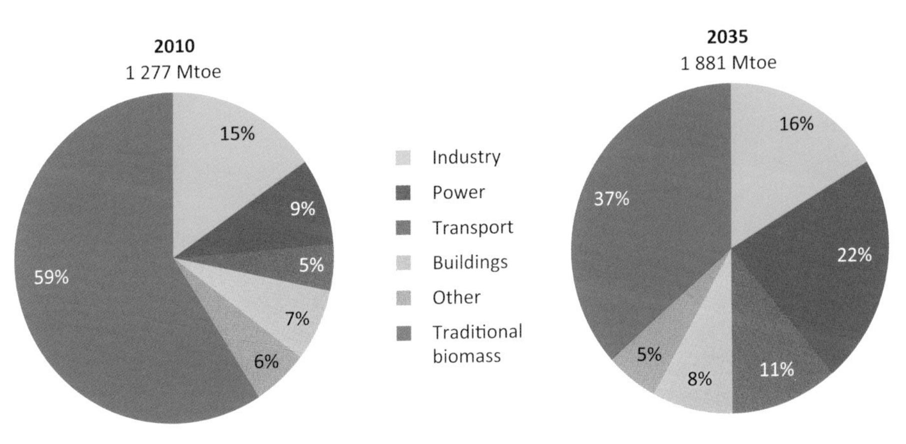

Excluding demand for traditional biomass, primary energy demand for bioenergy is largest in the European Union, rising from 130 Mtoe in 2010 to about 230 Mtoe by 2035, with industrial and residential heat accounting for nearly half of this demand. The United States has the second-highest demand, reaching about 210 Mtoe by 2035, driven mainly by

4. The term bioenergy refers to the energy content in solid, liquid and gaseous products derived from biomass feedstocks and biogas. This includes biofuels for transport and products (*e.g.* wood chips, pellets, black liquor) to produce electricity and heat. Municipal solid waste and industrial waste are also included. Refer to Annex C for further descriptions.

increases in the transport and power sectors. Brazil is also projected to have a thriving biofuel industry over the projection period, bioenergy primary demand reaching about 140 Mtoe in 2035. Bioenergy demand, excluding traditional biomass, in both China and India reaches about half that of the European Union by 2035.

From 2000 to 2010, global electricity generation from bioenergy grew by 6.9% per year, with larger increases in the OECD than in non-OECD countries. The increase in absolute terms was about half that of wind power but more than five times that of solar PV. By 2010, generation from bioenergy reached 331 TWh globally, accounting for over 40% of global non-hydro renewables generation. In the New Policies Scenario, bioenergy generation soars to 1 487 TWh in 2035. Currently, the European Union, United States, Brazil and Japan generate the most electricity from bioenergy. Over the projection period, China surges well ahead of all other regions, generating 325 TWh by 2035. The United States and the European Union generate 259 TWh and 272 TWh from bioenergy respectively in 2035, with India the only other region generating more than 100 TWh.

Bioenergy is also used to produce heat in combined heat and power facilities (such as in industrial co-generation) or stand-alone boilers (in the industrial, residential and services sectors). Global bioenergy consumption for heat in the final consumption sectors grows in the New Policies Scenario from 294 Mtoe in 2010 to about 480 Mtoe in 2035, with industrial demand maintaining its share of about two-thirds of total bioenergy demand for heat throughout the projection period. Opportunities for expanding bioenergy for heat production in non-OECD countries are larger than in the OECD because of their rapid energy demand growth.

The growth of demand for bioenergy for power generation and heat production is driven largely by government policy. In regions that establish a carbon price, some proven bioenergy power generation technologies become competitive with fossil fuel-based power plants during the projection period, particularly combined heat and power, co-firing with coal and waste to energy. Policy interventions, including renewable energy standards and subsidies, contribute to the growth in the demand for bioenergy for power and heat.

In the New Policies Scenario, biofuels increasingly displace oil in transport and start making inroads in aviation over the *Outlook* period. Ethanol continues to be the main biofuel, accounting for about three-quarters of biofuels supply throughout the *Outlook* period, as consumption rises from 1.0 mboe/d in 2010 to 3.4 mboe/d in 2035 (Table 7.4). This increase is due largely by blending mandates for passenger light-duty vehicles (PLDVs). Biodiesel supply increases from 0.3 mboe/d in 2010 to 1.1 mboe/d in 2035. Biodiesel holds potential, particularly in heavy freight transport, where options to replace oil are much more limited than for PLDVs (see Chapter 3); but its use is less widely supported by policy at present and the development of advanced biodiesel is making only slow progress.

The United States remains the largest market for biofuels, with demand rising from 0.6 mboe/d in 2010 to 1.7 mboe/d by 2035. Correspondingly, the share of biofuels in road transport grows from 5% to 19% over the same periods. These increases are driven by both supply- and demand-side policies: a production target of 136 billion litres of biofuel by 2022 and blending mandates contained in the Renewable Fuel Standard.

Table 7.4 ▷ Ethanol and biodiesel consumption by region in the New Policies Scenario (mboe/d)

	Ethanol		Biodiesel		Total biofuels		Share of road transport	
	2010	2035	2010	2035	2010	2035	2010	2035
OECD	**0.6**	**1.7**	**0.2**	**0.8**	**0.8**	**2.5**	**4%**	**13%**
Americas	0.6	1.5	0.0	0.2	0.6	1.7	4%	15%
United States	0.6	1.4	0.0	0.2	0.6	1.7	5%	19%
Europe	0.0	0.2	0.2	0.5	0.2	0.7	4%	13%
Non-OECD	**0.4**	**1.7**	**0.1**	**0.3**	**0.5**	**2.0**	**3%**	**6%**
E. Europe/Eurasia	0.0	0.0	0.0	0.0	0.0	0.1	1%	2%
Asia	0.0	0.7	0.0	0.1	0.1	0.8	1%	4%
China	0.0	0.5	0.0	0.0	0.0	0.5	1%	5%
India	0.0	0.2	0.0	0.0	0.0	0.2	0%	5%
Latin America	0.3	0.9	0.1	0.1	0.4	1.0	12%	22%
Brazil	0.3	0.8	0.0	0.1	0.3	0.9	22%	37%
World	**1.0**	**3.4**	**0.3**	**1.1**	**1.3**	**4.5**	**3%**	**8%**
European Union	0.0	0.2	0.2	0.6	0.2	0.8	4%	16%

Brazil maintains the highest share of renewables in transport in the world through to 2035, reaching about one-third by 2035, following wider adoption of flex-fuel vehicles that can use either gasoline or ethanol. In the European Union, biofuels meet 10% of road transport fuel demand by 2020 (in line with the target set in the renewable directive) and 16% by 2035 – up from 4% in 2010. The United States, European Union and Brazil together accounted for about 90% of global biofuel consumption in 2010, but new markets are expected to emerge over the *Outlook* period, notably China and India, where biofuels meet around 5% of road transport fuel demand in 2035.

Generally, blending rates are expected to increase over time, although not without first overcoming additional challenges. For example, a recent regulation in Germany to move from 5% ethanol blended with gasoline (E5) to 10% (E10) has been met with resistance by consumers, who are concerned about the impact on their car engines. In the United States, the recent decision by the EPA to allow retailers to sell 15% ethanol blended with gasoline (E15) has been met with similar criticism. Across all regions, ethanol remains the main type of biofuel to 2035, although passenger vehicle efficiency improvements – which reduce the overall fuel consumption of the car – have a moderating impact on ethanol demand. This is less of an issue for biodiesel demand, as efficiency improvements in trucks – the main users of biodiesel – are less significant (see Chapter 3).

Growth in biofuels will largely continue to depend on policy support. The limited potential for greenhouse-gas emission savings from some conventional biofuels has raised questions about the benefits of including biofuels in climate mitigation policies, though their potential to reduce oil imports provides another important justification for support in some cases. We assume that the European Union meets its 2020 target by allowing only biofuels that substantially reduce emissions, relative to fossil fuels, to be blended with oil-based fuels. Such restrictions are already in place in the United States, where emissions thresholds have been defined in advance for various types of biofuels.

The pace of development of advanced biofuel technologies also adds uncertainty to biofuels prospects, as targets and accompanying measures are largely absent to date. The United States is currently the only country in the world to have a clear target for the amount of advanced biofuels to be produced. Financial support will be essential to attract investment in technologies to produce advanced biofuels, which are assumed to become commercially available (though not yet competitive with conventional fuels) around 2020 in the New Policies Scenario. By 2035, advanced biofuels make up 18% of total biofuel production. They are mostly used in OECD countries, where they account for 27% of all biofuel use by 2035.

Supply and Trade

Our assessment indicates that global bioenergy resources are more than sufficient to meet projected demand in the New Policies Scenario without competing with food production. In 2035, primary energy demand for bioenergy is nearly 1 900 Mtoe, while we estimate the potential supply to be an order of magnitude higher, similar to other estimates (IPCC, 2011; IEA, 2011a; IIASA, 2012). Potential bioenergy resources are not evenly distributed across regions. Some of the regions with the largest resource potentials are Latin America (especially Brazil), the United States and China. Government policies will be needed to minimise or avoid direct and indirect land use change as a result of expanding biomass feedstock production (Box 7.3).

The global bioenergy supply potential is the aggregate of the supply potentials for several types of feedstocks. Energy crops – those grown specifically for energy purposes, including sugar and starch feedstocks for ethanol (corn, sugarcane and sugar beet), vegetable-oil feedstocks for biodiesel (rapeseed, soybean and oil palm fruit) and lignocellulosic material (switchgrass, poplar and miscanthus) for advanced biofuels – make up the vast majority of this potential.[5] Residues – the leftover materials from harvesting crops and forestry activities, such as corn stover, bagasse from sugarcane and scraps from logging – have the potential to provide over 600 Mtoe of bioenergy, depending largely on the sustainable portion that must remain in the field to replenish soils and maintain future crop yields. Forestry products, grown specifically for energy purposes, contribute somewhat less to the overall bioenergy supply potential.

5. In this assessment, land demand for food crops is given priority and such land is subtracted from total available land before considering use for energy crops.

Box 7.1 ▷ Improvements to the World Energy Model: the bioenergy supply and trade module

A new module was added to the IEA's World Energy Model this year to analyse the supply and trade of bioenergy.[6] It includes 25 regions with detailed representation of bioenergy supply potentials and conversion technology costs for the power sector and biofuel production. In order to meet demand for bioenergy in each sector and region, domestic resources are given priority (after taking account of existing trade) and compete with each other on the basis of conversion costs (including feedstock prices). Regional resources are treated as "bioenergy available for energy purposes", where agricultural demands are met before supplying the energy sector. If domestic bioenergy resources cannot satisfy all demands in a given region, supplementary supplies are obtained on the global market. Regions with available resources beyond food and domestic energy needs supply the global market. The model uses a global trade matrix for ethanol, biodiesel and solid biomass pellets to match unsatisfied demand with available supply on a least-cost basis, including transportation costs. The transition from conventional to advanced biofuels occurs in the model as the economics improve through technological advances and learning, and policies raise demand.

In the New Policies Scenario, all sources of solid biomass supply increase to meet significantly higher demand in the power and transport sectors, however, the shares provided by different feedstocks change over time. Energy crops (sugar-, starch- or oil-based and lignocellulosic crops) provide the largest share of supply until late in the projection period, when residues (forestry and agricultural) provide a slightly larger share (Figure 7.4). This is due to bioenergy demand in the power sector far surpassing demand for biofuels and technology development enabling residues to make inroads as a feedstock for advanced biofuels. By 2035 in the New Policies Scenario, about one-third of the estimated maximum sustainable potential for residues is consumed. The supply of bioenergy from forestry products grows substantially over time, continuing to provide about 10% of solid biomass supply through 2035. In total, solid biomass feedstocks meet about three-quarters of demand for bioenergy in the power and transport sectors throughout the *Outlook*, with the remaining portion met by biogas and waste products.

International trade of biofuels and solid biomass for power generation, usually in the form of pellets, accounted for about 7% of demand in the power and transport sectors in 2010.[7] In the New Policies Scenario, trade expands to over 10% of supply, as policy goals exceed some regions' capacity to meet demand with domestic resources, particularly in the European Union and India (Figure 7.5). An increasing share of supply for bioenergy in the

6. For more information on the World Energy Model and this module, see *www.worldenergyoutlook.org*.

7. Biomass pellets, a high-density uniform product, can be made from residues and other feedstocks to facilitate transport over long distances and increase performance in certain applications, such as co-firing.

power sector is met from imports, with inter-regional trade increasing from 6 Mtoe in 2010 to about 40 Mtoe in 2035, or about one-tenth of bioenergy supply in the power sector. The European Union and Japan are projected to be the largest pellet importers, with the United States, Canada and Russia expected to be the major pellet exporters. While playing limited roles in the world market, China, the European Union, India and Brazil also produce large amounts of bioenergy for domestic use in the power sector.

Figure 7.4 ▷ Share of solid biomass supply for biofuels and power generation by feedstock in the New Policies Scenario

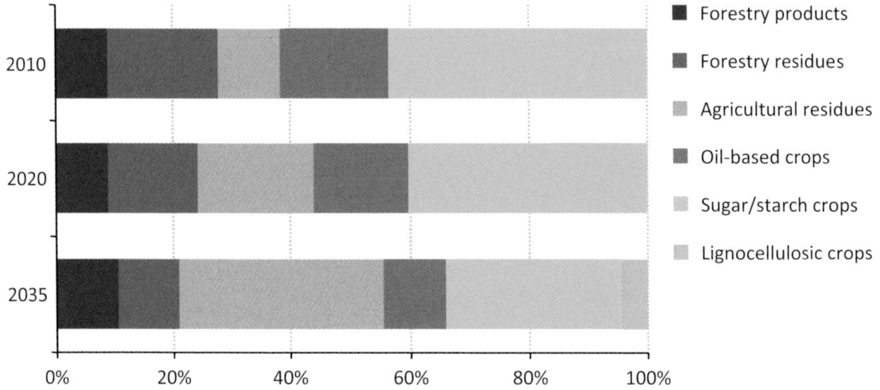

Trade of biofuels expands rapidly through the *Outlook* period, from 0.2 mboe/d in 2011 to about 0.9 mboe/d in 2035, or about one-fifth of total biofuel demand. In the New Policies Scenario, the European Union and India are far and away the largest importers of biofuels through 2035. Brazil is set to be the largest exporter of biofuels (mainly ethanol) in the world, with exports approaching 0.2 mboe/d in 2035. Other Latin American countries combined supply a similar volume of biofuels (with more emphasis on biodiesel) to the world market. Indonesia, the ASEAN region, and other non-OECD developing countries in Asia are also set to become large biofuels exporters. Brazil and the United States continue to be the largest producers of biofuels, though nearly all of US production is consumed domestically by the end of the period. China becomes a major biofuels producer over time as production increases sharply to meet growing domestic demand.

The projected growth in bioenergy trade is in line with recent trends. Solid biomass trade, in the form of wood pellets, increased six-fold between 2000 and 2010, with European Union countries accounting for two-thirds of the trade in 2010 (Lamers, 2012). Ethanol and biodiesel were traded mainly between European Union countries in 2011 (REN21, 2012). Ethanol imports to the European Union from countries outside the region are currently limited, but imports of biodiesel are more important, coming mainly from Latin America and Indonesia. Another emerging trend that is set to continue is the growing supply gap for ethanol in Asia (F.O. Licht, 2012). The United States is a major importer of ethanol from Brazil and other countries in Latin America, though it is also an exporter – net imports represent less than 1% of its total ethanol consumption.

Figure 7.5 ▷ Share of bioenergy demand for biofuels and power generation from domestic production and imports, 2010 and 2035

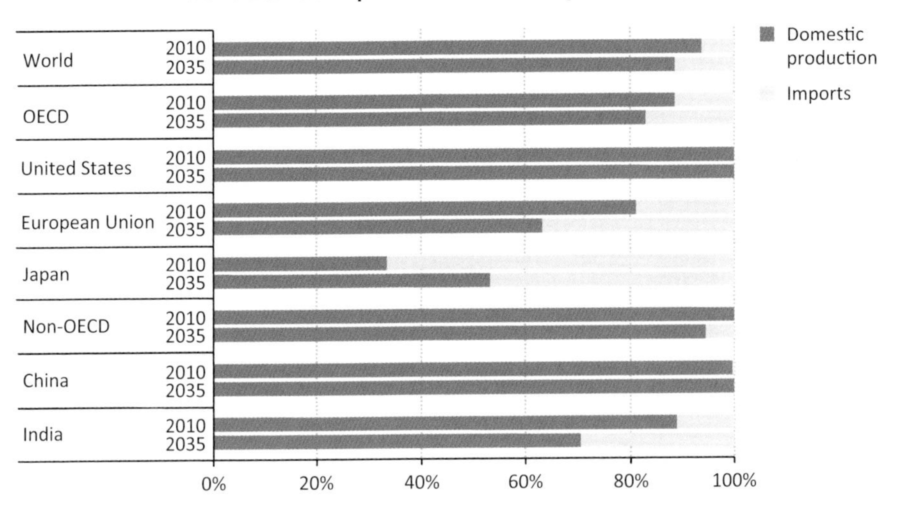

Notes: Trade within *WEO* regions is excluded. Aggregates present averages of the relevant *WEO* regions.

Economic and technical factors tend to limit demand for internationally traded bioenergy. Import and export tariffs, which apply mainly to biofuels, are one factor currently limiting ethanol and biodiesel trade. Lack of infrastructure for handling and transporting biomass, such as processing facilities, port infrastructure or ships, is another important barrier, which could prevent imports and exports increasing fast enough to meet demand. Technical standards need to be developed further, as well as sustainability criteria and certification schemes.

Hydropower

Hydro is currently the largest renewable source for power generation in the world, producing 3 431 TWh and meeting 16% of global electricity needs in 2010. It remains so over the projection period in the New Policies Scenario, with generation reaching 5 677 TWh in 2035, its share in total electricity generation dropping marginally to 15%. Projected growth in hydropower production in OECD countries – where the best resources have already been exploited – is limited. Nearly 90% of the increase in production between 2010 and 2035 is in non-OECD countries, where the remaining potential is higher and electricity demand growth is strongest. Most incremental hydro output is in Asia and Latin America, notably China, India and Brazil.

Global hydropower capacity is projected to increase from 1 067 GW in 2011 to over 1 680 GW in 2035. China's capacity almost doubles, to 420 GW, bringing its total installed hydropower capacity in 2035 close to that of the entire OECD in 2011 (Figure 7.6). Capacity jumps from 42 GW to 115 GW in India, from 89 GW to over 130 GW in Brazil and Africa continues to develop some of its vast hydro potential (Box 7.2).

Figure 7.6 ▷ **Installed hydropower capacity in selected regions in the New Policies Scenario**

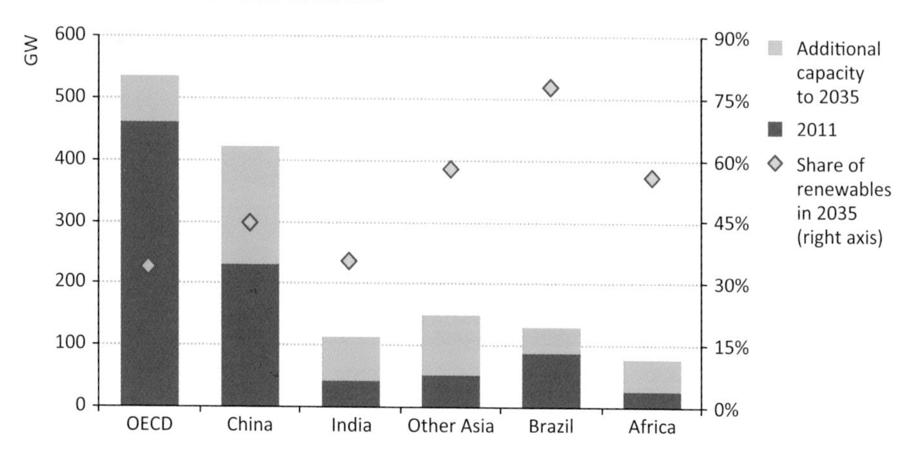

Box 7.2 ▷ **Hydropower prospects in Africa**

In 2010, 27 GW of installed hydropower capacity in Africa generated 105 TWh, supplying 16% of the continent's electricity. However, only a small fraction of Africa's hydropower potential has been developed (UNEP, 2010; IPCC, 2011). The technical potential has been estimated to exceed 1 800 TWh, located largely in the Republic of Congo, Ethiopia and Cameroon (WEC, 2010). In the New Policies Scenario, hydropower capacity rises to almost 80 GW by 2035, including several projects currently under construction, accounting for over 20% of the continent's total electricity generation.

Several challenges threaten the development of hydropower in Africa, particularly the availability of funding. Political and market risks, as well as local environmental considerations are barriers to securing the large initial investments required. However, opportunities for funding are enhanced by several international programmes, including the Clean Development Mechanism under the Kyoto protocol and a recent G20 initiative promoting investment in developing countries, which identified the Grand Inga project on the Congo River as a possible candidate for funding. Africa's energy needs are huge: 590 million of its people still lack access to electricity (see Chapter 18). Hydropower, both large and small scale, is an abundant source of clean energy that can make a major contribution to providing energy for all.

Wind power

Wind power is set to continue to expand rapidly as it becomes more cost-competitive with conventional sources of electricity generation, driven to a large degree by supportive government policies. In the New Policies Scenario, incremental electricity output from wind is greater than that of any other renewable source. Global generation from wind increases dramatically from 342 TWh in 2010 to around 2 680 TWh in 2035, pushing up its share in total electricity generation from 1.6% to 7.3%. Wind achieves the highest level of market penetration in the European Union, where it accounts for almost one-fifth of electricity generated in 2035, compared with less than 5% in 2010. Growth is also strong in the United States, China and India, in each of which wind reaches a share of 6-8% of electricity supply by 2035.

Wind power capacity worldwide increases from 238 GW in 2011 to almost 1 100 GW in 2035 (Table 7.5). Onshore wind makes up four-fifths of this growth. Offshore wind capacity expands rapidly, from 4 GW in 2011 to 175 GW by 2035, its deployment being underpinned by government support. There are still significant uncertainties about the achievement of cost savings on the required scale through deployment. Despite the cost reductions per unit of electricity produced by 2035, offshore wind costs remain well above wholesale electricity prices in most countries. By then, the European Union and China combined account for two-thirds of installed offshore wind capacity. The growing role of wind (and other variable renewables) underlines the importance of upgrading networks and adding flexible capacity into the power mix in order to maintain the overall reliability of supply.

Table 7.5 ▷ Installed onshore and offshore wind power capacity by region in the New Policies Scenario (GW)

	Wind Onshore			Wind Offshore			Total Wind		
	2011	2020	2035	2011	2020	2035	2011	2020	2035
OECD	**150**	**285**	**441**	**4**	**31**	**113**	**154**	**315**	**555**
Americas	53	107	175	-	4	26	53	112	202
United States	47	90	143	-	3	18	47	93	161
Europe	91	161	231	4	24	72	95	184	304
Asia Oceania	6	16	34	0	3	14	6	19	49
Japan	3	8	16	0	2	9	3	9	25
Non-OECD	**84**	**262**	**482**	**0**	**9**	**62**	**85**	**271**	**544**
E. Europe/Eurasia	2	6	16	-	0	3	2	6	19
Asia	79	239	411	0	9	53	79	248	464
China	62	191	280	0	9	46	62	200	326
India	16	44	93	-	-	5	16	44	97
Middle East	0	2	21	-	-	2	0	2	23
Africa	1	4	15	-	-	1	1	4	16
Latin America	2	11	19	-	-	3	2	11	22
World	**234**	**546**	**923**	**4**	**40**	**175**	**238**	**586**	**1 098**
European Union	90	159	218	4	23	70	94	182	288

Solar photovoltaics

Solar PV produced only a small fraction of the world's total electricity in 2010, but installed solar PV capacity has grown rapidly in recent years and is expected to continue to do so in the future. In the New Policies Scenario, electricity generation from solar PV in 2035 is over 26-fold that of 2010, increasing from 32 TWh to 846 TWh. Its share in total generation rises to just over 2% in 2035. Installed solar PV capacity increases from 67 GW in 2011 to just over 600 GW in 2035, thanks to continuing cost reductions and government support (Figure 7.7).

Figure 7.7 ▷ Solar PV gross capacity additions, average unit cost, and resulting investment requirements in the New Policies Scenario

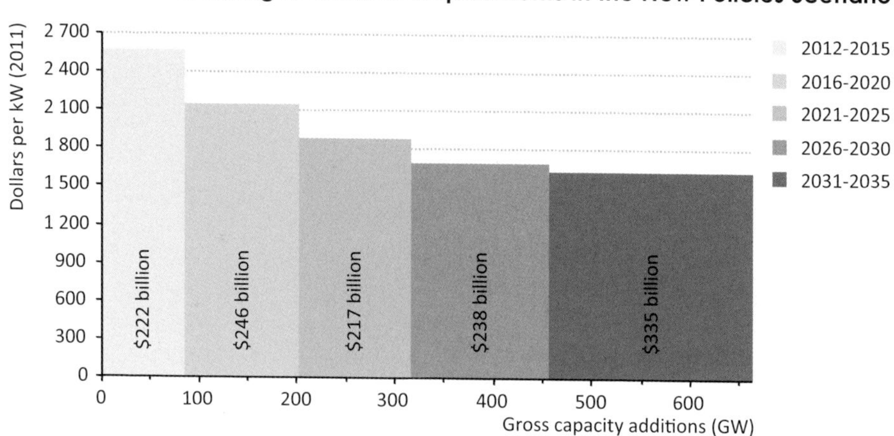

Notes: Gross capacity additions, which include replacements, are shown. In 2035, total global installed capacity of solar photovoltaics reaches some 600 GW. Total investment in each period is indicated vertically on each of the columns (capacity addition times cost per GW). Unit investment costs represent the weighted average of costs throughout all regions, for both large and rooftop installations.

This extremely rapid expansion is in line with recent experience – global solar PV capacity was just 1 GW in 2000. Over the course of 2011, solar PV capacity increased by about 30 GW, a 75% increase. Around 60% of the additions were in Germany and Italy, the world leaders in solar PV, with 25 GW and 13 GW of installed capacity respectively at the end of 2011. The European Union accounts for over three-quarters of global solar PV capacity in 2011. Over the *Outlook* period, EU capacity increases to some 146 GW, accounting for 5% of its electricity generation in 2035 (up from 1% in 2010). In the United States, capacity increases from 4 GW in 2011 to 68 GW in 2035. Other countries with large amounts of solar PV capacity in 2035 are China (113 GW), India (85 GW) and Japan (54 GW).

Investment in solar PV installations has been encouraged in recent years by substantial falls in solar PV costs, which resulted largely from widespread deployment and substantial oversupply (Spotlight). Between the first quarter of 2010 and the first quarter of 2012, solar PV generating costs fell by 44% (Frankfurt School UNEP Collaborating Centre and Bloomberg New Energy Finance, 2012). Solar PV costs continue to fall over the projection period, although at lower rates as the oversupply situation is corrected.

The increase in solar PV installations in European Union countries is thanks largely to feed-in tariffs, which considerably reduce project risk as returns are guaranteed, typically for periods of 10-20 years. These tariffs have been very generous in some cases and were not adjusted quickly enough to reflect the rapidly falling costs of solar PV. As a result, the returns offered were closer to those typically associated with high-risk investments and led to massive investment in solar PV installations. In some countries, governments responded quickly by reducing feed-in tariffs to levels that better reflected costs. As the costs of feed-in tariffs are passed on to consumers in most cases, it is essential to design incentives which attract sufficient investment while yet permitting adjustment of subsidies for new capacity additions as technology costs fall, to avoid unnecessary increases in electricity prices and maintain public acceptance.

SPOTLIGHT

Beyond the solar PV bubble

Solar PV cell manufacturing capacity has grown rapidly in response to booming global demand, initially in OECD countries, where demand first matured, and then in China, which expanded manufacturing capacity massively to support exports. In recent years, manufacturing capacity has expanded much more quickly than actual demand for solar PV panels. By 2011, estimated solar cell production capacity was around 20 GW higher than production, two-thirds higher than the new capacity installed worldwide that year. Since 2008, there has also been a very sharp fall in the cost of purified silicon, a key input for manufacture. Along with cost reductions from technological learning, these two factors have driven down the cost of PV systems sharply (IEA, 2011b).

Installers of solar PV systems and final electricity consumers have benefited greatly from falling solar PV prices, but solar PV manufacturers around the world, and particularly those in the United States and Europe, have suffered large financial losses. A wave of consolidation has been triggered within the industry, with a view to reducing costs and becoming more competitive. Several large companies have already gone bankrupt, such as Germany's Q-Cells – the largest solar cell manufacturer in Europe – in April 2012. Trade tensions have arisen between the United States, Europe and China, resulting in the imposition of import tariffs by the United States in 2012 on solar panels from China.

Difficulties are likely to persist in the short term, while the imbalance between supply and demand endures. How quickly the balance is restored depends largely on the rate of growth of demand for solar PV. China represents a large potential market, but its demand for solar PV in the short term is uncertain. In the New Policies Scenario, the oversupply continues over the short term.

Other renewables for electricity and heat

In the New Policies Scenario, electricity generation from concentrating solar power (CSP) plants soars from 1.6 TWh to about 280 TWh and capacity from 1.3 GW to 72 GW between 2010 and 2035. The majority of new projects in the near term are in the United States and Spain, but there are also developments in a number of other regions in the later years of the *Outlook* period, including North Africa, the European Union, India, Australia and South Africa. In 2035, CSP capacity is highest in China, followed by the Middle East. Most plants currently in operation or under construction are based on parabolic trough technology, the most mature of the CSP technologies. Further technology improvements and significant cost reductions are necessary to make CSP plants competitive on a large scale. The average capacity factor of CSP plants increases over the period, because of increasing use of CSP technologies with thermal storage.

Global geothermal electricity generation increases from 68 TWh to more than 300 TWh and capacity from 11 GW to over 40 GW between 2010 and 2035. Most of these projected increases occur in the United States, Japan and in Asia (Philippines and Indonesia). African countries, particularly North Africa, also increase their use of geothermal for electricity generation.

Electricity generation from marine energy, which includes tidal and wave power, increases from less than 1 TWh to almost 60 TWh between 2010 and 2035, with capacity growing from less than 1 GW to 15 GW. Tidal power is limited to select sites due to economic considerations, requiring a large tidal range and proximity to existing transmission lines to be considered viable. Wave power has notable potential to contribute to meeting electricity demand, but the relevant technologies are still in their infancy, requiring significant improvements to reduce costs.

Solar heat, used mainly in buildings to provide hot water, grows from 19 Mtoe to over 70 Mtoe worldwide between 2010 and 2035. China accounted for 68% of global solar heating capacity in 2011 (REN21, 2012). The country's use of solar heat is projected to expand substantially during the *Outlook* period, at about 30 Mtoe in 2035. US production also increases strongly, from 1.4 Mtoe to about 7 Mtoe. Geothermal heat is used mainly in buildings in the European Union and United States. In the New Policies Scenario, final consumption of geothermal heat increases from 3 Mtoe to almost 20 Mtoe between 2010 and 2035, coming mainly from the European Union, United States, China and Japan.

Costs of renewables

Investment

The projected increase in global renewables-based electricity generation capacity in the New Policies Scenario requires cumulative investment of $6.0 trillion (in year-2011 dollars), with annual investment increasing to over $300 billion by 2035. Renewables account for 62% of total investment in power generation capacity from 2012-2035, reaching almost 70% in 2035. Investment in wind power is higher than for any other source, at $2.1 trillion, representing 35% of total investment in renewables capacity (Figure 7.8). Investment in hydropower totals $1.5 trillion and solar PV $1.3 trillion over the projection period. OECD countries invest more than non-OECD in all sources, except hydropower and CSP. OECD countries invest $2.9 trillion in total renewables capacity, of which almost two-thirds goes to wind power and solar PV, while non-OECD countries invest $3.1 trillion, with over one-third going to hydro.

Figure 7.8 ▷ **Cumulative investment in renewables-based electricity generation by region and type in the New Policies Scenario, 2012-2035**

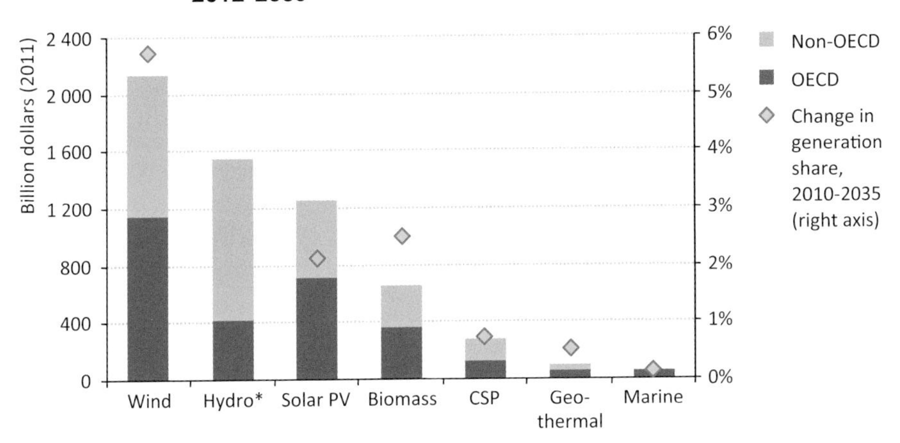

* The share of hydropower in total generation declines by half a percentage point in 2010-2035, starting from a share of 16% in 2010. All other renewable technologies start from a share of less than 2% in 2010.

To accommodate more renewables-based capacity, often in remote locations to capture the best renewable energy sources, additional transmission lines will need to be built and some existing transmission and distribution networks reinforced. Those additional investments are estimated at just below $230 billion from 2012-2035, or 3.2% of the total investment in electricity networks (see Chapter 6). This share is significantly higher for transmission lines, accounting for almost 10%. In the European Union and Japan, where the deployment of renewables continues to grow strongly, this share increases to around one-quarter. Extensive deployment of distributed renewables, such as solar PV in buildings, can reduce future transmission investment needs, though it would require additional investment in the reinforcement of distribution networks to accommodate electricity flowing from residential customers to the grid.

The projected demand for biofuels in the New Policies Scenario calls for a total of around $360 billion to be invested in bio-refineries worldwide. Almost two-thirds of this sum goes to conventional ethanol plants. Just over one-fifth of the total, $78 billion, goes to advanced biofuels (ethanol and biodiesel). Investment in conventional biodiesel amounts to $43 billion (12% of the total) and in aviation fuels around $12 billion (3%). Most biofuels investment is in OECD countries, where much of global demand is concentrated.

Production costs

In the electricity sector, the fall in investment costs (on a per-kilowatt basis) of many renewable energy technologies in recent years has resulted in lower production costs. However, based on these costs, most renewable technologies are not yet competitive with fossil fuel-based technologies. Solar PV generating costs have declined the most of any renewable energy technology over the past two decades, especially in the last few years (Spotlight) and are expected to achieve the largest reduction of any generating technology over the *Outlook* period as well, falling by between 40% and 60% in most regions compared to costs in 2011. However, this is not enough to decrease average solar PV generation costs below the wholesale price in any of the *WEO* regions during the projection period.

The cost of onshore wind power has also declined over time to where it is competitive with fossil fuel-based generation today in a few countries, and close in several others. Its costs are expected to continue to fall in the future, as a result of technological progress and economies of scale. Between 2011 and 2035, the levelised cost of energy generation from wind falls below wholesale prices in the European Union around 2020 and in China in the early 2030s, at which point it becomes competitive in the market without government support.[8] However, wind does not become competitive in the United States, as wholesale prices remain at low levels due to low gas prices and the absence of a CO_2 price.

The electricity production costs for offshore wind, CSP and marine fall throughout the *Outlook* period, but remain well above a competitive level. Those of bioenergy power plants, by contrast, see little reduction over time, as the technologies are already mature and their costs depend greatly on biomass feedstock prices, which are not expected to decrease through 2035.

Figure 7.9 ▷ Indicative biofuels production costs and spot oil prices

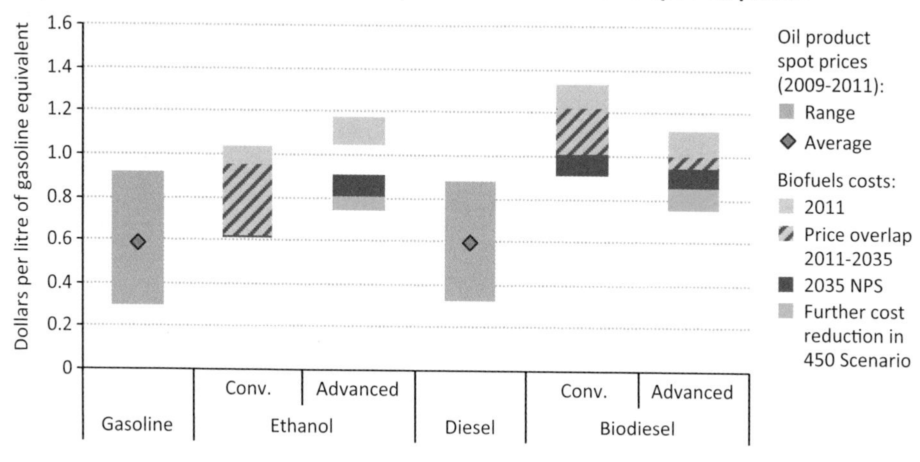

Notes: NPS = New Policies Scenario; Conv. = Conventional. The range of gasoline and diesel spot prices is taken from the monthly average spot price in the United States, Singapore and Rotterdam from 2009 to 2011. Biofuels costs are not adjusted for subsidies; cost variations can be even larger than depicted here, depending on feedstock and region.

8. The quality of available resources can vary widely across a region, resulting in differences in the generation costs of renewables. For example, sites with sub-optimal wind quality characteristics reduce wind turbine capacity factors, increasing the levelised costs of energy generation. Within a region, the best sites may become competitive earlier, with others becoming competitive later.

The costs to produce biofuels are higher than conventional fossil fuels, with a few exceptions, *e.g.* Brazil (Figure 7.9). Biofuel costs vary greatly by region, depending mainly on the feedstock, technology, land characteristics and climatic factors. In Brazil, ethanol derived from sugarcane is often cheaper to produce than gasoline due to favourable soil and climate conditions for the high yield crop. Ethanol produced from corn or sugar beets, such as in the United States and Europe, is generally more expensive. Biodiesel produced from soybeans and rapeseed, the most common feedstocks, currently cost significantly more than conventional diesel. Over the *Outlook* period, while the investment costs for conventional biofuel production processes are expected to fall, biofuel feedstock costs are not expected to decrease significantly, resulting in only small reductions in conventional biofuel production costs by 2035. Advanced biofuel technologies have higher potential cost reductions, decreasing by 10-20% between 2010 and 2035 in the New Policies Scenario. Increasing oil prices over the *Outlook* period help conventional and advanced biofuels to become more competitive.

Subsidies to renewables

To foster the deployment of renewable energy, governments use subsidies to lower the cost of renewables or raise their revenues, helping them compete with fossil fuel technologies. The justification is that imperfections in the market fail to factor in externalities (such as environmental costs attributable to other fuels) or deny nascent technologies the opportunity to mature without support. The ultimate goal is to help renewable energy technologies to achieve sufficient cost reductions to enable them to compete on their own merits with conventional technologies. At that point, any support should, accordingly, cease to be awarded to additional capacity.

Most of the current support mechanisms for renewables apply to electricity produced by capacity installed in a specific year, and for a fixed duration, which is typically 20 years. As cost reductions for renewable technologies are achieved, the level of support provided for new capacity installations needs to decline to avoid excessive and unnecessary increases in the cost of energy services. This structure means that even after the costs for new capacity of a renewable technology become competitive with fossil-fuel technologies, the payments related to the capacity installed in previous years will continue for the fixed duration.

Subsidies to renewables are generally paid to producers. They can be direct or indirect. Direct support includes tax credits for production and investment, price premiums and preferential buy-back rates (or feed-in tariffs). Indirect subsidies arise from mandates, quotas and portfolio standards, which support the uptake of renewables at higher costs to the economy or the consumer. The costs may be met either through government budgets (for example, tax credits) or by end-users collectively.

To ensure sustained deployment of renewables, it is critical to maintain investors' confidence through consistent policies. Repeated expirations of the production tax credit for wind power over the past fifteen years in the United States demonstrate the effects of inconsistent policies, as the wind industry has experienced boom and bust cycles over this period. Governments need to monitor market developments closely, set clear rules for calculating subsidies and make a credible commitment not to enact retrospective changes.

In 2011, renewables excluding large hydro received an estimated $88 billion in subsidies in various forms, up 24% from 2010, of which $64 billion went to electricity and the remainder to biofuels (Figure 7.10). Solar PV received more than any other renewable energy technology for electricity generation ($25 billion), followed by wind ($21 billion) and bioenergy ($15 billion). In the New Policies Scenario, total subsidies to renewables grow to about $185 billion in 2020 and reach almost $240 billion per year by 2035.[9] Support provided to bioenergy for power generation continues to grow over time, reaching $69 billion in 2035, exceeding that received by any other technology. The amount received by solar PV grows rapidly in the medium term, reaching $77 billion in 2027, before falling to $58 billion in 2035, as retired installations are replaced by new, less expensive capacity. Onshore wind power receives more support each year until around 2020, before falling to $14 billion by 2035, as this technology becomes increasingly competitive. Biofuels receive $24 billion in 2011, increasing to $46 billion in 2020 and $59 billion in 2035, with the vast majority going to conventional biofuels in 2035.

Figure 7.10 ▷ Global renewable energy subsidies by source in the New Policies Scenario

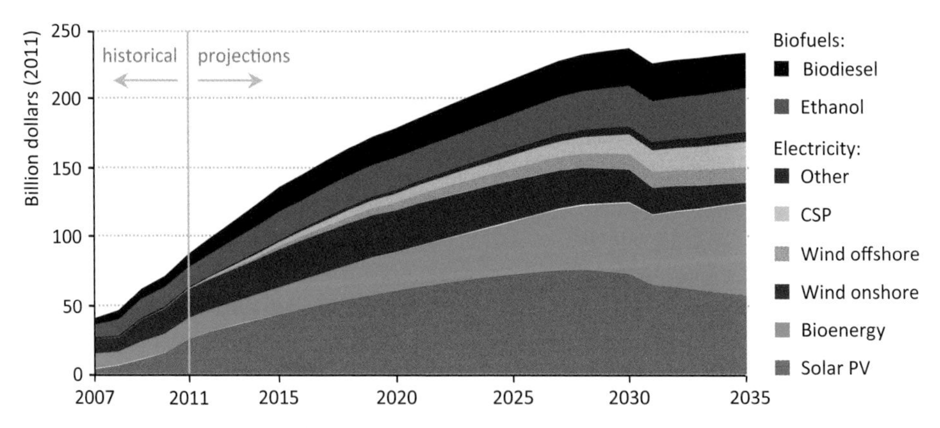

Notes: Other includes geothermal, marine and small hydro. CSP = Concentrating solar power.

While subsidies to renewables in the power sector increase in total, they decline on a per-unit basis as the costs of renewable energy technologies fall and electricity prices increase, mainly due to higher fossil fuel prices and the introduction – in some regions –

9. Projected subsidies to renewables are calculated by taking the difference between the average cost of electricity generated by the renewable energy technology and the regional wholesale electricity price. This level of subsidy is paid for each unit of electricity generated by the installed capacity over its lifetime. For biofuels, they are calculated by multiplying the volumes consumed by the difference of their cost to the reference price of the comparable oil-based products.

of a carbon price. They fall significantly for solar PV through 2035, due to continued cost reductions. However, outside of limited niche applications, solar PV continues to require subsidies through 2035. As onshore wind becomes more competitive, growing amounts of unsubsidised electricity are generated (Figure 7.11). Offshore wind, CSP and marine continue to require support through 2035. Subsidies per unit of biofuel also decline over time, with technological advances lowering the costs while oil prices increase.

Figure 7.11 ▷ Subsidised and unsubsidised renewables-based electricity generation by type in the New Policies Scenario

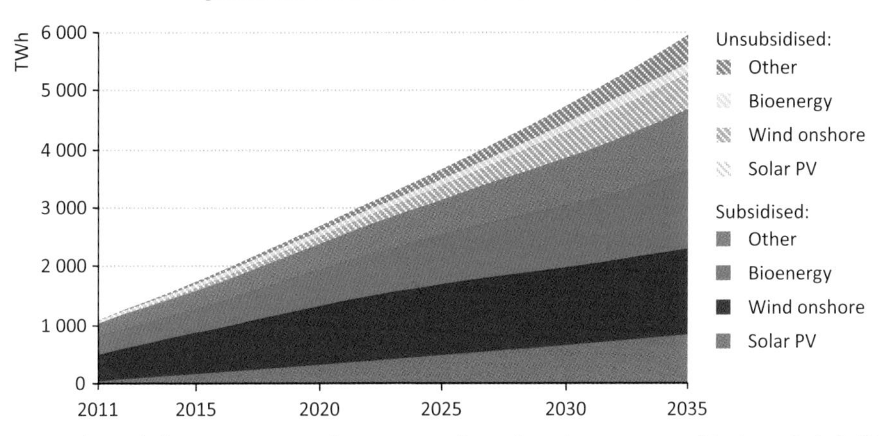

Note: Other includes concentrating solar power, geothermal, marine energy, small hydro and wind offshore.

In 2011, the European Union provided the highest level of total renewable energy support in the world, almost $50 billion, followed by the United States at $21 billion (Figure 7.12). Subsidies to biofuels were also the highest in the European Union, at $11 billion, the bulk of them going to biodiesel. In the United States, $8 billion in 2011 went to biofuels, mainly targeting ethanol.

Figure 7.12 ▷ Global subsidies to renewables-based electricity generation and biofuels by region in the New Policies Scenario

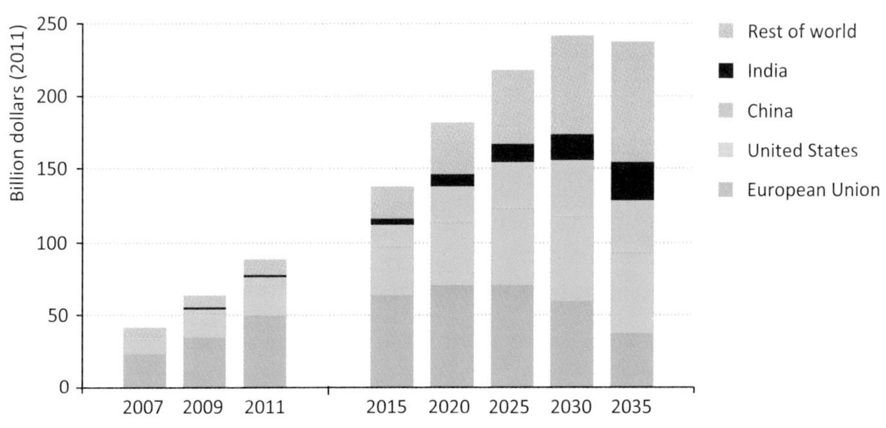

In the New Policies Scenario, subsidies to renewable energy in the European Union reach a plateau of around $70 billion in the 2020s, before declining to about half that level by 2035, as commitments to support higher cost capacity recently installed expire. In the United States, they increase until around 2030, peaking at $58 billion. In China, subsidies for renewables stabilise at above $35 billion during the late 2020s, while they keep increasing in India, reaching about $26 billion by the end of the projection period.

Subsidies to renewables-based electricity amount to a total of $3.5 trillion over 2012-2035. Of this, the capacity installed up to 2011 receives almost $1.0 trillion, continuing to receive payments until the early 2030s (Figure 7.13). The capacity added through to 2020 – mostly due to current targets – receives an additional $1.6 trillion. The remainder is paid for capacity built after 2020, with a part of the cost being paid beyond the time horizon of this *Outlook* for all the capacity built after 2015. Despite lower unit costs for most renewables over time, they are deployed in such large amounts after 2020 that they account for close to $1.0 trillion in subsidies from 2020 to 2035.

Figure 7.13 ▷ Subsidy lock-in of renewables-based electricity generation in the New Policies Scenario

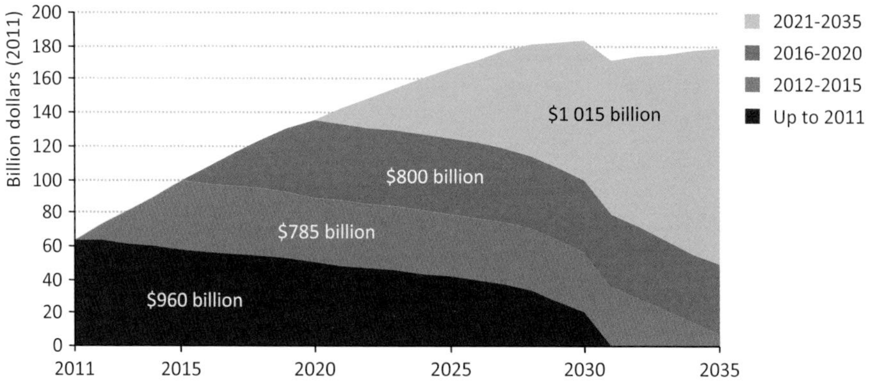

Note: Generation refers to all subsidised renewables-based electricity generation (excluding large hydro).

Integration of variable renewables into the electricity system

Electricity suppliers have always had to deal with weather-related demand volatility and often with capacity restrictions – for example, hydropower can be limited in dry years with low water inflows. However, the availability of wind and solar power is more sensitive to short-term changes in weather conditions, as wind conditions and cloud cover can change significantly in a matter of hours in a manner that is difficult to predict. Handling the natural variability of these and other renewable energy sources to ensure security of supply will become increasingly important as their share of overall capacity expands.

Several options exist or are being developed that can contribute to better management of the variability of the electricity system, including increasing interconnections, electricity storage, demand response and smart grids. Smart grids employ advanced technologies to monitor and manage the transmission of electricity from the generation point to the end-user, and can facilitate the integration of grids spanning large areas, allowing for a more efficient use of remotely located renewable resources. Demand response measures, shifting end-user demand to achieve a better distribution of load, can also help accommodate variable renewables. Electricity storage technologies can help smooth the supply of energy over short periods (minutes to hours), as well as allowing electricity produced during low demand periods (typically over periods of hours) to be available during times of peak demand. Improved forecasting techniques, over periods of minutes or hours, would also allow for better utilisation of variable renewable energy sources.

Until these measures are deployed on a larger scale, flexible capacity, with readily controllable electricity generation, will be required to accommodate variable renewables. It is important for flexible capacity to be able to start up quickly, ramp up to maximum output rapidly and operate at partial levels of output. Reservoir-based hydropower generally offers the best combination of these attributes, but costs and geography are limiting factors. Low investment costs, a wide range of available capacities and rapidly adjustable output often make gas-fired power plants the next best choice for flexible capacity. New coal-fired power plant designs can also provide considerable operational flexibility.

Today, there is adequate flexibility within most power systems to accommodate some further expansion of variable renewables. As larger amounts of variable renewables are deployed, additional capacity will be needed to ensure system adequacy. We estimate the additional global flexible capacity needs in the New Policies Scenario at 300 GW by 2035. In all *WEO* scenarios, natural gas dominates new flexible capacity. As the electricity sector decarbonises, the shares of both renewables and gas are expected to grow. While renewables may compete with gas in some cases, the two can be mutually beneficial, providing low-carbon electricity while maintaining the security of electricity systems.

The additional flexible capacity needed is calculated as the difference between the average annual power output of a variable renewable technology and the level of its output that can be relied on during times of peak demand (its capacity credit). The capacity credit, measuring their contribution to the adequacy of a system, depends on the renewable energy source and is typically lower than the average power output. In 2035, the average capacity credit of wind and solar PV (the two largest variable renewable technologies in terms of installed capacity), taken together, ranges between 5-20% depending on the region (Figure 7.14), against the average power output of 24% of their capacity. The capacity value is relatively low in the European Union, largely because of solar PV installed in countries where solar PV output is minimal or zero when electricity demand is at its peak. However, it is higher in India and Japan, where there is good correlation between peak electricity demand and solar PV output.

The costs of integrating variable renewables can be grouped into three broad categories: adequacy, balancing and grid integration, with total costs typically from $6-$25 per megawatt-hour (MWh) of variable renewable electricity generation. Adequacy costs arise from additional flexible capacity needed at times of peak demand and are on the order of

$3-$5/MWh. Balancing costs cover additional services to match supply and demand on a short-term basis and range from $1-$7/MWh. Grid integration costs add an additional $2-$13/MWh of variable renewables generation, including provision for transmission extensions for renewables located far from demand centres and reinforcements of existing transmission and distribution grids.

Figure 7.14 ▷ Installed wind and solar PV capacity and their contribution to system adequacy in the New Policies Scenario

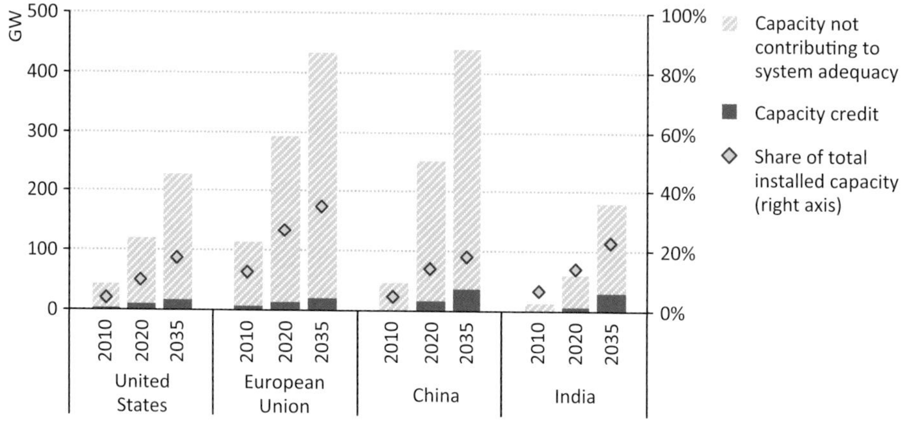

Note: Capacity credit is the amount of capacity that contributes to system adequacy, *i.e.* can be relied upon at times of peak demand.

Benefits of renewables

Several benefits are associated with the deployment of renewable energy technologies, including very low or no greenhouse-gas emissions, making them a key component in any climate change mitigation strategy (IPCC, 2011). In the New Policies Scenario, total CO_2 savings across all sectors from renewables are 4.1 gigatonnes (Gt) in 2035. In the power sector, renewables-based generation reduces emissions when it displaces power generation from the combustion of fossil fuels. Relative to the emissions that would be generated if the growing electricity demand of the New Policies Scenario were to be supplied using the electricity generation mix of 2010, renewables help to reduce CO_2 emissions in the power sector by 3.6 Gt in 2035 (Figure 7.15). These savings represent some 10% of the level of emissions reached in 2035, with more than 40% of the savings coming from increased wind generation. Although hydropower generation increases by about two-thirds over the projection period, its share of total generation declines in many regions, therefore its contribution to CO_2 savings is more limited than other renewables. Heat produced from renewable sources, as in wood pellet or solar heat boilers for example, reduce CO_2 emissions by 150 million tonnes in 2035 by displacing heat from boilers using coal, oil or gas.

Figure 7.15 ▷ **CO_2 emissions savings from greater use of renewables, relative to 2010 fuel mix* in the New Policies Scenario**

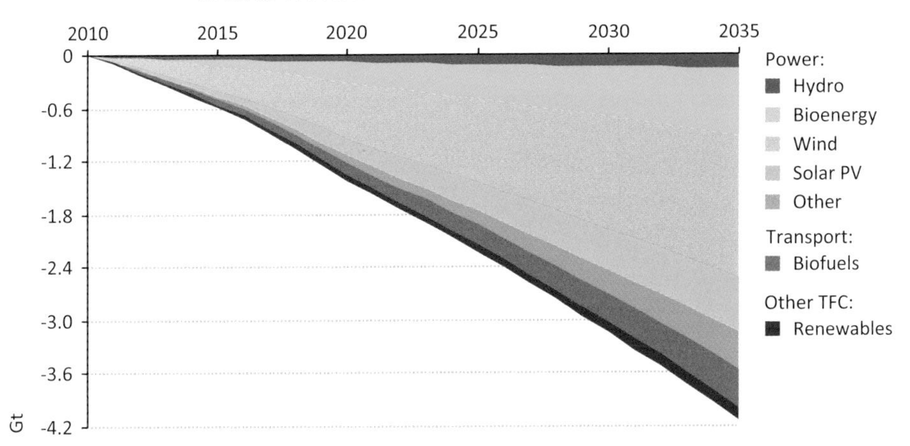

* The emissions savings compared with the emissions that would have been generated for the projected level of electricity generation in the New Policies Scenario were there no change in the mix of fuels and technologies and no change in the efficiency of thermal generating plants after 2010.

Notes: Other includes concentrating solar power, geothermal and marine energy. TFC = total final consumption.

Biofuels reduce emissions from oil in the transport sector by an estimated 0.4 Gt in 2035, but only so long as their production does not result in increases in emissions from direct or indirect land-use changes (Box 7.3). This aspect of biofuels has come under close scrutiny in recent years. Sugarcane ethanol and advanced biofuels have the highest potential to reduce emissions (IEA, 2011a). Biofuels have also been criticised for competing with food supply and contributing to deforestation. The negative impacts of biofuels, however, can be minimised or avoided if the right policies are established and enforced.

Renewable energy is largely a domestic source of energy (although some proportion of biofuels and other bioenergy is traded internationally). When it displaces imported fuels, it contributes to greater national energy security and directly reduces import bills, which represent a fairly significant percentage of gross domestic product (GDP) in many importing countries and often contribute to a trade deficit. Biofuels have the potential to reduce these effects significantly. Moreover, greater use of renewables could indirectly put downward pressure on oil and gas prices and reduce price volatility. In the electricity sector, renewables mainly reduce the need to import gas or coal, as oil use is limited in this sector.

The use of fossil fuels gives rise to several pollutants that worsen ambient air quality and have a negative impact on human health. Two of the most important of these pollutants are sulphur dioxide (SO_2) and nitrogen oxides (NO_x); SO_2 coming mainly from burning coal but also from diesel fuel, while NO_x come from burning all types of fossil fuels. They cause a number of environmental problems, such as acid rain and ground-level ozone formation.

Their impact is local and regional. Air pollution is a major problem in several large cities in non-OECD countries (and in some cities in the OECD). Integration of air quality and renewable energy policies can be more effective than separate actions.

Box 7.3 ▷ Indirect land-use change and the European Union's biofuels policy

The European Union's 2009 directive on renewable energy requires that 10% of transport demand in 2020 come from renewable sources, a target largely expected to be met through increased use of biofuels. The directive set sustainability criteria for biofuels, taking into account direct land-use changes, and established thresholds for greenhouse-gas emissions savings from biofuels, starting at 35% and rising to 50% in 2017 and to 60% in 2018. It also mandated the European Commission to review the impact of indirect land use changes on emission savings from biofuels. Direct land-use changes occur when, for example, forests and grasslands are converted to cropland to produce biofuels. Indirect land-use changes (ILUC) may occur when growing crops for biofuels in one area displaces previous agricultural or forest production to other areas. Studies have shown that ILUC can significantly reduce the greenhouse-gas savings potential of biofuels or even lead to increased emissions. Sugar-based ethanol has the highest savings potential and biodiesel some of the lowest (IFPRI and CEPII, 2010). These new findings have made it clear that ILUC emissions need to be taken into account in formulating biofuels policies, alongside other benefits of biofuels, notably reduced oil import dependence. No decision has been taken yet at the European Union level regarding the impact of these findings on future biofuels policy.

Several types of renewable energy technologies for electricity generation require significantly less water for their operation than fossil fuel-based and nuclear power plants. Solar PV and wind power do not use water to produce electricity and require only small amounts for cleaning purposes (see Chapter 17).[10] If the significant use of water during the extraction of fossil fuels and uranium is taken into consideration, the differences in water use are even greater. Use of solar PV and wind power also avoids thermal pollution and contamination that may be caused by the discharge of cooling water for thermal power plants.

10. Bioenergy, concentrating solar and geothermal power plants use water for cooling purposes, at levels close to those of fossil fuel-based and nuclear power plants. Bioenergy also requires water to grow the feedstocks.

Climate change mitigation and the 450 Scenario

Can energy efficiency help to avoid carbon lock-in?

Highlights

- Global energy-related CO_2 emissions in 2011 increased by 3.2% to reach a record high of 31.2 Gt. In the New Policies Scenario, our central scenario, CO_2 emissions increase to 37 Gt in 2035, corresponding to a 50% probability of limiting the long-term average global temperature increase to 3.6 °C relative to pre-industrial levels, and a 6% probability of limiting it to 2 °C.

- The 450 Scenario assumes strong policy action globally to put greenhouse-gas emissions on a long-term trajectory that will ultimately limit the global average temperature increase to 2 °C. It results in global energy-related CO_2 emissions peaking before 2020 at 32.4 Gt and declining to 22.1 Gt in 2035. To achieve this emissions trajectory, the share of fossil fuels in total primary energy demand declines from 82% in 2011 to 63% in 2035.

- The transformation of the global energy system in the 450 Scenario requires additional cumulative investment of $16 trillion compared with the New Policies Scenario, but delivers significant co-benefits in terms of reduced fossil-fuel import bills and local air pollution. Energy efficiency contributes more than half of all emissions savings, while renewables account for 21%, CCS for 12% and nuclear for 8%. Correspondingly, the majority of investment is directed to energy efficiency.

- Some 81% of the total CO_2 emissions allowable over the *Outlook* period in the 450 Scenario is already locked-in with the existing energy infrastructure. The scope for reaching the 2 °C goal is accordingly severely constrained. If global co-ordinated action to reduce CO_2 emissions is not taken before 2017, the infrastructure existing at that time will account for all the remaining CO_2 emissions allowable up to 2035 in the 450 Scenario.

- Rapid deployment of energy-efficient technologies could delay the lock-in of CO_2 emissions allowable in the 450 Scenario until 2022, leaving the door to a 2 °C world open for an extra five years. In addition to avoiding the severe effects of climate change, such investments are economically justified in their own right.

- About two-thirds of today's total carbon reserves of 2 860 Gt CO_2 – fossil-fuel reserves expressed in terms of their equivalent CO_2 emissions when combusted – are concentrated in only four regions: North America, the Middle East, China and Russia. Of these carbon reserves, 74% are publicly-owned. Less than 900 Gt CO_2 can be emitted up to 2050 in a 2 °C world, meaning that, in the absence of significant deployment of CCS, more than two-thirds of the current fossil-fuel reserves could not be commercialised before 2050.

Introduction

This chapter, after first setting the scene by reviewing recent developments in relation to emissions trends, focuses on the 450 Scenario. This scenario sets out a global energy pathway compatible with a near 50% chance of limiting the long-term increase in average global temperature to two degrees Celsius (2 °C) above pre-industrial levels. Unlike the other scenarios presented in this *Outlook*, the 450 Scenario imposes the assumptions necessary to achieve this specific outcome. It requires long-term stabilisation of the atmospheric concentration of greenhouse gases at below 450 parts per million (ppm) of carbon-dioxide equivalent (CO_2-eq). To achieve this target, we assume that significant further policy engagements are made in order to meet the specified emissions trajectory, while also trying to keep costs as low as possible (although it is possible to achieve the same target in other ways). Comparisons to other *WEO-2012* scenarios pathways are given at intervals in order to illustrate the extent of the additional efforts that would be necessary to achieve the 2 °C temperature goal.

Recent developments

The United Nations Framework Convention on Climate Change (UNFCCC) 17th Conference of the Parties, meeting in Durban, South Africa (COP-17) in December 2011 launched a process to develop "a protocol, another legal instrument or an agreed outcome with legal force applicable to all countries" related to global climate change mitigation. It is to be completed no later than 2015 so it may be adopted and come into effect from 2020. In addition, it was decided to extend the Kyoto Protocol to 2017 or 2020, but imposing reduction targets on only the European Union (EU), several other European countries, with Australia and New Zealand not having decided yet. Russia and Japan decided not to commit, while Canada withdrew altogether from the Protocol. Taking account of these developments and national climate policies, roughly 15% of global energy-related carbon-dioxide (CO_2) emissions will be subject to a carbon price, in one form or another, in 2015. Adequate investment in low-carbon energy technologies is dependent on greater certainty about long-term climate policies – the outcome of the Durban climate talks cannot be said to provide these conditions.

At the national level, there have been both positive and negative developments. In the United States, although there was a fall in emissions in 2011, policy development has been hesitant. There is still substantial uncertainty about the extension of current national support measures for renewables beyond 2012, *e.g.* the wind production tax credit (see Chapter 7), while planned regulation of greenhouse-gas emissions by the US Environmental Protection Agency faces political challenges. In the European Union, at the time of writing, the CO_2 price in the European Emissions Trading System had not exceeded the modest level of €10 per tonne of CO_2 in 2012, as a consequence of the weak economic situation and a steady supply of carbon credits from the Clean Development Mechanism (CDM). In Japan, following the accident at the Fukushima Daiichi nuclear power station in March 2011, the switch by power generators to natural gas and, to a lesser extent heavy fuel oil, pushed up CO_2 emissions. The government has now released the Innovative Strategy for Energy and the Environment, which provides for a greater focus on renewables and improving energy efficiency (see Chapter 9).

Some other signs of progress were seen in 2012. At the beginning of the year, the US state of California started to regulate greenhouse-gas emissions, by means of a cap-and-trade scheme, which will become binding from 2013. Australia introduced an emissions trading scheme in July 2012, starting with a fixed price of A$23 per tonne CO_2 and moving to full trading in July 2015. Korea has plans to join New Zealand and Australia to become the third Asia-Pacific country to adopt a cap-and-trade scheme to reduce greenhouse-gas emissions. Korea's scheme will come into effect in 2015 and will help to achieve the recently set target to reduce emissions by 4% in 2020, compared with 2005 emissions levels. The United States started an initiative to reduce short-lived greenhouse gases, including methane and hydrofluorocarbons, though it is not yet clear how this will be implemented.

Most recent emissions trends have not been encouraging. In 2011, energy-related CO_2 emissions increased by 3.2% – a rate of growth above the already rapid average increase over the last ten years (2.5%) – reaching a record high of 31.2 gigatonnes (Gt). A significant portion of the emissions growth was due to increased coal demand in China and India. As in previous years, coal accounted for the bulk (71%) of additional global CO_2 emissions, followed by oil (17%) and natural gas (12%) (Figure 8.1). In addition, the last three years have not demonstrated the decoupling of CO_2 emissions from economic growth that will be necessary if we are to mitigate the detrimental effects of climate change. Sluggish economic growth, high unemployment and limited access to cheap capital pose immediate hurdles to decisive action on climate change mitigation.

Figure 8.1 ▷ Annual change in global energy-related CO_2 emissions by fuel and global GDP growth

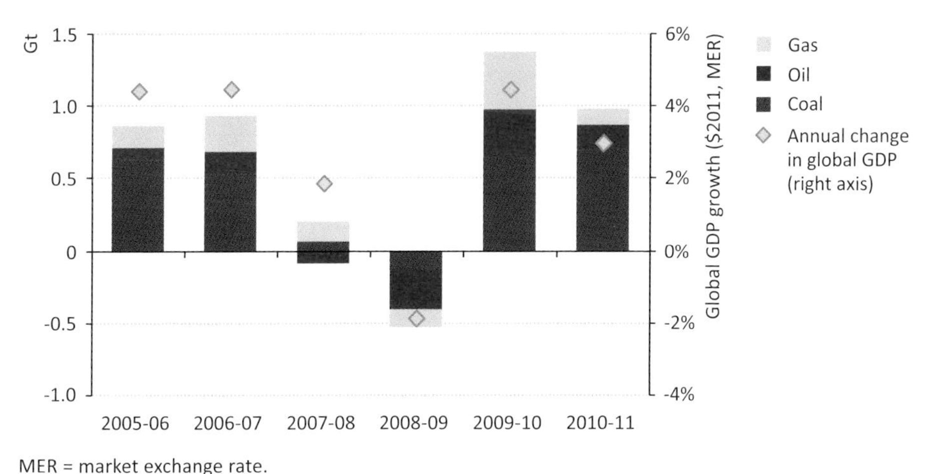

MER = market exchange rate.

CO_2 emissions in non-OECD countries rose by slightly more than 1 Gt (or 6%) in 2011, bringing their share of the global total to 59%. This increase was partially offset by a reduction of 50 million tonnes (Mt) (or 0.4%) in OECD countries. China, the world's biggest emitter, made the single largest contribution to the global increase, its emissions rising by 650 Mt, or 9%. To put this into perspective, the addition to China's emissions last year were roughly equivalent to total combined energy-related CO_2 emissions from France and Spain in 2011. Nevertheless, China's CO_2 intensity fell by 21% over 2005-2011, which is in line with its commitment to achieve a 40-45% reduction in 2020 with respect to 2005 levels.

In India, emissions rose by 140 Mt CO_2 in 2011, or 8%, the second-largest increment of any region. This moved India ahead of Russia to become the world's fourth-largest emitter behind China, the United States and the European Union. Russia's emissions increased by 2% in 2011, driven mainly by higher natural gas consumption.

CO_2 emissions in the United States fell by 80 Mt, or 1.5%, in 2011, primarily because of increased electricity generation from wind turbines, switching from coal to natural gas in power generation and an exceptionally mild winter and a wet year, which reduced demand for space heating and increased hydropower output (displacing fossil-fuelled power generation). CO_2 emissions in the European Union in 2011 were lower by 60 Mt, or 2%, as a relatively warm winter reduced heating needs. Had there not been a sharp increase in electricity generation from coal-fired power plants in the European Union in 2011 – caused by low CO_2 prices and relatively high gas prices – the reduction in emissions would have been twice as high. Despite rigorous actions by the government to promote energy efficiency, Japan's emissions increased by 50 Mt, or 4%, as a result of the substantial increase in fossil-fuelled power generation after the Fukushima Daiichi accident.

Figure 8.2 ▷ Energy-related CO_2 emissions per capita by region, 2011

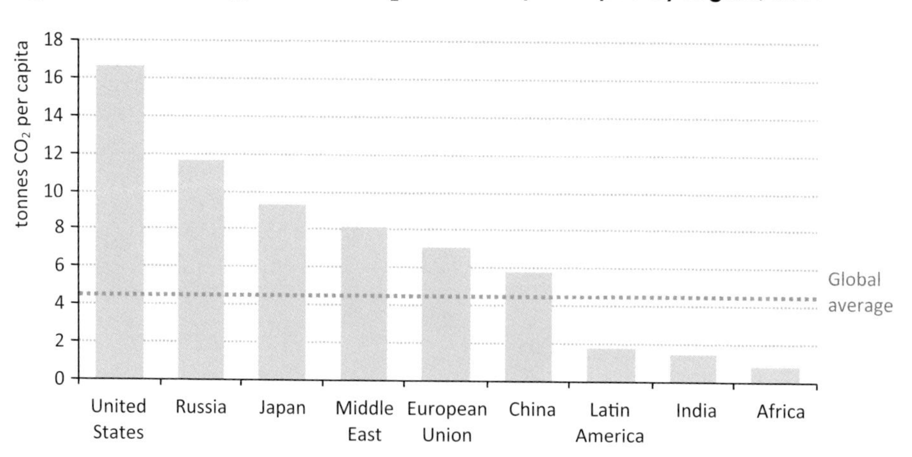

Notwithstanding the absolute increase in energy-related emissions over the previous years, China's current emissions per capita are 59% of the OECD average, while the figure for India is only 15%. However, per-capita emissions in China are growing quickly: since 2006, its per-capita CO_2 emissions have been above the global average and they are now 17% below the level of the European Union (Figure 8.2).[1] Among the largest CO_2 emitters, the United States still has by far the highest per-capita emissions, at 16.6 tonnes CO_2, more than three times the world average.

Taken together, current pledges to reduce CO_2 emissions are far from enough to enable the trajectory of the 450 Scenario to be followed. Individual countries' pledges leave a gap of 6-11 Gt CO_2-eq in 2020 on the pathway consistent with limiting global warming to 2 °C (UNEP, 2011; Höhne, *et al.*, 2012).

The world of the 450 Scenario

Comparison to other scenarios

The 450 Scenario sets out an emissions pathway consistent with limiting global warming to 2 °C above pre-industrial levels, considered the threshold for preventing dangerous anthropogenic interference with the climate system. Neither the Current Policies Scenario nor the New Policies Scenarios would achieve this goal. Figure 8.3 details the different emissions trajectories and splits the emissions savings between OECD and non-OECD countries. Emissions in the Current Policies Scenario increase from 31.2 Gt CO_2 in 2011 to 44.1 Gt CO_2 in 2035. The annual rate of growth in CO_2 emissions in the Current Policies Scenario is 1.5%, more than double that in the New Policies Scenario of 0.7%, which sees emissions of 37 Gt CO_2 in 2035. The Efficient World Scenario, which assumes that the long-term economic potential of energy efficiency is realised by 2035 (Box 8.2 and see Chapter 10), has a similar emissions trajectory to the 450 Scenario up to 2020, while the later decline is less steep and results in an emission level of 30.5 Gt CO_2 in 2035. In the 450 Scenario, emissions peak before 2020 at 32.4 Gt CO_2 and then decline to 22.1 Gt CO_2 in 2035. Almost two-thirds of emissions come from non-OECD countries in the New Policies Scenario; emissions avoidance in these countries makes up the majority of abatement in the 450 Scenario.

As in previous editions of the *WEO*, we have projected non-CO_2 greenhouse-gas emissions using the OECD ENV-Linkages model (Table 8.1). This includes non-energy related CO_2 emissions (such as process emissions from cement and emissions from land-use change), methane, nitrous oxide and F-gases. Projections for emissions from land use, land-use change and forestry are taken from the Baseline Scenario in OECD (2012) and have therefore an indicative character. They are kept constant across the scenarios, but decline over time. Total greenhouse gases increase to 64.7 Gt CO_2-eq in 2035 in the Current Policies Scenario, while emissions are just over half that level in the 450 Scenario in the same year.

1. A recent report by Oliver, Janssens-Maehnhout and Peters (2012) puts China's per-capita CO_2 emissions merely 4% lower than EU's per-capita emissions including sources beyond fossil-fuel combustion.

Figure 8.3 ▷ **Global energy-related CO_2 emissions by scenario**

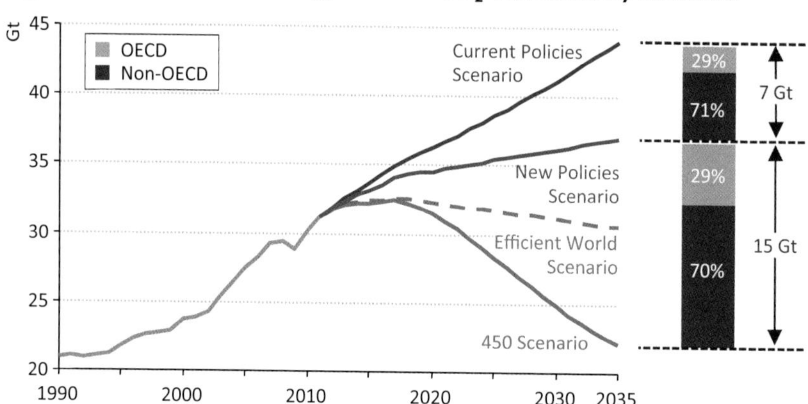

Note: There is also some abatement of inter-regional (bunker) emissions which, at less than 2% of the difference between scenarios, is not visible in the 2035 shares.

Based on the emissions trajectories shown, levels of the atmospheric concentration of greenhouse gases have been calculated (Figure 8.4). To establish the trajectory of atmospheric concentrations, we project that greenhouse-gas emissions peak around 2070 in the Current Policies Scenario and 2050 in the New Policies Scenario, steadily declining thereafter. The atmospheric concentration of greenhouse gases stabilises at around 950 ppm CO_2-eq in the Current Policies Scenario and 660 ppm CO_2-eq in the New Policies Scenario. The 450 Scenario sees stabilisation at 450 ppm CO_2-eq. These values can be translated into a temperature range, via radiative forcing, once greenhouse-gas concentration levels are stabilised in the long term.[2] The stabilisation level in the Current Policies Scenario results in a 50% likelihood of limiting the long-term temperature increase above pre-industrial levels to 5.3 °C. The New Policies Scenario puts us on a less emission-intensive pathway, but nonetheless leads to a median temperature increase of 3.6 °C.

Table 8.1 ▷ **World anthropogenic greenhouse-gas emissions by scenario** ($Gt\ CO_2$-eq)

		450 Scenario		New Policies		Current Policies	
	2010	2020	2035	2020	2035	2020	2035
CO_2: energy	30.2	31.4	22.1	34.6	37.0	36.3	44.1
CO_2: other	1.5	1.0	0.9	1.3	1.2	1.8	2.0
CH_4	7.6	6.2	5.2	7.3	7.7	8.8	10.1
N_2O	3.1	2.7	2.5	3.1	3.1	3.6	4.0
F-gases	0.9	0.5	0.6	0.6	0.8	1.5	2.6
LULUCF	5.3	4.0	1.9	4.0	1.9	4.0	1.9
Total	48.7	45.9	33.2	50.8	51.8	56.0	64.7

Notes: F-gases include hydrofluorocarbons (HFCs), perfluorocarbons (PFCs) and sulphur hexafluoride (SF_6) from several sectors, mainly industry. CO_2 other = CO_2 from industrial processes. LULUCF = land use, land-use change and forestry. Peat emissions are not included.

Sources: IEA-OECD analysis using the IEA World Energy Model and OECD ENV-Linkages model.

2. Irreversible effects on the climate from overshooting the long-term stabilisation level are not considered.

Figure 8.4 ▷ **Greenhouse-gas concentration pathways (left) and probability distribution of equilibrium temperature increase above pre-industrial levels (right)**

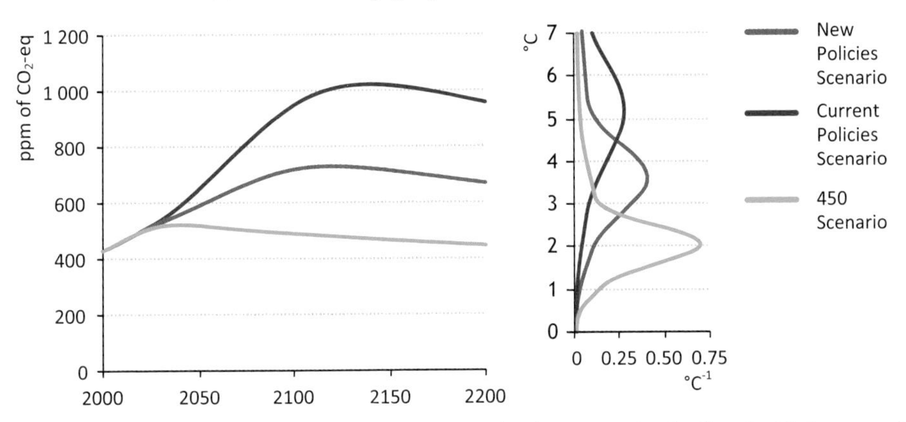

Notes: The median of the temperature distribution in the three scenarios is aligned with the respective greenhouse-gas concentration levels in 2200, where levels are almost stabilised. The probability distribution function for the temperature range was derived based on the equilibrium climate sensitivity distribution given in Rogelj, Meinshausen and Knutti (2012). PPM = parts per million.

Sources: IEA analysis using the MAGICC (version 5.3v2) and OECD ENV-Linkages models.

Based on the temperature distribution in a long-term equilibrium,[3] we calculated the likelihood of staying below a specified temperature threshold in each of the main *WEO* scenarios (Table 8.2). The likelihood of staying below 2 °C increases from 2% to 45% when shifting from the Current Policies Scenario to the 450 Scenario. The probability of exceeding 4 °C is 7% in the 450 Scenario, and this probability increases drastically in the New Policies Scenario to 37% and to 83% in the Current Policies Scenario, increasing the likelihood of extreme weather events (see Box 8.1).

Table 8.2 ▷ **Probability of staying below a specified temperature threshold by scenario**

	Current Policies	New Policies	450 Scenario
1 °C	0%	1%	4%
2 °C	2%	6%	45%
3 °C	7%	23%	85%
4 °C	17%	63%	93%
5 °C	40%	83%	97%
6 °C	66%	89%	98%

Note: The probabilities are derived based on the equilibrium climate sensitivity distribution given in Rogelj, Meinshausen and Knutti (2012).

3. The equilibrium climate sensitivity distribution is the average of 10 000 possible outcomes that span the range in the Intergovernmental Panel on Climate Change's 4th Assessment Report. This distribution is uncertain and different shapes are possible, which could lead to different temperature ranges. These sensitivities generally do not account for the effect of all climate feedbacks, which could lead to higher equilibrium temperatures.

Box 8.1 ▷ Extreme weather events and the energy sector

Extreme weather events have impacts on both energy supply and demand, creating significant challenges for meeting energy demand. The frequency and severity of extreme weather events has increased in recent years and it is expected to increase with a growing concentration of greenhouse gases (WMO, 2012). According to the latest research by the Intergovernmental Panel on Climate Change a warming world will likely be a world with more weather extremes (IPCC, 2012). The report finds it is either likely (or very likely) that there will be an increase in:

- The length, frequency and/or intensity of heat waves and droughts.
- Average and extreme sea levels.
- The average maximum wind speed of tropical cyclones.
- The frequency of heavy precipitation events.

The Current Policies Scenario implies the highest increase in extreme weather events, and therefore the greatest need for resilient energy systems in the future. This need will be strongest in developing countries, where the frequency of extreme weather events is accelerating fastest and where their cost threatens to be higher.

Fossil fuel-based electricity production can be adversely affected by air and water temperatures. Thermal and nuclear plants have reduced efficiencies in hotter temperatures, while the cooling processes for these plant types are constrained by regulations on river levels and the maximum allowable temperature for return water (see Chapter 17). During recent summers, cooling water restrictions forced several nuclear and thermal plants to reduce production. The Current Policies Scenario results in an increase in thermal (fossil and nuclear) capacity from 2011 to 2035 of around 2 200 GW, while capacity stays roughly constant in the 450 Scenario. Oil and gas supply, in particular offshore production operations, but also import operations, *e.g.* liquefied natural gas, will need to be adapted and upgraded to maintain supply security.

The supply of renewable energy sources is also vulnerable to an increase in extreme weather events. Hydropower will be affected by increased variability of rainfall and increased evaporation. Changes in wind patterns or insolation will affect the variability of wind- and solar-based electricity generation.

The energy sector can experience both increased demand and diminished supply during extreme weather events. For example, during summer heat waves, higher electricity demand for cooling coincides with hydropower shortfall, reduced efficiency of thermal plants and cooling constraints at thermal plants. The intensity of such conflicts is lower in a world where global warming is limited to 2 °C, in which a significant amount of electricity will be generated from wind, geothermal and solar PV. Many energy systems designed to mitigate emissions could prove more resilient to an increase in extreme weather events, as they entail more fuel diversity and more interconnection.

Primary energy demand and electricity generation in the 450 Scenario

In the 450 Scenario, global demand for primary energy increases by 16% from 2010 to reach 14 800 million tonnes of oil equivalent (Mtoe) in 2035. This overall figure masks a significant slowdown in energy demand growth during the *Outlook* period: while energy demand grows at 1.1% annually through 2020, growth slows to 0.3% annually from 2020 to 2035. This is largely a result of energy demand decoupling from economic growth, due to large-scale improvements in energy efficiency and structural changes in the economy.

The composition of energy demand changes significantly over the next 25 years in the 450 Scenario (Figure 8.5). Coal consumption peaks before 2020 and declines to under 3 400 million tonnes of coal equivalent (Mtce) in 2035, 33% lower than in 2010. Global demand for oil peaks before 2020 and declines to under 80 million barrels a day (mb/d) in 2035, or 10% less than in 2011. Thanks to its availability and the fact that it is the least carbon-intensive fossil fuel, natural gas in the 450 Scenario experiences an increase in demand from about 3 300 billion cubic metres (bcm) in 2010 to almost 4 000 bcm in 2035.

Figure 8.5 ▷ **Primary energy demand in the 450 Scenario by fuel**

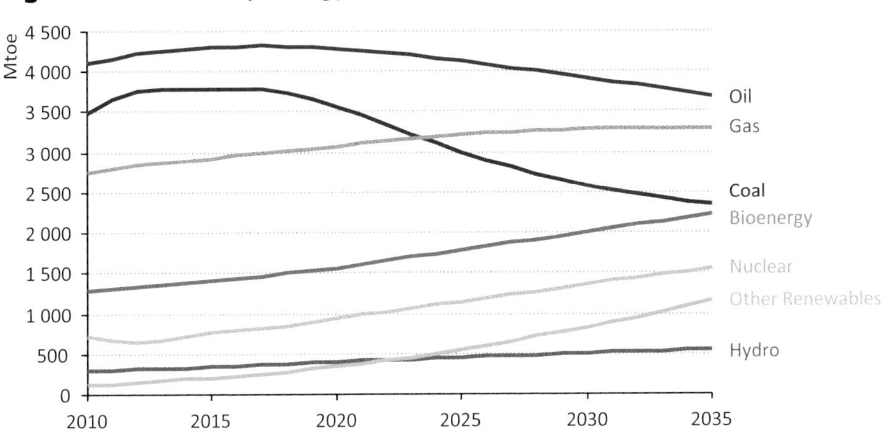

To stay within strict emissions limits, the overall share of low-carbon energy sources (including renewables, carbon capture and storage, and nuclear) increases from 19% in 2010 to 42% in 2035. Demand for bioenergy increases by 75% over 2010-2035, both because of strong growth in the use of bioenergy for electricity generation and rising biofuels production. The growth of low-carbon energy sources is particularly strong in electricity generation. On a global scale, total electricity generation increases in the 450 Scenario from around 21 400 terawatt-hours (TWh) in 2010 to 31 750 TWh in 2035. This is equivalent to an annual growth rate of 1.6%, which is significantly lower than the 2.2% seen in the New Policies Scenario. In 2010, a third of electricity generation was provided from low-carbon sources, with hydropower and nuclear representing the largest shares. This share grows to nearly 80% in 2035 in the 450 Scenario (Figure 8.6).

Figure 8.6 ▷ **Electricity generation from low-carbon technologies and share by scenario, 2010 and 2035**

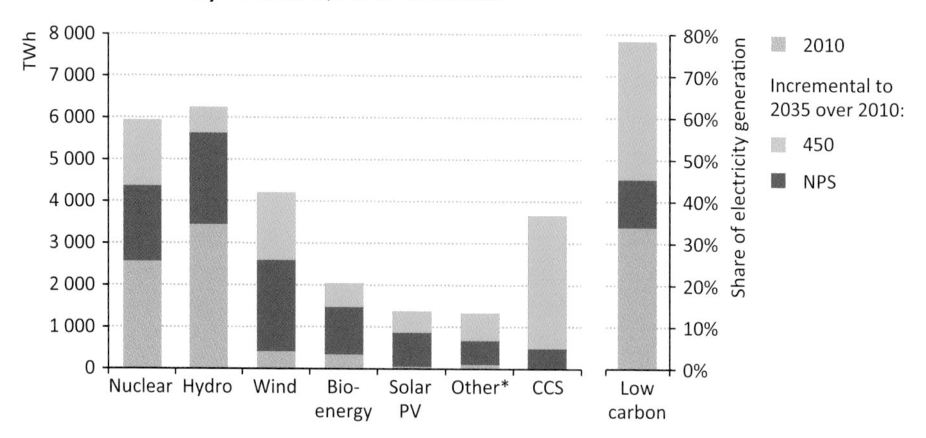

* Other includes geothermal, concentrating solar power and marine.

Note: 450 = 450 Scenario; NPS = New Policies Scenario.

Electricity generated from low-carbon sources already increases 2.3 times from almost 7 000 TWh in 2011 to 16 200 TWh in 2035 in the New Policies Scenario, mainly due to higher generation from wind, hydropower and nuclear. In the 450 Scenario, the amount of electricity generated from low-carbon sources rises further, to a total of over 24 900 TWh, which is 3.6 times higher than 2010 levels. Electricity generated from fossil-fuel plants with carbon capture and storage (CCS), nuclear and wind see the largest incremental gains in the 450 Scenario relative to the New Policies Scenario. Among these, CCS achieves the largest growth in electricity generation in absolute terms, compared to the New Policies Scenario, since it relies the most on carbon pricing. The growth in nuclear power in the 450 Scenario is, to a large extent, driven by the rapid growth in the number of nuclear power plants in China. Overall, nuclear's share in global electricity generation increases from 13% in 2010 to 19% in 2035 in the 450 Scenario. Two-thirds of the additional electricity generation from wind is in China, the European Union and the United States. Policy support and cost reductions encourage greater deployment of offshore wind, resulting in it providing 29% of all wind power by 2035. Since large wind turbines, especially those used in offshore wind installations, currently depend on rare earth permanent magnets to reduce the size and weight of the generators, the limited availability of these minerals may constrain this growth (Spotlight).

SPOTLIGHT

What is the role of rare earths in climate change mitigation?

Rare earths are seventeen distinct chemical elements (such as neodymium and dysprosium) and many are used in technologies that span much of the energy sector, from oil refineries and power generation to end-use equipment. They are used most prominently in permanent magnets, advanced batteries and phosphors. Permanent magnets play an important role in light-weight motors and generators used in hybrid and electric vehicles, as well as in certain types of wind turbines. Advanced batteries are necessary for hybrid and electric vehicles. Phosphors are a critical component of fluorescent and light-emitting diode (LED) lighting.

Attaining the trajectory of the 450 Scenario in our analysis means increasing wind capacity seven times from 2011, 650 million more hybrid and electric vehicles on the road and a substantial increase in energy-efficient lighting. If the use of rare earths in these technologies remained constant, there would be massive growth in the need for rare earths, which would be very challenging to meet.

Despite their name, rare earths are found all around the world. However, they are often found in low concentrations making the economics of extraction unattractive at current prices. In addition, rare earth elements are often found co-located with radioactive materials, increasing the cost and challenges of extraction. In recent years, China has provided nearly all of the supply of rare earth elements. Stringent export quotas have recently tightened the markets and increased prices dramatically. The result has been a push to develop new rare earth mines around the world, most notably in the United States, Australia, Vietnam and South Africa.

However, as deployment levels increase, technology is likely to improve, including reducing the dependency on rare earths. Research and development in this area is already underway both in the public and private sectors. Achieving the reductions in use thought possible today would drastically reduce demand, often by 50% or more by 2035 (US DOE, 2010). This would greatly alleviate future supply concerns, requiring only moderate continued expansion of mining capacity over time.

Beyond rare earth elements, indium, gallium, tellurium, cobalt and lithium play important roles in technologies that are important to the mitigation of climate change. Indium, gallium and tellurium are used in thin films for solar photovoltaic cells, while cobalt and lithium are used in advanced batteries for hybrid and electric vehicles. In much the same way as rare earths, use of these materials needs to be more efficient in the 450 Scenario to eliminate potential obstacles to the deployment of the necessary technologies.

Energy-related emissions and abatement

In the 450 Scenario, global energy-related CO_2 emissions peak at 32.4 Gt before 2020, *i.e.* only 1.2 Gt above the level in 2011. After 2020, emissions decline significantly to reach 22.1 Gt CO_2 in 2035, *i.e.* only 1.1 Gt higher than in 1990, under the impetus of strong and decisive action. In total, $16 trillion of additional investment is necessary, compared to the New Policies Scenario, an amount equivalent to the current gross domestic product (GDP) of the United States. The transport sector accounts for a major part of the additional investments ($6.3 trillion), followed by buildings ($4.4 trillion), power generation ($3.2 trillion), industry ($1.5 trillion) and biofuels ($0.6 trillion).

Over the *Outlook* period, energy intensity – energy demand per unit of GDP – declines by 2.4% per year and CO_2 intensity – CO_2 emissions per unit of energy use – falls by 1.8%. When compared to recent trends, the challenge of the 450 Scenario becomes clear. Over the last ten years, energy intensity declined by only 0.5% per year, while CO_2 intensity grew by 0.1% per year.

Figure 8.7 details the contribution made by specific mitigation measures, in our analysis, to a reduction in CO_2 emissions. In 2020, almost three-quarters of the emissions saved originate from energy efficiency, including electricity savings, end-use efficiency and power plant efficiency. The largest savings in the short term accrue from reduced electricity demand. While global electricity demand is higher in transport in the 450 Scenario compared with the New Policies Scenario, it is lower by over 800 TWh (or 7%) in buildings in 2020, due to more efficient appliances, space heating equipment and lighting. In industry, electricity demand is 600 TWh (or 6%) lower, mainly due to more efficient motor systems. CO_2 abatement from energy efficiency increases in absolute terms from 2.2 Gt CO_2 in 2020 to 6.4 Gt CO_2 in 2035, but the share of efficiency in CO_2 abatement declines towards the end of the projection period as renewables and CCS are used more widely.

The industry sector accounts for about half of the savings in end-use efficiency up to 2020, with the rest mostly shared between buildings and transport. Energy-efficiency related savings are important in industry because there are numerous opportunities to phase out inefficient old infrastructure. Emissions abatement in the buildings sector is mainly related to efficiency improvement of oil-and gas-fired boilers for space and water heating, as well as fossil fuel-based cooking. Energy efficiency improvements in transport take longer to materialise because fuel-economy standards apply only to new vehicles and it takes significant time for fuel-efficient vehicles to penetrate markets. Efficiency-related savings in power and heat increase up to the end of the 2020s through the introduction of efficient coal and gas power plants and less efficient plants being either mothballed, retired early or operated less. From then on, efficiency-related savings in the power sector decline because many of the newly-built power plants are retrofitted with CCS in the 450 Scenario as opposed to the introduction of very efficient coal without CCS and gas-fired power plants in the New Policies Scenario.

Figure 8.7 ▷ Global energy-related CO_2 emissions abatement in the 450 Scenario relative to the New Policies Scenario

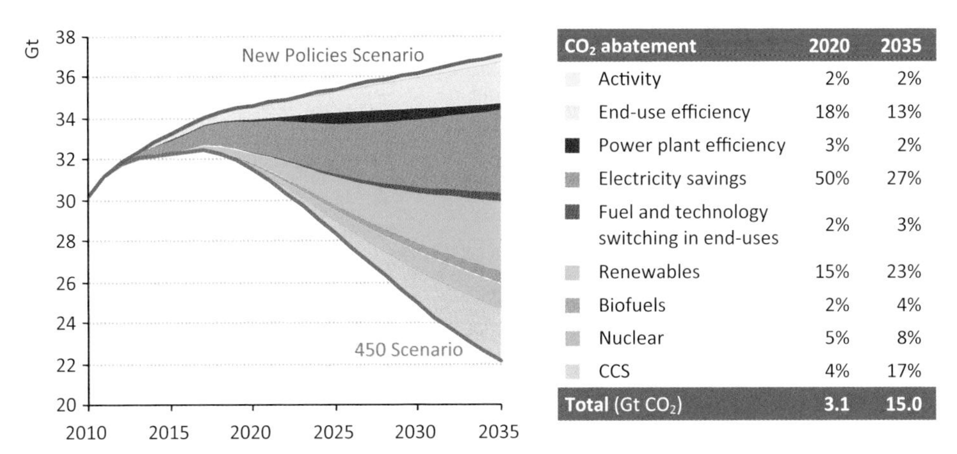

CO_2 abatement	2020	2035
Activity	2%	2%
End-use efficiency	18%	13%
Power plant efficiency	3%	2%
Electricity savings	50%	27%
Fuel and technology switching in end-uses	2%	3%
Renewables	15%	23%
Biofuels	2%	4%
Nuclear	5%	8%
CCS	4%	17%
Total (Gt CO_2)	**3.1**	**15.0**

Notes: Activity describes changes in the demand for energy services, such as lighting or transport services, due to price responses. Power plant efficiency includes emissions savings from coal-to-gas switching. For more detail on the decomposition technique used, see Box 9.4 in Chapter 9.

The second most important abatement measure are renewables (excluding biofuels), with increased use not only in power generation but also in buildings (for space and water heating) and in industry (as an alternative fuel source). Among renewables in the power sector, wind, hydro, and biomass are the most important sources of CO_2 abatement. Next to renewables, CCS saves 2.5 Gt CO_2 in 2035, becoming a significant source of mitigation from 2020 onwards. In several countries, such as China and the United States, very efficient coal-fired power stations are built up to 2020 and are retrofitted with carbon capture and storage in the following years as a consequence of a rising CO_2 price.

Transport is the end-use sector that has seen – by far – the most rapid increase in emissions over the last twenty years. CO_2 emissions in the sector increased by 2.2 Gt CO_2 from 1991 to 2011, or by almost 50%. Reducing emissions in transport thus forms a crucial element for any comprehensive strategy to reduce global CO_2 emissions. Road transport accounts for about three-quarters of global transport emissions and a diverse set of mitigation measures will be required in this sub-sector (Figure 8.8). Up to 2020, lower vehicle usage, fuel efficiency gains and an increase in the use of biofuels dominate abatement. Over the longer term, improvements in vehicle fuel economy represent the most important abatement measures, accounting for 51% of cumulative savings in the transport sector from 2011 to 2035.

Higher fuel prices lead to a lower vehicle usage in the 450 Scenario. The increase in the 450 Scenario is due to the assumed removal of subsidies in developing countries and an increase in fuel duty in OECD countries, which increases end-use prices and limits the rebound effect from more efficient vehicles. The tax level corresponds to an increase in the fuel duty for gasoline of \$0.43 per litre (l) in the European Union and \$0.34/l (\$1.29 per gallon) in the United States.

Figure 8.8 ▷ Global CO_2 emissions reduction in the 450 Scenario relative to the New Policies Scenario in global road transport

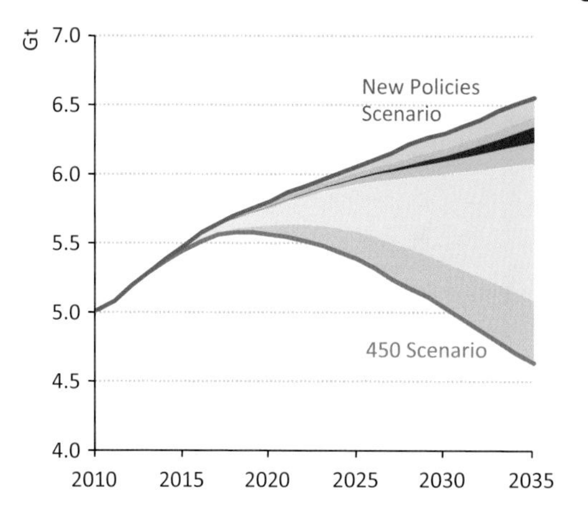

CO_2 abatement	2020	2035
Lower vehicle usage	4%	7%
Hybrid vehicles	10%	4%
Electric vehicles	1%	6%
Plug-in hybrid vehicles	5%	8%
Fuel economy	54%	51%
Biofuels	26%	24%
Total (Gt CO_2)	**0.2**	**1.9**

None of the mentioned abatement measures – lower vehicle usage, fuel economy and greater biofuels usage – require a change in the existing fuel infrastructure. They are therefore quickly deployable. On the other hand, while fuel-economy standards can have an immediate impact on new vehicles, these only gradually become dominant in the entire fleet. Overall, improvements in fuel economy and greater biofuels use[4] are the most important mitigation measures in the transport sector in 2035. An increase in the consumption of ethanol and biodiesel reduces emissions by 0.25 and 0.2 Gt CO_2, respectively, in 2035. The contribution from plug-in hybrid and full electric vehicles increases significantly, contributing 14% of the total emissions abatement in 2035.

Regional energy-related CO_2 emissions and abatement

In the 450 Scenario, it is assumed that the mechanisms in place under a climate agreement ensure an efficient allocation of emissions reduction across countries. Accordingly, energy-related CO_2 emissions in the OECD decrease from 12.3 Gt in 2011 to 6.1 Gt in 2035 in the 450 Scenario. This represents a reduction of about 50% and reduces the share of OECD emissions in the world total from 39% in 2011 to 28% in 2035. In non-OECD countries, CO_2 emissions decrease from 17.8 Gt in 2011 to 14.7 Gt – a reduction of 17%. In terms of emissions reduction relative to the New Policies Scenario, non-OECD countries deliver 70% of the abatement in 2035.

China's energy-related CO_2 emissions peak before 2020 at 8.7 Gt and decline to 4.9 Gt in 2035. Consequently, China's per-capita emissions exceed those of the European Union prior to 2020. However, towards the end of the *Outlook* period, Chinese per-capita emissions fall and stay only marginally above EU per-capita emissions, while still remaining significantly below the level of the United States and Japan (Figure 8.9). Despite the reduction in annual

4. For more detail on the implications on indirect land-use change, see Box 7.4 in Chapter 7.

emissions, China remains the largest emitter up to 2035. Compared with the New Policies Scenario, China achieves cumulative emissions reductions of 53.9 Gt CO_2, more than any other country. The trajectory of this reduction is consistent with reducing energy-related CO_2 emissions per unit of GDP by 57% in 2020 compared with the 2005 level, above China's Copenhagen pledge of 40-45%. This is based on the assumption that additional reductions are achieved through market-based mechanisms. The largest source of emissions abatement is efficiency, contributing 56%, followed by a higher share of renewables, 18%, and wide deployment of CCS in power and industry, 12%.

Figure 8.9 ▷ Per-capita emissions and total emissions by region

a) 2011

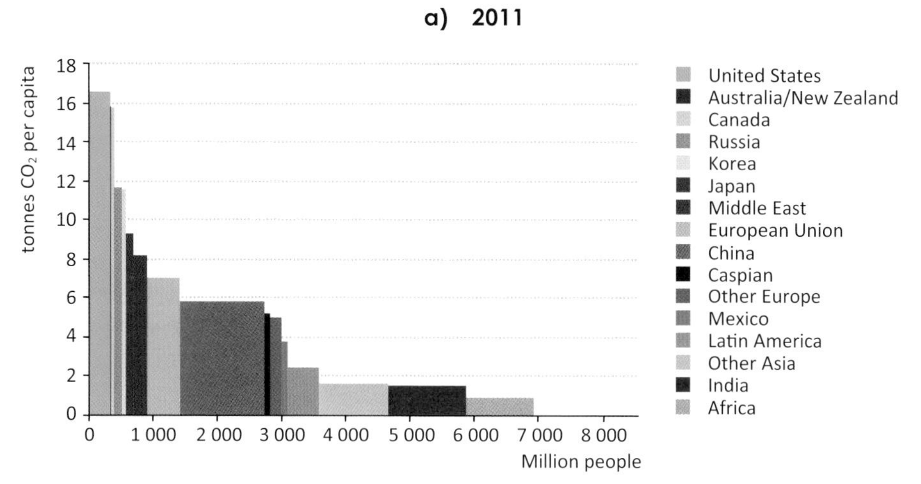

b) 450 Scenario, 2035

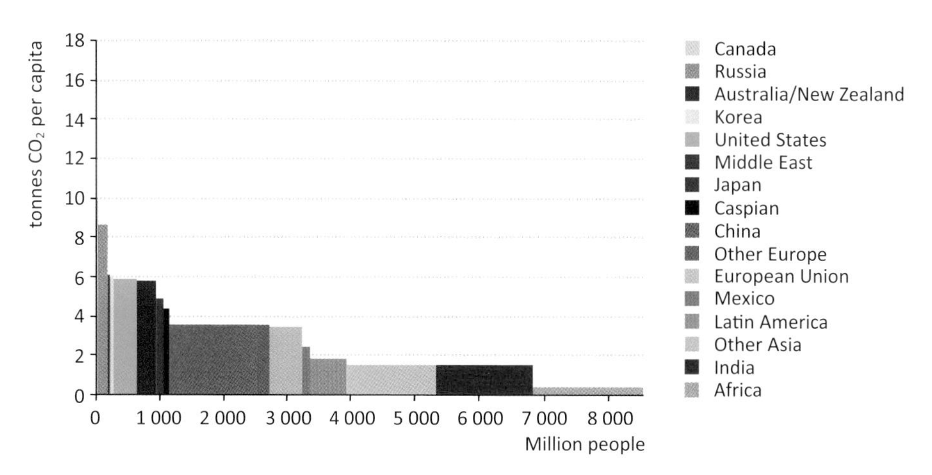

Note: Area in figures is equal to total emissions.

8

The United States, presently the second-largest global emitter, reduces domestic CO_2 emissions in 2035 by 3 Gt below 2011 levels, which represents a reduction of 58%. Per-capita emissions are projected to fall from 16.6 tonnes CO_2 to 5.9 tonnes CO_2 (above the global average of 4.5 tonnes CO_2 in 2011). This pathway implies that the United States reduces domestic energy-related CO_2 emissions by 16% in 2020 compared with 2005. The most important emissions savings are achieved by use of CCS and renewables. CCS saves 0.7 Gt CO_2 and renewables 0.6 Gt CO_2 in 2035 in the 450 Scenario, compared with the New Policies Scenario. An implicit carbon price rising above $75/tonne CO_2 causes almost all coal-fired power plants installed after 2012 to be retrofitted with CCS from the end of the 2020s. Prior to that, a switch from coal to gas in the power sector and the installation of very efficient fossil fuel power plants account for around 18% of emissions savings. From 2011 to 2035, wind is the largest abatement source within renewables, followed by concentrating solar power and solar PV. In end-use sectors, transport saves the most CO_2 emissions (0.3 Gt) in 2035, as a result of the higher use of biofuels and improved vehicle fuel economy.

The European Union nearly halves its CO_2 emissions in the 450 Scenario, from 3.5 Gt in 2011 to 1.8 Gt in 2035. In 2020, European Union domestic energy-related CO_2 emissions are down 27% compared with 1990, requiring it to buy another 3% on international carbon markets to meet the conditional offer of reducing emissions by 30%. Electricity savings and energy efficiency account for 45% of the abatement, and even more in the short term. The second most important abatement measure is the shift in the power sector away from fossil-fuel power plants without CCS towards renewable energy sources, of which wind power is the most important, followed by bioenergy and hydro. Road transport in the European Union saves about 2 Gt CO_2 emissions from 2011 to 2035, compared with the New Policies Scenario, more fuel efficient vehicles contributing 1 Gt CO_2, biofuels 0.7 Gt CO_2 and plug-in vehicles 0.2 Gt CO_2.

Energy-related CO_2 emissions in India increase from 1.8 Gt in 2011 to 2.2 Gt in 2035, an average annual increase of 1%. Compared with the New Policies Scenario, India saves 1.6 Gt CO_2 in 2035. India's Copenhagen pledge specifies a 20-25% reduction in the CO_2 intensity of GDP by 2020, compared with the 2005 level. In the 450 Scenario, India's CO_2 intensity is reduced by 39% in 2020, with the surplus being traded in carbon markets. With 59% of the cumulative emissions savings, efficiency is the most important source of reduction, yielding additional benefits in terms of economic growth and energy access (see Chapter 10). This saving of 59% is split into 12% from end-use efficiency, 34% from electricity savings and 13% from more efficient fossil fuel power plants. Among the end-use sectors, industry is the sector that saves the most CO_2 emissions, 2.3 Gt CO_2 over the entire *Outlook* period. This abatement is achieved nearly equally through a substantial increase in energy efficiency and the widespread adoption of CCS.

Russia, currently the fifth-largest emitter, reduces energy-related CO_2 emissions from 1.7 Gt in 2011 to 1.1 Gt in 2035 in the 450 Scenario. This means that Russia's per-capita emissions are reduced from 11.7 tonnes CO_2 to 8.6 tonnes CO_2 in 2035, still among the highest in the world. Compared with the New Policies Scenario, emissions are 0.7 Gt CO_2 lower in 2035. Energy efficiency contributes most to abatement, with 59% of cumulative savings, driven by reduced electricity consumption. The vast majority of the remaining savings are achieved in the power sector through the deployment of CCS, renewable energy sources – particularly hydro and biomass – and nuclear.

Japan's emissions are halved from 1.2 Gt CO_2 in 2011 to 0.6 Gt CO_2 in 2035 in the 450 Scenario. Compared to the New Policies Scenario, this represents a further reduction of 0.4 Gt CO_2 in 2035. In the 450 Scenario, Japan reduces energy-related CO_2 emissions by 10% in 2020 compared with 1990 levels, requiring Japan to buy a significant amount of CO_2 certificates on international markets in order to meet the target it submitted as part of the Cancun Agreement. Energy efficiency plays the most important role in reducing emissions over the *Outlook* period contributing 36% of cumulative savings, mainly in the form of reduced electricity consumption. Renewables form another important pillar of Japan's pathway to a low-carbon energy system, mainly in the form of an increase in the installed capacity of wind, hydro and solar PV.

Benefits of the 450 Scenario

Reduced fossil fuel use not only lowers energy-related CO_2 emissions but reduces fossil-fuel import bills and limits damage to human health from air pollutants. In the 450 Scenario, import bills in the five major importing regions – the European Union, the United States, Japan, China and India – are significantly lower, compared with the New Policies Scenario (Figure 8.10). The policy framework of the 450 Scenario favours faster deployment of low-carbon technologies and spurs the uptake of energy efficiency. As a result, demand for coal, oil and gas is reduced and the energy mix is more diverse relative to the New Policies Scenario. In the United States, China and India, import bills fall also as a consequence of higher domestic fossil fuels production, in particular natural gas, and lower fossil fuel prices, which result from less demand for fossil fuels. Compared with the New Policies Scenario, the global oil price in the 450 Scenario in 2035 is $25/barrel lower and the coal price almost 40% lower. The price for natural gas falls by 23% in Europe and 4% in North America.[5]

Fossil fuel-import bills for the five largest importers in 2035 are lowered from just over $2 trillion in the New Policies Scenario to slightly above $1.2 trillion in the 450 Scenario. This difference is mainly attributable to lower oil-import bills, which peak before 2015 in Japan, the United States and the European Union and then steadily decline towards 2035. In Japan, the oil-import bill in 2035 is 46% below the 2011 level. In the 450 Scenario, China starts having the highest fossil-fuel import bills in the world from around 2030, overtaking the European Union. The import bills of India and China in 2035 are lower than in the New Policies Scenario, but higher than today, mainly due to their higher oil demand.

5. Lower levels of international fossil-fuel prices do not necessarily translate into lower end-user prices compared to the New Policies Scenario. Subsidies removal, the introduction of CO_2 prices, and the shift towards more costly electricity generation options counteract lower fossil-fuel prices at the end-user level.

Figure 8.10 ▷ Fossil fuel-import bills in selected regions by scenario

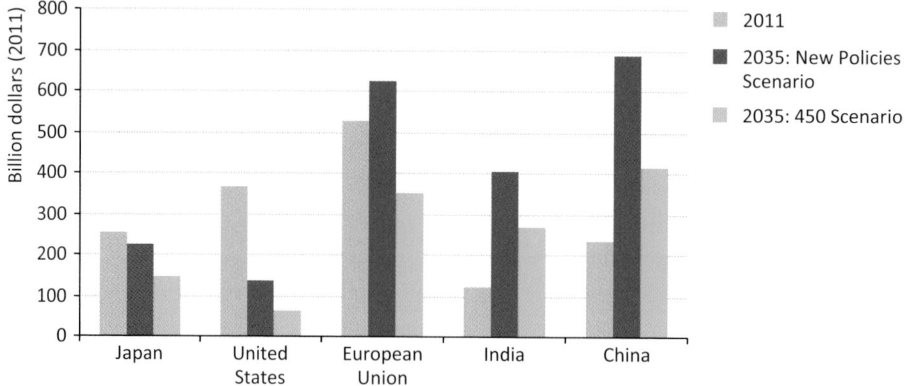

Reduced demand for fossil fuels in the 450 Scenario, relative to the other scenarios, also reduces local air pollution (Table 8.3). Emissions of sulphur dioxide (SO_2), nitrogen oxides (NO_X) and particulate matter ($PM_{2.5}$) at any level increase respiratory diseases and shorten life expectancy. In the Current Policies Scenario, NO_X, SO_2 and $PM_{2.5}$ emissions fall in OECD countries, while SO_2 and $PM_{2.5}$ emissions stabilise and NO_X emissions increase by 36% in the developing world. In the New Policies Scenario, emissions of all three types of local air pollutants decrease in OECD and non-OECD countries, with the exception of NO_x in non-OECD countries, where emissions increase by about 20%.

In the 450 Scenario, emissions of SO_2 in 2035 are a quarter lower than in the New Policies Scenario. NO_X emissions are lower by 24% and those of $PM_{2.5}$ by 7%. This has direct consequences for the number of years of life lost, which is obtained by multiplying the number of people exposed to anthropogenic $PM_{2.5}$ emissions by its impact in terms of reduced life expectancy, measured in years. In China, this number declines from 1.4 billion life years in 2035 in the New Policies Scenario to 1.2 billion life years in the 450 Scenario. This corresponds to an improvement of 15%, while for India years of life lost are reduced by 23%.

Table 8.3 ▷ Global air pollution by scenario (Mt)

		Current Policies		New Policies		450 Scenario	
	2010	2020	2035	2020	2035	2020	2035
SO_2 emissions	86.3	79.1	84.2	75.1	72.4	69.3	54.1
NO_X emissions	84.7	80.2	92.2	77.7	82.5	73.0	62.9
$PM_{2.5}$ emissions	42.8	42.8	43.0	42.2	41.2	41.3	38.2

Source: IIASA (2012).

Carbon in energy reserves and energy infrastructure

Potential CO_2 emissions in fossil-fuel reserves

In order to reduce the probability of average global warming exceeding 2 °C to 50%, CO_2 emissions from fossil fuels and land-use change in the first half of this century need to be kept below 1 440 Gt (Meinshausen, *et al.*, 2009). Since a total of 420 Gt CO_2 have already been emitted between 2000 and 2011 (Oliver, Janssens-Maenhout and Peters, 2012) and we estimate that 136 Gt CO_2 will be emitted from non-energy related sources[6] in the period up to 2050, a maximum of 884 Gt CO_2 more can be emitted by the energy sector from 2012 to 2050.

To put this number into context, we have calculated the carbon reserves – fossil-fuel reserves, expressed as CO_2 emissions when combusted – of global fossil-fuel reserves in 2012. Almost two-thirds of carbon reserves are coal, while oil accounts for 22% and natural gas for 15% of the carbon reserves. In energy-equivalent terms, coal accounts for 53% of all fossil-fuel reserves and it has the highest emissions factor. Coal when combusted, emits 68% more CO_2 than natural gas for the same energy-equivalent amount of fuel, while this emissions ratio is 42% more for coal relative to oil.

8

We estimate that total potential emissions from fossil-fuel reserves in 2012 amount to 2 860 Gt CO_2.[7] Seen in relation to the CO_2 emissions from other sources detailed above, this is equivalent to saying that, without a significant deployment of CCS, more than two-thirds of current proven fossil-fuel reserves cannot be commercialised in a 2 °C world before 2050.[8] Countries with high carbon reserves – fossil-fuel reserves expressed as CO_2 emissions when combusted – stand to be particularly affected by a decarbonisation of the global energy system. In the absence of widespread CCS adoption, the profits of public and private companies in fossil-fuel rich countries could be cut and state income from taxes and royalties reduced. Because coal, natural gas and oil reserves are not distributed equally across the globe, some regions would be more affected than others (Figure 8.11).

Two-thirds of potential CO_2 emissions from fossil-fuel reserves are concentrated in only four regions: North America, the Middle East, China and Russia. In the case of the United States and China, 93% and 96% of carbon reserves are coal. The situation is different for the Middle East, where oil accounts for 66% and natural gas for 33% of carbon reserves. Russia possesses significant reserves of all three fossil fuels, coal accounting for 65%, natural gas for 24% and oil for 11%. Other countries that hold a significant amount of uncombusted carbon in the form of coal are India and Australia. On a global scale the European Union has very limited fossil-fuel reserves, accounting for 3.6% of all carbon reserves, and Japan virtually none at all. Similarly, all African countries, except South Africa, collectively hold 83 Gt carbon reserves, equivalent to 2.9% of the global total.

6. Process-related CO_2 emissions and emissions from land use, land-use change and forestry

7. Unconventional gas and oil reserves are included. For more detail see Figure 3.14 in Chapter 3 and Box 1.1 in IEA (2012).

8. According to the IPCC's RCP 2.6 scenario (Vuuren, *et al.*, 2011), which is broadly consistent with the 450 Scenario, less than 50 Gt CO_2 could be emitted in the second half of the 21st century with CO_2 emissions turning negative from 2070.

Figure 8.11 ▷ **Potential CO_2 emissions from proven fossil-fuel reserves at the end of 2011 by region** (Gt CO_2)

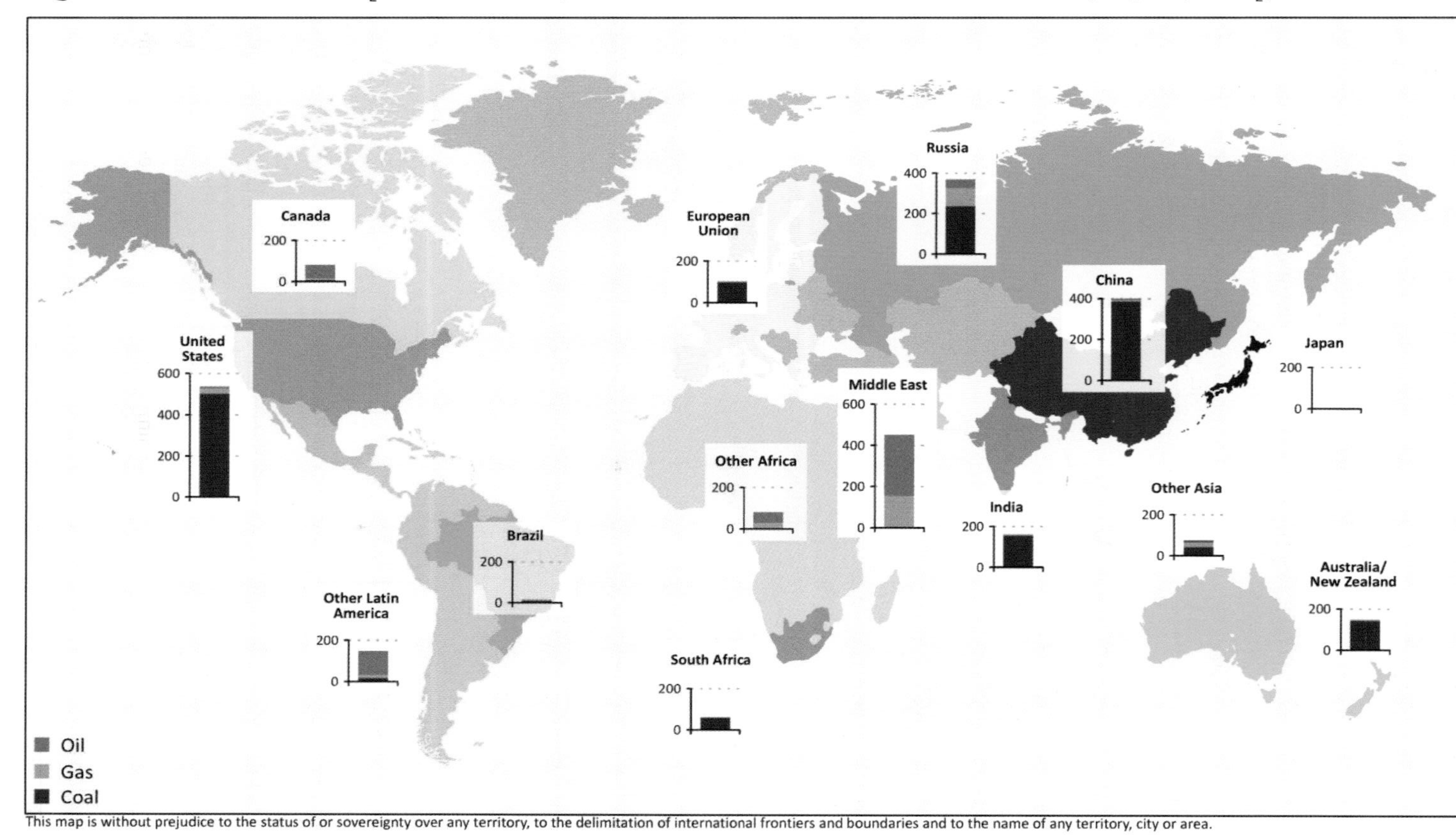

This map is without prejudice to the status of or sovereignty over any territory, to the delimitation of international frontiers and boundaries and to the name of any territory, city or area.

We estimate that 74% of all carbon reserves are held by government-owned companies and the rest by private companies (Figure 8.12). In non-OECD countries, 88% of carbon reserves are in public hands. In OECD countries, the share of carbon reserves owned by government-owned companies is 40% on average. The vast majority of oil and natural gas reserves in the Middle East are held by national oil and gas companies.

Figure 8.12 ▷ Potential CO_2 emissions from remaining fossil-fuel reserves by fuel type

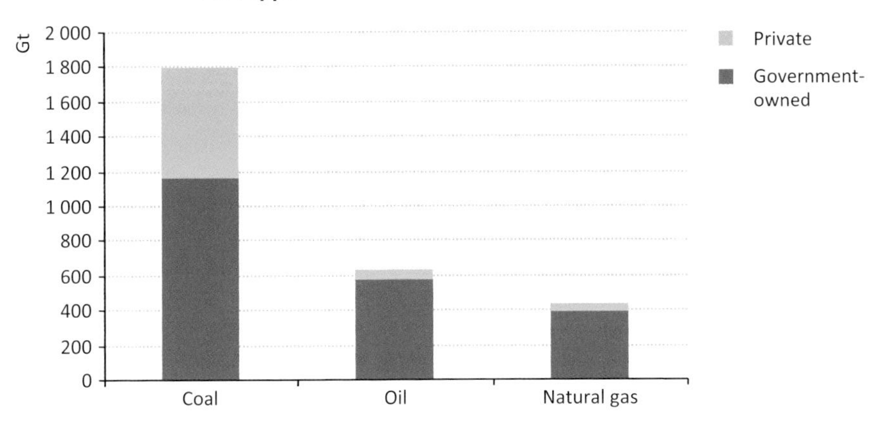

Investing in accelerated development and deployment of CCS represents an important hedging response for fossil-fuel rich regions, just as it offers a promising way forward for major users. This option promises to preserve the economic value of the reserves in a world undertaking significant action to mitigate climate change. This consideration is of relevance not only to the owners of coal and gas reserves, the fuel types used in the application of CCS, but also to oil reserve owners, since oil could then take a larger share of the remaining carbon budget. Investors are already weighing these considerations in their valuations of private companies.

Emissions lock-in

Potential CO_2 emissions are stored not only underground in the form of coal, oil and natural gas, but are also implicit in the nature of existing infrastructure. Emissions that will come from the normal use of infrastructure currently in place and under construction are locked-in for many years, as the average lifetime of energy infrastructure is long, as shown in *WEO-2011*. Barring dramatic shifts in relative fuel prices and technological breakthroughs, emissions from existing infrastructure cannot be avoided without decisive policy action that entails premature retirements, costly refurbishments or leaving capacity idle. In the absence of a global agreement on climate change mitigation, emissions-intensive infrastructure continues to be built, locking in emissions for decades to come. *WEO-2011* presented analysis on the implications of delaying action to limit the temperature increase to 2 °C (IEA, 2011). This year, we expose the extent of CO_2 emissions lock-in by OECD and non-OECD countries and analyse the effect of pursuing stringent energy efficiency policies on locked-in emissions (Box 8.2 compares the 450 and Efficient World Scenarios).

A detailed analysis of the global capital stock in place in all energy sectors – power generation, industry, transport, buildings, non-energy use and agriculture – shows that the infrastructure that either exists today or is under construction will, in normal use, emit 550 Gt CO_2 on a cumulative basis up to 2035. This amount is 81% of the cumulative emissions allowed over the period in the 450 Scenario, leaving a very limited amount available for additional facilities. This calculation assumes that infrastructure is allowed to operate for its full technical lifetime during the *Outlook* period, as in the New Policies Scenario.

In OECD countries, 79% of emissions allowed in the 450 Scenario between now and 2035 are already implicit in the current infrastructure. In 2020, emissions from the infrastructure existing in 2011 account for 81% of the CO_2 emissions permitted in that year in the 450 Scenario, while in 2035 it represents 71% of the emissions (Figure 8.13).

Figure 8.13 ▷ Energy-related CO_2 emissions from locked-in infrastructure in 2011 and in the 450 Scenario in OECD countries

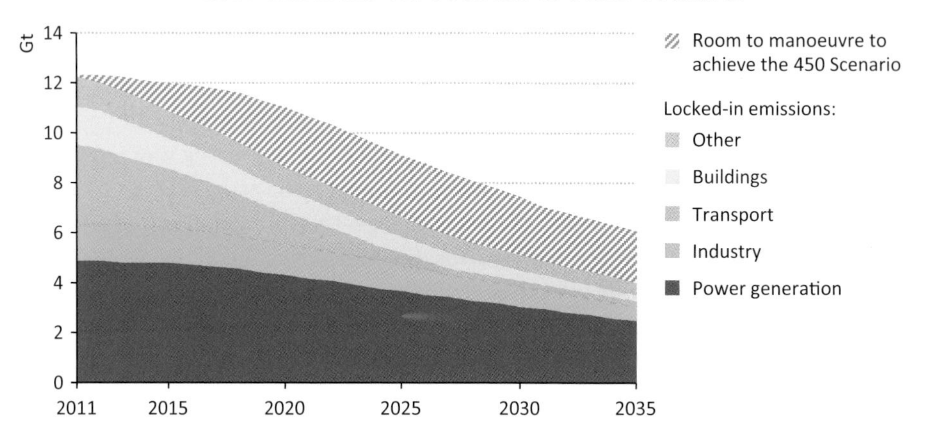

Most of the locked-in emissions in OECD countries originate from the power sector and industry, because the lifetime of plants is typically several decades. These plants will continue to operate, and emit CO_2, unless very strong policies are adopted that either force their premature retirement or make such early retirement economic. Power plants account for 4.9 Gt CO_2 of emissions in 2011, or 39% of the total. By far the greater part of these emissions comes from coal-fired power plants, while gas-fired power plants make up most of the rest. In 2035, only 2.5 Gt CO_2 will arise from power plants that exist today, since 54% of the generation capacity will be retired in the next 23 years. Industrial facilities often have a lifetime of around 30 years (though in some cases it up to 50 years). CO_2 emissions associated with industry made up only 12% of total emissions in the OECD in 2011. Because the facilities of energy-intensive industries in OECD countries are relatively old, only 41% of emissions from currently installed industrial infrastructure are locked in through 2035.

Transport is the second most important sector in terms of CO_2 emissions, its share in 2011 being 27%. Emissions come primarily from road transport, mainly passenger light-duty vehicles, which have a lifetime of around fifteen years. Thus there is scope to change much of the road transport vehicle fleet by 2035 and locked-in emissions decline from 3.3 Gt in 2012 towards 0.2 Gt in 2035. After 2030, locked-in emissions in transport are dominated by maritime transport and aviation emissions. Transport emissions are also partly locked-in by the infrastructure for fuel distribution and manufacturing, but this is assumed in this analysis to change as required by the changing vehicle fleet.

Emissions in the buildings sectors account for only 12% of CO_2 emissions in 2011, because emissions associated with the use of electricity are accounted for in power generation. Equipment for space and water heating have a lifetime of up to 25 years, while the lifetime for electric appliances is much shorter (IEA, 2011). Natural gas distribution networks have a much longer lifetime (like fuel distribution infrastructure in transport), but are also not treated in this analysis as a constraint on change.

In non-OECD countries, current infrastructure will last substantially longer than in OECD countries (Figure 8.14). While in OECD countries 33% of the facilities emitting CO_2 in 2011 will still be emitting CO_2 emissions in 2035, this share is 62% for non-OECD countries. This difference is attributable to the much younger age of existing power plants and industrial facilities in non-OECD countries, the significantly larger amount of infrastructure under construction and higher vehicle lifetimes. Emissions in non-OECD countries in 2011 were 17.8 Gt CO_2. Due to recent and planned investments, emissions from facilities existing or planned in 2011 in non-OECD countries increase over the next few years and then decrease to 11 Gt CO_2 in 2035. In non-OECD countries, 363 GW of fossil-fuel plants are currently under construction, reflecting strong electricity demand growth.

8

Figure 8.14 ▷ Energy-related CO_2 emissions from locked-in infrastructure in 2011 and in the 450 Scenario in non-OECD countries

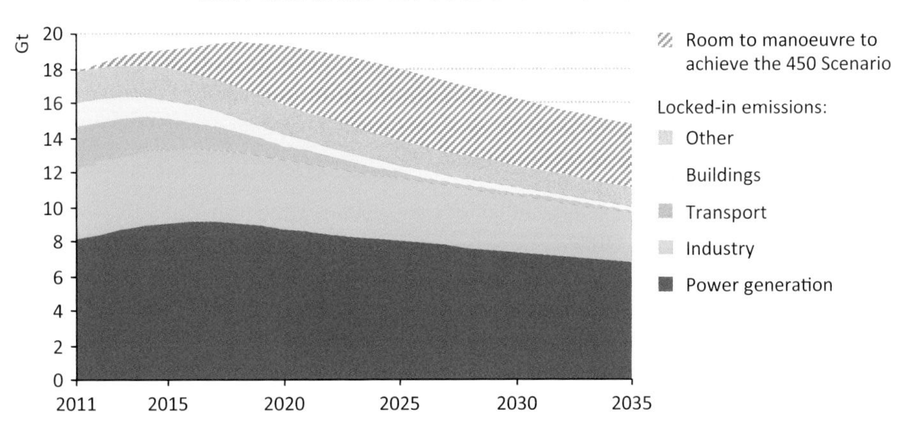

Box 8.2 ▷ The relationship between the 450 and Efficient World Scenarios

The 450 Scenario is designed to illustrate the energy world which results from taking plausible actions to limit the likely global temperature increase to 2 °C. The Efficient World Scenario, for its part, explains the implications for the economy, the environment and energy markets of doing no more than exploit energy efficiency opportunities which justify themselves in economic terms. The average global temperature increase in the Efficient World Scenario is an outcome of the modelling, not a pre-determined target.

In the 450 Scenario, we assume that an international climate agreement is put in place of sufficient stringency to ensure governments adopt the policies necessary rapidly to decarbonise the energy system. Such policies include a substantial price for CO_2 emissions and support for low-carbon technologies. OECD countries take action as early as 2015, adopting CO_2 prices that rise to as high as $120/tonne CO_2 in 2035. Non-OECD countries join the efforts by 2020, phasing out fossil fuel subsidies and, in major emitting countries, introducing a CO_2 price. Carbon pricing enables, in addition to national measures, a fast uptake of energy efficiency across all sectors.

The underlying assumptions in the Efficient World Scenario are that energy efficiency investments will be made as long as they are economically viable with the market prices prevalent in this scenario and that market barriers obstructing their realisation having been removed. The energy efficiency potentials are determined for each sector and region following a thorough review of energy efficiency options and their associated payback periods.

In general, measures that decrease CO_2 emissions also save energy. As energy efficiency is often the most economic way to reduce CO_2 emissions, the CO_2 trajectories in the 450 Scenario and the Efficient World Scenario are very similar up to 2020. However, in the 450 Scenario beyond 2020, additional measures are necessary, especially a large increase in the use of renewables and uptake of CCS. As a result, emissions in the 450 Scenario fall below their level in the Efficient World Scenario. CCS and higher use of biomass reduce CO_2 emissions, but increase the energy requirements in most cases. Therefore, the Efficient World Scenario includes neither a large increase in the use of biomass nor the deployment of CCS.

The CO_2 savings attributed to energy efficiency are similar in the 450 Scenario and the Efficient World Scenario. However, the policy measures employed are not necessarily the same. The Efficient World Scenario sees a focus on the removal of market barriers, end-user prices consequently fall lower than in New Policies Scenario, as less demand depresses international fossil-fuel prices (albeit with a partial rebound in the demand for energy). In the 450 Scenario, mandatory standards and international sectoral agreements are assumed to be widespread. Electricity prices are high, as the implicit price of CO_2 is passed through to consumers and end-user prices reflect the removal of subsidies; in transport, a tax is applied to compensate for lower international oil prices.

Like the power sector, the non-OECD industrial sector has seen a significant amount of capacity added in recent years, as increasing demand for industrial products, such as steel and cement has spurred industrial growth. As a result, emissions in 2035 from industrial plants already built are 70% of the 2011 level, far higher than the OECD figure at 41%. The continuing rapid expansion of industry infrastructure and buildings stock over the next few years in non-OECD countries presents a crucial window of opportunity to reduce locked-in emissions by installing efficient and low-carbon infrastructure.

On a global basis, if infrastructure investments, particularly in industry facilities and power plants, continue in line with the New Policies Scenario and are operated as in that scenario, infrastructure in existence in 2017 and expected to continue to operate through to 2035 would emit all the cumulative emissions allowed in the 450 Scenario (Figure 8.15). In other words, after 2017 any new power plants, industrial plants, new buildings, road vehicles or water boilers that consume fossil fuels could be built only if existing infrastructure were retired early to the extent necessary to offset emissions from the additional infrastructure.

Figure 8.15 ▷ Global energy-related CO_2 emissions in the 450 Scenario and from locked-in infrastructure

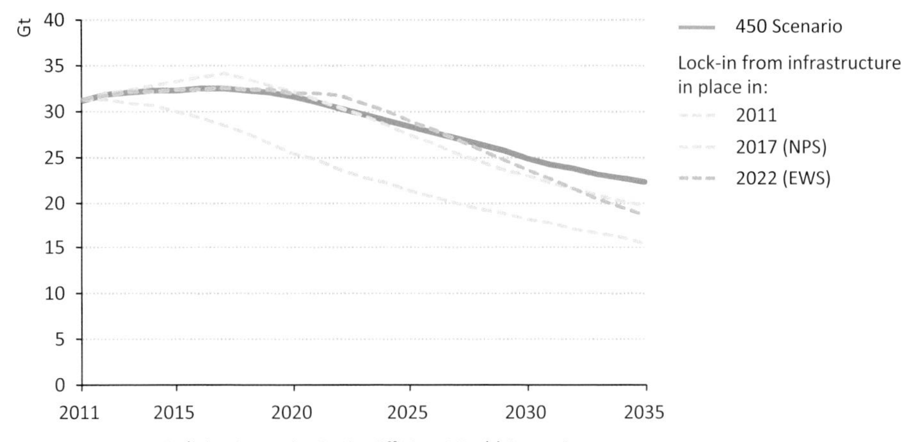

Note: NPS = New Policies Scenario, EWS = Efficient World Scenario.

In the absence of a global climate agreement, measures taken solely in pursuit of greater energy efficiency could significantly delay emissions lock-in from existing infrastructure, if vigorously pursued. If infrastructure investments were made in line with the Efficient World Scenario, new plants and facilities could continue to be built up to 2022 before the entire emissions budget of the 450 Scenario became locked-in, as against the date of 2017 in the New Policies Scenario (Figure 8.15). In other words, the Efficient World Scenario can buy another five years grace in the effort to achieve a 2 °C target. CO_2 emissions of the 450 Scenario would initially be exceeded, but would later fall substantially to regain the earlier excess emissions. Chapter 10 describes in detail what needs to be done to realise the energy savings in the Efficient World Scenario.

PART B
FOCUS ON ENERGY EFFICIENCY

PREFACE

Part B of this *WEO* (Chapters 9-12) analyses in depth the outlook for energy efficiency in final uses and in energy production and transformation activities.

Chapter 9 reviews recent energy efficiency trends – whether global energy use is, or is not, becoming more efficient per unit of output – and the changes that have taken place recently on the policy front. It discusses key barriers to change, an essential precursor to discussing how energy efficiency might be improved in the future. Energy efficiency gains in the New Policies Scenario are then presented, together with their implications and the level of investment in energy efficiency needed to attain those gains.

Chapters 10-12 break new ground, presenting an Efficient World Scenario. The scenario makes no bold assumptions about technical breakthroughs, but instead shows the extent of benefits that could be achieved if known best technologies and practices to improve energy efficiency were systematically adopted. Technologies implemented in the Efficient World Scenario are subject to a stringent test of their economic viability, expressed as the acceptable payback period for each class of investment. Necessary policies to realise the scenario are discussed in the context of categories of energy use and government actions needed to eliminate barriers presently obstructing the uptake of energy efficiency. Chapter 10 presents quantitative analysis of the Efficient World Scenario for the global energy economy and examines its implications. Chapter 11 looks at the scenario sector-by-sector. Mainly through figures and tables, Chapter 12 presents results for the world and five key regions and countries.

Energy efficiency: the current state of play

Are we putting enough energy into improving efficiency?

Highlights

- Energy efficiency curbs demand growth, reduces energy imports and mitigates pollution. In the last year, all major energy-consuming countries introduced new legislation on energy efficiency, making provisions for a 16% reduction in energy intensity by 2015 in China, new fuel-economy standards in the United States and a cut of 20% in energy demand in the European Union in 2020. Japan also aims to achieve a 10% reduction in electricity demand by 2030 in its new energy strategy.

- Implementation of those policies and of those under discussion in many other countries, at the level assumed in our New Policies Scenario, would result in annual improvements in energy intensity of 1.8% over 2010-2035, a very significant increase compared with 1.0% per year achieved over 1980-2010. In the absence of those gains, global energy demand in 2010 would have been 35% higher, almost equivalent to the combined energy use of the United States and China.

- In the New Policies Scenario, efficiency accounts for about 70% of the reduction in projected global energy demand in 2035, compared with the Current Policies Scenario. China, the United States, the European Union and Japan account for more than half of the savings, reflecting their dominance in global energy use and the emphasis placed on energy efficiency in these regions. Additional investment of $3.8 trillion to improve energy efficiency in end-use sectors is needed over 2012-2035, an average of $158 billion per year. Energy efficiency measures in the New Policies Scenario account for 68% of the cumulative global savings in CO_2 emissions relative to the Current Policies Scenario.

- The payback periods of the efficiency measures included in the New Policies Scenario are short, ranging from as low as two years for electrical equipment to eight years for space and water heating; but non-technical barriers remain a major obstacle. Monetising those barriers significantly increases payback periods and renders energy efficiency investments less attractive, especially for the buildings sector. These are the barriers governments have to tackle.

- Despite the vital role that energy efficiency plays in cutting demand in the New Policies Scenario, only a small part of its economic potential is exploited. Over the projection period, four-fifths of the potential in the buildings sector and more than half in industry still remain untapped. Much stronger policies could realise the full potential of energy efficiency and deliver significant economic, environmental and energy security gains.

Introduction

Policy makers confronted with the twin challenges of ensuring reliable and affordable energy supplies and dealing with climate change have consistently identified energy efficiency as an essential means of moving to a more sustainable energy future. Energy and economic analyses, including in previous editions of the *World Energy Outlook*, point to the same conclusion: improving energy efficiency in energy-importing countries reduces import needs or slows their growth; measures can be implemented quickly compared with often lengthy projects to expand production; it is among the cheapest of the large-scale carbon dioxide (CO_2) abatement options; and it can play a role in spurring economic growth and reducing energy bills, both of particular importance during this period of economic uncertainty and persistently high energy prices.

Box 9.1 ▷ What are we including when measuring energy efficiency?

Improving energy efficiency can be defined as using less energy to provide the same level of service. For example, when a compact florescent light (CFL) uses less electricity than an incandescent bulb to produce the same amount of light, the CFL is considered to be more energy efficient. But energy savings can arise from more than just switching to more energy-efficient technology. Fuel switching can also reduce primary energy needs. For example, switching away from a gas boiler for space heating to the use of heat pumps can substantially reduce energy needs per unit of heat produced. Energy consumption is also dependent on behavioural factors, such as the chosen level of thermal comfort or the preference to use a private car for personal mobility. In many cases, savings that arise from behavioural changes are classified as energy conservation, rather than energy efficiency. The main difference between the two is that reducing the absolute level of energy demand is the primary goal of energy conservation, if necessary, at the expense of personal comfort or satisfaction, while improved energy efficiency aims to reduce the energy consumed in delivering a given energy service (IEA, 2012a).

In this *Outlook*, we measure energy efficiency improvements as the savings that arise exclusively from improvements in technology (without a change of fuel). We separately quantify changes in energy demand arising from fuel switching (including changes in technology) and from price-driven behavioural changes. This decomposition approach allows us to understand the relative importance each factor is set to play in curbing future energy demand.

While energy efficiency clearly has many merits, it is challenging to measure: unlike primary fuels, it does not appear in national energy balances, and it is neither traded nor priced, except in a few countries. Energy efficiency can be measured at a micro-level, for example, by quantifying the reduction in the volume of fuel needed to drive a certain distance if a particular efficiency measure is adopted. Far more challenging is to understand the contribution made by a multitude of energy efficiency improvements to aggregate energy savings at a national level. Often the challenge relates to calculating what might have been used, under the same conditions without improved technology and practices. Many associated effects need to be disaggregated before the specific energy efficiency effect on energy consumption can be estimated. The detailed country-by-country data needed to separate out these effects are available only for a small number of OECD countries (IEA, 2011a).

For the purpose of this *Outlook*, in order to analyse the role energy efficiency is set to play in the future, we use decomposition analysis. This decomposes changes in energy demand into changes in efficiency, fuel and technology switching, and activity (Box 9.4). To compare the role energy efficiency has played historically in tempering growth in energy demand, we use energy intensity as the best available proxy, as in most cases the data that would be needed to disaggregate the savings into their various contributing factors – including efficiency – are not available. The shortcoming of energy intensity as an indicator of energy efficiency nonetheless needs to be recognised, as it fails to distinguish the effects of factors such as changes in the structure of a country's economy or its climate. For example, service-oriented and temperate countries typically have lower energy intensities than manufacturing-based and colder countries, regardless of their energy efficiency. Moreover, countries with a high proportion of energy-intensive industries, such as Korea, may be extremely efficient in the way they use energy, but still have high energy intensities.

The current status of energy efficiency

Global energy intensity, expressed as the amount of energy used to produce a unit of gross domestic product (GDP),[1] has fallen over the last several decades, primarily as a result of efficiency improvements in the power and end-use sectors and a transition away from energy-intensive industries. But the rate of decline in energy intensity has slowed considerably: from 1.2% per year on average between 1980 and 2000, to only 0.5% per year between 2000 and 2010 (Figure 9.1). This slowdown can largely be explained by an ongoing shift of global economic activity towards countries in developing Asia which have relatively high energy intensities due to their heavy reliance on energy-intensive industries and on coal-fired power generation, which is typically less efficient than other power generation options. In 2009 and 2010, global energy intensity actually increased – bucking the long-term downward trend – due to colder or hotter than average weather in some

1. Energy intensity is measured using GDP at market exchange rate (MER). It can be measured also with GDP expressed in terms of purchasing power parity (PPP), which enables differences in price levels among countries to be taken into account. However, it is misleading to do so when comparing long-term projections of energy intensity. This is because the PPP factors are likely to change in the future as countries become richer or poorer relative to the world average. Thus, using PPP-adjusted GDP projections tends to overstate the future relative importance of emerging economies, as it is likely that their PPP factors will be adjusted down in the future as they become richer (though it is possible that the applicable currency exchange rates may also rise, offsetting this effect to some degree). Therefore, we use GDP at market exchange rates in order to provide a coherent set of information for past and future trends.

regions, lower energy prices and economic contraction in 2009.[2] Preliminary data point to a slight improvement in energy intensity in 2011 of 0.6%, meaning that the long-running trend has been restored after the outliers of 2009 and 2010.

Figure 9.1 ▷ Global energy intensity average annual growth rates, 1971-2010

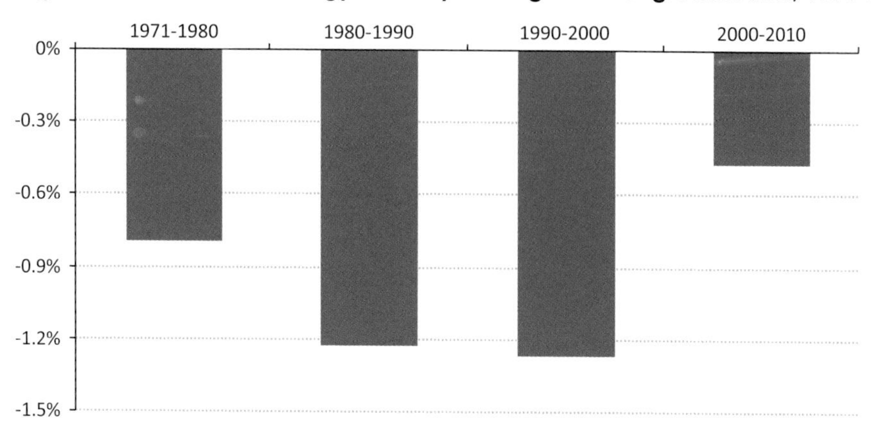

Note: Energy intensity is measured using GDP at market exchange rate (MER) in year-2011 dollars.

Global energy demand grew from 7 234 million tonnes of oil equivalent (Mtoe) in 1980 to 12 730 Mtoe in 2010, a 76% increase. Over this same period, global GDP expanded by 137%. Without the improvements in energy intensity that were realised over the period, global energy demand in 2010 would have been 35% higher. Energy prices have an important influence on energy intensity, a lesson learned in the 1970s when oil price spikes spurred a wave of energy efficiency and conservation measures that contributed to significant savings in energy demand in the 1980s. Energy intensity improvements in the 1990s cut growth in energy demand over the decade by more than 50%, while in the 2000s they resulted in savings of almost 20%.

Energy intensities tend to be much higher in developing countries than in the OECD, although there has been progressive convergence over the past three decades: the ratio among the highest and lowest values has declined from an average factor of nine in the 1980s to just under five currently (Figure 9.2). This has arisen primarily from globalisation, which has made similar technologies available at similar costs in different parts of the world, and from countries sharing best practices in energy efficiency policy and management techniques. Although energy efficiency is now getting more policy attention than in the past in some parts of the Middle East, the region's energy intensity has been increasing since the 1980s, in large part driven by artificially low energy prices that discourage the deployment of energy-efficient technologies (Figure 9.3). Africa's economic development drove energy

2. Running an industrial facility below full capacity increases its energy intensity. This is because energy consumption declines by a smaller amount than economic output, as a result of an element of fixed energy consumption that is independent from the output level, *e.g.* space heating, ventilation and lighting.

intensity up until around 1995, but it has since been on a downward path, with incremental energy demand (mainly for traditional biomass for cooking) growing at a slower rate than GDP. Latin America as a whole has seen economic output per unit of energy input remain relatively flat, with energy-producing countries' relatively high level of energy intensity offsetting less energy-intensive use in importing economies. High energy prices, energy efficiency measures, CO_2 abatement policies, the move towards more service activities, increasing productivity and, in some countries, a switch away from traditional biomass explain the declining trends in other regions.

Figure 9.2 ▷ **Energy intensities by regions, 1980 and 2010**

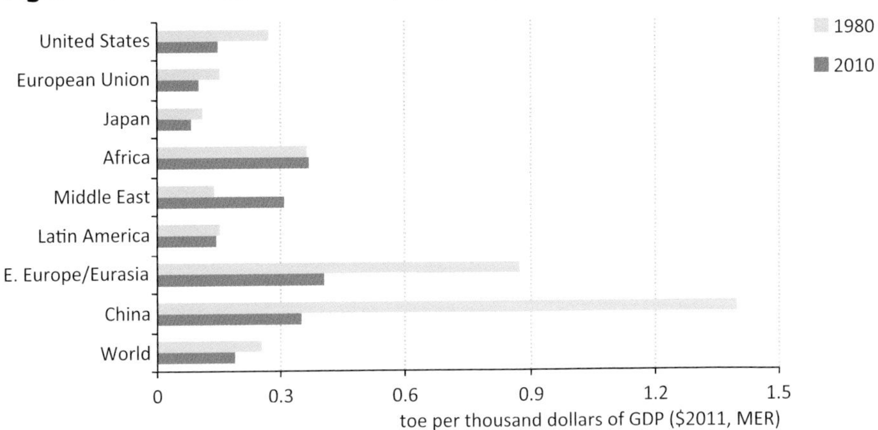

Among OECD countries, the United States has achieved the biggest improvement in energy intensity in recent decades, albeit from relatively high levels. Its energy intensity declined at an average rate of 2% per year from 1980 to 2010. In recent years, policy efforts to further improve energy efficiency have been reinforced. The 2009 economic stimulus package in the United States included new energy efficiency initiatives and substantial additional funding for existing programmes, while, in 2011, fuel-economy standards were introduced for heavy-duty vehicles, alongside a tightening of standards for light-duty vehicles.

Japan and the European Union are the regions with the lowest energy intensity. This reflects a high general level of technical efficiency, spurred by relatively high energy prices and policies to raise awareness among citizens and companies, and to promote energy-efficient technologies. Japan experienced an average improvement of 0.9% per year from 1980 through 2010; while the European Union achieved 1.4% per year, benefiting from the significant potential in some member countries to further improve efficiency.

Figure 9.3 ▷ **Energy intensity trends by region, 1980-2010**

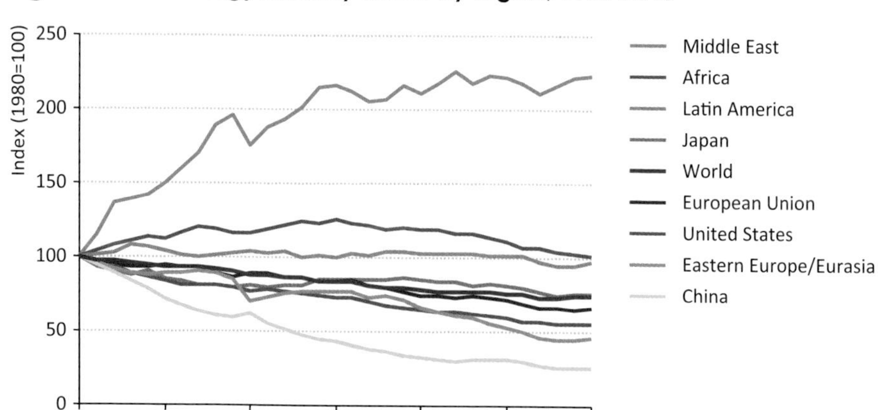

Existing policies

An increasing number of countries have made energy efficiency a key pillar of their energy strategies, viewing it as a means of realising gains in multiple areas of public policy. The European Union is in the process of adopting an energy efficiency directive – to complement its carbon and renewables policies – that envisages a 20% reduction in energy demand by 2020 against a business-as-usual approach.[3] Even though Japan is already a leader in energy efficiency, in September 2012 the government released the Innovative Strategy for Energy and the Environment, which calls for a greater focus on improving energy efficiency and includes a target to cut electricity demand by 10% in 2030, compared with 2010 (Box 9.2). The United States extended the US Corporate Average Fuel Economy standards to 2025, one element in an emerging trend towards much lower oil-import needs (see Chapter 2). Among the emerging economies, China has adopted ambitious energy intensity targets in its 11th and 12th Five-Year Plans, underpinned by a broad set of policies, including policies that aim to facilitate structural changes in the economy. In India, continuing the persistent efforts of the past few years, preparatory work for the 12th national plan identifies energy efficiency as an important cornerstone of a secure energy future. The range of policies being adopted by governments to improve energy efficiency is wide: regulations, market-based instruments, financial instruments, and information and awareness measures all feature, often brought together in an overarching strategy or framework (Table 9.1).

Cross-sectoral policies typically form the framework for energy efficiency measures across individual sectors. They often include targets, in terms of energy efficiency, energy intensity improvements or energy savings. In the European Union, for example, the Energy Efficiency

3. The European Parliament and Council reached an agreement in June 2012 on the text of the Energy Efficiency Directive, which specifies that primary energy demand should not exceed 1 474 Mtoe, or final energy demand exceed 1 078 Mtoe in 2020. The European Parliament voted in favour of the Directive in September 2012, and it will be submitted to Council vote before the end of the year.

Directive requires member states by April 2013 to set national targets for primary energy savings for 2020 and to achieve end-use energy savings of 1.5% per year; member states are required to report on these targets and develop National Energy Efficiency Action Plans. Other important instruments with wide application include energy price and tax policies, as well as financial mechanisms to incentivise and facilitate investment in energy efficiency. For example, China has provided $21 billion of tax incentives for industrial energy efficiency investments. Australia has recently introduced a carbon price that will incentivise investment in energy efficiency through increased electricity prices, as well as the Energy Efficiency Opportunities programme, which seeks to induce all large energy-users to improve their energy efficiency. Many countries have provided grants and tax incentives for energy efficiency investments and are now increasingly developing longer-term financial mechanisms to catalyse private sector investments in energy efficiency. However, subsidies to consumer prices for energy still remain a major barrier to improving energy efficiency, limiting the incentive for consumers to reduce their energy use (see Chapter 2).

Countries are increasingly introducing policies to improve the energy performance of buildings, which made up one-third of global final energy demand in 2010. Most OECD countries have mandatory energy codes for new and existing buildings, albeit with different levels of stringency and enforcement. In non-OECD countries, where new buildings represent a much higher share of the stock, energy efficiency codes for buildings are increasingly being introduced (including in China), although sufficient attention is not always being given to enforcement. Some countries, such as Germany, have introduced financing options to encourage private investors to improve energy efficiency in buildings. Energy labelling is an important way of informing consumers of the energy use of buildings, equipment and appliances at the time of purchase or rental. It is now mandatory for buildings and some appliances in a growing number of regions, including across the European Union. Some countries have set targets for passive or zero-energy buildings.[4]

In the transport sector, fuel economy policies are increasingly being implemented: most of the major car markets in the world – which collectively accounted for two-thirds of global passenger vehicle oil demand in 2010 – now have fuel-economy standards in place. An area which is only recently starting to attract more attention is freight transport – though it accounted for around one-third of road transport oil demand in 2010, fuel-economy standards for the freight fleet are in force or under discussion in only a few countries. Fuel economy labelling and related fiscal measures are also widely used for passenger vehicles, but less so for trucks. They are important especially in countries without the technical capacity to introduce (or enforce) fuel economy regulations (IEA, 2012b). Aside from improving the fuel economy of road vehicles, policies to reduce private vehicle use by improving public transport and land-use planning are also becoming more widespread.

4. Buildings that have net zero-energy consumption, achieved with extremely efficient building shells and heating/cooling systems combined with energy producing technologies, like solar (BPIE, 2011).

Table 9.1 ▷ Overview of key energy efficiency policies that are currently in place by country/region and sector

	United States	Japan	European Union
Cross-sectoral			
Energy efficiency strategy or target	None	None	EU Energy Efficiency Directive agreed; National Energy Efficiency Action Plans required; EU-level target to reduce primary energy consumption by 20% in 2020; EU ETS.
Buildings, appliances, equipment and lighting			
Building energy performance requirements	Mandatory energy requirements in building codes in some states.[5]	Voluntary guidelines in place.	Building energy performance requirements for new buildings (zero-energy buildings by 2021) and for existing buildings when extensively renovated; 3% renovation rate of central government buildings.
Energy labelling	Voluntary buildings labelling; mandatory and voluntary labelling for some appliances and equipment.	Voluntary buildings labelling; national voluntary equipment labelling programmes.	Labelling mandatory for sale or rental of all buildings and some appliances, lighting and equipment.
Equipment energy performance requirements	45 products covered.	Top Runner: 23 products covered.	15 product groups in EcoDesign Directive, further product groups planned end-2012; phase-out of incandescent light bulbs.
Transport			
Fuel-economy and GHG standards	PLDV: 34.5 mpg by 2016 (6.8 l/100 km), 54.5 mpg by 2025 (4.3 l/100 km); trucks: starting MY 2014.	PLDV: 16.8 km/l by 2015 (5.95 l/100 km); trucks: 12.2% target by MY 2015.	PLDV: 130 g CO_2 /km by 2015 (5.2 l/100 km); 95 g CO_2/km (3.8 l/100 km) planned by 2020; LCV: 147 g CO_2/km (5.9 l/100 km) by 2020; no trucks yet.
Fuel economy labelling	PLDV: yes; trucks: none.	PLDV: yes; trucks: yes.	PLDV: yes; trucks: none.
Fiscal incentives	Gas guzzler tax; EV tax credit; rebates in many states for EVs.	Registration taxes by CO_2 emissions and fuel economy.	Taxes are lower for vehicles with low average CO_2 emissions in most countries.
Industry			
Energy management programmes	Voluntary energy management programme for the implementation of ISO-50001.	Periodic energy audits and nationally certified energy managers for large industries.	Voluntary agreements in place or planned in many countries.
MEPs for electric motors	Premium efficiency (IE3) MEPs for 3-phase induction motors.	Adding 3-phase induction MEPs to Top Runner programme.	IE3 for 3-phase induction motors < 7.5 kW by 2015; all IE3 (IE2 + variable speed drive) in 2017.

5. For residential buildings in 18 states, based on 2009 code, and in 24 states for public and commercial buildings, based on 2007 code. One state adopted 2012 code for residential and 2010 for commercial buildings.

Table 9.1 ▷ Overview of key energy efficiency policies that are currently in place by country/region and sector (continued)

Russia	China	India	Brazil
Cross-sectoral			
2009 Federal Law No. 261-FZ on energy saving and improving energy efficiency; reduce energy intensity by 40% by 2020.	12th Five Year Plan (2011-2015): target to reduce energy intensity by 16% by 2015.	11th Five-Year plan (2007-2012): target to improve energy efficiency by 20%; 12th Five-Year plan forthcoming.	2011 National Energy Efficiency Plan; reduce projected power consumption by 10% by 2030.
Buildings, appliances, equipment and lighting			
Mandatory building codes (but not yet fully implemented).	Mandatory codes for all new large residential buildings in big cities.	Energy Conservation Building Code (2007), with voluntary requirements for commercial and residential buildings.	Voluntary guidelines in place.
Information on energy efficiency classes for appliances required since January 2011.	Labelling mandatory for new, large commercial and governmental buildings in big cities.	Voluntary Star Ratings for office buildings.	Voluntary for residential and commercial buildings.
Phase-out of incandescent >100 Watt light bulbs.	46 products covered by labelling schemes.	Mandatory S&L for room air conditioners and refrigerators, voluntary for 5 other products.	13 products covered by voluntary labels.
Transport			
PLDV: none; trucks: none.	PLDV: 6.9l/100 km by 2015, 5.0 l/100 km by 2020; trucks: proposed MY 2015.	PLDV: under development; trucks: none.	None
PLDV: none; trucks: none.	PLDV: yes; trucks: none.	PLDV: none; trucks: none.	None
None	Acquisition tax based on engine size.	Registration taxes by vehicle and engine size, sales incentives for advanced vehicles.	None
Industry			
Periodic energy audits required for some industries.	Top 10 000 programme setting energy savings targets by 2015 for the largest 10 000 industrial consumers.	PAT (Perform, Achieve, Trade) in force since 2011. Audits mandated for designated consumers.	None
None	High-efficiency (IE2) MEPs for 3-phase induction motors in place.	None	High-efficiency (IE2) MEPs for 3-phase induction motors in place.

Notes: ETS = emissions trading system; PLDV = passenger light-duty vehicle; trucks = road freight trucks > 3.5 tonnes; EV = electric vehicle; LCV = light commercial vehicle; mpg = miles per gallon; MY = model year; l/100 km = litres per 100 kilometres; g CO_2/km = grammes of CO_2 per kilometre; MEPs = minimum energy performance standards; IE = international efficiency classes for motors (IE1 = standard efficiency; IE2 = high efficiency; IE3 = premium efficiency); S&L = standards and labels.

Box 9.2 ▷ Lessons from Japan's energy-saving "Setsuden" campaign

Following the devastating earthquake and resulting tsunami that hit Japan in March 2011, many power plants were shut down or damaged, leading to a substantial loss of generating capacity. For example, Tokyo Electric Power (TEPCO), which was supplying 42 million people and entities responsible for 40% of Japan's GDP, lost 40% of its capacity (METI, 2011a). Steps were quickly taken to restore supplies, including by repairing damaged plants, bringing emergency generators into service, purchasing power from independent and private producers and using pumped-storage power stations.

Despite these efforts, major power shortages were anticipated, prompting the government to launch Japan's energy-saving "Setsuden" campaign. In May 2011, a target was set to reduce peak power demand in the east of the country by 15% (METI, 2011b). Mandatory demand restrictions were applied, for the first time since 1974, to all large businesses, while small businesses were encouraged to take voluntary measures. Industries reduced or changed their working hours. Diesel generators were installed. In some cases, production lines were moved to factories in west Japan or overseas. In the public sector, lights were removed, dimmed or switched off, air-conditioning temperatures were raised, and trains and metros ran less frequently. Households were encouraged to use electric fans instead of air conditioners, to use blinds to reduce heat from sunlight and to disconnect electric appliances not in use.

As a result of the Setsuden campaign, and thanks to a fairly mild summer, summer peak power demand in east Japan was cut by over 15% (equal to one-fifth of the output from nuclear plants in 2010) and unplanned blackouts were avoided (METI, 2011c). This success can be attributed to the collective efforts of the Japanese people and the government's leadership in identifying the efficiency potential by sector and providing advice about how it might be achieved. But not all of the measures applied would be appropriate during ordinary times. Workers cannot routinely be asked to shift working hours or delay or bring forward their holidays. Some industrial processes cannot abruptly cut demand without prejudicing their competitiveness. Saving power is an important aim, but not at the expense of damaging economic performance.

Based on experience gained from the Setsuden campaign, Japan plans to further improve its energy efficiency, even though it is already a global leader in the field. Steps being taken include amending the energy conservation regulations to favour measures to reduce peak demand and adding building materials to the existing energy efficiency target programme.

Since the 1970s, countries have introduced numerous policies and measures to promote energy efficiency in the industry sector, with varying levels of success. The most common measures include incentives (in the form of subsidies or energy taxes), emissions trading schemes, equipment performance standards, energy management programmes and funding of research and technology development. However, while most countries have some type of energy efficiency policies targeting industry, progress in integrating energy efficiency consideration into business decision making has been limited. This may be due, in part, to a lack of measures such as support for capacity building and training, and facilitating access to energy efficiency service providers and financing; but it is also due to the fact that energy consumption makes up only a small part of costs in most industries (UNIDO, 2011). The introduction of energy efficiency policies is just the first step in ensuring that energy savings materialise; of equal importance is monitoring their implementation (Box 9.3).

Box 9.3 ▷ The importance of effective implementation

To support governments in their implementation of energy efficiency, the IEA recommended to the G-8 summits in 2006, 2007 and 2008 the adoption of specific energy efficiency policy measures. The consolidated set of recommendations covers 25 fields of action across seven priority areas: cross-sectoral activity, buildings, appliances, lighting, transport, industry and energy utilities (IEA, 2009). These recommendations were updated and endorsed by IEA member countries in 2011. However, while proposing or putting in place a set of policies is a necessary requirement to tap energy efficiency potential, it is not sufficient to deliver the intended savings. Verification of effective implementation is essential. Evaluations of member countries in 2009 and 2011 revealed mixed progress in implementation of the IEA 25 Energy Efficiency Policy Recommendations (IEA, 2009, 2011a, 2011b and 2012c). In 2011, only 40% of them were either fully or substantially implemented (IEA, 2012c).

Three key issues need to be addressed to increase effective implementation:

- *Market surveillance and enforcement.* Non-compliance with minimum energy performance standards (MEPS) for equipment can be as high as 20-50% in the absence of enforcement regimes (IEA, 2010). For example, although building codes apply in theory in many countries, their enforcement remains a challenge in both OECD and non-OECD countries. Without a credible enforcement strategy, mandatory regulations or standards lose much of their force, representing a significant policy and economic failure for the implementing country.
- *Institutional arrangements.* Agencies tasked with establishing and delivering energy efficiency policies must co-ordinate a number of very diverse tasks, ranging from policy analysis, through project design and management, marketing, programme evaluation and many other elements. Ensuring effective co-ordination horizontally (within a single level of government, for example between national-level institutions) and vertically (between different levels of government, for example national to regional) is essential to effective delivery of energy efficiency policy outcomes.
- *Energy efficiency data.* Monitoring progress in energy efficiency requires detailed end-use sectoral data on energy consumption at regular intervals, which means adequate resources must be dedicated to this effort. Only a few OECD countries include energy efficiency data in their energy sector statistics.

Barriers to energy efficiency deployment

While investment in many energy-efficient technologies and practices appear to make good economic sense, the level of their deployment is often much lower than expected. This is due to the existence of a number of barriers that discourage decision makers, such as households and firms, from making the best economic choices (Table 9.2).

Table 9.2 ▷ Key barriers and remedial policy tools

	Barrier	Effect	Remedial policy tools
Visibility	Energy efficiency is not measured.	Opportunity not known to exist and so not acted upon.	Test procedures/measurement protocols/efficiency metrics.
	Efficiency is measured but not made visible to decision makers.	Opportunity not visible to decision makers and so not acted upon.	Ratings/labels/disclosure/ benchmarking/audits/real-time measurement and reporting.
Priority	Low awareness of the value of efficiency.	Energy efficiency is undervalued.	Awareness raising and communication efforts.
	Efficiency investments are bundled with all other investment decisions.	Efficiency investments can appear to be a low priority.	Regulation, mechanisms to decouple efficiency actions from other concerns.
Economy	Split incentives.[6]	Costs and benefits are not taken into account fully and energy efficiency is undervalued.	Regulation, financing mechanisms that incentivise investment in efficiency.
	Insufficient finance available or competing needs.	Under-investment in efficiency.	Stimulation of capital supply for efficiency investments, support of new efficiency business and financing models.
	Energy consumption subsidies.	Market conditions do not encourage efficiency.	Removal of subsidies.
	Unfavourable perception or treatment of risks.	Financing cost of efficiency projects is inflated, or energy price risk is under-estimated.	Better information on project and energy price risks, mechanisms to reduce efficiency project risk.
Capacity	Limited know-how on implementing energy efficiency measures.	Energy efficiency implementation is constrained.	Capacity building programmes.
	Limited government resources to support implementation.	Barriers addressed more slowly.	Shift government resources toward efficiency goals.
Fragmentation	Energy consumption is split among diverse range of end-uses and users.	Efficiency is more difficult to implement collectively.	Targeting regulations and other policies toward high-impact groups.
	Business models focused on either energy supply or energy demand.	Energy supply often favoured over energy service.	Regulations that reward overall energy service provision rather than just energy supply.
	Fragmented and under-developed supply chains.	Efficiency opportunities are more limited and more difficult to implement.	Programmes aimed at better market integration and overall economies.

6. A split incentive, sometimes referred to as a principal-agent problem, refers to the potential difficulties in motivating one party to act in the best interests of another when they may have different goals and/or different levels of information.

A major barrier to greater deployment of energy-efficient technologies and practices is a lack of awareness, or "visibility", by a decision maker. This may either be because the energy efficiency opportunity has not been measured, and therefore is not even known to exist, or because a decision maker has not been made aware of the fact that it exists. In both cases, it means that the decision maker is taking an investment decision based on imperfect information.

Another significant barrier is the fact that the costs and benefits of an energy efficiency investment decision can often fall on different actors. This is sometimes referred to as "split incentives". These split incentives can occur in many ways, such as:

- Between landlords and tenants, where building owners have little incentive to invest in improving energy efficiency because they do not pay the energy bill (the tenant does).
- Between buyers and operators, where the person in a firm that is responsible for purchasing a piece of capital equipment is not the person responsible for its operation and maintenance budget. In this case, the equipment with the lowest purchase cost may be chosen even if higher running costs mean it is more expensive over its operational life.
- Where the payback period[7] for investing in an energy efficiency improvement is longer than the length of time the buyer intends to own the asset (or thinks they might own it). This is a particularly serious concern for long-lived assets, such as buildings, where the payback period on an investment may be longer, but can also apply to relatively short-lived energy-using equipment, such as appliances.
- Where the benefits of improved energy efficiency are valued by a collective group, or society as a whole, rather than by the individual making their own investment decision.

The barriers to energy efficiency deployment can be compounded by the fact that energy consumption is divided across a multitude of diverse end-uses and users, suppliers and business models (fragmentation). This fragmentation complicates the design and adoption of widely-applicable solutions, increasing the complexity and cost of developing effective energy efficiency intervention programmes.

Whenever money is paid to overcome these barriers there is a direct transaction cost and, even when no money is paid directly, there may be considerable implicit costs, such as the time spent searching for information. These costs need to be factored into any energy efficiency investment decision. One technique that is used to value implicit transaction costs is to apply a monetary value to the time spent in overcoming the barrier, *e.g.* by assigning a value for time equal to the average hourly wage.

When transaction costs are included, the average payback period of an energy efficiency measure can increase to the point that some potential investments become unattractive. Studies suggest that the transaction costs for energy efficiency projects undertaken by industry can increase costs by between 9% and 40%, with smaller projects typically

7. The payback period is the length of time required to recover the initial cost of an investment. In this *Outlook*, the payback period is measured in years.

experiencing the largest increase (Mundaca and Neij, 2006). Transaction costs for households are often higher still (Joskow and Marron, 1992; EBRD, 2011), as obstacles to behavioural change, and the costs associated with finding or acquiring relevant information can be higher than for firms (Sathaye and Murtishaw, 2004).

It is important to recognise that there are a variety of energy efficiency policy instruments that are capable of removing most of the barriers and the associated transaction costs. Government action to develop and implement energy efficiency policies and programmes is crucial to ensure that they are overcome. For example, there is an important role to simply help raise awareness about the costs and benefits of potential investment decisions. Split incentives can be tackled through a mix of regulation and by devising business and financial solutions that share benefits equitably between different parties. Capacity and fragmentation barriers can often be reduced through stronger governance and resource allocations, regulation and capacity building efforts. Experience suggests that policy measures which target the removal of barriers affecting a concentrated and influential set of actors, such as specific energy-intensive industries, often produce the fastest results at the lowest cost.

The outlook for energy efficiency

Energy efficiency delivers the single largest share of energy savings in achieving the New Policies Scenario and in moving beyond it, reflecting the large amount of cost-effective potential that exists. Efficiency accounts for about 70%, or 1 060 Mtoe, of the reduction in projected global energy demand in 2035 in the New Policies Scenario, compared with the Current Policies Scenario (Figure 9.4). Energy demand in the New Policies Scenario still grows by 35% in the period 2010-2035, but without the implementation of the assumed efficiency measures the growth would be 43%. As a result, global energy intensity in the New Policies Scenario declines at 1.9% per year on average over the period 2011-2035, a significant improvement on the trend over the past decade.

Figure 9.4 ▷ Change in global primary energy demand by measure and by scenario

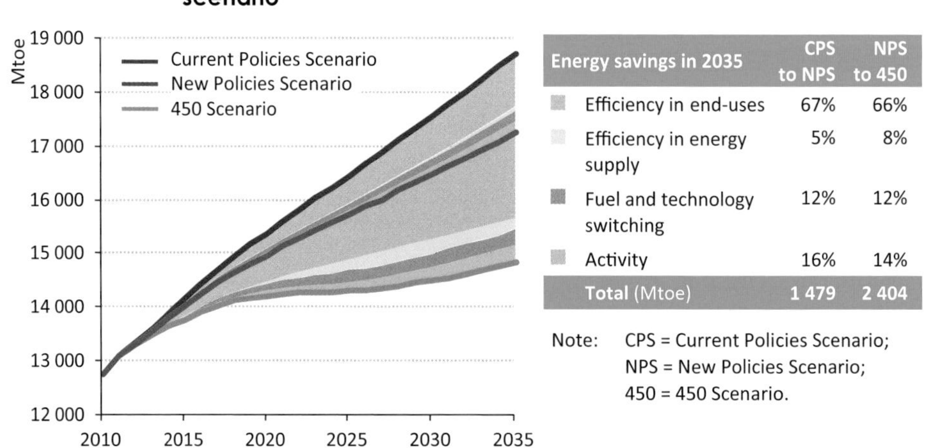

Energy savings in 2035	CPS to NPS	NPS to 450
Efficiency in end-uses	67%	66%
Efficiency in energy supply	5%	8%
Fuel and technology switching	12%	12%
Activity	16%	14%
Total (Mtoe)	**1 479**	**2 404**

Note: CPS = Current Policies Scenario; NPS = New Policies Scenario; 450 = 450 Scenario.

Box 9.4 ▷ **Decomposing the role of energy efficiency in curbing energy demand**

To analyse the role energy efficiency is set to play in the future, this *Outlook* uses a decomposition technique, which quantifies the relative importance of the different factors shaping future energy demand.[8] This has enabled us to attribute differences in projected energy demand across the *WEO-2012* scenarios to: efficiency effects; fuel and technology switching effects; or activity effects:

- *Efficiency effects.* Without changing the fuel used, the same level of energy service can be achieved with a lower level of energy use. Energy efficiency can be improved, for example, by changing from a conventional to a condensing boiler, by using a light-emitting diodes (LEDs) lamp instead of an incandescent one, or through better energy management of industrial processes.
- *Fuel and technology switching effects.* The same energy service can be provided by different fuels and technologies. For example, an oil-fired boiler or a heat pump can provide space heating in buildings; an electric arc furnace or a basic oxygen furnace can produce steel; and a car with an internal combustion engine or an electric motor can provide a transport service from one place to another. In each case, both technologies serve the same energy service demand, but with different associated levels of energy efficiency. This effect captures more than pure fuel switching as it also includes technology switching.
- *Activity effects*. Demand for energy services is influenced by a variety of factors, including economic and population growth and the level of end-user prices. As we assume the same GDP and population growth rates across our main scenarios, the primary reason for changes in activity between them is change in end-user prices.

We have applied this decomposition technique to all energy sectors – transport, residential, commercial, industry and power – using the best available information on sectoral energy demand, energy consumption and technologies (see Chapter 10, Figure 10.1). As we have focussed on changes in energy demand between scenarios, the results are not influenced by other factors – such as economic structure, non-price related changes in human behaviour, capacity utilisation in industry and weather patterns – that will no doubt change over time, but are the same across our scenarios.

Between 2011 and 2035, improvement in energy efficiency in end-use sectors (buildings, transport and industry), accounts for about two-thirds of the reduction in primary energy demand in the New Policies Scenario, compared with the Current Policies Scenario – by far the largest component. Energy efficiency improvements in the power sector, refineries, and transmission and distribution networks play a much smaller, yet still significant, role. The response to end-use energy price increases, through subsidies removal, CO_2 pricing and the relative increase of more expensive electricity producing technologies in the power

8. For the decomposition, we use the Logarithmic Mean Divisia Index I (Ang, Zhang and Choi, 1998; Kesicki, 2012). For more detail, see *www.worldenergyoutlook.org*.

mix, accounts for savings of 16% in 2035. Fuel and technology switching play a similar role in reducing primary energy demand in 2035.

Energy efficiency is also the largest contributor to energy savings in the 450 Scenario, compared with the New Policies Scenario, representing almost three-quarters, or around 1 800 Mtoe, of the reduction in energy use in 2035. In the 450 Scenario, energy savings due to improved efficiency on the supply side is more pronounced, as the power sector becomes more efficient under the impact of more widespread and higher CO_2 prices. In the 450 Scenario, energy prices to final consumers are significantly higher than in the New Policies Scenario, in some regions because of the introduction of CO_2 pricing, in others because of further reductions to fossil-fuel subsidies. Higher prices provide additional incentives to improve energy efficiency, while also encouraging energy conservation.

Trends by region

Energy efficiency policies in the European Union, China, the United States and Japan account for 53% of the reduction in global energy demand in the New Policies Scenario, compared with the Current Policies Scenario. This result reflects the sheer size of their energy markets and the emphasis these countries are placing on energy efficiency in their policies (Table 9.3).

The European Union has established a comprehensive energy efficiency policy framework with targets for 2020, notably a 20% reduction in energy demand in 2020 *vis-à-vis* their baseline. A key regulation of the EU energy efficiency directive focuses on engaging energy providers to help consumers – industry and household – to increase their investment in energy efficiency. The EU target is missed in the New Policies Scenario, with energy savings reaching just 14% in 2020 on the assumption of implementation delays in some member countries. However, necessary policies are fully implemented over time, a key factor in our projection that EU energy demand in 2035 is 2.5% lower than in 2010 and almost 90 Mtoe less than in the Current Policies Scenario (Figure 9.5). Without implementing the assumed efficiency measures, EU energy demand in 2035 would be 3% higher compared with 2010.

China has established a comprehensive framework to improve energy efficiency, with an overall goal of reducing energy intensity by 16% between 2011 and 2015. A key pillar of its efforts is an ongoing restructuring of its economy, which is expected to bring about significant savings in energy consumption per unit of GDP. Other key elements include innovation and energy savings in companies identified by the government in a list of "ten thousand energy-saving and low-carbon industrial enterprises", which collectively make up 37% of the targeted energy savings by 2015. The longer-term impact of the policies that China is pursuing, coupled with greater availability of natural gas, promises to change the country's energy landscape, drastically cutting coal demand growth post-2015. In the New Policies Scenario, China's energy demand grows by 60% over 2010-2035, however, without the implementation of the assumed efficiency measures, growth would be 72%. In our projections, China successfully meets its 2015 target for energy intensity.

Table 9.3 ▷ **Key energy efficiency assumptions in major countries/regions in the New Policies and 450 Scenarios**

	New Policies Scenario	450 Scenario
United States	Fuel-economy standards for new PLDVs at 54.5 mpg (4.3 l/100 km) in 2025. Fuel-economy standards for new trucks (up to 21% by 2017/2018, depending on type). Increased state and utility budgets for energy efficiency.	Fuel-economy standards for new PLDVs at 67 mpg (3.5 l/100 km) in 2035. Fuel economy improvement for new trucks up to 45% by 2035 (depending on type). Mandatory building codes; zero-energy buildings initiative. Extension of grants for energy efficiency.
European Union	Partial implementation of the Energy Efficiency Directive. Fuel-economy standards for PLDVs at 95 g CO_2/km in 2020 (3.8 l/100 km); LCV at 147 g CO_2/km (5.9 l/100 km) in 2020.	Full implementation of the Energy Efficiency Directive. Fuel-economy standards for PLDVs at 80 g CO_2/km in 2035 (3.2 l/100 km)*; LCV at 100 g CO_2/km (4.0 l/100 km). Zero-energy buildings in new construction by 2020. Ecodesign.
Japan	Fuel-economy standards for PLDVs at 20.3 km/l (4.9 l/100 km) in 2020. Measures to contain electricity demand growth.	Fuel-economy standards for PLDVs at 32 km/l (3.1 l/100 km) in 2035. More stringent measures to contain electricity demand growth.
China	Strategies to meet the target in the 12th Five-Year Plan (2011-2015) to cut energy intensity by 16% including: - structural changes to the economy. - top 10 000 companies. - innovation. Fuel-economy standards for PLDVs at 5.0 l/100 km in 2020.	Fuel-economy standards for PLDVs 3.1 l/100 km in 2035. Introduction of zero-energy buildings for all new construction, starting 2025.
India	Full implementation and extension of the National Mission on Enhanced Energy Efficiency. Fuel-economy standards for PLDVs: assumed annual improvement of 1.3% between 2010 and 2020. Compact florescent lamps financed through CDM. Average annual improvement of PLDVs fuel economy by 1.3% until 2020.	Minimum performance standards for new coal plants (38% in 2035). Reduction in transmission and distribution losses. Mandatory standards for appliances and increased penetration of energy-efficient lighting. Fuel-economy standards for PLDVs at 3.0 l/100 km.
Middle East		Fossil-fuel subsidy rates decline to a maximum of 20% by 2035.

* Fuel-economy standards in the 450 Scenario that are presented in g CO_2/km are related only to improvements in efficiency. Additional reductions in average g CO_2/km can occur through the use of alternative fuels such as biofuels, electricity or natural gas. All fuel-economy standards refer to test-cycle fuel consumption.

Figure 9.5 ▷ **Savings in primary energy due to energy efficiency in the New Policies Scenario compared with the Current Policies Scenario by region, 2035**

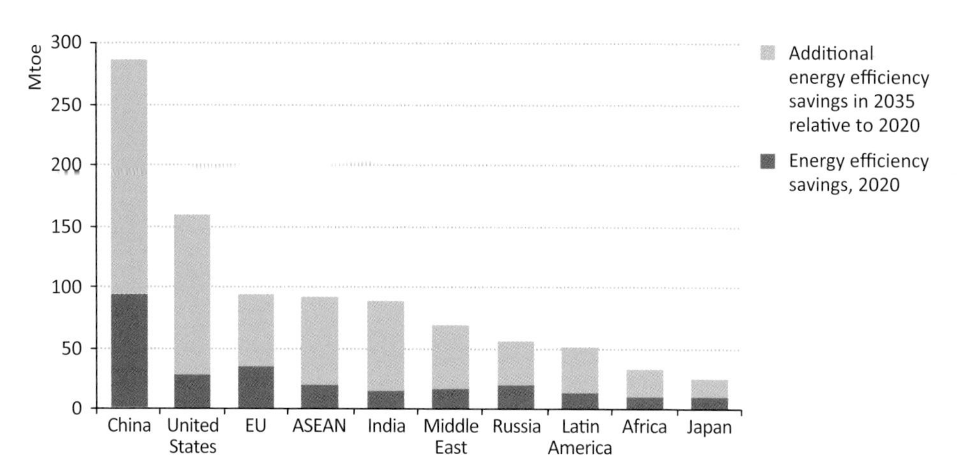

Note: EU = European Union.

The United States is in the process of revising its minimum energy performance standards (MEPS) for appliances and equipment, a policy area in which the country has been very active since 1978. Twenty-four states have adopted long-term energy savings targets, which are driving utility investments in energy efficiency. State budgets for electricity efficiency programmes increased to $4.5 billion in 2010 from $3.4 billion in 2009 (ACEEE, 2011). While the United States does not have an economy-wide energy efficiency target, it is focusing increasingly on improving energy efficiency in road transport, long recognised as having significant potential for improvement at low cost. The new 2025 fuel economy target for passenger cars of 54 miles per gallon (mpg), compared with around 35 mpg today, would exploit much of the known technical potential of conventional vehicles, assuming there is no significant change in the average size and power of the fleet over the period. In the New Policies Scenario, this policy alone allows the United States to cut oil consumption by some 2.1 mb/d of oil by 2035, with significant economic benefits due to reduced spending on imports. In terms of total demand, efficiency gains result in a 1% decline in energy demand by 2035, compared with current levels, versus 6% growth in the absence of the fuel-economy measures.

Japan's new Innovative Strategy for Energy and the Environment, released in September 2012, includes a major focus on energy efficiency. One target is to reduce electricity demand in 2030 by 10%, compared with 2010 levels. This is expected to be backed-up by measures to incentivise the introduction of more efficient technologies in the residential sector and, to a lesser extent, the industrial sector. In the New Policies Scenario, the new measures temper electricity demand growth significantly; though by 2030 it is still 6% higher than in 2010. Japan's primary energy demand is projected, in the New Policies Scenario, to be 10% lower in 2035 than in 2010, with the bulk of the reduction coming from efficiency measures. In the absence of these measures, we estimate demand in 2035 would still be lower than in 2010, but only by 5%.

The centrepiece of India's efforts to save energy is its innovative Perform, Achieve and Trade (PAT) mechanism for energy efficiency, which is aimed at large energy-intensive industries. Other countries in developing Asia, particularly in the ASEAN region, have been identified as having large potential to improve energy efficiency in industry (IEA, 2011c). In the New Policies Scenario, we assume that governments facilitate the introduction of energy service companies (ESCOs) and tap a portion of this new market. In the New Policies Scenario, energy demand in developing Asia, excluding China, increases by 95% between 2010 and 2035; but in the absence of the assumed efficiency measures, the rise would be 109%.

Due to plentiful energy resources and low energy prices, improving energy efficiency has historically not been a key priority throughout much of the Middle East. With the exception of a few countries, subsidised prices have significantly hampered the uptake of efficient technologies in the power sector, road transport and buildings. For example, the average efficiency of fossil fuel power generation in the region is currently just 33%, nine percentage points below the OECD average. In the New Policies Scenario, very few measures specifically targeted at improving energy efficiency are assumed to be implemented in the region. Nonetheless, efficiency gains cut projected Middle Eastern energy demand by 7% in 2035, compared with the Current Policies Scenario. This can largely be attributed to the spill-over effect as the implementation of energy efficiency policies in other parts of the world makes more efficient industrial and transport capital stock available in the international market.

In much of Africa, with the exception of South Africa and a few countries in North Africa, providing access to basic energy services and increasing the availability of energy to underpin economic growth have been given considerably more attention by policy makers than energy efficiency. While improving energy access is fundamental to fulfilling development aspirations, integrating energy efficiency strategies at the very start of such programmes could enable access to be provided to more people in a shorter timeframe (see Chapter 18). During the projection period, efficiency measures make up 78%, or over 360 Mtoe, of the 2% reduction in energy demand in Africa in the New Policies Scenario, compared with the Current Policies Scenario. The bulk of the savings are made in South Africa. Elsewhere in the continent savings stem from the more widespread use of improved biomass cook stoves. As is the case in the Middle East, spill-over effects from the improved efficiency of energy-using equipment traded in the world markets play an important role in raising efficiency levels in Africa.

Trends by sector

In comparing the New Policies Scenario with the Current Policies Scenario, 44% of the final energy savings that result from efficiency improvements come from the industry sector, as rising energy prices strengthen the economic case for making energy efficiency investments in energy-intensive industries (Table 9.4). The share of energy in total industry input costs in large developing countries has been estimated at 10% for chemicals and basic metals, and 6.5% for non-metallic minerals (UNIDO, 2011). These shares tend to be higher in small developing countries, where access to technology and capital is often limited, and lower in OECD countries. In the New Policies Scenario, an increasing share of industrial output comes from emerging economies, where most of the new capacity is added. The uptake of more efficient technologies is strong in OECD countries and China, because of the introduction of minimum energy performance standards and increased energy prices, due to the introduction of CO_2 prices.

Table 9.4 ▷ Energy demand and savings due to efficiency measures in the New Policies Scenario compared with the Current Policies Scenario by end-use sector (Mtoe)

	Energy demand in the New Policies Scenario			Cumulative energy savings due to energy efficiency
	2010	2020	2035	2011-2035
Industry	2 421	3 035	3 497	3 221
Transport	2 377	2 778	3 272	2 510
Buildings	2 910	3 302	3 748	1 138
Other	970	1 107	1 232	465
Total	8 678	10 223	11 750	7 334

Note: Other includes agriculture and non-energy use.

In the New Policies Scenario, increasingly stringent regulation delivers substantial energy savings in the transport sector, making it the second-largest contributor to the reduction in energy demand over the projection period due to efficiency measures. Several countries are discussing the introduction of ambitious fuel-economy standards, often in a bid to slow or cut oil imports or to reduce local pollution. Tighter standards are expected to increase the rate of innovation in the automotive industry, which will have spill-over effects for efficiency levels globally. As energy prices are lower in the New Policies Scenario than in the Current Policies Scenario, some of the energy efficiency savings are offset by the rebound effect, *i.e.* the increase in the demand for energy services that occurs when their overall cost declines (included in Figure 9.6 under activity).

In the New Policies Scenario, energy savings in the buildings sector are relatively small, compared with other sectors, because of the high transaction costs. Most of the energy efficiency savings occur in commercial buildings, where regulation is easier to apply. Some demand reduction also occurs in the residential sector thanks to the assumed reduction in fossil-fuel subsidies in some countries, including India, Russia and parts of the Caspian region.

Figure 9.6 ▷ **Decomposition of the change in final energy consumption in the New Policies Scenario compared with the Current Policies Scenario by sector in 2035**

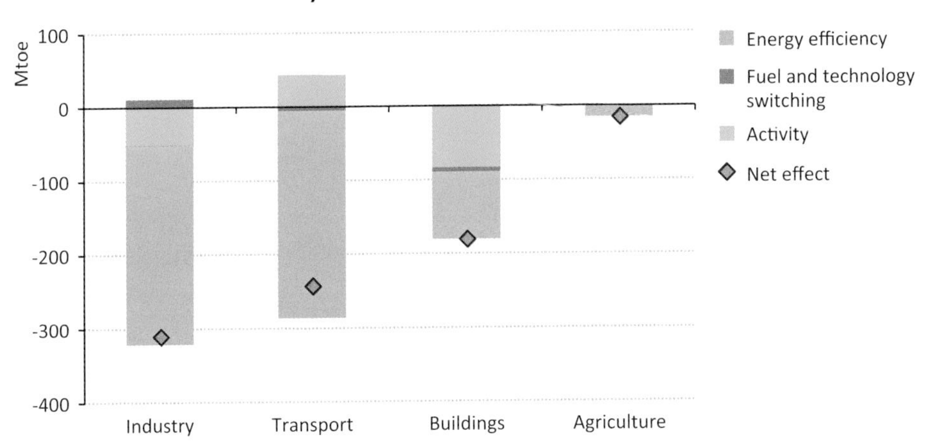

Role in reducing CO_2 emissions

Energy efficiency measures in the New Policies Scenario reduce CO_2 emissions by 4.6 gigatonnes (Gt) in 2035, compared with the Current Policies Scenario. This saving accounts for 65% of total savings in 2035 (Figure 9.7). The share of energy efficiency in total savings declines over time, as energy efficiency is cheaper than other abatement options and is among the first options used. Lower electricity demand from more efficient appliances, industrial motors and buildings reduces fuel input to the power sector and is the largest factor in CO_2 reduction through efficiency measures (2.9 Gt of savings in 2035). Fuel savings achieved through more efficient vehicles, industrial processes and heating applications save an additional 1.3 Gt CO_2 in 2035. Higher power generation efficiency accounts for an additional 0.3 Gt of savings in the New Policies Scenario, less significant than the contribution from increased renewables (1.6 Gt) or nuclear (0.4 Gt).

Figure 9.7 ▷ **Contribution of change in CO_2 emissions by policies in the New Policies Scenario compared with the Current Policies Scenario**

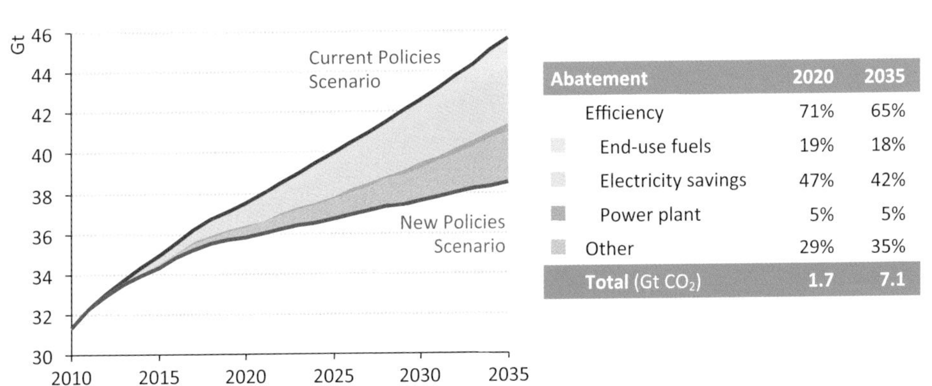

Abatement	2020	2035
Efficiency	71%	65%
End-use fuels	19%	18%
Electricity savings	47%	42%
Power plant	5%	5%
Other	29%	35%
Total (Gt CO_2)	**1.7**	**7.1**

The policies that we assume to be implemented, particularly subsidies to renewable energy, the introduction of carbon pricing and coal to gas switching in the power sector, underpin a decline in carbon intensity[9] of 0.4% per year on average (Figure 9.8). This trend would represent a shift with respect to the last decade, when global carbon intensity remained flat. Nonetheless, the declining energy and carbon intensities in the New Policies Scenario are not enough to limit the long-term increase in the global mean temperature to two degrees Celsius: the 450 Scenario requires annual average improvements of 2.4% in energy intensity and 1.8% in carbon intensity (Chapter 8).

Figure 9.8 ▷ Trends in energy and CO_2 intensity by scenario, 2010-2035

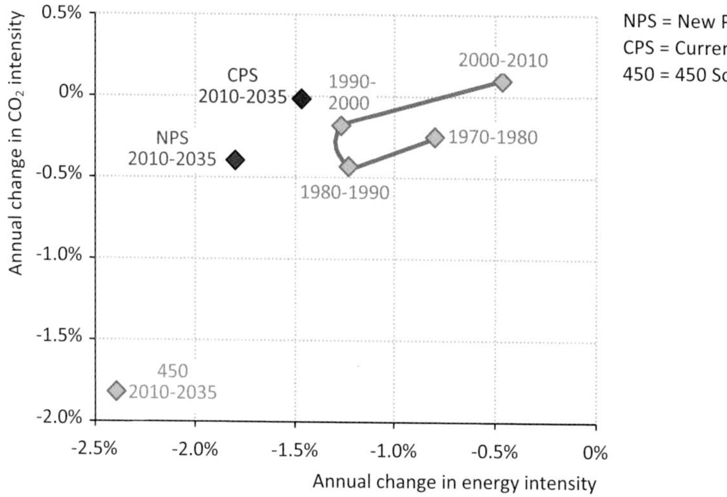

Untapped economically viable potential in the New Policies Scenario

While the New Policies Scenario represents a significant improvement over current trends, our detailed sector-by-sector analysis shows that it leaves significant potential for energy efficiency untapped. The Efficient World Scenario, presented in Chapter 10, examines the implications of realising this full economic potential. Comparing the results of the two scenarios demonstrates that the share of economic potential realised in the New Policies Scenario ranges from less than 20% in buildings to up to 44% in industry (Figure 9.9).

In the New Policies Scenario, regions including the European Union, the United States and China introduce fuel-economy standards. Globally, these measures capture 37% of the full potential for energy efficiency improvements in the transport sector. The assumed fuel-economy standards for passenger light-duty vehicles, coupled with end-user pricing policy, lead to wider deployment of hybrids and very efficient internal combustion engines (ICEs) (IEA, 2011c). Driven mostly by higher energy prices, targets and incentives, the use of more efficient technologies in energy-intensive industries in non-OECD countries (where

9. In this context, carbon intensity is the amount of CO_2 emissions per unit of primary energy consumed.

almost all new additions occur) exploits around 44% of the global efficiency potential in the industrial sector.[10]

Figure 9.9 ▷ **Utilised long-term energy efficiency economic potential in the New Policies Scenario, 2011-2035**

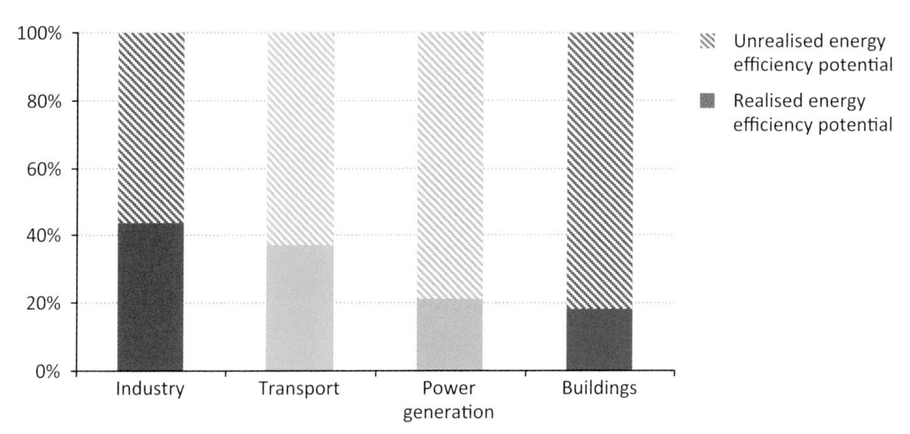

Thus far, energy efficiency policies in buildings have received less attention from policy makers, as they are in general more difficult to implement, enforce and verify, and are sometimes more costly. For these reasons, the buildings sector only achieves one-fifth of the economically available potential in the New Policies Scenario (See Chapter 10). Buildings and their associated energy technologies may have very different time scales. For example, a building shell may be in place for a structure's entire life and, if it is inefficient when first constructed, it will impose a high energy load over its full lifetime, unless it is retrofitted. Energy retrofits are usually economically justified when included in routine technical renovation, provided that the associated costs can be amortised over the remaining life of the building. However, the case may be less compelling when the owner is not the occupier and is not responsible to pay the energy bill (a case of split incentives); or if the tenure of the owner-occupiers is not known and may be shorter than the remaining life of the building. There are few commercial solutions to these market barriers. This, in addition to the absence of long-term incentive schemes for retrofitting in most countries creates a significant lock-in effect that, unless overcome through very strong policy intervention, will continue to deter improvement of the energy performance of existing buildings. Few countries are considering such incentives.

The potential for raising the average efficiency of thermal power plants depends largely on the relative price of fossil fuels and their competitiveness, policies in place to stimulate or reduce the deployment of specific technologies and the overall level of electricity demand. Moreover, the absence of a widespread price on carbon provides little incentive to increase the efficiency of such plants. Lower electricity demand requires fewer new

10. A study undertaken in Australia found that 40% of the potential to improve energy efficiency in the industrial sector (compared with the efficiency that could be achieved through implementing the policies under consideration) had been realised (ClimateWorks, 2012).

additions to the power system, therefore limiting the scope for increasing the weighted-average efficiency of all fossil fuel plants. This is one of the underlying factors for why power plant average efficiency in the New Policies Scenario is sometimes only marginally higher than in the Current Policies Scenario. More generally, it is often cheaper to refurbish an old, inefficient plant, than to invest in a new very efficient one, and this may be decisive where uncertainties about policies affecting future operations persist. For these reasons, the power sector achieves no more than 21% of the energy efficiency potential in the New Policies Scenario.

SPOTLIGHT

Do energy efficiency measures "undermine" carbon markets?

As most energy efficiency measures are less costly than other climate change mitigation measures, they should be among the first steps taken by countries seeking to rein in CO_2 emissions. While energy prices that include value for carbon externalities provide an additional economic incentive for energy efficiency investment, some barriers to their uptake reflect market failures that are not addressed by a carbon price by definition (IEA, 2011d).

Where carbon pricing does provide an additional incentive for efficiency improvements, the converse is also true – specific energy efficiency measures are also needed to complement CO_2 prices to ensure that cost-effective potential energy savings are realised. Energy efficiency and carbon-pricing policies interact and need to be aligned. Superposing energy efficiency measures on existing carbon-pricing systems may reduce the CO_2 price to a level at which there is insufficient incentive to undertake long-term low-carbon investments, locking in more carbon-intensive infrastructure and causing higher mitigation costs in the future. For example, energy efficiency measures tend to reduce electricity demand and therefore wholesale electricity prices. This means that technologies such as renewables need greater support per unit of capacity installed. An appropriate level of carbon pricing, which increases wholesale prices, would render renewables more competitive.

Investment in energy efficiency

To achieve the ongoing efficiency gains projected in the New Policies Scenario, investment in energy efficiency (beyond that in the Current Policies Scenario) needs to increase steadily from \$117 billion in 2020 to \$290 billion in 2035 (Figure 9.10).[11] This includes additional investment in each of the end-use sectors: transport, residential, industry and services.

11. Energy efficiency investment is used to denote expenditure on a physical good or service which leads to future energy savings, compared with the energy demand expected otherwise. This section focuses on energy efficiency investment in end-use sectors – transport, residential, industry and services – as this is where most of the savings and additional investment occur. Over the *Outlook* period, power sector investment is reduced by \$0.5 trillion, due to lower electricity demand, while the additional energy efficiency investment in the power sector is \$350 billion in total.

Over the *Outlook* period, cumulative additional investment in the New Policies Scenario amounts to $3.8 trillion.[12]

Figure 9.10 ▷ **Average annual increase in energy efficiency investment in the New Policies Scenario compared with the Current Policies Scenario**

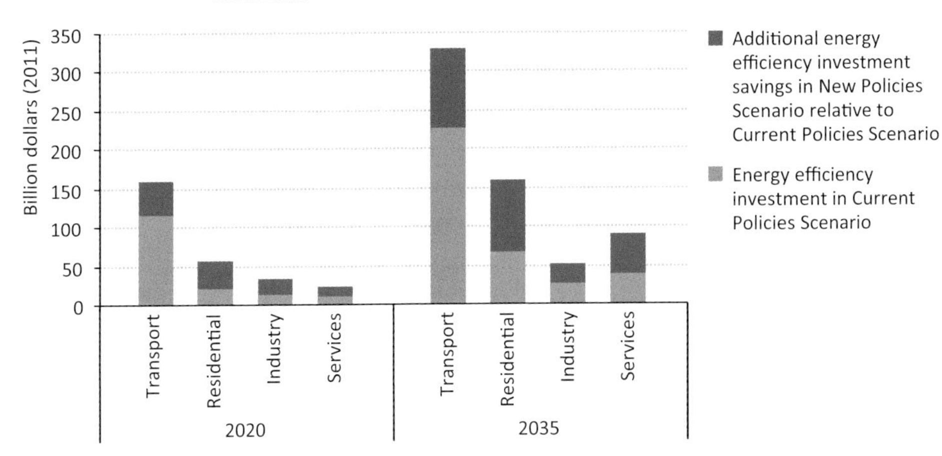

Investment in transport increases by $1.6 trillion, 41% of the total additional investment for all sectors worldwide. OECD countries account for two-thirds of the incremental transport investment. This high share is a function of the higher cost of increasing fuel economy in the OECD countries on one hand, as they already tend to be at higher levels, and, on the other, the fact that most OECD countries have adopted fuel economy policy targets, while only China has adopted such measures among non-OECD countries. Fuel economy improvements are achieved in road transport through improvements in internal combustion engine technology, such as changes to the thermodynamic cycle or reducing engine friction, as well as weight reductions and improvements to auxiliary systems. In addition, energy savings result from greater use of hybrid cars and alternative fuel vehicles and more rapid market penetration of light weight materials. Such technological advances come at a cost: in 2035, the average additional cost per vehicle is $550 in non-OECD countries and around $1 400 in OECD countries, compared with the Current Policies Scenario. In the United States and Europe, the additional costs are around $1 600 per vehicle, given the stringency of fuel-economy standards, while in China the additional cost is around $810 per vehicle. But this adds only a few percent to the price of a vehicle. Improving vehicle efficiency is, of course, generally cheaper in countries with a larger share of inefficient vehicles in the existing fleet.

12. The estimates of capital costs for end-use technology used in this analysis are based on the results of work carried out in co-operation with a number of organisations, including the Cement Sustainability Initiative, and Global Buildings Performance Network, and have been verified with a number of industrial sources for the iron and steel sector, petrochemicals, cement, road and freight vehicles, buildings and energy management systems. Several independent sources were used to check consistency.

About one-third of the additional investment in residential and services sector buildings is in improving the efficiency of electrical equipment (appliances and lighting), totalling $745 billion over the *Outlook* period. Building energy management, often a cheap option per unit of energy saved, is assumed not to be widespread in the residential sector but is increasingly used in commercial and public buildings.[13] In the New Policies Scenario, improvements to insulation and heating systems are mostly achieved in new buildings. Such investments often do not increase property value and have payback periods that exceed the average length of ownership, creating substantial barriers to their being undertaken without a revised policy framework. Investment in retrofits, insulation and thermal efficiency total $1.0 trillion over the *Outlook* period.

In the New Policies Scenario, additional investment in improving energy efficiency in industry amounts to $450 billion between 2011 and 2035, compared with the Current Policies Scenario. Average annual investment increases over time as cheaper options are tapped in the early years and the number of projects undertaken to increase efficiency increases. About two-thirds of the additional investment in industry is in improving the efficiency of heat systems, where much unrealised potential exists (GEA, 2012). The remainder of the investment is in electrical equipment, mostly industrial motors. Improved motor systems have been available on the market for some years, but their uptake has been slow, especially in developing countries (UNIDO, 2011). This is one of the most striking examples that, even when energy efficiency investments are economic and pay back very quickly, other barriers can be serious impediments. In the New Policies Scenario, increasing electricity prices make the economic case stronger.

The payback periods of the energy efficiency measures that are assumed to be adopted in the New Policies Scenario are short, with the exception of buildings in OECD countries where retrofits are being implemented.[14] In road transport, for example, fuel standards adopted in OECD countries pay back in around four years, even allowing for transaction costs. For cars and trucks such transaction costs are in fact small, as fuel economy is one of the main parameters taken into consideration when making the investment. Additionally, information on vehicle fuel economy is readily available to consumers in most OECD countries. Energy efficiency measures in industry in OECD countries have a payback of less than five years (two-and-a-half years for motors), while the payback period for investment in industry in non-OECD countries is below two years (Figure 9.11). Including transaction costs averaging 20% of the investment cost does not greatly change the payback period.

13. Buildings energy management, or active control, comprises measures that through automation and control ensure that equipment works in an optimised way, *i.e.* only when needed and not more. Automation ensures turning off devices when not needed and regulating motors or heating at the optimised level.

14. To calculate the payback period of energy efficiency measures we have associated the additional investment in each year to the savings in fuel expenditures that such investment will entail over the lifetime of the capital stock. Savings in fuel expenditures are a function of the energy savings resulting from the mix of the efficiency measures taken and end-use fuel prices. The payback is calculated as an average for the investment occurring over the *Outlook* period, accounting for savings that occur even beyond the period.

The average payback period of efficiency measures undertaken in buildings in OECD countries is relatively long as building retrofits, the most expensive measure in this sector, are undertaken in some countries. Including transaction costs more significantly increases the average payback period for investment in space heating and cooling, and water heating efficiency measures in the residential sector. In OECD countries the payback period increases to over fourteen years and in non-OECD it reaches almost nine years, almost bordering the lifetime of some of the capital stock for which the investment is made.

Figure 9.11 ▷ Average payback periods for energy efficiency measures in the New Policies Scenario by sector including and excluding transaction costs

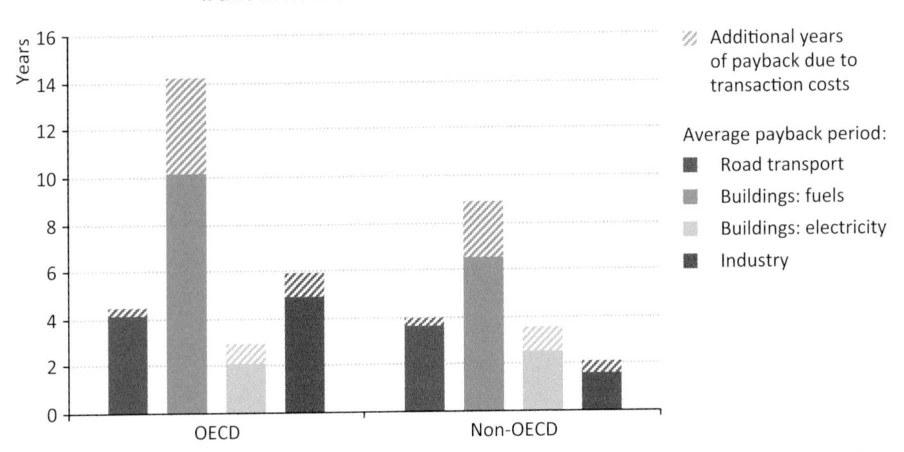

Note: Transaction costs are assumed to be equivalent to 9% of the investment cost of the energy efficiency measure for transport, 20% for industry and 40% for buildings.

Measures to address some of the barriers to energy efficiency implementation are adopted in the Current Policies Scenario, though the effort falls far short of what is needed to remove them. More is done in the New Policies Scenario but, even here, the level of policy intervention aimed at addressing energy efficiency barriers is such that large transaction costs remain. The Efficient World Scenario, presented in Chapters 10-12, offers a much more robust policy framework that largely eliminates the implicit transaction costs.

Box 9.5 ▷ How much money is currently flowing into energy efficiency?

Despite the increasing interest of policy makers, financial institutions and other stakeholders, investments in energy efficiency are seldom tracked systematically and no comprehensive estimate is available of current global investment in energy efficiency. This is due to the fact that energy efficiency investments are undertaken by a multitude of agents, households and firms, often using their own funds. Moreover, there is no standard definition for what constitutes an energy efficiency investment.

To overcome this information gap, for this *Outlook* we have made a first attempt to estimate energy efficiency investment through a country-by-country survey. We have found that many countries track investment in improving energy efficiency in buildings and industry, but data for the transport and power sectors are more difficult to obtain or estimate. Given these limitations, we have used the following approach:

- Use country sources and estimates, wherever available. This proved possible for larger countries.
- Infer energy efficiency investment from data from multilateral development banks and other sources which detail public funding invested in energy efficiency projects to which a multiplier is applied, based on the economic circumstances and practices of the individual country (AGF, 2010).

Using this methodology, we estimate that global investment in projects aimed principally at improving energy efficiency amounted to $180 billion in 2011 (Figure 9.12). This is significantly lower than the investment in expanding or maintaining fossil fuel supply (nearly $600 billion). About two-thirds of the estimated investment in energy efficiency in 2011 was undertaken in OECD countries. Investment in the European Union reached approximately $76 billion, while investment in the United States reached $20 billion (Capital E, 2012). Investment in China reached $31 billion, 85% of it undertaken by companies (CPI, 2012). There is increasing activity in energy efficiency in India as a result of government programmes to promote energy efficiency through regulation and capacity building, estimated to have amounted to $9.5 billion, split between the buildings and industrial sectors (HSBC, 2011). There is also significant energy efficiency investment in Russia by the government and also by the European Bank for Reconstruction and Development; in total we estimate it reached $5.7 billion in 2011.

Figure 9.12 ▷ Investment in energy efficiency by country and region, 2011

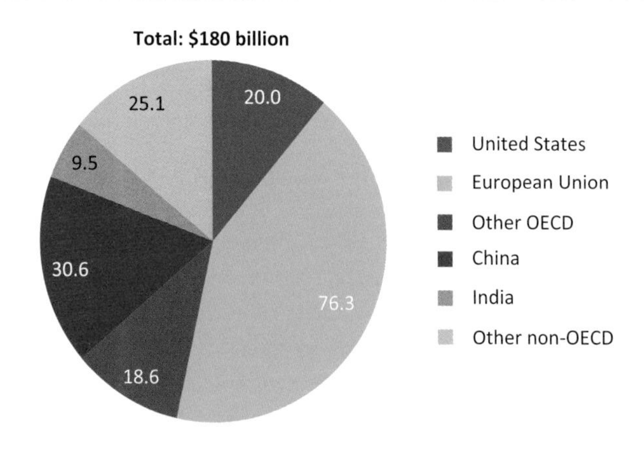

A blueprint for an energy-efficient world

Six steps toward a hat-trick of energy goals

Highlights

- The Efficient World Scenario offers a blueprint to realise the economically viable potential of energy efficiency. We set out the policies that governments need to enact to lower market barriers, thereby minimising transaction costs and enabling the necessary energy efficiency investments. Those investments pay back well before the end of the lifetime of the energy capital stock and result in huge gains for the economy, energy security and the environment.

- The Efficient World Scenario results in a more efficient allocation of resources, boosting cumulative economic output through 2035 by $18 trillion – equivalent to the current size of the economies of the United States, Canada, Mexico and Chile combined. GDP gains in 2035 are greatest in India (3.0%), China (2.1%), the United States (1.7%) and OECD Europe (1.1%). Additional investment of $11.8 trillion in more efficient end-use technologies is needed, but is more than offset by a $17.5 trillion reduction in fuel expenditures and $5.9 trillion lower supply-side investment.

- Growth in global primary energy demand over 2010-2035 is halved, relative to the New Policies Scenario, and energy intensity improves at 2.6 times the rate of the last 25 years. Oil demand peaks at 91 mb/d before 2020 and declines to 87 mb/d in 2035. Coal demand is lower in 2035 than today, while natural gas demand still rises. We estimate that the rebound effect (increased energy demand due to higher GDP and lower prices) approaches 10% of the cumulative savings from efficiency improvements, but this can be cut by more than half with end-user pricing policies.

- Oil demand is reduced by 12.7 mb/d in 2035, compared with the New Policies Scenario, equal to the current production of Russia and Norway combined. This eases pressure for new discoveries and developments. Oil-import bills in the five largest importers are cut by 25%. Though Middle East oil-export revenues are lower than in the New Policies Scenario, they still increase significantly relative to today.

- Energy-related CO_2 emissions peak before 2020 and decline to 30.5 Gt in 2035, pointing to a temperature increase of 3 °C; in addition to energy efficiency, low-carbon technologies will be needed to achieve the 2 °C goal. Emissions of local pollutants are cut sharply, bringing benefits to China and India, in particular.

- We propose six categories of policy action, which, if widely implemented, can turn the Efficient World Scenario into reality. Key steps include: strengthening the measurement and disclosure of energy efficiency, to make the gains more visible to consumers; regulations to prevent the sale of inefficient technologies; and financing instruments.

The Efficient World Scenario

Introduction

Energy efficiency policies already in place and many of those currently under discussion are set to accelerate the rate of efficiency improvement in the global energy economy. These improvements are reflected, respectively, in the Current Policies Scenario and the New Policies Scenario. Yet those scenarios see the exploitation of only a small portion of the available energy efficiency potential, as many barriers still exist to the implementation of the necessary measures, even when they are fully viable in economic terms.

In the following three chapters, we present an Efficient World Scenario. This quantifies the implications of dismantling the impediments, so that a wave of investment is released for measures to improve energy efficiency. All of these investments satisfy stringent tests in terms of their payback period. The objective is to understand the transformation such measures would achieve – how far they would take us towards a sustainable energy economy – and how the change might be achieved. Our analysis is based on detailed scrutiny of the known opportunities for action to realise this economic potential and of the policies required to do so. The Efficient World Scenario provides a comprehensive global picture – region-by-region and sector-by-sector.

The chapter starts by defining the approach and key assumptions that underpin the Efficient World Scenario and shows how its realisation would represent a triple win, cutting energy demand, improving the global economy and delivering environmental benefits. While the primary goal of energy efficiency policy is to realize energy savings, it is important to take these associated benefits into account, as they can be very significant, bolstering the case for policy intervention. Some of them can readily be given a monetary value, but for others this is much more difficult. At the end of the chapter, we summarise the government action that would be needed to obtain these benefits.

Methodology and assumptions

The core assumption in the Efficient World Scenario is that policies are put in place to allow the market to realise the potential of all known energy efficiency measures which are economically viable. To calculate the economic potential, which varies by sector and by region, two key steps were undertaken.

First, the technical potentials were determined, identifying key technologies and measures to improve energy efficiency by sector, in the period through to 2035. This process involved analysis, over a number of sub-sectors and technologies, of a huge amount of data and information from varied sources (Figure 10.1). For the industry, power and transport sectors, we undertook detailed surveys of companies, with operations across the world, to ascertain the efficiencies of the best technologies and practices available now and how these are likely to evolve based on the efficiency and costs of technologies that are in the process of being developed and demonstrated. For the buildings sector[1], we consulted

1. The buildings sector comprises energy demand from residential and services and, hence, includes energy consumed by appliances as well as lighting.

Figure 10.1 ▷ Representation of energy efficiency by end-use sector* in the World Energy Model as considered in the Efficient World Scenario

Industry

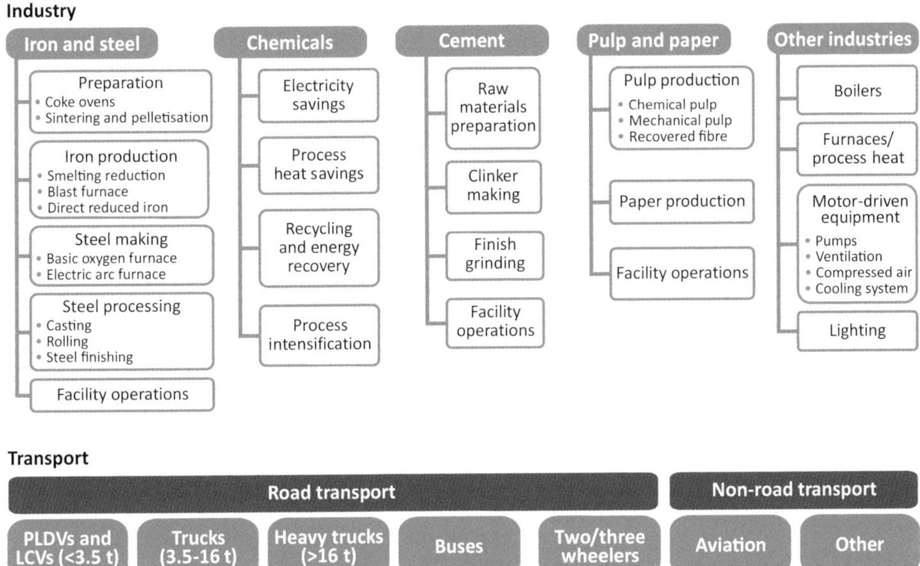

Transport

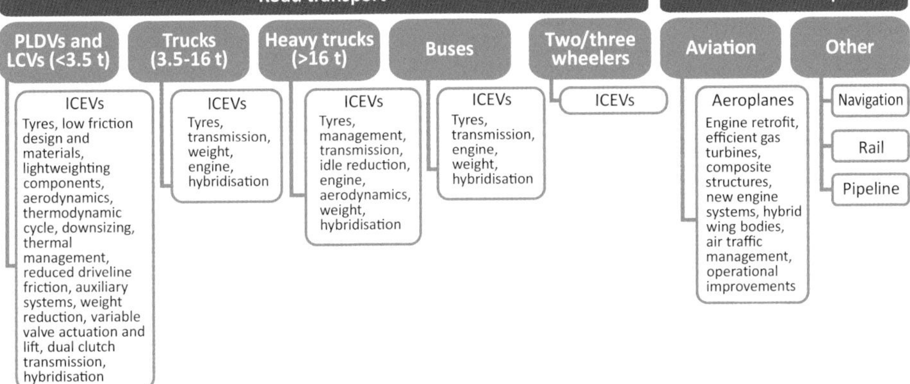

Buildings

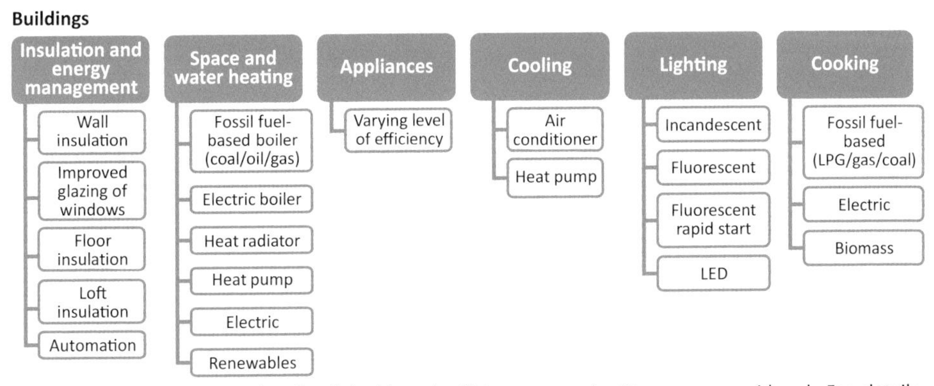

* In the power sector, twelve fossil fuel-based efficiency opportunities were considered. For details, see *www.worldenergyoutlook.org/weomodel/*. Note: PLDV = passenger light-duty vehicles; LCV = light commercial vehicle; ICEV = internal combustion engine vehicle; t = tonnes.

with a large number of companies, experts and research institutions at national and international levels. We also conducted an extensive literature search to catalogue the technologies that are now in use in different parts of the world and judge their probable evolution. The Efficient World Scenario assumes neither major or unexpected technological breakthroughs, nor more holistic concepts (such as prioritising energy efficiency at all levels of urban planning), nor changes in consumer behaviour (except where induced by lower energy prices). While such measures might well be cost-effective, measuring their cost and impact at global level is speculative, in that they represent a significant departure from current practices and, therefore, data for the quantification of their potential are limited. If adopted at scale, however, they could achieve reductions of energy demand beyond what is achieved in the Efficient World Scenario. The scenario is, rather, based on a bottom-up analysis of currently available technologies and practices, and considers incremental changes to the level of energy efficiency deployed.

In a second step, we identified those energy efficiency measures which are economically viable. All practicable measures of what is economically viable are imperfect, not least because of data difficulties and the challenge of arriving at an acceptable average discount rate across many different communities or circumstances. The criterion we adopted was the amount of time an investor might be reasonably willing to wait to recover the cost of an energy efficiency investment (or the additional cost, where appropriate) through the value of undiscounted fuel savings (Figure 10.2).[2] Acceptable payback periods were calculated as averages over the *Outlook* period and take account of regional and sector-specific considerations. For example, the payback period adopted in relation to non-OECD industry is shorter than for the OECD, since energy efficiency investments are generally considered attractive outside the OECD only if the investments can be recovered quickly, whereas within the OECD there is more pressure on industry to demonstrate action to improve energy efficiency, so that a longer acceptable payback period is allowed.[3]

The payback periods that result are, in some cases, longer than what is required today by some lending institutions, households or firms; but they are always considerably shorter than the technical lifetime of the individual assets. The periods chosen are in line with prevailing judgements in the literature and have been deemed acceptable under the policy assumptions of the Efficient World Scenario by stakeholders who have been consulted in the course of the work.

2. Using a 5% social discount rate as a typical proxy for a societal perspective would have increased payback periods of building fuels (21 years in OECD, 12 years in non-OECD countries), power generation (22 years in OECD, 9 years in non-OECD countries) and transport (9 years). All of these payback periods are still within an acceptable limit from the perspective of asset lifetimes, but they require specific policies in some cases directed at overcoming initial deployment hurdles.

3. Payback periods can be much longer, *e.g.* in the case of the closure of small industrial facilities in China. These are, though, exceptions and have little impact on the average payback periods as there are many energy efficiency options with very short payback periods in non-OECD countries.

Figure 10.2 ▷ **Efficient World Scenario methodology**

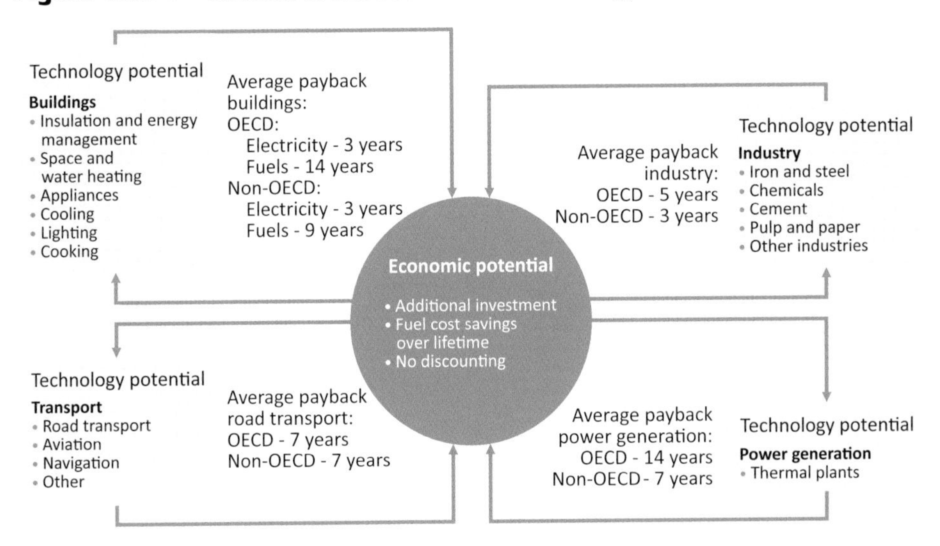

Notes: Payback periods are specified by region and sub-sector. They refer to investment additional to the New Policies Scenario over 2012-2035, but savings are also accounted for beyond the projection period when the lifetime of the capital stock exceeds the *Outlook* period.

The payback periods adopted do not take into account the transaction costs associated with overcoming the present barriers to investment (see Chapter 9). The Efficient World Scenario is posited on the basis that these barriers will be overcome by a bundle of targeted policy measures, so eliminating or, at least, minimising, transaction costs (see end of this chapter for a discussion of such policy measures). Of course, action of this kind will entail a cost; for example, the cost of enforcement of minimum required standards. Estimating this cost is fraught with difficulties; but it can confidently be stated that it is much less than the economic benefits which will ensue (see later section on implications for the global economy).

Moreover, the payback calculation does not take into account the co-benefits to society associated with energy efficiency.[4] These benefits are reported separately in our analysis in order to leave decision makers free to assign weights to them in accordance with their national standards and priorities. The use of relatively straightforward payback periods is widespread among investors and lending institutions, and so will be a familiar methodology to those taking investment decisions in the market place today.

Policies in areas other than efficiency are assumed to be the same as in the New Policies Scenario. In the case of renewables, one consequence could be lower deployment in regions where policy targets are expressed as a share of total generation or total fuel consumption (as in the case of biofuels-blending mandates). In countries with carbon

4. Other metrics such as the global cost method, life-cycle costs analysis and benefits calculations take co-benefits into account and are also adopted in policy making, for example in the European Union.

pricing, carbon dioxide (CO_2) prices are lower than in the New Policies Scenario, as energy efficiency measures contribute to targeted emissions reductions. Fossil-fuel subsidies are phased out by 2035 at the latest in all regions except the Middle East, where they are reduced to a maximum rate of 20% by 2035.

Energy markets in the Efficient World Scenario

In the Efficient World Scenario, the implementation of energy efficiency measures in both energy transformation and consumption reduces the growth in primary energy demand significantly (Table 10.1). World primary energy demand reaches over 14 800 million tonnes of oil equivalent (Mtoe) in 2035 – a reduction of 14% relative to the New Policies Scenario (equivalent to 18% of global energy use in 2010) and 21% relative to the Current Policies Scenario.[5] Global energy demand still grows, but at an average annual rate of 0.6%, compared with 1.2% in the New Policies Scenario. The energy savings are less marked in the period to 2020, due to the relatively low capital stock turnover in the energy sector in this period, but they are far from negligible: in 2020, demand is 6% lower than in the New Policies Scenario.

Table 10.1 ▷ World primary energy demand in the Efficient World Scenario by fuel (Mtoe)

	Total primary energy demand			Change versus NPS		CAAGR*
	2010	2020	2035	2020	2035	2010-2035
Coal	3 474	3 648	3 274	-11%	-22%	-0.2%
Oil	4 113	4 311	4 061	-3%	-13%	-0.1%
Gas	2 740	3 070	3 541	-6%	-14%	1.0%
Nuclear	719	887	1 094	-1%	-4%	1.7%
Hydro	295	382	476	-2%	-3%	1.9%
Bioenergy	1 277	1 502	1 749	-2%	-7%	1.3%
Other renewables	112	293	650	-2%	-8%	7.3%
Total	12 730	14 093	14 845	-6%	-14%	0.6%

* Compound average annual growth rate. Note: NPS = New Policies Scenario.

The Efficient World Scenario sees the adoption of more efficient equipment and processes in industry, improvements in building shells, windows, insulation, lighting and appliances, the use of more efficient vehicles and aeroplanes, and improvements in efficiency in power generation and grids. The use of all fuels is affected, fossil and non-fossil, though the reduction is greatest for fossil fuels (in both absolute and relative terms), as reducing their consumption is the target of most efficiency measures. Their share of primary energy consumption falls from 81% in 2010 to 74% in 2035, as demand for both oil and coal peaks before 2020 and then declines through 2035 (Figure 10.3).

5. In the course of this chapter, the Efficient World Scenario is generally compared against the New Policies Scenario, the central scenario of this year's *Outlook*.

Figure 10.3 ▷ **World primary energy demand in the Efficient World Scenario by fuel** (Mtoe)

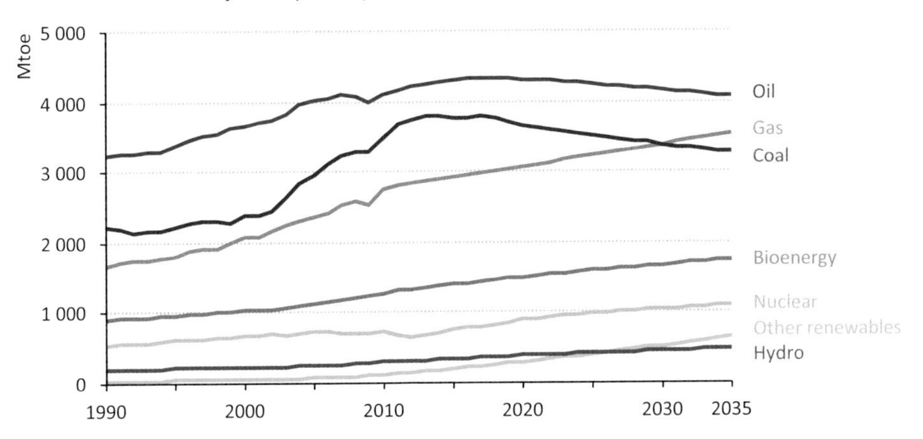

All of the growth in global energy demand in the Efficient World Scenario takes place in non-OECD countries. Collectively, their demand grows at an average annual rate of 1.3% between 2010 and 2035, reaching close to 9 600 Mtoe by 2035, which is 14% lower than in the New Policies Scenario. Energy demand peaks around 2015 in the OECD and then gradually declines to about 4 900 Mtoe in 2035, which is 13% lower than in the New Policies Scenario and 10% lower than in 2010. China, the United States and India, currently the world's three largest energy-consuming countries, combined account for 50% of the energy savings in 2035, compared with the New Policies Scenario. The largest reduction in energy demand in percentage terms occurs in the Middle East, where the combination of energy efficiency policies and the phase-out of fossil-fuel subsidies leads to a 19% reduction in primary energy demand in 2035, compared with the New Policies Scenario.

Trends by fuel

In the Efficient World Scenario, oil demand peaks at 91 million barrels per day (mb/d) before 2020 and then declines to 87.1 mb/d in 2035 (Figure 10.4). The reduction of 12.7 mb/d in 2035, compared with the New Policies Scenario, is comparable to the total oil production today of Russia and Norway combined. The transport sector makes the biggest contribution to this reduction, accounting for more than 60% of the cumulative savings over the *Outlook* period. This is thanks to increased efficiency across all sub-sectors and the phase-out of fossil-fuel subsidies, notably in the Middle East and Africa (see Chapter 11 for trends at the sectoral level in the Efficient World Scenario). Improved efficiency for space and water heating in the buildings sector is responsible for some 16% of cumulative oil savings, while the remainder arises through improvements in efficiency in industrial and transformation processes, as well as in power generation.

Non-OECD countries realise two-thirds (or 8.4 mb/d) of the 12.7 mb/d reduction in oil demand in 2035, compared with the New Policies Scenario. China sees the largest savings in absolute terms, with a reduction of 2.1 mb/d, followed by the Middle East (2.0 mb/d), and India (1.0 mb/d). Oil demand in OECD countries is lower by 3.4 mb/d in 2035, with the biggest reductions occurring in OECD Europe (1.3 mb/d) and the United States (1.0 mb/d). Savings in OECD countries would be much greater but for the savings already projected in the New Policies Scenario. The remaining savings are from international bunkers.

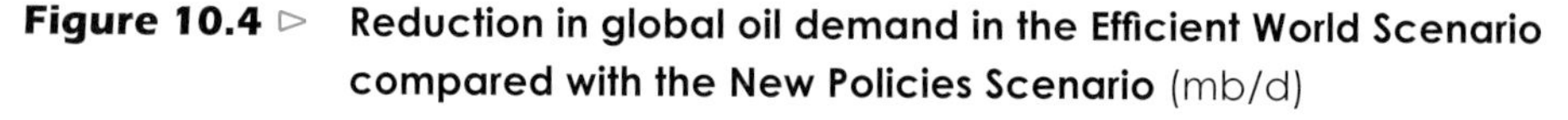

Figure 10.4 ▷ Reduction in global oil demand in the Efficient World Scenario compared with the New Policies Scenario (mb/d)

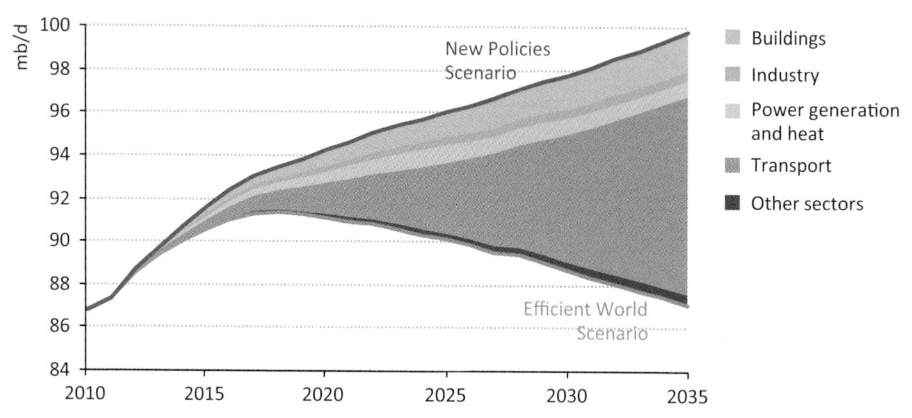

Prospects for coal and gas, currently the dominant fuels in the power sector, are closely linked to electricity demand, which increases at just 1.6% per year on average between 2010 and 2035 in the Efficient World Scenario, compared with 2.2% in the New Policies Scenario (Figure 10.5). Electricity demand growth slows in both OECD and non-OECD regions, reflecting the significant potential that currently exists worldwide to deploy more energy-efficient lighting, appliances, air-conditioning and motor systems, or to use them more intelligently through the use of automation, *i.e.* through active controls. The deployment of such equipment could significantly temper near-term demand growth, as their relatively short operating lifetimes allows for a quick replacement of existing inefficient stock, and as they can be brought into service at significant volumes quickly. This is particularly the case in developing countries, where ownership levels are still low but growing very quickly.[6]

Lower electricity demand growth diminishes the need to expand power generation capacity. Coal is the fuel that is most affected, partly because a large share of the savings arises in countries that are heavily dependent on coal for electricity generation. The faster deployment of more efficient coal-fired power plants, as well as more efficient processes in industry, further reduces coal use. Global coal demand peaks before 2020, at around 5 400 million tonnes of coal equivalent (Mtce), before dropping to about 4 700 Mtce in 2035, which is 22% lower than in the New Policies Scenario (Figure 10.6). China drives this global trend, with some 21% lower coal demand at the end of the *Outlook* period compared

6. The widespread application of technologies that could imply a countervailing trend, such as a large-scale deployment of electric vehicles, is largely beyond the policy framework adopted in the Efficient World Scenario.

with the New Policies Scenario. Demand in India still increases throughout the period, but is 29% lower in 2035 than in the New Policies Scenario. OECD coal demand declines at an average annual rate of 2.3%, reaching 870 Mtce in 2035, or 26% lower than in the New Policies Scenario. The United States is responsible for more than 50% of the cumulative reduction in OECD coal demand, with Europe accounting for a further 33%. Global coal-fired power capacity additions are 35 gigawatts (GW) per year on average between 2012 and 2035, 23% lower than in the New Policies Scenario.

Figure 10.5 ▷ Reduction in electricity demand by region in the Efficient World Scenario compared with the New Policies Scenario

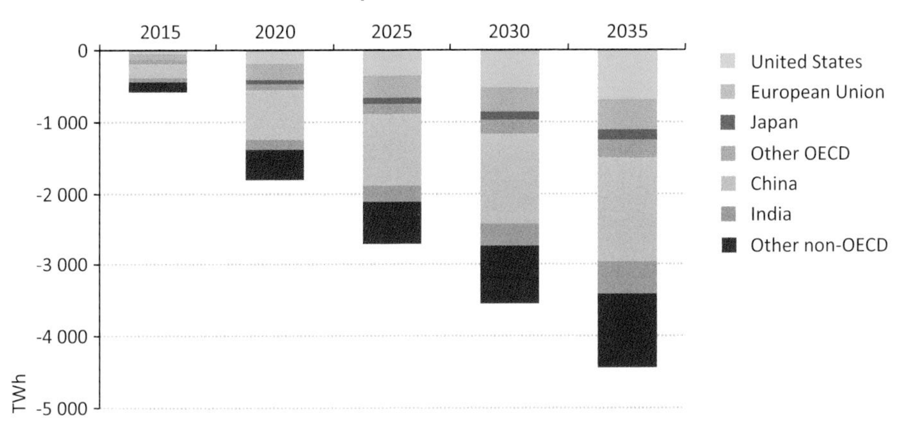

Unlike for the other fossil fuels, global demand for natural gas still increases in the Efficient World Scenario, as it remains an important fuel in the power, industry and buildings sectors. Total demand reaches 3 700 billion cubic metres (bcm) in 2020 and almost 4 300 bcm in 2035 (Figure 10.7). Nonetheless, demand is 14% (or 680 bcm) lower in 2035 than in the New Policies Scenario, which is roughly equivalent to US natural gas demand in 2010. Global demand grows at an average annual rate of 1.0%, compared with 1.6% in the New Policies Scenario, and gas overtakes coal in the early 2030s to become the second most important fuel in the energy mix after oil. Around 30% of the cumulative savings occur in the buildings sector, where gas is used much more efficiently. Gas remains an important fuel for residential space and water heating, as well as cooking. In power generation, which contributes almost 50% of cumulative savings, gas demand in 2035 is 16% lower than in the New Policies Scenario, mostly due to lower electricity demand, but also higher-efficiency plants. Average additions of gas-fired power capacity are 50 GW per year between 2012 and 2035, 13% lower than in the New Policies Scenario. Non-OECD countries account for almost 60% of the difference in global gas demand in 2035, relative to the New Policies Scenario. The Middle East sees the biggest reduction in demand, at 115 bcm, driven by reduced electricity demand and increased efficiency in the power and buildings sectors. The second-largest reduction in 2035 in non-OECD countries occurs in Russia, at 60 bcm, followed by China, at around 40 bcm. In OECD countries, the bulk of the savings arise in the buildings sector, through buildings renovation and the uptake of more energy-efficient space and water heating equipment, and in power generation.

Figure 10.6 ▷ **Reduction in global coal demand in the Efficient World Scenario compared with the New Policies Scenario**

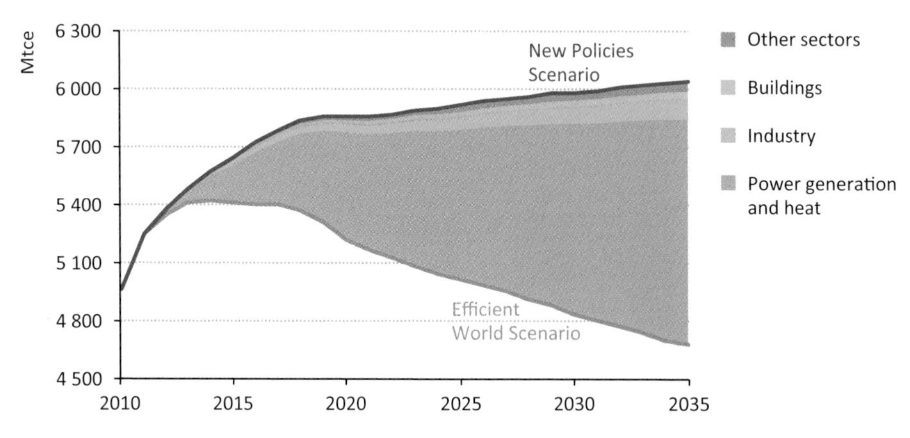

The outlook for non-fossil fuels in the Efficient World Scenario is closely linked to that of electricity demand, as policy assumptions related to subsidies for renewables are unchanged from the New Policies Scenario. With global electricity demand 14% lower in 2035 in the Efficient World Scenario, as a result of demand-side efficiency measures, there is a reduction in electricity generation from other renewables (-9%), bioenergy (-9%), nuclear (-4%) and hydro (-3%). The share of all non-fossil fuels in the generation mix still increases from 33% in 2010 to 48% in 2035 in the Efficient World Scenario. Average additions of renewables and nuclear generating capacity are 129 GW per year between 2012 and 2035, 7% lower than in the New Policies Scenario, with wind power accounting for almost 40% of the reduction.

Figure 10.7 ▷ **Reduction in global natural gas demand in the Efficient World Scenario compared with the New Policies Scenario**

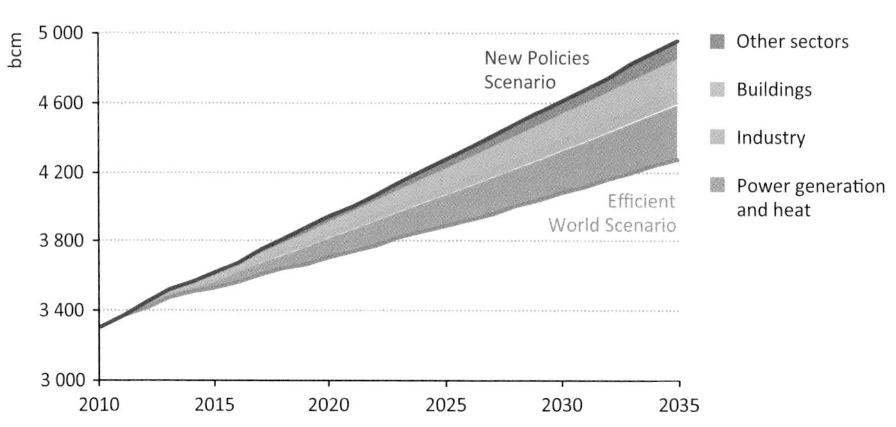

Energy intensity

Global energy intensity – the amount of energy used to produce a unit of gross domestic product (GDP) – declines at an average annual rate of 2.4% in the Efficient World Scenario over the *Outlook* period, representing a dramatic improvement on past trends (Figure 10.8). By 2035 global energy intensity is 45% lower than in 2010. In OECD countries, energy intensity falls by 2.4% per year (almost twice as much as over the past 25 years) and is 45% lower in 2035 than in 2010. The decline is led by the United States, where energy intensity falls at an average annual rate of 3.0%. Japan's energy intensity falls at a slower rate than the OECD average, as it is already among the lowest in the world and there is less scope for improvement, but it is still around 40% lower in 2035 than in 2010. In non-OECD regions, energy intensity falls at an average annual rate of 3.2%, compared with 1.7% over the last 25 years, reducing the intensity by more than half over the *Outlook* period. China is the country that changes most, with an average annual improvement of 4.2%, followed by India at 3.7% and Russia at 3.0%.

Figure 10.8 ▷ Average annual change in global energy intensity in the Efficient World Scenario

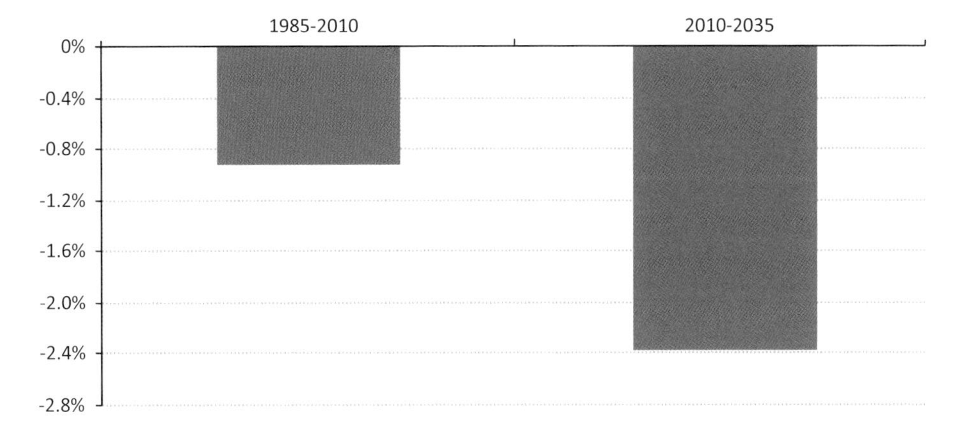

Energy prices

The policy measures that drive efficiency improvements in the Efficient World Scenario have important repercussions on energy prices, both domestically and on international markets. In real terms, the IEA crude oil import price needed to balance supply and demand reaches \$116/barrel (in year-2011 dollars) in 2020 and declines to \$109/barrel in 2035. The oil price in 2035 is \$16/barrel lower than in the New Policies Scenario. Likewise, coal and natural gas prices are lower in the Efficient World Scenario (Table 10.2). These new equilibrium prices lead to a "rebound effect" – *i.e.* part of the savings in energy consumption that would be achieved (but only part) is offset as consumers respond to lower prices by using more energy (Box 10.2).

Table 10.2 ▷ Fossil-fuel import prices in the Efficient World Scenario (year-2011 dollars per unit)

	Unit	2011	2020	2035	Change from NPS in 2035
IEA crude oil imports	barrel	108	116	109	-13%
Natural gas					
United States	MBtu	4.1	5.4	7.7	-4%
Europe imports	MBtu	9.6	11.1	10.6	-15%
Japan imports	MBtu	14.8	13.8	13.0	-12%
OECD steam coal imports	tonne	123	106	100	-13%

Note: NPS = New Policies Scenario; MBtu = million British thermal units.

Energy trade

The introduction of more efficient capital stock in energy production and consumption has implications for energy bills and trade. Energy-importing countries import less oil, natural gas and coal, compared with the New Policies Scenario. However, certain importing countries that are also producers experience a lower reduction in import needs in absolute terms than their demand savings. This is because lower international prices alter supply too, cutting production disproportionately in regions with higher production costs.

Although oil demand peaks and then declines in the Efficient World Scenario, oil trade (between *WEO* regions) still increases by 5% over the level of 2011, due to a mismatch of the locations of demand and supply. The growth in international trade is, nevertheless, about three-quarters lower than in the New Policies Scenario. Oil imports in OECD countries drop sharply over the *Outlook* period, but they continue to rise in other oil-importing regions. Imports in developing Asia rise from 10.6 mb/d in 2011 to 22.0 mb/d in 2035, a reduction of 3.5 mb/d, compared with the New Policies Scenario. Energy efficiency policies in China and India, mostly directed at their fast-expanding car and truck fleets, reduce imports by 2.0 mb/d and 1.0 mb/d respectively in 2035, compared with the New Policies Scenario. Between OECD regions, Europe sees the biggest savings in 2035, compared with the New Policies Scenario. Imports into the European Union drop from 9.8 mb/d in 2011 to 7.0 mb/d in 2035, a reduction of 1.1 mb/d in 2035, compared with the New Policies Scenario. All other OECD importing countries also see big drops in their oil-import requirements, compared with 2011 levels. US imports in 2035 are 0.5 mb/d (or 14%) lower than in the New Policies Scenario, as the additional reduction in US demand that takes place in the Efficient World Scenario is partly offset by lower domestic oil production, due to lower international oil prices. The United States is already projected to cut its oil demand significantly in the New Policies Scenario, due to recent changes in policy, thereby limiting the scope for further reductions in the Efficient World Scenario (Chapter 3).

Natural gas trade (between *WEO* regions) grows substantially over the projection period, from 675 bcm in 2010 to almost 990 bcm in 2035, but this is a reduction of almost 20% compared with the New Policies Scenario. EU imports grow from 335 bcm in 2010 to about 420 bcm in 2035, a reduction of over 100 bcm, compared with the New Policies Scenario. Japan's imports remain at close to the levels of 2010, compared with an increase of more than 20% in the New Policies Scenario. Import requirements in China and India grow over time, but at a slower pace. On the supply side, exports from Russia and the Middle East are reduced significantly, compared with the New Policies Scenario.

Trade in hard coal (coking and steam coal) between *WEO* regions mirrors the demand trends, peaking soon after 2015, at around 970 Mtce, and declining slightly thereafter. India, the largest importer of coal in 2035, remains a net importer throughout the *Outlook* period, but imports are cut by almost one-third in 2035, compared with the New Policies Scenario. By 2035, Japan and the European Union also see their imports of coal reduced, reaching a level of around 120 Mtce (26% lower than in 2010) and almost 110 Mtce (31% lower than in 2010) respectively. China, which is expected to become the world's largest coal importing country in 2012, also imports much less. Exports from Indonesia and Australia are significantly curtailed relative to the New Policies Scenario, due to the lower demand in the Asia-Pacific region. In the United States, production is reduced, in response to lower prices and lower domestic demand. It remains a net exporter, though the level is about 20% lower in 2035 than in the New Policies Scenario.

As a result of reduced import needs and lower international fossil fuel prices, importing countries see their spending on energy imports reduced significantly. The energy efficiency measures implemented cut aggregate oil, gas and coal import bills in 2035 for the major importing regions by almost $570 billion, compared with the New Policies Scenario (Figure 10.9). Almost 70% of these savings accrue from lower oil import bills. The United States, the European Union and Japan all spend less on oil imports in 2035 than today and less than projected in the New Policies Scenario. Spending by China and India increases, compared with current levels, but remains significantly lower than in the New Policies Scenario. The oil-import bills of the five largest importers are cut by 25%. Spending on natural gas imports increases over current levels in all major importing countries, with the exception of Japan.

Oil and gas revenues earned by producers in the Middle East increase to $930 billion in 2035. Although this represents a 26% reduction on their revenues compared with the New Policies Scenario, there are other offsetting economic benefits, as the assumed partial phase-out of fossil fuel subsidies slows the increase in domestic energy consumption, thereby increasing the availability of oil and gas for export over the longer term, to earn vital state revenue streams (Spotlight).

Figure 10.9 ▷ **Fuel import bills in selected countries by fuel and scenario**

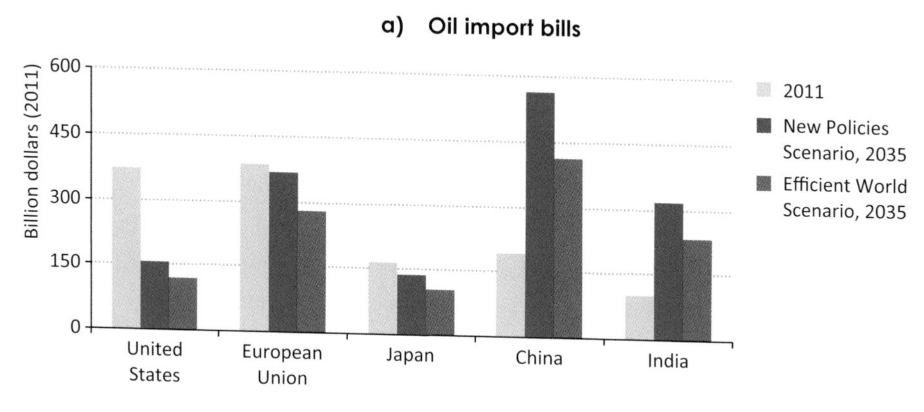

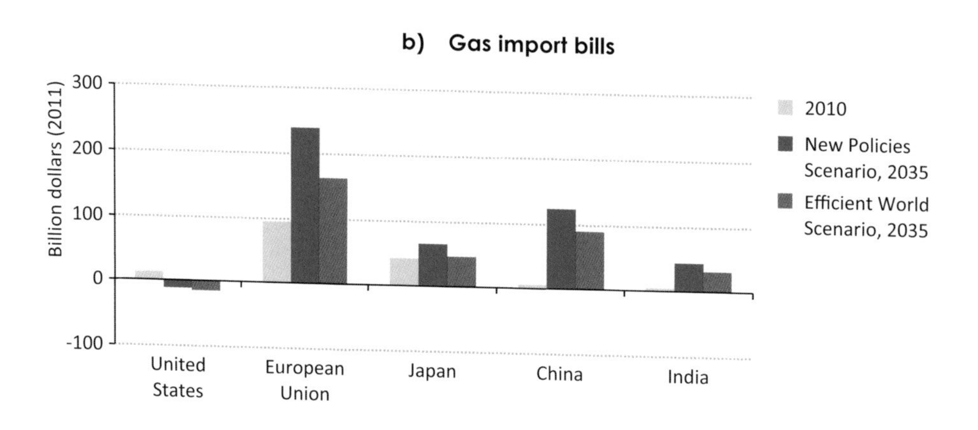

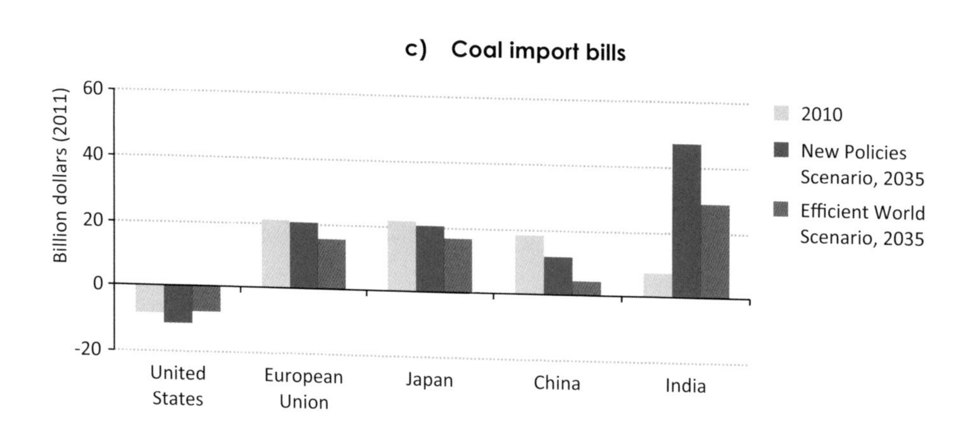

Note: Negative numbers refer to export revenues.

SPOTLIGHT

Will improved energy efficiency be good for the oil-exporting countries?

Historically, energy efficiency has not been a focus of attention in many of the oil-rich countries in the Middle East: unlike all the other regions in the world, energy intensity in the Middle East has actually increased since the 1980s (see Chapter 9). But the attention of policy makers in the region is increasingly shifting towards more efficient uses of energy, in particular in Gulf countries. Saudi Arabia established an energy efficiency centre in 2012, while the United Arab Emirates has launched a national energy efficiency and conservation programme, with the aim of improving efficiency in buildings.

A number of factors have led to this shift in policy focus: around 70% of electricity consumption in most Gulf states comes from the use of air-conditioning and ventilation devices, boosting peak electricity demand and affecting network stability. Significant losses in transmission and distribution networks often add to the problem, while the use of fuel oil and the direct combustion of crude oil for generating electricity have led to the region having an extremely low average power plant efficiency of 33%, much lower than typical thermal efficiencies in OECD countries. In the transport sector, a key area of concern is the much higher average fuel consumption of passenger cars, which is around 70% higher than the OECD average. The consequences of this low efficiency in domestic energy use are sharp domestic demand growth in the short term and reduced availability of oil for export in the longer term (Chapter 3, Box 3.1).

The main barrier to the adoption of more energy-efficient technologies in the Middle East lies in the subsidies to fossil fuels and electricity, which undermine the economic incentive for consumers to invest in energy-efficient technologies and encourage wasteful consumption. For example, due to extremely low petrol prices in Saudi Arabia, the payback period for a car which consumes half the petrol per 100 kilometres than the average car today is close to 20 years (compared with five years in the United States or seven years in China). This is far too long to incentivise any such investment. Similarly, subsidised electricity prices decrease the individual's incentive to consider more efficient air-conditioning and encourages wasteful use.

If the Middle East makes additional investments in energy efficiency of about $500 billion, compared with the New Policies Scenario, the reduction of domestic demand achieved increases the scope for Middle East countries to export. In 2035, an additional 2 mb/d of oil and 115 bcm of gas could be exported with cumulative revenues of $1.6 trillion over the *Outlook* period, if global demand was the same as in the New Policies Scenario.

Investment and fuel savings[7]

In the Efficient World Scenario, the implementation of policies to overcome barriers to the deployment of energy efficiency drives a steady stream of additional energy efficiency investments in the end-use sectors, compared with the New Policies Scenario. In the case of buildings, these additional investments rise from around $165 billion (in year-2011 dollars) in 2020 to more than $330 billion in 2035, amounting to a cumulative addition of $4.6 trillion over the *Outlook* period. However, this is more than offset by total fuel bill savings of $7.6 trillion over the same period (Table 10.3). The cumulative additional investments in industry reach $1.1 trillion by 2035, giving rise to $3.3 trillion in energy bill savings over the same time frame. For transport (excluding international bunkers), additional annual investment rises from about $90 billion in 2020 to around $575 billion in 2035, making a cumulative total of $4.8 trillion.[8] Once again, this investment is more than outweighed by global fuel cost savings, in this case of $5.7 trillion over the period to 2035, compared with the New Policies Scenario. The lifetime of the capital stock extends beyond the projection period in all sectors, so additional fuel cost savings accrue post 2035.

Table 10.3 ▷ Investment in energy efficiency, energy savings and fuel cost savings by end-use sector in the Efficient World Scenario compared with the New Policies Scenario, 2012-2035

	OECD			Non-OECD		
	Additional investment ($ trillion)	Energy savings (Mtoe)	Fuel cost savings ($ trillion)	Additional investment ($ trillion)	Energy savings (Mtoe)	Fuel cost savings ($ trillion)
Industry	0.4	668	1.2	0.7	3 482	2.2
Transport	1.6	1 121	3.0	3.2	2 731	2.7
Buildings	3.2	3 478	5.9	1.4	3 704	1.7
Total	5.3	5 267	10.0	5.2	9 917	6.6

Note: Early retirement of industrial facilities before the end of the technical lifetime by five years is assumed in the Efficient World Scenario and is included in the investment figures.

Global primary energy demand in the Efficient World Scenario is a cumulative 28 000 Mtoe lower than in the New Policies Scenario between 2010 and 2035, and electricity demand falls by a cumulative 56 000 terawatt-hours (TWh), or 8%. This reduces the need for investment in energy supply infrastructure along the entire value chain (Figure 10.10). Investment in energy supply in the Efficient World Scenario, including coal, oil and gas extraction and transportation, biofuels production and electricity generation, transmission and distribution (T&D), drops by a cumulative total of $5.9 trillion from 2012 to 2035, or 16%, compared with the New Policies Scenario. This sum offsets half of the additional investment on the demand side. Investment in the supply of fossil fuels falls by $3.4 trillion

7. Detailed analysis of the Efficient World Scenario at the sectoral level is presented in Chapter 11.

8. Additional investment occurs in international aviation and marine bunkers and amounts to an additional $1.3 trillion.

and investment in the power sector by $2.5 trillion across the whole electricity supply value chain, as average annual capacity additions are 28 GW lower than in the New Policies Scenario. This reduces cumulative investment requirements in new power generation capacity by $1.3 trillion, compared with the New Policies Scenario, and more than offsets the additional cumulative investment of $350 billion required for more efficient power plants. Investment in T&D in the power sector is reduced by $1.6 trillion, due to lower electricity demand and reduced losses.

Figure 10.10 ▷ Change in investment across the electricity value chain in the Efficient World Scenario, compared with the New Policies Scenario, 2012-2035

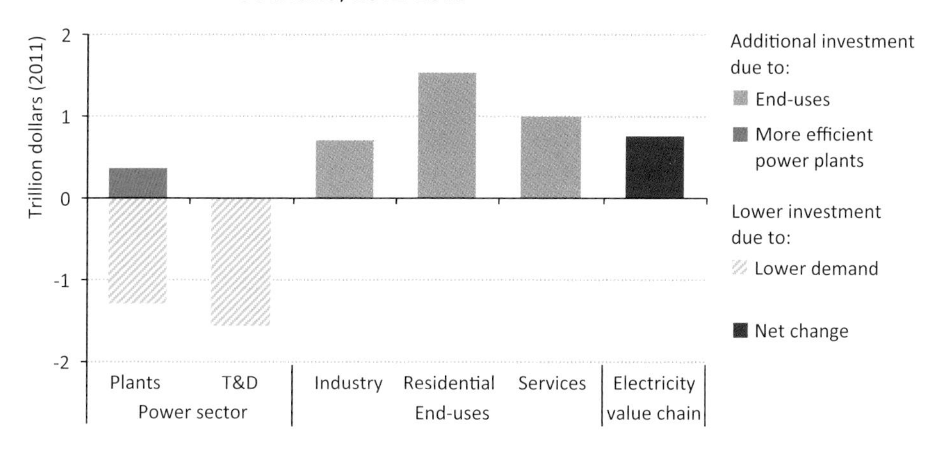

Implications for the global economy

Achieving the Efficient World Scenario would give a boost to the global economy of $18 trillion over the *Outlook* period, with a 0.4% higher global GDP in 2035 than in the New Policies Scenario (Box 10.1). This reflects a gradual reorientation of the global economy, as the production and consumption of less energy-intensive goods and services frees up resources to be allocated more efficiently. The reduction in energy use and the resulting savings in energy expenditures increase disposable income and encourage additional spending elsewhere in the economy.

While the global economy benefits overall, the impact differs across countries (Figure 10.11). In OECD countries, household consumption tends to account for a large share of GDP and means that the policies of the Efficient World Scenario are seen most clearly in increased household demand for more energy efficient goods and services. In 2035, the economy of the United States is 1.7% larger in the Efficient World Scenario, with $450 billion more economic output, most of it in the form of services. Europe's GDP is more than 1% larger, and, in addition to services, Europe sees a notable increase in domestic production and purchase of more energy-efficient road vehicles.

Box 10.1 ▷ **Assessing the impact of the Efficient World Scenario on global economic growth**

As well as the direct impact of energy efficiency policies on energy demand and energy-related investment, it is important to assess their broader impact on the global economy. Our analysis seeks to do this by linking the IEA's World Energy Model (WEM), a partial equilibrium model designed to replicate how energy markets function, to the OECD ENV-Linkages general equilibrium model, an economic model that describes how economic activities are linked to each other across sectors and regions (OECD, 2008). Specifically, the ENV-Linkages model was calibrated using the outputs of the WEM for energy demand, investment in energy consuming and producing equipment and fuel savings (OECD, 2012).

Our analysis shows that the economic impact of the Efficient World Scenario would feed through a number of channels. In general, the policies included in the Efficient World Scenario would encourage firms and households to shift their spending patterns towards more energy-efficient capital goods, which, in turn, reduces their expenditure on energy consumption. This change in the balance of spending, and therefore supply and demand, has a cascade effect on the relative price of all goods and factors of production in the economy.

Firms producing less energy-intensive goods and services are faced with increased demand and react by trying to maximise profits. By contrast, demand for more energy-intensive goods and services declines. At a household level, the move towards less energy-intensive goods and services results in a reduction in energy expenditure, which boosts disposable income and increases spending elsewhere. In addition, trade flows between countries respond to changes in relative prices between regions. For instance, if steel becomes relatively cheaper to produce in China because of increased energy efficiency, Chinese firms gain market share.

Figure 10.11 ▷ **Change in real GDP in the Efficient World Scenario compared with the New Policies Scenario**

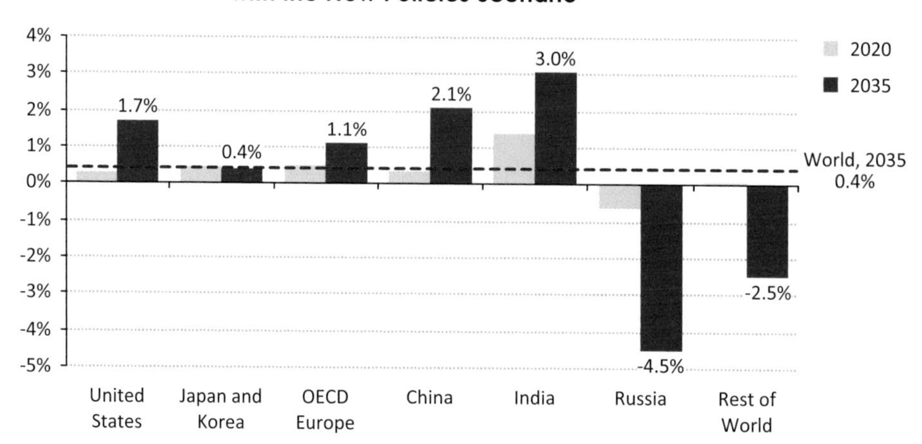

In many non-OECD countries, investment and exports play a larger role in the economy, relative to household consumption. This means that, in addition to the shift toward consumption of domestically produced goods and services, there is a more significant impact observed in manufacturing, construction and energy-intensive industries. India and China receive the largest relative and absolute boost in the Efficient World Scenario, with their economies being 3% and 2.1% larger respectively. In contrast to most other countries, the economies of the largest oil and gas exporters, such as Russia, experience lower levels of economic growth, mainly as a result of lower growth in oil and gas export revenues (due to reduced demand and prices).

The services sector ($700 billion) and transport services sector ($320 billion) experience the largest net growth of value-added in the Efficient World Scenario in 2035 (Figure 10.12), relative to the New Policies Scenario. Service-related sectors experience particularly strong growth in nearly all countries, especially the United States and European countries. The transport sector, which includes freight and public transportation, sees particularly strong growth in Europe driven, in part, by the enforcement of stringent fuel-economy standards and the rapid uptake of energy-efficient vehicles. Manufacturing grows more than in the New Policies Scenario in most countries and is focused on more energy efficient products, such as more efficient cars and electrical appliances. Overall, the construction sector sees slightly increased activity in the Efficient World Scenario, as inefficient buildings are refurbished and new buildings are required to comply with stringent energy efficiency standards. Globally, energy-intensive industries experience sizeable interregional reallocations of production but the net effect on the value-added is virtually zero. The chemical industry in the United States and the iron, steel and cement sectors in China gain most market shares.

Figure 10.12 ▷ Changes in value-added by sector and region in the Efficient World Scenario compared with the New Policies Scenario in 2035

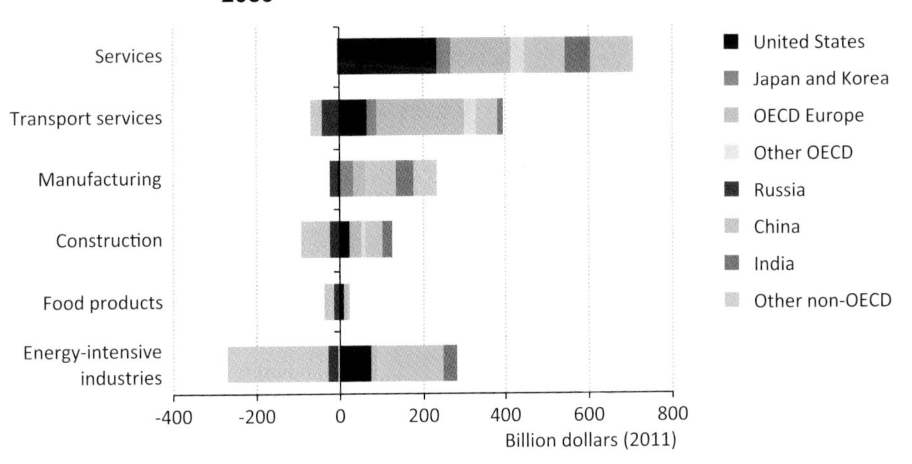

Note: In the OECD ENV-Linkages model, the manufacturing sector includes the manufacture of electronic devices and machinery, motor vehicles and trailers, transport equipment and clothing products.

While the global economy is larger in 2035, global trade is actually more than 2% lower in the Efficient World Scenario, equivalent to $0.5 trillion in goods and services (Figure 10.13). This stems from a move to less energy-intensive goods and services, which implies that a greater proportion of all goods and services are being produced and consumed domestically. Indeed, service-related sectors in OECD countries are strongly stimulated by the Efficient World Policy and capture two-thirds of additional investments. Additionally, fewer cargoes of heavily traded commodities and energy-intensive goods are transported around the world. As a result, many regions, such as Europe, China and India, improve their trade balance.

Box 10.2 ▷ How large is the rebound effect?

Increased energy efficiency does not always deliver the full energy savings predicted by engineering analysis. The undesirable side-effects are commonly referred to as the "rebound effect".

Where does the rebound effect originate? Increased efficiency of a product or facility, saving operating costs, may lead to increased use, such as when the owner of a more efficient car starts driving it more often. This is usually referred to as a direct rebound effect. An indirect effect also occurs, as a result of the increase of disposable income due to reduced energy expenditures by households and firms, which may lead to spending the available money on other energy-consuming products. This pushes up energy demand, particularly in developing countries (Bergh, 2011). Lower energy prices as a result of lower energy demand have a similar effect.

How significant is the rebound effect? Despite increased attention to the problem and extensive academic debate, large uncertainties remain about the actual size of the rebound effect and its various components (Sorrell, 2007). More recently, even the definition and methodology of calculating the rebound effect has come under scrutiny (Turner, 2012). Since the rebound effect is related to income levels and to the degree of energy service saturation in a particular country, every assessment is usually country-specific. Generalising, it can be said that, depending on the country or the consumption sector at stake, the direct rebound effect is generally small, ranging from 0-10% (see Nadel, 2012 for a discussion of the case of the United States). Estimates of the indirect rebound effect range from small to very large, with some studies suggesting 100% or more. Those figures are highly controversial.

Our estimate of the overall rebound effect is 9% of the savings that are achieved in the Efficient World Scenario, compared with the New Policies Scenario. A significant portion of this could be avoided by appropriate pricing policy. Keeping end-user prices at the same level as in the New Policies Scenario, for example, would reduce the rebound effect by more than 50%.

Figure 10.13 ▷ **Change in global trade flows for selected sectors in the Efficient World Scenario compared with the New Policies Scenario in 2035**

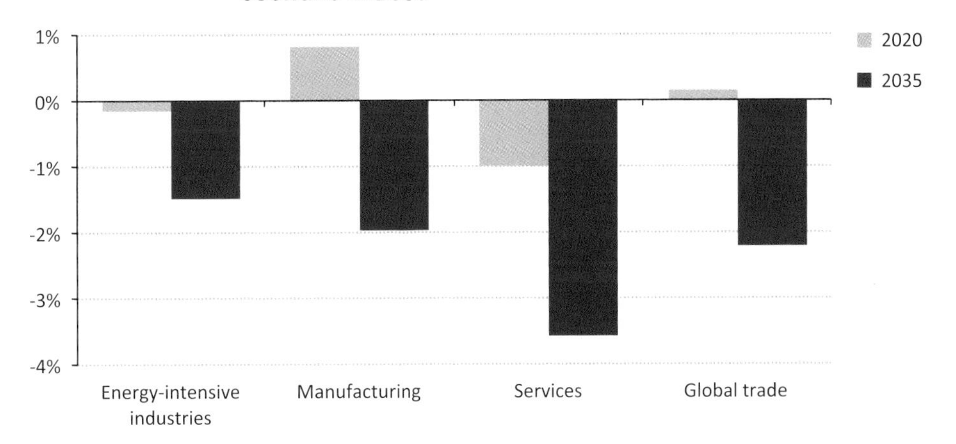

Source: IEA-OECD analysis using OECD ENV-Linkages model.

Environmental implications

Energy-related CO_2 emissions

Energy-related CO_2 emissions in the Efficient World Scenario peak before 2020, at 32.4 gigatonnes (Gt), and decline steadily from then on, to 30.5 Gt in 2035. Due to the faster deployment of energy-efficient technologies, emissions in 2035 are 6.5 Gt lower than in the New Policies Scenario (Figure 10.14). Energy efficiency, including end-use efficiency, electricity savings and efficiency gains in power plants, is responsible for 95% of the reduction in CO_2 emissions in the Efficient World Scenario, compared with the New Policies Scenario in 2035. The remainder comes from technology and fuel switching in the end-use sectors, mainly higher use of natural gas and heat pumps.

The reduction due to energy efficiency can be separated into direct and indirect savings. Direct emission savings arise from the use of less fossil fuel for the same unit of energy service provided. The transport sector is responsible for 45% of the cumulative energy efficiency-related direct CO_2 savings, followed by industry with 30%, buildings with 17% and the power sector with 7%.

Indirect emissions savings from energy efficiency arise from avoided emissions in power generation due to lower electricity and heat demand in the end-use sectors. These savings result from reduced demand in the buildings sector and in the industry sector.[9] Appliances account for roughly 40% of the electricity savings in residential buildings, with lighting and space heating contributing additional savings in OECD countries and space cooling in

9. In the New Policies Scenario, electricity contributes less than 2% to energy demand in the global transport sector in 2035. Hence, the potential for indirect emission savings in this sector is very small.

non-OECD countries. A key factor in reducing electricity-related emissions in industry is the deployment of more efficient motor systems (see Chapter 11).

Figure 10.14 ▷ Energy-related CO_2 emissions by scenario and abatement measures

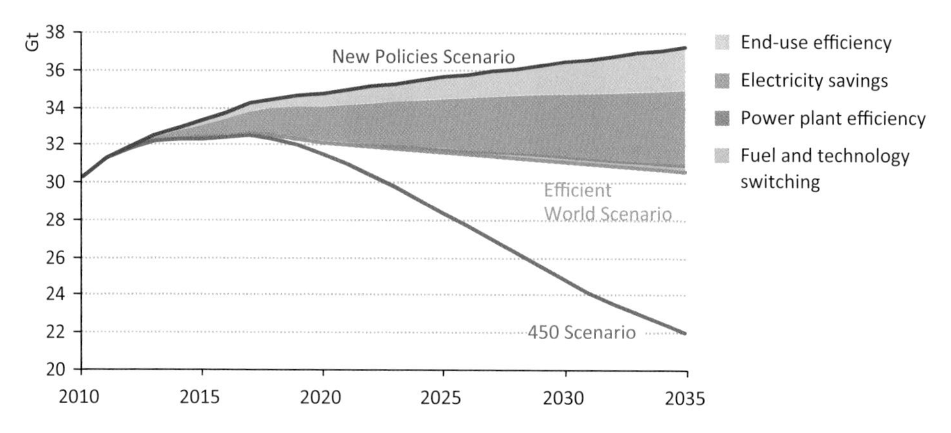

Indirect savings account for more than 60% of the total CO_2 emissions savings from energy efficiency achieved in the Efficient World Scenario, relative to the New Policies Scenario (Figure 10.15). Non-OECD countries are responsible for almost three-quarters of indirect savings as a result of their more carbon-intensive electricity generation on the one hand, with a higher share of fossil fuels and lower average efficiency levels. On the other hand, the potential to improve the efficiency of the use of electricity is higher in non-OECD countries, in particular in industry, as their average efficiency levels are comparatively low at present and as they represent the bulk of the growth in energy demand.

Figure 10.15 ▷ Cumulative efficiency-related CO_2 emissions savings in the Efficient World Scenario relative to New Policies Scenario by sector and region

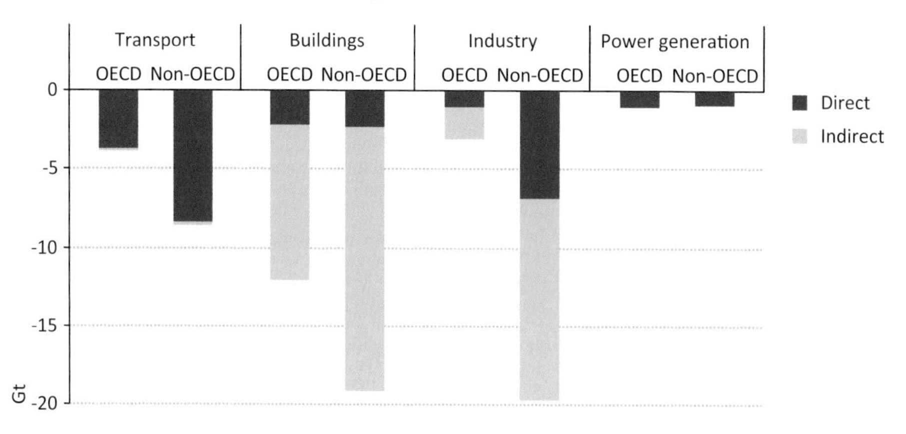

The Efficient World Scenario puts CO_2 emissions on a long-term trajectory consistent with stabilising the atmospheric concentration of greenhouse-gas emissions at around 550 parts per million. This trajectory is consistent with a 50% probability of staying below a 3 degrees Celsius (°C) temperature increase above pre-industrial levels in the long term, compared with 3.6 °C in the New Policies Scenario. This emphasises that, while energy efficiency is an indispensable element of any decarbonisation pathway, additional measures would be needed to achieve the international goal of limiting the temperature increase to 2 °C. Compared with the 450 Scenario, energy-related CO_2 emissions are 8.5 Gt higher in 2035, despite a similar level of energy consumption (see Chapter 8 for a discussion of the implications of the Efficient World Scenario on the level of emissions locked-in over time).

Local pollution

More than two million people die each year from indoor and outdoor air pollution (WHO, 2011). The rapid deployment of energy-efficient technologies that is assumed in the Efficient World Scenario would not only reduce energy consumption and CO_2 emissions, but could also save thousands of lives every year. China and India are responsible for almost half of global sulphur dioxide (SO_2) emissions, the main source of acid rain. In China, SO_2 emissions decreased over the last decade because of the installation of desulphurisation units on power plants, while India's emissions increased, in the absence of strict emission limits. In the Efficient World Scenario, China reduces its SO_2 emissions by 37% over the *Outlook* period, which results in SO_2 emissions being 12% lower in 2035 compared with the New Policies Scenario. In India, annual growth in SO_2 emissions slows from 2.6% in the New Policies Scenario to 1.4% in the Efficient World Scenario. On a wider level, SO_2 emissions are reduced by slightly more than 11% in OECD countries and almost 15% in non-OECD countries in 2035 in the Efficient World Scenario, compared with the New Policies Scenario, thanks to higher efficiency levels in power generation and industry (Table 10.4).

The largest sources of nitrogen oxides (NO_x) emissions are road transport and power generation, even though there have been major reductions in NO_x emissions from road transport in OECD countries over the past few years. NO_x emissions and particulate matter ($PM_{2.5}$) are the primary causes of smog in urban areas and can significantly damage the human respiratory system. While NO_x emissions are reduced over the *Outlook* period in the New Policies Scenario, in the Efficient World Scenario they are cut by a further 13% in 2035 in the OECD and 16% in non-OECD regions respectively, due to air pollution controls in vehicles and more efficient processes in power generation and industry. $PM_{2.5}$ emissions cause a range of health problems, including asthma and lung cancer, and are responsible for a significant number of premature deaths. Burning of traditional biomass and industrial processes cause the majority of $PM_{2.5}$ emissions in developing countries. Such emissions are reduced in the Efficient World Scenario through partial replacement of traditional biomass by more efficient cooking equipment. Compared with the New Policies Scenario, the largest reduction in $PM_{2.5}$ emissions in absolute terms is achieved in China and India, which are responsible for two-thirds of the global reduction in $PM_{2.5}$ emissions by 2035.

Table 10.4 ▷ **Air pollution by region and sector** (million tonnes)

	2005	2010	Efficient World Scenario 2020	Efficient World Scenario 2035	Change in EWS vs NPS 2020	Change in EWS vs NPS 2035
Sulphur dioxide (SO_2)						
OECD countries	29.2	18.5	12.8	10.3	-5.4%	-11.3%
Power generation	18.2	9.4	4.3	2.2	-11.6%	-30.5%
Buildings	1.4	1.2	0.9	0.7	-9.3%	-18.1%
Industry*	8.6	7.5	7.2	7.2	-0.9%	-2.1%
Road transport	0.3	0.1	0.0	0.0	-2.0%	-13.8%
Other**	0.8	0.5	0.4	0.2	-2.5%	-10.6%
Non-OECD countries	68.1	67.7	56.9	51.8	-7.6%	-14.9%
Power generation	38.0	33.5	24.0	18.8	-13.5%	-26.3%
Buildings	4.4	4.7	4.5	3.4	-6.1%	-16.3%
Industry*	23.6	28.0	26.9	28.0	-2.3%	-4.8%
Road transport	0.9	0.4	0.3	0.4	-1.7%	-14.9%
Other**	1.2	1.1	1.2	1.1	-3.5%	-12.5%
Nitrogen oxides (NO_x)						
OECD countries	38.4	29.5	18.7	13.9	-4.6%	-12.7%
Power generation	9.8	7.2	4.3	3.0	-9.4%	-21.3%
Buildings	2.0	1.9	1.8	1.6	-8.0%	-20.1%
Industry*	5.4	5.0	5.1	5.1	-1.5%	-3.4%
Road transport	15.0	10.3	3.7	1.7	-2.3%	-13.2%
Other mobile sources	6.1	4.9	3.7	2.4	-3.9%	-13.6%
Other**	0.1	0.1	0.1	0.1	0.0%	0.0%
Non-OECD countries	48.9	55.2	54.7	56.0	-5.9%	-15.8%
Power generation	12.7	14.8	14.3	13.8	-13.2%	-24.5%
Buildings	3.0	3.2	3.3	3.1	-3.4%	-8.9%
Industry*	9.8	12.4	13.6	14.5	-2.2%	-4.7%
Road transport	14.5	15.0	13.2	13.4	-3.6%	-19.6%
Other mobile sources	8.5	9.3	9.8	10.7	-3.2%	-14.0%
Other**	0.6	0.5	0.6	0.6	0.0%	0.0%
Particulate matter ($PM_{2.5}$)						
OECD countries	4.5	4.2	3.8	3.6	-3.2%	-6.0%
Power generation	0.3	0.2	0.2	0.1	-12.5%	-30.7%
Buildings	1.3	1.3	1.3	1.3	-6.3%	-9.7%
Industry*	0.9	0.8	0.8	0.9	-0.3%	-0.7%
Road transport	0.6	0.4	0.2	0.2	-2.1%	-12.8%
Other mobile sources	0.4	0.4	0.2	0.2	-3.8%	-12.8%
Other**	1.0	1.0	1.1	1.1	0.0%	0.0%
Non-OECD countries	36.2	38.6	37.5	35.3	-2.2%	-5.6%
Power generation	2.0	2.3	2.5	2.3	-14.8%	-27.4%
Buildings	16.1	17.1	16.8	14.7	-1.7%	-5.0%
Industry*	11.1	12.2	11.2	10.8	-0.5%	-1.2%
Road transport	1.0	0.9	0.6	0.7	-3.4%	-18.9%
Other mobile sources	0.7	0.8	0.8	0.9	-3.4%	-14.5%
Other**	5.2	5.4	5.7	5.9	0.0%	0.0%

* Includes industrial processes. ** Other includes waste management, agriculture, other mobile sources (if not separately shown) and fuel extraction. Note: NPS = New Policies Scenario; EWS = Efficient World Scenario.

Source: IIASA (2012).

The role of energy efficiency in increasing energy access

Access to adequate energy services is fundamental to pulling communities out of poverty, as it is a vital input for social and economic development. Currently 1.3 billion people do not have access to electricity worldwide and 2.6 billion people rely on biomass as their primary source of fuel for cooking. Despite the action being undertaken, such as the efforts recently announced under the umbrella of the UN Sustainable Energy for All initiative, these numbers are not reduced significantly by 2035 in the *Outlook* (see Chapter 18).

Countries with low levels of access to modern energy services rarely concentrate on energy efficiency, as they face other, more pressing, challenges. Energy efficiency policies also demand a high level of sophisticated co-ordination between government agencies, which is difficult to achieve even in OECD countries. Governments with limited energy resources often first look to improve the efficiency of the supply side, in order to extend their capacity to provide energy to households. The next step is often to address technical losses in generation and distribution systems and then to seek to cut demand in end-uses, in order to free up additional resources for use by others. By improving energy affordability, energy efficiency can make it easier for lower income households to pay energy bills, freeing up funds for other priorities.

Figure 10.16 ▷ Additional electricity needed to achieve energy for all in India compared with savings in the Efficient World Scenario, 2011-2030

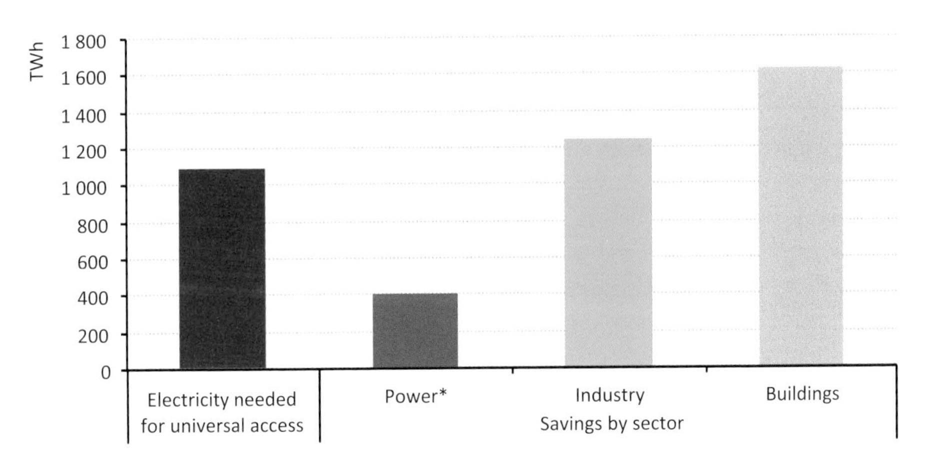

*Efficiency improvements in fossil-fuel power plants and system grids.

A number of developing countries have already identified the important role that energy efficiency can play in improving energy access. India, for example, announced the National Mission for Enhanced Energy Efficiency, which identifies increased electricity access and improved reliability as important co-benefits of improved efficiency. Our analysis shows that in certain countries, including India, the energy saved in the Efficient World Scenario is more than enough to provide the additional electricity needed over the level of the New

Policies Scenario to serve basic energy needs of the entire population by 2030.[10] Although, the challenges associated with grid extension and the development of off-grid solutions would also need to be met (Figure 10.16). In other cases, while the savings are not by themselves sufficient to satisfy the basic needs universally, they could enable electrification programmes to expand at a much faster rate by freeing up financial resources to be devoted to this purpose.

Building the Efficient World Scenario: a blueprint for savings

The discussion on barriers to energy efficiency in Chapter 9 makes it plain: the energy savings identified in the Efficient World Scenario will not happen if market actors are left to their own devices. To seize the opportunity, the Efficient World Scenario rests on the foundation of a raft of concrete, forceful and complementary policy measures taken to overcome these barriers. These stimulate private and public sector actions that generate the energy savings and co-benefits of the Efficient World Scenario. As the nature of the barriers to energy efficiency are manifold and divergent, depending on the circumstances of the end-use and economy considered, a portfolio of measures is needed. But, whatever the specifics of the sector or economy being addressed, certain key principles need to be adhered to. Implementation of the Efficient World Scenario envisages the prior adoption of policy measures in line with the public policy framework, or blueprint, which follows. While much can be achieved by individual countries or regions, full realisation of the benefits of the blueprint is likely to depend on a formal global commitment, to raise energy efficiency and report results regularly, using mutually agreed verification mechanisms.

Make it visible

The energy performance of each energy end-use and service needs to be made visible to the market. Under the Efficient World Scenario, it is envisaged that governments take the lead, in partnership with private sector agents, to ensure that the energy performance of all major energy services and end-uses is measured and reported to consumers, clients and statistical authorities in a consistent, accessible, timely and reliable manner. Governments need to frame the market for energy services in this manner so as to ensure that energy efficiency options can compete on a level playing field with energy supply options. A key element of this is to ensure that the relative energy efficiency of different products and services is visible in the market place. Policy makers need to ensure that, beyond the basic measurement of consumption per unit of output, common, agreed measurement test procedures and/or protocols are developed to measure energy efficiency. They need to ensure also that the resulting information is routinely available, in a readily comprehensible form, to all those considering the procurement of energy using assets or equipment or the

10. This timeframe is chosen to be consistent with meeting the UN goal of full electrification by 2030, which we analyse in detail in Chapter 18 in the Energy for All Case.

optimisation of an existing system.[11] Such increased visibility lowers information costs, an important element of transaction costs.

Make it a priority

The profile and importance of energy efficiency needs to be raised. Visibility stimulates market actors to consider energy efficiency, but is often not enough to motivate them to demand it. Government needs to take additional steps to ensure the full value of higher energy efficiency is clear to the individual and to society at large. Available measures include: regulatory requirements, such as the obligations China and Japan place on industry to implement energy efficiency measures, and those some European and North American regulators place on utilities; market transformation programmes; measures to obligate companies to address questions of energy performance at board level, such as corporate social responsibility reporting requirements; and awareness raising and promotional activities.

Make it affordable

Create and support business models, financing vehicles and incentives to ensure investors in energy efficiency reap an appropriate share of the rewards. Tailored financing instruments are needed to address the various split incentive barriers to energy efficiency; for example, where the asset ownership period is shorter than the payback period, such as for the retrofit in buildings. These mechanisms may need to be structured to encourage the redeployment of long-term capital, ordinarily targeting energy supply-side investments, into investments aimed at energy-demand reduction through efficiency. In any case, they need to remove the risks to asset owners of potential asset sale before the return on investment is accrued. Examples of these instruments include utility-operated or funded energy efficiency finance schemes, typically tied to demand-side management or utility energy efficiency obligation schemes; pay-as-you-save schemes; and supportive frameworks for the energy services industry. Other instruments can be deployed to help increase the attractiveness of energy efficiency investments when they are competing for capital with alternative investment opportunities that are perceived to have a higher rate of return. These include soft loans, grants, credit lines, loan guarantees, and special funds. Equally, when appropriate, fiscal and financial incentives can help increase the attractiveness of energy efficiency investment. Such incentives can often be temporary, designed to bridge the initial cost gap. They allow market volumes to grow and the cost gap to reduce, due to economy of scale effects.

11. Examples of policy instruments used to address the measurement of energy efficiency are: energy performance test procedures, energy efficiency measurement metrics; energy system and sub-system metering and energy auditing. Examples of policy instruments used to ensure that energy efficiency is reported to end-users and procurers are: energy labelling, rating and energy performance disclosure schemes, energy performance benchmarking, smart metering and performance feedback systems, such as continuous commissioning (*i.e.* constant monitoring and tuning of building equipment to ensure it is operating optimally without wasting energy).

Perceptions of financial risk are another barrier to energy efficiency investment. They can be tackled by measures to lower the risk premiums applied to lending for energy efficiency projects, combined with measures to alert end-users to the value of energy efficiency investment because of the potential volatility of energy prices. These measures include risk guarantees, credit lines, mechanisms to standardise and bundle project types, and awareness and capacity building efforts among the finance community. The type and scale of these instruments, as in the Efficient World Scenario, varies by sector and economy, but needs to be of sufficient scale to address many of the primary affordability barriers to energy efficiency investment by dramatically reducing actual or implicit transaction costs. It is important to note that many of these financing instruments essentially operate to re-deploy supply-side capital investments into demand-side efficiency improvement investments, which has the effect of changing energy supply businesses or financiers into energy service (supply and demand) businesses and financiers. The success of these measures in the Efficient World Scenario helps to lower bills through reducing demand (despite the recycling of a part of these savings to finance the energy efficiency measures that results in a slight increase in energy prices). Overall, however, the reduction in demand lowers both bills and prices and gives rise to a modest rebound effect. This, in turn, is offset in certain economies by the removal of energy subsidies by 2035. The removal of subsidies not only creates a level playing field for energy efficiency investment, helps improve the viability of energy businesses and reduces energy security risks, but is one of the more important direct drivers of savings under the Efficient World Scenario.

Make it normal

Energy efficiency needs to be normalised if it is to endure. Once a high-efficiency technology or service solution has been widely adopted, there is rarely a step backwards: the old less-efficient technology or approach is rapidly forgotten. Usually the cost differentials for higher-efficiency technologies and services decline substantially as adoption rates increase. Accordingly, policies which build the market are helpful and necessary to "normalise" energy efficiency. Under the Efficient World Scenario, a mix of regulations is deployed to prohibit the least-efficient approaches and impose minimum energy performance standards for equipment, vehicles, buildings and power plants. This is, indeed, the single most important category of policy mechanisms in the Efficient World Scenario. In some instances, regulatory requirements are also placed on industry – to develop, implement, monitor and report effective energy saving programmes – and utilities – to finance and implement energy efficiency schemes for their customer base. The scale of this activity lowers the transaction costs. Complementary measures to ensure that the energy-efficient solution becomes the normal solution include efforts to boost the supply of new, higher efficiency technologies and services into the market. Resulting benefits from learning and economies of scale help make the most energy-efficient option the normal solution.

Make it real

Monitoring, verification and enforcement activities are needed to verify claimed energy efficiency. Without such efforts, experience has shown that savings will turn out to be less than expected and the overall policy objectives be undermined. Under the Efficient World Scenario, there is a substantial increase in the scale of such activities. Verification builds confidence in claimed performance and outcomes. Enforcement is necessary to secure compliance (for example, with vehicle efficiency standards or the application of buildings codes). Monitoring provides the principal inputs for evaluation, which is essential to ensure energy efficiency programmes are delivering the expected outcomes and to facilitate any necessary adjustments.

Make it realisable

Achieving the supply and widespread adoption of energy efficient goods and services depends on an adequate body of skilled practitioners in government and industry. The Efficient World Scenario foresees the adoption of systematic programmes to develop and sustain a body of skilled energy efficiency workers. The required skills extend to policy development and implementation, product and service development, monitoring, verification and enforcement, fostering innovative business models and the implementation of quality assurance efforts to ensure there is no loss of service or satisfaction through the adoption of the efficient option. The buildings sector illustrates the scale of the challenge. The diverse nature of the sector (the "fragmentation barrier", see Chapter 9, Table 9.2) requires sustained and extensive capacity building for the necessary skills, once developed, to be widely transferred.

Investment in governance and programme development, and implementation is required on a scale considerably beyond current practice. Governance of energy efficiency is intrinsically more complex than the governance of energy supply, because there are many more energy end-uses and services, which are often major industries in their own right and which have unique characteristics that require targeted policy measures. While a substantial increase in administrative funding is required, the sums involved are a very small fraction of the value of the savings and greater economic efficiency they facilitate. One of the world's best resourced and most successful equipment standards programmes reports that the cost of administering the programme is just $1 for every $650 of the value of the energy savings it produces (US DOE, 2012).

Unlocking energy efficiency at the sectoral level

What is needed and where?

Highlights

- In the Efficient World Scenario, primary energy demand is reduced by 2 350 Mtoe in 2035 compared with the New Policies Scenario, mostly occurring in the power sector. However, 85% of these savings are the result of energy efficiency measures in end-uses. If the savings are attributed to the end-use sectors, buildings account for 41% of the reduction in primary energy demand in 2035, followed by industry (23%), transport (21%) and power (8%).
- Energy used in the buildings sector grows at an average annual rate of 0.4% between 2010 and 2035 in the Efficient World Scenario, a substantially slower rate than the 1% of the New Policies Scenario. The savings are driven by faster uptake of efficient lighting, appliances and equipment in all regions, retrofitting of existing buildings in OECD countries, and more efficient new-build, technology switching and energy price reforms in non-OECD countries.
- Energy demand growth in the industry sector falls to 1.1% per year on average in 2010-2035 in the Efficient World Scenario, from 1.5% in the New Policies Scenario. Despite an increase of 113% in industrial sector activity, energy use increases by only 31% over the period. Savings arise from faster adoption of more efficient technologies, phasing out older facilities, process change and system optimisation, including of electric motor driven systems.
- Efficiency gains in the transport sector reduce oil demand in 2035 by 9.1 mb/d compared with the New Policies Scenario. The global average fuel economy of new sales (test-cycle) for passenger light-duty vehicles reaches 3.5 litres per 100 kilometres (l/100 km) in 2035, down from 7.6 l/100 km in 2010. The on-road global average fuel consumption of heavy trucks in 2035 is around 45% lower than what it was in 2010. Key policy instruments that help to achieve the fuel economy improvements include stringent standards, fuel economy labelling, tax breaks and other incentives.
- Electricity demand growth in 2010-2035 is reduced by one-third, compared with the New Policies Scenario, mainly as a result of higher efficiency in the equipment used in buildings and industry. Almost 200 Mtoe of savings stem from increased efficiency in the power generation sector. The overall efficiency of fossil-fuel power generation rises to 49% in 2035, six percentage points above 2010 levels. This increase is 2.5 percentage points higher than in the New Policies Scenario, and is achieved mainly thanks to efficiency and emission standards that prevent the construction of inefficient plants and the refurbishment of old ones.

The balance of sectoral opportunities

The discussion of the Efficient World Scenario in Chapter 10 illustrated the benefits available from exploiting known opportunities for economic investment in energy efficiency. This chapter concentrates on the opportunities by sector. The measures included on a sectoral level are detailed in Table 11.1. Primary energy demand in 2035 is reduced by some 2 350 million tonnes of oil equivalent (Mtoe), or 14%, compared with the New Policies Scenario, the majority of it in the power sector (1 263 Mtoe). However, 85% of the savings in the power sector are the result of demand-side savings in other sectors, especially buildings and industry. If those savings are attributed to the end-use sectors where the demand reduction occurs, the buildings sector accounts for almost 41% of the savings (Figure 11.1), mostly due to improvements in the energy efficiency of building shells and electrical equipment. In 2035, within industry and the buildings sector, almost two-thirds of the energy savings are in the form of electricity and heat. By contrast, savings in the transport sector are dominated by a reduction in oil demand, mainly driven by improved fuel efficiency in road transport.

Figure 11.1 ▷ **Energy savings in 2035 by fuel and sector in the Efficient World Scenario compared to the New Policies Scenario**

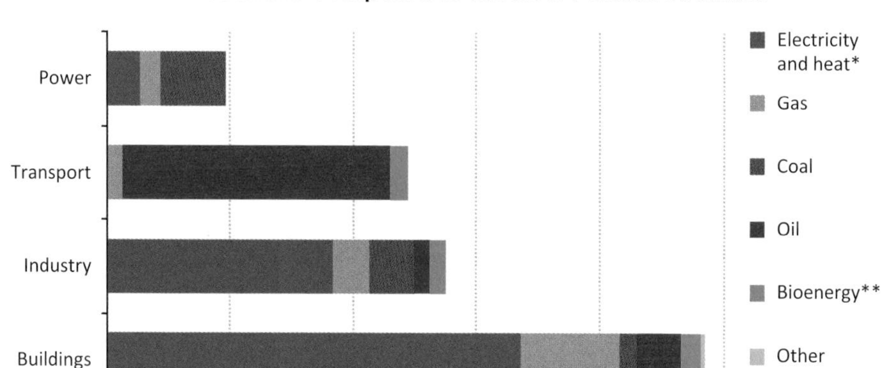

* Electricity demand savings in end-use sectors are converted into equivalent primary energy savings and attributed to each end-use. The savings allocated to the power sector arise solely from the increased efficiency of the plant, the grid and system management. ** Bioenergy includes waste.

Note: The figure excludes savings in non-energy use, other energy sectors and agriculture, which together account for 152 Mtoe.

Table 11.1 ▷ **Summary of key policies by sector in the Efficient World Scenario**

Sector	Policy framework in the Efficient World Scenario	Policies beyond the scope of the Efficient World Scenario
Buildings	• Stringent building energy codes for new buildings and those undergoing renovation implemented by 2015 and enhanced by 2020. • Retrofits, beyond the level of the New Policies Scenario, in existing buildings (in OECD countries). • Minimum energy performance standards (MEPS) for all major appliances and equipment, implemented/enhanced by 2015. • Building energy management systems in all new construction in OECD from 2015 and in non-OECD from 2020.	• Changes in urban design (horizontal versus vertical cities). • Architectural improvements, such as reduction in per-capita floor space requirements through better layout design. • Increased access to electricity. • Support for distributed renewable energy generation. • Energy conservation induced by behavioural change beyond that induced by price.
Industry	• All new equipment having efficiency levels matching best available technology by 2015. • Early retirement of inefficient existing facilities by five years. • Process change, when applicable to local conditions. • Implementation of process control and energy management systems. • Adoption of high-efficiency electric motor systems.	• Deployment of carbon capture and storage. • Support for low carbon energy. • Structural changes in the economy beyond those included in the New Policies Scenario.
Transport	• Deployment of the most efficient vehicle options in road transport by 2035, driven by mandatory fuel-economy standards, fuel-economy labelling, tax breaks and incentives. • International sectoral agreement in the aviation and maritime sectors.	• Fuel switch beyond the level of the New Policies Scenario. • Integrated transport and land-use planning. • Modal shift policies. • Demand management strategies (car-pooling, teleworking, etc). • Behavioural changes beyond those induced by price.
Power generation and grids	• Efficiency standards on existing fossil fuel plant, reducing refurbishment and lifetime of inefficient plant. • Efficiency standards on new fossil fuel plants, reducing or prohibiting the construction of coal subcritical or gas steam power plants. • Support for smart grids and efficiency standards for power networks.	• Introduction of CO_2 pricing beyond the countries assumed in the New Policies Scenario. • Enhanced support for renewables. • Stronger support and penetration of CCS technology. • Stronger support for nuclear power plants.

Buildings

Techno-economic potential and policy framework

There is a very large technical potential to improve the energy efficiency of the building stock and the equipment used within it. New buildings can be constructed to use less than 10% of the energy of typical designs and can be net zero-energy or even net positive-energy contributors if on-site distributed generation is used (NREL, 2007).[1] Holistic retrofits can save up to 90% of the thermal energy use in existing buildings (ECEEE, 2011); while the technical savings potential from the use of energy efficient equipment and appliances can range between 5-90%, depending on the end-use.[2] But a range of barriers exist which discourage realisation of these savings. While the New Policies Scenario already includes some measures targeting the buildings sector, these address only part of the economically viable savings potential.

The Efficient World Scenario addresses most of the remaining gaps through the adoption and implementation of a raft of strong policy measures. In buildings, it assumes that all energy efficiency policies now under consideration for the buildings sector are fully implemented, reinforced and strengthened and that their breadth and scope is extended. In particular, stronger measures are adopted to overcome the factors which deter individual building owners and developers from implementing energy efficiency measures which, in themselves, are fully economic. For the building shell and structure these measures include stringent building energy codes that apply to both new and existing buildings and that progress at the fastest rate the local building industry is capable of meeting; more effective code compliance; building energy labelling and performance disclosure; linking the permission to plan and build to attainment of building energy performance objectives; access to financing through mechanisms such as dedicated energy efficiency credit facilities and pay-as-you-save schemes; strong fiscal and financial incentives; capacity building, training and awareness; and research and development. These policy measures are structured to ensure that new buildings advance as rapidly as is realistically achievable towards net zero-energy consumption levels, while also ensuring the rate and depth of energy efficient retrofit of the existing building stock is substantially increased through target-based holistic renovation programmes (especially for OECD countries). However, only energy efficiency investments that are repaid within strictly-defined payback periods are considered under the Efficient World Scenario (see Chapter 10).

For building components, such as windows and insulation, and energy using equipment, a complementary set of policies is applied. This includes mandatory labelling schemes and minimum energy performance standards (MEPS) for all significant energy end-uses and also for energy-related equipment, such as windows, showerheads, faucets and insulation. The stringency of these policies is increased to take better account of the true improvement

1. Such deep reductions are not necessarily achievable for all building types or climate zones, and are often uneconomic for new construction.

2. 90% is for standby power and the 5% is for cooking appliances. All other equipment and appliances have savings potentials in between these two limits (IEA, 2009a).

potential and technology learning curves for energy using and related equipment. Much greater use is also made of supporting policies, such as utility demand-side management (DSM) programmes and incentives to accelerate extensive renovations and the replacement of less efficient technologies by more efficient alternatives. Additionally, policy support is given to energy management systems, like automation, active controls, smart metering and monitoring systems (for consumer feedback). The incentive to adopt higher energy efficiency measures is further enhanced by the assumed partial removal of fossil-fuel subsidies (see Chapter 10).

Of these policies, adoption of progressively more stringent building energy codes and minimum energy performance requirements for all significant energy-using equipment is the key policy for the buildings sector. Equipment efficiency standards can be readily applied in all markets, but building codes are more difficult to implement in markets with a greater share of informal construction and need to be phased in more gradually. Strengthening compliance is essential for policies in both areas to work, and the necessary investment by the authorities in market monitoring, verification and enforcement activities is assumed. For the existing building stock, ambitious programmes need to be implemented, including mandatory annual renovation rates, under the stimulus of appropriate incentives.

Outlook

Energy demand in buildings in the Efficient World Scenario grows at an average annual rate of 0.4% from 2010 to 2035 to reach almost 3 200 Mtoe, or 15% less than in the New Policies Scenario (Table 11.2). There is a substantial reduction in buildings sector energy intensity: the rate of growth in energy use is appreciably slower than growth in building floor area, which increases at an average annual rate of 1.7%. This is attributable to the much wider adoption of measures to improve the thermal performance of the building stock and raise the efficiency of equipment and appliances.

Table 11.2 ▷ Global buildings energy demand by fuel and energy-related CO_2 emissions in the Efficient World Scenario

	2010	2020	2035	CAAGR* 2010-35	Change versus New Policies 2020	Change versus New Policies 2035
Total (Mtoe)	2 910	3 080	3 193	0.4%	-7%	-15%
Coal	124	111	69	-2.3%	-10%	-27%
Oil	329	299	233	-1.4%	-9%	-22%
Gas	616	655	697	0.5%	-8%	-19%
Electricity	831	960	1 161	1.3%	-10%	-17%
Heat	148	150	145	-0.1%	-6%	-15%
Bioenergy	841	866	809	-0.2%	-1%	-4%
Other renewables**	21	39	78	5.5%	-6%	-8%
CO_2 emissions (Gt)	2.9	2.8	2.6	-0.5%	-8%	-21%

*Compound average annual growth rate. ** Includes solar and geothermal.

The use of fossil fuels in the sector reduces gradually, as growth in demand for natural gas is more than offset by a decline in demand for coal and oil. Coal exhibits the largest reduction in demand (27%) in 2035 relative to the New Policies Scenario, followed by oil (22%) and then electricity and gas (both at around 18%). The share of electricity in building energy use continues to grow strongly, as it does in the New Policies Scenario, rising from 29% in 2010 to 36% in 2035. Despite the strong growth in electricity demand in the Efficient World Scenario, the additional savings, compared with the New Policies Scenario, reach more than 240 Mtoe by 2035, largely due to the improved efficiency of appliances and equipment. These electricity savings are expressed in terms of final energy consumption; but if the primary energy needed to produce electricity and heat are factored in the savings amount to about 675 Mtoe in 2035.

Although the cumulative energy savings over the outlook period in buildings, relative to the New Policies Scenario, are almost equally divided between the OECD and non-OECD regions, the reduction comes from very different sets of options. In OECD countries, where new construction activity is estimated to be as low as 1% of the building stock each year and demolitions 0.3-0.5%, the biggest potential savings are made in existing buildings (IEA, 2009b).[3] On the other hand, non-OECD countries generally have much higher new construction activity than OECD countries, especially in emerging economies, where the rate of renewal is estimated to be around 5% of existing stock for residential buildings and 10% for commercial buildings (IEA, 2009b). Consequently, the biggest saving potential in non-OECD countries is in new buildings, through the penetration of more efficient appliances and equipments.

Accordingly, the majority of savings in buildings in the residential and services sectors in the OECD region is attributable to the reduction in energy use for space heating and cooling, due to refurbishment of existing buildings and, especially, to improvements in the insulation of building shells for both. Recent case studies show that the renovation of building shells and openings, combined with installation and appropriate operation of heat control and measuring devices, can improve building energy efficiency by up to 60% (IEA, 2009b). In the Efficient World Scenario, energy demand in OECD buildings declines at 0.15% per year on average. Fossil-fuel consumption declines, while use of modern biomass and other renewables grows. There is a substantial reduction in the use of natural gas and oil, due to improved insulation and operation, and a shift towards heat pumps for space heating. In spite of this shift, the consumption of electricity in buildings is reduced sharply relative to the New Policies Scenario, as a result of the broader implementation of mandatory MEPS and labelling, which stimulate the adoption of more efficient appliances and equipment, and insulation and retrofits in building shells, which lower space heating and cooling service demand. Overall, OECD building electricity consumption falls by 16% in 2035, compared with the New Policies Scenario.

3. Except for Japan, where the demolition rate is much higher.

Energy use in buildings within non-OECD countries rises from 1 658 Mtoe in 2010 to 1 985 Mtoe in 2035 in the Efficient World Scenario: 12% less than in the New Policies Scenario. Most of the major non-OECD countries are considering the introduction of legislation to set mandatory building energy codes for new buildings and MEPS and labelling for appliances, lighting and equipment. In the Efficient World Scenario, these measures are fully implemented and strengthened. They are further complemented by capacity building, training, demonstration projects, awareness campaigns and the provision of financial incentives. The partial removal of fossil-fuel subsidies gives a further stimulus to efficiency improvements and so reduces energy demand.

China has the world's largest building stock, absorbing 16% of global energy consumption in buildings in 2010. The total floor area of all buildings in China is about 48 billion square metres currently and it is expected to reach 60 billion square metres by 2035.[4] This expansion is largely driven by housing demand in urban areas, where per-capita floor space has been increasing by one square metre per year in the recent past (China Daily, 2008). Although significant efforts have been made to implement energy conservation measures, such that most new urban construction now complies with building codes, there is still potential to increase code compliance in rural areas and to improve the energy efficiency of the existing building stock (LBNL, 2010). In the Efficient World Scenario, a wide range of policies and measures are assumed to be adopted in China, producing energy savings in buildings of 18% by 2035, compared with the New Policies Scenario.

Most existing buildings in Russia and Eastern Europe have very high energy intensities, with losses estimated to be up to 40% of supplied energy (IEA, 2009b). There is large potential to reduce thermal energy use through the refurbishment of the building stock and the heat networks which supply it (IEA, 2011a). Although energy efficiency is already becoming a priority in Russia, the policy package adopted within the Efficient World Scenario is much more comprehensive than the policies currently under consideration. It entails the full implementation of legislation providing subsidies for retrofits and energy efficiency technologies and fines for owners of buildings that fail to respect the defined standards. These measures are combined with reform of the heat markets and imposition of more stringent building energy codes and equipment efficiency standards. These measures result in energy savings of 15% in 2035, compared with the New Policies Scenario.

In India, total energy demand in buildings in 2035 is reduced by 14%, compared with the New Policies Scenario. India is expected to construct more buildings in the period 2012-2035 than the total stock existing in 2010. In addition to the broader adoption and enforcement of building energy codes, much of the savings in the near term come from technology switching, such as the adoption of liquefied petroleum gas (LPG) stoves and light-emitting diodes (LED) lighting, triggered by subsidy removal, and from the adoption of high-efficiency equipment, stimulated by more stringent and comprehensive MEPS.

4. GBPN and CEU (2012); IEA analysis.

Trends by sub-sector

Residential sector buildings

Energy savings in the buildings sector stem from reductions in heating and cooling demand, resulting from greater insulation; efficiency savings from higher-efficiency equipment and technology switching (*e.g.* from gas-fired boiler space heating to the use of heat pumps); and reduction in overall demand due to the removal of energy subsidies (Figure 11.2). Retrofits play a greater role for OECD countries, where the stock turnover is not very high.

Figure 11.2 ▷ Savings in residential energy demand in the Efficient World Scenario relative to the New Policies Scenario by contributing factor

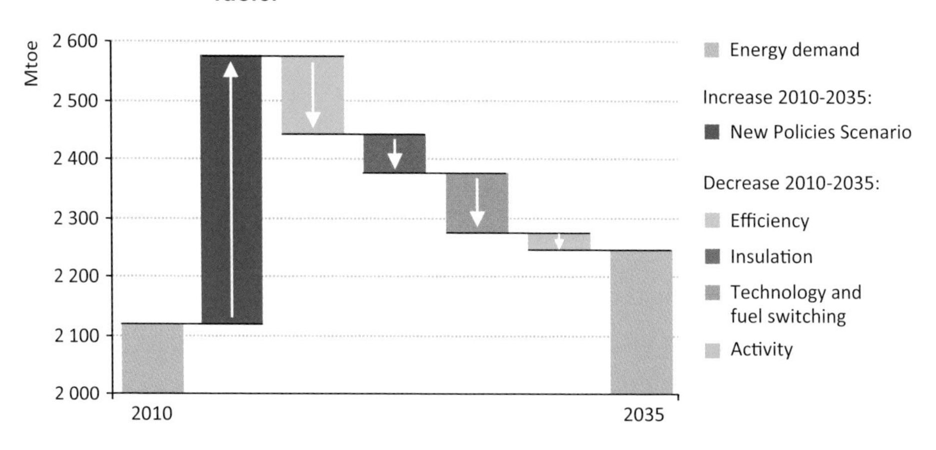

Note: Details on decomposition analysis can be found in Chapter 9, Box 9.4.

Standard end-uses in residential buildings include space and water heating, appliances, lighting, cooking and cooling. In the Efficient World Scenario, reductions in residential energy use in space and water heating account for 56% of the cumulative savings at the world level to 2035, compared with the New Policies Scenario. In the OECD, this accounts for two-thirds of cumulative savings (Figure 11.3). Appliances, lighting and cooling make up almost another 20% of the reduction. Unlike in OECD regions, space cooling is a key priority in non-OECD countries, due to climate conditions. Interestingly, out of the 50 largest metropolitan areas in the world today, the vast majority with high annual cooling degree-days are in non-OECD countries (Sivak, 2009).[5] In those countries, appliances, cooling and lighting achieve significant energy savings, attributable to the spread of mandatory labelling schemes and MEPS. Cooking is also a key end-use sector in terms of energy savings potential, due to present widespread use of inefficient cooking technologies, most using traditional biomass.

5. Cooling degree-days are the number of degrees per day that the daily average temperature is above a given "comfort" temperature.

Figure 11.3 ▷ **Change in energy demand in the residential sector in the Efficient World Scenario and the New Policies Scenario from 2010 to 2035 by end-use**

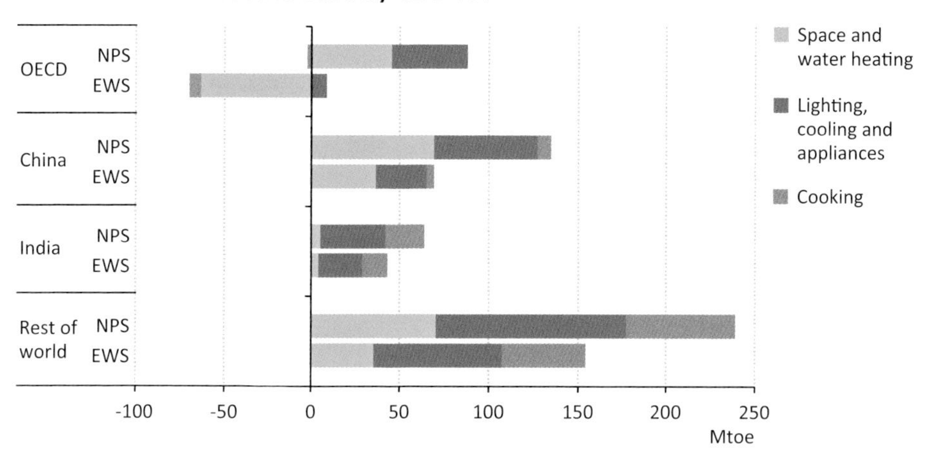

Note: NPS = New Policies Scenario; EWS = Efficient World Scenario.

Service sector buildings

In the Efficient World Scenario, energy consumption in the services sector is 19% lower in 2035, about two-thirds of the reduction coming from more efficient heating and cooling and better insulation (including automated building energy management systems). Deployment of higher-efficiency solutions increases in the service sector, as in the residential sector. The fabric of existing and new buildings incorporates integrated design, better insulation and shading to optimise thermal efficiency and use of daylight. Heating, cooling and ventilation systems are controlled better and are more efficient, taking advantage of natural ventilation and cooling, heat pump technology, heat recovery and/or cool storage and other energy saving techniques, when appropriate. New lighting systems are much more efficient and provide for greater user control, while exploiting the energy benefits of day-lighting when possible. The adoption of advanced LED and plasma lighting technology is accelerated to reduce lighting energy consumption in lighting of both high and low intensity. Fixed speed pumps give way to variable-speed systems, with more appropriate sizing for the required task. Commercial information technology and refrigeration systems are also improved to exploit a greater proportion of the technical savings potential available from efficient equipment.

Space heating

In the Efficient World Scenario, energy savings in space heating come about by the greater use of insulation and high-efficiency glazing in both new build and retrofits, coupled with increased rates of retrofitting. Savings also arise from the increased use of more efficient heating equipment, such as condensing boilers, micro combined heat and power (CHP) plants and heat pumps, and from improved control via intelligent thermostats. New building designs take greater advantage of passive and integrated design techniques to make better use of ambient energy flows, such as solar gains, while minimising overheating. The overall energy savings are attributable to a blend of efficiency gains, due to the use of more efficient technologies for the same fuel, *e.g.* efficient gas boilers, and change from gas- or oil-based boilers to more efficient heat pumps. Space heating savings are larger in the OECD than the non-OECD region throughout the projection period, because the former has greater demand for space heating and correspondingly greater scope for reductions. This greater demand for space heating in the OECD mainly arises from the climate in the region having more heating degree-days and a demand for higher average thermal comfort levels.

In many cases, the higher-efficiency technologies bring substantial improvements in the quality of service and important benefits beyond simple energy, economic and emissions savings. Use of integrated design and improved insulation gives better thermal comfort than reliance on thermo-mechanical heating systems as it helps to even out the differences in radiative temperature to which the human body is highly sensitive. Efficient glazing, using selective radiative coatings to reflect heat back into a room, inert gas-filled cavities, or even evacuated cavities, together with low conductivity frames not only saves energy but also makes a particularly strong contribution to overall thermal comfort. Additionally, it helps reduce noise, compared with standard single glazing. Insulation also eliminates thermal bridges and mould growth and thus improves indoor air quality and health. In many markets, these factors are valued highly.

Equipment

There are favourable prospects for improving the efficiency of energy-using equipment, although there still remain many barriers and also lock-in effects. Some building equipment, such as heating systems and plumbing, may be long-lasting and relatively difficult to change. Others, such as consumer electronic devices, may be very short-lived. For such equipment, the most efficient step is to ensure the equipment meets the highest attainable standards when it is bought and installed (Box 11.1).

Lighting and appliances are the end-use sectors that offer the fastest energy savings. In the case of residential sector lighting, incandescent lamps, which are typically replaced annually, due to their short lifetimes, are replaced initially by compact fluorescent lamps (CFLs) that last six times as long and use a quarter of the energy. Over the medium term, solid state lighting, such as LEDs, gain a substantial part of the market, while average LED efficiency levels continue to rise.

In the case of electrical appliances, energy efficiency levels continue to be driven upwards, in line with established technology learning curves. In OECD countries, the most efficient refrigerators today consume 20% of the energy of the average refrigerator on the market in the mid-1990s.[6] It is assumed that the current most efficient technology gradually becomes

6. For the European Union, A+++ appliances use 20% of the D to E class average, which is the mean threshold for the labelling scheme established in 1995.

the global norm, while technology improvement continues to occur. Similar potential for improvement exists in air conditioners, where global average efficiency levels are roughly a third of the levels attained by the most efficient technologies on today's market and yet further technical improvements are possible (Econoler, *et al.*, 2011). Other major electricity using end-use equipments in households, such as televisions, clothes washers, dryers, dishwashers, rice cookers, and information and communication technologies (ICT), all have strong savings potential (Waide, 2011) above and beyond the levels within the New Policies Scenario. Additional savings are achieved by smart solutions for demand management, including automation and active controls.

Box 11.1 ▷ Determining the cost-effective efficiency potential of appliances in the Efficient World Scenario

The Efficient World Scenario combines the same growth in electrical appliances as that projected in the New Policies Scenario, with greater efficiency wherever this is attainable cost effectively. The efficiency levels reached have been determined using the BUENAS (Bottom-Up Energy Analysis System) model, an international appliance policy model developed by Lawrence Berkeley National Laboratory (LBNL). BUENAS covers thirteen economies, which together account for 77% of global energy consumption, and twelve different end-uses, including air conditioning, lighting, refrigerators and industrial motors (LBNL, 2012).

This type of model is particularly important in assessing the impact of minimum energy performance standards (MEPS) and energy labelling, which are the key policies driving equipment energy efficiency gains and their associated electricity and fossil-fuel savings in the buildings sector. In the Efficient World Scenario, we assume that these policies are fully enacted where already in place and are introduced where not in place yet. In addition, in cases where the equipment coverage of such policies is only partial, the coverage is assumed to be extended so that almost all energy used by end-use equipment in buildings is covered by these policies. Consequently, the share of equipment energy use that is subject to MEPS increases from typical current levels of 30-60% in OECD economies, 55% in China and from 0-10% in other countries, to 95% in all countries in 2035 under the Efficient World Scenario.

Despite the very significant savings these policies produce, the modelling of the cost-effective savings potential is conservative, as it only includes currently-available technologies and applies current equipment prices. In reality, the price of high-efficiency equipment is likely to go down over time, as manufacturers find ways to lower costs through economies of scale, process improvement and other innovations. The US Department of Energy has recently allowed for this effect in its equipment rulemakings.

With such improvements in building shell and end-use equipment, there is a significant reduction in the energy intensity of buildings. For the residential sector, energy demand grows at an average annual rate of 0.23% from 2010-2035, much slower than the growth in building floor area, which increases by an average annual rate of 1.7%. This decoupling of energy demand from floor space is achieved by a near 30% reduction in the energy required per square metre of residential floor space over the *Outlook* period.

Industry

Techno-economic potential and policy framework

In 2010, industry was responsible for 28% of global final energy use and 32% of energy-related carbon dioxide (CO_2) emissions (including indirect emissions).[7] Energy-intensive industries, such as iron and steel, cement, chemicals, and pulp and paper, currently account for roughly half of total final industrial energy consumption. In our projections, most of the increase in industrial production through to 2035 occurs in non-OECD countries (Figure 11.4).

Figure 11.4 ▷ Cumulative new industry capacity as a share of currently installed global capacity in the Efficient World Scenario

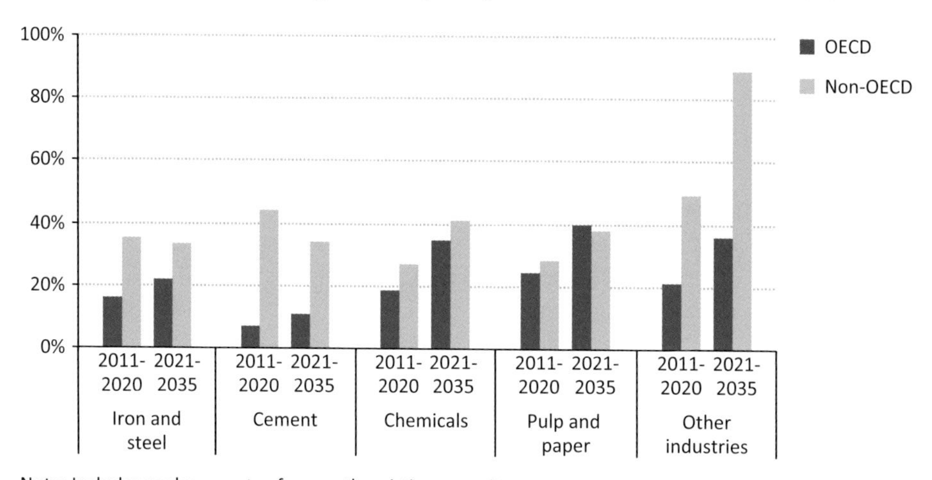

Note: Includes replacements of currently existing capacity.

Source: IEA analysis.

The potential for energy efficiency improvements in industry varies across sub-sectors. While in many OECD countries large energy-intensive industries already use efficient technologies, further improvements can be realised by replacing older facilities, optimising processes or through enhanced energy management practices. Untapped potential

7. Industry sector energy demand is calculated in accordance with IEA energy balances, *i.e.* neither demand from coke ovens, blast furnaces, nor petrochemical feedstocks is included.

also remains in the non-energy-intensive industry sector. In non-OECD countries, new manufacturing facilities in energy-intensive industries are often equipped with the latest efficient technologies. These new plants are often large scale and therefore more energy efficient, since production size has a strong influence on specific energy consumption (energy consumption per unit of output). However, older infrastructure in non-OECD regions is in most cases less efficient and accelerating the closure of plants with outdated technology can produce significant energy savings. Pure technological changes can achieve only a part of the energy savings; the rest requires systems optimisation and wider process changes (Box 11.2).

Box 11.2 ▷ Types of energy efficiency improvements in industry

Energy efficiency improvements in industry can be classified into three main categories:

- *Better equipment and technology*. It is estimated that the accelerated adoption of best available technology (BAT) could cut global industrial energy use by almost a third (IEA, 2012a). Replacing technologies such as inefficient compressors, which often lose up to 80% of input energy as heat, could contribute to radical energy cuts.
- *Managing energy and optimising operations*. Efficiency improvements through systems optimisation can, in some cases, achieve additional savings, up to 20% (UNIDO, 2011). Systems optimisation means going beyond component replacement towards integrated system design and operation. Optimisation of electric motor systems, such as fans, pumps, compressors and drives, has potential for particularly large and cost-effective savings in all industry sectors (IEA, 2011b).
- *Holistically transforming production systems*. More radical reductions in industrial energy use require an integrated approach to the management of resources and waste over the whole industrial process and consumption chain. Strategies for transforming production systems include increased use of recycled or waste materials and energy, sharing resources among industries and dematerialisation.

There are significant barriers to the implementation of energy efficiency measures in industry and these are often hard to overcome. They include the requirement for short payback periods, in some cases lack of awareness and know-how, and concern that time spent on efficiency improvement is a distraction from core business and that change could interrupt production or affect reliability. Government intervention can address these barriers, creating incentives for companies and ensuring that enabling and supporting systems are in place.

Since the 1970s, industrial energy efficiency policies have been implemented in many countries around the world. Key measures include the funding of research and technology development, incentives in the form of subsidies or energy taxes, emissions trading schemes, equipment performance requirements and energy management programmes. In addition, a variety of supporting measures, such as capacity building, provision of training, facilitating access to energy efficiency service providers and sources of finance are used to promote the uptake of energy efficient technologies and practices (IIP, 2012).

However, there are still gaps that existing policies and policies currently under discussion will not close. The Efficient World Scenario assumes a substantial extension and increase in the scale of the policy efforts that underlie the New Policies Scenario (see Chapter 10). In particular, it assumes accelerated deployment and further development of existing policy instruments, such as energy efficiency targets, benchmarking, energy audits and energy management requirements. These are complemented by supportive measures, like training, capacity building, the provision of information and guidance. The efficiency of industrial equipment and systems is promoted by the development of progressive energy performance requirements. Tools, guidance and information measures also help to promote the deployment of energy efficient systems and assist systems optimisation. In addition, new policy measures that go beyond the energy sector are developed, to promote the use of recycled materials, waste heat, and materials and processes that reduce manufacturing energy requirements. Due to the significant investments required and their relatively long payback periods, fiscal and financial incentives, as well as effective financing mechanisms, play an important role. Importantly, the verification processes for systems measurement and energy savings are improved so that energy efficiency benefits can be confidently assessed, contributing to easier access to finance.

In OECD countries, policy measures are taken to increase the rate of energy efficiency refurbishment and systems optimisation in existing facilities. In emerging and developing economies, greater emphasis is placed on establishing an efficient industrial base by ensuring that the most efficient technologies are used when designing and commissioning new facilities, and that there is an acceleration in the closure, or comprehensive retrofit of facilities with obsolete technology. Technology and knowledge transfer to developing countries is increased together with experience exchange on effective policy making.

Outlook

In the Efficient World Scenario, demand for final energy in the industry sector increases by 31% between 2010-2035, compared with a rise of 44% in the New Policies Scenario. Global energy consumption continues to grow in all sub-sectors, as the annual intensity improvements achieved (ranging from 0.5-1.6%) are unable to counteract the rapid growth in industrial production (Figure 11.5).

Figure 11.5 ▷ **Average annual change in industrial activity, efficiency and energy demand by industrial sub-sector and scenario, 2010-2035**

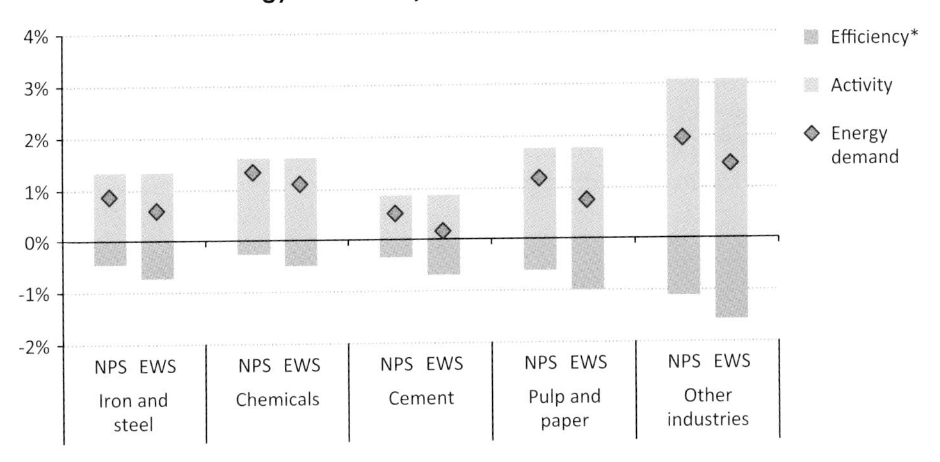

* Negative values for efficiency represent improvements.

Note: NPS= New Policies Scenario; EWS = Efficient World Scenario.

Most of the cumulative final energy savings in industry, with respect to the New Policies Scenario, come from reduced use of electricity (40%), followed by lower use of coal (23%) and natural gas (18%). Demand for oil remains broadly flat, while demand for gas, electricity and biomass increases significantly. China accounts for 39% of the cumulative energy savings and India for 14%. Only 16% of savings arise in OECD countries. The extensive deployment of energy-efficient technologies contributes to climate mitigation objectives by slowing growth in energy-related CO_2 emissions from the industry sector.

Table 11.3 ▷ **Global industry energy demand by fuel and energy-related CO_2 emissions in the Efficient World Scenario**

	2010	2020	2035	CAAGR* 2010-35	Change versus New Policies	
					2020	2035
Total (Mtoe)	2 421	2 901	3 171	1.1%	-4%	-9%
Coal	676	769	748	0.4%	-4%	-9%
Oil	321	343	330	0.1%	-4%	-7%
Gas	463	577	688	1.6%	-4%	-8%
Electricity	638	838	999	1.8%	-6%	-12%
Heat	126	133	121	-0.2%	-4%	-8%
Bioenergy**	197	242	285	1.5%	-4%	-8%
CO_2 emissions (Gt)	9.8	10.9	10.5	0.3%	-7%	-15%

* Compound average annual growth rate. ** Includes other renewables.

Note: CO_2 emissions include indirect emissions from electricity and heat.

Trends by sub-sector

Iron and steel

Currently, some 70% of world steel is produced via the blast furnace/basic oxygen furnace route (World Steel Association, 2012). With blast furnaces accounting for by far the largest part of energy consumption, a particular focus in the past has been on reducing their energy consumption. In the Efficient World Scenario, widespread adoption of top pressure recovery turbines and blast furnace gas recovery takes place. Pulverised coal injection is increased, to reduce coke demand, and combined-cycle gas turbines are used in place of steam turbines, to increase the thermal efficiency of power generation from blast furnace gas.

When electric arc furnaces (EAF) are used for steel making, direct current arc furnaces can significantly reduce energy intensity; but this technology is applicable only to furnaces above a certain production size. In the Efficient World Scenario, we assume a higher proportion of scrap metal being recycled in some economies, resulting in major energy savings. We also assume a higher share of EAFs, which results in higher overall electricity consumption, but lower fuel consumption. Both process changes – a higher use of scrap metal and a higher share of EAFs – account for more than a third of all the energy savings in the iron and steel sector. Gas-based direct reduced iron (DRI) is another option for less energy-intensive iron and steel making, as emphasised by the DRI facilities planned in Iran and under construction in Louisiana in the United States. However, the future development of DRI is uncertain, partly due to the uncertainty about the future development of gas prices. In the Efficient World Scenario, the combination of the above changes decreases the fuel intensity of iron and steel production between 2010-2035 by 11% in OECD countries and 19% in non-OECD countries. Energy savings in iron and steel in 2035, compared with the New Policies Scenario, are 35 Mtoe, or 6%.

Chemicals

The chemical sector is very diverse and so are the technological options to save energy. Significant energy savings are possible from the recovery and use of waste heat, co-generation, efficiency gains in steam crackers, increasingly selective catalysts, and through increasing the size of crackers and furnaces. Additional savings can be realised from process intensification and the co-ordination of energy use with neighbouring plants. Moreover, the integration of petrochemical and refinery plants can result in not only energy savings, but also lower transport costs, lower storage requirements and increased feedstock flexibility. In the Efficient World Scenario, the wider deployment of these technologies and organisational measures reduces the sub-sector's energy use in 2035 by 5%, or 28 Mtoe, compared with the New Policies Scenario.

Cement

The energy intensity of cement production is largely dependent on the type of kiln technology employed for clinker production. Dry kilns with pre-heaters and a precalciner are significantly more efficient at clinker production than shaft kilns, which are still common in China and India, or wet/semi-dry/dry kilns which are commonly used in the European Union, Russia and the United States. Important savings can be achieved by implementing heat recovery. In the Efficient World Scenario, it is assumed that there is a complete transition by 2035 to dry kilns with preheaters and precalciners in North America and the European Union, while shaft kilns are completely phased out in India and China.

Energy savings are realised in raw materials preparation and grinding by the introduction of high-efficiency classifiers and by the use of vertical roller mills (CSI and ECRA, 2009). Compared to today, additional efforts are made to replace clinker with alternatives, such as fly ash, blast furnace slag, limestone and pozzolana, which yield substantial energy savings. The reduction of the clinker-to-cement ratio accounts for roughly a fifth of overall energy savings in the cement sector. Globally, the measures adopted reduce energy demand in cement manufacturing in 2035 by 8%, or 24 Mtoe, compared with the New Policies Scenario.

Pulp and paper

In pulp and paper production, the chemical pulping process is the most energy-intensive step. Black liquor gasification has the potential to save a significant amount of energy in this step, although its use is currently limited. In the mechanical pulp production process, the use of high-efficiency grinding, efficient refiners and pre-treatment of wood chips can reduce energy consumption substantially, compared with conventional processes. However, by far the greatest potential for savings is from higher use of recycled fibre. Much of this potential has already been realised in some economies, such as in the European Union, but the use of recycled paper in pulp production can be further increased, especially in many non-OECD countries. At the global level, 50% of waste paper is currently recycled (IEA, 2010). The use of recycled paper as an input to paper production is driven not only by energy considerations, but also by factors such as availability and product quality specifications. Technologies to reduce energy consumption in paper production include shoe press, heat recovery and new efficient drying techniques. Systems optimisation, in the form of improved process control, monitoring, and management can help to reduce energy consumption beyond the improvement achievable by single equipment components. The deployment of all of these options is increased in the Efficient World Scenario, reducing energy demand in pulp and paper in 2035 by 10%, or 19 Mtoe, compared with the New Policies Scenario.

Other industries

The category "other industries" includes the remaining industry sub-sectors, which generally are not energy intensive. This category includes a wide range of very different sub-sectors. The largest energy consumers are food and tobacco, machinery, non-ferrous metals, mining and quarrying, and textiles. In total, this category accounted for 49% of total industrial energy use in 2010, but in the Efficient World Scenario it achieves 65% of the total cumulative energy savings in industry from 2011-2035. This is because energy-intensive sectors have, in the past, made significant energy savings, so that the largest potential for additional energy savings now lies in non-energy-intensive sub-sectors, where the share of energy costs in total production costs rarely exceeds 5% (UNIDO, 2011). Roughly half of all savings in other industries is in the form of electricity, and it is estimated that 70% of all electricity used in industry is related to electric motor systems that are used for ventilation, pumps, compressed air and mechanical movement (IEA, 2011b). The introduction of variable-speed drives and the proper sizing of motors achieve significant savings, since electric motors operate more efficiently at full power. Further areas for energy improvements include boilers, furnaces and specific process technologies. The overall effect is an energy reduction in 2035, when compared with the New Policies Scenario, of 220 Mtoe or 11%.

Transport

Techno-economic potential and policy framework

There are substantial opportunities to improve energy efficiency across all transport sectors (road, aviation, maritime, rail and other), mainly through increased deployment of energy-efficient technologies, but also by improving the efficiency of transportation systems overall. In road transport, vehicles powered primarily by internal combustion engines (conventional and hybridised) are set to continue to dominate the passenger light-duty vehicle (PLDV) market through to 2035 (IEA, 2011a). We estimate that the fuel economy improvement potential of PLDVs over that period ranges from 40% to 67% (including hybridisation), compared with an average vehicle today, depending on the technology type and the region, and that this can be achieved with current technologies (IEA, 2012b). Improvements in vehicle fuel economy typically entail engine downsizing, weight reduction and changes in the thermodynamic cycle, but there is additional potential for savings by using dual clutch transmissions and improving auxiliary systems, aerodynamics, the rolling resistance of tyres, etc. A reduction of average vehicle size is another plausible option for reducing average fuel consumption by vehicle, but represents a change of consumer preferences and is therefore not taken into account in the Efficient World Scenario.

There is less scope for freight trucks to improve fuel efficiency, as they mostly use diesel engines, which are already better optimised for fuel consumption. The potential to reduce fuel consumption in trucks by 2035 is in the range of 30-50%, compared with today's vehicles (IEA, 2012b). Additional reductions of 5-15% are possible by educating drivers, as in the case of PLDVs (IEA, 2012c). However, there is less certainty surrounding the efficiency gains that are possible and not all will always be available. The full potential of hybridisation, for example, can be realised only when driving in stop-and-go situations, such as in urban areas or in regions with lower level speed variations, such as Europe; long haul trucks drive in a continuous manner, with less potential for regenerative braking. Furthermore, other policy objectives, such as reducing air pollution, are not always consistent with improving vehicle fuel economy.

The International Air Transport Association (IATA) estimates that fuel efficiency in new aeroplanes could be improved by up to 50% by 2035, due to new engine systems and hybrid wing bodies (IATA, 2009). Existing aeroplanes could be made 7-13% more fuel efficient, by retrofitting engines and deploying more efficient gas turbines or composite secondary structures, while additional fuel economy potential lies in improvements in air traffic management (12%) and operational improvements (6%). For maritime transport, improvements can come from improving vessel design, engines, propulsion systems or operational strategies, such as reducing ship speed (Crist, 2012). For rail, the technology opportunities encompass the scrappage of old inefficient trains, hybridisation of diesel rail, switching lines to electric rail and optimising operation.

While most of the technologies needed to achieve significant fuel economy improvements are already available, policy intervention is necessary to increase their deployment. Governments are focusing on improving the fuel economy of road vehicles, in passenger light-duty vehicles in particular (see Chapter 9, Table 9.1). Fuel-economy standards for PLDVs are already widely deployed in many OECD countries, while only China has adopted such standards among non-OECD countries, although India is discussing their adoption. Additional policy measures targeting the efficient fuel use of PLDVs, including fuel economy labelling, are also widely adopted. For road freight trucks, only Japan and the United States have adopted fuel-economy standards so far; the European Union, Canada and China are in the process of setting standards.[8] Information on fuel economy is often limited for trucks. There is currently no policy framework that explicitly aims at improving fuel efficiency in non-road transport sectors, even though energy efficiency guidelines have been adopted by different governmental bodies, such as the Energy Efficiency Design Index by the International Maritime Organization in early 2012.

Under the Efficient World Scenario, the policy framework is significantly strengthened. Stringent standards become mandatory for road vehicles in all countries and are progressively raised, such that the efficiency level of new vehicles reaches its maximum potential by 2035, provided the required payback period stays within certain limits (see Chapter 10, Figure 10.1). Fuel-economy standards need to be set at levels that are ambitious enough to accelerate the rate of improvement in the fuel economy of vehicles and this requires the standards setting process to take long-term technology learning curves into account. This is especially important, given the long lead times in the regulatory process and the time required by manufacturers to adapt their production. Stringent fuel-economy standards are also set in the Efficient World Scenario for freight vehicles, aviation and the maritime sector in all economies. Stringent and consistent fuel economy labelling schemes, and vehicle and fuel price signals, including the phase-out of fossil-fuel subsidies, are adopted to encourage fuel-efficient vehicle purchase and operation. They play a key role in driving the market towards fuel-efficient vehicles and in overcoming initial deployment hurdles. The further harmonisation of vehicle testing systems across regions

8. Road freight comprises light commercial vehicles (gross vehicle weight less than 3.5 tonnes), trucks (gross vehicle weight more than 3.5 tonnes) and heavy freight trucks (gross vehicle weight more than 16 tonnes).

helps reduce test facility investment in individual countries and facilitates the transfer of fuel-economy standards and labelling among countries. Improved information exchange among countries also supports improved efforts to raise compliance with standards and labelling requirements across jurisdictions. Other measures adopted include tyre rolling resistance labelling and incentives to assist feedback to drivers on fuel efficient techniques.

Outlook

The Efficient World Scenario has important long-term implications for energy demand in the transport sector. Total final energy demand from transport grows at an average 1.3% per year in 2010-2020 and then plateaus, reaching 2 780 Mtoe in 2035, or 15% lower than in the New Policies Scenario (Table 11.4). Only 5% of the cumulative energy savings in the transport sector over the *Outlook* period occur prior to 2020, as it takes time for more efficient technologies to have an impact on the entire vehicle fleet and because the New Policies Scenario already adopts numerous policies that lead to improvements in fuel economy in the period to 2020. Over the entire time horizon, however, growth in energy demand averages 0.6% per year, less than half the rate in the New Policies Scenario. In 2035, the sector's CO_2 emissions are 15% lower than in the New Policies Scenario.

Table 11.4 ▷ Global transport energy demand by fuel and energy-related CO_2 emissions in the Efficient World Scenario

	2010	2020	2035	CAAGR* 2010-35	Change versus New Policies	
					2020	2035
Total (Mtoe)	2 377	2 704	2 780	0.6%	-3%	-15%
Oil	2 201	2 449	2 414	0.4%	-3%	-15%
Gas	90	111	134	1.6%	-3%	-16%
Biofuels	59	108	176	4.5%	-2%	-15%
Electricity	24	35	56	3.5%	-1%	-2%
Other	3.4	0.2	0.2	-10.5%	0%	0%
CO_2 emissions (Gt)	6.8	7.6	7.5	0.4%	-3%	-15%

* Compound average annual growth rate.

By 2035, transport oil demand is 9.1 million barrels per day (mb/d) lower than in the New Policies Scenario, representing 72% of the total oil savings achieved across all sectors in the Efficient World Scenario (Figure 11.6). The use of other fuels is also reduced by 1.2 million barrels of oil equivalent per day (mboe/d), as the Efficient World Scenario retains the same policy assumptions for alternative fuels and electric vehicles as the New Policies Scenario. More than half of these savings are biofuels and most of the remainder are natural gas.

An important consideration in realising the full potential of fuel economy improvements is how a fall in fuel prices will affect driver behaviour. The potential "rebound effect" can be around 20% for passenger cars in the OECD, *i.e.* a fall in price of 10% increases the kilometres driven by 2%, and in non-OECD countries the change is potentially larger. In the

Efficient World Scenario, international oil prices in 2035 are $16 per barrel lower than in the New Policies Scenario, which increases passenger-kilometres driven in most countries and dampens the energy savings realised. Were end-user prices prevented from falling, *e.g.* through increased taxation, the savings would be higher (see Chapter 10, Box 10.2). There are, however, notable exceptions to this general rule in countries in the Middle East and Africa, where the assumption of partial removal of fossil-fuel subsidies offsets the effect of lower international oil prices and reduces the passenger-kilometres driven.

Figure 11.6 ▷ **Fuel savings in the transport sector in the Efficient World Scenario**

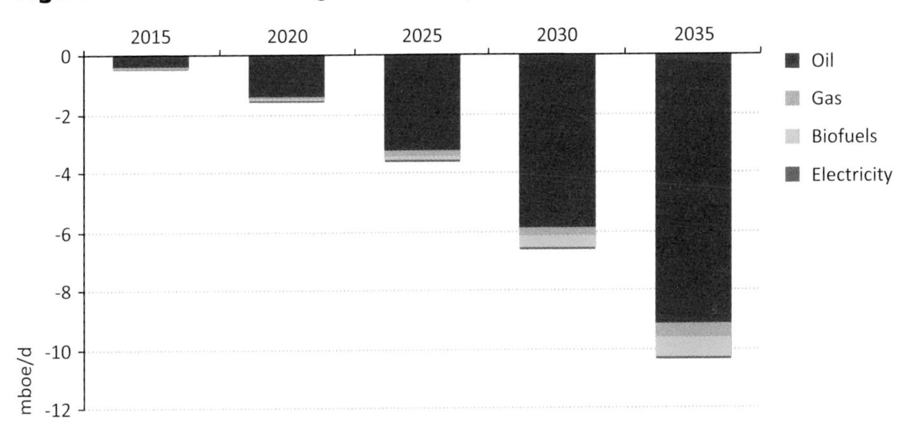

Today, almost 60% of energy demand in the transport sector is in OECD countries, but this share is set to shrink, as demand for mobility is growing rapidly in non-OECD countries, where the level of motorisation is still very low (IEA, 2011a). At the same time, many OECD countries and China are making persistent efforts further to reduce transport fuel demand, particularly oil use in PLDVs.

11

In the Efficient World Scenario, non-OECD countries are responsible for more than two-thirds of cumulative energy savings in transport (Figure 11.7). Almost one-fifth of global savings are made in China, where the full implementation of the fuel-economy standards for PLDVs that are already planned and their extension to achieve a tested average of 3.4 litres per 100 kilometres (l/100 km) in 2035 (compared with 4.8 l/100 km in the New Policies Scenario) helps reduce total transport oil demand by 1.7 mb/d in 2035, compared with the New Policies Scenario. Almost 15% of cumulative global savings are achieved in the Middle East, where the adoption of fuel-economy standards and the reduction of fossil-fuel subsidies lowers oil demand by 1.2 mb/d in 2035, compared with the New Policies Scenario. India's oil demand is cut by 0.6 mb/d in 2035.

Despite the fact that both the United States and the European Union already have fuel-economy standards for 2025 and 2020, respectively (that are assumed to be implemented in the New Policies Scenario), they are responsible in the Efficient World Scenario for almost 20% of cumulative fuel demand savings, relative to the New Policies Scenario, because of the large size of their markets. By 2035, oil demand in the transport sector is 0.7 mb/d lower in both the United States and the European Union. In the United States, the savings are driven by further tightening of PLDV standards, to over 60 miles-per-gallon (mpg), around 3.8 l/100 km by 2035, increased sales of hybrid vehicles and more stringent standards for heavy trucks, reducing average on-road fuel consumption by about 45%, compared with today.

Box 11.3 ▷ Modal shift and behavioural change in transport energy efficiency

About 60% of the world's population is projected to live in urban areas by 2035, up from 50% today, creating an opportunity for holistic transport concepts targeting energy efficiency in urban planning. One option is to encourage passengers to shift to less energy-intensive transportation modes, such as rail, buses (*e.g.* bus rapid transit systems), trams, cycling and water-based transit. The use of bus rapid transit systems shows great potential as a cost-effective way to reduce the use of passenger cars (IEA, 2012a). Intelligent transport systems and improved logistics can greatly improve the energy efficiency of road and rail freight systems.

Other options that can help increase overall system efficiency are improved planning and operation of urban transport networks and systems, and sustainable mobility measures. These options include the optimisation of traffic signals and network flows, dedicated lanes for low energy-intensity, high carrier-capacity traffic, restrictions on parking, especially when this impedes traffic flows, congestion charging, car-pooling, and park-and-ride schemes.

However, these measures are assumed to be deployed to no greater extent in the Efficient World Scenario than in the New Policies Scenario. The savings realised in the Efficient World Scenario are, accordingly, entirely due to the adoption of more efficient vehicle technology. Nonetheless the energy savings potential from mode shifting and other transport system measures is considerable, particularly for individual mobility. Emerging economies with quickly-growing demand for mobility and cities that are undergoing rapid development have scope to integrate traffic optimisation and modal shift into urban planning policy. There is less scope for mode shifting in freight as local delivery, in particular, will continue to be dominated by road transport.

Quantifying the global potential to improve energy efficiency through modal shift and behavioural change is very difficult, as it depends heavily on local characteristics. But, for the purpose of illustration, we estimate that a global reduction in individual PLDV travel by 10% in 2035, in favour of electric rail, would cut projected oil demand in our New Policies Scenario by 2.3 mb/d.

Figure 11.7 ▷ **Energy savings in the transport sector by region* in the Efficient World Scenario relative to the New Policies Scenario**

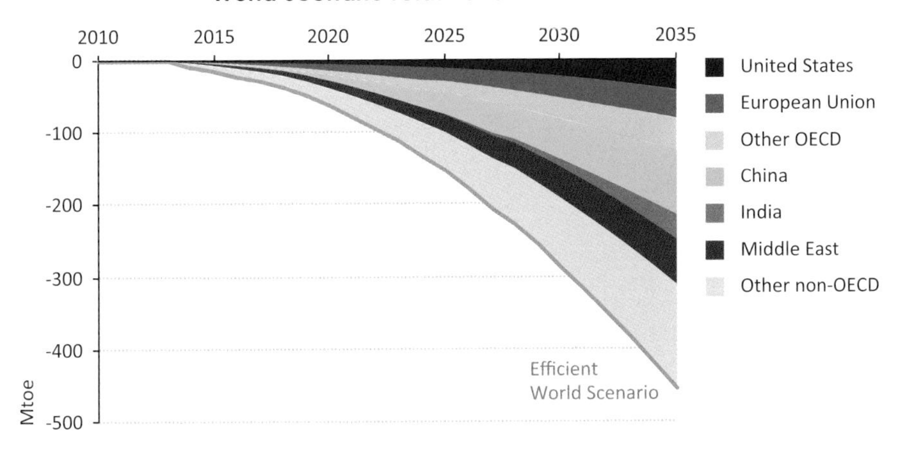

* Excludes international bunkers.

Trends by sub-sector

Road transport accounts for more than 85% of the 10.2 mboe/d of energy that is saved in the transport sector by 2035 under the Efficient World Scenario (Figure 11.8). The remaining savings are concentrated in aviation and maritime, even though much of the potential in these areas is already realised in the New Policies Scenario, driven by increasing international oil prices. Nonetheless, by 2035 a new aeroplane in the Efficient World Scenario is about five percentage points more efficient than in the New Policies Scenario.

Figure 11.8 ▷ **Energy savings in the transport sector in the Efficient World Scenario relative to the New Policies Scenario by sub-sector**

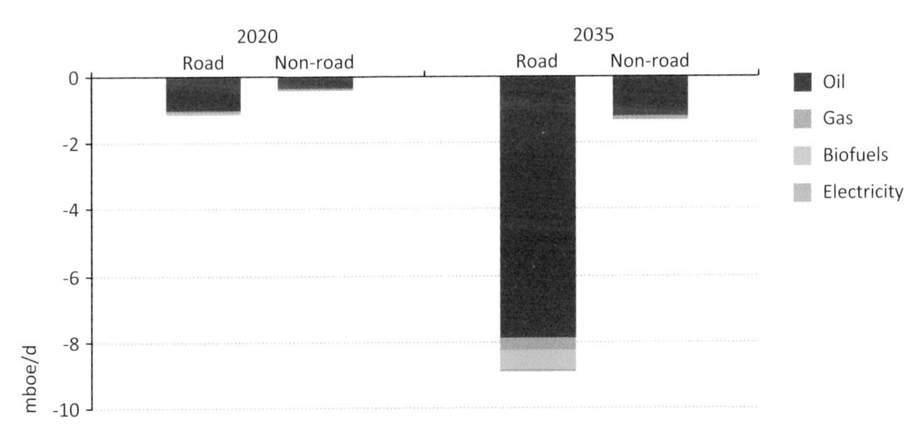

The dominant contribution road transport makes to the sector's total energy savings is due to its sheer size (relative to the other sub-sectors) and the significant potential to improve its efficiency that goes unrealised in the New Policies Scenario. Around three-quarters of all fuels used in the transport sector today are consumed in road transport, and due to currently low motorisation rates in non-OECD countries, there is significant scope for additional growth. By 2035, the stock of PLDVs is projected to be more than twice as large as today, at 1.7 billion vehicles. In the New Policies Scenario, oil demand from PLDVs grows by 1.2% per year on average and in 2035 it constitutes almost half of global transport oil demand. The largest growth in transport oil demand, however, comes from road freight traffic. It accounts for about one-third of global road fuel demand today, and this share is set to rise in the New Policies Scenario with increasing movement of goods (see Chapter 3). Only Japan and the United States currently have fuel-economy standards for heavy freight vehicles (both of which are considered in the New Policies Scenario), and since most truck operators require payback periods below eighteen months before committing to fleet upgrades (which is shorter than many current technologies deliver), the realisation of the large technical potential is far from certain without government intervention.

In the Efficient World Scenario, around 95% of cumulative fuel savings in road transport arise in PLDVs and road freight, split fairly evenly between the two (Figure 11.9). The average global tested fuel economy of new PLDVs reaches 3.5 l/100 km in 2035, compared with 4.9 l/100 km in the New Policies Scenario and around 7.6 l/100 km in 2010. The disproportionately large savings achieved in road freight reflect the current dearth of plans to improve efficiency in this sub-sector.

Figure 11.9 ▷ Energy savings in road transport in the Efficient World Scenario relative to the New Policies Scenario

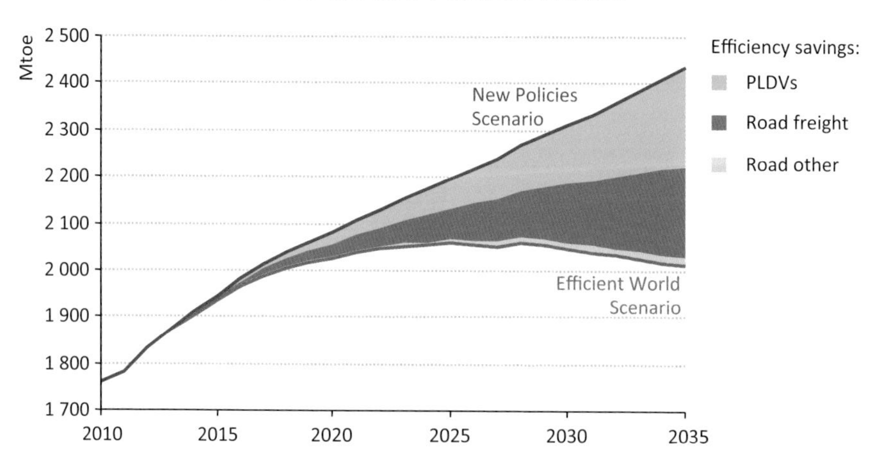

We assume that reducing fuel consumption is given at least as much attention as other policy objectives, such as improving safety and air quality, and that policies are put in place to overcome the initial deployment hurdles associated with the purchase of more efficient trucks. The average on-road fuel economy of heavy trucks (gross vehicle weight in excess of 18 tonnes) reaches 19 l/100 km in 2035, compared with 28 l/100 km in the New Policies Scenario and 36 l/100 km in 2010.

Power generation and electricity demand

Techno-economic potential and policy framework

Just as there are opportunities to save energy through the deployment of more energy efficient demand-side technologies, savings may also be made through improvements in supply-side efficiency in the power sector. These include the faster introduction of efficient electricity generation plants and the phasing out of less efficient ones; the adoption of higher-efficiency transformers to reduce transmission and distribution (T&D) losses; the adoption of smart-grid technologies; and the greater use of modern, high efficiency, combined heat and power generators. Fuel substitution can also contribute to efficiency improvements, because gas-fired generation from combined-cycle gas turbines (CCGTs) is considerably more efficient than coal-fired power generation. Globally, energy demand in the power sector was 4 839 Mtoe in 2010, two-thirds more than the buildings sector and double the transportation sector. The power sector is also the largest sector in terms of carbon emissions, representing 41% of global energy-related CO_2 emissions in 2010.

Currently there are many inefficient power plants in operation around the world and many more are planned or under construction. In many OECD countries, ageing plants, with low efficiencies, continue to run because the incremental cost of generation can be very low. In several non-OECD countries, another important factor behind the high share of inefficient plants is that fuel prices are often low due to fossil-fuel subsidies or abundant low-cost domestic fossil-fuel supplies. In many countries, transmission and distribution losses of electricity are extremely high, due to the poor quality of the network infrastructure or the manner in which the grid is operated.

A number of measures are currently being used or considered specifically to improve efficiency in the power sector. For generation, they can take the form of minimum efficiency standards for new plants or specific incentives and targets for CHP. For electricity networks, regulations aimed at reducing losses from power grids create direct incentives for the adoption of more energy-efficient technologies and practices.

In addition to explicitly efficiency-orientated measures, a number of policies being applied to the power sector for other purposes have the secondary effect of improving efficiency. Measures that raise the cost of fossil fuels, such as removing subsidies or establishing a CO_2 price, increase the value of energy efficiency in generation. Policies designed to reduce pollution from power plants can also have the secondary effect of closing older, less efficient power plants.

In the Efficient World Scenario, supply-side energy efficiency options are more widely deployed than in the New Policies Scenario. This is achieved through a co-ordinated set of government policies to encourage more efficient use of fossil fuels in the generation and dispatch of electricity and in its delivery to end-users.

These policies include:

- Setting minimum energy efficiency standards for new thermal power plants that prohibit the construction of new subcritical coal plants and have the effect of replacing them with supercritical, ultra-supercritical or integrated gasification combined-cycle (IGCC) plants.
- Setting minimum energy efficiency standards for new gas and oil plants that lead to increased capacity additions of combined-cycle gas turbines (CCGTs).
- Supportive regulatory and incentive measures for the deployment of combined heat and power (CHP) plants, in place of heat-only plants, in conjunction with the imposition of efficiency standards to improve average CHP plant efficiency.
- The introduction of efficiency and emission standards on existing coal, natural gas and oil plants, which have the effect of reducing refurbishments of old plants and of shortening plant lifetimes, thereby increasing the proportion of new, higher-efficiency, capacity additions.
- Support mechanisms for the deployment of smart grid technologies, to enhance the ability to optimise power flows and reduce transmission and distribution losses.
- Setting and strengthening minimum efficiency requirements for transformers, to reduce transmission and distribution losses in the power network.

Outlook

In the Efficient World Scenario, electricity demand growth from 2010-2035 is reduced by one-third, compared with the New Policies Scenario. As a result, electricity demand is 14% lower than in the New Policies Scenario by 2035. This translates into total savings in gross electricity generation of some 5 460 terawatt-hours (TWh) in 2035, which is larger than the combined power output of China and India in 2010. By 2035, fossil fuel-fired generation in the Efficient World Scenario falls by 22%, compared with the New Policies Scenario, peaking around 2015 and declining thereafter. Nuclear and renewables-based electricity generation are affected less than fossil fuels, as their expansion is policy-driven in many countries. Their output is 4% and 6% lower than the New Policies Scenario respectively by 2035, principally due to lower electricity demand, particularly in the period after 2020. While the majority of savings in the power sector in the Efficient World Scenario, relative to the New Policies Scenario, stem from demand-side measures, improvements in supply-side efficiency reduce primary energy consumption in 2035 by almost an additional 195 Mtoe, or 15% of the overall power sector savings (Figure 11.10). These savings can be split into two categories: improved power plant efficiencies and reduced losses in transmission and distribution. The overall efficiency of fossil-fuel generation gradually rises to 49% in 2035 – an improvement of six percentage points compared with 2010. This is a 70% larger increase in efficiency than that seen in the New Policies Scenario, where overall efficiency increases by three and a half percentage points. Transmission and distribution losses decline, reaching 7.6% in 2035 compared with 8.3% in the New Policies Scenario (Box 11.4).

Figure 11.10 ▷ Power generation energy savings by measure in the Efficient World Scenario

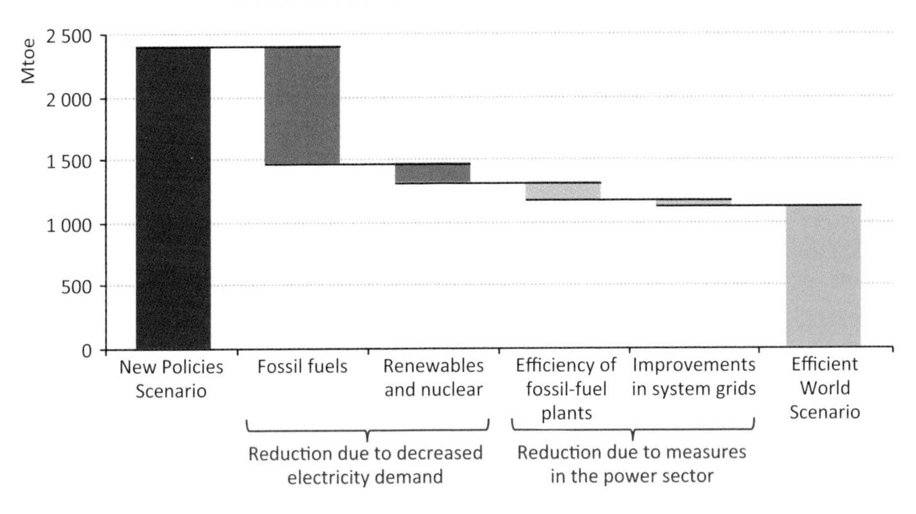

Box 11.4 ▷ The value of smart grids

Electricity grids play a vital role in providing reliable electricity supplies to consumers. The fundamental building blocks of network infrastructure have changed little in over a century, but developments in electronics, information and communication technology over recent decades are changing the way in which this infrastructure is operated. These "smart grid" technologies have several advantages. For example, using information and communication technologies to remotely detect where and why a fault has occurred can reduce the amount of time required to restore supply to consumers (IEA, 2011c).

Smart grid technologies also have the potential to yield improvements in efficiency. Technologies that monitor networks and reduce power flows at peak times also have the effect of reducing losses. Smart meters allow consumers to monitor their energy use in real time and adjust their consumption accordingly, especially in response to time dependent tariffs. This could allow consumers to reduce their electricity bills and at the same time reduce the need for expenditure on fuel and generation capacity in the power sector, resulting in benefits for the economy as a whole. Smart grids may also indirectly enable greater integration of distributed generation, including renewables and combined heat and power. The degree to which this potential is realised will depend on how quickly and broadly smart grid technologies are adopted worldwide, and – in the case of smart meters – on the degree to which consumers will change their behaviour in response to the information provided by these new technologies.

In the Efficient World Scenario, we assume an improvement in grid efficiency, relative to the New Policies Scenario. This reflects investment in more efficient grid infrastructure and grid operation, including smart grid technologies.

At a regional level, the reduced growth in electricity demand in the Efficient World Scenario, relative to the New Policies Scenario, is especially marked in non-OECD countries. In China, the energy consumed by the power sector increases by almost 100% over the *Outlook* period in the New Policies Scenario, but in the Efficient World Scenario it grows by 60%. The equivalent increase in India is 43% lower in the Efficient World Scenario than in the New Policies Scenario. In the OECD, the primary energy used by the power sector increases by 10% in the New Policies Scenario, but falls by 7% in the Efficient World Scenario. The overall reductions in the use of fossil fuels in power generation in OECD and non-OECD countries are similar in relative terms. The share of the fossil-fuel reduction accounted for by coal is greater in the non-OECD, however, while gas takes more of the reduction in the OECD countries. Nuclear power experiences a larger relative reduction in the OECD countries.

Box 11.5 ▷ The potential of combined heat and power

The global average efficiency of power plants that generate only electricity is 41%. Almost three-fifths of the primary energy used in these plants becomes waste heat, of no economic value. Combined heat and power (CHP) generation allows some of this energy to be used, either for industrial processes or for space and water heating in residential and commercial buildings. This leads to a significant improvement in overall efficiency, CHP units having a global average efficiency of 62% when the useful electricity and heat produced are both taken into account. New CHP plants can achieve efficiencies of over 85%. A number of countries have adopted policies to support the use of CHP. The US federal government has adopted a goal of deploying 40 gigawatts (GW) of new industrial CHP by the end of 2020 (US White House, 2012), and many states offer incentives for CHP projects (US EPA, 2012). In Europe, the EU's Cogeneration Directive of 2004 obliged member states to take steps designed to promote and monitor the development of CHP in their countries and to remove barriers to its deployment: it is intended that these provisions will be superseded by new requirements in the forthcoming Energy Efficiency Directive (see Chapter 9). At a national level, European countries have used a number of policy tools, including green certificates and feed-in tariffs, to support CHP.

A key constraint on the deployment of CHP is the difficulty of distributing heat over long distances. Because of this, CHP units must be located close to demand, potentially increasing costs. Lack of data makes analysis of CHP on a global level difficult. In the IEA's statistics, the heat produced by CHP installations is measured only if the heat is sold by the producer to another entity. Heat produced in an industrial CHP facility and used by the same firm is not reported and only the corresponding fuel consumption is accounted for. This makes it difficult to analyse the current extent of CHP use globally, and to model its future development.

Table 11.5 ▷ **Installed capacity, fuel consumption and electricity generation in the Efficient World Scenario**

	Capacity (GW)				Consumption (Mtoe)				Generation (TWh)			
	2010	2020	2035	Δ NPS*	2010	2020	2035	Δ NPS*	2010	2020	2035	Δ NPS*
OECD	2 718	3 059	3 269	-11%	2 267	2 202	2 102	-16%	10 848	11 246	11 567	-13%
Coal	669	599	374	-25%	876	720	394	-35%	3 747	3 217	1 998	-28%
Gas	819	955	1 042	-13%	471	472	506	-17%	2 544	2 672	2 975	-15%
Oil	218	123	58	-14%	71	32	13	-35%	309	139	58	-36%
Nuclear	327	321	317	-3%	596	603	621	-3%	2 288	2 315	2 382	-3%
Renewables	685	1 062	1 478	-6%	253	375	568	-8%	1 960	2 904	4 155	-6%
Non-OECD	2 465	3 756	5 029	-12%	2 572	3 183	3 862	-18%	10 560	14 793	19 612	-16%
Coal	980	1 346	1 444	-21%	1 373	1 572	1 556	-28%	4 940	6 197	6 669	-27%
Gas	532	788	1 103	-10%	632	713	887	-15%	2 216	2 979	4 263	-14%
Oil	217	208	152	-15%	204	158	95	-24%	691	554	345	-26%
Nuclear	68	148	241	-5%	122	284	473	-5%	468	1 087	1 814	-5%
Renewables	668	1 267	2 090	-5%	241	457	852	-6%	2 245	3 975	6 522	-6%
World	5 183	6 815	8 299	-11%	4 839	5 385	5 964	-17%	21 408	26 039	31 180	-15%
Coal	1 649	1 945	1 817	-22%	2 249	2 292	1 950	-29%	8 687	9 413	8 667	-27%
Gas	1 351	1 742	2 145	-11%	1 102	1 184	1 393	-16%	4 760	5 651	7 237	-15%
Oil	435	331	209	-14%	275	190	108	-26%	1 000	693	403	-28%
Nuclear	394	468	558	-4%	719	887	1 094	-4%	2 756	3 402	4 196	-4%
Renewables	1 354	2 329	3 568	-5%	494	832	1 420	-7%	4 206	6 879	10 677	-6%

* Change in Efficient World Scenario relative to New Policies Scenario.

In the Efficient World Scenario, global installed power capacity in 2035 is around 8 300 GW, some 1 050 GW lower than in the New Policies Scenario (Table 11.5). More than three-quarters of the reduction in capacity is in fossil-fuel plants, with the biggest contribution coming from coal (49%), followed by gas (26%). A reduction in wind power capacity accounts for more than one-third of the remaining capacity difference. Retirements of power plants over the *Outlook* period amount to above 2 350 GW, an increase of 19% compared with the New Policies Scenario. Gross capacity additions are over 5 200 GW in the Efficient World Scenario, a decrease of 11% from the New Policies Scenario.

The increase in the efficiency of coal-fired generation is more pronounced in the Efficient World Scenario than in the New Policies Scenario, as older plants are mothballed or decommissioned earlier and there is a greater uptake of newer, more efficient plants. There is a marked shift away from subcritical coal plants in all regions, particularly in the OECD, where their share of coal generation drops from 63% in 2011, to 51% in 2020 and to 16% in 2035 (Figure 11.11). Subcritical plants are displaced by more efficient coal technologies, such as high pressure and temperature ultra-supercritical plants or IGCC plants. Furthermore, a combination of new build and retirements means that after a period of growth in the early years of the *Outlook*, coal generation peaks before 2020 and declines almost to the 2010 level by 2035 in the Efficient World Scenario.

Figure 11.11 ▷ Coal-fired capacity by technology in the Efficient World Scenario

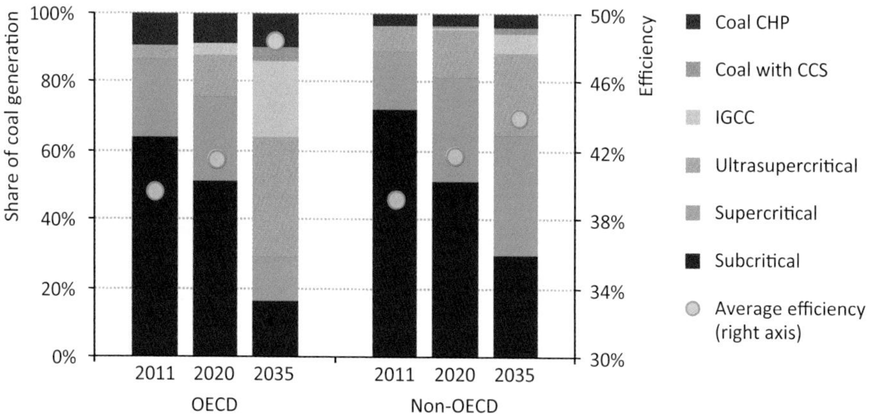

Note: Data are for electricity-only plants.

Compared with the New Policies Scenario, the average improvement in the efficiency of fossil fuel-based generation by 2035 is 3.6 percentage points higher in the OECD in the Efficient World Scenario and around 2 percentage points higher in the non-OECD region. The bigger improvement in the OECD arises from a larger reduction in the share of coal-fired generation in total generation than in non-OECD countries and a greater increase in the share of gas-fired generation. In addition, the coal plants that are in operation later in the projection period are more efficient. This is especially true for the United States, where constructing highly-efficient coal plants instead of refurbishing old inefficient plants raises the average coal plant efficiency by six percentage points.

Pathways to energy efficiency
Country and regional profiles

What is included in the profiles?

This chapter presents the results of the Efficient World Scenario in profiles for the world and five major countries and regions: the United States, the European Union, Japan, China and India. The regions covered represented nearly 60% of global energy demand in 2010. The profiles are aimed at providing decision makers with a data-rich set of information on the potential, costs and benefits of achieving a high energy efficiency pathway. The policy, technology and economic assumptions that underlie the Efficient World Scenario, and the results presented in the profiles, are described in Chapter 10 (A blueprint for an energy-efficient world) and Chapter 11 (Unlocking energy efficiency at the sectoral level).

The presentation is primarily in the form of figures and tables. For each country and region, we show the potential for energy savings, including the change in total primary energy demand and the efficiency gains achievable in different sectors, given our assumptions about technology improvements and adopted policies; key energy efficiency indicators (energy intensity,[1] energy demand per capita and sectoral indicators) are also provided. Costs and benefits associated with the Efficient World Scenario are shown in the form of the additional investment required, the impact on fossil-fuel import bills and economic and emissions indicators. A set of country/region-specific policy opportunities are listed to show, based on bottom-up analysis, what actions might appropriately be adopted to achieve the energy efficiency gains. These are ordered by sector, according to where the energy savings potential is greatest.

How to read the profiles

The country and regional profiles that follow are presented in a consistent format. Each contains a set of figures and tables corresponding to the categories described below. Major findings are drawn from these and listed in Highlights at the opening of the profiles; at their conclusion, we identify policy opportunities[2] for realising energy efficiency gains.

Primary energy demand by scenario (Figures 12.1a to 12.6a)

These charts show the change in total primary energy demand by fuel in the Efficient World Scenario relative to the New Policies Scenario between 2010 and 2035. The tables that accompany these charts show total primary energy demand by source in 2010 and 2035 in the Efficient World Scenario (EWS). The category "other renewables" includes wind, solar photovoltaic (PV), geothermal, concentrating solar power and marine.

1. See Chapter 9 for an explanation of the choice of this indicator.

2. For the buildings sector, many such opportunities include the adoption of minimum energy performance standards (MEPS).

Key indicators (Tables 12.1a to 12.6a)

These tables show historical data for energy efficiency indicators and projections in 2020 and 2035 in the New Policies Scenario (NPS) and the Efficient World Scenario (EWS). Economy-wide energy intensity is the ratio of primary energy demand (in tonnes of oil equivalent [toe]) to gross domestic product (GDP), measured in market exchange rate (MER) terms and in year-2011 dollars. Energy demand per capita is the ratio of primary energy demand (in toe) to population. Energy intensity at the sector level is measured in units that relate to the service provided:

- Residential energy intensity is the energy consumed (in kilowatt-hours [kWh]) per unit of floor space (in square metres [m^2]). Energy consumed includes all household end-uses.[3]
- Services energy intensity is the energy consumed per unit of economic value-added (VA) of the services sector (in MER terms and in thousands of year-2011 dollars, [$1 000 VA]).
- Fuel consumption in transport is the average volume of gasoline-equivalent liquid fuel consumed (in litres [l]) (across all vehicle sales) per unit of distance travelled (in 100 kilometres [km]). Fuel consumption by passenger light-duty vehicles (PLDVs) is shown under test-cycle conditions, as this is the metric used in current policy making; on-road conditions are shown as the relevant metric for fuel consumption for heavy trucks.
- Other industries energy intensity is indexed to the year 2000.
- The average thermal efficiency of fossil-fuelled power plants is the indicator shown for the power sector.

Impact on world GDP (Figure 12.1b)

This chart shows the impact of the Efficient World Scenario on real world GDP and economic value-added by sector. Projections are provided by OECD (2012).

Fossil fuel net trade and import bills (Figures 12.2b to 12.6b)

These charts show fossil fuel net trade and import bills for oil, gas and coal in 2010 and in 2035 in the New Policies Scenario and Efficient World Scenario. Fossil fuel net trade is shown in million barrels per day (mb/d) for oil, billion cubic metres (bcm) for gas and million tonnes coal equivalent (Mtce) for coal. Negative values signify exports and export revenues.

Economic and environment benefits (Tables 12.1b to 12.6b)

These tables show historical data for GDP in purchasing power parity (PPP) terms, import bills, consumer expenditures on energy, and energy-related emissions and projections for

3. The indicator illustrates current intensity and future reduction potential, and should not be used for regional comparison due to the inherent variations, such as family size, average area of a dwelling and climatic conditions.

2020 and 2035 in the New Policies Scenario and Efficient World Scenario. Energy-related emissions are shown for carbon dioxide (CO_2) (in gigatonnes [Gt]) and the major air pollutants, sulphur dioxide (SO_2), nitrogen oxides (NO_X) and particulate matter ($PM_{2.5}$) (all in million tonnes [Mt]). Historical and projected emissions of major air pollutants are from IIASA (2012).

Additional investment (Figures 12.1c to 12.6c)

These charts show incremental investment needs in end-use sectors (industry, transport, residential and services) in the Efficient World Scenario, relative to the New Policies Scenario.

Change in energy consumption in energy-intensive industries (Figures 12.1d to 12.6d)

These charts show the change in energy consumption in the iron and steel, chemicals, cement, and pulp and paper industry from 2010 to 2035 in the Efficient World Scenario. The net change in energy consumption for each industry is the result of the difference in energy consumption due to changes in activity and energy savings due to energy efficiency improvements. An accompanying table shows consumption levels in the Efficient World Scenario.

Oil savings in the transport sector (Figures 12.1e to 12.6e)

These charts show oil savings (in mb/d) achieved in different transport sub-sectors in the Efficient World Scenario, relative to the New Policies Scenario. Other road includes light commercial vehicles, trucks less than 16 tonnes, buses and two/three wheelers.

Energy demand and savings in the residential sector (Figures 12.1f to 12.6f)

These charts show residential consumption in the New Policies Scenario and the Efficient World Scenario. Energy savings over the period 2010-2035 are shown according to change in activity[4] (*i.e.* the difference in energy service demand), efficiency measures like insulation (including retrofits and energy management systems) and technology improvements in space heating (including heat pumps), appliances, cooling, lighting and others (water heating and cooking). The tables that accompany the charts show fuel savings by change in activity and technology in million tonnes of oil equivalent (Mtoe) in 2020 and 2035, as well as their share of the overall savings achieved in the residential sector in those years.

Final energy consumption (Tables 12.1c to 12.6c)

These tables show historical data for final energy consumption by sector and energy source and projections for 2020 and 2035 in the New Policies Scenario (NPS) and the Efficient World Scenario (EWS). They are not balances.

4. Appears only at world level (Figure 12.1f), reflecting the reduction in demand due to partial subsidy removal in fossil fuel-net-exporting countries.

World

Highlights

- Primary energy savings achieved in the Efficient World Scenario in 2035 are equivalent to 18% of global energy demand in 2010.
- Efficiency gains boost GDP by 0.4% in 2035, relative to the New Policies Scenario.
- Additional investments required in end-use efficiency are $11.8 trillion over 2012-2035; this saves consumers $17.5 trillion in energy expenditures during that period.

Energy consumption

Figure 12.1a ▷ Global primary energy demand by scenario

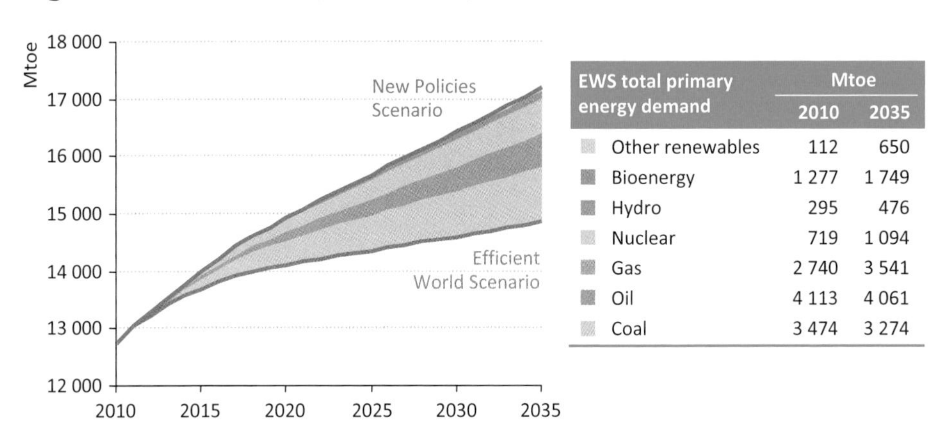

EWS total primary energy demand	Mtoe 2010	Mtoe 2035
Other renewables	112	650
Bioenergy	1 277	1 749
Hydro	295	476
Nuclear	719	1 094
Gas	2 740	3 541
Oil	4 113	4 061
Coal	3 474	3 274

Table 12.1a ▷ World key indicators

	2000	2010	2020		2035	
			NPS	EWS	NPS	EWS
Energy intensity (toe/million dollars, MER)	197	188	157	149	119	103
Energy demand per capita (toe/capita)	1.66	1.86	1.96	1.85	2.01	1.73
Residential energy intensity (kWh/m²)	155	138	125	118	113	98
Services energy intensity (kWh/$1 000 VA)	219	216	191	174	148	120
Fuel consumption new PLDVs test-cycle (l/100 km)	8.5	7.6	5.8	5.1	4.9	3.5
Fuel consumption new heavy trucks on-road (l/100 km)	37	36	29	27	28	19
Energy intensity of other industries (2000=100)	100	84	72	68	56	50
Fossil-fuel power plant efficiency (%)	32%	34%	37%	37%	39%	41%

Figure 12.1b ▷ Changes in global real GDP and value-added by sector in the Efficient World Scenario relative to the New Policies Scenario

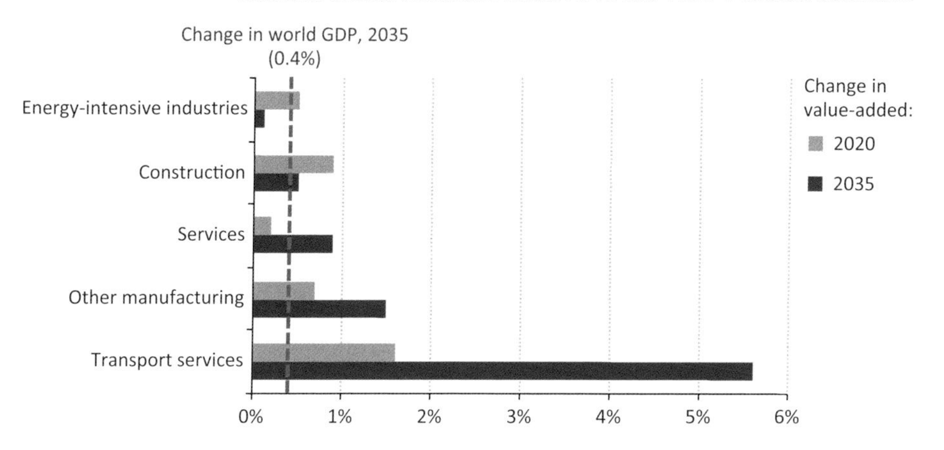

Table 12.1b ▷ Global economic and environmental benefits

	2000	2010	2020		2035	
			NPS	EWS	NPS	EWS
GDP ($2011 trillion, PPP)	64.4	76.2	113.0	113.4	181.5	182.2
Consumer expenditures ($2011 billion)	4 502	5 851	8 524	8 066	10 594	9 016
CO_2 emissions (Gt)	27.4	30.2	34.6	32.0	37.0	30.5
SO_2 emissions (Mt)	97.3	86.3	75.1	69.7	72.4	62.0
NO_x emissions (Mt)	87.3	84.7	77.7	73.4	82.5	69.9
$PM_{2.5}$ emissions (Mt)	40.8	42.8	42.2	41.3	41.2	38.9

Figure 12.1c ▷ Global additional investment by end-use sector in the Efficient World Scenario relative to the New Policies Scenario

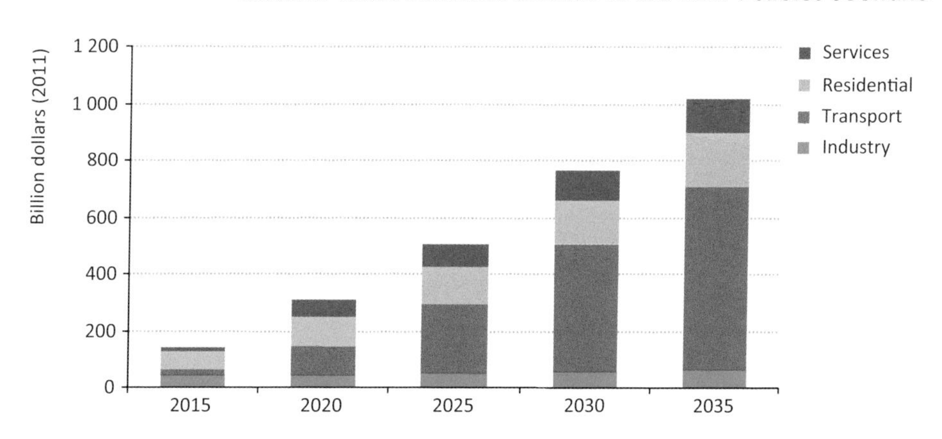

Figure 12.1d ▷ **Global change in energy consumption in energy-intensive industries in the Efficient World Scenario, 2010-2035**

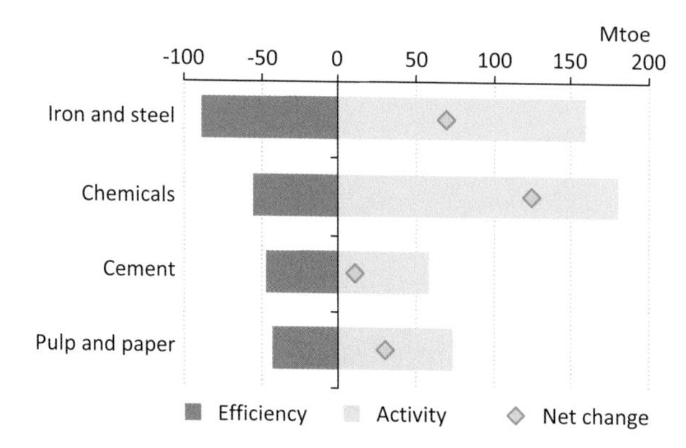

Energy demand	Mtoe	
	2010	2035
Iron and steel	436	506
Chemicals	393	518
Cement	247	258
Pulp and paper	150	181

Figure 12.1e ▷ **Global oil demand savings in the transport sector in the Efficient World Scenario relative to the New Policies Scenario**

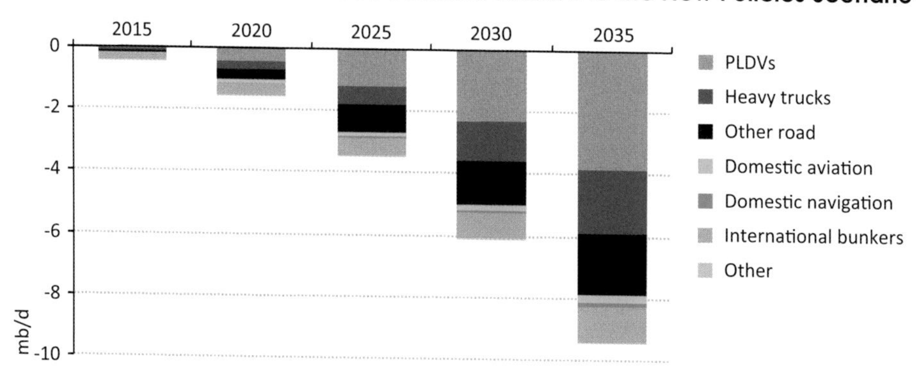

Figure 12.1f ▷ **Global energy demand and savings in the residential sector in the Efficient World Scenario relative to the New Policies Scenario**

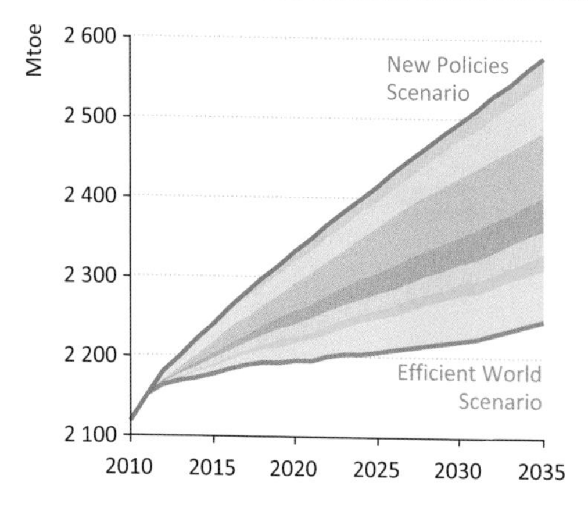

Fuel savings	Mtoe		Share	
	2020	2035	2020	2035
Activity	12	30	8%	9%
Insulation*	28	65	21%	20%
Space heating	34	78	25%	24%
Appliances	17	43	12%	13%
Cooling	14	29	10%	9%
Lighting	9	20	6%	6%
Others	24	67	17%	20%

* Includes energy management systems.

Table 12.1c ▷ World final energy consumption

	Energy demand (Mtoe)						% Change NPS to EWS	
			New Policies		Efficient World			
	2000	2010	2020	2035	2020	2035	2020	2035
Total final consumption	**7 078**	**8 678**	**10 223**	**11 750**	**9 789**	**10 363**	**-4**	**-12**
Coal	581	853	982	976	939	876	-4	-10
Oil	3 114	3 557	3 984	4 336	3 873	3 806	-3	-12
Gas	1 116	1 329	1 612	1 993	1 529	1 742	-5	-13
Electricity	1 091	1 537	2 047	2 676	1 893	2 297	-8	-14
Heat	247	278	303	305	287	270	-5	-12
Renewables	929	1 125	1 294	1 464	1 269	1 372	-2	-6
Industry	**1 907**	**2 421**	**3 035**	**3 497**	**2 901**	**3 171**	**-4**	**-9**
Coal	435	676	799	822	769	748	-4	-9
Oil	326	321	356	354	343	330	-4	-7
Gas	415	463	600	748	577	688	-4	-8
Electricity	459	638	890	1 133	838	999	-6	-12
Heat	100	126	138	131	133	121	-4	-8
Renewables	172	197	252	310	242	285	-4	-8
Transport	**1 952**	**2 377**	**2 778**	**3 272**	**2 704**	**2 780**	**-3**	**-15**
Oil	1 861	2 201	2 517	2 850	2 449	2 414	-3	-15
Electricity	19	24	36	57	35	56	-1	-2
Biofuels	10	59	111	206	108	176	-2	-15
Other fuels	62	93	115	159	112	134	-3	-16
Buildings	**2 451**	**2 910**	**3 302**	**3 748**	**3 080**	**3 193**	**-7**	**-15**
Coal	105	124	124	96	111	69	-10	-27
Oil	351	329	327	300	299	233	-9	-22
Gas	528	616	711	856	655	697	-8	-19
Electricity	583	831	1 062	1 402	960	1 161	-10	-17
Heat	143	148	160	170	150	145	-6	-15
Renewables	742	862	919	925	905	888	-1	-4
Other	**768**	**970**	**1 107**	**1 232**	**1 103**	**1 219**	**-0**	**-1**

United States

Highlights

- Primary energy savings achieved in the Efficient World Scenario in 2035 are equivalent to 15% of the country's energy demand in 2010.
- Efficiency gains boost GDP by 1.7% in 2035 relative to the New Policies Scenario.
- Additional investments required in end-use efficiency are $1.9 trillion over 2012-2035; this saves consumers $2.5 trillion in energy expenditures during that period.

Energy consumption

Figure 12.2a ▷ United States primary energy demand by scenario

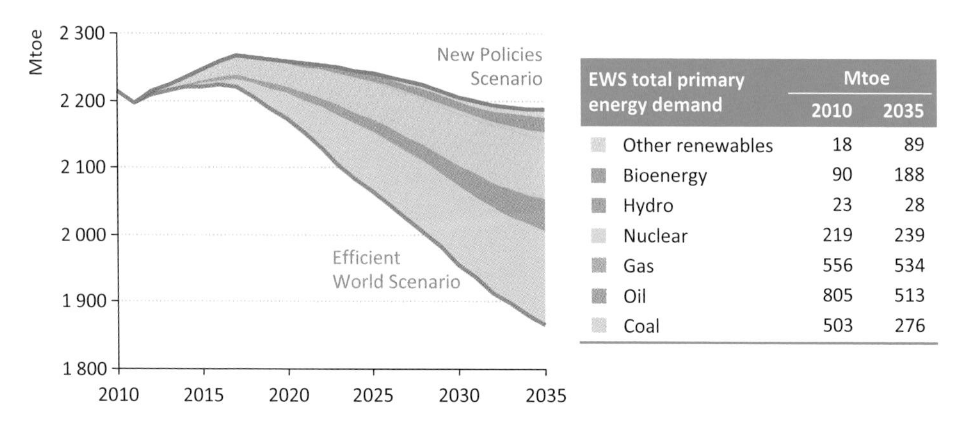

EWS total primary energy demand	Mtoe	
	2010	2035
Other renewables	18	89
Bioenergy	90	188
Hydro	23	28
Nuclear	219	239
Gas	556	534
Oil	805	513
Coal	503	276

Table 12.2a ▷ United States key indicators

	2000	2010	2020		2035	
			NPS	EWS	NPS	EWS
Energy intensity (toe/million dollars, MER)	179	149	118	113	82	70
Energy demand per capita (toe/capita)	7.92	7.04	6.62	6.37	5.80	4.95
Residential energy intensity (kWh/m^2)	159	137	123	115	111	89
Services energy intensity (kWh/$1 000 VA)	235	218	185	174	144	113
Fuel consumption new PLDVs test-cycle (l/100 km)	10.0	8.8	5.9	5.9	4.5	3.8
Fuel consumption new heavy trucks on-road (l/100 km)	40	39	31	29	28	21
Energy intensity of other industries (2000=100)	100	85	71	69	55	50
Fossil-fuel power plant efficiency (%)	37%	40%	42%	42%	45%	49%

Figure 12.2b ▷ United States fossil fuel net trade (current) and import bills in the Efficient World Scenario and the New Policies Scenario

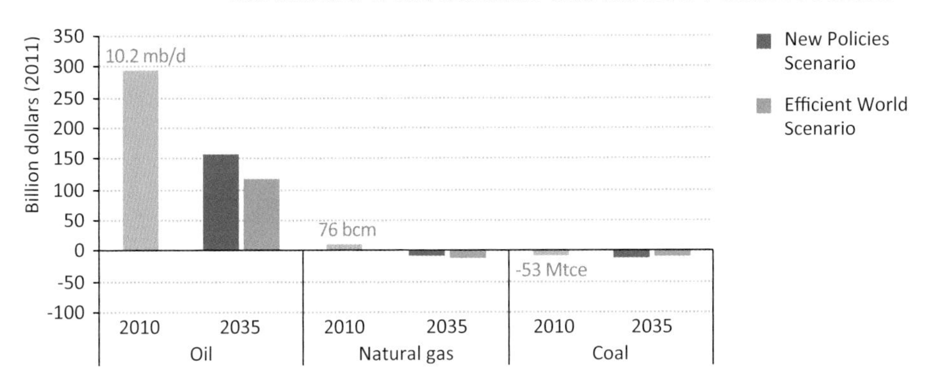

Table 12.2b ▷ United States economic and environmental benefits

	2000	2010	2020		2035	
			NPS	EWS	NPS	EWS
GDP ($2011 trillion, PPP)	14.3	14.8	19.2	19.3	26.6	27.1
Energy-import bills ($2011 billion)	307	308	242	241	157	118
Consumer expenditures ($2011 billion)	1 067	1 106	1 378	1 318	1 379	1 136
CO_2 emissions (Gt)	5.7	5.3	5.2	4.9	4.3	3.5
SO_2 emissions (Mt)	13.8	7.5	4.2	4.0	3.3	2.7
NO_x emissions (Mt)	18.0	12.7	7.7	7.4	6.1	5.2
$PM_{2.5}$ emissions (Mt)	1.2	1.0	0.9	0.9	0.9	0.8

Figure 12.2c ▷ United States additional investment by end-use sector in the Efficient World Scenario relative to the New Policies Scenario

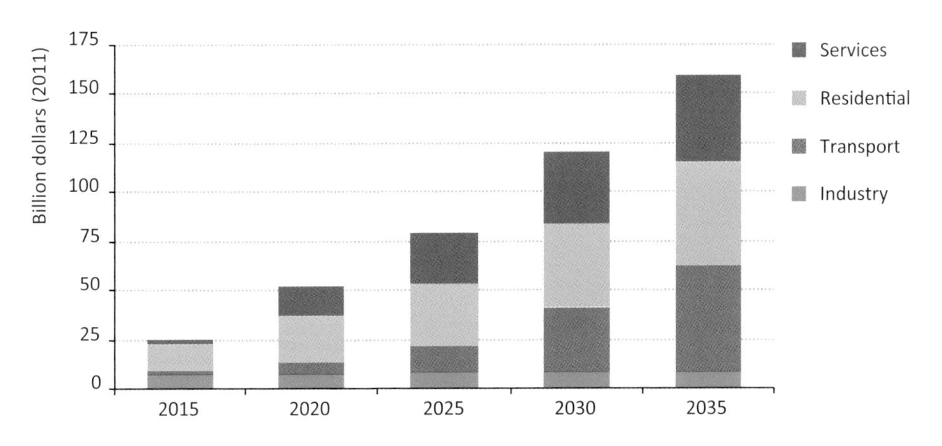

Figure 12.2d ▷ United States change in energy consumption in energy-intensive industries in the Efficient World Scenario, 2010-2035

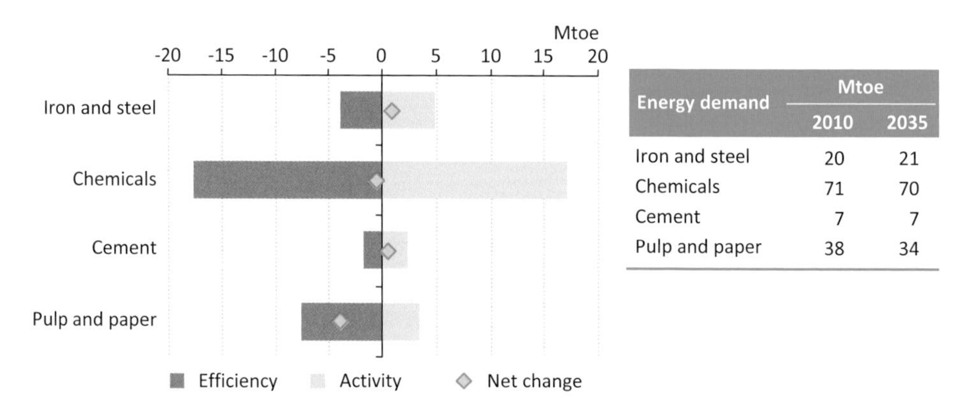

Energy demand	Mtoe	
	2010	2035
Iron and steel	20	21
Chemicals	71	70
Cement	7	7
Pulp and paper	38	34

Figure 12.2e ▷ United States oil demand savings in the transport sector in the Efficient World Scenario relative to the New Policies Scenario

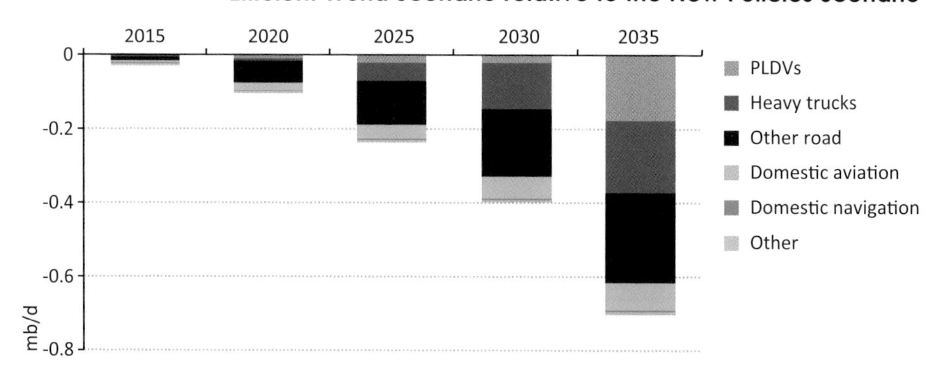

Figure 12.2f ▷ United States energy demand and savings in the residential sector in the Efficient World Scenario relative to the New Policies Scenario

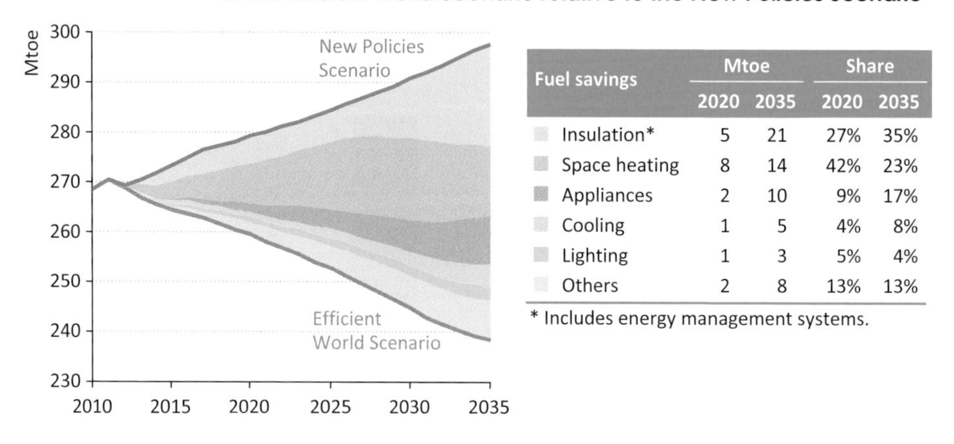

Fuel savings	Mtoe		Share	
	2020	2035	2020	2035
Insulation*	5	21	27%	35%
Space heating	8	14	42%	23%
Appliances	2	10	9%	17%
Cooling	1	5	4%	8%
Lighting	1	3	5%	4%
Others	2	8	13%	13%

* Includes energy management systems.

Table 12.2c ▷ United States final energy consumption

	Energy demand (Mtoe)						% Change NPS to EWS	
			New Policies		Efficient World			
	2000	2010	2020	2035	2020	2035	2020	2035
Total final consumption	**1 546**	**1 500**	**1 546**	**1 482**	**1 497**	**1 299**	**-3**	**-12**
Coal	33	27	27	22	26	20	-3	-8
Oil	793	749	716	543	708	499	-1	-8
Gas	360	319	342	349	322	288	-6	-18
Electricity	301	327	359	403	342	344	-5	-15
Heat	5	7	6	4	5	3	-4	-10
Renewables	54	72	97	161	94	145	-3	-10
Industry	**332**	**280**	**290**	**275**	**280**	**253**	**-3**	**-8**
Coal	30	25	26	21	25	19	-3	-7
Oil	26	30	26	19	25	18	-3	-6
Gas	138	111	115	105	112	97	-3	-8
Electricity	98	76	79	77	76	70	-4	-8
Heat	4	5	4	3	4	3	-3	-8
Renewables	36	32	40	51	38	46	-4	-10
Transport	**588**	**583**	**576**	**480**	**572**	**437**	**-1**	**-9**
Oil	569	541	518	371	514	339	-1	-9
Electricity	0	1	1	4	1	4	0	0
Biofuels	3	25	39	77	38	70	-0	-8
Other fuels	15	16	18	28	18	24	-2	-12
Buildings	**459**	**486**	**523**	**571**	**488**	**453**	**-7**	**-21**
Coal	2	2	1	1	1	0	-8	-49
Oil	49	37	26	12	23	3	-13	-74
Gas	189	182	197	203	181	153	-8	-25
Electricity	202	251	279	322	265	269	-5	-16
Heat	1	1	1	0	1	0	-6	-29
Renewables	15	14	18	33	17	28	-5	-15
Other	**167**	**151**	**157**	**156**	**157**	**155**	**-0**	**-1**

Policy opportunities

- Make building energy codes and energy labelling schemes mandatory for all new and existing buildings; require renovations of existing buildings to incorporate stringent energy requirements and provide resources necessary for effective implementation.
- Strengthen appliance and equipment MEPS to best available technology with appropriate labelling scheme and make regular updates.
- Extend fuel-economy standards for heavy trucks to make a 45% improvement by 2035 relative to 2010.
- Require small and medium-size enterprises to carry out regular energy audits, and promote research and development into energy-efficient technologies.
- Introduce energy performance requirements for existing coal-fired power plants.

European Union

Highlights

- Primary energy savings achieved in the Efficient World Scenario in 2035 are equivalent to 13% of the region's energy demand in 2010.
- Efficiency gains boost GDP by 1.1% in 2035 relative to the New Policies Scenario.[5]
- Additional investments required in end-use efficiency are $2.2 trillion over 2012-2035; this saves consumers $4.9 trillion in energy expenditures during that period.

Energy consumption

Figure 12.3a ▷ **European Union primary energy demand by scenario**

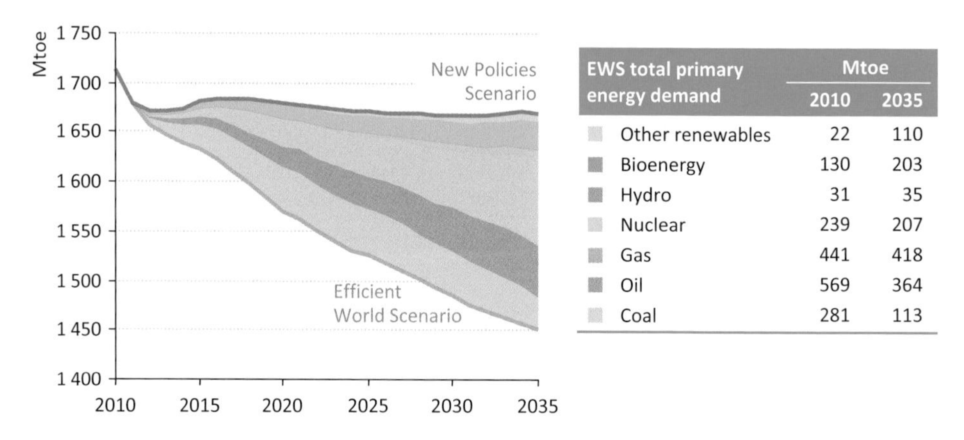

EWS total primary energy demand	Mtoe 2010	Mtoe 2035
Other renewables	22	110
Bioenergy	130	203
Hydro	31	35
Nuclear	239	207
Gas	441	418
Oil	569	364
Coal	281	113

Table 12.3a ▷ **European Union key indicators**

	2000	2010	2020 NPS	2020 EWS	2035 NPS	2035 EWS
Energy intensity (toe/million dollars, MER)	112	99	83	77	63	55
Energy demand per capita (toe/capita)	3.48	3.41	3.27	3.06	3.22	2.80
Residential energy intensity (kWh/m^2)	201	193	185	167	177	142
Services energy intensity (kWh/$1 000 VA)	145	151	141	127	122	98
Fuel consumption new PLDVs test-cycle (l/100 km)	7.3	6.2	4.4	3.8	3.9	3.4
Fuel consumption new heavy trucks on-road (l/100 km)	33	31	28	24	26	16
Energy intensity of other industries (2000=100)	100	83	73	71	62	59
Fossil-fuel power plant efficiency (%)	37%	38%	39%	39%	40%	41%

5. GDP gains are for OECD Europe.

Figure 12.3b ▷ European Union fossil fuel net trade (current) and import bills in the Efficient World Scenario and the New Policies Scenario

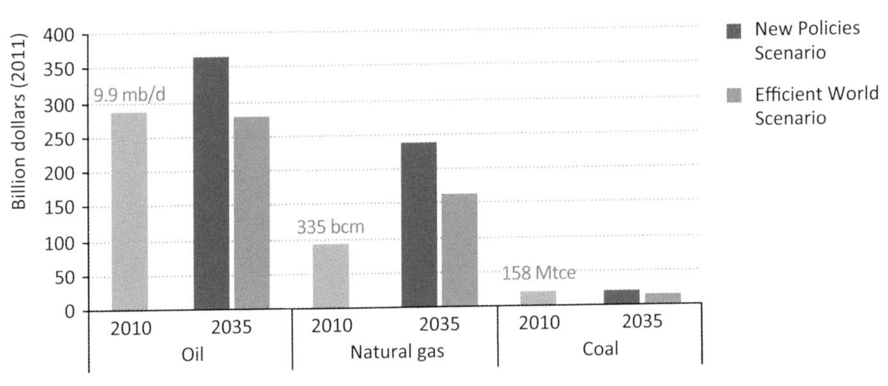

Table 12.3b ▷ European Union economic and environmental benefits

	2000	2010	2020		2035	
			NPS	EWS	NPS	EWS
GDP ($2011 trillion, PPP)[5]	16.0	16.9	20.2	20.3	26.6	26.9
Energy-import bills ($2011 billion)	304	402	597	540	626	455
Consumer expenditures ($2011 billion)	1 299	1 567	1 887	1 739	2 001	1 606
CO_2 emissions (Gt)	3.9	3.6	3.3	3.0	2.7	2.3
SO_2 emissions (Mt)	2.0	4.5	2.7	2.5	2.0	1.9
NO_x emissions (Mt)	5.5	8.8	5.4	5.0	3.7	3.3
$PM_{2.5}$ emissions (Mt)	0.7	1.6	1.5	1.4	1.4	1.3

Figure 12.3c ▷ European Union additional investment by end-use sector in the Efficient World Scenario relative to the New Policies Scenario

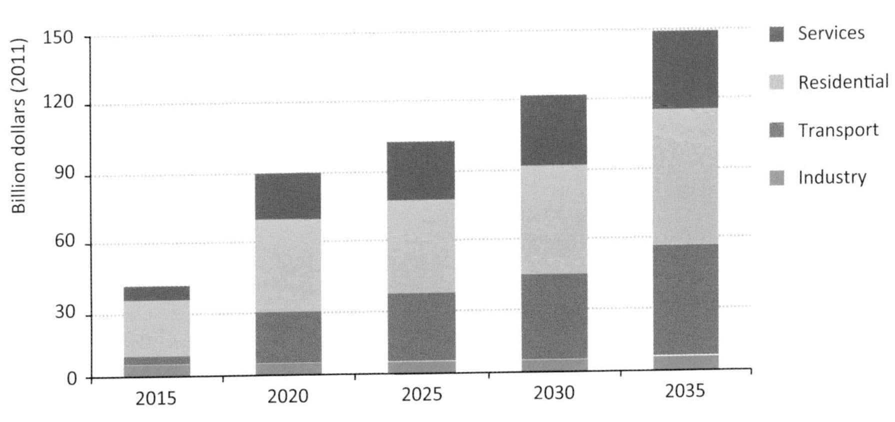

Figure 12.3d ▷ European Union change in energy consumption in energy-intensive industries in the Efficient World Scenario, 2010-2035

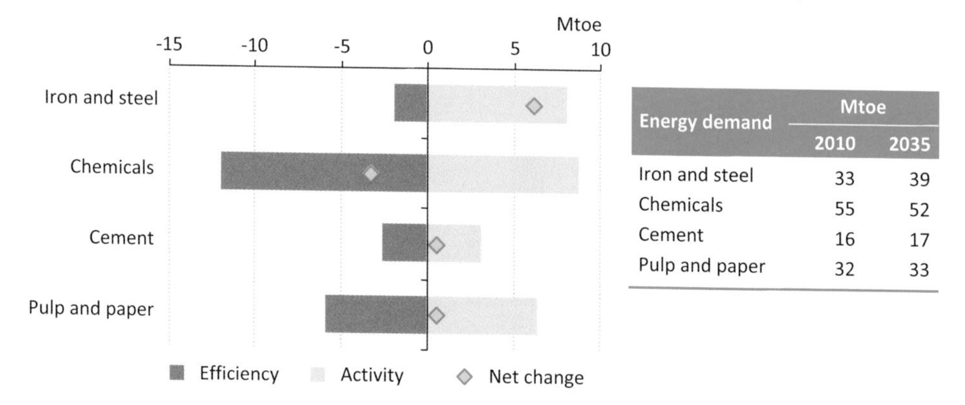

Energy demand	Mtoe	
	2010	2035
Iron and steel	33	39
Chemicals	55	52
Cement	16	17
Pulp and paper	32	33

Figure 12.3e ▷ European Union oil demand savings in the transport sector in the Efficient World Scenario relative to the New Policies Scenario

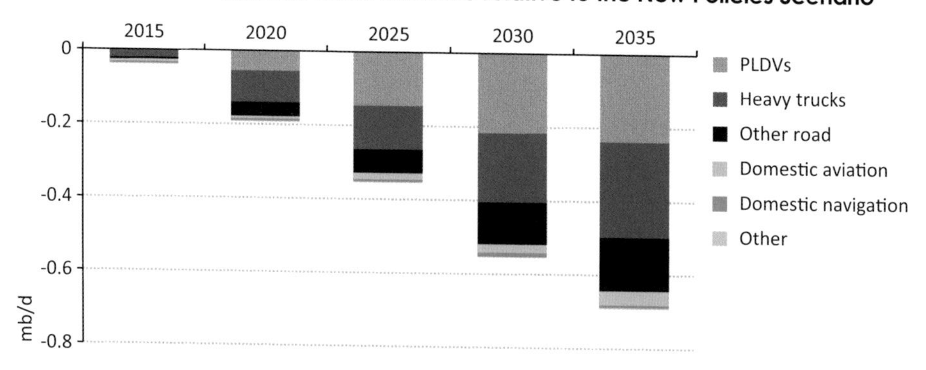

Figure 12.3f ▷ EU energy demand and savings in the residential sector in the Efficient World Scenario relative to the New Policies Scenario

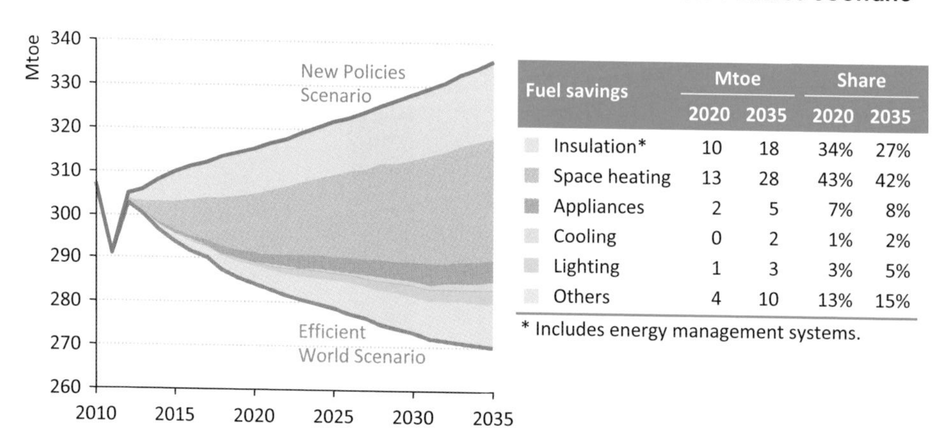

Fuel savings	Mtoe		Share	
	2020	2035	2020	2035
Insulation*	10	18	34%	27%
Space heating	13	28	43%	42%
Appliances	2	5	7%	8%
Cooling	0	2	1%	2%
Lighting	1	3	3%	5%
Others	4	10	13%	15%

* Includes energy management systems.

Table 12.3c ▷ European Union final energy consumption

	Energy demand (Mtoe)						% Change NPS to EWS	
			New Policies		Efficient World			
	2000	2010	2020	2035	2020	2035	2020	2035
Total final consumption	**1 169**	**1 194**	**1 201**	**1 227**	**1 134**	**1 068**	**-6**	**-13**
Coal	52	41	40	31	38	29	-5	-8
Oil	536	501	442	380	424	333	-4	-13
Gas	271	274	281	304	260	259	-7	-15
Electricity	217	244	262	288	245	252	-7	-13
Heat	44	53	58	62	55	52	-5	-17
Renewables	49	81	119	161	112	144	-6	-11
Industry	**306**	**273**	**287**	**290**	**279**	**277**	**-3**	**-4**
Coal	38	25	27	23	26	22	-1	-3
Oil	50	34	30	24	30	23	-2	-4
Gas	101	85	86	87	84	85	-2	-2
Electricity	91	89	97	99	93	93	-4	-5
Heat	10	16	16	16	15	15	-3	-7
Renewables	17	24	31	42	30	39	-3	-7
Transport	**303**	**319**	**296**	**270**	**286**	**231**	**-3**	**-14**
Oil	295	297	260	218	251	185	-4	-15
Electricity	6	6	7	10	7	10	0	1
Biofuels	1	13	26	38	25	32	-3	-15
Other fuels	1	2	3	4	3	4	-3	-10
Buildings	**421**	**470**	**495**	**544**	**446**	**437**	**-10**	**-20**
Coal	12	13	11	6	9	4	-15	-29
Oil	80	64	54	42	46	28	-14	-33
Gas	149	170	176	196	157	153	-11	-22
Electricity	115	145	154	176	141	145	-9	-18
Heat	34	37	42	47	40	37	-6	-20
Renewables	30	42	58	78	53	69	-9	-11
Other	**139**	**132**	**123**	**122**	**123**	**122**	**-0**	**0**

12

Policy opportunities

- Require deep renovations of existing buildings to incorporate stringent energy requirements and provide resources necessary for effective implementation.
- Strengthen appliance and equipment MEPS to best available technology with appropriate labelling; extend to more products and make regular updates.
- Implement fuel-economy standards for heavy trucks; extend and strengthen PLDV policy targets to 2035.
- Ensure the synergies of energy efficiency measures and the emissions trading system are exploited to mutually reinforce both policies and optimise their impact.

Japan

Highlights

- Primary energy savings achieved in the Efficient World Scenario in 2035 are equivalent to 10% of the country's energy demand in 2010.
- Efficiency gains boost GDP by 0.4% in 2035 relative to the New Policies Scenario.[6]
- Additional investments required in end-use efficiency are $0.5 trillion over 2012-2035; this saves consumers $1.2 trillion in energy expenditures during that period.

Energy consumption

Figure 12.4a ▷ Japan's primary energy demand by scenario

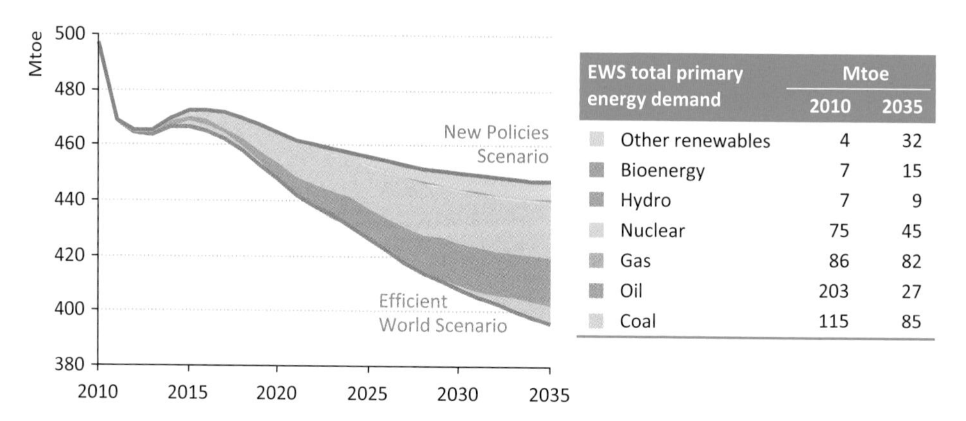

EWS total primary energy demand	Mtoe 2010	Mtoe 2035
Other renewables	4	32
Bioenergy	7	15
Hydro	7	9
Nuclear	75	45
Gas	86	82
Oil	203	27
Coal	115	85

Table 12.4a ▷ Japan's key indicators

	2000	2010	2020		2035	
			NPS	EWS	NPS	EWS
Energy intensity (toe/million dollars, MER)	94	84	70	67	56	50
Energy demand per capita (toe/capita)	4.09	3.90	3.70	3.57	3.78	3.35
Residential energy intensity (kWh/m^2)	139	125	111	103	101	82
Services energy intensity (kWh/$1 000 VA)	188	178	168	155	155	123
Fuel consumption new PLDVs test-cycle (l/100 km)	7.2	6.2	4.3	4.3	3.9	3.0
Fuel consumption new heavy trucks on-road (l/100 km)	32	27	24	21	20	14
Energy intensity of other industries (2000=100)	100	83	80	79	73	71
Fossil-fuel power plant efficiency (%)	43%	45%	50%	49%	52%	53%

6. GDP gains are for Japan and Korea.

Figure 12.4b ▷ Japan's fossil fuel net trade (current) and import bills in the Efficient World Scenario and the New Policies Scenario

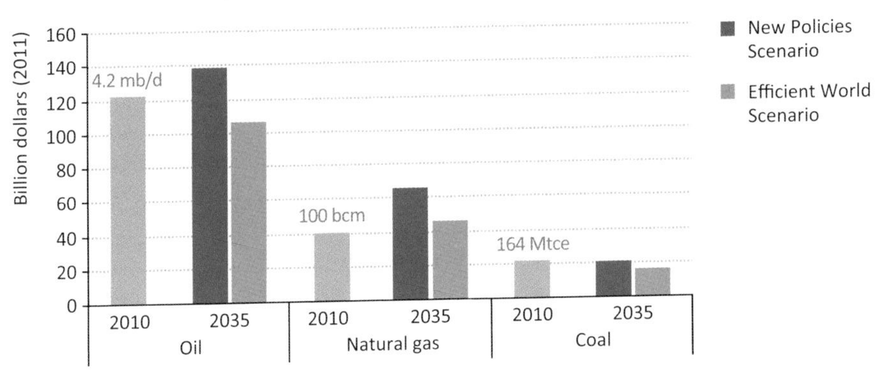

Table 12.4b ▷ Japan's economic and environmental benefits

	2000	2010	2020		2035	
			NPS	EWS	NPS	EWS
GDP ($2011 trillion, PPP)[6]	5.7	6.0	7.2	7.2	9.0	9.0
Energy-import bills ($2011 billion)	141	186	243	223	226	170
Consumer expenditures ($2011 billion)	379	445	509	476	503	406
CO_2 emissions (Gt)	1.2	1.1	1.0	1.0	0.9	0.8
SO_2 emissions (Mt)	0.8	0.6	0.5	0.5	0.5	0.4
NO_x emissions (Mt)	2.3	1.7	1.0	1.0	0.8	0.7
$PM_{2.5}$ emissions (Mt)	0.2	0.2	0.1	0.1	0.1	0.1

Figure 12.4c ▷ Japan's additional investment by end-use sector in the Efficient World Scenario relative to the New Policies Scenario

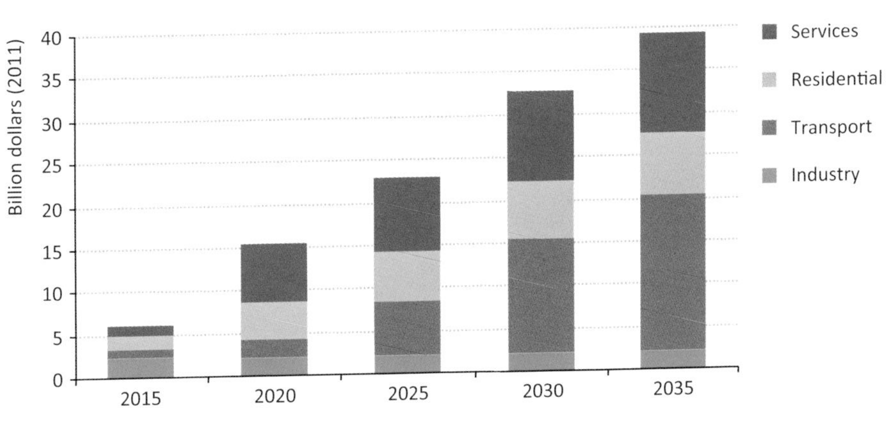

Figure 12.4d ▷ **Japan's change in energy consumption in energy-intensive industries in the Efficient World Scenario, 2010-2035**

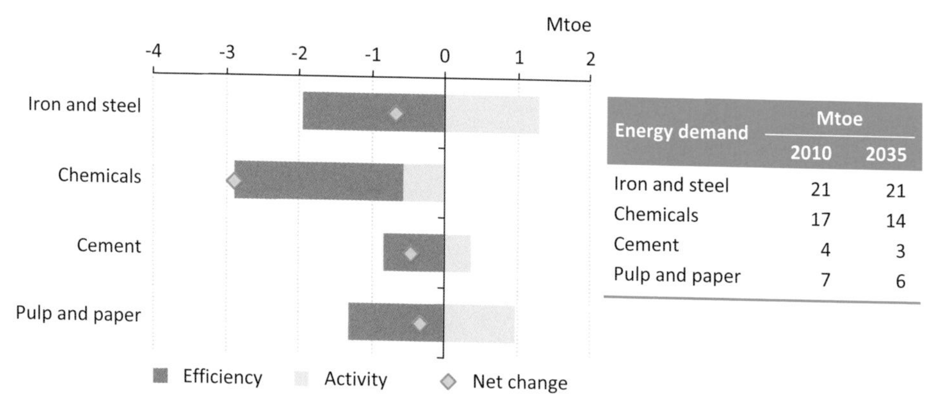

Energy demand	Mtoe	
	2010	2035
Iron and steel	21	21
Chemicals	17	14
Cement	4	3
Pulp and paper	7	6

Figure 12.4e ▷ **Japan's oil demand savings in the transport sector in the Efficient World Scenario relative to the New Policies Scenario**

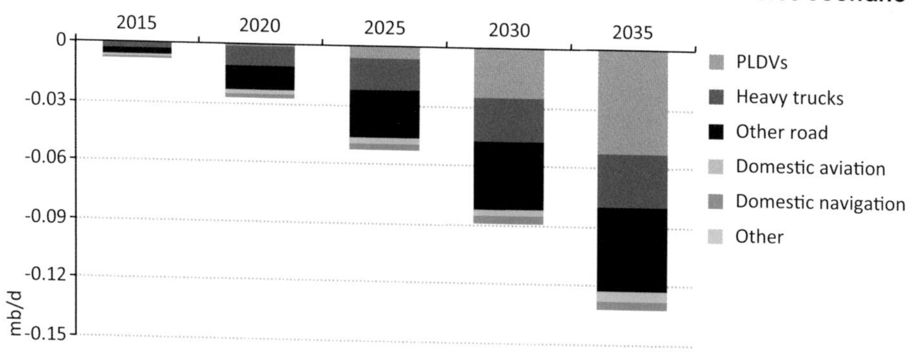

Figure 12.4f ▷ **Japan's energy demand and savings in the residential sector in the Efficient World Scenario relative to the New Policies Scenario**

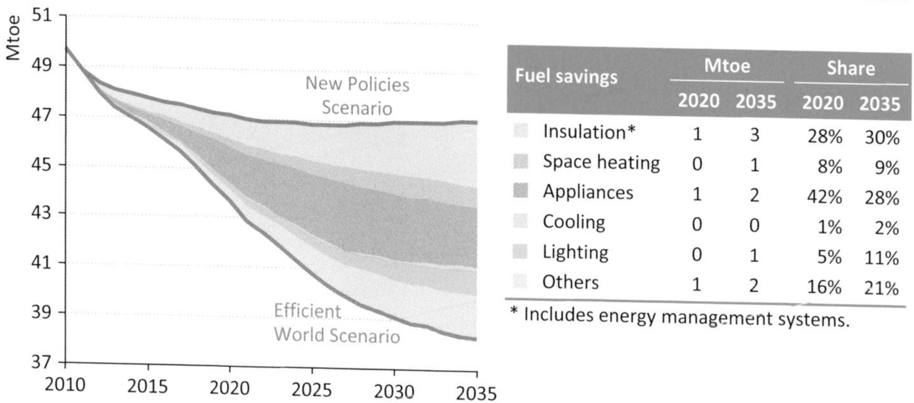

Fuel savings	Mtoe		Share	
	2020	2035	2020	2035
Insulation*	1	3	28%	30%
Space heating	0	1	8%	9%
Appliances	1	2	42%	28%
Cooling	0	0	1%	2%
Lighting	0	1	5%	11%
Others	1	2	16%	21%

* Includes energy management systems.

Table 12.4c ▷ Japan's final energy consumption

	Energy demand (Mtoe)						% Change NPS to EWS	
			New Policies		Efficient World			
	2000	2010	2020	2035	2020	2035	2020	2035
Total final consumption	**345**	**325**	**318**	**306**	**307**	**273**	**-4**	**-11**
Coal	27	29	28	27	28	27	-1	-1
Oil	210	171	158	135	155	121	-2	-10
Gas	23	34	39	43	36	36	-6	-17
Electricity	81	86	88	93	83	81	-6	-12
Heat	1	1	1	1	1	1	-6	-19
Renewables	4	3	4	7	4	7	-1	3
Industry	**100**	**90**	**92**	**91**	**91**	**89**	**-1**	**-2**
Coal	26	28	27	27	27	26	-1	-1
Oil	35	23	22	18	22	18	-1	-1
Gas	5	8	11	13	10	13	-1	0
Electricity	31	29	29	30	29	28	-2	-4
Heat	-	-	-	-	-	-	-	-
Renewables	3	3	3	4	3	4	-3	-6
Transport	**88**	**77**	**67**	**51**	**66**	**45**	**-2**	**-13**
Oil	86	75	65	49	64	42	-2	-13
Electricity	2	2	2	3	2	2	-7	-16
Biofuels	-	-	-	-	-	-	-	-
Other fuels	-	-	0	0	0	0	-8	-22
Buildings	**111**	**114**	**116**	**124**	**108**	**100**	**-8**	**-20**
Coal	1	1	1	1	0	0	-5	-16
Oil	42	31	30	30	28	22	-7	-26
Gas	18	26	28	30	25	23	-8	-25
Electricity	48	56	57	60	52	51	-8	-16
Heat	1	1	1	1	1	1	-6	-19
Renewables	1	1	1	3	1	3	8	18
Other	**47**	**43**	**42**	**40**	**42**	**40**	**-0**	**0**

Policy opportunities

- Implement rigorous and mandatory building energy codes, with stringent energy requirements for existing buildings.
- Develop a strategy towards net zero-energy buildings in new construction and encourage the introduction of energy management systems in buildings.
- Extend and strengthen PLDV fuel-economy standards. Extend fuel-economy standards for heavy trucks to make a 45% improvement in 2035 relative to 2010.
- Introduce pricing mechanism to improve energy management and reporting for large industry and power generation.

China

Highlights

- Primary energy savings achieved in the Efficient World Scenario in 2035 are equivalent to 25% of the country's energy demand in 2010.
- Efficiency gains boost GDP by 2.1% in 2035 relative to the New Policies Scenario.
- Additional investments required in end-use efficiency are $2.4 trillion over 2012-2035; this saves consumers $4.9 trillion in energy expenditures during that period.

Energy consumption

Figure 12.5a ▷ China's primary energy demand by scenario

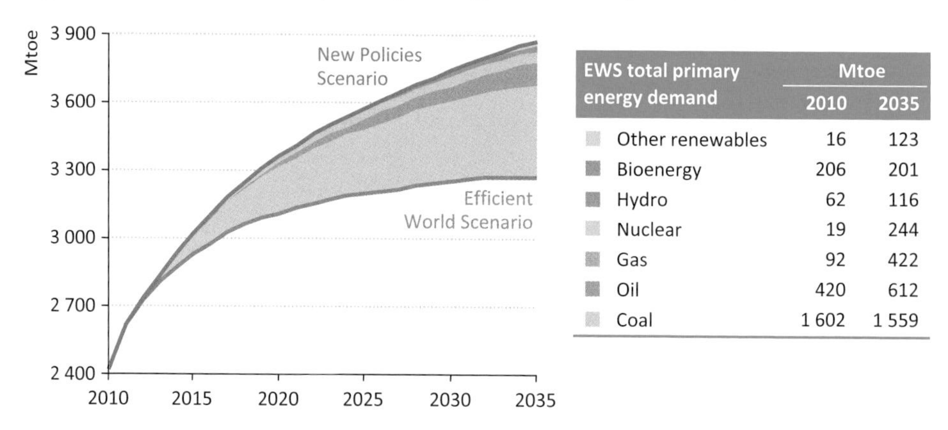

EWS total primary energy demand	Mtoe 2010	Mtoe 2035
Other renewables	16	123
Bioenergy	206	201
Hydro	62	116
Nuclear	19	244
Gas	92	422
Oil	420	612
Coal	1 602	1 559

Table 12.5a ▷ China's key indicators

	2000	2010	2020		2035	
			NPS	EWS	NPS	EWS
Energy intensity (toe/million dollars, MER)	456	350	228	211	141	119
Energy demand per capita (toe/capita)	0.94	1.80	2.41	2.23	2.79	2.36
Residential energy intensity (kWh/m²)	132	109	98	90	85	71
Services energy intensity (kWh/$1 000 VA)	336	270	208	183	134	108
Fuel consumption new PLDVs test-cycle (l/100 km)	8	8	6.7	5.0	4.8	3.4
Fuel consumption new heavy trucks on-road (l/100 km)	43	40	30	30	29	21
Energy intensity of other industries (2000=100)	100	70	56	51	40	33
Fossil-fuel power plant efficiency (%)	28%	32%	35%	35%	38%	38%

Figure 12.5b ▷ China's fossil fuel net trade (current) and import bills in the Efficient World Scenario and the New Policies Scenario

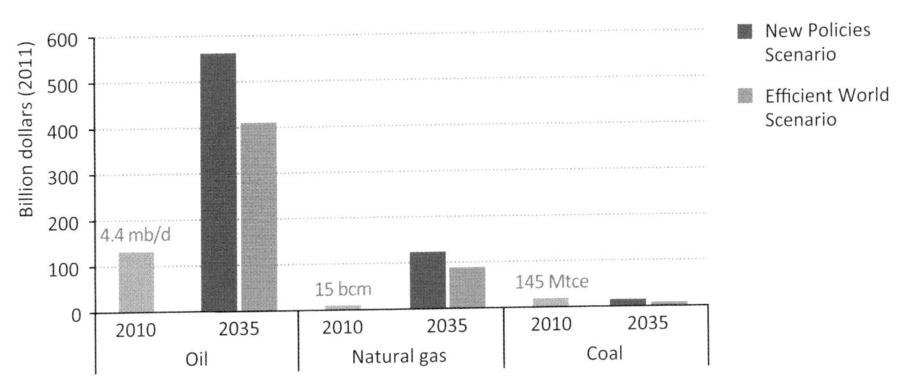

Table 12.5b ▷ China's economic and environmental benefits

	2000	2010	2020		2035	
			NPS	EWS	NPS	EWS
GDP ($2011 trillion, PPP)	6.4	10.7	22.8	22.9	42.6	43.5
Energy-import bills ($2011 billion)	61	154	454	412	695	503
Consumer expenditures ($2011 billion)	267	586	1 264	1 135	1 884	1 441
CO_2 emissions (Gt)	5.4	7.2	9.5	8.6	10.2	8.3
SO_2 emissions (Mt)	32.4	29.5	26.1	24.3	21.0	18.5
NO_x emissions (Mt)	16.4	21.1	22.3	20.8	22.0	18.4
$PM_{2.5}$ emissions (Mt)	12.9	14.8	12.8	12.4	10.9	10.0

Figure 12.5c ▷ China's additional investment by end-use sector in the Efficient World Scenario relative to the New Policies Scenario

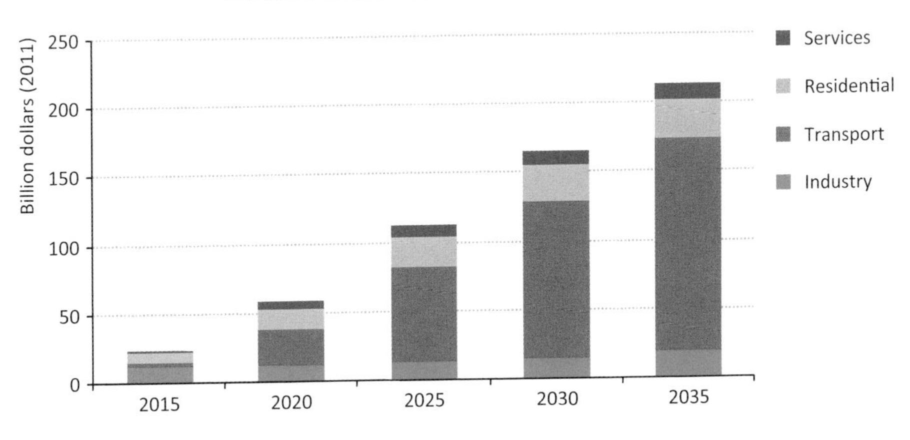

Figure 12.5d ▷ **China's change in energy consumption in energy-intensive industries in the Efficient World Scenario, 2010-2035**

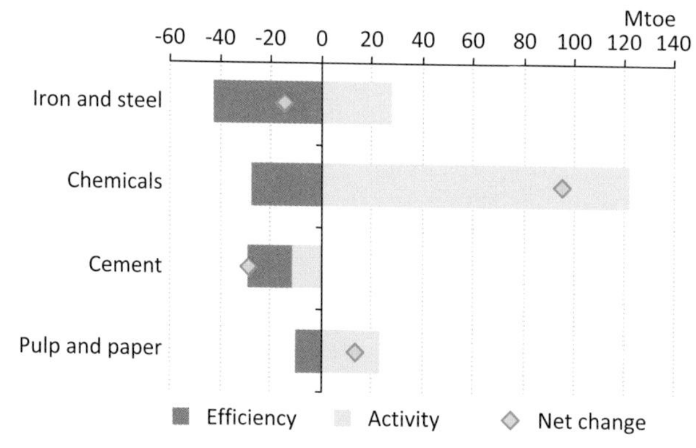

Energy demand	Mtoe	
	2010	2035
Iron and steel	210	195
Chemicals	110	204
Cement	136	107
Pulp and paper	26	40

Figure 12.5e ▷ **China's oil demand savings in the transport sector in the Efficient World Scenario relative to the New Policies Scenario**

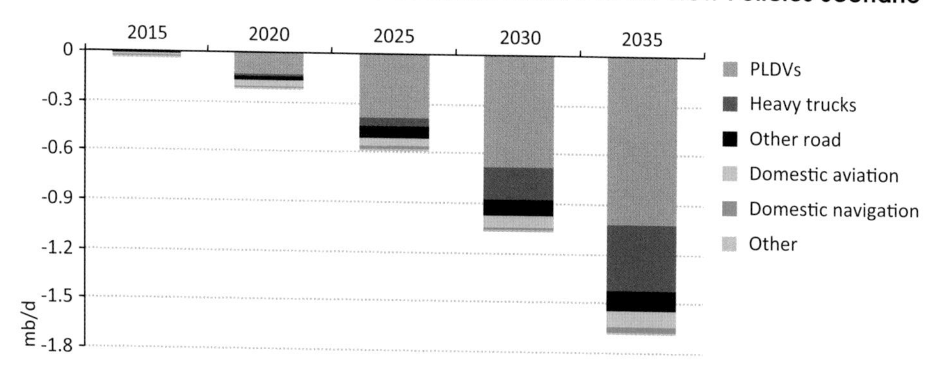

Figure 12.5f ▷ **China's energy demand and savings in the residential sector in the Efficient World Scenario relative to the New Policies Scenario**

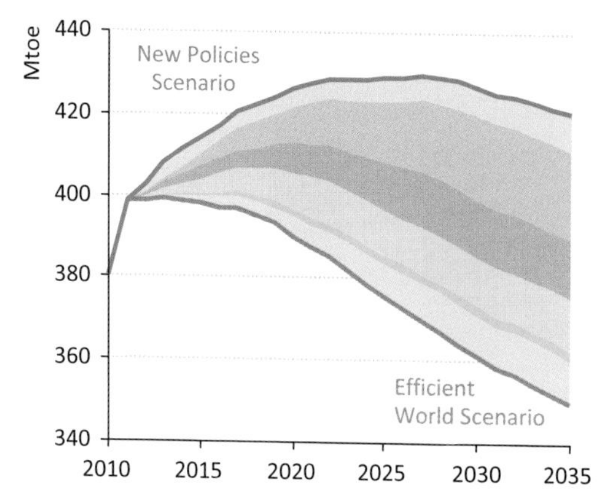

Fuel savings	Mtoe		Share	
	2020	2035	2020	2035
Insulation*	4	9	12%	12%
Space heating	9	22	25%	31%
Appliances	7	14	20%	20%
Cooling	9	13	26%	18%
Lighting	1	2	4%	3%
Others	5	11	13%	15%

* Includes energy management systems.

Table 12.5c ▷ China's final energy consumption

	Energy demand (Mtoe)						% Change NPS to EWS	
			New Policies		Efficient World			
	2000	2010	2020	2035	2020	2035	2020	2035
Total final consumption	**824**	**1 506**	**2 099**	**2 402**	**1 983**	**2 087**	**-6**	**-13**
Coal	305	514	564	478	541	432	-4	-10
Oil	184	357	554	677	538	583	-3	-14
Gas	12	57	161	269	151	241	-6	-10
Electricity	92	300	544	736	484	611	-11	-17
Heat	25	64	80	74	75	66	-6	-11
Renewables	205	213	198	168	194	154	-2	-8
Industry	**331**	**714**	**1 001**	**1 090**	**945**	**968**	**-6**	**-11**
Coal	216	401	447	385	431	356	-3	-7
Oil	32	48	63	63	60	58	-6	-7
Gas	5	16	79	120	74	111	-6	-8
Electricity	60	203	356	472	327	396	-8	-16
Heat	19	45	56	50	53	45	-6	-10
Renewables	0	0	0	0	0	0	-	-
Transport	**87**	**184**	**351**	**517**	**342**	**430**	**-3**	**-17**
Oil	82	169	323	460	315	381	-3	-17
Electricity	1	3	10	19	10	18	-0	-2
Biofuels	-	1	6	22	5	18	-3	-19
Other fuels	4	10	12	16	12	12	-4	-21
Buildings	**323**	**452**	**552**	**586**	**501**	**483**	**-9**	**-18**
Coal	59	68	67	44	59	27	-12	-39
Oil	25	49	53	39	49	29	-7	-25
Gas	3	24	56	110	51	94	-8	-14
Electricity	26	81	162	226	132	178	-19	-21
Heat	7	19	23	24	22	21	-6	-12
Renewables	205	211	191	144	187	135	-2	-7
Other	**82**	**156**	**196**	**208**	**195**	**207**	**-0**	**-1**

Policy opportunities

- Strengthen industry-wide programmes aimed at energy conservation, including sectoral targets for energy intensity improvement, and ease access to finance for energy efficiency projects.
- Continue closing inefficient industry facilities; incentivise recycling of waste materials.
- Broaden and strengthen MEPS and labelling schemes to include all major categories of appliances and equipment.
- Strengthen building energy codes and extend to new and existing buildings in small and medium cities.
- Extend and strengthen PLDV fuel-economy standards to 2035 and implement heavy truck fuel-economy standards.

India

Highlights

- Primary energy savings achieved in the Efficient World Scenario in 2035 are equivalent to 40% of the country's energy demand in 2010.
- Efficiency gains boost GDP by 3.0% in 2035 relative to the New Policies Scenario.
- Additional investments required in end-use efficiency are $0.6 trillion over 2012-2035; this saves consumers $1.1 trillion in energy expenditures during that period.

Energy consumption

Figure 12.6a ▷ India's primary energy demand by scenario

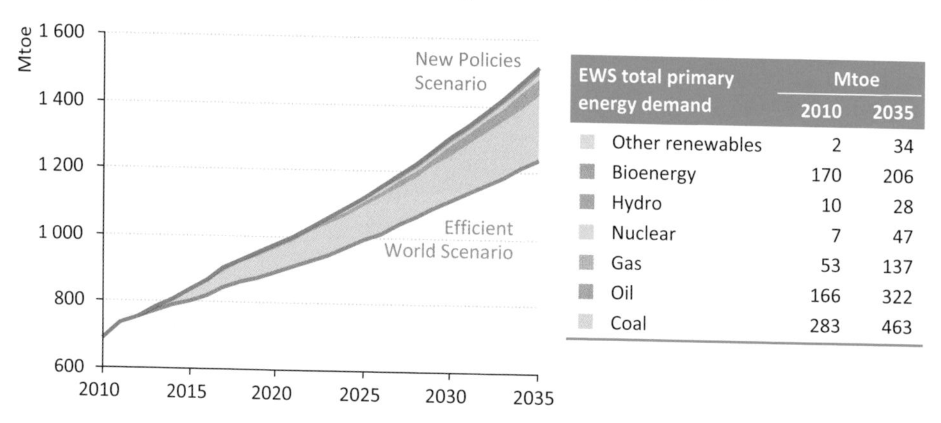

EWS total primary energy demand	Mtoe 2010	Mtoe 2035
Other renewables	2	34
Bioenergy	170	206
Hydro	10	28
Nuclear	7	47
Gas	53	137
Oil	166	322
Coal	283	463

Table 12.6a ▷ India's key indicators

	2000	2010	2020 NPS	2020 EWS	2035 NPS	2035 EWS
Energy intensity (toe/million dollars, MER)	601	442	313	286	210	171
Energy demand per capita (toe/capita)	0.45	0.59	0.73	0.67	1.00	0.82
Residential energy intensity (kWh/m^2)	220	188	154	147	124	108
Services energy intensity (kWh/$1 000 VA)	478	350	247	213	143	117
Fuel consumption new PLDVs test-cycle (l/100 km)	7	6	5.4	5.4	5.1	3.2
Fuel consumption new heavy trucks on-road (l/100 km)	44	42	33	33	31	22
Energy intensity of other industries (2000=100)	100	86	65	61	46	39
Fossil-fuel power plant efficiency (%)	29%	29%	33%	34%	38%	41%

Figure 12.6b ▷ **India's fossil fuel net trade (current) and import bills in the Efficient World Scenario and the New Policies Scenario**

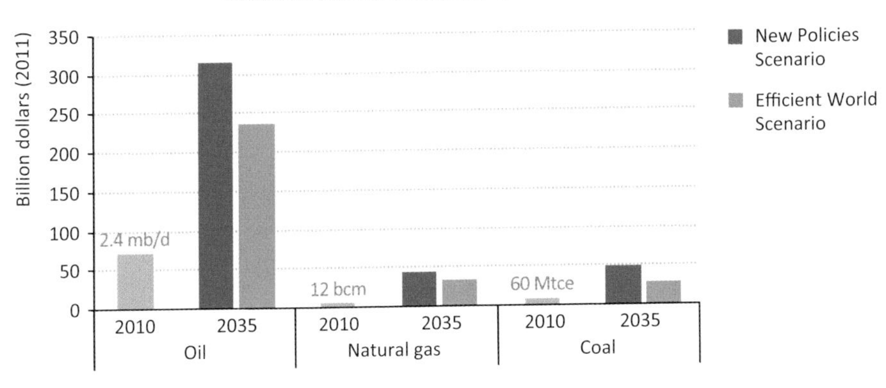

Table 12.6b ▷ **India's economic and environmental benefits**

	2000	2010	2020		2035	
			NPS	EWS	NPS	EWS
GDP ($2011 trillion, PPP)	2.8	4.2	8.3	8.4	19.2	19.8
Energy-import bills ($2011 billion)	42	83	200	180	406	296
Consumer expenditures ($2011 billion)	117	168	327	304	650	511
CO_2 emissions (Gt)	1.2	1.6	2.4	2.1	3.8	2.9
SO_2 emissions (Mt)	5.9	8.1	11.4	9.8	15.4	11.5
NO_x emissions (Mt)	4.2	5.5	6.8	6.3	11.8	9.5
$PM_{2.5}$ emissions (Mt)	6.0	6.1	6.6	6.3	6.8	6.1

Figure 12.6c ▷ **India's additional investment by end-use sector in the Efficient World Scenario relative to the New Policies Scenario**

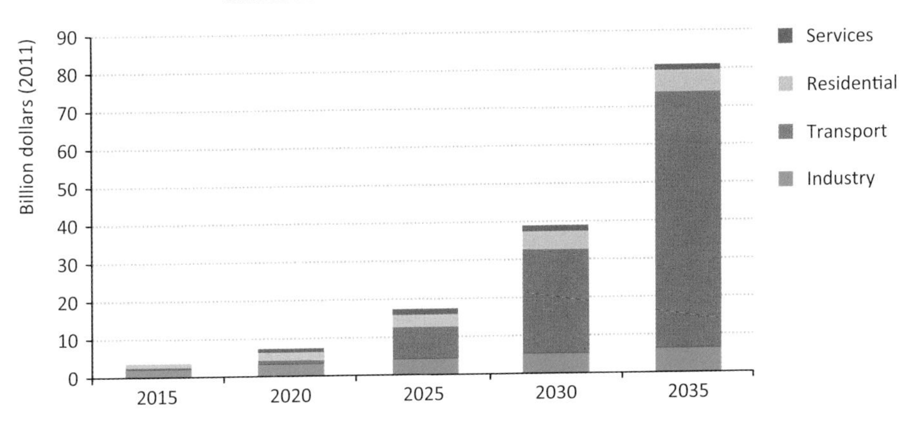

Figure 12.6d ▷ **India's change in energy consumption in energy-intensive industries in the Efficient World Scenario, 2010-2035**

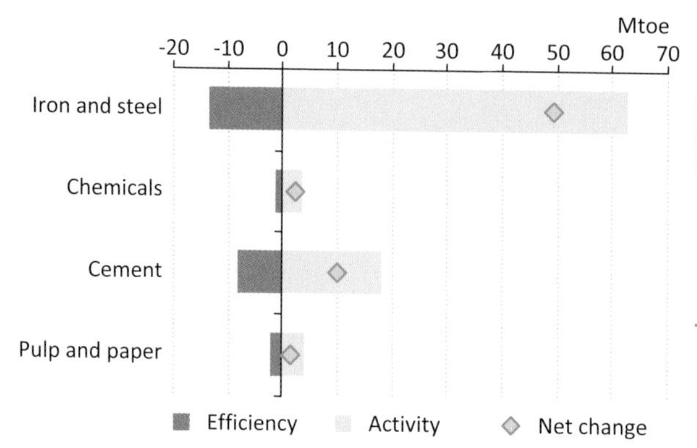

Energy demand	Mtoe	
	2010	2035
Iron and steel	31	80
Chemicals	5	7
Cement	14	24
Pulp and paper	4	6

Figure 12.6e ▷ **India's oil demand savings in the transport sector in the Efficient World Scenario relative to the New Policies Scenario**

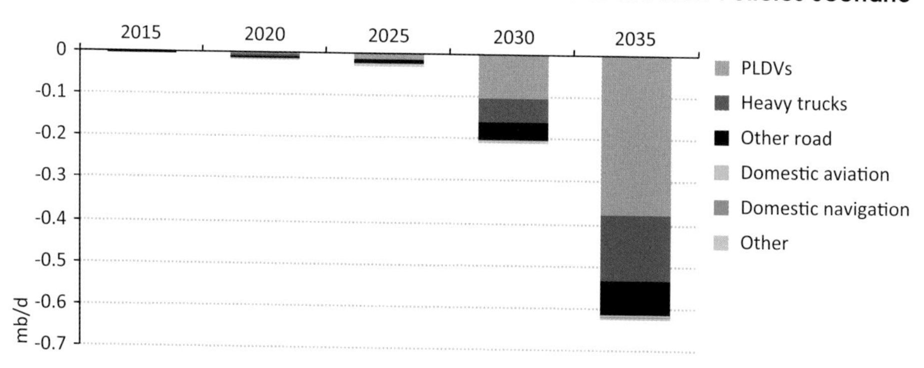

Figure 12.6f ▷ **India's energy demand and savings in the residential sector in the Efficient World Scenario relative to the New Policies Scenario**

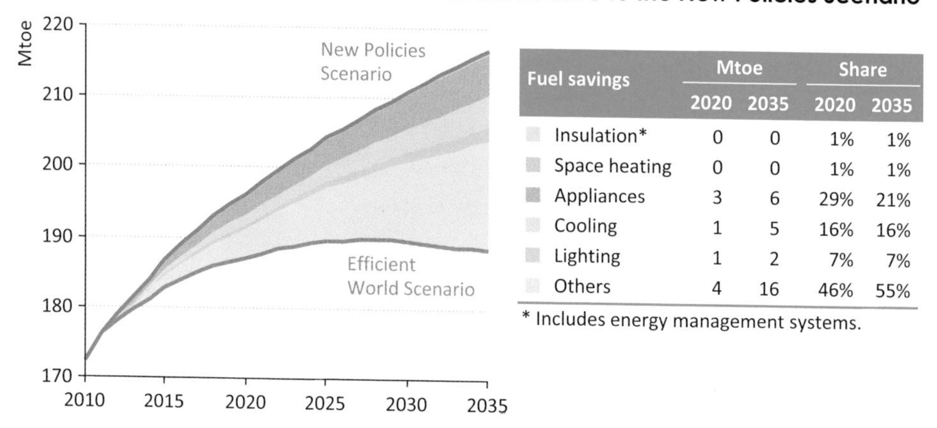

Fuel savings	Mtoe		Share	
	2020	2035	2020	2035
Insulation*	0	0	1%	1%
Space heating	0	0	1%	1%
Appliances	3	6	29%	21%
Cooling	1	5	16%	16%
Lighting	1	2	7%	7%
Others	4	16	46%	55%

* Includes energy management systems.

Table 12.6c ▷ India's final energy consumption

	Energy demand (Mtoe)						% Change NPS to EWS	
			New Policies		Efficient World			
	2000	2010	2020	2035	2020	2035	2020	2035
Total final consumption	**319**	**462**	**634**	**978**	**605**	**850**	**-5**	**-13**
Coal	33	76	122	188	113	158	-8	-16
Oil	96	132	185	340	179	296	-3	-13
Gas	10	23	29	51	29	47	1	-8
Electricity	32	61	111	212	98	175	-11	-17
Heat	-	-	-	-	-	-	-	-
Renewables	148	169	187	186	185	175	-1	-6
Industry	**87**	**152**	**236**	**375**	**220**	**322**	**-6**	**-14**
Coal	25	62	106	175	99	148	-7	-15
Oil	22	26	32	40	30	35	-6	-12
Gas	0	7	9	16	8	14	-6	-12
Electricity	14	28	51	98	47	80	-9	-19
Heat	-	-	-	-	-	-	-	-
Renewables	26	29	37	47	37	45	-2	-4
Transport	**32**	**55**	**85**	**225**	**85**	**190**	**1**	**-16**
Oil	31	52	76	200	77	170	1	-15
Electricity	1	1	2	2	2	2	-0	-0
Biofuels	0	0	2	10	2	8	0	-18
Other fuels	0	2	4	13	4	10	1	-23
Buildings	**159**	**198**	**234**	**272**	**220**	**233**	**-6**	**-14**
Coal	8	14	16	14	14	10	-12	-28
Oil	18	23	30	40	26	30	-13	-24
Gas	0	0	1	5	1	6	101	17
Electricity	10	21	40	83	33	66	-19	-21
Heat	-	-	-	-	-	-	-	-
Renewables	122	140	147	130	146	122	-1	-6
Other	**40**	**56**	**80**	**106**	**79**	**105**	**-0**	**-0**

Policy opportunities

- Expand innovative programmes for industry, such as Perform, Achieve and Trade, offering market-based incentives for meeting improvement targets, and facilitate energy efficiency project financing by clustering small- and medium-size businesses.
- Implement rigorous and mandatory building energy codes; extend MEPS and labelling, with stringent energy requirements, to more energy-using appliances and equipment.
- Ensure that the efficiency savings continue to facilitate improved energy access, while electricity generation capacity is constrained in the near term.
- Implement PLDV fuel-economy standards.
- Continue support to the National Biomass Cookstove Initiative and extend it to cover all households by 2020.
- Direct investments in coal-fired power plants to modern and efficient technology.

PART C
IRAQ ENERGY OUTLOOK

PREFACE

Part C of this *WEO* (Chapters 13-16) continues the past practice of examining in depth the prospects of a country of special significance to the global energy outlook. The spotlight falls this time on Iraq.

Chapter 13 surveys the situation as it is today. It also explains the basis for the projections which follow and how the analytical approach differs, where necessary, from that in the earlier chapters.

Chapter 14 provides a detailed analysis of Iraq's oil and gas resources. It shows what will be involved in their future exploitation, including the requirements for supporting transportation and export infrastructure, and the provision of water supplies. The scale of necessary investment is assessed.

Chapter 15 focuses on Iraq's internal requirements, projecting energy demand growth and how it can be met, including the growing potential role for natural gas. Repairing, renewing and expanding the electricity system is an early priority.

Chapter 16 brings the threads together, examining the implications of the projections both for the development of Iraq's economy and for the global oil market, in which Iraq is already a major player and is set to become increasingly important.

Iraq today: energy and the economy

Old country, new start?

Highlights

- Iraq's prosperity will depend on its energy sector. It is estimated to have the fifth-largest proven oil reserves and the 13th-largest proven gas reserves in the world, as well as vast potential for further discoveries. These resources can fuel its social and economic development.
- Energy is already the cornerstone of Iraq's economy, with oil exports accounting for 95% of government revenues and equal to over 70% of GDP in 2011. The pace of Iraq's rehabilitation depends to a significant degree on the oil sector: how quickly production and exports are increased and how effectively revenues are managed. Iraq's oil production is now above 3 mb/d and it is the third-largest oil exporter in the world, with an increasing share of exports going to Asia. Even conservative projections of future oil production imply profound effects on the Iraqi economy.
- A key obstacle to Iraq's development is the lack of reliable electricity supply. Power stations produce more electricity than ever before, but supply is still insufficient to meet demand; power cuts are a daily occurrence and the use of back-up diesel generators is widespread. Building a modern electricity system, with sufficient generation capacity and supplies of fuel, is recognised as an immediate priority.
- Oil makes up more than 80% of Iraq's primary energy mix, compared with less than 50% in the rest of the Middle East. Despite the significant economic advantages of using natural gas instead of liquid fuels, particularly in electricity generation, almost 60% of gas production in Iraq was flared in 2011, as the facilities were not in place to gather it and make it available for productive uses.
- Iraq has ambitious investment plans for its energy sector, supported by an increasing number of international energy companies. In the last five years, oil production and exports have both increased by more than 40% and grid-based electricity supply around 70%. However, progress is still patchy and the state of Iraq's energy transport, storage and export infrastructure, while improving, continues to be a serious constraint.
- Meeting Iraq's energy policy objectives will require substantial progress across a wide front: improved institutions and human capacity, better co-ordination of decision making, a strengthened and unambiguous legal and regulatory framework, enhanced conditions to support participation by the private and financial sectors, and a broader political consensus on the direction of future policy. Success will make a huge difference to the future outlook for Iraq's energy sector and for its economy more generally, potentially putting it on a path to becoming a powerhouse of the regional and global energy system.

Iraq's energy sector[1]

The energy sector is the cornerstone of Iraq's economy and the key to its future as it recovers from three decades punctuated by conflict (Figure 13.1). Three wars (with Iran from 1980 to 1988 and with US-led coalition forces in 1991 and 2003), international sanctions and internal instability have taken a severe toll. Living standards have fallen sharply, as Iraq's per capita gross domestic product (GDP) declined by more than one-fifth in real terms between 1980 and 2011, leaving this indicator as one of the lowest in the Middle East (Figure 13.2). Yet Iraq is also a country of immense potential. It has the fifth-largest proven oil reserves in the world and the 13th-largest proven gas reserves (and significant scope for further discoveries); the development of these resources can and should underpin Iraq's reconstruction and its social and economic development.

Figure 13.1 ▷ Iraq hydrocarbon resources and infrastructure

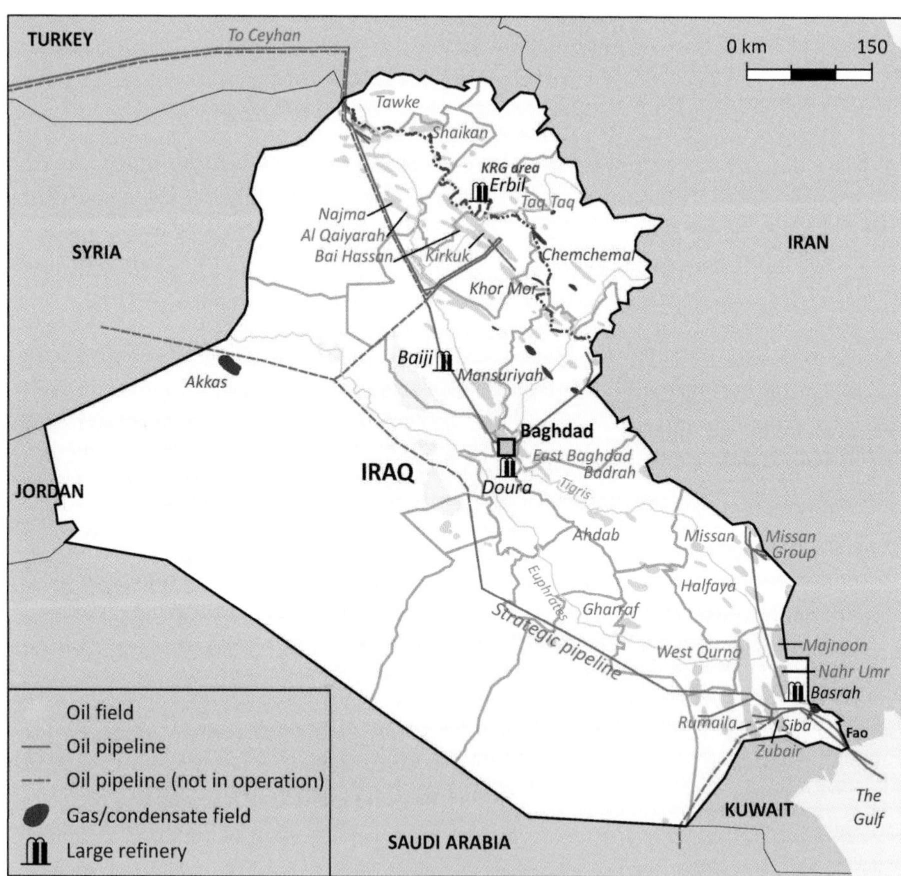

This map is without prejudice to the status of or sovereignty over any territory, to the delimitation of international frontiers and boundaries and to the name of any territory, city or area.

1. This analysis benefited from a workshop held by the IEA in Istanbul, Turkey on 4 May 2012, and was first published as a *World Energy Outlook* Special Report titled "*Iraq Energy Outlook*" on 9 October 2012.

Figure 13.2 ▷ **GDP per capita for selected countries, 2011**

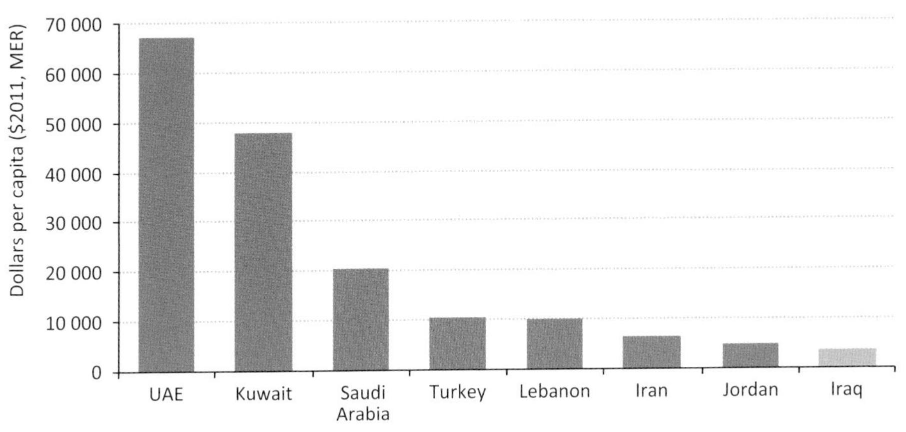

Note: GDP is measured at market exchange rates (MER) in year-2011 dollars.

Sources: International Monetary Fund (IMF) databases; IEA analysis.

The pace of Iraq's revival in the coming decades depends very heavily on the oil sector: how quickly production and export are increased and how the resulting revenues are managed and spent. By mid-2012, oil output was above 3 million barrels per day (mb/d), of which around 2.4 mb/d was exported. Iraq's improving stability, its huge resource base and contracts concluded with international companies to develop the country's major fields provide the foundation for a rapid increase in oil production in the coming years. Iraq will need to overcome a set of challenges relating to investment in infrastructure, institutional reform and the legal framework for the hydrocarbons sector, enhance human capacity and consolidate political stability and security.

Iraq's oil production over the coming years, even when estimated on a conservative basis, constitutes the core of its economy, as revenue from exports accounts for an overwhelming share of national wealth. In 2011, oil revenue accounted for around 95% of government income and was equivalent to more than 70% of Iraq's GDP. These figures are high even by the standards of other resource-rich countries in the Middle East (Figure 13.3). Sustained and growing export revenues are essential to meet the imperatives of reconstruction, as Iraq is still struggling to provide basic services, such as electricity and clean water. Over the long term, a central challenge for Iraq is to use its oil wealth to create a more diversified economy. Translating growth in oil receipts into prosperity for the people of Iraq is the promise and challenge.

Figure 13.3 ▷ Hydrocarbon revenue relative to GDP in selected countries, 2011

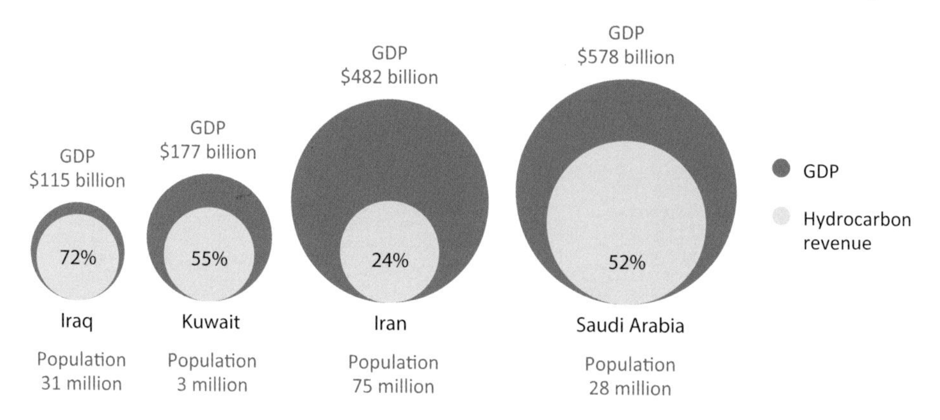

Note: GDP is measured at market exchange rates (MER) in year-2011 dollars.

Sources: IMF, UNPD, World Bank and national government databases; IEA analysis.

Overview of energy supply

Crude oil

The trajectory of Iraq's oil supply since 1980 confirms the malign influence of multiple conflicts, as production peaks have been followed by steep declines in every decade (Figure 13.4). As a result, Iraq has yet to surpass its historical peak of production in 1979, an average of 3.5 mb/d, and remains well below the 5.3% of global oil output that this represented (Box 13.1). Iraq's historical production record is dominated by just two super-giant fields[2]: the Kirkuk field in the north of the country, which has been producing since the 1920s, and the Rumaila field in the south, which began operation in the 1950s. Together, these two fields have produced around 28 billion barrels of oil, 80% of Iraq's cumulative oil production.

A snapshot of Iraq's oil supply for the month of June 2012 shows daily production at an average of 3 mb/d. During this period around 670 thousand barrels per day (kb/d) were delivered to domestic refineries, around 70 kb/d were used directly for power generation and over 2.4 mb/d were exported (Figure 13.5). Most of the export flows were made by tanker through offshore terminals and mooring systems in the south, but a smaller share was exported via the northern pipeline to the Turkish Mediterranean port of Ceyhan. This picture highlights the continued pre-eminence of the southern Rumaila field, with 1.3 mb/d of oil production, as well as the significant contributions from the nearby West Qurna (Phase I) and Zubair fields. Kirkuk makes a notable contribution to production in the north of Iraq, although its output has fallen back in recent times and is currently around 270 kb/d.

2. Super-giant fields are defined as those with ultimately recoverable resources greater than 5 billion barrels.

Figure 13.4 ▷ Iraq oil production

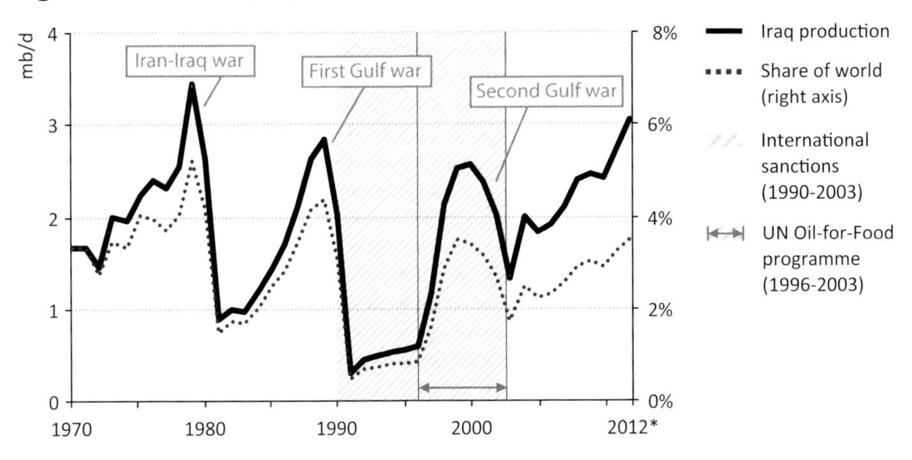

* Based on first five months.

Box 13.1 ▷ Breaking through the historical ceiling on Iraq's oil supply

Iraq's oil production and exports have consistently lagged behind the potential implied by its resources. In the early years, Iraq's oil output was determined by the investment choices of the Iraq Petroleum Company (IPC), made up exclusively of international companies, which held concessionary rights to oil production covering almost the entire country. The position of IPC went into decline after the Iraqi revolution of 1958, but the process of establishing greater national control over Iraq's resource wealth was a prolonged one, which lasted well into the 1970s. Following the creation of the Iraq National Oil Company (INOC) in 1964, Iraq started to develop the national expertise to operate and expand the country's oil output, with foreign companies eventually assisting as contractors. But the process was slow and the net result was that Iraq missed out on the region's oil boom that started in the late 1960s. In the mid-1960s, Iraq's oil output (1.3 mb/d) did not diverge widely from that of Saudi Arabia (2.2 mb/d). But Iraq had reached only 2 mb/d in 1973, by which time Saudi production was close to 8 mb/d.

Iraq had some success in the 1970s, bringing oil production towards the strategic goal formulated at that time of creating production capacity of 5.5 mb/d by 1983. But this goal was never realised, because of the Iran-Iraq war. A subsequent plan to raise capacity to 6 mb/d by the mid-1990s was, again, not realised, this time because of the 1991 Gulf war. National ambitions now extend beyond the levels foreseen in earlier plans, based on the availability of new technologies and a willingness to bring in international expertise.

Figure 13.5 ▷ **Iraq average daily oil production and transportation, June 2012**

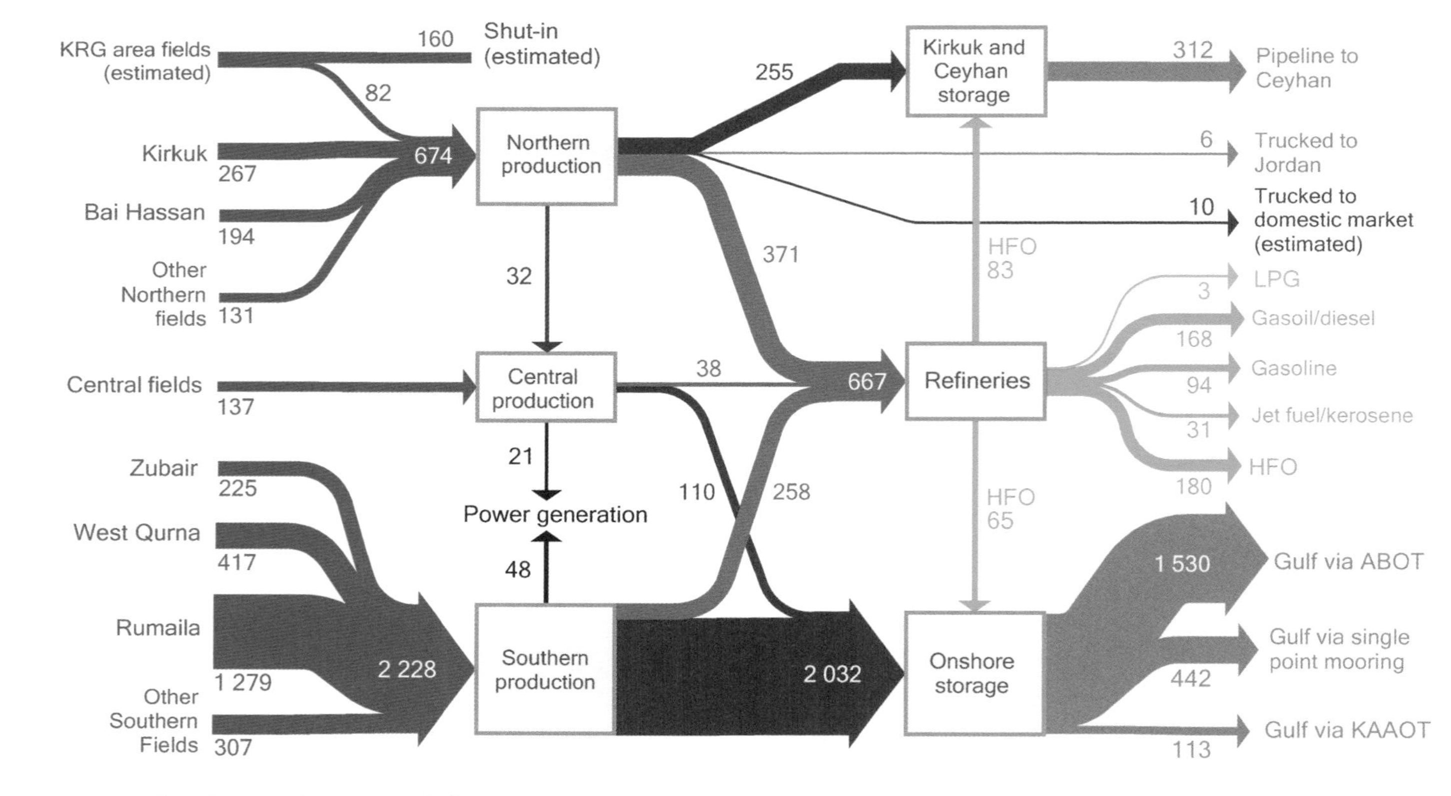

Notes: Numbers shown on diagram are in kb/d. Changes in storage levels account for flow imbalances. Northern, central and southern production refers to regions in Iraq as shown on the map in Figure 14.1. NGL production is included. ABOT = Al-Basrah Oil Terminal; HFO = heavy fuel oil; KAAOT = Khor al-Amaya Oil Terminal.

Sources: Based on direct communication with Iraq's Ministry of Oil and company reports; IEA analysis.

More than 70% of Iraq's oil production comes from fields that are being operated by international oil companies under technical service contracts (see the later section, legal and institutional framework). Nineteen of these contracts have been awarded by the federal authorities, one in 2008 and the reminder as a result of four national licensing rounds since 2009. These contracts cover all the main southern fields as well as smaller oil and gas projects elsewhere in Iraq (but not Kirkuk). If all of these contracts deliver on their oil plateau production commitments, the result would be dramatic growth in Iraq's productive capacity in the coming years, with potential output of more than 12 mb/d from these projects alone before the end of the decade. In addition, the Kurdistan Regional Government (KRG) in the north of Iraq has concluded around 50 contracts, almost all of which are production-sharing contracts, covering areas in which there has hitherto been no substantial commercial production.[3] The legitimacy of these contracts has been contested by the federal authorities, but exploration activities are, in many cases, well underway and some production capacity is already in place.

Oil products

Oil supply to domestic refineries of around 670 kb/d in June 2012 was slightly higher than the 630 kb/d average delivered in 2011. The current nameplate capacity of Iraq's refineries stands at around 960 kb/d, but we estimate that only about 770 kb/d of this capacity is operational, with the country's three largest refineries at Baiji, Doura and Basrah accounting for around 70% of the total. These are supplemented by a large number of small topping plants, but these are unable to produce high-quality petroleum products. The downstream sector, like much of Iraq's essential infrastructure, has suffered the effects of long periods of under-investment.

The range of oil products produced by Iraq's refineries falls well short of its domestic needs and of the possibilities afforded by modern, more complex refineries (Figure 13.6). Around 45% of the products coming out of Iraqi refineries are heavy fuel oil, with gasoline accounting for less than 15% of the total. This product mix means that Iraq has to import around 8.5 million litres per day of gasoline and 2.6 million litres per day of diesel to meet demand.[4] It also has a large surplus of heavy fuel oil for which it has no domestic use or export possibilities. In 2011, Iraq blended an average of 150 kb/d of heavy fuel oil into the exported stream of crude oil, lowering its quality and price. With product yields similar to those of an average refinery in the United States, Iraq could have avoided the shortfall of gasoline and diesel and eliminated its surplus of heavy fuel oil.

3. Actual oil output from the KRG area, as shown in Figure 13.5, was significantly below production capacity due to a dispute with the federal authorities that halted exports to international markets from this region in April 2012. Export shipments restarted in early August 2012 and continue as of the time of writing, but a number of underlying issues are yet to be resolved (see later section on legal and institutional framework).

4. Preliminary data for the first half of 2012 show a sizeable increase in diesel imports compared with 2011.

Figure 13.6 ▷ Average refinery product slate in Iraq compared to the United States, 2011

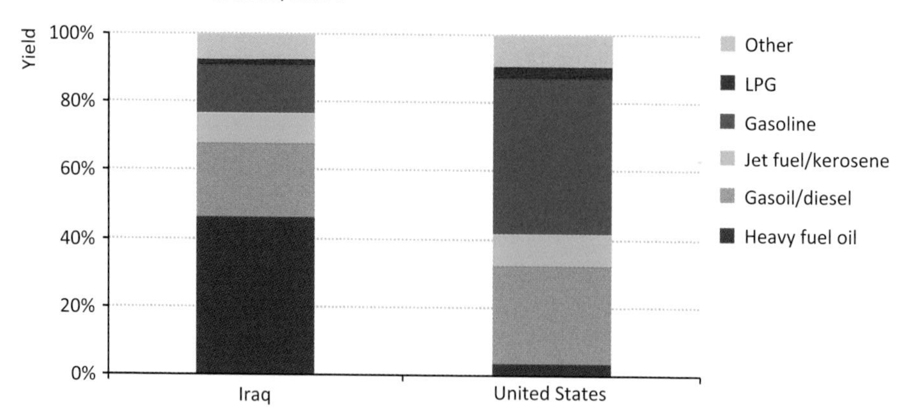

Sources: Iraq Ministry of Oil; US Energy Information Administration (EIA); IEA analysis.

Natural gas

Iraq's gas production is dominated by associated gas and has therefore followed the rollercoaster profile of oil output. Historically, much of this gas was flared: Iraq began to invest in large-scale gas processing facilities only in the 1980s and maintenance and expansion of these facilities has not kept pace with the volumes produced. In June 2012, nearly 2 billion cubic metres (bcm) of gas were produced, with around 55% coming from southern oilfields (Figure 13.7). However, we estimate that, due to the lack of gas processing capacity, more than half of the gas produced was flared (rather than marketed and consumed productively). This monthly estimate is consistent with our estimates for 2011, where total production was around 20 bcm, of which around 12 bcm was flared. Flaring is hugely wasteful given the continuing shortfall in electricity supply in Iraq and has damaging environmental effects. Putting gas gathering and processing facilities in place, developing the gas transmission network and bringing online new gas-fired power plants are therefore urgent priorities for the authorities.[5]

Transport, storage and export infrastructure

The condition of Iraq's transport, storage and export infrastructure, although improving, has been a serious constraint on progress in the energy sector. Processing facilities, pumping stations, storage tanks and pipelines, where not destroyed outright, saw their condition deteriorate substantially in the 1990s, as international sanctions interfered with proper maintenance. Some spare parts and equipment were brought in under the terms of the United Nations Oil-for-Food programme and inventive and improvised solutions were often found to keep equipment running, but Iraq's effective oil export capacity is still below the levels reached in 1979.

5. Iraq has also joined the Global Gas Flaring Reduction Partnership, which supports efforts to reduce flaring.

Figure 13.7 ▷ Iraq monthly gas production and transportation, June 2012

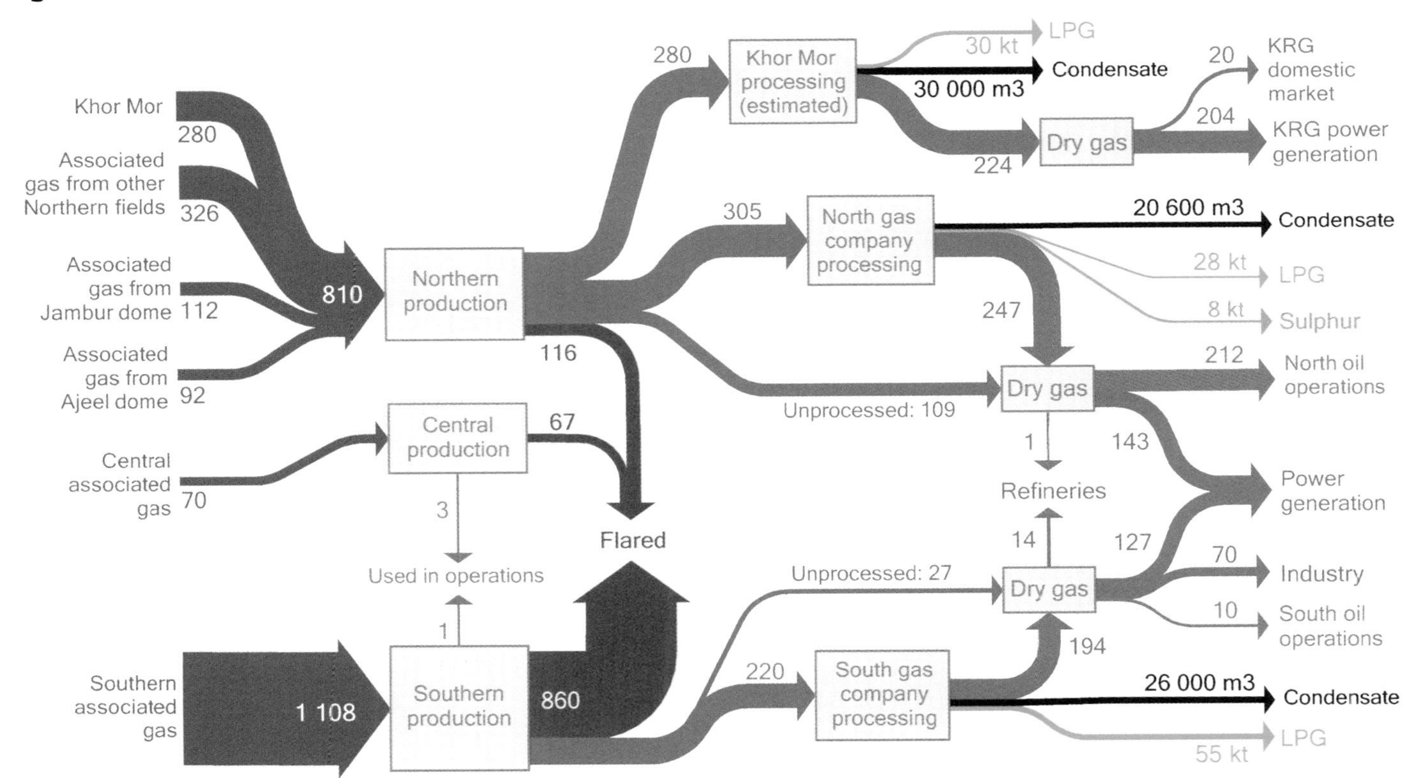

Notes: Total production for the month of June 2012. All numbers shown on diagram are in million cubic metres (mcm) unless otherwise stated. Northern, central and southern production refers to regions in Iraq as shown on the map in Figure 14.1.

Sources: Based on direct communication with Iraq's Ministry of Oil and company reports; IEA analysis.

13

The lack of sufficient storage in the south of Iraq is a particular problem, as it means that any delays or weather-related interruptions to loading tankers at the offshore facilities can lead directly to a shut-down of production. In 2011, production at southern oil fields was curtailed regularly in response to infrastructure constraints. There are signs that this is improving, with new storage capacity being built at the main export depot at Fao and fewer curtailments observed so far in 2012; but further timely expansion of transportation and storage capacity is essential for Iraq's oil output growth.

Several projects are underway and southern export capacity was expanded in early 2012, when the commissioning of new offshore crude loading facilities, called single-point mooring systems (SPMs), brought nominal export capacity up to nearly 3.4 mb/d. However, not all of this can yet be utilised due to bottlenecks elsewhere. For the month of June 2012, around 442 kb/d of crude oil was exported by tanker via these new SPMs, using just over one-quarter of their capacity. The remainder of Iraq's sea-borne exports went through the existing offshore terminals, Al-Basrah Oil Terminal (ABOT) and Khor al-Amaya Oil Terminal (KAAOT). Iraq's other main export route is the northern pipeline route to Ceyhan, but only one of the twin pipelines is operational. The maximum available capacity along this route, according to the Ministry of Oil, is 600 kb/d, much less than the nameplate capacity of 1.6 mb/d. Actual flows averaged just over 300 kb/d in June 2012.

The southern fields are connected to central and northern Iraq via the Strategic Pipeline, a reversible, domestic link, which was intended to give Iraq the option of pumping Kirkuk crude southwards for export via the Gulf or of pumping southern crude northwards, for export via Turkey or to refineries around Baghdad. Damage to parts of the pipeline mean that it is currently used only to transport oil for domestic purposes, principally from the South Oil Company to the refinery at Doura, near Baghdad. The Strategic Pipeline is designed to carry 850 kb/d, but operating capacity is estimated to be much lower: actual flows in June 2012 were below 200 kb/d. Attempts to diversify oil export options have largely fallen victim to conflict and regional politics: a pipeline from southern Iraq to the Saudi Arabian port of Yanbu on the Red Sea was commissioned in 1990, but closed after Iraq's invasion of Kuwait and later expropriated by Saudi Arabia. Another westward export system, built to the Mediterranean ports of Banias in Syria and Tripoli in Lebanon, likewise ran into political difficulties and has been largely inoperative since 1982.

Despite the problems with infrastructure, Iraq has been an increasingly important supplier of oil to global markets. Oil exports have rebounded in recent years, making Iraq the third-largest oil exporter in the world (after Saudi Arabia and Russia). An increasing share of these exports has been directed to fast-growing Asian markets, rising from 32% in 2008 to 52% by 2011 (Figure 13.8), while the shares going to North America and to Europe fell over the same period to 26% and 22% respectively.

Figure 13.8 ▷ Iraq crude oil exports by destination market

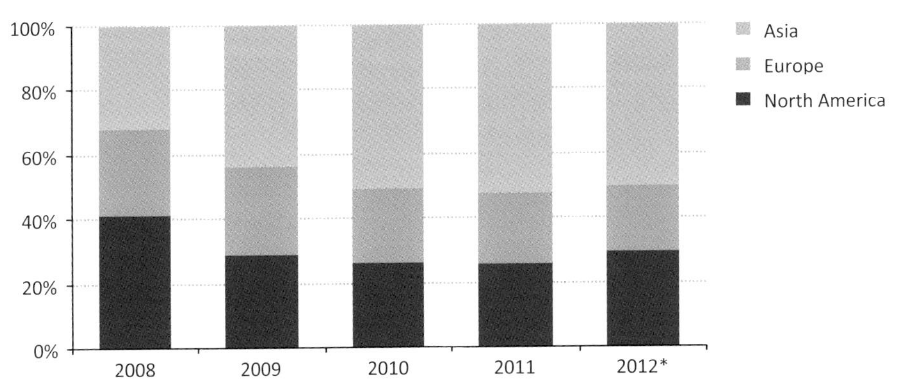

* Based on first five months.

Sources: Direct communication with Iraq's State Oil Marketing Organization; IEA analysis.

Overview of energy demand

Energy consumption in Iraq has nearly quadrupled over the last three decades. But the rate of growth has been much less strong than elsewhere in the Middle East and the pattern of energy use has been frequently disrupted. As of 2010, primary energy demand for the country as a whole was 38 million tonnes of oil equivalent (Mtoe), or 1.3 tonnes of oil equivalent (toe) per capita. The per capita figure is lower than the global average of 1.9 toe and only a little more than one-third of the level in the rest of the Middle East (Figure 13.9). However, relative to the size of Iraq's economy, low productivity and efficiency means that national energy consumption is actually high by global standards: Iraq uses 0.4 toe to produce each $1 000 of national output (based on market exchange rates in year-2011 dollars), nearly one-fifth higher than the Middle East average and twice the global average.

Figure 13.9 ▷ Primary energy demand per capita in Iraq and the Middle East

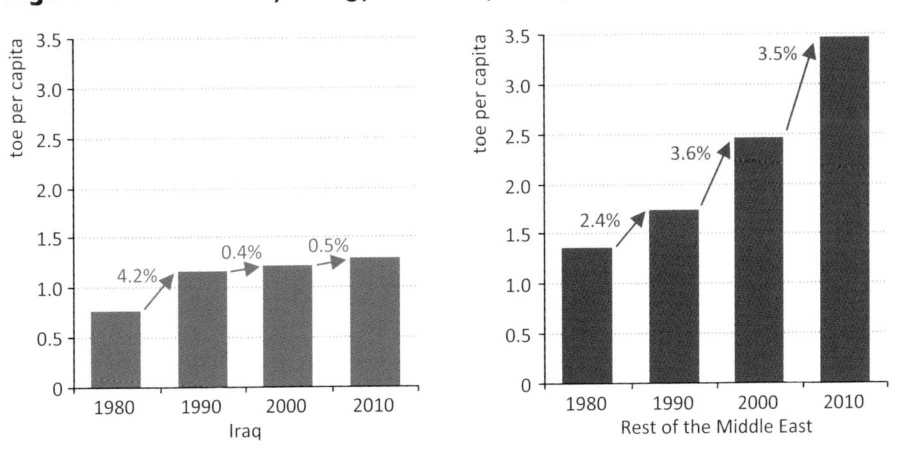

Note: The percentage figures are compound average annual growth rates.

Iraq's domestic energy consumption is dominated by fossil fuels. Over time, the general trend in the Middle East has been towards greater consumption of gas in the energy mix, gas often replacing oil for power generation as well as taking a greater share of industrial energy use. By 2010, the share of oil in the Middle East's primary energy demand (excluding Iraq) had fallen below 50% (Figure 13.10). Although Iraq has plans to increase gas production and use, oil still accounts for around 80% of primary energy demand. As elsewhere in the region, the share of non-fossil fuels in Iraq's primary energy mix is small, although there is some electricity generated from hydropower plants in the north of Iraq. These plants have gross installed capacity of 2.3 gigawatts (GW), but operating capacity is estimated at less than 1.5 GW, due to a combination of low water levels in reservoirs upstream, constraints imposed by the need to match irrigation flows and safety concerns (as in the case of Iraq's largest facility at Mosul).

Figure 13.10 ▷ Evolution of the energy mix in Iraq and the Middle East

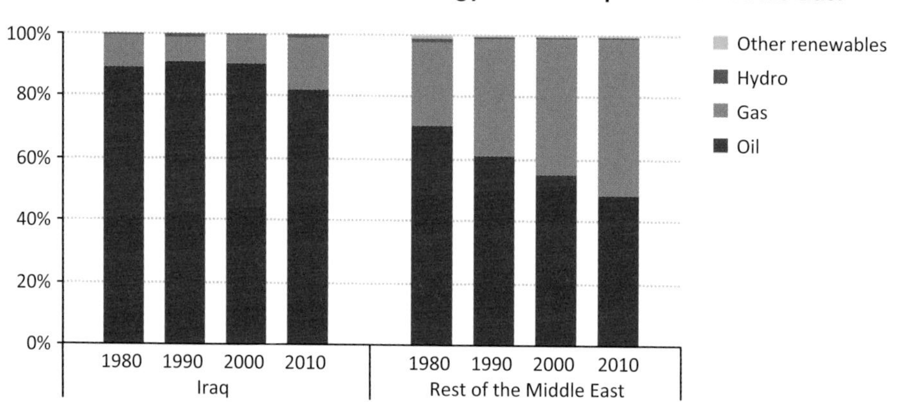

On the basis of the available data (which have limitations), we have constructed a domestic energy balance for Iraq in 2010 (Figure 13.11). This shows that the sector consuming the largest amount of oil is transport, followed by power generation and buildings. Energy use in transport, which accounts for around 60% of total final consumption, is dominated by road transport. Iraq has seen strong growth in demand for gasoline as car ownership levels have surged in recent years; the country now has an estimated three million passenger light-duty vehicles, more than half of which have been imported since 2003. Most of the oil consumed in the main grid-connected power stations is heavy fuel oil (although some crude oil and gasoil is also used), while substantial amounts of gasoil are used to fuel private generators and large grid-connected diesel units. Oil use in buildings consists largely of liquid petroleum gas (LPG), which is widely used as a fuel for cooking, and kerosene for winter heating. Oil products are subsidised in Iraq, with the domestic price of gasoline being 28% of the price paid for imports and that of diesel 11%. The government took steps to reduce oil product subsidies in 2007 and this appears to have helped dampen demand, but we estimate that the value of subsidies to Iraq's oil consumption, based on the prices available on international markets, was around $20 billion in 2011 – equivalent to around thirteen weeks of oil export revenue.

Figure 13.11 ▷ Iraq domestic energy balance*, 2010 (Mtoe)

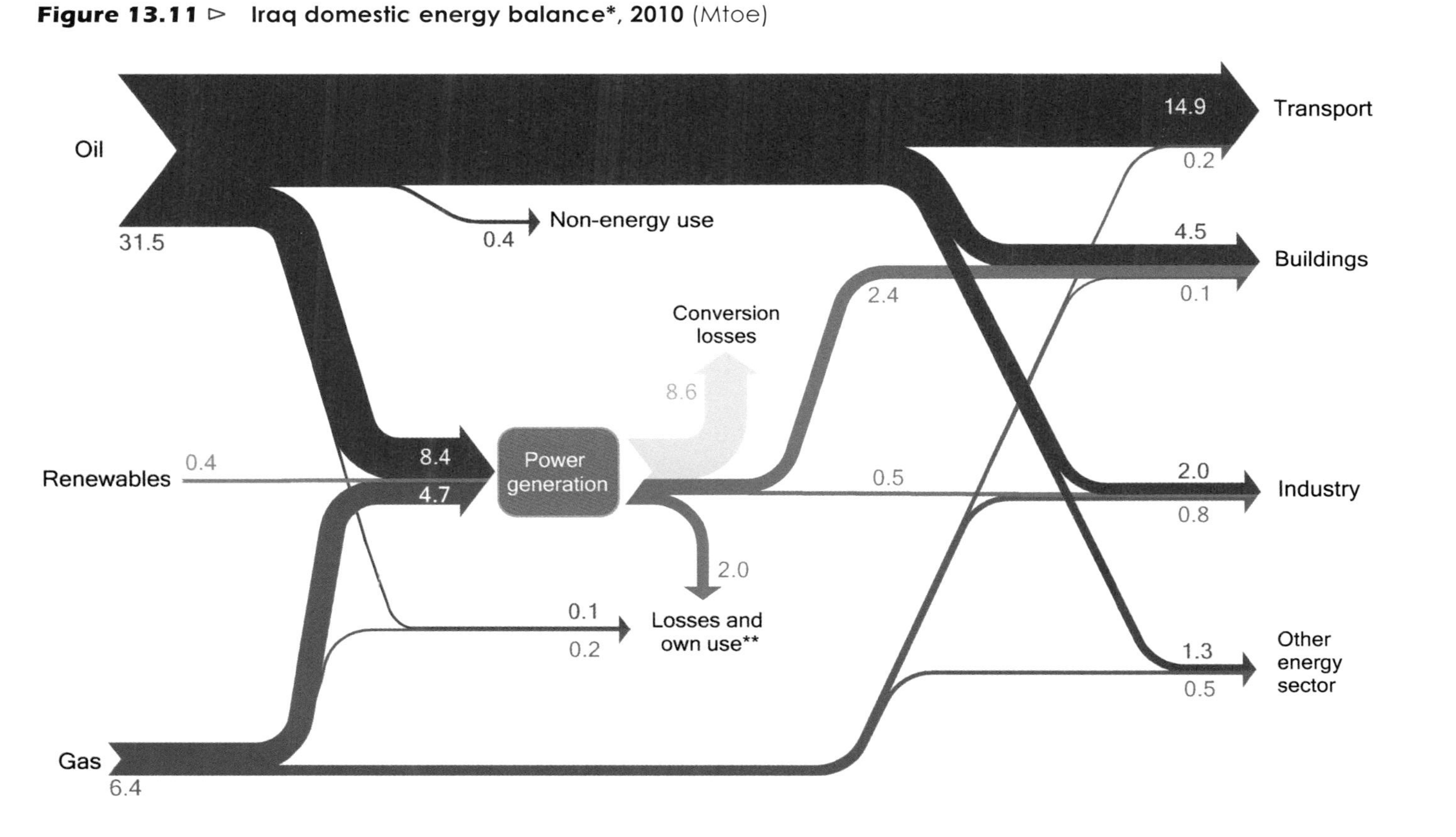

* Oil exports, oil product exports/imports and electricity imports not shown. Gas flaring not shown. ** Includes losses and fuel consumed in oil and gas production, generation lost or consumed in the process of electricity production, and transmission and distribution losses.

Industrial energy use is very low in Iraq at just over 3 Mtoe and has been stagnant at around these levels since the early 1990s. Iraq's industrial base consists largely of state-owned assets across a range of sectors, from heavy industry (including iron and steel, fertiliser, chemicals, petrochemicals, cement, glass and ceramics) to light industry (textiles, leather, furniture, dairy products, etc.), but with outdated technology that makes them ill-equipped to compete in a market environment. Many of these enterprises are operating at well below their nominal capacity (or not at all). The government has tried to attract investment to rehabilitate (and in many cases reopen) facilities and modernise equipment, but with limited success so far.

Electricity

One of the main obstacles to Iraq's economic and social development is the lack of reliable electricity supply. Despite a significant increase in grid-based electricity capacity in recent years (peak net daily production in 2011 was around 70% higher than 2006), it is still far from being sufficient to meet demand. We estimate that the net capacity available at peak in 2011 was around 9 GW while the estimated net capacity required to meet peak demand was 15 GW, resulting in a need for around 6 GW more available capacity – an increase of around 70% (Figure 13.12). This is before taking account of the increase in demand likely to occur as the electricity supply becomes more reliable. Building additional generation capacity, and ensuring that it has adequate supplies of fuel, is the immediate priority for the power sector in Iraq.

Figure 13.12 ▷ Iraq difference between gross installed generation capacity and available peak capacity, 2011

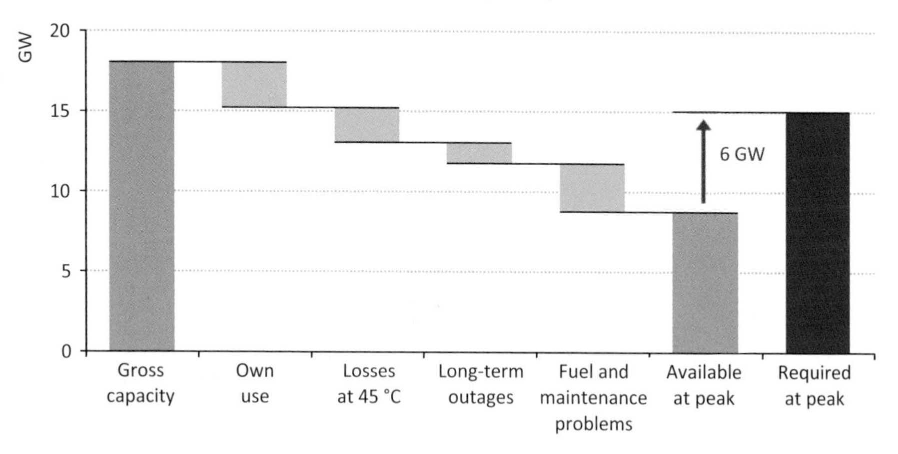

Notes: At high ambient temperatures, the maximum achievable output from thermal power plants declines compared to their production under normal conditions. In Iraq, peak demand occurs on hot summer days, when the ambient temperature is typically around 45 °C. Because of this, the theoretically available capacity of power plants has to be reduced at times of peak demand to reflect their output at this temperature.

To try and fill some of the electricity supply gap, around 90% of Iraqi households supplement the public network with private generators, either a private household generator or a shared generator operating at neighbourhood level (IKN, 2012). The generation provided from such sources is difficult to quantify, but we estimate that in 2011 shared generators accounted for 3 terawatt-hours (TWh), on top of the 37 TWh of consumption that came from the grid. In central Baghdad alone, a 2009 survey estimated that approximately 900 megawatts (MW) of private generation was available for use (Parsons Brinckerhoff, 2009a). Private generators currently play an important role in reducing Iraq's shortfall in electricity supply (helping to reduce the number of blackouts) and also to bring benefits in terms of flexibility and providing electricity access to rural areas. However, even though private generators receive subsidised fuel from the government, the price of the electricity they provide to consumers is considerably higher than grid electricity: the same 2009 survey suggested that residential customers were paying ten to fifteen times more for the electricity supplied from private generation than from the grid. As well as being an expensive way to provide electricity, diesel generators also contribute to local air pollution.

Even with the use of non-grid generation, the average availability of electricity to end-users (from all sources) was limited in 2011 to around eleven to nineteen hours per day, varying across the country (Figure 13.13). Electricity supply in the Kurdistan Regional Government (KRG) area is increasingly reliable, although cut-offs still occur at times of peak demand. Reliability is also higher than average in some southern governorates, notably in Basrah. At the other end of the scale, electricity supply from all sources was reported to be twelve hours per day or less in six governorates, accounting for 30% of Iraq's population. An additional challenge for Iraq is that electricity demand is seasonal, with the highest peak occurring in the summer months as a result of very high temperatures in much of the country. During the summer, peak hourly electricity demand could be expected to reach levels around 50% above the average demand level, increasing the gap between grid-based electricity supply (operating at capacity) and demand.

Existing generation, distribution and transmission infrastructure is in need of rehabilitation and upgrading, as well as rapid expansion, to catch up with and meet growing demand for electricity. Just over half of Iraq's existing nominal generation capacity pre-dates 1990; a further 47% has been added since 2000; a mere 2% (250 MW) was added during the entire 1990s, during which time the state of the existing generation stock also deteriorated significantly. Recent capacity additions have helped to improve the overall efficiency of the sector, but it is still low by international standards and conversion losses represent a significant item in Iraq's energy balance (Figure 13.11). The efficiency of gas-fired plants in particular is currently 31% (compared with around 55% achievable for new combined-cycle gas turbines under good operating conditions). There are largely separate electricity grids for the KRG area and the rest of Iraq, with distinct strategies for their development. There are also different policy approaches to providing grid electricity within Iraq, with independent power producer (IPP) contracts having played a useful role in increasing electricity supply from the grid in the KRG area in recent years.

Figure 13.13 ▷ Iraq source and reliability of electricity supply by governorate, 2011

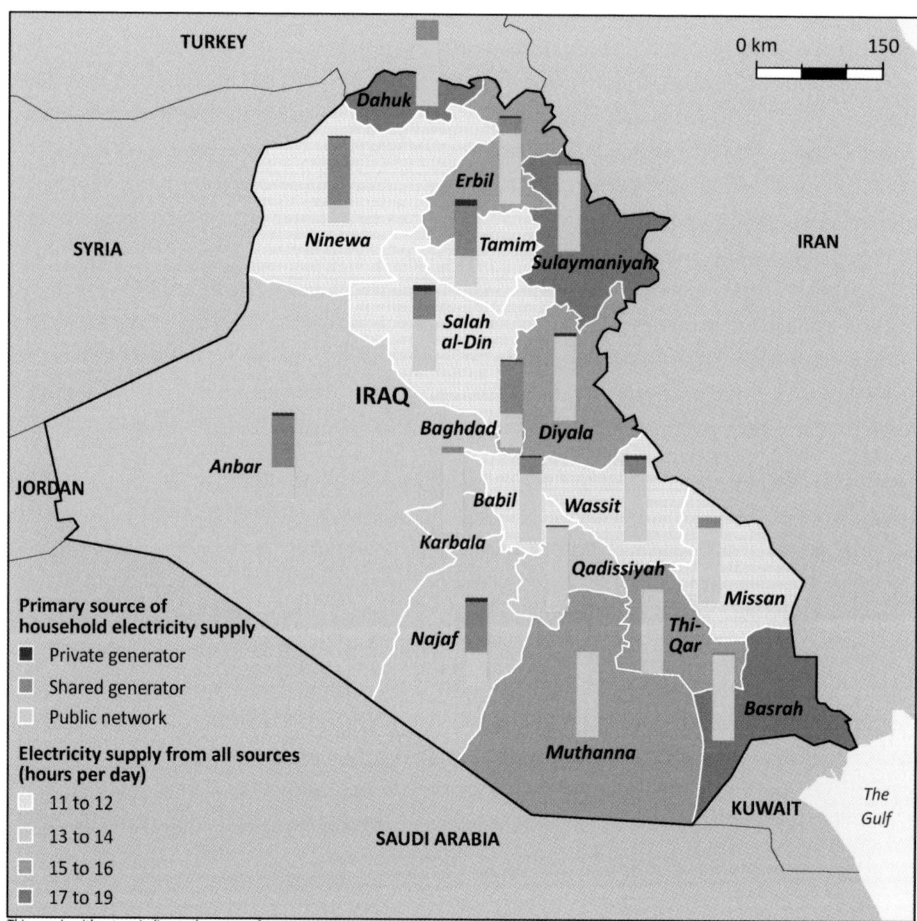

This map is without prejudice to the status of or sovereignty over any territory, to the delimitation of international frontiers and boundaries and to the name of any territory, city or area.

Sources: IKN (2012); IEA analysis.

Electricity transmission and distribution losses in Iraq are the highest in the Middle East region (Figure 13.14), in large part because of damage sustained during the 1991 Gulf war, subsequent sabotage and a lack of maintenance. The performance of the high-voltage transmission system remains relatively robust, with the network of 400 kilovolt (kV) transmission lines in reasonable working order in terms of functioning and reliability, though not in terms of attaining modern levels of efficiency. Cross-border connections with Iran and Turkey allow for some electricity imports, which amounted to almost 6 TWh in 2010. However, the distribution network, which is designed to connect 98% of Iraqi households to the grid, has been severely degraded.

Figure 13.14 ▷ Electricity losses by type in selected countries, 2010

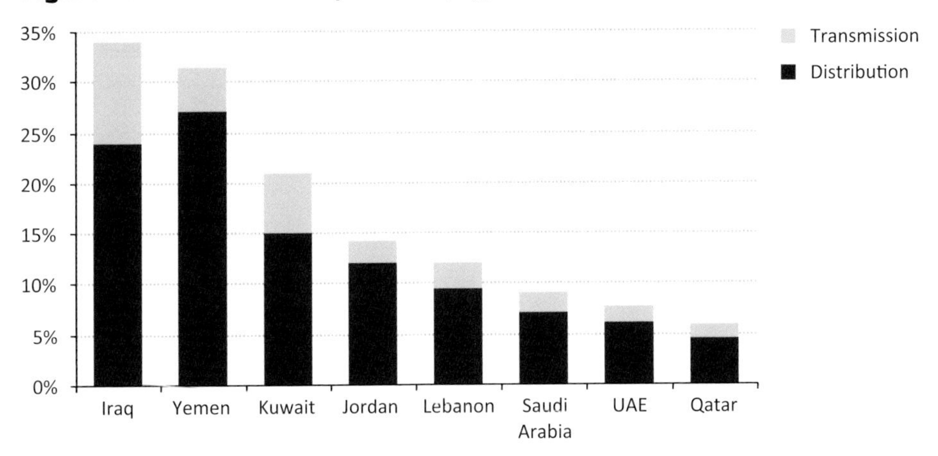

Note: The split between transmission and distribution losses is estimated for Saudi Arabia and the UAE.
Sources: Arab Union of Electricity (2010); IEA databases and analysis.

Prices for electricity from the grid are heavily subsidised in Iraq, reflecting historical pricing patterns and the political impossibility of raising prices while the quality of service is so low. Electricity provided from the grid to household consumers is charged on a sliding scale, starting from $0.017 per kilowatt-hour (kWh) for consumption up to 1 000 kWh per month, with higher prices for consumption above this threshold. Average per household consumption in Iraq's residential sector in 2010 (the last year for which we have data) was only 800 kWh per year, so most households consume all or most of their electricity at the lowest price band. This price is almost ten times lower than the OECD average household electricity price in 2010 of $0.16 per kWh. Since grid-based electricity is relatively cheap compared with electricity sourced from private and shared generators, end-users have an incentive to concentrate their consumption, so far as possible, at times when grid-based electricity is available, exacerbating problems with system reliability. Payment discipline is also relatively weak – it is estimated that tariff revenue is collected only on around one-third of the electricity that enters the distribution network – and there are many illegal network connections. The high proportion of residential consumers who use grid electricity without paying at all means that the effective rate of subsidy is even higher than is implied by the low tariff level. The tariff for industrial users is a flat rate of $0.10 per kWh, much higher than for the residential sector.

The context for Iraq's energy development

The economy

The heavy dependence of Iraq's economic development on the success of its oil sector makes the economy highly dependent on the prevailing conditions in the global oil market. Monthly oil export revenues are subject to sharp swings (Figure 13.15); price falls in late-

2008 and early-2009 cut Iraq's revenue sharply, forcing the federal government to cut back quickly on planned expenditure. Oil revenue volatility is typically transmitted to the non-oil economy through fiscal policy, which means that the sound management of national finances is essential to mitigating this risk. Figure 13.15 also shows the high dependence of Iraq's economy on its sea-borne exports in the south.

Figure 13.15 ▷ Iraq monthly oil export revenues by route

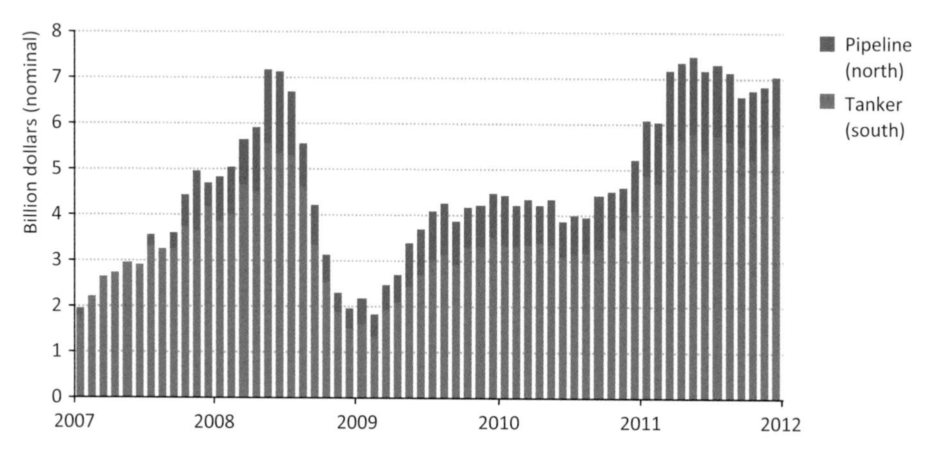

Note: Iraq also receives additional revenue from exporting a small amount by truck.

Sources: Direct communication with Iraq's State Oil Marketing Organization; IEA analysis.

As in many oil exporting countries, the oil price that Iraq requires in order to match oil revenues to planned government expenditure – its fiscal breakeven price – has increased in recent years and is getting closer to the actual market prices. This increases its vulnerability to a decline in oil prices. Governments with large financial reserves (such as central bank reserves, used primarily for backing a national currency, or sovereign wealth funds, often used to serve a broader range of government objectives) can potentially use these resources to help smooth out mismatches between government revenues and expenditures, but Iraq's ability to do this is relatively limited. Based on 2011 data, Iraq's central bank reserves were less than oil revenues in the same year, whereas Saudi Arabia had closer to double the coverage, Kuwait more than three times and the United Arab Emirates seven times (Figure 13.16).

While oil export revenues are equivalent to more than 70% of GDP, the oil and gas sectors account directly for less than 2% of total employment. Although there are many jobs created indirectly (via suppliers, service companies and so on), the largest impact that oil has on the broader labour market is via the revenues that enable Iraq to maintain one of the largest public sectors in the world relative to its population. The public sector workforce has more than doubled since 2003 and the associated wage bill has increased as a share of GDP, from around 15% in the mid-2000s to 30% in 2012. Sustainable and balanced growth in the economy requires increased diversification away from oil, into other industrial sectors and services, as well as the improved performance of labour-intensive sectors such as agriculture. Policies to facilitate growth in the private sector, particularly small and medium enterprises are required. However, the business environment in Iraq, as it stands today, puts a brake on entrepreneurial activity: the costs of doing business are high relative to those in many other countries, mainly as a result of security concerns, the unreliable provision of essential services (such as electricity), regulatory obstacles, bureaucratic delays and poor access to finance. Iraq was ranked 164 out of 183 countries in the *Doing Business 2012* survey (World Bank, 2012).

Figure 13.16 ▷ Number of years cover for oil revenues from financial reserves in selected countries, 2011

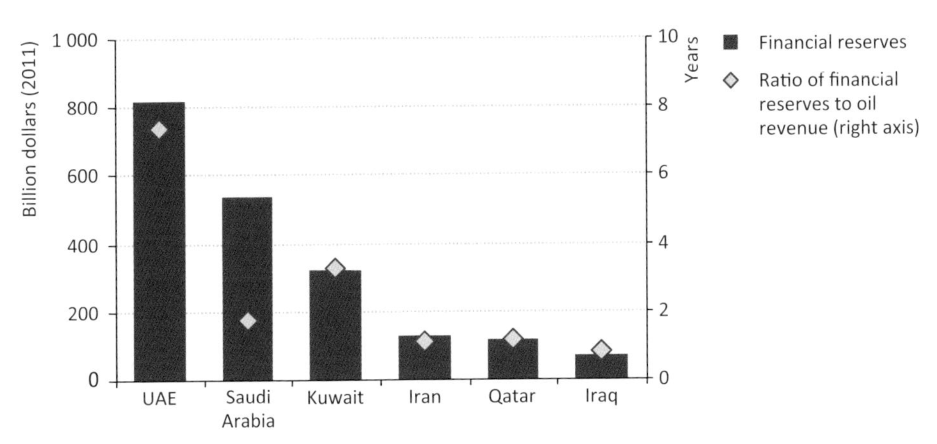

Notes: Financial reserves estimates include central bank reserves (used primarily for backing a national currency) and, where appropriate, the estimated value of sovereign wealth funds (often used to serve a broader range of government objectives). While Qatar has a relatively low number of years cover, it also has the lowest fiscal breakeven oil price of the countries shown.

Sources: IMF, SWF Institute and Development Fund for Iraq databases; IEA analysis.

Legal and institutional framework

The legal framework for the energy sector operates within the overall boundaries established by the 2005 constitution, which establishes a federal democratic system of governance, with the KRG area accorded the status of a federal region. According to the constitution, oil and gas resources are "owned by all the people of Iraq in all the regions and governorates". The federal government, with the producing governorates and regions, is tasked with the management of oil and gas extracted from present fields and with the necessary strategic policies to develop Iraq's oil and gas wealth in a way that achieves the highest benefit to the people. However, the constitution does not explicitly cover the question of jurisdiction over hydrocarbon exploration and development, and has been subject to various interpretations. In the case of electricity production and distribution, the environment and water resources, the constitution shares responsibility between the federal and regional authorities.

Various drafts of new federal hydrocarbons legislation have been under discussion since 2006.[6] Under these drafts, co-ordination between the federal level and the regions would be entrusted to a Federal Oil and Gas Council, but no consensus has yet been reached on the Council's composition and competences.[7] The complex debate over control and decision-making authority in the hydrocarbons sector has become intertwined with broader issues, notably question of revenue sharing between the centre and the regions. The package of hydrocarbon laws under consideration also includes possible reforms to the sector's institutional structure, notably the (re-)creation of an Iraq National Oil Company (INOC) as a state-owned company managing all of the state interests in current and new fields and operating with a degree of financial and administrative autonomy. The delay in passing new hydrocarbon laws means that for the moment a federal system of resource development (based on technical service contracts) co-exists uneasily with the approach followed by the Kurdistan Regional Government (based on production-sharing contracts consistent with its own legislation from 2007), whose legitimacy has been contested by the federal government (Box 13.2).

Box 13.2 ▷ Technical service contracts and production-sharing contracts

Technical service contracts and production-sharing contracts are among the instruments used by major resource holders to develop their oil and gas. Direct comparisons between them are complicated by the variety of terms and the impact of key variables, notably the oil price and the size of field. But, at their core, the difference is that technical service contracts provide a fee per barrel of oil produced to the contractor as remuneration for work done (although the timing and nature of this remuneration can vary considerably), while production-sharing contracts provide returns based on the value of the oil/gas found or produced. The production-sharing contract holder therefore sees both the upside benefit and downside risk of oil price fluctuation, as well as geological, technical and other market risks. In both types of contract, the financing of exploration and development costs is borne by the outside investors (state investors are generally "carried" by the outside investors) until the resource is brought to market, at which point developers receive remuneration and cost recovery in the form of either cash or oil production (payment in kind). Resource holders with discovered fields of low geological risk often favour technical service contracts. Offering fixed remuneration technical service contracts for exploration is less common (but not unknown), as it can be difficult to strike the right balance between risk and reward without a clear view of the size and quality of potential resources.

6. In the absence of new legislation, a range of laws that predate the constitution (some of which date back decades) are also still applicable.

7. In August 2012, a special parliamentary committee was established in Iraq with the intention of trying to overcome the impasse in discussions over new federal hydrocarbon laws.

Nineteen technical service contracts have been awarded by the federal government to date, each to a consortium led by an international operating company (Table 13.1). One of Iraq's state-owned operating companies is present, with a 25% stake, in each of these consortia. Key parameters of the technical service contracts are the initial production target ("initial target" in Table 13.1), the plateau production commitment ("plateau target") and the maximum remuneration fee per barrel ("max. fee"). The initial target is important because it provides the trigger for the reimbursement of costs and the payment of fees, incentivising operators to reach this level as quickly as possible. Four projects (Ahdab, Rumaila, West Qurna Phase I and Zubair) have already reached this initial threshold. The maximum remuneration (in dollars per barrel of oil equivalent) is fixed in each contract, with the level falling as the overall profitability of the contract increases (measured in terms of the ratio of cumulative income to cumulative expenditure). The plateau production commitment is the amount that the consortium has agreed to produce, although the amounts that have been contracted are considered in many cases to be higher than those likely to be achieved in practice: the remuneration fee can be reduced in the event that production during the plateau production period is lower than agreed.

Table 13.1 ▷ Contracts awarded by federal authorities for hydrocarbon exploration and development

				Production*			
Bid round	Project or licensing block	Operator	Type	Initial Target	June 2012	Plateau target	Max. fee**
2008	Ahdab	Petrochina	Oil	25	129	140	6.00
One (2009)	Rumaila	BP	Oil	1 173	1 279	2 850	2.00
	West Qurna (I)	ExxonMobil	Oil	268	417	2 825	1.90
	Zubair	Eni	Oil	201	225	1 200	2.00
	Missan Group	CNOOC	Oil	97	91	450	2.30
Two (2009)	West Qurna (II)	Lukoil	Oil	120	-	1 800	1.15
	Majnoon	Shell	Oil	175	21	1 800	1.39
	Halfaya	Petrochina	Oil	70	34	535	1.40
	Gharraf	Petronas	Oil	35	-	230	1.49
	Badra	GazpromNeft	Oil	15	-	170	5.50
	Qairayah	Sonangol	Heavy oil	30	2	120	5.00
	Najmah	Sonangol	Heavy oil	20	-	110	6.00
Three (2010)	Akkas	KOGAS	Gas	1.03	-	4.1	5.50
	Mansuriyah	TPAO	Gas	0.78	-	3.1	7.00
	Siba	Kuwait Energy	Gas	0.26	-	1.0	7.50
Four (2012)	Block 8	Pakistan Petroleum	Gas-prone	n/a	-	n/a	5.38
	Block 9	Kuwait Energy	Oil-prone	n/a	-	n/a	6.24
	Block 10	Lukoil	Oil-prone	n/a	-	n/a	5.99
	Block 12	Bashneft	Oil-prone	n/a	-	n/a	5.00

* Production figures are in kb/d for oil projects and bcm per year for gas projects. ** The maximum remuneration fee (Max. fee) is in US dollars per barrel of oil equivalent.

Commentators largely agree that the federal authorities have driven a hard bargain with international oil companies in these contracts. The agreed maximum remuneration fees are at low levels per barrel of oil produced, meaning that the overwhelming share of the revenue generated is retained by the government; but this is offset, in part, by the prospect of high volumes, the expectation that the fields are of such size and quality that there is little technical risk and the consideration that companies are not taking on price risk or exploration risk. In most cases, the bidding process and hard bargaining has been followed by contract implementation, but some companies have explicitly or implicitly reviewed their positions in the south. Statoil has sold its stake in the West Qurna Phase II project. ExxonMobil, operator of the West Qurna Phase I project, has engaged to pursue exploration opportunities in the KRG area, as have Total (part of the consortium for the Halfaya field with Petrochina) and GazpromNeft. The federal government has made it clear that no company with activities in the KRG area is allowed to bid in the national licensing rounds for projects in the rest of Iraq, but the implications for companies with existing contracts are not clear.

The KRG has awarded around 50 contracts for hydrocarbon exploration and development, mostly to medium-size international companies.[8] The bulk of these are production-sharing contracts for exploration blocks, and the terms reflect the possibility that no commercial hydrocarbons will be found and that the contract holder will carry the burden of all costs with no remuneration. The subsequent terms, after a commercial discovery has been made, are generally assessed as generous, but not unprecedented.[9] The KRG has also awarded service contracts for the development of the Khor Mor and Chemchemal gas fields, the former of which was producing at the time the contract was signed.

There are international examples of both types of contractual arrangement (technical service contracts and production-sharing contracts) within national approaches to resource development. This in itself is a reason to imagine that accord should be attainable on the appropriate contract form (or forms). However, arguments about the details of the contractual arrangements are secondary to more fundamental questions about which entities should have the power to authorise and conclude such contracts in the first place, and the consequences that arise for Iraq from the absence of a consistent country-wide policy in such a strategic sector.

The areas of difference between the federal government and the KRG have been thrown into sharper relief as oil production from the KRG area has grown, as this has raised the issue of the destination market for this production (production in the KRG area exceeds local demand) and also the payment of recoverable costs under the KRG contracts. The federal government controls the export of hydrocarbons and the associated revenue

8. Since 2010, some large international oil companies have started to acquire licences in KRG: Murphy Oil and Marathon in 2010, followed by Hess, Repsol and ExxonMobil in 2011, and stakes in existing licences acquired by Chevron, Total and GazpromNeft in 2012.

9. At an oil price of $90 per barrel, the government take on a typical KRG contract has been estimated at 83% (Genel Energy, 2011).

through the State Oil Marketing Organization (SOMO). Agreements in early-2011 resulted in some claims for costs from KRG producers being paid by the federal government and oil from the KRG area being exported through the pipeline to Ceyhan (the volumes were scheduled to be up to 100 kb/d in 2011 and up to 175 kb/d in 2012). These arrangements came under strain from continued disagreements over the appropriate mechanism for auditing costs and broader questions of revenue sharing and, as a result, exports from the KRG area via SOMO channels were suspended in April 2012. These re-started in early August and agreement was reached in September that could see export volumes from the region rise to 200 kb/d for the rest of 2012, accompanied by a resumption of cost recovery payments. However, the underlying issues relating to hydrocarbon governance remain to be fully resolved. Efforts by the KRG to market oil or gas directly to international buyers would run contrary to the federal claim to sole authority over hydrocarbon exports.

Another difficult issue that is yet to be resolved is that of the disputed territories. These straddle the borders of the "Green Line", a boundary relating to KRG-controlled territories that were recognised in the 2004 Transitional Administrative Law (the precursor to the constitution). The Kurdish presence extends beyond the Green Line, with parts of these territories thought to hold considerable oil and gas resources but, with uncertainty over who administers them, projects here could face higher legal risk.

Institutions

The key institutions for the energy sector in Iraq are the Ministry of Oil (for the hydrocarbons sector) and the Ministry of Electricity (for the power sector), which both report to the Deputy Prime Minister for Energy (Figure 13.17). These ministries combine policy making, regulatory and operational functions, including – in the case of the Ministry of Oil – direction and supervision of regional state-owned oil production companies (the South Oil Company in Basrah, the North Oil Company in Kirkuk, the Missan Oil Company in Missan, the Midland Oil Company in Baghdad), the State Oil Marketing Organization, which manages the exports and imports of crude oil and oil products, and the South and North Gas Companies, which process gas. The institutional picture could be affected substantially by the adoption of new federal hydrocarbons legislation. At a regional level, the KRG has its own Ministry of Natural Resources and Ministry of Electricity.

The need to improve Iraq's institutional capacity has been recognised as an important priority by the Iraqi government. It is closely linked to the broader issue of human resource development in Iraq. In the 1970s, Iraq had one of the highest per capita levels of post-graduate qualifications in the world, but the quality of the education system, in common with other Iraqi institutions, has deteriorated since. Moreover, many highly-trained Iraqis left the country during the years of conflict and instability. A generational gap is evident, with little parity between the senior professionals, educated before the onset of the multiple wars, and those educated since. The government is trying to remedy this situation, but it will take considerable time. In Iraq's energy institutions, there is a scarcity of skills across a number of areas, including procurement and project management, which are

Figure 13.17 ▷ Iraq main government institutions related to the energy sector

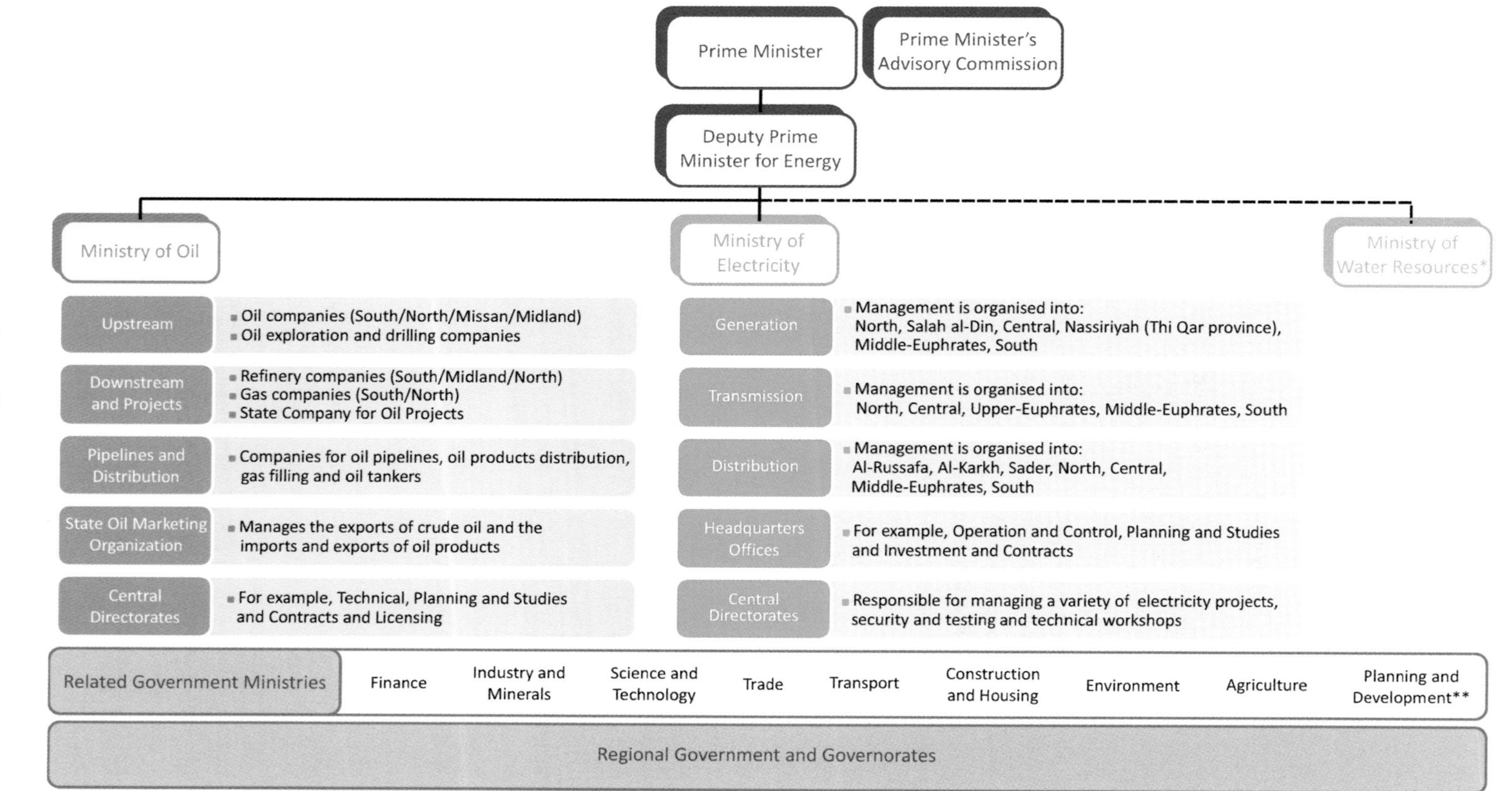

* The responsibility of the Deputy Prime Minister for Energy, but with a broader remit than energy. ** Includes Central Statistics Organization.

essential to the repair and expansion of Iraq's energy infrastructure. This can result in decision making, even on relatively minor issues, being concentrated in the hands of a small number of senior officials, a situation which can lead to procedural and project delays. Policy making is also hampered by occasionally inconsistent and incomplete energy data (Box 13.3).

Box 13.3 ▷ Iraq energy data

Obtaining robust data on energy supply and demand is an essential starting point for competent energy sector analysis, but the quality of Iraq's energy data varies widely by sector. Our research reveals that data on oil production and export are relatively reliable, though there are still some inconsistencies in the crude and product balances. Data on gas production and consumption is less readily available, particularly outside the power sector. Information on final energy consumption is generally hard to find. A further complication is that data are often collected and published separately by the KRG, risking differences in methodology and coverage which can make it difficult to establish with confidence a single dataset for the whole of Iraq.

The situation is improving thanks to the efforts of the Iraqi authorities and to the increasing number of surveys, such as household expenditure surveys and enterprise surveys, and one-off analytical studies that have been completed as inputs to policy making: examples of the latter are detailed master plans for the electricity sector (Parsons Brinckerhoff, 2009b and 2010), and master plans that are now being prepared for water and transport. Thanks to excellent co-operation from the Iraqi government, we have had access to a wealth of information for this report, which has allowed us to reconcile various anomalies and to put into the public domain a consolidated picture of Iraq's energy production and use. Continued efforts to improve the consistency and reliability of Iraq's energy data will be essential for effective policy making and the implementation of an integrated energy strategy.

Security

The security situation has been, and will continue to be, crucial to the development of both Iraq's energy sector and the economy overall. In general, the number of security incidents in Iraq has declined markedly in recent years but the risk of violence remains an important concern for companies working in the energy sector, necessitating close attention and considerable expense (Figure 13.18). The security risks are by no means uniform across Iraq, with the number of incidents in the key oil-producing province of Basrah and, even more so, in the KRG area, being lower than in and around the capital. There is also a legacy of landmines and unexploded ordnance in Iraq. Under the terms of contracts signed with the federal government, international oil companies are obliged to survey and clear any areas relevant to oil exploration and production that are suspected of being contaminated (UN Inter-Agency Information and Analysis Unit, 2012).

Figure 13.18 ▷ Total number of attacks in Iraq (weekly)

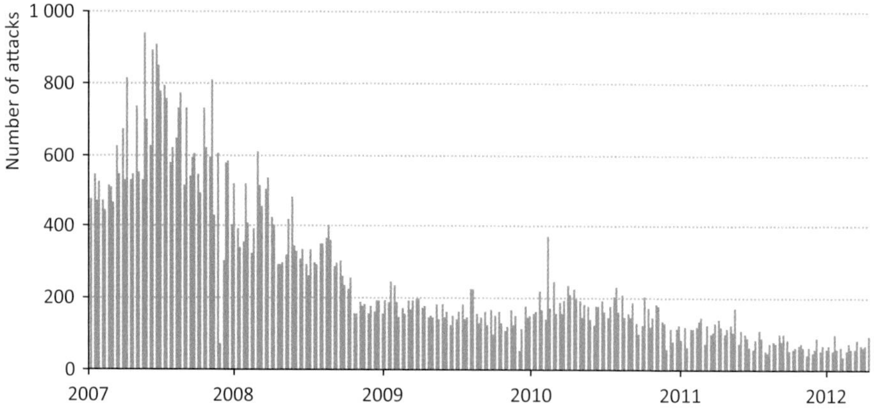

Source: UN Inter-Agency Information and Analysis Unit.

Environment and water

The natural environment has not been spared the degradation observed elsewhere in Iraq. Changing weather patterns, deforestation, poor farming and water management practices and an under-developed institutional/legal framework have all had negative impacts. The consequences are seen in such things as desertification (the decline of arid but productive land until it becomes barren), reduced supplies of clean water, reduced agricultural output and the greater prevalence of disruption from dust and sandstorms. Transport congestion, inefficient industrial plants and extensive use of diesel generators all contribute to local pollution in urban areas. Population growth, internal migration, a growing economy and the expected growth in the exploitation of Iraq's natural resources all have the potential to exacerbate these trends.

The energy sector contributes directly to the environmental challenge, be it from water/soil contamination and gas flaring in the course of oil and gas production, the extensive burning of crude oil and petroleum products in power generation (including local generation) or leaded fuels of low quality consumed by an ageing transport fleet. We estimate Iraq's energy-related CO_2 emissions to be about 100 million tonnes (Mt) in 2010 (excluding gas flaring). In addition, Iraq flared 11 bcm of gas in 2010, resulting in an estimated 25 Mt of CO_2 emissions, near equivalent to the annual emissions of 3.7 million cars. Despite steps in the right direction, including provisions in contracts with international oil companies to reduce flaring, environmental considerations remain subordinate in energy policy making.

Competition for limited resources makes water availability and use important issues in Iraq. Around 60% of Iraq's water resources come from outside its borders, in the form of the Tigris and Euphrates rivers, though water scarcity in Iraq is not attributable solely to the actions of countries further upstream: the amount of water available per capita in Iraq is still high by regional standards and this has contributed to inefficient methods and technologies for water use, exacerbated by the poor state of Iraq's existing water infrastructure and the effect of recent conflicts. Agriculture is the sector with by far the largest water needs, as well as the largest potential for more efficient use. There is no price on water for most agricultural users and inefficient irrigation methods are widespread. There are also shortages of water for households. Nearly two-thirds of households use the public network as their main drinking water source, yet 25% of these households receive less than two hours of water per day (IKN, 2012). Public network use is particularly low in the south of the country, which is downstream of all other activities and sees less rainfall. Although water requirements for the energy sector are small by comparison, Iraq's vulnerable situation, particularly in the south, could be exacerbated by the water injection needed to raise and maintain oil production and, to a lesser extent, by water use in the power sector, particularly if the water used is drawn from sources for which there are competing uses (see Chapter 14). In the case of the Kirkuk and Rumaila oil fields, fresh water is already being used for water injection.

Projecting future developments

Building on the data presented here, of the actual state of Iraq's energy economy, subsequent chapters project how the sector might evolve in the period to 2035. The basic analytical approach to those projections is similar to that taken in the rest of the *WEO-2012*, illustrating the range of possible outcomes which might derive from today's uncertainties and seeking to draw out the implications for policy makers.

The international context for the Iraq projections is the New Policies Scenario of *WEO-2012*.[10] For the Iraq analysis, the structure of our World Energy Model[11] has been adapted to accommodate some specific features of Iraq's energy situation and to allow for a more disaggregated view of oil and gas supply. New areas for quantitative analysis have also been developed, notably to derive assumptions about Iraq's GDP growth from different oil production trajectories and to assess when the gap of unmet electricity demand from the grid can be closed. The scope of the Iraq model has been narrowed in some areas where data is more limited, such as energy use in industry and households. On the supply side, the main resource-rich areas of Iraq have been modelled individually to allow for a more detailed consideration of Iraq's production outlook and potential.

We have labelled the main product of our analysis of Iraq the ***Central Scenario***; but we also develop two cases which illustrate outcomes falling either side of this scenario (Box 13.4).[12]

10. The New Policies Scenario is therefore used for all country/regional comparisons in Chapters 14-16.
11. For which a model for Iraq's energy system has been developed.
12. The Current Policies and 450 Scenarios are not developed at country level for Iraq; results for these two scenarios are presented for the Middle East as a whole.

Box 13.4 ▷ Overview of the Iraq Central Scenario and cases

The Iraq outlook sets out detailed projections for the period to 2035 (the *Outlook* period) for:

- A ***Central Scenario*** that reflects our judgement about a reasonable trajectory for Iraq's development, based on an assessment of current and announced policies and projects. In line with the assumptions underpinning the *WEO-2012* New Policies Scenario (of which the Central Scenario forms a part), we are cautious in assessing the prospects for full implementation of these policies and projects, bearing in mind the many institutional, political and economic difficulties that could arise.

In addition, we discuss at various points in the text:

- A ***High Case***, in which we take a more favourable view on the prospects for energy sector development. In this case, Iraqi oil production rises rapidly to surpass 9 mb/d by 2020 and increases further to a level around one-quarter higher than the Central Scenario by the end of the *Outlook* period. We analyse the implications of this case for investment in Iraq's energy supply infrastructure, for Iraq's domestic economic prospects and energy demand, and the potential impact on global energy markets.
- We also present findings of a ***Delayed Case***, in which investment in Iraq's energy sector rises only slowly from levels seen in 2011. For the projection period as a whole, investment is around 60% of the level in the Central Scenario, acting as a significant constraint on the pace at which the sector develops. We analyse the implications of this case for Iraq's domestic economic and energy demand prospects, and the potential impact on global energy markets.

GDP and population

Elsewhere in this *Outlook*, assumptions about GDP growth by country and region do not change between the main scenarios examined (see Chapter 1). For Iraq, however, variations in the assumptions about GDP are an integral part of our approach. The growth rates for GDP (Table 13.2) reflect the predominant role of revenues from the oil sector in the Iraqi economy and the way that different projections for oil supply correspondingly produce different outcomes for the national economy as a whole.

Table 13.2 ▷ Growth rates* for Iraqi oil output and GDP by scenario

		2000-2010	2010-2015	2010-2020	2010-2035
Central Scenario	Oil output	-1.3%	11.7%	9.6%	5.0%
	GDP	2.4%	10.0%	10.6%	6.9%
High Case	Oil output	-1.3%	20.1%	14.8%	6.4%
	GDP	2.4%	15.1%	13.8%	7.6%
Delayed Case	Oil output	-1.3%	4.6%	3.3%	1.7%
	GDP	2.4%	5.9%	5.9%	4.7%

* Compound average annual growth rates.

Iraq's troubled history over the past three decades has held back its economic development. If Iraq's economy had developed at the same rate as the Middle East average since 1980, then it would have been more than 50% larger in 2010; if it had developed over this period in line with the average for non-OECD countries, its economy would have been more than twice as large. The implication of our GDP assumptions for the Central Scenario and for the High Case is that Iraq makes up for some of this lost time (Figure 13.19). In the Central Scenario, Iraq catches up and, around 2017, overtakes its notional GDP path from 1980 based on Middle East average annual growth. In the High Case, Iraq gets back to a path closer to one based on average non-OECD growth since 1980. In the Delayed Case, however, Iraq remains short of these benchmarks.

Figure 13.19 ▷ Variations in Iraq's GDP growth by scenario

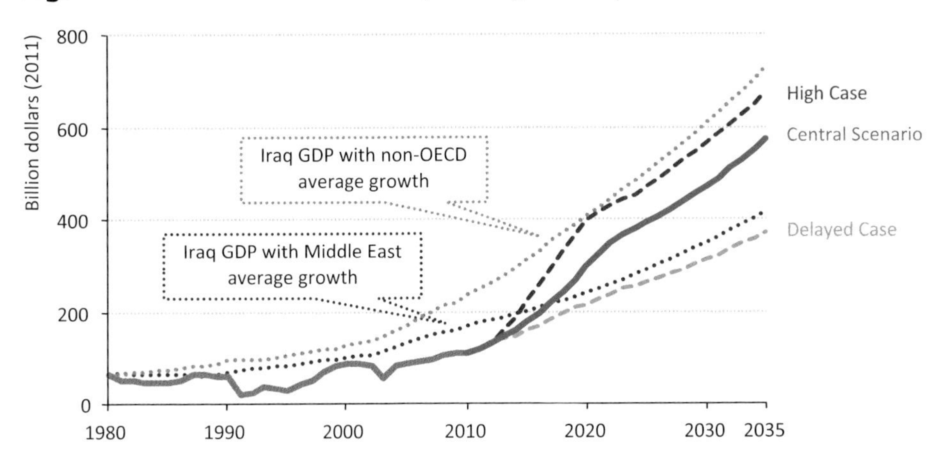

We assume in all cases that Iraq makes gradual progress with the diversification of its economy, although the oil sector continues to exercise a predominant role in GDP. The share of value added provided by services increases from current levels as Iraq's economy develops. The GDP outlooks presented here are contingent on Iraq consolidating and extending the gains achieved in recent years in terms of security and political stability. They also depend, particularly in the High Case, on Iraq managing effectively the risk that very rapid growth in oil receipts prejudices the development of other parts of the economy or of national institutions. This challenge is particularly important because of demographic trends in Iraq; the median age is eighteen (compared with a world average of 29) and more than 40% of the population is under fifteen years of age. The opportunities afforded this new generation of Iraqis will be critical. In our scenarios, we assume that the population of Iraq almost doubles over the projection period from around 30 million in 2010 to around 58 million by 2035. The population changes that Iraq has seen in recent decades (mass emigration and significant numbers of internally displaced persons) could have been the basis for varying our population assumption across different future cases. However, we have elected not to do so as this would make it more difficult to draw comparisons between the various projections.

Policies

The challenges of reconstruction are no less daunting than the scale of reconstruction itself. They are also wide ranging, covering issues such as ensuring security, effective economic management, structural reform of an economy that is dominated by the public sector and heavily centralised, the need for extensive legal and regulatory reform, institutional reform and effective implementation, and the need to overcome political differences. The extent to which each of these challenges is overcome is very uncertain, yet critical to Iraq's future.

The longer-term outlook for Iraq is difficult to assess when so much of the attention of policy makers has understandably been on short-term imperatives. However, evidence of rather longer-term thinking is emerging, both on national development priorities and,

Box 13.5 ▷ Strategic thinking for Iraq's energy sector

The Iraqi vision for the energy sector is that it should act as a motor for national development, providing revenues from export that can be used to provide secure and reliable domestic energy supply and to promote the diversification of the national economy and employment. The National Development Plan and the statements of senior Iraqi policy makers give some indication of the main strategic objectives for this sector:

- A rapid increase in oil production and export; there are different views as to what target should be adopted, but no divergence of views on the central importance of such an increase to Iraq's economic development.
- Sufficient supply of gas, via associated gas, development of non-associated gas resources and reductions in flaring, to provide for growing demand for power generation, for industrial development (such as petrochemicals) and for export.
- Efforts to replace and expand oil and gas reserves through increased exploration.
- Create reliable and adequate storage, transportation and export capacities in time to accommodate oil and gas development plans.
- Expand and modernise refining capacity to provide the range of oil products to support a growing Iraqi economy.
- Increase power generation capacity, and improve transmission and distribution, so as to meet increasing demand for electricity and provide reliable, high-quality service.

Among the key challenges in meeting these objectives will be to ensure the co-ordination and sequencing of investments in different parts of the energy sector, the mobilisation of the huge investments required, and the integration of social and environmental concerns. These have been central points for discussion in the development of Iraq's Integrated National Energy Strategy.

specifically, on the outlook for the energy sector. An important manifestation of this shift was the completion of a National Development Plan for 2010 to 2014 (Ministry of Planning of Iraq, 2010), which contains objectives for a range of policy areas, including a medium-term vision for the energy sector, and has been an important source of guidance for our policy assumptions. This plan has been followed by work on a more detailed Integrated National Energy Strategy, co-ordinated by the Prime Minister's Office, which had not been published at the time of writing (Box 13.5).

In the Central Scenario, we assume that Iraq makes substantial progress towards its declared objectives, taking into account additional assumptions in related policy areas (Table 13.3). We also consider a range of institutional, economic, logistical and regulatory factors that will affect the likelihood and speed of meeting these aims in full. The High Case retains the overall analytical framework, but takes a generally more favourable view about the extent and pace at which the various obstacles to a rapid transformation of the energy sector might be overcome. In the Delayed Case, cumulative investment in Iraq's energy supply is around 60% of the level foreseen in the Central Scenario. This could be attributable to a more difficult political or regulatory environment, poor co-ordination, or slow contracting and delivery of projects in the energy sector. As a consequence, the growth in oil supply is slower than in the Central Scenario, as is the increase in gas supply, and there is also an impact on the power sector.

Table 13.3 ▷ Main assumptions for Iraq in the Central Scenario

Policy area	Assumptions
Energy prices	The average IEA crude oil import price – a proxy for international oil prices – rises to \$120/barrel (in year-2011 dollars) in 2020 and \$125/barrel in 2035. Gradual progress towards a partially liberalised market for oil products. A gradual, partial removal of subsidies to electricity begins after full domestic electricity supply is ensured. The price of gas increases slowly in real terms, as the government maintains the rate of subsidy, seeking to incentivise its use in power generation and in the industrial/ transformation sectors.
Oil supply	Slower targeted ramp-up in production than implied by current contracts, accompanied by a review of the requirements for individual fields (see Chapter 14).
Gas supply	Priority is given to gas use for domestic power generation, followed by industrial use. Gas flaring is gradually reduced.
Power generation	Peak demand for electricity in 2011 was estimated at 70% higher than the amount actually supplied from the grid. The relative prices of gas and oil products mean that gas is used increasingly in the power sector as it becomes more available domestically, while the share of oil in the generation mix falls. Beginning in 2015, combined-cycle gas turbines enter the generation mix, either through conversions from existing gas turbines or as new plants.
Refining	New refinery capacity starts to become available from 2019.
Industry	An Iraqi petrochemicals industry is expanded from 2020, and industries such as fertilisers and cement grow in line with our general assumption of gradual economic diversification.
Transport	There is a gradual rehabilitation of the road and rail networks.

Iraq oil and gas resources and supply potential

How much is enough?

Highlights

- Iraq's ambition to expand its oil and gas output over the coming decades is not limited by the size of its hydrocarbon resources nor by the costs of producing them, which are among the lowest in the world. Contracts and field development plans imply an extraordinary increase in production. How this develops in practice will be determined by the speed at which impediments to investment are removed, clarity on how Iraq plans to derive long-term value from its hydrocarbon wealth, international market conditions and Iraq's success in consolidating political stability and developing its human resource base.

- In our Central Scenario, Iraq's oil production increases to more than 6 mb/d in 2020 and reaches 8.3 mb/d in 2035. In the High Case, production surpasses 9 mb/d already in 2020, before rising to 10.5 mb/d in 2035. Meeting these trajectories (the High Case in particular) will require rapid, co-ordinated progress in many areas to ensure the timely availability of rigs, sufficient water for injection to maintain reservoir pressure, and adequate storage, transportation and export capacity.

- Anticipated production increases are driven mainly by super-giant fields in the south, but here the requirements for water injection to support oil production are high, at 9 mb/d in the Central Scenario by 2035 and 11 mb/d in the High Case. Early investment is essential to bring water from the Gulf to the southern fields and reduce potential stress on freshwater resources.

- The KRG area in the north of Iraq is now one of the most active exploration regions in the world. Resolution of current differences between the regional and federal governments would open up the possibility for this area to deliver substantial output growth. In our projections, the outlook is for production of between 500-800 kb/d in 2020 and between 750 kb/d-1.2 mb/d in 2035.

- Iraq's production of natural gas is projected to grow to almost 90 bcm in 2035 in the Central Scenario and to almost 115 bcm in the High Case. The largest share is associated gas from southern oil fields around Basrah, bolstered by new gas processing capacity that reduces gas flaring; but we also project rising volumes of non-associated gas production, particularly in the north of Iraq.

- In the Central Scenario, cumulative oil and gas sector investment during the period 2012-2035 is almost $400 billion (an average of $16 billion per year); in the High Case this figure rises to $580 billion ($24 billion per year). This represents a large step up from the estimated $7 billion spent in 2011. If investment were to remain closer to 2011 levels, as in a Delayed Case, Iraq's oil output would rise much more slowly, reaching only 4 mb/d in 2020 and 5.3 mb/d in 2035.

Oil

Reserves and resources

The petroleum geology of Iraq is similar to that of the rest of the Middle East region: as Africa and the Arabian plate moved towards Eurasia over geological times, the ancient Tethys ocean was slowly closed, creating an exceptional combination of good source rocks, reservoir rocks of a variety of characteristics (both sandstones and carbonates) and strong cap rocks keeping the hydrocarbons in place. The movement of the Arabian plate towards Eurasia created the Zagros Mountains in Iran and, to the west/southwest of that, the three main hydrocarbon basins of Iraq:

- The Zagros foldbelt, immediately west of the mountains, with a series of gentle folds, including the super-giant Kirkuk reservoir and the fields in the Kurdistan Regional Government (KRG) area. It runs close to the Iranian border in the east and takes in most of the northern part of Iraq. In subsequent tables in this chapter, it is labelled Northern Zagros and referred to as "North". The foldbelt itself continues into Iran, where it is also a prolific hydrocarbon province, rich in both oil and gas.

- The Mesopotamian foredeep basin, which is where most of Iraq's super-giant fields have been found. Although not geologically separated, we split this basin between Southern Mesopotamia (referred to as "South"), the region near Basrah where most of the fields operated by international companies are located, and Central Mesopotamia ("Centre"), extending across the alluvial plains of the Euphrates and Tigris rivers around Baghdad. The geological basin extends into Iran, Kuwait and Saudi Arabia. It is generally more oil prone than gas prone.

- The Widyan Basin-Interior Platform, to the west, broadly overlaps with the Western Desert, extending into Saudi Arabia. It is the least explored of the three main basins and is thought to be more gas-rich than oil-rich. In this report, it is referred to as the Western Desert or "West".

Iraq's contribution to global oil supply over the coming decades is not limited by the size of its subsurface hydrocarbon potential. The country's proven reserves are already sufficient to support a major expansion in production and may represent only a small part of the overall hydrocarbons potential. Much of Iraq remains unexplored or, at least, greatly under-explored compared with other major oil-producing countries. In October 2010, Iraq's Ministry of Oil increased its figure for the country's proven reserves to 143 billion barrels, almost 25% more than the previous 115 billion barrels (proven reserves shown in Table 14.1 are further updated for production and new developments in the KRG area to the end of 2011). The new reserve estimate gives Iraq the fifth-largest proven oil reserves in the world and the third-largest conventional proven oil reserves after Saudi Arabia and Iran (O&GJ, 2011a; BP, 2012).

Figure 14.1 ▷ Iraq main hydrocarbon basins and fields

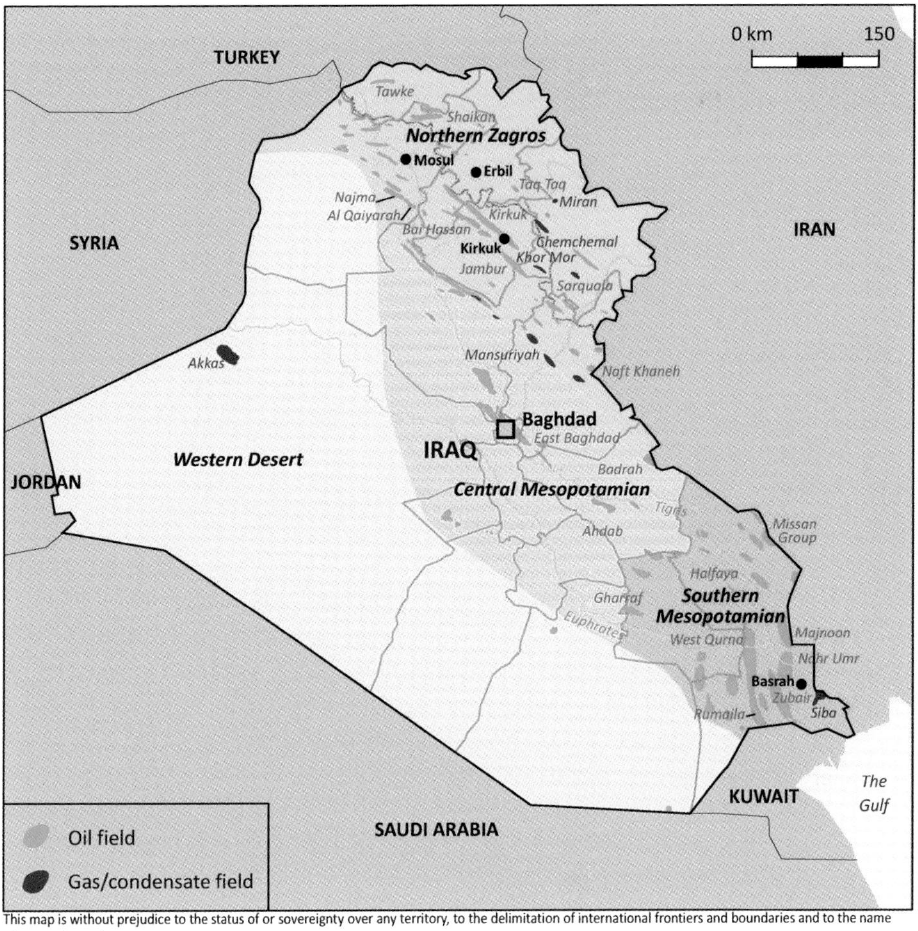

This map is without prejudice to the status of or sovereignty over any territory, to the delimitation of international frontiers and boundaries and to the name of any territory, city or area.

The extent of ultimately recoverable oil resources in Iraq is subject to a large degree of uncertainty. The figures presented in Table 14.1 are derived from the United States Geological Survey (USGS) 2000 assessment and subsequent updates; we use USGS data as the primary input to the *World Energy Outlook* series because it is the only source that covers many countries and regions using a consistent methodology. From this source, IEA analysis puts the level of ultimately recoverable resources at around 232 billion barrels (crude and natural gas liquids), of which 35 billion barrels had already been produced at the end of 2011. This leaves a figure for remaining recoverable resources of just under 200 billion barrels, almost three-quarters of which consists of proven reserves. The remainder, around 55 billion barrels, is made up of anticipated reserve growth (to the extent that this is not already included in the proven reserve number) and undiscovered resources.

Table 14.1 ▷ Iraq oil resources by region and super-giant field* (billion barrels)

	Proven** reserves, end-2011	Ultimately recoverable resources	Cumulative production, end-2011	Remaining recoverable resources	Remaining % of ultimately recoverable resources
Southern Mesopotamian	**107**	**135**	**18**	**116**	**86%**
West Qurna	43	44	1	43	98%
Rumaila	17	31	14	17	54%
Majnoon	12	12	0	12	99%
Zubair	8	10	2	8	80%
Nahr Umr	6	6	0	6	98%
Central Mesopotamian	**12**	**19**	**0**	**18**	**99%**
East Baghdad	8	8	0	8	100%
Northern Zagros Foldbelt	**24**	**66**	**17**	**49**	**75%**
Kirkuk	9	23	14	9	38%
Western Desert	**0**	**13**	**0**	**13**	**100%**
Total Iraq	**143**	**232**	**35**	**197**	**85%**

* Figures include crude oil and natural gas liquids (NGLs). Super-giant fields are defined as those with ultimately recoverable resources greater than 5 billion barrels. ** Proven reserves are approximately broken down by basin, based on information provided by the Iraqi Ministry of Oil, supplemented with company presentations for fields in the KRG area.

Sources: Data provided to the IEA by the US Geological Survey and the Iraqi Ministry of Oil; IEA databases and analysis.

Estimates from other sources of Iraq's undiscovered oil resources are considerably higher. Alongside its announcement of 143 billion barrels of proven reserves, the Ministry of Oil stated in 2010 that Iraq's undiscovered resources amounted to some 215 billion barrels. A detailed study by Petrolog, published in 1997, reached a similar figure and did not include the parts of northern Iraq in the KRG area or examination of the stratigraphic traps that are numerous in central and western regions of the country (O&GJ, 2011b). Even using the more conservative USGS figure, Iraq has thus far produced only 15% of its ultimately recoverable resources, compared with 23% for the Middle East as a whole.

Exploration efforts can be expected to add substantially to proven reserves over the coming decades. Of 530 potential hydrocarbon-bearing geological prospects identified by geophysical means in Iraq only 113 have been drilled, with oil being found in 73 of them (O&GJ, 2011b). Although it is reasonable to assume that the most promising structures were drilled first, it is also likely that new seismic data and more sophisticated analysis of historical data will increase the number of structures that may be considered hydrocarbon-bearing. Prior to the recent surge in exploration activity in the KRG area, more than half of the exploratory wells in Iraq had been drilled prior to 1962, a time when technical limits and a low oil price gave a much tighter definition of a commercially successful well than would be the case today.

For the past three decades, the world's reserves growth has come more from known fields than from new discoveries and most of these barrels were found in super-giant oilfields. As well as techniques for secondary and tertiary recovery that can increase the volume of oil that is ultimately recovered from reservoirs under development, drilling deeper into discovered fields offers opportunities to increase the reserves in Iraq: the country's currently known reserves are almost entirely found in rocks of the Cretaceous, Paleogene or Neogene periods and the deeper geological layers have seen only very limited exploration.

According to data from the Ministry of Oil, Iraq's proven reserves are spread across 66 fields, with total oil-in-place at these fields exceeding 500 billion barrels. This implies an average recovery factor of around 35% (Figure 14.2). Five super-giant fields in the south of Iraq – Rumaila, West Qurna, Zubair, Majnoon and Nahr Umr – account for 60% of the total proven reserves. The other main reserve-holding fields are East Baghdad, in the centre, and the longstanding producer, Kirkuk, in the north. Of the total remaining reserves, around two-thirds are in fields being operated by international companies. These include most of the larger fields with higher anticipated recovery factors,[1] a reasonable indication that these are the fields that are easiest to develop.

Figure 14.2 ▷ Iraq oil reserves by field, end-2011

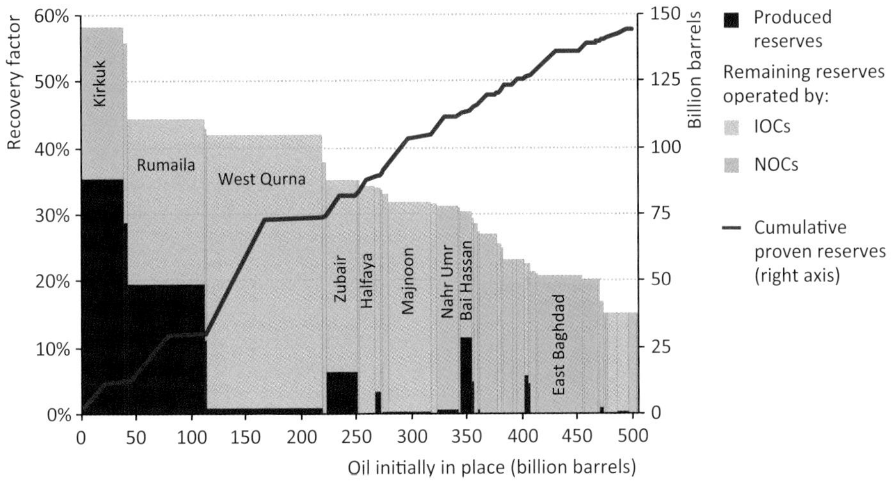

Notes: IOCs = international oil companies; NOCs = national oil companies. This figure includes data from 66 fields, but only the largest are indicated by name.

Sources: Data provided to the IEA by the Iraqi Ministry of Oil; IEA databases and analysis.

This chart includes only a small figure for proven reserves from recent discoveries in the KRG area, which we currently estimate at around 4 billion barrels. However, as a result of the contracts awarded by the regional government, this is now one of the most actively

1. The exceptions are Kirkuk, in which international oil companies have no stake, and the Najma and Qayara fields (two IOC-shaded bars to the right-hand side of the figure), which are smaller heavy oil fields.

explored regions in the world; the present, modest figure for proven reserves can be expected to increase substantially in the coming years. The Kurdistan Regional Government states that oil resources in the region amount to 45 billion barrels. This figure includes a large allowance for resources to be identified as a result of future exploration and arising from secondary recovery.

Production costs

The cost of developing Iraq's oil fields is very low by international standards. The super-giant fields being developed in the south are some of the largest in the world, bringing large economies of scale to their exploitation. Moreover, the geology is relatively uncomplicated when compared with major ongoing projects elsewhere in the world, for example the Kashagan field in Kazakhstan, where the reservoir is deep and at very high pressure, or deepwater pre-salt developments in offshore Brazil. Iraq's fields are all onshore and, as in the case of the fields around Basrah, are often located in relatively unpopulated and flat terrain, reducing the costs of wells, pipelines and other facilities. The oil produced is of a medium grade, requiring no specialist upgrading, and can be pumped and handled quite easily.

Table 14.2 ▷ Indicative oil development and production costs in selected countries

	Type of project	Scale of project (mb/d)	Capital cost* per barrel of capacity ($2011/bbl)	Operating cost** ($2011/bbl)
Iraq	Expansion super giant (south)	1.00	7 000-12 000	2
	New super-giant (south)	1.00	10 000-15 000	2
	Mid-size (north)	0.25	15 000-20 000	2-3
Saudi Arabia	Generic expansion	0.50	15 000	2-3
Brazil	Deepwater pre-salt	0.25	70 000-80 000	15-20
Kazakhstan	North Caspian offshore	0.25	70 000-80 000	15-20
Canada	Canadian oil sands with upgrading	0.25	100 000-120 000	25-30

* Capital cost per barrel of plateau rate production capacity. ** Operating cost includes all expenses incurred by the operator during day-to-day production operations but excludes taxes or royalties that might be levied by the government as well as other compensation to the operator, such as remuneration fees.

Source: IEA analysis.

The southern fields are all located within easy reach of coastal export facilities, keeping the primary export pipelines relatively short (whereas secondary export routes are much longer). Such proximity to an international port is an important consideration not only for getting the crude to market, but also for bringing in equipment, in this case via the Shatt Al-Arab waterway that extends up from the Gulf to beyond Basrah. This ease of access is in marked contrast to the logistical difficulty of bringing heavy equipment to many other parts of the world, for example to the Caspian region or to fields in northern Russia. There are also well-established fabrication yards and other industrial facilities within relatively easy reach.

Apart from the super-giant Kirkuk field, the oilfields in the north are smaller (although still large by international standards[2]), often containing 0.5 to 1.0 billion barrels of recoverable oil. The region's favourable geology is yielding wells with high initial oil production rates, in many cases higher than are reported elsewhere in Iraq; but the cost of bringing oil to market is nonetheless higher than in the south because of the need for additional expenditure on infrastructure and supply logistics, given the distance to the nearest ports.

Uncertainties above the ground, related to politics, security and the performance of Iraq's institutions, temper the geological and geographical advantages described above, and can add significantly to the cost and difficulty of executing projects. But this does not diminish the fact that, from a technical perspective, oil projects in Iraq are among the more straightforward and least-cost in the world, both in terms of the capital cost per unit of new production capacity and the operating expense.

Production

Development plans and contracts already in place imply an extraordinary increase in Iraq's oil production over the coming years. This increase would come from fields operated under the technical service contracts awarded in the national licensing rounds, from fields managed directly by the national oil companies[3] and from fields or prospective areas held under production-sharing contracts in the KRG area. On the basis of a detailed examination of 46 Iraqi fields,[4] we found that – if all of these proceed according to their currently envisaged schedules – oil production capacity in Iraq would reach 14.6 million barrels per day (mb/d) by 2020, almost five times the 3 mb/d being produced as of mid-2012. Given that the largest production commitments were volunteered by international companies during the licensing rounds and subsequently incorporated in binding technical service contracts (see Box 13.2 in Chapter 13), the Ministry of Oil has solid grounds for the formal position that, well before the end of the decade, Iraq should have more than 12 mb/d of production capacity from fields covered by technical services contracts alone.

Iraq's resource base can support an expansion of this magnitude; but there is a range of constraints – infrastructure, institutional, logistical and security – that cast doubt on whether an expansion at this pace is achievable. It is also reasonable to ask whether such a vertiginous path corresponds to Iraq's interest in deriving maximum value from its hydrocarbon wealth. It would mean, for example, that Iraq's major discovered fields (with the exception of Kirkuk and East Baghdad) not only reach peak production at the same time, but also that their post-peak declines broadly coincide. In order to provide a stable

2. The Shaikan field is currently being appraised and could come close to super-giant status.

3. These are the South, North, Midland and Missan oil companies, all of which are state-owned and part of the Ministry of Oil.

4. These include all the fields currently producing in Iraq, as well as all those that have been tested and have field development plans in place.

long-term export revenue stream, there are arguments in favour of sequencing the build-up from the various fields so as to reach a lower combined production plateau, but one that could be sustained for longer.

Alongside the above-ground factors within Iraq that might hold back output, the capacity of international markets to absorb the growth in Iraq's production could also be limited. There are arguments for allowing production capacity to run ahead of actual output, creating spare capacity that can play an important role for the stability of global markets; but, beyond a certain point, Iraq could face the possibility that spare capacity becomes redundant capacity, a particularly inefficient use of scarce investment capital in the case of Iraq. The role of Iraq's oil production in global markets is considered further in Chapter 16.

Given these uncertainties, the Iraqi authorities are considering optimal oil output paths and strategies for the future, taking into account reservoir characteristics and production economics at the various fields, infrastructure priorities and constraints, projections for market needs, as well as broader considerations such as the country's fiscal requirements, the need to maximise value from national resources and the contribution that the oil sector can make to economic development and employment. We anticipate that these deliberations will result in a slower targeted ramp-up in production than implied by current contracts and that this process will be accompanied by a review of the requirements for individual fields. Decisions on the detailed field development plans and enhanced redevelopment plans that are due to be submitted in 2012-2013 by operators under the technical service contracts will be important in determining future production profiles and resource management policies.

We have developed two detailed trajectories for Iraq's possible future oil production.[5] The Central Scenario is based on strong progress with capacity additions throughout the projection period, reflecting our judgement of the time required for projects to be executed in the developing circumstances of Iraq and taking account of any constraints arising from international market conditions. Achieving this trajectory requires a significant, co-ordinated growth in investment all along the energy supply chain: we analyse, in particular, the inter-linked requirements for increased upstream activity (drilling rigs and wells), for water that is injected to maintain reservoir pressure and for new transportation, storage and export capacity. In the High Case, we examine what would be required for Iraq to reach production in excess of 9 mb/d by 2020. Both the Central Scenario and the High Case see increases in oil output greater than anything achieved in the past by Iraq, with the Central Scenario well above the growth achieved in the 1970s (Figure 14.3). Achieving the High Case would match the highest sustained growth in the history of the industry – that of Saudi Arabia between 1966 and 1974.

5. The possibility of delays in investment, which would bring oil output below the levels anticipated in the Central Scenario, are discussed in the concluding section of this chapter.

Figure 14.3 ▷ Iraq production outlook to 2020 in context

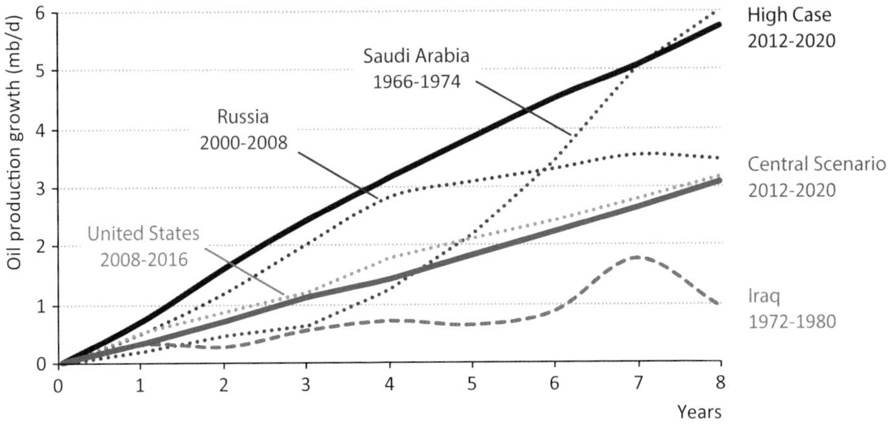

Note: The figures for the United States are actual data to 2011 and then projections (from *WEO-2012* New Policies Scenario) to 2016.

The production profile in both cases, broken down by producing basin, involves a rapid increase to 2020, then a slower rate of output growth thereafter to 2035 (Table 14.3). Given the volume of projects underway or under consideration, Iraq has the potential to meet this output growth in a variety of ways, with greater or lesser contributions from individual fields and regions. How this evolves in practice will depend on commercial negotiations and political decisions, the detailed outcome of which we have not attempted to anticipate. The breakdown by region and by field in our projections is simply proportional to the future ambitions expressed in existing contracts, company announcements and development plans, with the downside risks and upside potential discussed in the text. Existing discoveries are more than sufficient to support these oil production profiles, so we have not anticipated contributions from undiscovered fields except in the KRG area (included in "North" in Table 14.3), where there is a large amount of exploration underway and planned.[6]

Most of the increase anticipated in both scenarios comes from the large southern fields: the "Big 4" of Rumaila, West Qurna, Zubair and Majnoon contribute more than two-thirds of Iraq's projected production in 2035, slightly higher than their present share (most of which comes from Rumaila). A strategy to prioritise the lowest-cost sources of production could result in an even larger share for the major southern fields, but – because of the fast anticipated growth – this is also the area where constraints related to infrastructure, logistics and water availability could have the largest impact, potentially bringing down production levels below the range of our projections.

6. This means, for example, that we have not taken into consideration possible contributions to the oil balance from the exploration blocks included in the fourth bid round. There is, however, a larger contribution from yet-to-find fields to natural gas production and some volumes of natural gas liquids from these fields are included in the oil balance.

Among the major fields:

- **Rumaila** is well positioned over the next decade to maintain its pre-eminence in Iraq and its standing as the second-largest producing field in the world, after Saudi Arabia's Ghawar. It has a very productive main reservoir (for which the recovery factor could be well in excess of the 44% estimate from the Ministry of Oil that is used in Figure 14.2), and at least two other oil-bearing formations that have been identified but not yet developed. Its production history means that its capacities and potential are well understood and its access to export infrastructure well established. It also has more natural water support than other major fields in the region. The maximum remuneration fee per barrel, at $2 per barrel, is – along with the fee agreed for Zubair – the highest for the large southern fields. In our projections, output ranges between 1.7-2.2 mb/d in 2020 and a gradual decline starts from the mid-2020s.
- **West Qurna** is, in geological terms, a northward extension of the Rumaila field. It is split, for development purposes, by the Euphrates River, with the southern part (Phase I) contracted to a consortium led by ExxonMobil and the northern part (Phase II) by Lukoil. Thus far, there has been production only from the southern area, currently at 400 kb/d, but the combined plateau production commitments of the two phases amount to more than 4.5 mb/d. Our projections are more modest initially, at 1.5-2.5 mb/d in 2020 before reaching 3.2-4.6 mb/d in 2035, the wide range reflecting our concerns over the availability of transportation infrastructure and water for injection. There is a risk that production growth could be slowed by Statoil's exit from the consortium operating Phase II and by uncertainty over ExxonMobil's position in the Phase I project.
- **Zubair** was one of the first major discoveries in the region but it has remained in the shadow of its larger neighbour, Rumaila, with historical production rarely exceeding 200 thousand barrels per day (kb/d). A technical service contract for the field was awarded to a consortium led by Eni in 2009. Our projections for 2020 range from 500-750 kb/d and around 700 kb/d for 2035.
- **Majnoon** has seen only very limited production since its discovery (by Petrobras) in the 1970s. Development is complicated by the large amount of unexploded ordnance in the area, a legacy of the 1980s war with Iran. A consortium led by Shell is now working to raise production, initially to 175 kb/d (the threshold for cost-recovery payments). Shell is reportedly in negotiations with the Ministry about lowering the plateau production commitment for the field from its current 1.8 mb/d. Our projections are between 550-950 kb/d in 2020 and 0.7-1 mb/d in the 2030s.

After these four super-giants, the next largest contribution in the south comes from the Halfaya field, which is located in the Missan governorate and operated by a consortium led by PetroChina. Production has started from newly built facilities, ahead of the original project schedule. We project that output from Halfaya reaches between 150-275 kb/d in 2020 and 200-325 kb/d in 2035 although, given the rapid progress thus far, these projections could well be exceeded if the connecting infrastructure to export facilities is available on time.

Production from the central region of Iraq is expected to remain under 350 kb/d, with the major contributors being the Ahdab field, operated by PetroChina (which increased production rapidly in 2012 to reach its target plateau of 115 kb/d), and the Badra field, operated by GazpromNeft (where first production is expected in 2013). Further growth from this region is held back as we do not anticipate major progress with the region's super-giant East Baghdad field that underlies the capital. It was offered in the second national licensing round but did not receive any bids. Output from the west of the country is marginal, consisting only of some natural gas liquids produced later in the projection period as part of new natural gas developments.

Table 14.3 ▷ Iraq oil production by region in the Central Scenario and the High Case (mb/d)

Central Scenario	2011	2015	2020	2025	2030	2035
South	2.0	3.2	4.8	5.4	5.8	6.4
Big 4	1.8	2.8	4.2	4.7	5.1	5.6
Centre	0.0	0.2	0.2	0.2	0.2	0.3
North	0.7	0.8	1.1	1.3	1.4	1.6
West	0	0	0.01	0.01	0.02	0.02
Total	**2.7**	**4.2**	**6.1**	**6.9**	**7.5**	**8.3**
High Case	**2011**	**2015**	**2020**	**2025**	**2030**	**2035**
South	2.0	4.6	7.3	7.6	7.8	8.1
Big 4	1.8	4.0	6.4	6.6	6.8	7.1
Centre	0.0	0.2	0.3	0.3	0.3	0.3
North	0.7	1.0	1.6	1.7	1.9	2.1
West	0	0.01	0.02	0.02	0.03	0.03
Total	**2.7**	**5.9**	**9.2**	**9.6**	**10.0**	**10.5**

Notes: The figures provided and discussed are for actual oil production and not for production capacity. The "Big 4" includes production from Rumaila, West Qurna, Zubair and Majnoon.

Around a quarter of Iraq's production in 2011 came from the north of the country (Figure 14.4), primarily from the Kirkuk and Bai Hassan fields. This northern share declines to one-fifth by the end of the projection period, but the geography of Iraq's northern output changes substantially in our projections with the rise in production under contracts awarded by the KRG (Box 14.1). Production from other northern fields grows at a slower pace. The super-giant Kirkuk field includes three large production areas, two of which are operated by the North Oil Company, with the third (the Khurmala dome) being developed under a contract awarded by the KRG. In the absence of firm development plans, we project that production from the areas operated by the North Oil Company rises only slightly from current levels.

Box 14.1 ▷ Outlook for oil production in the KRG area

With 23 rigs drilling exploration wells in mid-2012 (more than double the number from early-2011), the KRG area is now one of the most intensive areas for oil and gas exploration in the world, reflecting high expectations of significant discoveries in the heavily folded and faulted subsurface of the northern Zagros foldbelt. The regional government has awarded around 50 contracts with international companies to explore for and produce oil, and has stated its ambition to raise the region's production to 1 mb/d by 2015, based on existing discoveries, and to 2 mb/d by 2019, based on existing and expected discoveries (our projections are more conservative). The largest discovery to date has been the Shaikan field, which could have as much as 12-15 billion barrels of oil in place, although further drilling will be required in order confidently to assess the resources. There are currently five fields that are producing under contracts awarded by the KRG: Tawke, Taq Taq, the Khurmala dome of the Kirkuk field, Shaikan (early production) and condensates from the Khor Mor gas field (Figure 14.4). The Sarqala field is undergoing extended well tests and five additional fields are expected to be in early production tests by the end of 2012.[7]

Our projections for oil production from the KRG area are contingent on resolution of the current deadlock between the regional government and Baghdad over governance of the hydrocarbons sector, which have generated disputes over the modalities and payment for cost recovery and export (see Chapter 13). A harmonious resolution would open up the possibility for the KRG area to deliver substantial output growth in the longer term, not least because it would facilitate the arrival and operation of larger international companies. A stable framework for export, an area in which the federal government claims exclusive authority, is essential to mobilise the multi-million dollar investments that are required. In the case of the Shaikan field, contract-holder Gulf Keystone estimates that full field development may cost up to $10 billion (they hope to reach initial production of 100 kb/d through investment of around $500 million). As long as there is no certainty that future output can be converted into a reliable revenue stream, investment will, at best, be made only in small increments. Our projections for oil from the KRG area (included in "North" in Table 14.3) see production rising to between 500-800 kb/d in 2020 and 750 kb/d-1.2 mb/d in 2035.

Aside from the complex political considerations, the outlook for the KRG area is also constrained by the time needed to put in place the necessary production and transportation infrastructure. With the exception of a small-capacity link between the Tawke field and the Iraq-Turkey export pipeline, there is very limited pipeline infrastructure in the region and most oil is transported by truck. A number of proposals exist to tie in existing fields to the main Iraq-Turkey pipeline, either through connections around the Kirkuk area (a link from the Taq Taq field is under construction) or, alternatively, through new connections that would bring oil directly to the main pumping station at Fish Khabur near the Turkish border.

7. Production capacity in the KRG area as of mid-2012 is estimated at 250 kb/d.

Figure 14.4 ▷ Northern Iraq oil and gas fields and infrastructure

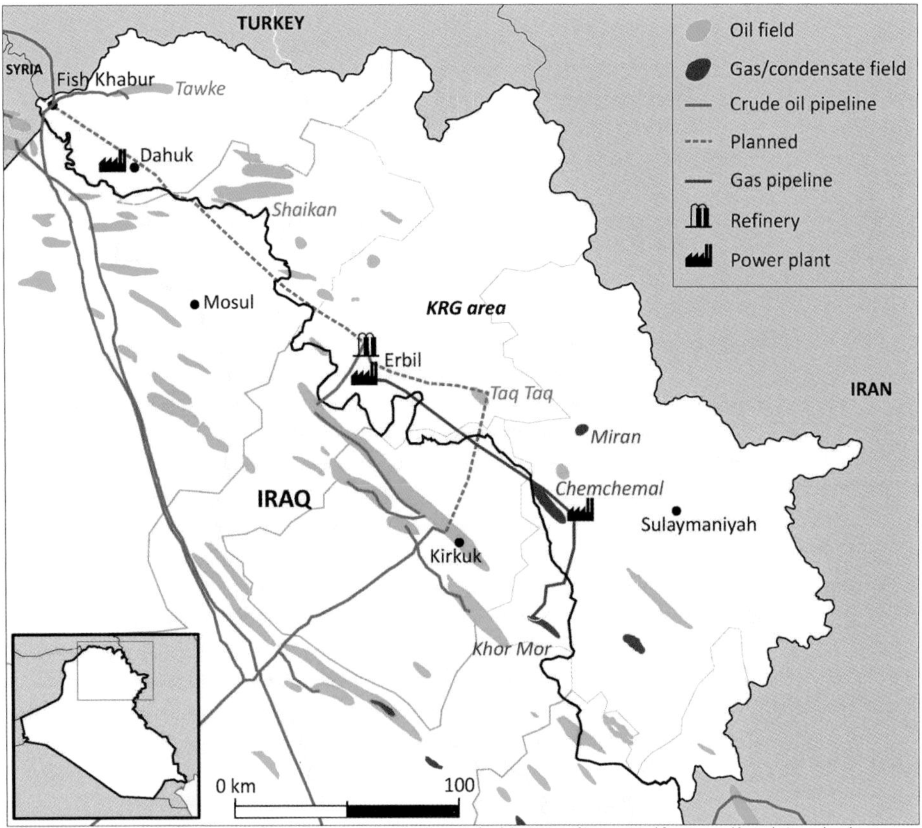

This map is without prejudice to the status of or sovereignty over any territory, to the delimitation of international frontiers and boundaries and to the name of any territory, city or area.

Wells and other upstream facilities

Bringing on new production and compensating for the natural decline in output from existing facilities requires new wells and surface equipment to gather the produced hydrocarbons and to separate out natural gas from the crude oil. To meet the production profile of the Central Scenario, Iraq requires, on average, 500 kb/d of additional capacity every year to 2020 while in the High Case, the average annual requirement is close to 900 kb/d. After 2020 the pace of overall output growth is lower, but more investment is required each year to combat natural decline and maintain production levels.

The number of producing wells in Iraq has risen sharply in recent years: in 2010 Iraq recorded 1 526 producing wells and just a year later 1 695 wells, an increase of 169 (OPEC, 2012). This is a testament to the improving security situation and the surge in development activity that followed the signing of the technical service contracts in 2009 and 2010. Substantial repair work was immediately undertaken as operators worked urgently to reach the initial production threshold, after which they could recover costs.

Large numbers of wells are required, not only for oil and gas production but also for water injection at oil fields, as well as for exploration activities (Figure 14.5). In the Central Scenario, the number of wells needed continues to rise until around 2020, by which time the rig count in Iraq is projected to reach 135, compared with the current figure of around 80. After 2020, the flattening of the production profile brings a temporary reduction in the number of oil wells required, but in-country rigs could be re-deployed for exploration, for non-associated gas production (which expands rapidly in the 2020s), or to create spare oil production capacity. The addition of 55 drilling rigs by 2020, all requiring experienced crews and support services, would be a sizeable but still manageable task for Iraq: in the last ten years, the Middle East as a whole (excluding Iraq) has been adding around ten rigs per year. In the 2020s, the number of wells required to meet the Central Scenario starts to increase again as the average productivity of a new well is assumed to decline (from the current average, around 3 kb/d, to 2 kb/d) and the number of water injection wells, relative to producers, increases as the fields mature. However, Iraq manages with around 130 to 140 rigs until close to the end of the projection period.

Figure 14.5 ▷ Wells by type and rig count required in the Central Scenario

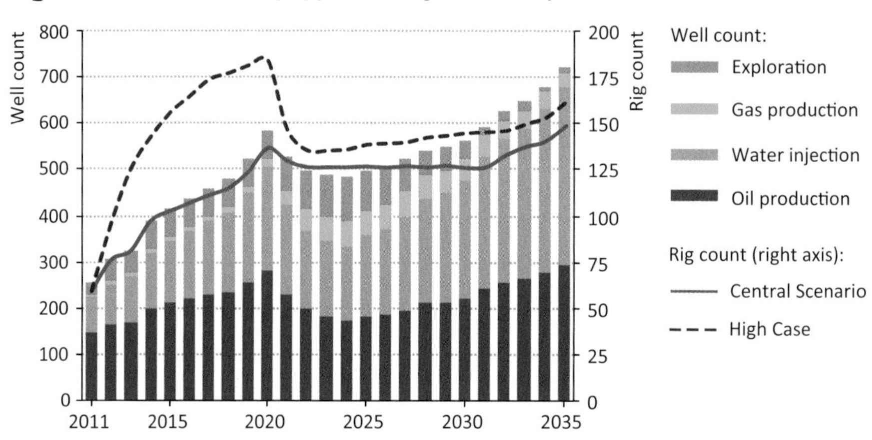

Notes: These figures do not include well workovers, *i.e.* major maintenance or remedial work on existing wells, or the rigs required for these operations. There are currently around 30 workover rigs in Iraq.

In the High Case, the total annual requirement for wells rises much more quickly in the coming years to over 800 by 2020, drops back sharply as output growth slows in the early 2020s, before growing again towards 800 in 2035. To achieve this, the rig count would need to more than double to 180 by 2020, before falling back. Sourcing this number of additional rigs and drilling crews in the coming years would be a considerable challenge, particularly for the short-lived peak in drilling towards 2020.

Crude oil conversion and bringing oil to market

A part of Iraq's new oil infrastructure is required to cater for the anticipated rise in domestic demand, with the largest share going to the rehabilitation of Iraq's refining sector. Thus far the government's attempts to attract large-scale private investment have not been successful: Iraq's three main refineries in Doura (near Baghdad), Basrah and Baiji remain in urgent need of upgrading. The main addition in recent years has been the 40 kb/d refinery in Erbil (now being expanded to 100 kb/d), but the bulk of investment made has been in even smaller-capacity topping plants, which can be built quickly but have not made a dent into Iraq's deficit of some key oil products, particularly gasoline, at a time when domestic demand is growing rapidly.

In addition to projects aimed at upgrading and de-bottlenecking the existing refineries, design work is at various stages for large projects in Karbala (140 kb/d), Nassiriyah (300 kb/d)[8], Kirkuk (150 kb/d) and Missan (150 kb/d) that would increase refinery capacity by almost 750 kb/d. Legislation offers investors the possibility of both a 5% discount on the average crude export price and 50-year operating licences, while the government is also considering a system of remuneration to investors based on a fixed fee per barrel. As yet there has been no firm commitment to the projects on offer. In order to provide the margins that could justify these projects, potential investors have indicated that they favour larger complexes than those proposed by the government and freedom to export petroleum products (given that oil product prices on the domestic market remain subsidised).

Figure 14.6 ▷ Iraq refinery capacity in the Central Scenario

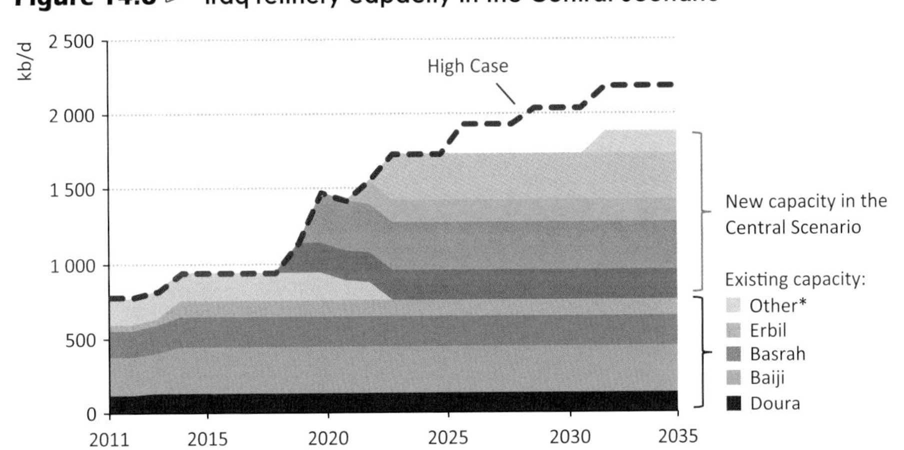

* Refers to smaller-volume refinery capacity, *i.e.* topping plants.

8. The Ministry of Oil has announced plans to include construction of this refinery as part of the Nassiriyah oil field development project, which is expected to be offered in an upcoming licensing round.

In our projections, we assume that this investment does go ahead, but later than currently envisaged by the authorities. In both the Central Scenario and the High Case, the first of Iraq's new refineries becomes operational in 2019, setting the scene for the start of decommissioning of Iraq's smaller, topping refineries (Figure 14.6). In the High Case, capacity continues to grow throughout the 2020s, reflecting the higher domestic demand for oil. The modernisation and expansion of Iraq's refinery capacity in our projections brings a significant improvement in the product slate, increasing the share of gasoline produced relative to heavy fuel oil (see Chapter 13). This is sufficient to reduce the country's reliance on imports of gasoline and diesel, but does not allow for significant export of oil products.[9]

Bringing Iraq's crude oil production to international markets is another major challenge. Our projections suggest that Iraqi exports will soon exceed the historical export peak of 3.3 mb/d reached in 1979, but this will depend on substantial additional investment to remove potential infrastructure bottlenecks, particularly in the south. There are three links along the supply chain that could constrain export from the main southern fields and hence production: the network of pipelines, storage facilities and pumping stations that provide a connection between the fields and the main export depot at Fao, the available pumping and storage capacity at Fao itself and the offshore export facilities for loading tankers (Figure 14.7). All of these constraints are being addressed, but not at the same pace.

Expansion of offshore crude loading capacity has made the most progress, with the commissioning of two single-point mooring systems in early 2012. The Ministry of Oil plans to add another three mooring systems and to expand the capacity of the existing offshore terminals at Khor al-Amaya (KAAOT) and Al-Basrah (ABOT). Once complete, with new or expanded pipelines linking them to onshore facilities, these projects would bring Iraq's sea-borne export capacity to 8 mb/d, a level sufficient to accommodate the exports projected in both the Central Scenario and the High Case.

The Fao terminal is a critical link between the onshore fields and the offshore loading facilities, but longstanding plans to rebuild storage and pumping capacity at this depot are only now coming to fruition. Twenty-four new storage tanks are envisaged (each of 58 thousand cubic metres, for a total capacity of more than 8 million barrels). The first set of eight tanks has been constructed, but the facility is set to begin large-scale operation only in 2013 once pumps and other ancillary equipment have been installed. The delay has necessitated the construction, for the new single point mooring systems, of interim pipelines bypassing the terminal. As a consequence, oil is pumped onto tankers directly from more than 100 km onshore and any halt to offshore loading (which can often be weather-related) can force a reduction in production at the fields.

9. For the moment, there are no plans to create a large export-oriented refining capacity of the sort that exists elsewhere in the Middle East. The logical place to site such a facility would be around Basra, but it would require dedicated export infrastructure for oil products; this would be only a longer-term possibility given other infrastructure priorities.

Figure 14.7 ▷ Southern Iraq oil and gas fields and infrastructure

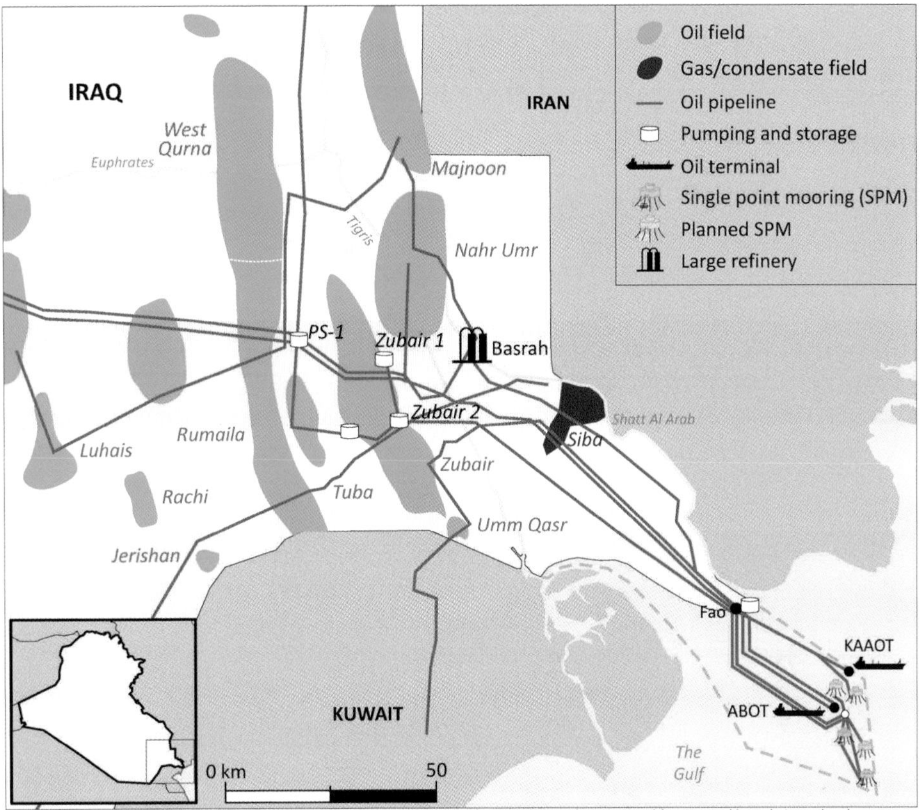

This map is without prejudice to the status of or sovereignty over any territory, to the delimitation of international frontiers and boundaries and to the name of any territory, city or area.

Infrastructure challenges further onshore include pipelines and equipment that are in a poor state of repair and inadequate in capacity, some missing links to accommodate output from new fields like Halfaya and Majnoon, and a shortage of storage and pumping capacity. Individually, none of these projects is technically complex or difficult to implement; but collectively they demand a high level of integrated planning and timely execution in order to avoid limiting the productive potential of the major southern fields (see concluding discussion in this chapter on institutions and project lead times).

The second historical axis for oil export from Iraq has been the northern route to the Mediterranean port of Ceyhan in Turkey. Rehabilitation of this route and raising its capacity from the current 600 kb/d to 1 mb/d is included in the Ministry of Oil's plan for export infrastructure to 2014; but this date is likely to be pushed back, as detailed feasibility work on Iraq's pipeline export options started only in 2011. Iraq is interested in developing a range of routes to market, with enhanced linkages between the north and south of the country (Box 14.2), but it is not yet certain which routes will be developed. Options on the table include not only the development of the existing route to Ceyhan, but also rehabilitated or new capacity to the Mediterranean via Syria or, possibly, across Jordan (or Saudi Arabia) to the Red Sea. The infrastructure choices facing Iraq need to be shaped by the range of crudes produced and by Iraq's evolving oil marketing strategy (see Box 16.2 in Chapter 16).

Box 14.2 ▷ The Strategic Pipeline

The anticipated need for a large increase in southern export capacity over the coming years brings with it the risk of over-reliance on the Straits of Hormuz for access to world markets. Development of alternative routes to market makes sense both for commercial and security reasons and the strategic position of the government is that Iraq should have at least three independent outlets to market for its crude exports: a major route through the Gulf and two pipeline routes (either both to the Mediterranean, or one to the Mediterranean and one to the Red Sea). Accompanying this expansion of export routes, a sizeable capacity (of up to 2.5 mb/d) should be kept available to shift oil from south to north, or vice versa. A large-capacity Strategic Pipeline, separate from the domestic transportation network, has been proposed as a way to link the different systems. Currently, some oil from the south is refined in central Iraq, but there is no functioning link with the northern export route. A new Strategic Pipeline, though it would cost around $1.5 billion to construct (with capacity of 2.5 mb/d over a distance of 650 km), would give Iraq valuable flexibility in marketing its oil for export. In the event of disruption to export along an existing route, the Strategic Pipeline, if it allowed Iraq to sustain exports of 2 mb/d by switching routes, would generate revenue equal to the initial investment in little more than a week.

Water requirements

The medium-term prospects for Iraq's oil production, from its southern fields in particular, are closely tied to the availability of sufficient water for injection to support reservoir pressure.[10] Both the historical mainstays of Iraq's oil production, Kirkuk and Rumaila, have required water injection. In the case of Kirkuk, reservoir pressure at the field dropped significantly after production of only around 5% of the oil-in-place (Figure 14.8). After an initial experiment with gas injection, engineers implemented a water injection project that reversed the decline in reservoir pressure and allowed production to continue at a high rate. Rumaila had produced more than a quarter of its oil-in-place before water injection was required, because its main reservoir formation (at least its southern part) is connected to a very large natural aquifer which has helped to push the oil out of the reservoir.

10. As oil is produced, the space it leaves behind in the reservoir rock needs to be filled, otherwise well flow rates will decline, gas will come out of solution from the oil and the ultimate recovery factor, *i.e.* the percentage of oil initially in place that will ultimately be produced, can be reduced significantly. If pressure support is not available naturally, for example from an aquifer connected to the reservoir, it needs to be provided by injecting gas or, more commonly in Iraq and around the world, by injecting water.

Figure 14.8 ▷ **Reservoir pressure and recovery factors prior to and after the start of water injection at the Rumaila and Kirkuk fields**

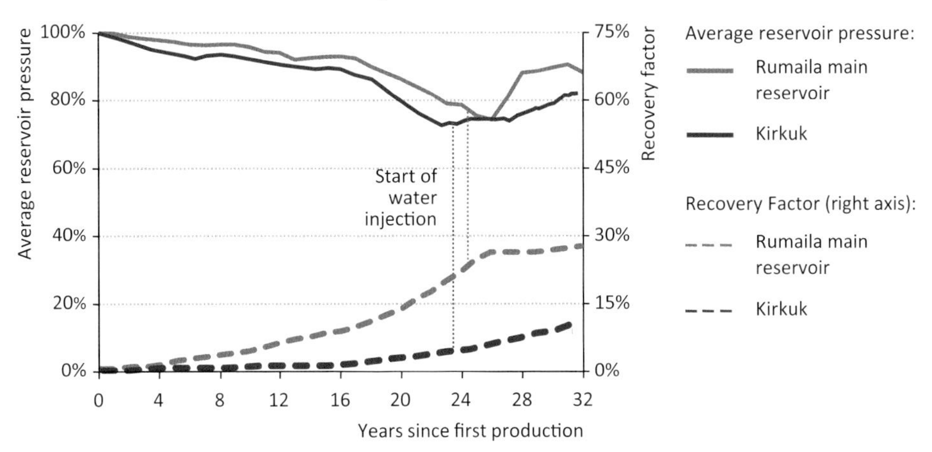

Note: Recovery factor is the percentage of oil initially in place that has been produced; the recovery factor shown for Rumaila is for the main reservoir and therefore differs from the figure for the field as a whole.

Sources: Al-Naqib, *et al.* (1971); Mohammed, *et al.* (2010); IEA analysis.

Water requirements for most of Iraq's oil fields will fall between these two cases. To meet the oil production levels in our Central Scenario, we estimate that Iraq's net water injection requirements will increase from 1.6 mb/d in 2011 to more than 12 mb/d in 2035.[11] Water needs for oil field injection are highest in southern Iraq (where water resources are also most strained), where they grow to about 9 mb/d at the end of the *Outlook* period (Figure 14.9). In the High Case, the net water requirements are correspondingly greater, rising to almost 16 mb/d for the country as a whole in 2035 and surpassing 11 mb/d for the fields in the south.

Future water injection plans in southern Iraq involve the construction of a Common Seawater Supply Facility (CSSF), which would treat seawater from the Gulf and pump it more than 100 km inland for use in the oil production areas. This solution is strongly favoured over other water sourcing alternatives: it provides a secure water supply, independent of future water availability; it reduces stress on freshwater resources, freeing them for other uses; and it achieves economies of scale, through the construction and operation of a single facility to provide the bulk of southern oilfield water needs. Iraq's total water injection needs in 2035 equate to around 2% of the combined average flows of the Tigris

11. This estimate is based on production and water injection data from Iraq's southern fields provided by the Ministry of Oil and by operators. The average replacement ratio is 1.5, *i.e.* 1.5 barrels of water must be injected to fill the "space" in the reservoir created by the production of 1 barrel of oil. This is applied across all oil production in Iraq. This estimate is net of produced water, *i.e.* we assume that the most economically advantageous approach is to treat and re-inject all the water that is produced along with the oil. If the produced water is not re-injected, then, in most cases, it will require additional treatment to remove oil and salts before being suitable for reuse (for agriculture or other purposes). To the extent that produced water is put to other uses, the net water injection requirement will, obviously, rise.

and Euphrates rivers (current flows measured in central Iraq), or 6% of their combined flow during the low season. While withdrawals at these levels might at first sight appear manageable, these rivers also have to satisfy other, much larger, end-use sectors, including agriculture (where water use is currently very inefficient).

Figure 14.9 ▷ Net water requirement* in southern Iraq oil fields by source in the Central Scenario

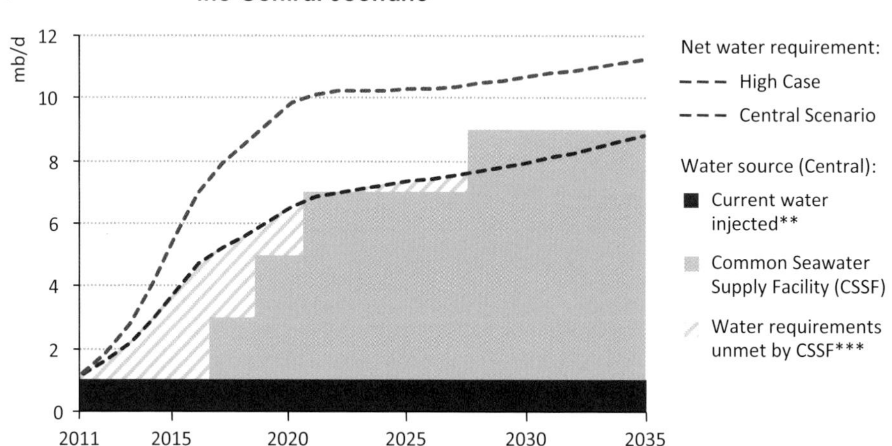

* Represents water brought in from sources external to the oil field, *i.e.* produced water that is re-injected is excluded. ** The Karmat Ali facility is the primary current source of water supply. The facility produces 0.9 mb/d from the Shatt al-Arab waterway and there are plans in place to raise its capacity to 2.2 mb/d in 2013. *** Sources may include aquifers or surface water.

The anticipated capacity of the CSSF is 10-12 mb/d, to be built in phases; but the start date for the operation of the facility remains uncertain. ExxonMobil withdrew from its role as co-ordinator of the project in early 2012 and, although a new project management contract is set to be awarded by the government before the end of 2012, the time required for detailed design and engineering work and project implementation is likely to be long. Saudi Aramco's Qurayyah Seawater Plant Expansion provides a reference for the potential timeline: this was a 2 mb/d expansion of an existing facility and took nearly four years from the awarding of the front-end engineering and design contract (May 2005) to the time that water first began to flow (early 2009). Given the importance attached by the Iraqi authorities to the CSSF, we anticipate relatively rapid progress with this project, but estimate that a likely start-date for an initial 2 mb/d phase of the CSSF is 2017, considerably later than initially planned.

In the Central Scenario, 8 mb/d of CSSF capacity is required, with subsequent phases (after the initial 2 mb/d) assumed to take two years to complete and the last tranche being added in the late 2020s. In the High Case, the overall capacity required would be 10 mb/d and the timeline for building to this level would extend into the 2020s, lagging well behind the higher water requirements in this case. Water that is not supplied from the CSSF would need to be sourced from aquifers and surface water (potentially made available by expanding existing facilities, such as the Karmat Ali facility in the south). Amounts up to 5 mb/d would be required from these sources in the High Case, considerably more than in the Central Scenario. Delay in planning and implementing the CSSF, beyond that already anticipated here, would quickly create additional demands on alternative sources of water.

To the extent that water injection needs are not met, reservoir pressure in unsupported fields falls (how quickly depends on the response of individual fields), thereby causing well flow rates to decline and making it more difficult to raise oil output. Other consequences might include higher gas production as a result of falling reservoir pressure (which would, if no facilities are in place to gather and process the gas, mean an increase in flaring) and, if water flooding is ignored to the point that reservoir damage occurs, a potential reduction in the volume of oil that can ultimately be recovered from the fields.

Natural gas

Reserves and resources

Iraq's proven reserves of conventional natural gas amount to 3.4 trillion cubic metres (tcm), or about 1.5% of the world total (Table 14.4), placing Iraq 13th among global reserve-holders. Around three-quarters of these proven reserves consist of associated gas, with the rest in a small number of non-associated fields. Iraq did not revise its figure for proven gas reserves in 2010 at the time of the upward revision of proven oil reserves, although it would have been reasonable to have done so, given the high share of associated gas. Geographically, Iraq's proven gas reserves are concentrated in the south, mostly as the large associated gas reserves in the super-giant fields of Rumaila, West Qurna, Majnoon, Nahr Umr and Zubair. The composition of the gas associated with oil production varies considerably between the north and the south of the country. Associated gas in the south has a relatively high content of natural gas liquids (NGLs).[12] The gas produced in the north is somewhat drier, but also requires treatment in order to make the gas marketable.

Ultimately recoverable resources are estimated to be considerably larger, at 7.9 tcm, of which around 30% is thought to be in the form of non-associated gas. This means that almost 40% of the resources yet to be found are expected to be in non-associated gas fields. The breakdown of our estimated figure for ultimately recoverable resources between the main Iraq's regions shows the continued predominance of the southern region, but also the potential of non-associated gas resources in both the northern and western parts of Iraq. Exploration and appraisal of these resources is at a very early stage and production is limited (in the north) and non-existent (in the west), but both these areas are considered strong prospects. Due to the higher revenues earned by the oil sector, gas has historically been a secondary consideration for the government; but attention to gas is growing as domestic demand increases, in particular for power generation.

12. According to data provided by the South Gas Company, the unprocessed associated gas consists, on average, of 70% methane and 30% NGLs, of which 15% is ethane and 8% propane.

Table 14.4 ▷ Iraq gas resources by region and super-giant field (bcm)

	Proven* reserves, end-2011	Ultimately recoverable resources**	Cumulative production, end-2011	Remaining recoverable resources	Remaining % of ultimately recoverable resources
Southern Mesopotamian	**2 202**	**4 298**	**351**	**3 947**	**92%**
West Qurna	780	1 139	18	1 121	98%
Rumaila	332	838	288	550	66%
Majnoon	203	388	2	386	99%
Zubair	156	334	39	295	88%
Nahr Umr	193	379	4	375	99%
Central Mesopotamian	**179**	**700**	**1**	**700**	**100%**
East Baghdad	126	367	1	367	100%
Northern Zagros Foldbelt	**993**	**2 027**	**158**	**1 869**	**92%**
Kirkuk	248	256	154	102	40%
Western Desert	**60**	**906**	**0**	**906**	**100%**
Total Iraq	**3 435**	**7 932**	**510**	**7 422**	**94%**
Associated	*2 558*	*5 279*	*505*	*4 773*	*90%*
Non-associated	*876*	*2 653*	*5*	*2 649*	*100%*

* Proven reserves are broken down approximately by basin, based on information provided by the Iraqi Ministry of Oil. ** Ultimately recoverable resources (URR) for associated gas is derived from oil URR and known gas-oil-ratios. This gives higher associated gas URR than the USGS analysis. As the total URR is derived from the USGS data, this results in a lower non-associated gas URR, which may be underestimated by several hundred billion cubic metres (bcm).

Sources: USGS; data provided to the IEA by the US Geological Survey and the Iraqi Ministry of Oil; IEA databases and analysis.

Production

Iraq's declared aim for the gas sector is to utilise a valuable domestic resource in support of its economic development, with the power sector a strong priority for gas use, followed by domestic industry. Iraq also aims to become an exporter of natural gas. In our projections, Iraq's marketed gas production (net of flaring, venting and reinjection) is expected to increase significantly over the projection period, from less than 10 billion cubic metres (bcm) in 2010 to almost 90 bcm by 2035 in the Central Scenario, and close to 115 bcm in the High Case (Table 14.5). These projections depend on Iraq putting in place the gas infrastructure to capture and process the rising volumes of associated gas, mainly from the southern oil fields, and successfully developing non-associated gas fields. At the oil production levels anticipated in the Central Scenario and the High Case, the volumes of associated gas, alone, are not sufficient to realise Iraq's ambitions.

Gas production in the south of Iraq is closely aligned with the outlook for oil production, with an additional contribution expected from the Siba field, which is the only non-associated field currently appraised in the area. The share of the south in Iraq's gas production rises to a high point of 70% in 2020 in the Central Scenario, before falling back later in the projection period as non-associated gas production in other regions rises. The major sources of non-associated gas are in the north of Iraq, particularly in the KRG area. There is also a modest expansion in the west, where resources are thought to be large, but large-scale production is held back in our projections by the difficulty, thus far, to attract investors to exploration blocks and the distance from major demand centres.

Table 14.5 ▷ Iraq gas production by region in the Central Scenario and the High Case (bcm)

Central Scenario	2010	2015	2020	2025	2030	2035
South	3	7	29	40	43	47
Centre	-	-	0.5	0.5	1	1
West	-	-	0.3	2	6	9
North	4	7	12	30	32	31
Total	**7**	**13**	**41**	**73**	**82**	**89**
Of which associated	*5*	*10*	*32*	*42*	*46*	*51*
High Case	**2010**	**2015**	**2020**	**2025**	**2030**	**2035**
South	3	10	45	53	55	58
Centre	-	-	0.5	1	4	6
West	-	-	0.3	4	7	10
North	4	9	17	35	39	40
Total	**7**	**18**	**63**	**92**	**105**	**114**
Of which associated	*5*	*13*	*49*	*55*	*59*	*62*

Investment in Iraq in new or rehabilitated gas processing facilities has not kept pace with increases in oil output and, as of 2012, the country has the capacity to process around 8 bcm of gas per year. There is only a very limited national distribution network. Our projections point to the need, over the projection period, for processing capacity to increase by more than ten times in the Central Scenario, and by more than thirteen times in the High Case. They also require the timely development of a transport network to the power plants and industrial facilities that will account for the bulk of Iraq's domestic consumption.

The implication of these projections is that natural gas ceases to be an occasionally useful by-product of oil production, as in the past, and becomes a more pivotal and autonomous part of Iraq's energy strategy. This will require integrated planning from the government to ensure that the production, capture and processing of gas proceed in a coherent way, that processing plants are well sized and located and that the richer components of the natural gas stream – condensate, liquefied petroleum gas (LPG) and, once there is a market for it, also ethane – are separated and used productively in the national economy.

Figure 14.10 ▷ **Iraq gas production in the Central Scenario**

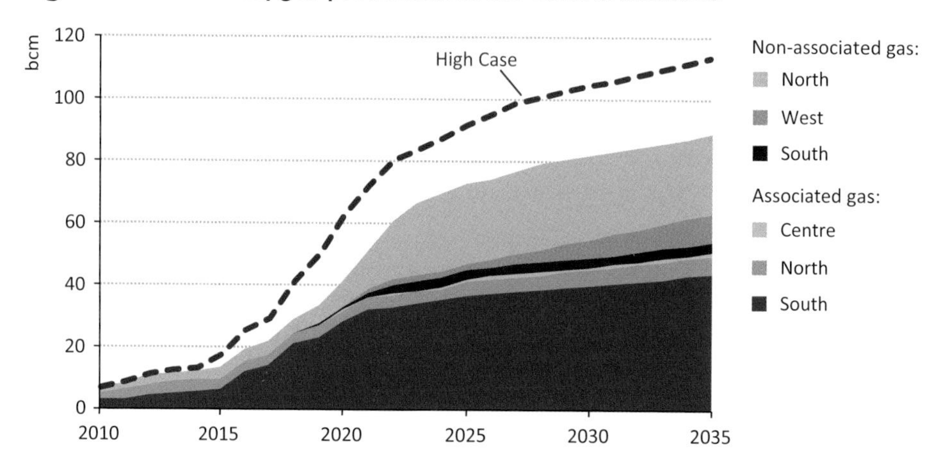

Associated gas

Iraq does not exploit all the associated gas that is currently being produced (much of which is flared), and faces a major task ahead to gather and process all the associated gas that will become available as oil production ramps up. The speed at which Iraq adds gas gathering and processing capacity will determine the pace at which it can realise its ambition to reduce gas flaring (Figure 14.11). In the Central Scenario, the projected addition of almost 30 bcm of annual processing capacity over the coming decade brings the volume of gas flared down below 4 bcm by the early 2020s, with flared volumes peaking in 2015 at around 17 bcm. In the High Case, more than 40 bcm of gas processing capacity is assumed to be added by 2020 but, because of higher projected oil output, the volumes of flared gas are higher, reaching a peak of 25 bcm in 2015.

Figure 14.11 ▷ **Iraq associated gas processing capacity additions and flaring reduction in the Central Scenario**

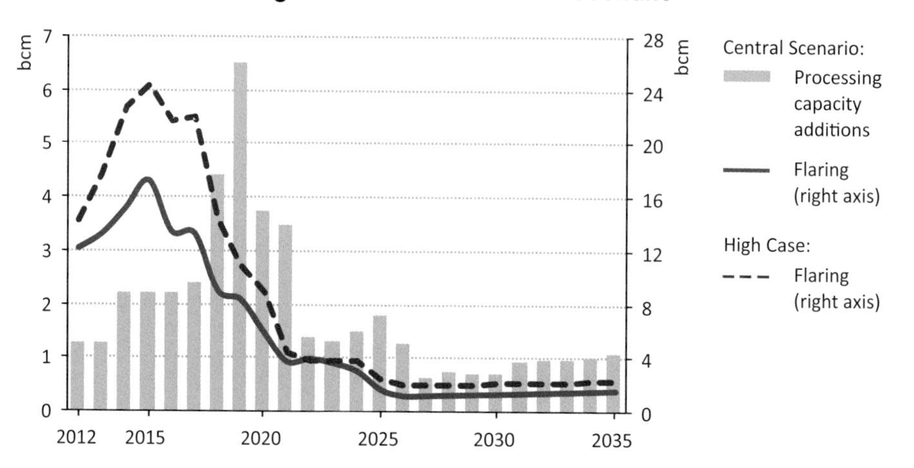

Box 14.3 ▷ Basrah Gas Company

The Basrah Gas Company (BGC), a joint venture established between the South Gas Company (51%), Shell (44%) and Mitsubishi (5%), is designed to provide an answer to a part of Iraq's gas needs by gathering and processing the associated gas from three southern fields (Rumaila, Zubair and West Qurna Phase I), much of which is currently flared. The short-term objective is to rehabilitate 30 existing facilities and two major processing plants in North Rumaila and Khor al Zubair previously operated by South Gas Company; but, as oil production increases, investment in new capacity will be essential to reach the planned throughput of 20 bcm of gas per year.[13] Total planned investment is $13 billion, with the possibility of an additional $4 billion for an LNG export plant if there is sufficient gas available beyond that required to meet Iraq's domestic needs.

A significant challenge for the BGC is to align the interests of the different parties involved. The technical service contracts for the fields initially supplying gas to the BGC do not include a remuneration fee for gas production, so the contractors have little commercial incentive to make gas readily available (albeit a strong public image incentive to reduce flaring). Moreover, while there is a flow of flared gas ready to be captured today, BGC's longer term investment decisions have to allow for the uncertainty over the timing and volumes of future associated gas output, both heavily dependent upon the trajectory of oil production.

The economic challenges BGC faces are similarly complex: low domestic gas prices (that are fixed at just over $1 per million British thermal units (MBtu) do not in themselves offer sufficient value to underpin the investment required. Instead, the company must rely on the returns it can earn from the higher value of the natural gas liquids extracted from the raw gas (condensate and LPG) and on a set of formulae, linked to the price of fuel oil, that determine the net price South Gas Company pays for the gas processed by BGC.[14] This price is higher than the domestic gas price (at an oil price of $100 per barrel, South Gas Company pays more than $2/MBtu), but the net value to Iraq is still substantial, primarily because gas displaces oil in power generation, freeing the oil for export at world market prices. Although BGC does not itself market any gas within Iraq, it remains dependent on how quickly demand develops – in particular how quickly gas-fired power capacity will be built and when the connecting infrastructure will be put in place.

13. This figure could be increased if additional gas volumes become available. We assume that BGC continues to process all gas from the three southern fields once their associated gas production exceeds this amount (soon after 2020 in our Central Scenario) and that additional gas processing capacity is developed to cater for other southern fields (whether within BGC or as part of a separate arrangement).

14. As well as having a 51% stake in BGC, the South Gas Company is also the entity that sells raw gas to BGC and that purchases processed gas from it.

The cumulative volume of gas flared during the years 2012-2020 in the Central Scenario is about 110 bcm; in the High Case, it is close to 160 bcm. If this gas were to be substituted for oil in Iraq's power generation, allowing the oil to be exported, the implied value of this flared gas is $70 billion in the Central Scenario and more than $100 billion in the High Case. This lost value provides a compelling reason for Iraq to move as quickly as it can to make early additions to gas processing capacity. This is the objective of the Basrah Gas Company (Box 14.3), although the area of operation of this company, as currently envisaged, does not cover all of Iraq's southern fields. Additional regulatory and institutional measures will be needed to encourage and require productive use of all Iraq's associated gas.

Non-associated gas

Iraq's non-associated gas production rises from a very low base to almost 40 bcm in the Central Scenario and to more than 50 bcm in the High Case (Figure 14.12). The bulk of this output comes from the north of the country, primarily from the KRG area: by 2035, total gas production from fields awarded by the KRG reaches 20 bcm in the Central Scenario and 29 bcm in the High Case, the overwhelming majority of which is non-associated.[15] The main discovered fields that are anticipated to contribute to production over this time are Khor Mor, Chemchemal and Miran in the KRG area, the Mansuriyah field (also included in the north in our projections), the Akkas field in the west, and the Siba field in the south. All of these are contracted to international operators. We also make allowance for production from fields that are yet to be found, in particular from the KRG area, given the high level of exploration activity underway.

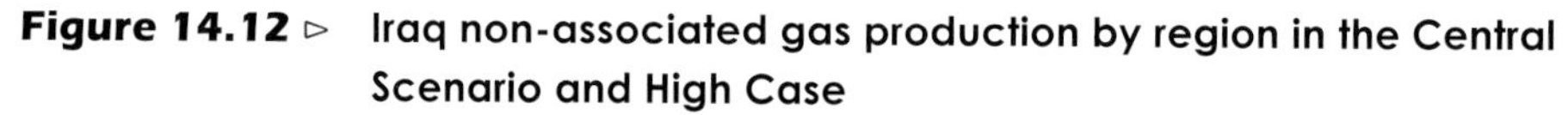

Figure 14.12 ▷ Iraq non-associated gas production by region in the Central Scenario and High Case

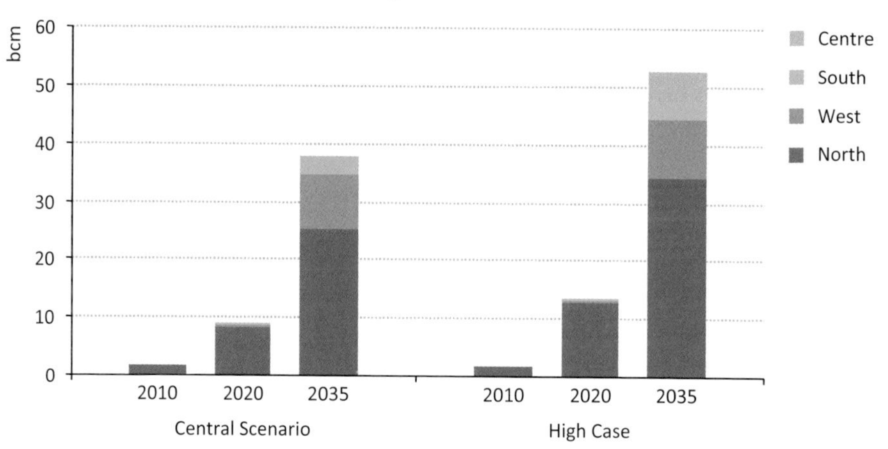

15. As in the case of oil production, the projection for the KRG area is highly contingent on the resolution of outstanding issues with the federal authorities regarding hydrocarbon governance and access to markets.

Thus far, the offer of gas exploration and development opportunities through the national licensing rounds has enjoyed only mixed success. The Akkas and Mansuriyah gas fields were included in the first national licensing round, but were awarded (together with Siba) only after being re-offered in the third round, with slightly improved terms.[16] In the fourth licensing round in 2012, seven exploration blocks considered gas-prone were on offer, only one of which was awarded (in central Iraq to a consortium led by Pakistan Petroleum). None of the six exploration blocks in the west of Iraq received bids. More attractive conditions may be proposed in a future licensing round.[17] The authorities may need to undertake prior exploration activities in prospective areas so that the risks and opportunities for potential investors are better understood. Over time, a more market-oriented electricity sector and a higher domestic gas price would provide stronger incentives for investment in gas for domestic use.

In the meantime, the main impetus to develop Iraq's non-associated gas resources has come from companies with an eye on the higher value offered by international markets (see Chapter 16). This has certainly been the case in the KRG area, where the region's gas resources have been proposed as a way to meet Turkey's gas demand and to fill pipelines onwards to southeast Europe. Even in the absence of clarity over the legal prospects for export, there are provisional proposals on the table to build gas export infrastructure from the Chemchemal, Khor Mor and Miran fields to the Turkish border. In other parts of Iraq, the possibility of export also appears to have been an important consideration for bidders, even though the structure of the technical service contracts on offer theoretically gives operators less of a stake in the eventual destination of the gas produced (as their remuneration is fixed). An example is the award of the contract for the Siba field in the south to a consortium led by Kuwait Energy, which has argued in favour of exporting this gas to Kuwait using a short cross-border gas pipeline from Rumaila that was built in the mid-1980s (but unused since 1990). While the evident interest in gas export (and in the valuable natural gas liquids that come with the produced gas) can be harnessed by policy makers to secure investment in Iraq's gas resources, they will need to safeguard adequate volumes for domestic use. More broadly, a national strategy for the gas sector – bringing together the anticipated levels of domestic demand and a vision for supply and infrastructure development, would help to clear away some of the uncertainties that impede large-scale investment.

Oil and gas investment

The projections in the Central Scenario and, even more so, in the High Case require very substantial levels of investment. In the Central Scenario, cumulative oil and gas sector investment during the period 2012-2035 amounts to almost $400 billion (in year-2011 dollars); in the High Case, this figure rises to almost $600 billion (Figure 14.13). A

16. The removal of signature bonuses and changes in the valuation of petroleum and gas offtake arrangements in the third round helped to attract interest in these fields.

17. Government officials are reportedly considering a fifth national licensing round.

combination of the speed at which production increases to 2020 and the incremental spending that is required to modernise infrastructure and refining capacity means that the investment requirement is spread unevenly over the projection period. The peak comes in the years between 2015 and 2020, when the required level of annual average investment in oil and gas is more than $20 billion in the Central Scenario. This effect is even more pronounced in the High Case, where the annual level of spending required between 2015 and 2020 is more than $30 billion. These levels of investment represent a large step up from oil and gas spending in 2011, which was an estimated $7 billion.

Figure 14.13 ▷ Iraq oil and gas investment in the Central Scenario

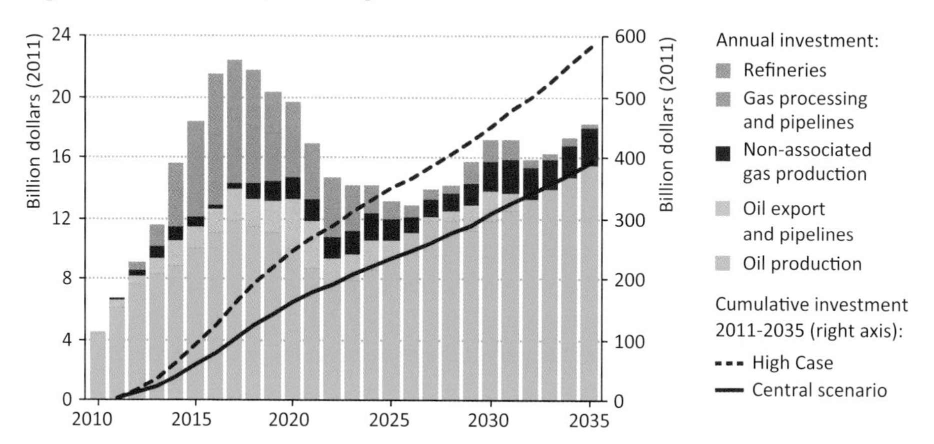

In our estimation, more than 90% of future investment in oil and gas will ultimately be paid for by the Iraqi treasury, either directly or via the cost recovery and fee arrangements made with the various companies operating in the Iraqi upstream (the main potential exception to this is the refining sector, where Iraq is looking for private investment). However, the responsibility for managing and executing projects at each stage of the value chain varies. In the upstream, this rests with the companies and consortia developing or exploring the various fields. Water supply is likewise the responsibility of individual operators, although in the south the economies of scale mean that a common approach has been chosen. Implementing transportation, storage and export projects is ordinarily taken on by entities responsible to the Iraqi Ministry of Oil.[18]

This division of responsibilities is not unusual in itself, but – if Iraq is to achieve a rapid increase in oil production and export capacity – it puts a premium on good performance by a range of public and private actors and on effective government decision making and co-ordination. The projected rate of increase in investment may stretch to the limit existing capacities within the various state entities that are responsible for project and contract management, many of which do not have a large body of officials with the necessary experience or training. There is a complementary risk that the companies charged with the

18. The exceptions are small tie-ins from individual fields to the main infrastructure and larger-scale privately built pipelines planned in the KRG area.

implementation of projects may encounter a range of bureaucratic obstacles, leading them to reconsider whether their ratio of reward to risk is attractive enough, compared with other investment opportunities.

To understand the scale of the task over the coming years, we undertook a detailed analysis of project preparation and completion timelines from across the Middle East, involving 216 large-scale investment projects that have been tendered since 2000 in Saudi Arabia, Qatar, the United Arab Emirates and Kuwait. We also examined the status of more than 140 projects at various stages of preparation in Iraq itself. For the oil and gas projects examined, we found that oil or gas upstream projects in the Middle East might take around two years from the first feasibility study to the final investment decision and then another two to six years to complete in full, with the average completion time being between three to four years. The timeline for refinery projects in the region was similar, but with the projects surveyed taking, on average, under four years to complete. Projects in domestic oil and gas transportation and storage were generally quicker to implement (Figure 14.14). In Iraq, a significant number of projects were re-tendered or cancelled at the project preparation stage but, where projects reached final investment decisions, the lead times were not untypical for the region. The sample size in each area was not large enough to draw firm conclusions about project completion times.

Figure 14.14 ▷ Average oil and gas project preparation and completion times in the Middle East

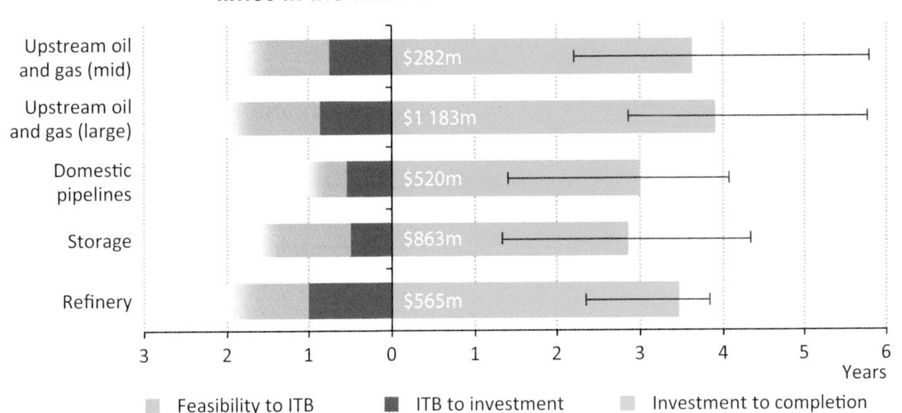

Notes: The periods measured are from the start of feasibility studies to the invitation for contractors or suppliers to bid on a defined project (the ITB or invitation-to-bid); then from the ITB to a final investment decision; then from a final investment decision to the completion of a project. The top and bottom 5% of the sample were excluded and the error bars show the range of completion times for the remainder. The figures represent the average size of the projects considered in each sample.

Sources: IEA analysis; Zawya database.

The implication of this analysis is that, as of mid-2012, the bulk of the projects that will determine Iraq's oil output to 2015 should already have reached a final investment decision or should be in the final stages of contract award. In addition, a sizeable contingent of the projects that will determine the supply trajectory during the latter part of the current decade should be at various stages of feasibility planning. The current signs are that, while progress is being made, there are important projects that have yet to be sanctioned. The Common Seawater Supply Facility is an important example. In the upstream, our analysis of recent contract awards shows that, so far in 2012, investment in 880 kb/d of new upstream facilities has been sanctioned, for the West Qurna (Phase II), Zubair and Badra fields. These facilities are expected to start to come on-stream in late 2013 and contribute to a further increase in production in 2014 and 2015. The number and scale of upstream project awards thus far in 2012 is higher than the levels seen in previous years. Overall, the scale of upstream projects awarded over the period 2010-2012 is broadly consistent with the levels of output anticipated in the Central Scenario. If the spending commitments seen thus far in 2012 are maintained (*i.e.* annual commitments to support new capacity of 900 kb/d), this would bring medium-term upstream capacity additions to the levels anticipated in the High Case.

Realising sustained growth in investment will, nonetheless, be a constant battle. The rate at which oil and gas investment will grow in the coming years will be affected by a wide range of factors, from questions of politics and security to everyday logistical and administrative concerns. A large number of these factors need to be moving in a positive direction in order to achieve a rapid increase in oil and gas production, since any missing elements or uncertainties in the regulatory framework will entail delays and a preference for smaller (often sub-optimal) tranches of investment.

If institutional, political or other obstacles prevent Iraq from moving oil and gas investment significantly beyond the levels seen in 2011, the anticipated growth in oil and gas production will not materialise at anything like the expected speed. To illustrate the downside risks, we examined the implications of a Delayed Case, in which annual oil and gas investment initially remains close to the levels attained in 2011, *i.e.* around $7 billion per year, and rises only gradually over the period to 2035. In this case, oil production continues to grow, but reaches only 4 mb/d in 2020 and 5.3 mb/d in 2035, 3 mb/d lower than in the Central Scenario by the end of the projection period. The implications of this case for Iraq's domestic energy balance and for international markets are covered in subsequent chapters.[19]

To mitigate the possibility of delay, a key priority for Iraq is to build up its capacity to manage investment projects (including enhanced capabilities in such areas as planning, budgeting and auditing, financing, procurement and environmental assessment) and to ensure that projects have a supportive environment for implementation (including some degree of insulation from macroeconomic risks – see Chapter 4). Further progress with defining a strategic vision and legislative framework for the hydrocarbons sector is also essential to promote co-ordination and common endeavour among the range of public and private actors shaping the future of Iraq's oil and gas.

19. Investment in other parts of the energy sector, including for power generation, is also assumed to occur later in the Delayed Case.

Iraq: fuelling future reconstruction and growth

An economy reaching full power, but when?

Highlights

- Iraq's domestic energy demand is projected to rise rapidly in the coming decades on the back of strong economic growth. In the Central Scenario, Iraq's energy demand is more than four times higher than 2010 by 2035, reaching 160 Mtoe, growing from being slightly greater than that of Kuwait to that of Italy today.

- Natural gas moves from sideshow to centre stage in Iraq's energy mix, with demand reaching more than 70 bcm in 2035. It becomes the main fuel for power generation, replacing oil and freeing up valuable resources for export. Fossil fuels dominate the energy mix, with hydropower and other renewables playing a small supporting role.

- An extensive effort enables installed gross power generation capacity in Iraq to almost quadruple by 2020 and reach more than 80 GW by 2035. Gas-fired power plants emerge as the lowest cost generation technology and, over time, Iraq moves from a predominantly oil-fired power generation mix to major use of gas, both in gas turbines and more efficient combined-cycle gas turbines (CCGTs). Without this transition, domestic oil demand would be around 1.2 mb/d higher in 2035 and Iraq would forego around $520 billion in cumulative oil export revenues.

- If planned new capacity is delivered on time, grid-based electricity generation will catch up with currently estimated peak demand around 2015, but it will take longer to build the necessary capacity buffer to allow for maintenance and unplanned outages without disruption.

- Iraq's power sector requires investment of more than $6 billion per year on average to 2035, but half is needed before 2020. Most of the investment is required in new generation capacity, but just under 40% goes to improving the transmission and distribution network, where efficiency gains can save the equivalent of the annual fuel input to six CCGT power plants.

- End-use energy consumption in Iraq more than doubles by 2020 and almost quadruples by 2035. A five-fold increase in the number of passenger cars in Iraq, from around three million today, underlies a rise in total oil demand to around 1.7 mb/d in 2020 and 2 mb/d in 2035. Pent-up demand for housing and household appliances bolsters residential energy use, while energy-intensive industries, such as cement, fertiliser and petrochemicals progressively attract increased investment.

- We estimate that, if action is not taken to phase out fossil fuel consumption subsidies in Iraq, they will triple to nearly $65 billion in 2035.

Overview of energy demand trends

This chapter presents energy demand projections for Iraq to 2035. It begins with an overview of energy demand trends, based on our Central Scenario, but noting also contrasting results in our High Case, where Iraq experiences a more rapid increase in oil and gas production, and in our Delayed Case, where we assume that lower levels of investment in Iraq's energy sector act as a significant constraint on the pace at which the economy develops.[1] Concentrating then on the Central Scenario, the chapter continues by examining the critical issue of the outlook for the power sector in Iraq and considers prospects in each of the main end-use energy sectors, transport, industry and buildings. An examination follows of the environmental outlook associated with the Central Scenario in terms of energy-related emissions and water use. The chapter concludes by briefly examining the most significant variations in sector-level energy demand trends and energy-related emissions in the High Case and the Delayed Case respectively.

Iraq's total primary energy demand moves into a new and prolonged phase of strong growth in the Central Scenario, increasing by nearly 6% per year to reach 160 million tonnes of oil equivalent (Mtoe) in 2035 – more than four times higher than in 2010 (Figure 15.1). Energy demand in the current decade alone grows by more than two-and-a-half times, reflecting rapid growth of the economy, fuelled by the proceeds of swift growth in hydrocarbon supply and a growing population. Energy demand growth moderates to an extent in the second half of the *Outlook*, as growth in the economy throttles back slightly as the pace of growth of oil production begins to diminish. Energy demand growth in Iraq outstrips that of its neighbours, resulting in its share of the Middle East total going from 6% in 2010 to 16% in 2035. Iraq's total energy demand grows from being slightly greater than that of Kuwait (with 3 million people) to that of Italy today (60 million people).

Figure 15.1 ▷ Iraq total primary energy demand by scenario

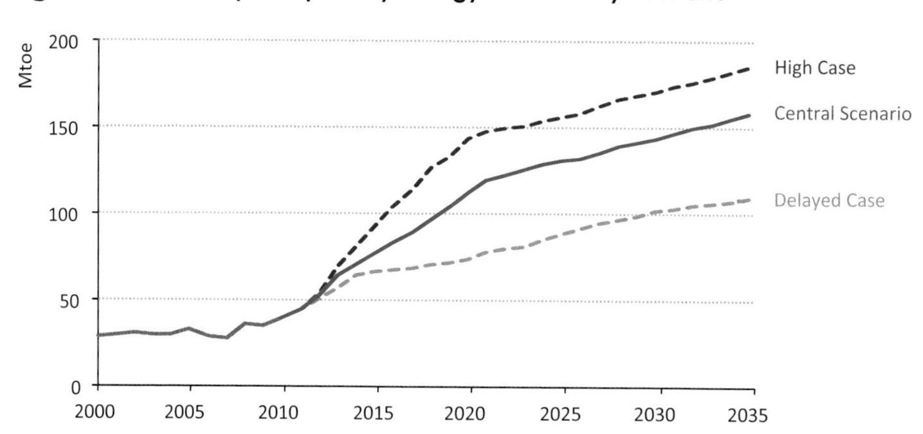

1. Chapter 13 defines the Central Scenario, High Case and Delayed Case in more detail.

In our High Case, faster oil production growth pushes economic growth higher and energy demand more than trebles over the current decade, reaching 187 Mtoe by 2035 – 17% higher than in the Central Scenario. Additional oil and gas export revenues stimulate higher government spending, and public and private consumption, resulting in the greater energy consumption. By contrast, in the Delayed Case, the flatter oil production growth profile results in slower economic growth and energy demand and Iraq's fuel mix also changes more gradually, as new power generation and gas processing facilities are built more slowly.

A common theme in the Central Scenario and the alternative cases examined is that fossil fuels continue overwhelmingly to dominate Iraq's energy economy between now and 2035 (Table 15.1). In the Central Scenario, fossil fuels still account for 99% of Iraq's energy mix in 2035 (58% oil, 41% natural gas), compared to 95% in the rest of the Middle East and 75% globally. Iraq has some limited electricity generation from hydropower, focused in the north of the country, and a small contribution from other renewables.

Table 15.1 ▷ Iraq primary energy demand by fuel and scenario (Mtoe)

			Central Scenario			High Case	Delayed Case
	1980	2010	2020	2035	2010-35*	2035	2035
Oil	9	32	75	92	4%	114	69
Gas	1	6	37	66	10%	71	39
Hydro	0.1	0.4	0.5	1.2	4%	1.0	1.3
Biomass	0.0	0.0	0.1	0.1	8%	0.1	0.1
Other renewables	0.0	0.0	0.1	0.3	n.a.	0.3	0.2
Total	**10**	**38**	**113**	**160**	**6%**	**187**	**110**

* Compound average annual growth rate.

The dominance of fossil fuels comes as no surprise in this hydrocarbon-rich country, but there are some shifting patterns of consumption over the projection period. In the Central Scenario, oil demand in Iraq more than doubles in the next ten years, to around 1.7 million barrels per day (mb/d) by 2020, and goes on to exceed 2 mb/d in 2035. In the period to 2020, demand growth is driven by transport and the power sector. Oil demand growth moderates to just over 1% per year after 2020, reflecting contrasting trends: a large and rapid decline in oil consumption in power generation (as natural gas availability increases), offset by a less rapid, but still significant, increase in consumption in end-use sectors, particularly transport. In the High Case, oil demand reaches 2 mb/d in 2020 and 2.5 mb/d in 2035, while it reaches only 1.6 mb/d in 2035 in the Delayed Case but retains a larger share of Iraq's overall energy mix.

In all scenarios, natural gas becomes a major pillar of the domestic energy economy. In the Central Scenario, gas demand increases by around 10% per year on average (Figure 15.2), reaching 39 billion cubic metres (bcm) in 2020 and 72 bcm in 2035. Natural gas accounts for around half of all energy demand growth in Iraq over the *Outlook* period, its share of Iraq's primary energy mix growing – at the expense of oil – from less than 20% in 2010 to more than 40% in 2035. Within these overall figures, demand varies across the country and through the year: there is a large summer peak in electricity demand (air conditioning), a smaller winter peak (heating) and variations from the north to the south, reflecting different climate conditions.

Figure 15.2 ▷ Iraq total primary energy demand by fuel in the Central Scenario

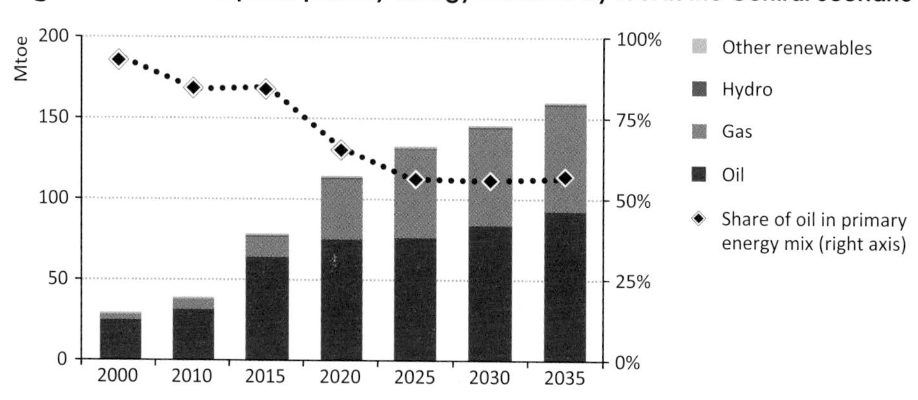

While the outlook for natural gas demand is strong, much depends on the extent and timing of adequate domestic supply (see Chapter 14). Delays on the supply side would have important implications for other fuels. If, for example, Iraq were not to achieve the switch to natural gas in its power sector posited in our Central Scenario, instead continuing to burn heavy fuel oil and crude oil at similar levels of efficiency to today, then this would come at a considerable cost in terms of lost oil export revenue over the *Outlook* period. In the High Case, natural gas demand reaches 55 bcm in 2020 and around 78 bcm in 2035. Gas use increases much more slowly in the Delayed Case, rising to 42 bcm in 2035, primarily as a result of lower availability of associated gas (due to the flatter oil production profile) and less rapid development of Iraq's non-associated gas resources. The switch to gas use in power generation is much less marked in this case.

In the Central Scenario, total primary energy demand per capita more than doubles in Iraq over the next decade, rapidly overtaking the world average in the process but stabilising at a level below that of the rest of the Middle East (Figure 15.3). The rapid increase is attributable to the combination of fast demand growth generally (including meeting currently unsatisfied demand) and intensive use of relatively-inefficient oil-fired power generation plant. As these trends weaken, and more efficient power plants are adopted, per capita energy demand stabilises at around one-third above the global average.

Figure 15.3 ▷ Total primary energy demand per capita in selected countries in the Central Scenario

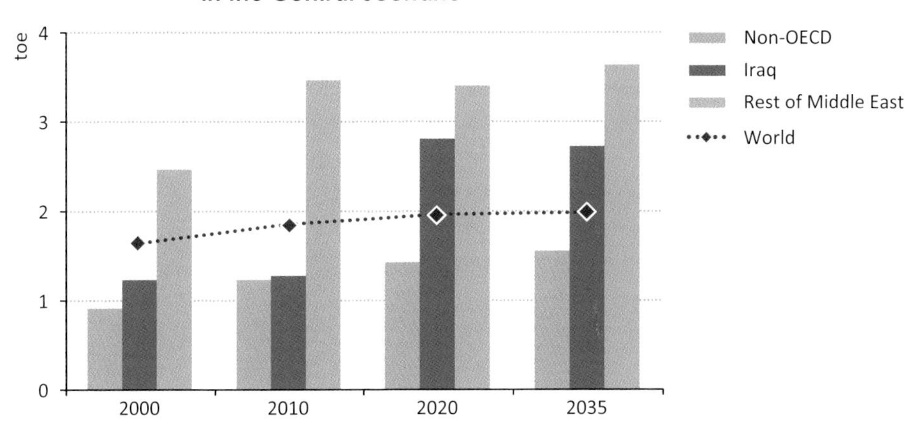

The energy intensity[2] of Iraq's economy was nearly double the global level in 2010. While its level was comparable with that of China, energy intensity in China is on a continuing improving trend, while Iraq's worsens early in the *Outlook* period, as it catches up with the existing overhang of unmet energy demand, before then improving gradually to 2035 (Figure 15.4). Even then, Iraq's energy intensity is more than double the projected global average at that time. In the High Case, additional economic growth outpaces the increase in energy demand early in the *Outlook* period, but reaching a very similar level to the Central Scenario by 2035. In the Delayed Case, the slower pace of GDP growth and of structural change in energy use means it does not even achieve the level of improvements in energy intensity seen in the Central Scenario.

Figure 15.4 ▷ Energy intensity in selected countries in the Central Scenario

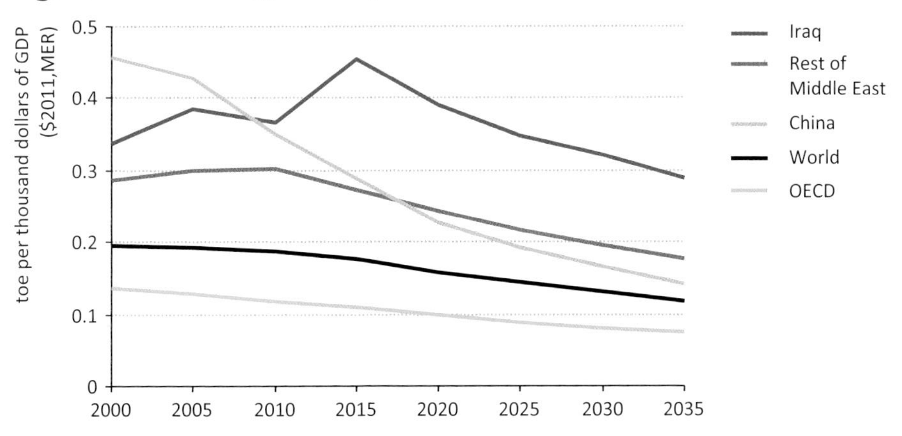

Note: MER = market exchange rates.

2. Energy intensity is an (imperfect) indicator of the energy efficiency of an economy (see Chapter 9). It is measured here as energy consumed (tonnes of oil equivalent) per thousand dollars of GDP, measured at market exchange rates (MER) in year-2011 dollars.

Outlook for the power sector in the Central Scenario

Developments in the power sector are of central importance to the outlook for Iraq. In the near term, the challenge is to provide an adequate supply of grid electricity to households and businesses, eliminating the considerable shortfall in generation relative to demand that has resulted from under-investment and war damage limiting supply, while economic and population growth have boosted demand (see Chapter 13). The solution adopted needs to develop Iraq's power system in a way that best supports its future economic and social development, in particular, by making sound choices over the technologies and fuels used to generate power, by improving network infrastructure and by reforming the electricity market.

Electricity demand

We estimate that electricity demand in Iraq was around 57 terawatt-hours (TWh) in 2010, but that the installed electricity generation capacity was able to satisfy only 58% of that level, around 33 TWh. Final electricity consumption grows five-fold in our Central Scenario, reaching just under 170 TWh in 2035 (Figure 15.5). Consumption growth averages more than 6.5% per year over the *Outlook* period, but the rate of growth is twice this level to 2020; responding to economic growth, population growth and the closing of the gap between demand and available supply. Of the three end-use sectors, industry's consumption increases fastest. From a relatively low starting point, it reaches seven times the level of 2010 by 2035. The residential sector retains the largest share of consumption over the period. It accounts for more than 45% of final consumption in 2035.

Figure 15.5 ▷ Iraq total final electricity consumption by sector and estimated unmet electricity demand in the Central Scenario

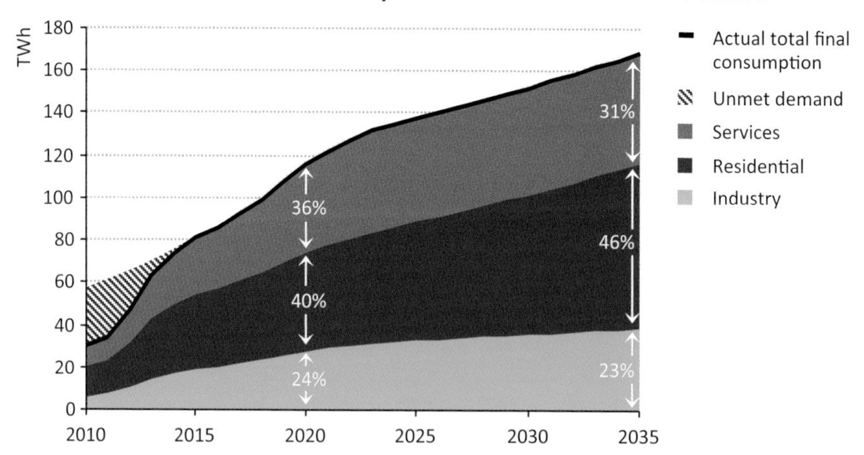

Meeting demand for electricity with reliable and continuous supply is the main immediate concern for Iraq's electricity sector. As a result, extensive activity is underway to build new generating capacity. If the planned new capacity is not delayed, we estimate that grid-based electricity generation will catch up with peak demand around 2015 in the Central Scenario (Figure 15.6), though this date is subject to uncertainties on both the supply and demand sides. Because peak demand is not currently met, it cannot be measured and can only be estimated. If demand proved to be 10% lower than is assumed in the Central Scenario, available capacity would exceed peak demand in 2014, all other things being equal; but if it proved to be 10% higher, equilibrium would not be reached until 2016. On the supply side, the timing of the commissioning of new plants (especially relative to the summer peak) is an important factor, as is the availability of fuel and any unexpected technical problems at existing plants. Whereas most electricity systems have a margin of 10-20% of spare capacity in order to guarantee stable supply, in the Central Scenario, Iraq achieves a 15% capacity margin only in 2017. Achieving the planned improvement in grid electricity would have a considerable impact on private diesel generators: their use declines rapidly once full grid supply is achieved.

Figure 15.6 ▷ Iraq net generation capacity available at peak and capacity required to meet peak demand in the Central Scenario

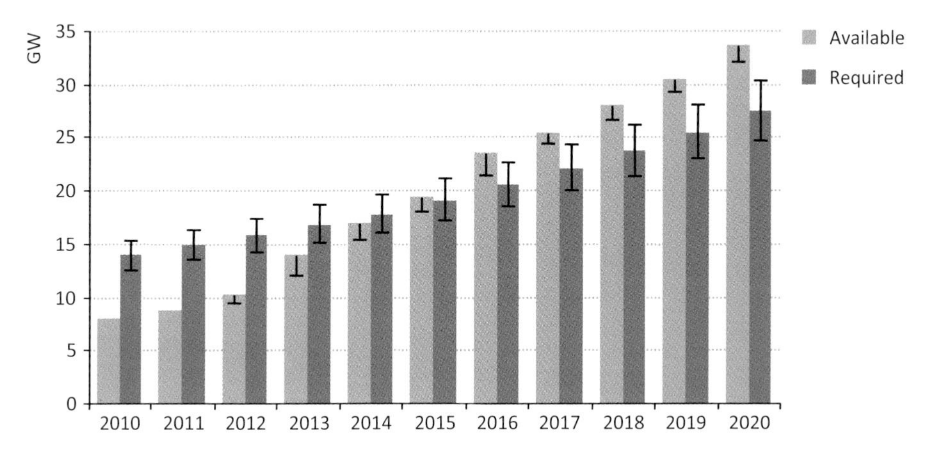

Notes: Error bars for "Required" represent ±10% of Central Case demand. Error bars on "Available" assume half the new capacity added comes online after the summer peak.

Iraq has two largely separate transmission systems: one in the Kurdistan Regional Government (KRG) area and the other covering the rest of the country. The electricity supply-demand balance in the KRG area is evolving differently to the rest of the country. Peak demand in the KRG area was estimated to be 2.9 gigawatts (GW) in 2011 (compared with 15.2 GW for the rest of Iraq). In the same year, peak supply in the KRG area was 2.3 GW, or over 80% of the estimated peak demand. In the rest of Iraq, peak supply in 2011 was 45% of estimated demand. Capacity additions are continuing in the KRG area and, as a result, it is likely that a continuous electricity supply will be achieved there sooner than in many other parts of Iraq.

Electricity generation

Generation capacity

In the Central Scenario, the gross capacity of grid-connected power plants grows quickly from 16 GW in 2010 to 60 GW in 2020, and then to 83 GW in 2035 (Figure 15.7). There is also a considerable shift in the technologies used. Early in the *Outlook* period, capacity from gas turbines (GTs)[3] grows quickly, reflecting projects currently under construction, projects for which contracts have been signed and the short-term government plans currently in place (Box 15.1 and Figure 15.8): in the period 2011-2015, two-thirds of new capacity additions are GTs burning crude oil, gasoil or heavy fuel oil (HFO).

Figure 15.7 ▷ Iraq grid-connected gross generation capacity in the Central Scenario

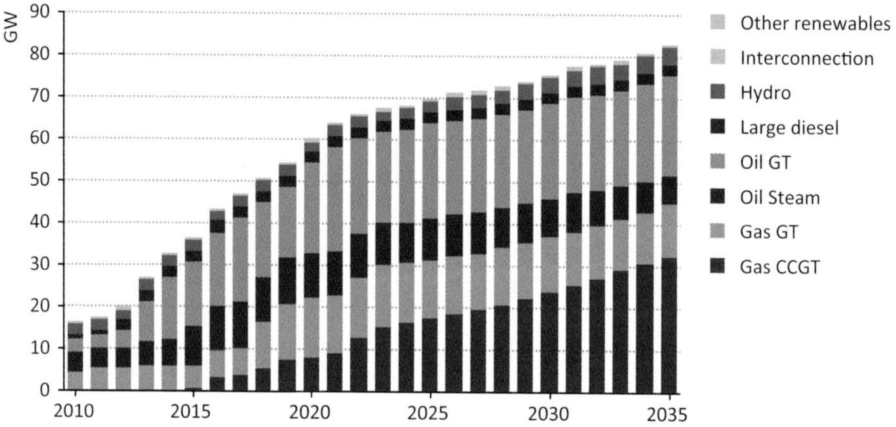

Notes: Large diesel refers to grid-connected diesel generators rather than small private or community generators. Despite their name, large diesel generators typically burn gasoil or HFO. Gas GTs include plants converted from oil GTs; gas CCGTs include plants converted from gas GTs. Interconnection refers to connections between Iraq and neighbouring countries.

GTs burning oil possess a number of characteristics that make them attractive to meet Iraq's immediate needs. They are modular technologies with relatively short construction times and low capital costs, which means that a large amount of capacity can be added quickly while minimising the call on the government's budgetary resources (Table 15.2). HFO, gasoil and crude oil in power generation also have advantages as fuels in the short term, given the lack of availability of natural gas and the relatively plentiful supply of HFO from refineries, crude from oil fields and, as required, gasoil from imports.

3. Throughout the power sector outlook for Iraq, there is an important distinction between the generation technology used and the fuel it consumes, which may change over time. Gas turbines (GTs) can burn natural gas and liquid fuels, such as HFO and crude oil. In this text, we therefore distinguish between gas GTs and oil GTs. Similarly, diesel generators are capable of burning a number of fuels other than diesel: in Iraq they use gasoil and HFO.

Box 15.1 ▷ Iraq near-term surge in power provision

Several different approaches have been used in order to achieve a rapid build up in Iraq's power generation capacity.

In 2008, the Ministry of Electricity procured a large number of gas turbines under so-called "Mega Deals" from GE and Siemens. For the power plants included in these deals, a separate contracting process was then undertaken to award engineering, procurement and construction (EPC) contracts to other companies for the installation of the turbines. Administrative delays in this second phase have added considerably to the time required to install new generation capacity: in some cases the time taken to award the EPC contracts was longer than the construction time for the plant itself.

Another approach, the independent power producer (IPP) model, has been used successfully as a commercial structure for power plants in the KRG area, with the government retaining responsibility for fuel supply and offering a specific rate of return on capital to the owner of the plant. The IPP model has been explored in the rest of Iraq, but not adopted successfully.

Figure 15.8 ▷ Expected completion time of selected power plants in Iraq

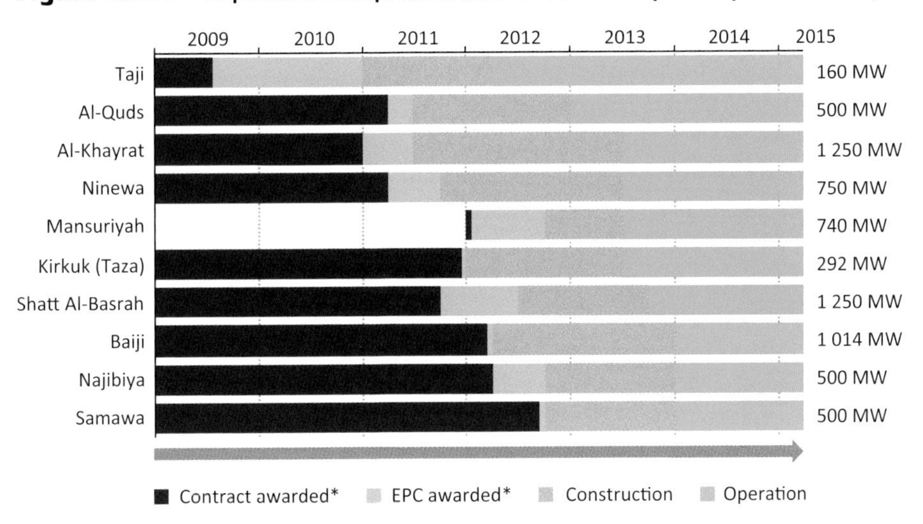

15

* Time elapsed from date of award.

Notes: EPC is an engineering, procurement and construction contract. The contract for Mansuryiah was awarded separately from the Mega Deals and an EPC contract was signed directly with the manufacturer, Alstom. All plants are gas turbines.

Sources: Zawya and Middle East Economic Digest databases; Special Inspector General for Iraq Reconstruction annual reports; IEA analysis.

Table 15.2 ▷ Capital costs, efficiency, and construction times for the main types of new generation technologies in Iraq

	Capital cost ($/kW)	Efficiency	Construction time
Gas turbine	900	36%	1.5 years
Combined-cycle gas turbine (CCGT)	1 200	57%	2.5 years
Steam turbine	1 900	39%	4 years
Diesel generator	1 800	38%	1 year
Hydro	3 700	n/a	4-7 years

Notes: Capital costs include interest during construction and "soft" costs such as legal expenses and EPC. Efficiencies are representative of the maximum currently attainable by each technology under standard conditions. Environmental factors, such as temperature, and operating conditions may mean that actual efficiencies achieved are lower.

Sources: Iraq Electricity Masterplan (Parsons Brinckerhoff, 2010); IEA analysis.

Beyond 2015, there is a marked shift in the generation technologies installed, driven primarily by the increased availability of natural gas to the power sector. This shift towards natural gas occurs partly through the construction of new gas-fired GTs and combined-cycle gas turbines (CCGTs), and partly as some oil-fired GTs are converted to burn natural gas instead of liquid fuels. In the Central Scenario, over 9 GW of GTs originally commissioned as oil-fired plants are converted by 2020 to use natural gas as a fuel. In addition to the increase in gas-fired GTs, the share of gas-fired CCGTs in the capacity mix increases considerably, growing from zero in 2010 to more than 32 GW – 40% of installed capacity – by 2035. Some of this increase is achieved by upgrading GTs. The first GT to CCGT conversion project, in Erbil, is underway, with another, in Sulaymaniyah, expected to start soon. Conversions from GTs to CCGTs entail additional capital costs but increase the efficiency of the plant considerably: for GTs installed in the period to 2015 in Iraq, this could result in a 44% increase in gross generation for the same fuel input. A small number of additional steam turbine and grid-connected diesel plants are added in the period to 2016, as a result of the government's short-term plans; beyond this there are no additions of these technologies given their high costs of generation relative to gas-fired technologies.

Generation by fuel

In the Central Scenario, Iraq's annual gross generation of electricity grows from around 50 TWh in 2010 to over 200 TWh in 2020, an increase in electricity output comparable to that expected in the European Union over the same period. As noted, there is an important accompanying shift in the fuel used. Iraq's current electricity generation comes primarily from liquid fuels: heavy fuel oil, crude oil and gasoil accounted for 57% of generation in 2010. Natural gas accounted for 33% of the mix. In the near term, most of the growth in electricity generation depends on liquid fuels: gross generation from crude oil and refined products more than triples to 100 TWh in 2015 – around 70% of the electricity output projected for that year (Figure 15.9).

This expansion of liquid fuels means that the share of natural gas in the power generation mix actually falls in the near term to just below 25%. However, in the period after 2015, this trend is reversed rapidly as an increasing amount of natural gas becomes available. The availability of gas as a fuel, at lower cost, means that it establishes itself as the dominant base-load and mid-load fuel by around 2020, with oil capacity used to meet demand only in the peak demand periods. The use of liquid fuels for power generation peaks around 2015 and then declines steadily. The share of gas increases to around 60% of the mix in 2020 and to nearly 85% by 2025, retaining this share for the rest of the period. The increase in the share of gas generation is accompanied by an increase in overall efficiency, which grows from 30% in 2010 to 42% in 2035, due largely to the introduction of CCGT technology. The marked reduction in HFO use after 2020 occurs as upgrades to Iraq's refining capacity come on stream, reducing the share of HFO in refining output.

Figure 15.9 ▷ Iraq electricity generation by fuel and overall efficiency of power generation in the Central Scenario

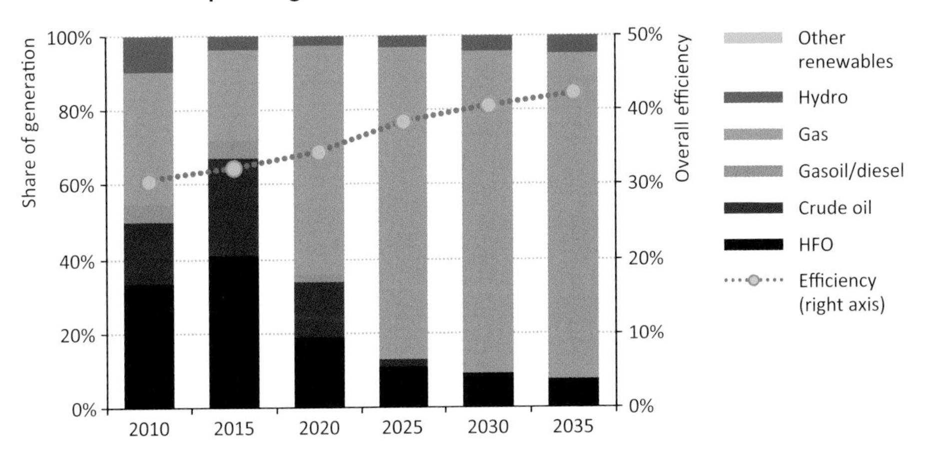

Electricity generated from hydropower made up nearly 10% of supply in 2010 (nearly 5 TWh). Capacity is projected to increase modestly over the *Outlook* period, and as a result hydro generation reaches 14 TWh in 2035, with a focus in the north of the country, where there is also potential for some small-scale hydro in more remote communities. This projection is sensitive to decisions taken in other countries that affect water flows entering Iraq. While Iraq has a large increase for non-hydro renewables, particularly solar, the costs of exploiting this for power generation remain high relative to alternative fossil fuel technologies. Due to this, and based on existing policies, there is only a small increase in non-hydro renewables, such as solar, over the *Outlook* period (Box 15.2).

The improvement in efficiency and the shift from oil to natural gas help to limit the environmental impact of electricity generation. In 2010, the CO_2 emissions-intensity of the Iraqi power sector was 700 grammes of carbon dioxide per kilowatt-hour (g CO_2/kWh), higher than the Middle East average, which stood at 680 g CO_2/kWh. By 2035, it reaches 442 g CO_2/kWh, a 37% reduction and a major decline, even by international standards.

Box 15.2 ▷ The role of renewables in Iraq

Renewable energy sources, almost all of which is hydro, play a small but important role in Iraq's existing electricity generation mix, accounting for nearly 5 TWh of generation in 2010.

A number of sites have already been identified as having potential for new hydroelectric plants, two of which – the Bekhma and Badoush dams on the Tigris River – saw the start of construction during the 1980s, but work was abandoned in the wake of the sanctions imposed in the 1990s. There is considerable potential for expanding hydroelectricity in the KRG area, as well as some opportunities in the rest of Iraq, but projects are very capital-intensive and water availability is constrained. Future dam construction will depend primarily on the extent and security of upstream water supply and the water management policies of the Ministry of Water Resources, with electricity generation being an associated benefit of projects undertaken primarily for other purposes. In the Central Scenario, an additional 2 GW of hydro capacity is installed over the *Outlook* period. The rapid growth of generation from other sources means that hydro's share of the generation mix falls to around 5% by 2035.

Iraq has very good solar resources. Even though the Middle East's best solar irradiance is farther south – in Saudi Arabia and Yemen for example – Iraq's average solar irradiance is similar to that in north Africa. Iraq also has some history of research into solar power (which was curtailed significantly during the decades of wars and sanctions). Today, solar research activities are sponsored by the Ministry of Electricity and the Ministry of Science and Technology. The Ministry of Electricity has a number of off-grid solar research stations, with capacity of a few tens of megawatts (MW). Despite the strength of the resource, grid-connected solar electricity generation – either through photovoltaics (PV) or concentrating solar power (CSP) – will remain a very high-cost option, compared to fossil fuels. Our Central Scenario assumes a small amount of solar PV capacity – less than 50 MW – is added by 2035.

Outside the electricity sector, solar water heating is likely to be a highly attractive option for buildings if subsidies for fossil-fuel alternatives are phased out. Wind speed in Iraq is relatively low and the biomass resource is moderate. Our Central Scenario does not assume any large-scale development of wind or biomass resources during the *Outlook* period, though international collaboration could help to change this picture. Iraq became a signatory in 2009 to the international convention creating the International Renewable Energy Agency (IRENA), paving the way to full membership of the organisation when the agreement is ratified.

Generation costs

Any fossil fuel used for power generation in Iraq is fuel that cannot be sold for export. The full cost of fossil-fuelled electricity generation to the Iraqi economy therefore depends on the international export value of the fuels used, rather than subsidised domestic fuel prices. In 2010, around 160 kb/d of oil was burned in power plants; at international market prices this had a value of over $4 billion. By 2015, 520 kb/d of oil is expected to be used in power generation, with an international market value of $22 billion. The planned shift to natural gas – which is already envisaged by the government and provided for in the Electricity Masterplans drawn up for the Iraqi power sector (Parsons Brinckerhoff, 2009a and 2010) – is driven by the relative costs of competing generation technologies, which are themselves strongly influenced by the relative value of oil-based fuels and natural gas in Iraq: per unit of energy, oil has a considerably higher export value than gas.

Combining the fixed costs, operational and maintenance costs, and fuel costs of new generation plants allows the total costs per unit to be compared. Using international oil and gas prices and considering the main generation technologies and fuels available in Iraq, gas-fired CCGTs emerge as the lowest cost base-load technology. Based on international prices in the Central Scenario, they have a levelised cost of $77 per megawatt-hour (MWh) in 2020 (Figure 15.10). For plants operating only at peak periods, gas-fired GTs are the most economic at international fuel prices, due to their lower capital costs.

Figure 15.10 ▷ Levelised cost of base-load plants with international fuel prices in 2020 (Central Scenario) and current domestic fuel prices in Iraq

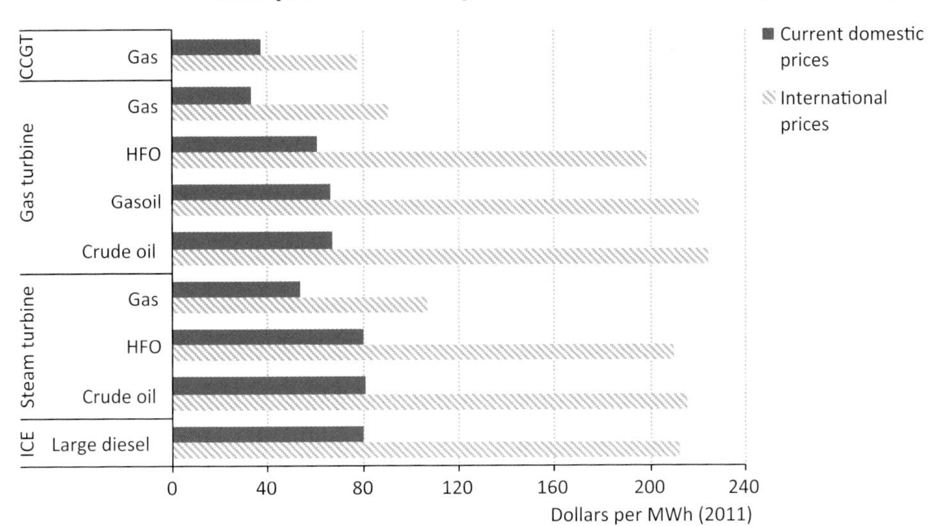

Note: CCGT = combined-cycle gas turbine; ICE = internal combustion engine.

Fuel prices for the power sector are currently set by the government at a level considerably below their international market price. Even at these subsidised fuel prices, gas-fired generation is more economic than oil as a base-load generation technology. In addition to the direct costs of oil, liquid fuels have several technical disadvantages relative to gas. Oil corrodes generating equipment more than gas, shortening the lifetime of the plants and increasing the cost and the amount of time per year that has to be spent on maintenance. For Iraqi plants, this may be expected to increase the duration of annual planned outages from 1-2 weeks to 5-6 weeks. The cost of replacement parts is also estimated to increase five-fold.

Transmission and distribution

Iraq's transmission network is divided into two. The network in the north is managed by the Ministry of Electricity in the Kurdistan Regional Government, while the rest of the country is served by another grid managed by the Ministry of Electricity of the federal government. Interconnection between the two transmission networks is very limited. Despite periodic attacks against them, Iraq's transmission networks are in relatively good working condition, so future investment is expected to be directed primarily to improving efficiency and reliability, and increasing capacity to accommodate new demand and generation. In order to have enough transmission capacity, with a reasonable degree of redundancy, Iraq is likely to need around an additional 28 000 kilometres (km) of 400 kilovolt (kV) and 132 kV lines between its main demand centres, such as Baghdad, Basrah and Mosul.

The distribution networks are seriously degraded, particularly outside the KRG area, suffering from poor design, lack of maintenance and theft of electricity. Losses in the distribution network are particularly high compared to most countries in the Middle East (see Chapter 13). As well as the waste of energy associated with such losses, the degraded condition of the distribution network means the supply to consumers is of poor quality, including low voltage levels and frequent disconnections. In addition to repairing the existing networks, the booming growth in electricity demand will require considerable expansion of both the transmission and distribution networks. Undertaking the repair and maintenance of the existing networks and expanding them to meet new demand will require a large number of well-trained staff as well as institutional reform. It has been recommended, for example, that a distribution code be adopted that sets uniform standards and procedures for upkeep of the distribution network (Parsons Brinckerhoff, 2010). Capacity building initiatives have the potential to play a positive role in the power sector and beyond, such as the Centre of Excellence for research and training on energy proposed within a joint declaration by the European Commission and Iraq on an enhanced strategic energy partnership.

In addition to the important wider economic benefits to be derived from a reliable supply of electricity, improving the design, maintenance and operation of the transmission and distribution networks should reduce system losses, thereby avoiding the construction of unneeded generation capacity and unnecessary use of fuel for generation. In the Central Scenario, we assume that measures to improve new and existing network infrastructure – which would include adoption of technical standards, institutional reforms and training of engineering staff – lead to a fall in network losses over the *Outlook* period from 34% to 29%. Although this figure is still considerably higher than the Middle East average, which reaches 16% by 2035, the associated saving in fuel is 3.3 Mtoe in 2035: equivalent to the annual fuel input to six CCGT power plants.

Investment in the power sector

Cumulative investment of over $140 billion will be required in the power sector over the *Outlook* period (Figure 15.11). This is 35% higher than Iraq's total GDP in 2010, but is only around 3% of its expected oil export revenues over the period. The need to make up for past under-investment and to meet rapidly growing demand means that half of the total investment needs to take place prior to 2020 – averaging $8.5 billion per year. This large commitment is vulnerable to increases in cost or decreases in revenue to fund it (see Chapter 16). Typically, customers' payments would be a key source of revenue to fund such investment, but the price of electricity in Iraq is low and payment enforcement is weak (Box 15.3).

Figure 15.11 ▷ Iraq annual average investment in power generation capacity and electricity networks in the Central Scenario

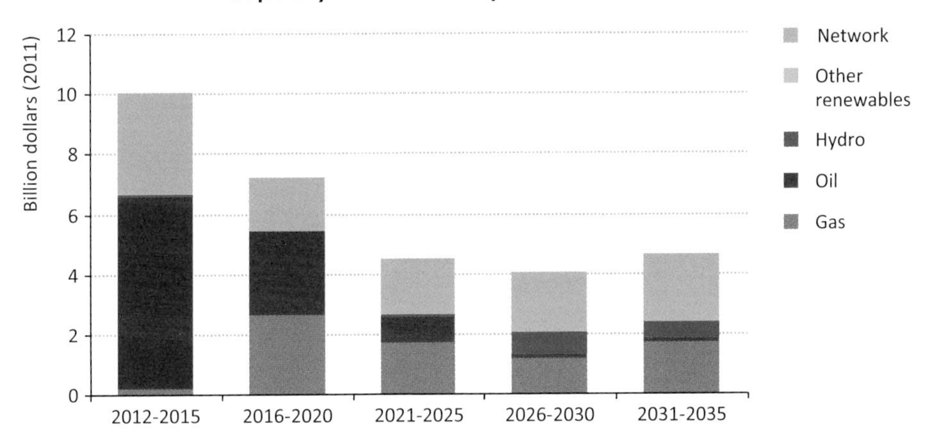

In the period to 2020, nearly three-quarters of investment in power plant capacity is in oil-fired installations, mostly oil-fired GTs. A significant amount of this capacity is later converted to use gas as a fuel, though some of it is retained as oil-fired peaking plants. A quarter of the investment goes to new dedicated natural gas plants, shared almost equally between GTs and CCGTs. From 2020, the pattern of investment changes, with almost two-thirds of power plant investment going towards gas-fired capacity, and an increasing share going to either new CCGTs or GT-to-CCGT conversions. Investment in hydropower is concentrated in the period 2026 to 2035, making up 30% of the investment in generation capacity during this period. Electricity networks represent a little under two-fifths of total power sector investment.

Box 15.3 ▷ Reforming the power sector

Beyond addressing the immediate need to add capacity quickly to the network, efforts are under way to put the electricity sector on a legal and regulatory footing that will facilitate its development over the coming decades. Two relevant laws have been drafted and are currently under consideration. The first is the Ministry of Electricity Law, which would restructure the Ministry of Electricity. The second is the Electricity Regulation Law, which would establish a regulatory office for the electricity sector, with responsibility for monitoring the sector, licensing participants, enforcing technical codes, and resolving disputes and consumer complaints. Other instruments of governance will also need to be developed and implemented. A positive step in this direction has been taken with the drafting of a grid code, setting out detailed technical standards for the use of the electricity network.

Electricity metering, billing and payment collection is an additional issue in need of urgent attention. Electricity consumption in many households and businesses is not metered, while readings for those that do have meters are taken infrequently. Even where customer use of electricity is metered, billing and collection of payments is not comprehensive, partly because of inadequate information technology systems for billing and tracking payments. It is too difficult at present to distinguish unofficial use of grid electricity from technical losses arising from problems with the network. It will be politically difficult to address the problem of payment collection fully and vigorously while electricity supply remains seriously inadequate, but plans need to be laid for robust action once circumstances permit.

Effective implementation will be needed across the board to establish a stable and well governed electricity sector, a process itself dependent on hiring and retaining well trained staff. Successful reform of the sector will create the investment climate necessary to encourage the private sector to shoulder more of the burden in future.

Measuring the value of changes in the power sector

Compared with a case in which the power generation mix does not change from 2015 to 2035, the collective impact of the changes in technology, fuel mix and efficiency in the Central Scenario, reinforced by reduced end-user subsidies, is to lower oil demand in the power sector by 1.2 mb/d by 2035. The cumulative export value of the fuel saved over the *Outlook* period is $520 billion (Figure 15.12). The savings come mainly from changes in the fuel mix, as the increased availability of natural gas means that the share of oil in electricity generation falls from over 70% in 2015 to under 10% in 2035. There are also efficiency gains (mainly from using CCGTs instead of GTs for generation) that increase the technical efficiency of the power sector from 30% to 42% in 2035; this improvement reduces overall fuel demand by more than one-fifth by 2035. Reducing transmission and distribution losses results in a 7% reduction in the amount of generation required by the end of the period. The reduction in end-user price subsidies for electricity results in additional fuel savings that yield additional export revenues (Spotlight). In addition, the amount of capital investment needed in generation capacity is reduced.

Figure 15.12 ▷ **Export value of fuel saved in the power sector as a result of changes from 2015**

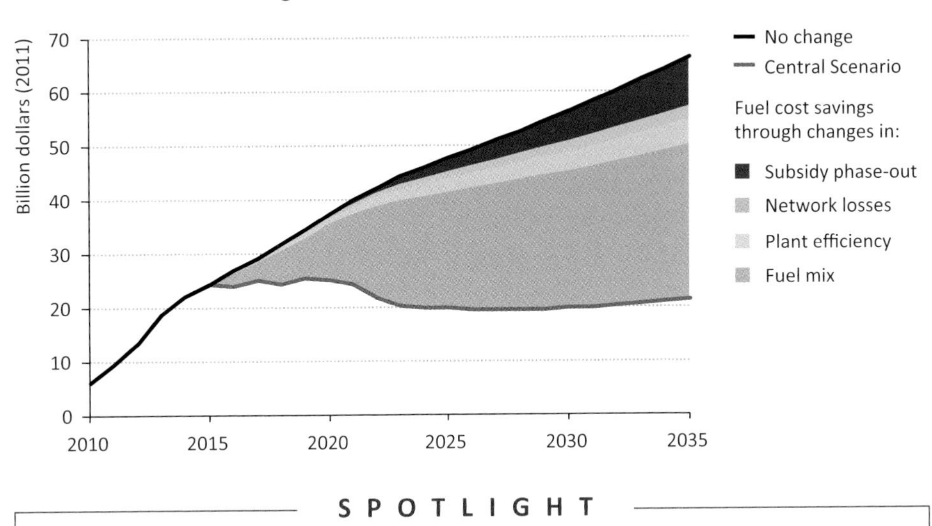

S P O T L I G H T

What is the potential cost of maintaining fossil-fuel subsidies in Iraq?

We estimate that the value of fossil-fuel subsidies in Iraq in 2011 was around $22 billion. While not the most pressing issue facing Iraq in the short term, the perspective to 2035 shows the enormous cost of maintaining subsidies at equivalent levels throughout the projection period. The experience of other countries in the Middle East (see Box 3.1) suggests that, without measures to moderate energy demand (with gradual moves to market-related pricing at the top of this list), there is a risk of runaway consumption growth, the associated subsidies cutting government revenues that could be allocated much more productively elsewhere.

In our Central Scenario, we assume a partial but incomplete subsidy phase-out in Iraq over the projection period, based on government statements and our assumptions for other oil exporters in the Middle East. The subsidy rate for electricity begins to decline from around 2020 to approximately half the level of today in 2035, with a similar pattern of reduction for oil subsidies. In the case of natural gas, we assume that the rate of subsidy remains unchanged, as the government seeks to encourage its use in power generation and in the industrial and energy transformation sectors. The partial phase-out means that the projected cost of subsidies in 2035 is $35 billion (around 6% of GDP), rather than nearly $65 billion (around 12% of GDP). The cumulative benefit to Iraq (fiscal and opportunity costs) is $214 billion over the *Outlook* period (Figure 15.13), although the value of these subsidies is nearly $1 trillion over the period as a whole.

There are a number of distinct trends underpinning these trajectories. Over the next decade, oil subsidies are seen to increase by the largest amount, as domestic demand grows strongly and the international oil price increases gradually. Electricity and natural gas subsidies are also expected to increase (gas subsidies increase to $5 billion in 2035), due to growing demand and increasing international fuel prices (which are used to derive our reference prices), even as the rate of subsidy remains fixed. Around 2020, the trends begin to change. Electricity demand is fully met, allowing currently very low prices to be increased gradually. New refinery capacity comes online in increments (reducing oil product imports), and products on the domestic retail market are priced, increasingly, at commercial rates. Natural gas largely replaces oil in the power generation mix, which lowers the subsidy burden because of the lower international reference price of gas relative to that of oil.[4] Overall, the total value of subsidies in our Central Scenario peaks before 2025 and then declines gradually.

Figure 15.13 ▷ Iraq fossil-fuel consumption subsidies in the Central Scenario

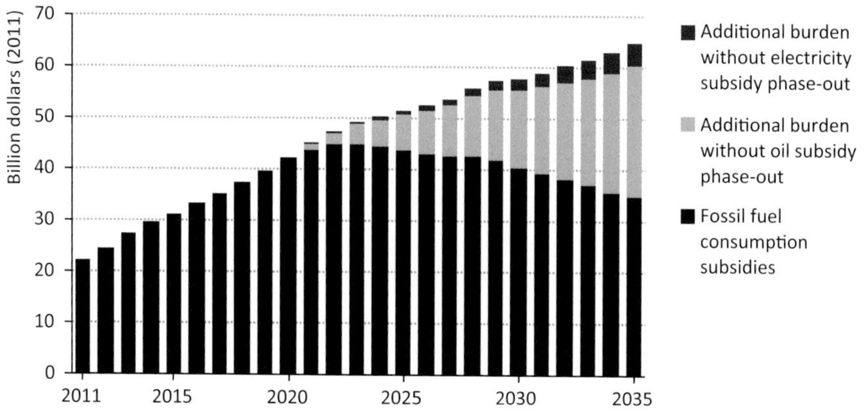

Outlook for end-use sectors

Transport

Oil demand grows more strongly in transport than in any other sector, increasing by 6% per year on average to reach 1.2 mb/d in 2035 (Figure 15.14). By 2035, Iraq's oil demand in transport is greater than that of Japan, even though it is projected to have significantly fewer vehicles. This reflects a lack of policies targeting fuel economy improvements and comparatively high average vehicle mileage. It also indicates the potential impact that new policies, aimed at increasing efficiency and promoting public transport, could have in restraining demand growth in this sector (potentially supporting increased oil exports).

The largest consumers of oil in Iraq's transport sector are passenger light-duty vehicles (PLDVs), of which there were an estimated three million on the road in Iraq in 2010.[5] Around 35% of households in Iraq have a private car, with higher shares in the

4. The savings identified as being due to subsidy phase-out in Figure 15.12 relate to lower volumes of fuel required, as a result of lower demand in the power sector. In Figure 15.13, they are captured as fossil-fuel subsidies because they represent a reduction in fossil fuel demand.

5. Direct communication with the Iraqi government.

KRG area (46%) and in Baghdad (40%), and an average of 30% in other governorates (Iraq Central Statistics Organization, 2012). The level of vehicle ownership is already quite high, at more than 100 vehicles per 1 000 inhabitants, just below the Middle East average, but more than double the level in China and nine times higher than in India. Iraq relies on vehicle imports – around 150 000 cars and pick-up trucks were imported in 2011 – and, while the picture is changing, the typical car imported at present is the Saipa Saba from Iran (a small vehicle with fuel consumption of up to 9 litres per 100 km, depending on the drive cycle) or a second-hand car.

Figure 15.14 ▷ Iraq PLDV ownership and oil demand in transport in the Central Scenario

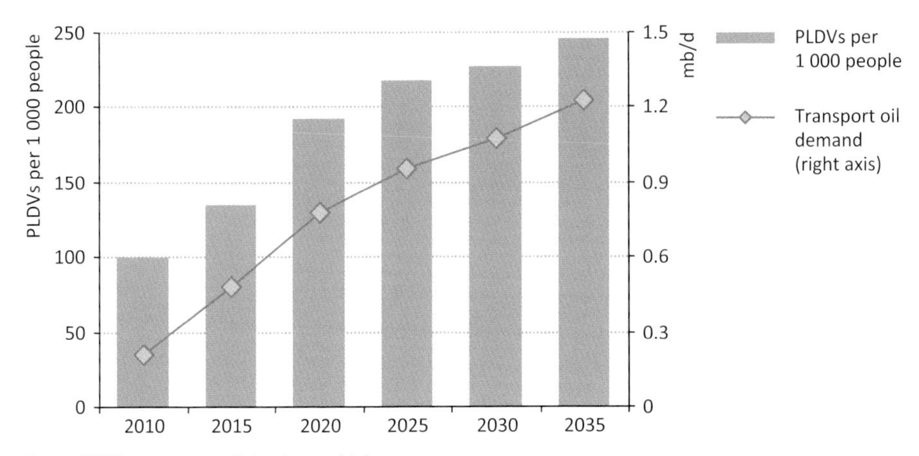

Note: PLDV = passenger light-duty vehicle.

In our Central Scenario, the number of PLDVs in Iraq grows strongly, reaching around 8 million in 2020 and over 14 million in 2035. Vehicle penetration also increases, overtaking the average level of the Middle East and reaching more than 190 PLDVs per 1 000 people in 2020 and nearly 250 per 1 000 people in 2035 (Figure 15.15). This increase is driven by high average GDP growth, of around 7% per year, limited alternative means of transport and an increase in the population of working age. Road freight activity (in million tonne-kilometres) is also set to more than triple to 2035. An important qualification to these projections arises from the fact that Iraq's road transport system is already in pressing need of rehabilitation and modernisation and its road density is relatively low, at around 0.1 km per square kilometre of land surface. Slow progress in re-building this network will constrain growth in road transport.

Iraq is expected to increasingly open up to regional and international transport, particularly aviation and road transport, in support of broader economic development. Its limited coastline will become increasingly busy with seaborne freight, oil exports in particular. Rail represents an opportunity for Iraq but its future is highly uncertain. Rail provided an important channel for personal mobility and freight traffic in Iraq until the 1980s, but the performance of the network has since collapsed and passenger usage is only a fraction of the previous level. Rehabilitation of the rail network offers an opportunity to facilitate economic growth, reduce the pressure on road transport and, potentially, to link into other rail projects in the region; but it would also require substantial investment. In our Central Scenario, we assume that the railway network expands only modestly, leaving most freight traffic and individual mobility dependent on road transport.

Figure 15.15 ▷ PLDV ownership and stocks in selected countries in the Central Scenario

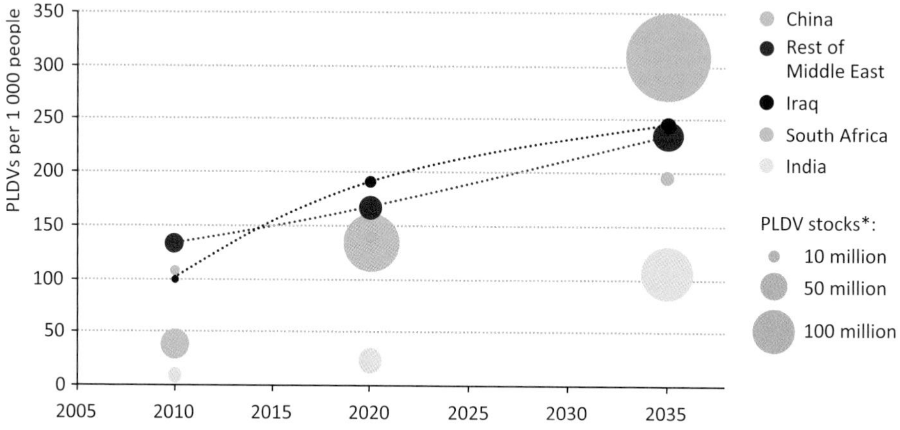

* Size of bubble reflects total stock of PLDVs.

Industry

Despite industry's habitual role as one of the foundations of an economy, in Iraq this sector consumed only slightly more than 3 Mtoe in 2010, around the same level as in 1990. This statistic reflects under-investment and disrepair that has left a legacy of under-utilised and largely uncompetitive assets. As a result, current assets do not provide any reliable guide to future activity. The level of public investment and how far Iraq goes to create a business climate conducive to private sector investment will be the critical determinants of future industrial energy demand.

In our projections, energy consumption in Iraq's industry sector increases by more than 6% per year, to reach just over 15 Mtoe in 2035 (Figure 15.16), with consumption being a mix of natural gas, natural gas liquids and electricity. The trend in industrial demand growth follows that of the economy more generally, being strong to 2020 and then slower for the remainder of the projection period. It is important to distinguish between the dominant oil industry in Iraq, which grows by 5% per year on average, and the much smaller non-oil industry, which sees growth of around 8% per year. Despite this stronger growth, the non-oil sector accounts for only one-quarter of total industry value-added in 2035. The trend towards private sector participation, already contributing towards growth in oil and gas production, is expected to spread gradually to other industrial sectors. The cement industry is a case in which private sector participation is beginning to occur, stimulated by an emerging boom in construction.

Figure 15.16 ▷ Iraq domestic energy balance* in the Central Scenario, 2035 (Mtoe)

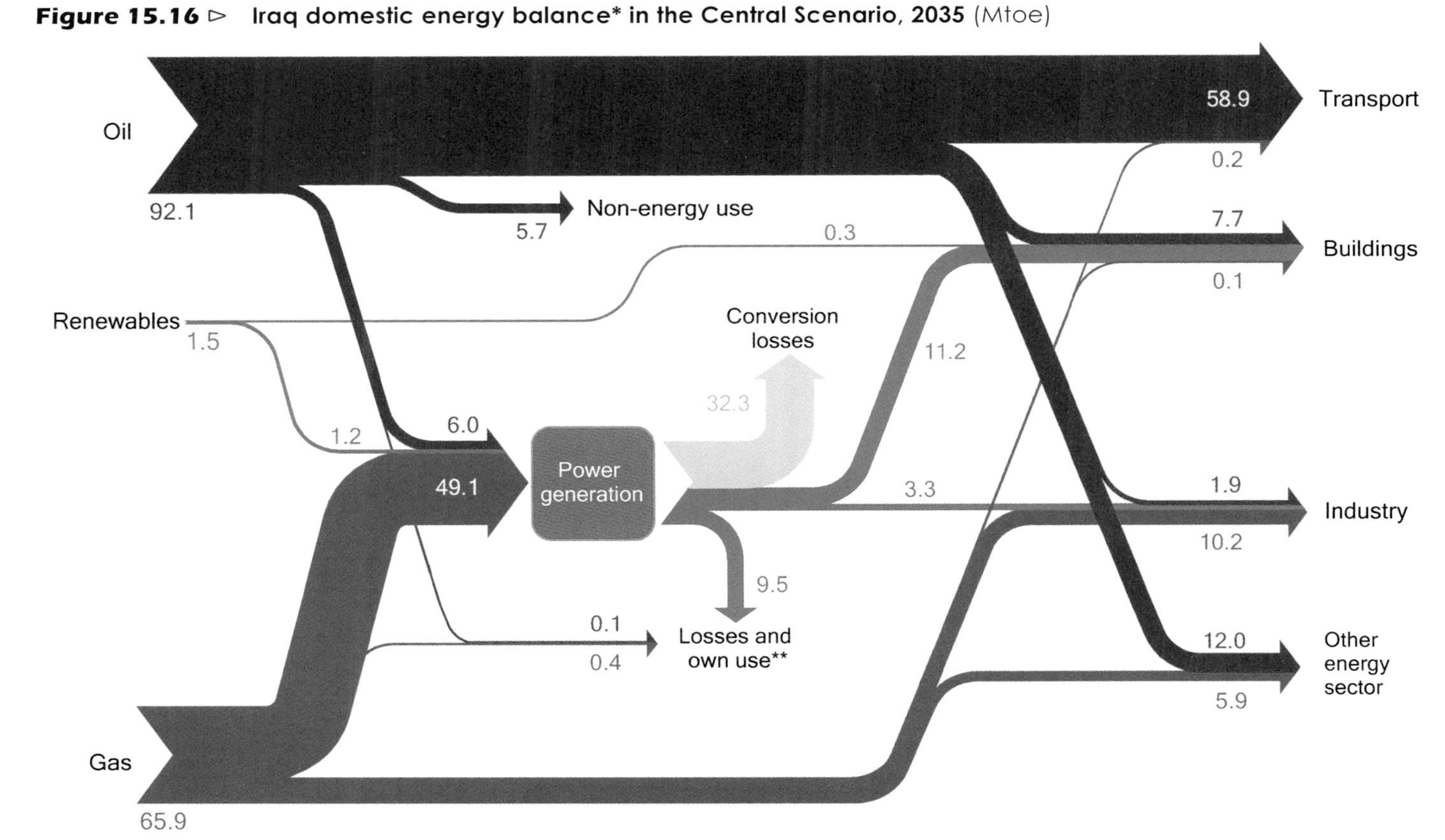

*Oil exports, oil product exports/imports and electricity imports not shown. Gas flaring not shown. ** Includes losses and fuel consumed in oil and gas production, generation lost or consumed in the process of electricity production, and transmission and distribution losses.

In the Central Scenario, demand for hydrocarbon feedstock for fertiliser and petrochemicals grows to nearly 5 Mtoe in 2035,[6] with growth concentrated in the period after 2020. Iraq has adopted the policy objective of building up industrial infrastructure that can use and add value to the country's hydrocarbon sector, seeking to derive comparative advantage from low input costs. Our projections suggest that, once processing facilities are in place, Iraq would have ample supplies of ethane fully capable of supporting a petrochemical industry selling particularly into Asian markets, where petrochemical and fertiliser demand is expected to grow. However, Iraq's existing petrochemicals sector is very small and will take time to develop, particularly given other investment priorities and the challenge of gaining a position in regional and global markets. As in many other sectors, a successful build up of the petrochemical industry in Iraq will depend on significant long-term investment and confidence as to the secure long-term availability of feedstock. The development of the Saudi petrochemical sector by SABIC is a good example of the challenges and opportunities, while also illustrating the long time required to develop higher value-added sectors and to move beyond basic commodities.

Buildings and other sectors

In our Central Scenario, the buildings sector (for this analysis, it includes the residential, services and agriculture sectors), sees demand increase from around 7 Mtoe in 2010 to over 19 Mtoe in 2035. The main factors underpinning the projected increase in consumption are population and economic growth, urbanisation, new housing supply and burgeoning appliance ownership. Iraq is currently short of between 1 million and 3.5 million housing units (Ministry of Planning, 2010). Several large-scale housing projects are either planned or under construction. This significant level of expected construction creates an important opportunity to incorporate high energy efficiency standards, rather than locking in inefficiency in capital stock that has a long lifetime. International standards already exist in this area. Consumers could then have lower electricity consumption and therefore lower bills, and less investment in power generation capacity would be needed.

In the Central Scenario, the share of electricity in total energy consumed in the buildings sector grows from 35% to 58%. Improved reliability of cheaper grid-based electricity gives a boost to demand, with households buying a greater range of appliances and using them more often. Air conditioning will remain a key component of consumption, along with other consumer goods; for example, the proportion of households owning a computer, now generally below 20% (IKN, 2012), is expected to increase. As with building standards, the introduction of energy performance standards for household appliances, especially air conditioning, is an obvious priority.

---

6. This implies investment in three large-scale crackers for ethylene production (running at 100% ethane), as well as additional midstream infrastructure (pipelines, fractionators) and downstream processing plants. This would bring Iraq's petrochemical output to a level greater than that of Qatar today.

Liquefied petroleum gas (LPG) is widely used as a cooking fuel in Iraq and we project LPG demand to remain strong, reaching more than 210 kb/d in 2035. Iraq has imported LPG for several years and this is projected to continue until around the end of this decade when, due to increased natural gas processing capacity, Iraq moves from being an LPG importer to a net exporter.

Environment

Energy-related emissions

Iraq's energy-related CO_2 emissions are projected to increase in the Central Scenario from around 100 million tonnes (Mt) in 2010 to just over 400 Mt in 2035 (Figure 15.17).[7] They grow to account for nearly 17% of Middle East energy-related CO_2 emissions in 2035 and a little more than 1% of global emissions – a high figure compared to Iraq's 0.3% of the global economy and 0.7% share of the global population. The increase in CO_2 emissions slows after 2020, as a result of lower economic growth, a move to more efficient power plants and a gradual reduction in fuel and electricity subsidies. Gas flaring reduces over time, coming down to below 4 bcm by the early 2020s, resulting in a reduction in related emissions (see Chapter 14).

Figure 15.17 ▷ Iraq energy-related CO_2 emissions by fuel and its share of Middle East emissions in the Central Scenario

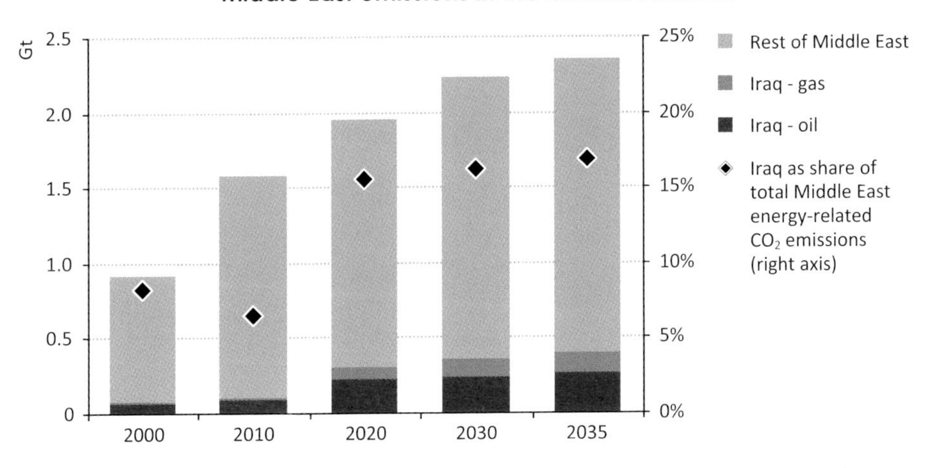

Reliance on fossil fuels, together with out-of-date technologies and a lack of energy efficiency standards, made Iraq one of the most carbon-intensive economies in the world in 2010. However, Iraq's carbon intensity is set to improve significantly, falling by a quarter by 2035 (Figure 15.18). This improvement is related mainly to the rapid growth in the size of Iraq's economy (based on investment in modern equipment), but changes in patterns of energy consumption also play a role. The most important of these is the shifting fuel mix

7. Figures for energy-related CO_2 emissions cover only the productive use of energy and so do not capture the effect from reduced flaring of natural gas that is anticipated in the early years of the projection period.

in electricity generation, as natural gas displaces liquid fuels. The projected reduction in private generation in the power mix also brings local pollution benefits, but a significant increase in road transport has an opposite effect. Substantially modifying the link between an expanding economy and increases in energy use and CO_2 emissions is an important challenge for Iraq, and one that will require policies directly aimed at the inefficient energy practices and technologies that are commonplace in the region – often the policies already mentioned, such as action to move towards end-user prices for energy that reflect the market value of the energy consumed, standards for energy-efficient appliances (in particular, for air conditioners, given their significant role for electricity demand) and fuel-efficiency standards for vehicles.

Figure 15.18 ▷ Carbon intensity in selected countries in the Central Scenario

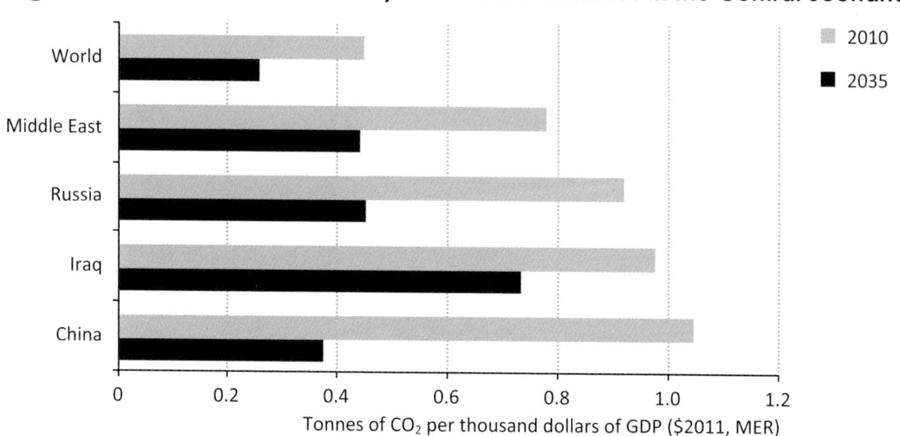

Water use

Water and oil are both important determinants of Iraq's future prosperity. A growing economy and population will result in an increasing overall requirement for water and energy. Available water supply has long been a concern, with water flow from the Tigris and Euphrates rivers expected to continue to fall over the projection period. However, water shortages facing Iraq are not attributable solely to developments further upstream, but also to inefficient methods used in agriculture and elsewhere.

Water requirements to support energy sector activities (oil and gas production and power generation) constitute just over 1% of Iraq's overall needs in 2015 (Figure 15.19) and so are not expected to be a determining factor in its water strategy. To take one example, growth in water requirements for cooling in power generation is not expected to match the pace of generation expansion, as CCGTs, in particular, have low water requirements relative to many other technology options, including the oil-fired power plants that they are often expected to replace in the Central Scenario. The development of hydroelectric capacity in Iraq's power sector will be strongly linked to water policy choices (see Box 15.2). However, water could still become an important factor in Iraq's energy strategy, especially in the south of the country, where water injection is needed to maintain oil production but freshwater supply is relatively scarce. To reduce excessive demand on freshwater in the south, early and ongoing investment is required to draw supplies of seawater from the Gulf for energy purposes (see Chapter 14). In the future, Iraq may also be expected to invest in some desalination capacity, mainly in the south, to help meet the growing demand for potable water, but the scale and timing is uncertain.

Figure 15.19 ▷ Iraq water requirements, 2015

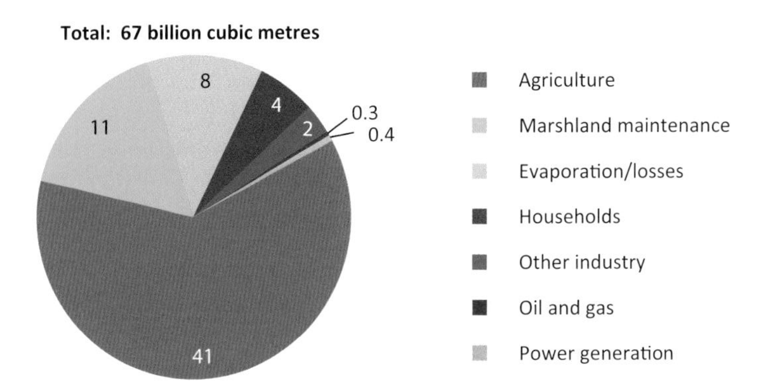

Note: The requirement for oil and gas is for the anticipated production in the Central Scenario.

Sources: Iraq Ministry of Water Resources; IEA analysis.

High Case

In our High Case, the rapid increase in oil production (see Chapter 14) feeds back into a significantly higher trajectory for GDP growth: Iraq's economy grows to well over five times its current size, with GDP being around $100 billion higher in 2035 than in the Central Scenario (Table 15.3). Iraq experiences GDP growth of over 14% per year to 2020 (much more rapid growth than that experienced by China in the last decade) and averages nearly 8% growth per year over the projection period as a whole.

Table 15.3 ▷ Iraq key domestic energy indicators by scenario

		Central Scenario		High Case		Delayed Case**	
	2011*	2020	2035	2020	2035	2020	2035
GDP (MER, $2011 billion)	115	289	552	384	649	186	331
Primary energy demand (Mtoe)	44	113	160	144	187	74	110
Oil demand (mb/d)	0.8	1.7	2.1	2.0	2.5	1.2	1.6
Gas demand (bcm)	8.5	39	72	55	78	18	42
Installed electrical capacity (GW)	17	60	83	69	86	48	63
Electricity generation (TWh)	55	202	277	233	301	142	208
CO_2 emissions (Mt)	123	304	402	380	477	202	277

* 2011 figures are estimates. ** For discussion of the Delayed Case, see the concluding section to this chapter.

The main domestic energy implications of the High Case are:

- More rapid economic growth driven by oil revenues sends energy demand even higher, particularly in the period to 2020. Total primary energy demand is over 140 Mtoe in 2020 – more than one-quarter higher than the Central Scenario – and nearly 190 Mtoe in 2035.
- Natural gas and oil are called upon to meet almost all of the additional energy demand, with a small additional contribution from hydropower. The increase in natural gas consumption is driven largely by power generation, but also by increased consumption in industry.
- Oil consumption is pushed higher by increased demand for transport, with PLDV ownership per 1 000 people reaching almost 300, nearly three times today's level. By 2035, Iraq's primary energy mix is 61% oil, 38% natural gas and 1% hydro and other renewables – a slightly higher reliance on oil than in the Central Scenario.
- Installed electricity generation capacity quadruples during the period to 2020, a huge increase. More GT capacity is installed early in the *Outlook* period, but there is also more rapid conversion to CCGT plants around 2020, resulting in a faster improvement in the average efficiency of the power sector.
- Additional generating capacity and increased electricity consumption largely offset one another in the near term, meaning that Iraq does not manage to fully meet electricity demand from the grid any more quickly than in the Central Scenario, still closing the demand gap around 2015.
- Almost all of the increase in energy demand is met from fossil fuels, meaning that – in the absence of additional policy changes – more rapid growth in the economy does not equate to a better outlook for renewables or CO_2 emissions, which are around 25% higher than the Central Scenario in 2020 and close to 20% higher in 2035.

The main uncertainty affecting the High Case is whether Iraq can succeed in turning a very rapid rise in export revenue into greater national prosperity and, importantly, whether much higher required levels of energy-sector investment can be achieved. The experience of other resource-rich countries suggests a flood of oil-generated wealth can, in some circumstances, offer substantial opportunities to improve social and economic welfare; but it can also have unintended and undesirable consequences for macroeconomic stability, for national institutions and governance and for the development of non-oil sectors of the economy (Chapter 16).

Delayed Case

In the Delayed Case, we look at the contrary and cautionary possibility that stubborn obstacles delay energy sector developments. It tests the consequences if investment in Iraq's energy sector were to rise only slowly from the levels seen in 2011. The reality of the risk is recognised by the Iraqi government, as noted in the foreword to the National Development Plan 2010-2014: "Over the past few years, Iraq has demonstrated the inability to manage the annual budget, allocate investments, and has struggled to create an economy that allows for progress and development" (Ministry of Planning, 2010). Delay in investment may occur because of a combination of factors, such as frustration of the government's efforts to modernise and reform Iraq's legal framework and institutions. These constraints could be progressively lifted, allowing Iraq to move back towards a trajectory consistent with the Central Scenario (or even the High Case) at a later time. But if the low investment assumption is extended throughout the projection period, almost half of the cumulative investment in the Central Scenario is lost, resulting in consistently slower growth in oil production (reaching 4 mb/d in 2020 and 5.3 mb/d in 2035) and slower progress in the construction of new electricity capacity. In this Delayed Case:

- Reduced oil export revenues dampen Iraq's GDP trajectory, with growth being around 2% lower per year on average than in the Central Scenario. As a result, total primary energy demand grows much more slowly, reaching 74 Mtoe in 2020 and 110 Mtoe in 2035.
- Iraq's energy mix is slower to change. Oil retains its predominant position for longer, with a 75% share of primary consumption in 2020 (compared with 67% in the Central Scenario) and a 63% share in 2035 (compared with 58%). This reflects the later appearance of additional gas supply and processing infrastructure.
- The slower development of gas infrastructure means that liquid fuels play a larger role in power generation for longer, which reduces their availability for export to 2.7 mb/d in 2020 and 3.8 mb/d in 2035. Delays to power generation projects mean that full electrification is achieved one year later than in the Central Scenario, despite the fact that electricity demand is lower.
- The delay in putting in place new gas processing facilities also means a more prolonged period of gas flaring. Over the projection period as a whole, total flared volumes are higher than in the Central Scenario, even though oil production (and therefore associated gas output) is considerably lower.

Overall, the cumulative impact on Iraq's economy is a loss of nearly $3 trillion relative to the Central Scenario, as export revenues are sharply lower, other industrial and services sectors fail to develop quickly and the power sector makes slower progress in moving to cheaper and more efficient generation.

Implications of Iraq's energy development

What does Iraq's energy mean for Iraq, and the world?

Highlights

- Iraq stands to gain almost $5 trillion in revenues from oil exports over the period to 2035 in the Central Scenario. Revenues of this magnitude offer a transformative opportunity to Iraq if they can be used to stimulate much-needed economic recovery and diversification.
- To realise this opportunity, Iraq will need strengthened institutions and human capacity, sound long-term strategies for the energy sector and for the economy, and ensure efficient and transparent management of revenues and spending. Direct employment in the hydrocarbons sector will generate only a small fraction of the jobs that Iraq needs for its youthful and growing population.
- Iraq will need cumulative investment of over $530 billion in new energy supply infrastructure to 2035. The bulk of this investment is in upstream oil, followed by the power sector. The annual requirement is highest in the current decade. Much of this investment is expected to be financed from the state budget: during the period to 2016, we estimate that Iraq will face a net financing requirement of $27 billion to meet the levels of spending required.
- Iraq adds 5.6 mb/d to global oil supply by 2035 in the Central Scenario, a larger contribution to global supply growth than any other producer. Iraq becomes the second-largest global exporter after Saudi Arabia and a key supplier to fast-growing markets in Asia. Without such supply growth from Iraq, oil markets would be set for very difficult times, characterised by higher and, in all probability, more volatile prices that would reach almost $140 per barrel in 2035 in real terms.
- In the High Case, Iraq's ambitions provide for near-term production to increase much more quickly. Growth in output from Iraq over the current decade would be enough to account for more than three-quarters of the growth in supply anticipated in this period in the Central Scenario. Non-OPEC oil supply is also expected to grow, at least in the medium term.
- Gas exports from Iraq start around 2020 in the Central Scenario and approach 20 bcm by 2035. The resources and market opportunities are there to expand exports further, as Iraq can potentially provide a very cost-competitive source of gas supply to neighbouring countries, to European markets and – via LNG from the south – to Asia.

Energy in Iraq's economic and social development

The Central Scenario

The anticipated increase in oil production in Iraq is set to generate exceptional growth in export revenues over the coming decades. In the Central Scenario, these revenues total almost $5 trillion in the period to 2035 (in year-2011 dollars), increasing from around $80 billion in 2011 to more than $200 billion per year in the 2020s and closing in on $300 billion by 2035 (Figure 16.1). Revenues of this magnitude create a transformative opportunity for Iraq, with the potential to support much-needed economic recovery and diversification. On a cumulative basis, $5 trillion is 10% of global revenues from oil trade to 2035; it is half the amount that China is projected to spend on oil imports over the same period and more than the oil import bill of India. This trajectory would mean that Iraq's share of total OPEC revenue is set to grow from the current 7% to around 18% by 2035. Iraq's gross domestic product (GDP) in 2035 would be five times larger (in real terms) than now, and its GDP per capita would rise to a level comparable with that of Brazil in 2010.

Figure 16.1 ▷ Iraq oil and gas export revenues in the Central Scenario

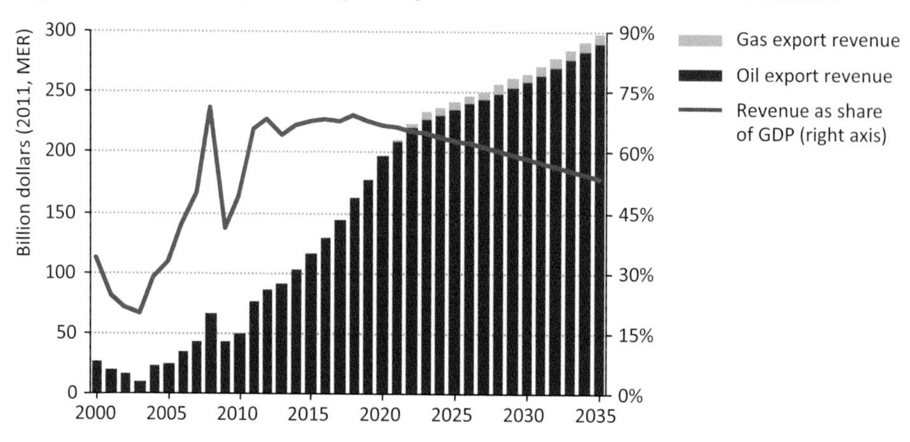

Note: GDP is measured at market exchange rates (MER) in year-2011 dollars.

But, even if realised, these revenues come with significant risks attached. They constitute a very large share of Iraq's national wealth, equivalent to around two-thirds of GDP in the early years. This figure is set to decline over the projection period, but the share of export revenue in GDP is projected to remain above 50%. This is very high by international standards: the figure for Iraq in 2035 is around the level of Kuwait in 2011 and well above the indicators for other major oil-exporting countries (Figure 16.2). With a large and growing population, Iraq needs to develop a self-sustaining and productive economy beyond the oil sector, a task that can actually be complicated by the scale of these oil revenues (quite apart from their dependence on potentially volatile international prices). There is a well-recognised threat to resource-rich countries, sometimes referred to as the "resource curse", the reality of which is evident in the economic performance of other countries relying heavily on wealth derived from natural resources, particularly those where institutional capacity is weak (Humphreys, *et al.*, 2007). Iraq will need to remain keenly alert to these pitfalls if it is to achieve successful and sustainable national development on the basis of its oil wealth.

Figure 16.2 ▷ Oil production and export revenue as a share of GDP in selected countries in the Central Scenario, 2011 and 2035

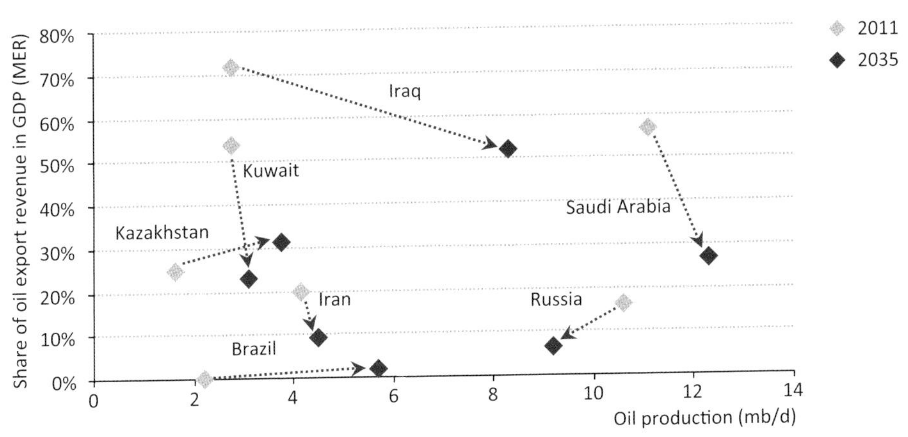

Note: Iraq's Central Scenario forms part of the *World Energy Outlook 2012* New Policies Scenario.

The investment required to raise production and generate the estimated revenues is concentrated in the early part of the projection period, at a time of simultaneous need to rehabilitate existing energy infrastructure that is often in a poor state of repair. In the Central Scenario, the cumulative investment requirement for the energy sector in the period to 2035 is $530 billion, more than $22 billion per year on average (Figure 16.3). Spending during the years between 2016 and 2020 is considerably higher, at $28 billion per year, and can be compared with our estimate of just over $9 billion invested in Iraq's energy sector in 2011 (oil, gas and electricity). The implication is that, during the period to 2020, a sum equivalent to more than one-fifth of Iraq's export revenues needs to be ploughed back into the energy sector each year. After 2020, the average ratio of energy investment to export revenues is lower – less than 10% – which is below the level observed in 2011.

Much of this investment is expected to be financed from the Iraqi treasury, whether directly or through re-payments of costs initially incurred by contractors. This is not the only way that Iraq could meet its infrastructure investment needs: an alternative would be to put in place a legal and policy framework that could encourage the private sector to assume more financial risk. Such a shift in policy has not been built into our scenarios, although the required strengthening of the business environment and the financial sector would serve the important supplementary purposes of contributing to wider economic diversification. On the present strategic approach, the adequacy and timing of investment in the energy sector depends to a large degree on the overall condition of public finances, with energy competing with other substantial claims on government spending: Iraq has a large and growing public sector, with a huge requirement for capital investment in other areas.

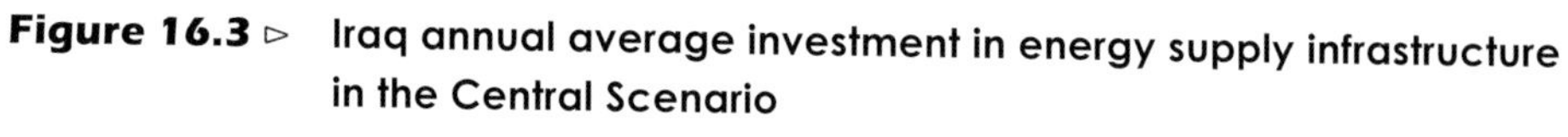

Figure 16.3 ▷ Iraq annual average investment in energy supply infrastructure in the Central Scenario

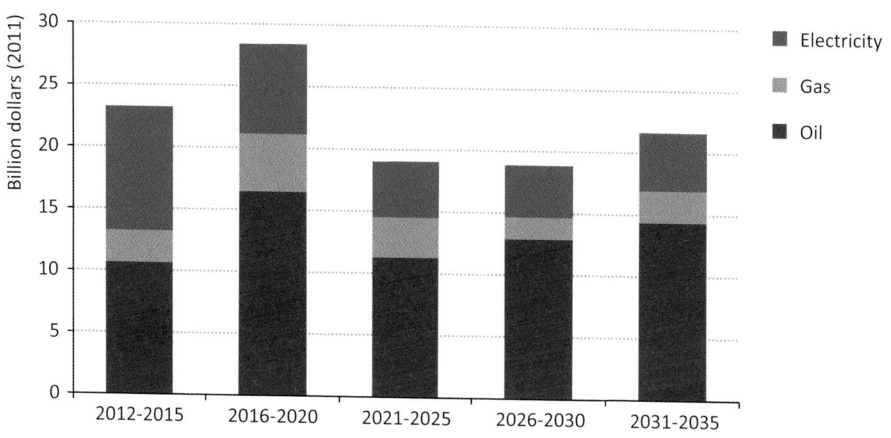

As noted above, economies that rely heavily on commodity exports to finance investment have to allow for the potential volatility of international prices: oil-export driven economic growth can be thrown off course by abrupt shifts in prices. If Iraq meets the overall level of capital spending implied by the Central Scenario (including allowance for the much higher investment in the energy sector) and keeps current government expenditure at today's levels, then it will face a net financing requirement over the period to 2016 of $27 billion. Such a net financing requirement should be well within Iraq's capabilities in the Central Scenario, though an oil price below the levels assumed in this scenario or one that is subject to sharp fluctuations could create difficulties. Swings in revenue can all too easily feed through into irregular capital spending, as there are even more constraints on cutting current expenditure, the bulk of which consists of public sector salaries.[1] In 2009, for example, a 31% decline in revenue, compared with 2008, led to a 12% fall in total Iraqi expenditure, with capital spending bearing the brunt of the reduction. If problems with the fiscal balance were to go so far as to cast doubt on uninterrupted operation of the mechanism for the recovery of costs under the technical service contracts, this could swiftly induce operators to hold back on their own investment or, at least, spend in smaller increments.

Another constraint is institutional; in recent years, Iraq's federal budget has consistently forecast a deficit but ended the year in surplus, in part because the federal government has

1. Public and state-owned enterprises account for around 40% of total employment and the associated wage bill represents two-thirds of the state budget.

spent just 60-70% of the capital expenditure that it had planned for.[2] This shortfall has been due to weaknesses in budget preparation, management and implementation in the various ministries. If this pattern were to be repeated in the coming years in the energy sector, the projected levels of energy investment in the Central Scenario would not be realised, investment moving instead towards the trajectory of the Delayed Case. The large and rapid build-up in the administrative capacity required to manage not only the expected surge in revenue but also its efficient expenditure is an urgent and major challenge for Iraq.

Though the hydrocarbons sector will generate enormous income for Iraq, it cannot generate anything like the employment opportunities that the country needs. As noted in Chapter 13, the oil and gas sectors currently employ no more than 2% of Iraq's workforce, with direct employment of around 125 000, of which around 100 000 work in state-owned companies engaged in production, refining, processing and distribution, and the rest in private sector operating or service companies. This compares with a current Iraqi workforce of just under 8 million. As the hydrocarbons sector is not labour-intensive, this ratio will continue and may even worsen as a youthful Iraqi population reaches working age over the projection period. Based on our population assumptions, the labour force will increase to over 11 million in 2020 and 20 million in 2035.[3] On average, Iraq's economy has to generate opportunities for half a million net additions to the labour force each year. In 2035, total direct employment in the oil and gas sectors may reach up to 200 000 (Figure 16.4). While the oil sector will support a much larger number of jobs in related construction, manufacturing and other service sectors, it is clear that the answer to Iraq's pressing social and economic needs will need to be found elsewhere.

Figure 16.4 ▷ Iraq labour force and estimated direct oil sector employment in the Central Scenario

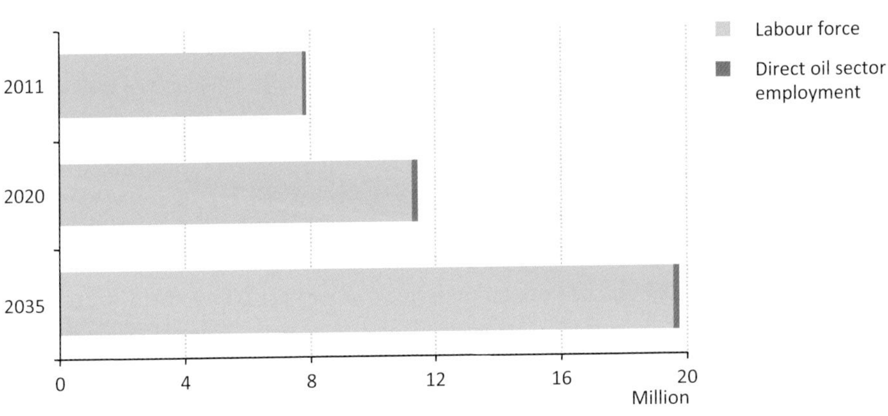

2. A contributing factor has also been (with the exception of 2009) a higher oil price, and consequently higher revenues, than had been assumed in budget calculations.

3. It is likely that a greater share of the working-age population will seek employment as education levels improve and more women enter the job market.

How can the energy sector help create wider opportunities elsewhere in the economy? The first and most important contribution is through the provision of modern, reliable energy services, particularly of electricity. In the Central Scenario, the shortfall in power supplies from the public network is eliminated around 2015, releasing the share of household and business spending that currently goes into more expensive local diesel generation. Second is through judicious use of Iraq's oil revenues, *i.e.* wise strategic choice of priorities for public spending. Alongside investment in functioning electricity and gas networks and a modern refining system, these revenues can develop other assets to underpin economic and social development, such as modern telecommunication and water networks, transport systems and public education. They would therefore contribute one important element of a supportive environment for the nascent private sector. Constant vigilance will be required to safeguard against the dangerous luxury of believing that such revenues make it possible to afford a very large public sector, risking the crowding-out of productive growth of the non-oil sectors of the economy.

As part of an overall economic strategy, the development of downstream oil and gas activities and energy-intensive industrial sectors can provide one element of economic diversification. This could be a viable path for Iraq, given the availability of energy inputs and feedstock at competitive prices. In the Central Scenario, we assume progress is made in this direction, notably with the development of a petrochemicals industry in the 2020s, but it is not without risk: Iraq is a relatively late mover and, in some cases, is looking to gain a foothold in markets with well-established regional players. Moreover, basic chemicals production offers relatively little in the way of employment. Realising a larger potential for job creation can be achieved by moving further down the product chain (into intermediates and plastics), but the manufacture of more sophisticated products also requires a more sophisticated policy environment to secure the necessary large-scale investment, as well as early efforts to ensure that sufficient skilled Iraq personnel are available.

At present, a large share of Iraq's energy spending goes on imported machinery, equipment and services. Given the speed at which Iraq expects to increase oil production and electricity generation, high reliance on imports is inevitable in the short- to medium-term as competitive alternatives are, in many cases, not available on the domestic market. Iraq's supply industry, in common with all parts of the economy, fell behind its regional and international peers during years of conflict and sanctions, losing a lot of expertise through emigration. Iraq can make up this lost ground, but the pace at which this happens will depend on how effectively government policy facilitates the development of local suppliers and how it chooses to promote a higher level of local content (Box 16.1).

Box 16.1 ▷ Gaining local benefits from Iraq's energy investment boom

The energy investment anticipated in Iraq requires a large number of components and inputs that could increasingly be supplied from within the country itself. Typically, up to half of investment spending in the oil and gas sectors consists of steel products and other materials, such as cement and aggregate; another fifth consists of equipment such as pumps, compressors and control systems; the rest is spent on construction labour, rigs, project management, and freight and shipping. Iraqi companies can expect to take on more in many of these areas, initially within Iraq itself but with the longer-term objective of expanding into the wider Middle Eastern market.

Iraq has long experience in oil field operations and steps to expand are already being taken: for example, the Iraq Drilling Company, a state-owned company that is part of the Ministry of Oil, has set up partnership agreements with an international oil services company. The process of developing capacity will require investment in training, technical and engineering education[4] and supportive conditions for new business development. There are some encouraging signs. In 2012, Iraq awarded two contracts, worth $280 million, for the construction of new factories in the south that will produce oil and gas pipe. If implemented effectively, the initiative announced in September 2012 to create near Basrah a specialised and secure free zone for the hydrocarbons sector could also help to attract businesses that manufacture, store or service oil and gas field equipment.

At present, there are no measures in place in Iraq that mandate a certain level of local content in energy sector projects. In some countries, measures of this kind have been introduced as a legislative requirement; but, in Iraq's circumstances in the oil and gas sectors, they could also be considered as conditions in future licensing rounds or field development plans. Any such requirement needs to be designed with care so that it acts as a genuine spur to local contracting, not running so far ahead of local capacity that it acts as an obstacle or cost barrier to project implementation. Where public authorities finance projects directly, their own procurement policies (which tend for the moment to favour state-owned enterprises) can provide an important means of building up independent domestic suppliers and contractors.

In our calculation of Iraq's GDP growth, which is based primarily on developments in the oil sector, we have assumed a gradual increase in the local content of investment spending, such that it reaches 45% by 2035. If this share were to rise more quickly, reaching 55% by the end of the projection period, around 1% would be added to Iraq's GDP, an extra $4 billion in national output in 2035.

4. Existing technical service contracts oblige contractors to support a Training, Technology and Scholarship Fund that is intended to provide education and training opportunities in petroleum engineering and operations to Iraqi nationals. Total payments to this fund are currently around $60 million per year.

Economic development in the High and the Delayed Cases

The High Case brings more rapid GDP growth to Iraq, but also implies a greater concentration of economic activity in the hydrocarbons sector until the early 2020s, thus actually reducing economic diversification until the latter part of the projection period. Oil export revenues rise to a level above 80% of GDP in 2015 and stay above 70% of GDP until the mid-2020s, when oil production growth tapers off. By the end of the projection period, the share of oil export revenue in GDP comes down to a level comparable to that of the Central Scenario.

In the High Case, oil exports generate an additional $610 billion in the period to 2020 and an additional $1.8 trillion over the projection period as a whole, compared with the Central Scenario (despite a slightly lower oil price assumption in this case, see next section). The potential positive impacts are clear, but the scale of this huge windfall could exacerbate problems with revenue management and the development of the non-oil sectors of the economy. There would be a higher bill for investment and the concentration of capacity additions in the period to 2020 could further increase unit costs. There would need to be a step-change in Iraq's institutional capacity in order to manage in an efficient and timely way the vastly increased capital budget: investment in the energy sector during the period between 2015 and 2020 would need to be at average levels of more than $40 billion each year, more than four times higher than our estimate of total energy investment in 2011.

Table 16.1 ▷ Key energy export, revenue and investment indicators from the projections for Iraq

		Central Scenario		High Case		Delayed Case	
	2011	2020	2035	2020	2035	2020	2035
GDP (MER, $ billion)	115	289	552	384	649	186	331
Oil production (mb/d)	2.7	6.1	8.3	9.2	10.5	4.0	5.3
Oil export (mb/d)	1.9	4.4	6.3	7.1	7.9	2.7	3.8
Gas production (bcm)	9	41	89	63	114	18	49
Gas export (bcm)	0	2	17	8	37	0	7
Billion dollars (2011)		**2012-2035**		**2012-2035**		**2012-2035**	
Oil sector investment		319		503		178	
Oil export revenue		4 880		6 644		3 288	
Gas sector investment		71		81		35	
Gas export revenue		106		211		17	
Power sector investment		142		154		103	

Notes: MER = market exchange rate; mb/d = million barrels per day; bcm = billion cubic metres.

Although not part of our full-scale projections, the High Case accentuates the possibility (which is present also in the Central Scenario) that developments in the oil sector might run ahead of those in the rest of the energy sector. The oil sector has stronger foundations for higher output – in the form of existing contracts with skilled international companies –

than might be achieved in other sectors (despite the various impediments, discussed in Chapter 14, could hold back oil output growth). Compared with gas and electricity, operations in an export-oriented oil sector are less dependent on domestic policy and the institutional environment, so long as reliable access to international markets is ensured. Rapid increases in oil output, with the associated growth in revenues, could even be the occasion for holding back investment in other parts of the energy economy (as well as the economy as a whole) by appearing to lessen the urgency of upgrading inefficient technologies or changing inefficient practices, such as the large-scale use of oil in power generation. If realised, the combination of higher oil production with a slower pace of change in other sectors would be a worrying development for Iraq, as the increasing dependency on a single commodity would be symptomatic of the onset of the "resource curse".

In a Delayed Case, oil export revenue is almost $1.6 trillion lower over the projection period than in the Central Scenario, despite higher international prices, resulting from tighter oil markets. Annual spending on energy investment rises compared with 2011 but remains well below the levels required for the Central Scenario, reaching average levels of $13 billion over the projection period as a whole. Overall, the power sector absorbs a slightly higher share of investment spending than in the Central Scenario, which creates the potential for stress on public finances as the balance in spending shifts away from the high revenue-generating oil sector. The share of oil export revenue in GDP, though, remains stubbornly high at around 60% (this indicator falls to a lower level by 2035 in both the Central Scenario and the High Case).

Iraq's impact on international oil markets

The international market context for Iraq in the Central Scenario and the other cases is that projected in the *WEO-2012* New Policies Scenario, as described in detail in Chapter 3. The speed and location of global oil consumption growth implies for Iraq a continued shift in orientation towards markets in the Asia-Pacific region, to which around 50% of Iraq's oil exports go at present. While recognising any exporter's interest in diversity of markets, Iraq's position as the major provider of additional barrels of oil to the world market means that it will naturally be drawn to the high-growth markets, notably China and India, where growth in global oil consumption will be concentrated.[5] In the same period, overall demand for oil in European and North American markets declines by around 1% per year and the demand for imported oil in North America falls much faster, because of growth in indigenous production.

5. Chinese oil companies are active in Iraq's upstream: CNPC is a partner in the consortium development of the Rumaila field and operator of Ahdab; Petrochina is operator of Halfaya and CNOOC of the Missan Group. These fields account for more than 2 mb/d of Iraq's anticipated production in 2020 in the Central Scenario. India's Reliance has two licences (Sarta and Rovi) in the KRG area.

Figure 16.5 ▷ **Growth in OPEC and non-OPEC supply in the Central Scenario**

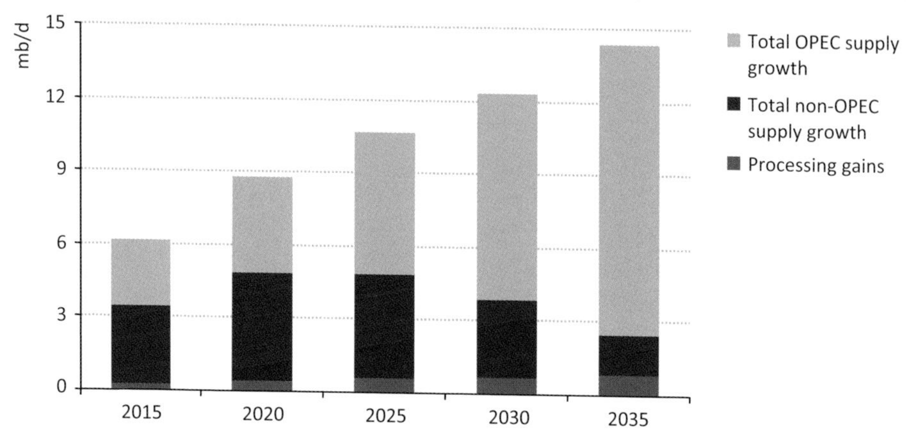

Note: 2010 is used as the base year for this chart to exclude the impact of temporary market exigencies in 2011, such as the loss of supply from Libya and the high level of non-OPEC outages.

On the supply side, the major trend of importance to Iraq in the near term is the rise in non-OPEC production, driven by the rise of unconventional oil, especially light tight oil,[6] in North America and anticipated increases in output from Brazil and, to a lesser extent, Kazakhstan. For the next few years, the projected growth in non-OPEC supply is sufficient to meet a significant part of incremental global demand. However, towards 2020, non-OPEC oil supply starts to tail off, requiring substantial growth in production of crude oil and natural gas liquids (NGLs) from the OPEC countries. Compared with a baseline of 2010, the requirement for additional production of crude and NGLs from OPEC countries rises substantially, starting from the latter part of this decade, and reaches around 4 million barrels per day (mb/d) in 2020 and almost 12 mb/d 2035 (Figure 16.5).

The Central Scenario

Over the period to 2035, Iraq is set to become one of the main pillars of global oil output. On the basis of the projections in the Central Scenario, Iraq, by some distance, plays the largest role in global oil supply growth over the period to 2035 (Figure 16.6), at the end of which only twelve countries in the world are projected to produce significantly more oil than they did in 2011.[7] The growth in Iraq's output of 5.6 mb/d over this period is more than 2 mb/d higher than that of Brazil and is double that of Canada.

6. Light tight oil is produced from shale, or other very low permeability rocks, with technologies similar to those used to produce shale gas.

7. The other producers with higher projected production in 2035 than in 2011 are Venezuela, Qatar, Kuwait, the United Arab Emirates, Iran and Libya, the latter's contribution amplified because of low production in the base year.

Figure 16.6 ▷ **Major contributors to global oil supply growth to 2035 in the Central Scenario**

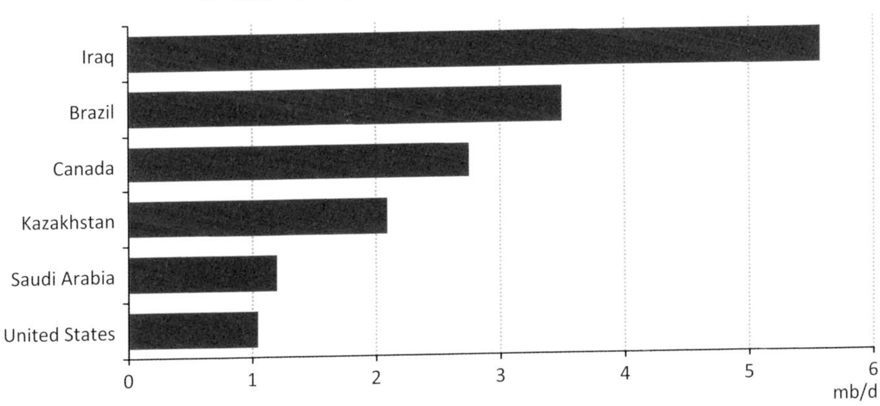

Offsetting projected internal demand in Iraq from projected supply in the Central Scenario allows Iraq's oil exports to increase substantially during the projection period, with the steepest period of growth starting around 2015 and continuing into the early 2020s, after which growth is more gradual. Iraq's exports rise to 4.4 mb/d in 2020 and 5.2 mb/d in 2025, finishing the projection period at 6.3 mb/d (Figure 16.7). In the period to 2015, a larger share of incremental oil production is absorbed on the domestic market, primarily for power generation, but this temporary increase slows (and is then reversed in the power sector) as larger volumes of natural gas become available.

Figure 16.7 ▷ **Iraq oil balance in the Central Scenario**

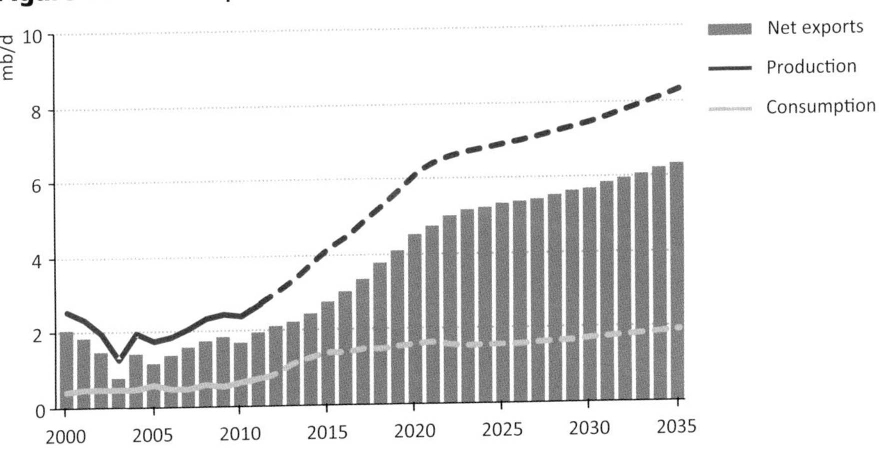

These projections also make Iraq the largest contributor to global oil export growth. As a result, Iraq is set to overtake Russia around 2030 to become the second-largest global oil exporter, behind Saudi Arabia. Virtually all of the oil exported at present from Iraq is crude oil and we anticipate that this will continue to be the case (Box 16.2).

Box 16.2 ▷ Crude quality and marketing options

The crude oil produced by Iraq varies in quality and density, meaning that the country's ability to combine crudes into a consistent blend is an important element in maximising export values. At present, Iraq exports two different crudes, Basrah Light (with a contractual American Petroleum Institute [API] gravity of 34°) by tanker from the south and Kirkuk Blend (which is actually lighter, with a contractual 36° API, than its southern counterpart) by pipeline to the north. However, it often has to offer discounts to compensate for the specification of the delivered oil being heavier than the contractual figures, as a result of heavier crudes and heavy fuel oil being blended into the export stream.

Over time, our production profile implies that a greater variety of crudes will be produced in Iraq, including a larger share of heavier crudes. Most production thus far has been from a geological formation (the Zubair, not to be confused with the field of the same name) containing an intermediate density crude around 35° API. But as this formation is gradually depleted, so new supplies are set to come in larger volumes from the shallower Mishrif layer, which generally contains heavier crude (with an API gravity in the mid to high 20s), together with more limited quantities from the deeper Yamama layer (which is lighter). The expected quality variation is not so great as to cause difficulties for buyers of Iraqi crude since we anticipate that much of Iraq's export crude will go to markets in Asia, where demand is growing most quickly and where the large, modern refineries can handle crudes with a range of specifications. Nonetheless, in order to ensure that it captures the maximum value from its exports, Iraq will need to anticipate in its marketing strategy these changes in the quality of its oil production. For the moment, it may be optimal to manage the heavier crude stream by blending it, but – as the share of heavier oil increases – it may become more advantageous to avoid blending the heavy oil, selling it separately, and maintaining the integrity and consistency of the distinct streams.

The strategy adopted by the government will have an impact on the delivery and surface facilities required. If the overriding objective is to diversify export routes, avoiding over-reliance on the Straits of Hormuz, this would point to the early addition to (or expansion of) overland export routes to the Mediterranean. However, our projections suggest that the highest market value for Iraq is to be found in the other direction, towards Asia. This would suggest giving a priority to establishing dedicated facilities to handle an additional crude export stream by tanker from the south. These options are by no means mutually exclusive, but Iraq needs to make a judgement between market and strategic considerations to determine priorities for infrastructure spending.

In the Central Scenario, Iraq can meet the lion's share of the required growth in OPEC supply of crude and NGLs during the period to 2020. After 2020, the incremental production is shared more broadly among OPEC countries, with Iraq accounting for just over a quarter of the increase. Iraq's claim to incremental OPEC supply over the current decade is buttressed by the need for export revenue to finance its recovery after decades of conflict and stagnation (a period during which the absence of Iraqi capacity contributed to the build-up of large financial reserves and wealth funds in many other producing countries). If Iraq's exports grow as foreshadowed in our projections in the Central Scenario, estimated revenues earned by the rest of OPEC remain well above $1 trillion per year throughout the projection period, rising steadily after 2020 to over $1.3 trillion in 2035.

Despite the inevitable uncertainty over future market conditions, it can be said with some confidence that the needs of the global market for growth in production from Iraq along the lines of the Central Scenario are in line with Iraq's own needs for export revenue to support its own reconstruction and development. Realising the projections in this scenario would represent a major contribution from Iraq to the stability of global oil markets.

Iraq's oil market impact in the High and Delayed Cases

In the High Case, the projected level of Iraqi oil production and exports is substantially higher than in the Central Scenario, particularly in the period to 2020 (Figure 16.8). As production rises to more than 9 mb/d in 2020, projected exports rise to above 4 mb/d already in 2015 (approximately the level reached in the Central Scenario in 2020), and are more than 7 mb/d in 2020. Production growth of this magnitude would represent a much larger share of global incremental oil supply, more than three-quarters of the total anticipated increase to 2020 in the Central Scenario. This is also a period during which there is significant anticipated growth in non-OPEC supply. If Iraq were to capture the incremental market share to this extent, in order for OPEC to keep its share of global production at the level anticipated in the Central Scenario the rest of OPEC would need to reduce production up to 2020 well below the level in 2010. In the High Case that we have modelled, part of the additional increase in Iraq's output (compared with the Central Scenario) is indeed accommodated by adjustments by other OPEC producers, but not all of it. As a result, there is a shift in the global balance between supply and demand, producing a lower trajectory for the international oil price than that assumed in the Central Scenario. In the High Case, the international price is around $115 per barrel in 2020, remaining around these levels to 2035. As a result of lower prices, oil consumption is slightly higher, reaching 101 mb/d by the end of the projection period.

Figure 16.8 ▷ **Iraq oil balance in the High Case**

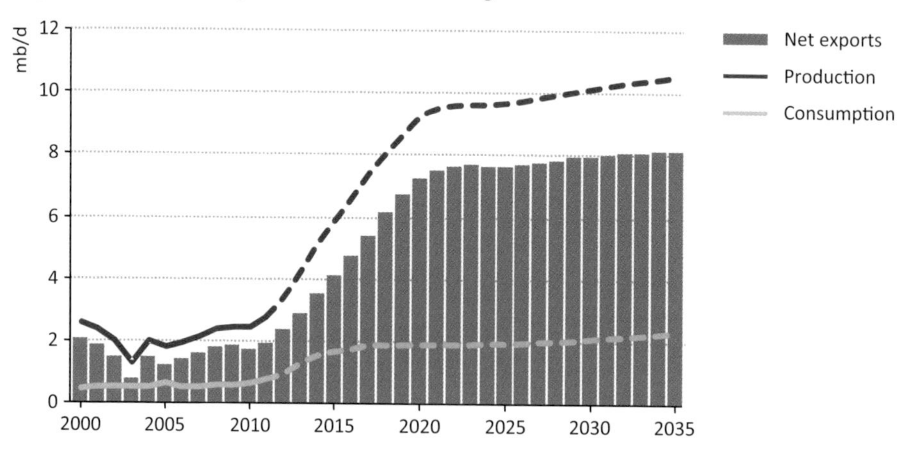

Global market uncertainties must be taken into account. To the extent that growth in global oil use is faster than we project – or growth in global supply from other producers lower than we anticipate – the requirement for incremental Iraqi oil would increase. These circumstances could result from faster than expected economic growth over the coming years, governments failing to implement announced policies to constrain the growth in oil consumption or a change in upstream prospects (as, for example, the moratorium on deepwater drilling following the Deepwater Horizon disaster in the Gulf of Mexico). Alternatively, the Iraqi authorities could seek to manage market impacts to some extent by opting to develop spare production capacity over and above actual output, reducing the volumes exported but retaining some flexibility to respond to short-term market needs (Spotlight). Deliberately creating spare production capacity is unlikely to be an early Iraqi priority but, if in the longer term the country's most pressing needs have been met, Iraq's aim to hold a reasonable reserve of spare capacity would bring new confidence to global markets.

The projections in the Delayed Case present a very different outlook for oil markets, which – without the growth in supply from Iraq projected in our Central Scenario – would be heading for troubled waters. In a Delayed Case, Iraq's oil production, though it continues to rise, reaches only 5.3 mb/d by 2035, 3 mb/d short of the level expected in the Central Scenario. Supply from Iraq at the levels projected in the Delayed Case would act to tighten oil markets considerably over the projection period, producing a significant increase in the international oil price. This is 3% higher than in the Central Scenario in 2020 and 11% higher by 2035 (almost $140 per barrel in real terms in 2035 or $240 per barrel in nominal terms). There is also likely to be a significant increase in price volatility.

SPOTLIGHT

Why invest in spare oil production capacity?

Iraq's contribution to global oil markets depends not only on its capacity to produce, but also on its potential to respond to market needs by adjusting output. This means retaining some spare oil production capacity, ready to be deployed as and when Iraq judges that responding to additional market needs best serves its national interest. At present, the major share of global spare oil production capacity rests with Saudi Arabia, but Iraq has set the strategic aim also to develop a degree of flexibility.[8]

Spare production capacity is an international public good that provides a buffer against sudden or unforeseen market developments and, like the oil stocks held by importing countries, some insurance against the impact of disruptions to supply. Avoiding price spikes and consequent demand destruction is demonstrably in the interest of oil producers as well as consumers, but spare production capacity can be difficult to justify since – by definition – there is no reliable revenue stream associated with this investment.

To the extent that a rise in Iraq's production displaces supply by other producers, Iraq could well have an indirect impact on global levels of spare capacity before it develops any of its own. The timing of any Iraqi decision deliberately to create its own spare capacity is uncertain: given Iraq's pressing needs in other sectors of the economy, this is unlikely to be a viable proposition in the short- to medium-term, though it could ultimately be an important opportunity. We estimate that a barrel per day of oil production capacity in Iraq costs, on average, $10 000-$15 000 (in year-2011 dollars). This means that the upstream investment required to build 1.5 mb/d of spare capacity would be between $15-$22.5 billion. For comparison, oil export revenue in the Central Scenario is projected to be around $200 billion in 2020.

Prices at these levels would have an impact on global oil demand in the Delayed Case, which reaches 94 mb/d by 2020 (0.5 mb/d lower than the Central Scenario) and 97 mb/d by 2035 (2.4 mb/d lower), primarily as a result of reduced growth in the transport sector. The response from consumers is a mix of conservation (for example via reduced driving distances), faster penetration of more efficient technologies and switching away from oil to alternative fuels. The shortfall in Iraq's supply, compared with the Central Scenario, means additional reliance on other producers to raise output, and this would be anticipated to come primarily from OPEC countries (an additional 700 thousand barrels per day [kb/d] by 2035). The strain on markets becomes increasingly evident in the period from 2020 onwards, as the global market starts to rely on a smaller number of producers for additional production.

8. The technical service contracts concluded by the federal authorities and the production-sharing contracts with the Kurdistan Regional Government include provisions allowing the authorities to curtail production. However, the modalities to compensate contractors for capacity built but not used are not clear, in some cases, and not yet tested in practice.

Figure 16.9 ▷ **Key oil market indicators in 2035 in the Delayed Case, relative to the Central Scenario**

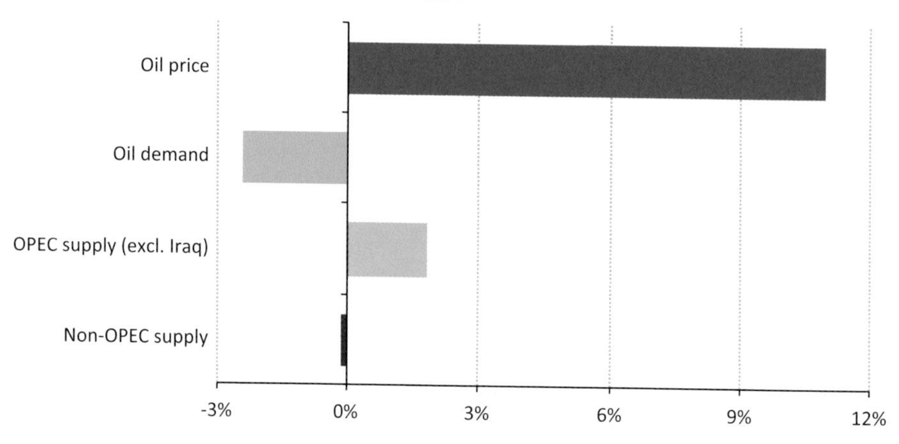

Iraq's impact on international gas markets

Natural gas plays an increasingly important role in Iraq's energy balance in all the scenarios and cases examined. Marketed gas production in Iraq in the Central Scenario rises to almost 90 billion cubic metres (bcm) in 2035, a full 80 bcm higher than in 2011. This rate of increase is one of the fastest in the world and only five countries achieve faster absolute growth over this period (Figure 16.10). Among major gas producers, none comes close to matching Iraq's ten-fold increase in marketable gas output (albeit from a very low base).[9] Iraq's projected production in 2035 consists of roughly equal shares of associated and non-associated gas. As noted in Chapter 15, gas provides an important substitute for oil use in the domestic market, freeing up the more valuable and more easily exported commodity.

Whereas in the oil sector domestic consumption remains small compared to anticipated levels of production and export volumes can be counted upon to grow fast, the perspectives for gas are less clear-cut. There is still significant uncertainty over the extent to which gas will be used in the domestic market for electricity and industrial uses and how quickly gas flaring will be reduced. Moreover, according to our projections, the associated gas that will be produced along with rising oil production will not be sufficient in itself to allow for gas export. In the Central Scenario, cumulative production of associated gas is enough to cover only around 70% of anticipated demand from within Iraq over the period to 2035. As discussed in Chapter 14, the development of Iraq's non-associated gas resources will therefore be the key to determining the prospects for, and extent of, potential gas export.

9. The ten-fold increase in marketable gas production should be seen in the context of the volumes of gas that are currently flared. Iraq's total gas production (including flaring) is today close to 20 bcm, so we are in practice projecting that total output will be about 4.5 times larger than today, but with a significant reduction in gas flaring.

Figure 16.10 ▷ Growth in natural gas production in selected countries in the Central Scenario

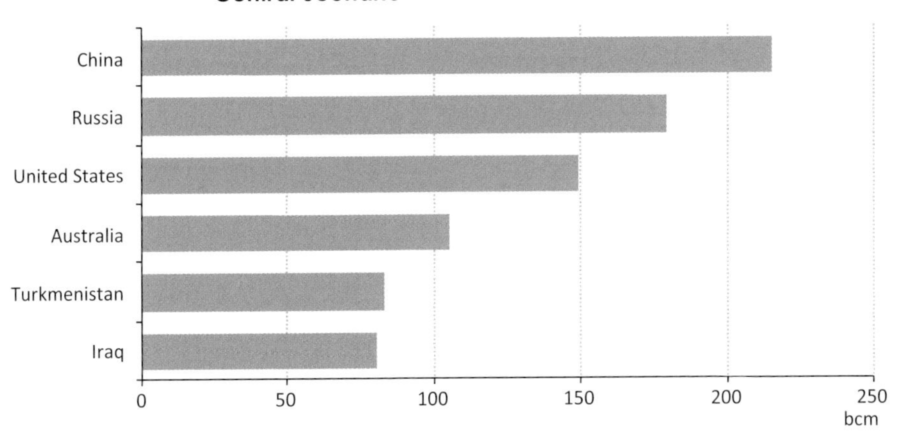

In the Central Scenario, rising gas production is accompanied by increased use of gas in power generation until the latter part of this decade, with a surplus in the national balance available for export from 2020.[10] Gas export is stable around 10-15 bcm during the 2020s, before rising again after 2030 to reach 17 bcm by the end of the projection period (Figure 16.11). The growth in gas processing capacity also means that Iraq will become a substantial producer of liquefied petroleum gas (LPG), which is widely used in Iraq and in the region as a fuel for cooking and heating and, increasingly, also within the petrochemicals sector. From its current position requiring LPG imports of around 33 kb/d in 2011, Iraq becomes a net exporter in 2016 and, by 2035, is expected to export around 220 kb/d.[11]

For policy makers and potential investors in the sector, a key question to be clarified is how Iraq will manage the trade-off between the low value of gas on the domestic market – which needs to be supplied first – and the more attractive returns available on export markets. This requires a clear strategic vision on how the gas supply-demand balance might best evolve (which, because of the importance of associated gas, therefore depends to a significant degree on the trajectory targeted for oil production) and on the incentives in place for non-associated gas production. There is also the question of how decisions

10. There are plans in place for gas imports to central Iraq via a new pipeline from Iran, primarily for use in power generation. Volumes of up to 25 million cubic metres per day have been agreed in principle for a period of five years, which would boost gas availability until domestic supplies become available (and free up some of the oil that would otherwise be used for power generation). The longer-term perspectives are not clear. If imports from Iran were to continue beyond the agreed five-year period, this could allow for a commensurate increase in Iraq's gas exports.

11. LPG (a mixture of propane and butane) is a product both of gas processing and oil refining. Exports of LPG – as well as condensate and other natural gas liquids – are included in the volumes and revenue calculations for oil. Iraq's projected output of 440 kb/d (15 million tonnes [Mt] per year) in 2035 would make it one of the world's leading LPG producers: global LPG production in 2010 was 7.7 mb/d (240 Mt), with Saudi Arabia the largest producer in the region at 700 kb/d (21 Mt).

regarding gas export will be taken, an aspect of the vexed issue of overall governance of the hydrocarbons sector and the differences between the federal authorities and the Kurdistan Regional Government (KRG). Clarity on these questions, already vital for successful oil exploitation, could bring forward and expand the prospects for gas export from Iraq, while continued uncertainty will push them back. A further possibility is that Iraq might use part of any gas surplus to generate electricity for export to neighbouring countries: over relatively short distances, exporting "gas-by-wire", *i.e.* gas-fired electricity, can be competitive with "gas-by-pipe".

Figure 16.11 ▷ Iraq gas balance in the Central Scenario

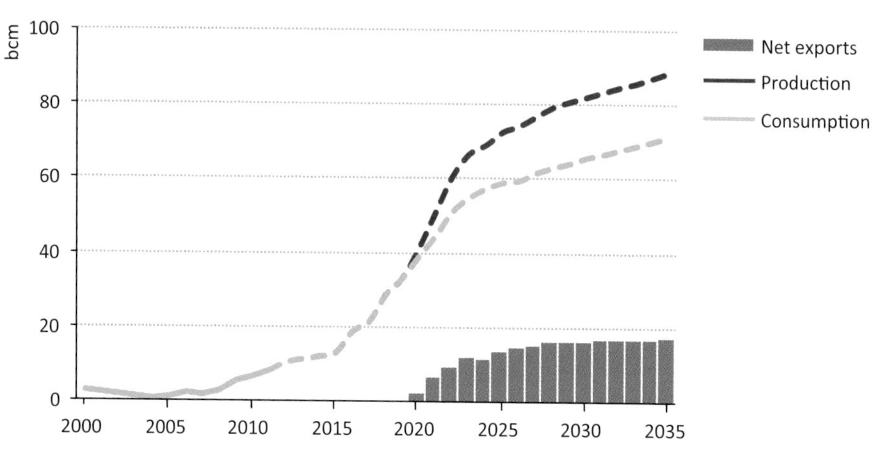

The geography of gas production within Iraq is important. In the south of the country, the overall perspectives are closely tied to the outlook for oil and associated gas production, how quickly the currently flared volumes are captured and processed and decisions on gas-consuming industrial development in the area around Basrah. As part of the investment plan for the Basrah Gas Company, Shell and Mitsubishi have proposed a southern export facility, with a capacity of 4 million tonnes (Mt) of LNG per year (around 5.5 bcm per year). Such a gas export project would require a large up-front investment in liquefaction and export facilities, provisionally estimated to cost more than $4 billion. There is also the southern non-associated Siba gas project, awarded in the third national licensing round to a consortium led by Kuwait Energy, some of the production from which could be allocated for export to nearby Kuwait.

In the west of Iraq, the main potential source of gas for export is the Akkas field, near the border with Syria, the development of which was awarded in the third national licensing round to a consortium led by Korea's KOGAS. The debate over the Akkas field illustrates well the potential for friction between the aims of providing gas for domestic supply or for export: responding to the possibility that gas from Akkas might be earmarked for supply to Syria, the authorities in Anbar province (where the field is located) argued that gas should be primarily used on the local market for power generation and to promote industrial development. The result (as reflected in the deal signed with KOGAS in 2011) was to prioritise supply of produced gas to Anbar's power plants, leaving scope for any surplus to be exported. For the moment, unrest in Syria has closed the debate about export, leaving a question mark over the source of demand for any uncommitted gas produced in the Western Desert. Our projections for this part of Iraq see gas production picking up only after 2020.

Although the resources in the west may be of comparable size, the main potential for non-associated gas production in Iraq over the projection period is in the north of the country, particularly in the KRG area. Turkey provides a large, proximate export market, with an increasing need for imported gas, as well as a conduit to markets in the rest of southeast Europe and, further, to the European Union. A report for the European Commission (Mott MacDonald, 2010) identified Iraq as a very promising source of gas supply to the European Union, suggesting trade volumes in 2030 at 15 bcm (in a base case) and 30 bcm (in an optimistic scenario), figures that are broadly in line with our own projections. Northern Iraq is seen as the most likely initial source of gas, with a second phase potentially involving also associated gas from southern fields.

The potential for gas production in the north is undoubtedly high and the possibility of export is generating strong interest from international investors in gas projects. The prospects and timing for export are though subject to considerable uncertainty. The Khor Mor and Chemchemal fields, for which contracts have been awarded by the KRG, represent a significant share of the region's gas potential in the medium term (and Khor Mor is currently producing). These are being developed under a technical service contract which provides for gas to be sold for local power generation at very low prices. Mobilising investment to develop the fields requires a perspective on monetising the natural gas liquids (an LPG plant was commissioned in 2011), a change in the pricing of gas for the local market, or clarity on the conditions for natural gas export. For the moment, however, the dispute that has meant only intermittent use of federal channels for oil export from the KRG area, due to the differences between the KRG and the federal authorities (who insist that they alone have constitutional authority for international energy trade), has cast a similar shadow over gas export. Moreover, raising the price paid for local consumption of gas involves a no less politically challenging decision to increase the price of electricity paid by end-users and would, in any case, be unlikely to raise the value of the gas to that available on international markets.

Potential gas export markets and routes

If and when Iraq does start to export its gas, it will have a wide range of available markets, with many of its neighbours dependent on imports of gas to meet their domestic needs. Four of Iraq's neighbours are currently reliant upon imports (Turkey, Jordan, Kuwait and Syria) and Saudi Arabia also faces a looming deficit of gas. Gas consumption in the

Middle East as a whole has been increasing at more than 8% per year over the last three decades and, although this annual growth is expected to slow as gas markets become more mature, gas use in the region as a whole is still projected to rise by more than 70% in the period to 2035. There are, therefore, robust drivers for regional trade, underpinned by the consideration that export would, in many cases, require only short cross-border interconnections.

There are, though, relatively few of these links in place and energy trade among the countries of the Middle East remains far below its potential. Many energy flows in the past have been interrupted by politics or conflict. Iraq's only existing gas export pipeline is a case in point: a link between Rumaila and Kuwait was completed in the mid-1980s to transport Iraqi gas to Kuwait, but it has been inoperative since Iraq's invasion in 1990. This legacy of political risk is likely to continue to hamper the development of energy trade (including electricity trade) within the region. A further potential constraint on export to Middle Eastern markets is the seasonality of electricity and gas use, with many countries – including Iraq – experiencing higher demand in the summer months. Where the export volumes are sourced in whole or in part from associated gas (in which case the operator has much less discretion to vary gas output), reliable export to these countries would require investment in sufficient storage to match the variations in demand.[12]

In the Central Scenario, Turkey and other European markets are expected to require substantial additional volumes of imported gas. The likelihood of new gas transportation capacity across Turkey to accommodate export volumes from Azerbaijan offers an opportunity for tie-ins from northern Iraq, and the possible extension of these pipeline routes into southeast Europe provides an opening for Iraq to become a supplier to European gas markets, which are projected to require almost 200 bcm in additional gas imports as demand rises and indigenous production falls. To the south, LNG export would provide Iraq with an entry to the fast-growing markets in the Asia-Pacific region.

Our analysis of the cost of delivering gas from Iraq to a variety of possible regional and international markets in 2020 suggests that it can be very competitive (Figure 16.12). The highest supplied cost is that of LNG to Asia-Pacific markets, primarily because of the large investment required in liquefaction capacity; this option does, though, offer flexibility in terms of destination and access to the higher prices assumed to be available in Asia-Pacific markets, compared with Europe. The analysis shows that the delivered cost of pipeline gas exported to any of the regional markets, including markets in southeast Europe (assumed here as the western border of Turkey with Bulgaria or Greece) would be well below the price assumed for these markets in 2020.

12. This storage could be either in the importing or the exporting country. At present, Iraq has no underground gas storage: a facility built in the 1980s in the Kirkuk region for LPG, which held a surplus during summer months for retrieval during periods of peak demand in the winter, is no longer in use.

Figure 16.12 ▷ Indicative delivered supply costs and prices for Iraq gas export, 2020

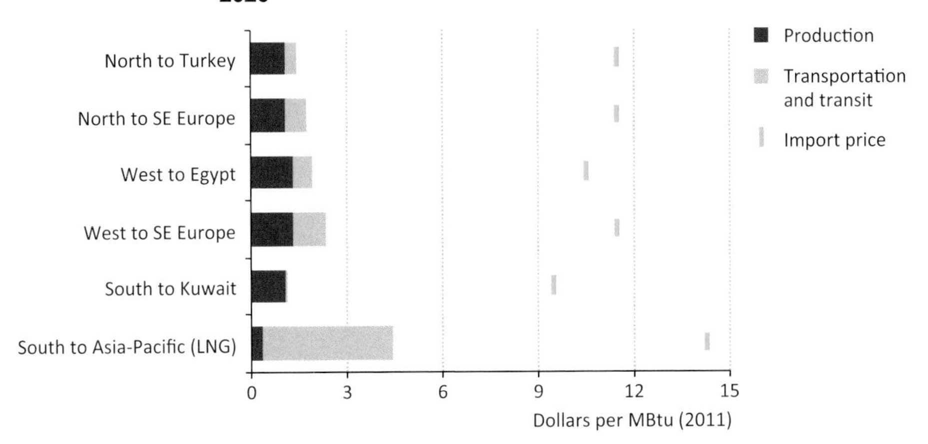

Notes: All costs are indicative, in year-2011 dollars. Dry gas costs are assumed to be low because gas production and processing in Iraq is generally accompanied by substantial output of natural gas liquids. Supply costs for individual projects could vary significantly, depending on their detailed design. The source of the gas is assumed to be non-associated gas fields, with the exception of southern production feeding into a possible LNG plant, which is assumed to be associated gas. Production costs do not include taxes or royalties. The cost of shipping LNG to the Asia-Pacific market is an average of the cost of transportation to the Indian and Chinese markets. The bulk of the cost, in any case, consists of the cost of facilities for LNG liquefaction. Import price assumptions for 2020 are from the Central Scenario. MBtu = million British thermal units.

Variations in the High Case and the Delayed Case

In the High Case, the volume of gas available for export is pushed higher as the increase in production (both from associated and non-associated gas) is larger than the projected increase in domestic demand (from higher GDP growth). Gas export also starts slightly earlier than in the Central Scenario, in 2019, and reaches 25 bcm by the mid-2020s and 37 bcm by the end of the projection period. By contrast, in the Delayed Case, associated gas is held back by lower oil production, while opportunities in non-associated gas are deemed unattractive or too risky because of prolonged uncertainty over the conditions for investment and export. In this case, production is initially able to keep pace with (lower) domestic gas demand, but a surplus emerges only after the mid-2020s, with projected exports reaching 7 bcm by 2035 (Figure 16.13).

Although attention will rightly be focused on oil output as a key indicator of Iraq's energy performance in the coming years, progress in developing the gas sector – both as a fuel for domestic use and eventually also for export – may in practice be a more reliable indicator of the quality of Iraq's institutions and policies. This sector highlights more than any other the challenges of policy co-ordination, requiring that gas availability match the growth of new sources of gas demand (in power generation and industry) and takes account of the growth of modern refining capacity (to absorb otherwise surplus heavy fuel oil). If Iraq does not succeed in managing these challenges, there will be a range of lost opportunities across the energy sector as a whole, likely to result in increased domestic oil demand (with implications also for oil exports and for Iraq's fiscal position). If Iraq does succeed in developing a viable domestic gas sector, this will be a substantial step not only in moving the country away from high direct dependence on oil but also a signal that the institutional capacity is there to take on other challenges facing the Iraqi energy sector and the economy as a whole.

Figure 16.13 ▷ Natural gas production and export in the High Case and the Delayed Case

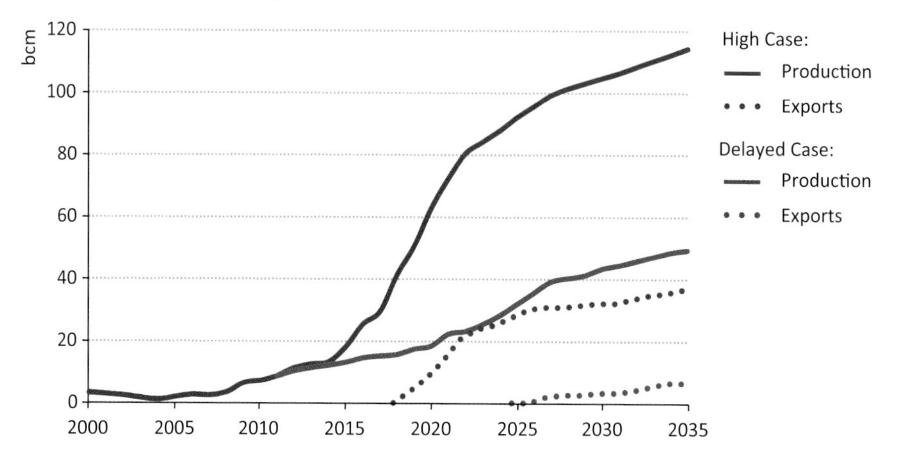

PART D
SPECIAL TOPICS

PREFACE

Two special studies form the final part of this year's *WEO.*

Chapter 17 analyses – for the first time in the *WEO* series – current and future freshwater requirements for producing energy. The chapter highlights the dependence of energy on water, and the particular vulnerabilities faced by the energy sector in a more water-constrained future. It projects energy-related water needs by scenario, energy source and region, and then identifies some of the key stress points that may constrain energy development.

Chapter 18 continues a long *WEO* tradition of analysing what needs to be done to bring basic modern energy services to a substantial part of the world's population. It reports on recent policy developments, presents projections of how the situation may develop in years to come and offers a new Energy Development Index to assist the evaluation of the effectiveness of increasing actions directed to the objective of ending global energy poverty.

Water for energy

Is energy becoming a thirstier resource?

Highlights

- Energy depends on water – for power generation, the extraction, transport and processing of fossil fuels, and the irrigation of biofuels feedstock crops – and is vulnerable to physical constraints on its availability and regulations that might limit access to it. A more water-constrained future, as population and the global economy grow and climate change looms, will impact energy sector reliability and costs.
- Global water withdrawals for energy production in 2010 were estimated at 583 billion cubic metres (bcm), or some 15% of the world's total water withdrawals. Of that, water consumption – the volume withdrawn but not returned to its source – was 66 bcm. In the New Policies Scenario, withdrawals increase by about 20% between 2010 and 2035, but consumption rises by a more dramatic 85%. These trends are driven by a shift towards higher efficiency power plants with more advanced cooling systems (that reduce withdrawals but increase consumption per unit of electricity produced) and by expanding biofuels production.
- The water requirements for fossil fuel-based and nuclear power plants – the largest users of water in the energy sector – can be reduced significantly with advanced cooling systems, although this entails higher capital costs and reduces plant efficiency. Future water needs for biofuels depend largely on whether feedstock crops come from irrigated or rain-fed lands and the extent to which advanced biofuels – whose feedstock crops tend to be less water-intensive – penetrate markets. Water requirements for fossil fuel production are comparably lower, though potential impacts on water quality are an important concern.
- Energy efficiency, wind and solar PV contribute to a low-carbon energy future without intensifying water demands significantly. Compared with 2010, withdrawals in the 450 Scenario rise by only 4% in 2035, though consumption doubles due to much higher biofuels production. Several low-carbon energy technologies – nuclear power, power plants fitted with carbon capture and storage equipment and certain types of concentrating solar power – can be highly water-intensive.
- Water is growing in importance as a criterion for assessing the physical, economic and environmental viability of energy projects. Among other examples, the availability of and access to water could become an increasingly serious issue for unconventional gas development and power generation in parts of China and the United States, India's large fleet of water-dependent power plants, Canadian oil sands production and maintaining reservoir pressures to support oil output in Iraq. Such vulnerabilities are manageable, in most cases, but will require deployment of better technology and greater integration of energy and water policies.

Introduction

Energy and water are valuable resources that underpin human prosperity and are, to a large extent, interdependent. Water is ubiquitous in energy production: in power generation; in the extraction, transport and processing of fossil fuels; and, increasingly, in irrigation to grow feedstock crops used to produce biofuels. Similarly, energy is vital to the provision of water, needed to power systems that collect, transport, distribute and treat it (Box 17.1). Each faces rising demands and constraints in many regions as a consequence of economic and population growth and climate change, which amplify the mutual vulnerability of energy and water. For the energy sector, constraints on water can challenge the reliability of existing operations and the viability of proposed projects, imposing additional costs for necessary adaptive measures.

This chapter addresses water for energy in the context of the *WEO-2012* energy scenarios. It provides information on future water requirements for energy production and identifies the particular water resource risks associated with our energy scenarios. The chapter begins with an overview of global water trends and energy sector vulnerabilities to water constraints. It discusses water use in the production of primary energy and electricity. Water use factors are applied to our energy projections to estimate water requirements by scenario, energy source and region. These results are then analysed with respect to water resources in key regions to identify several important stress points that might arise relating to the energy supply figures in our scenarios.

Facing a more water-constrained future

Water is a plentiful resource, but it is not always available for human use in the quantities or at the quality, time and place required. Only about 2.5% of the world's water is freshwater.[1] Of that, less than 1% is accessible via surface sources and aquifers – the rest is locked up in glaciers and ice caps, or is deep underground (Figure 17.1).

Figure 17.1 ▷ World water resources and human freshwater use

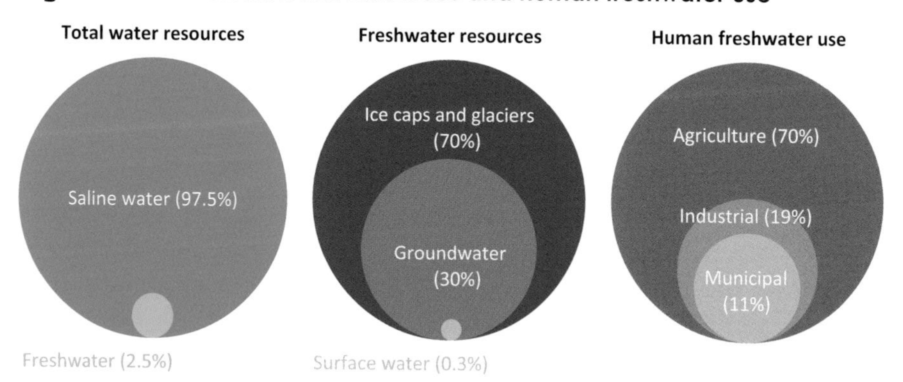

Sources: Shiklomanov (1993); UN FAO Aquastat database.

1. Unless otherwise noted, the term "water" in this chapter refers to accessible renewable freshwater.

Freshwater is generally renewable,[2] replenished naturally by the water cycle. Its availability, however, is uneven, due to wide differences in climatic patterns, geography and human use. Water resources are abundant in most parts of Brazil, for example, which sees relatively high rainfall. By contrast, many countries in the Middle East and North Africa face chronic water scarcity, receiving minimal rainfall and low water flow from outside their borders. They must therefore turn to other sources of water supply, such as non-renewable aquifers or desalination. The quantity and quality of water resources varies temporally – during summer months, for instance, surface water levels are lowest and temperatures highest. Furthermore, water resources can be adversely (and unexpectedly) affected by extreme events, such as droughts, heat waves and floods.

The extent to which water resources come under strain in a given country or region depends on how human use relates to supply. Globally, agriculture is the principal user of water, accounting for 70% of water use, followed by industry (including mining and power generation) at 19% and municipal networks, which serve the water needs of public and private users, at 11% (UN FAO, 2012). For the purpose of indicating present physical water scarcity, this chapter uses renewable water resources per capita, thus taking into account supply and a proxy for demand (population).[3] Significant tracts of almost every continent face water scarcity, stress or some degree of vulnerability (Figure 17.2). Even countries with seemingly ample water availability at a national level may face scarcity in particular regions, such as the southwest United States and non-coastal Australia. Water resources can vary significantly from one water basin to the next and can be located far from areas where demand is greatest.

Climate change coupled with population and economic growth portends a more water-constrained future in many regions. The water and climate cycles are inextricably linked: rising temperatures will accelerate the movement of water, increasing both evaporation and precipitation. Expected impacts include falling average surface water flows (glacier melt being an exception); higher surface water temperatures; a reduction of snowpack and change in the timing of the snowmelt season; sea level rise, which will contaminate freshwater supplies; and droughts, heat waves and floods that are more frequent and more severe (IPCC, 2008). Future water demand is expected to grow with rising populations, urbanisation, higher standards of living and higher food demand, particularly in non-OECD countries as direct water use increases with readier access and there is a dietary shift from plants to (more water-intensive) meat (WWAP, 2012).

2. A country's renewable water resources include exploitable internal resources, *i.e.* surface water and groundwater generated from endogenous precipitation, and external resources, *i.e.* water that enters a country from upstream countries via rivers or aquifers. Both are naturally renewed, but a few sources are non-renewable, *e.g.* some underground aquifers.

3. An index based on renewable water resources per capita (Falkenmark, *et al.*, 1989) is one of the most commonly used measures of water scarcity, cited by UNDP (2006) and UN FAO (2010). Some methodologies also evaluate economic water scarcity, which takes account of limitations by human, institutional and financial capital on water availability (IWMI, 2007).

Figure 17.2 ▷ Renewable water resources per capita in 2010

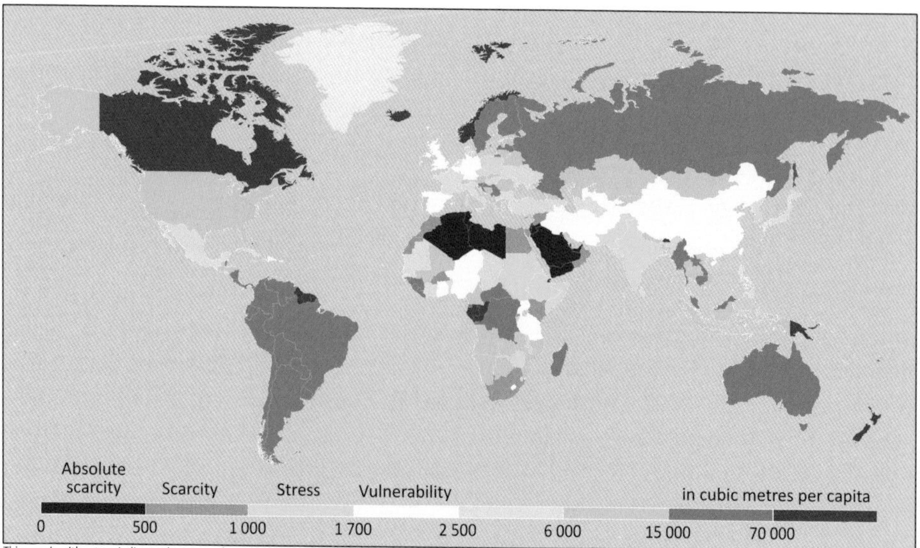

This map is without prejudice to the status of or sovereignty over any territory, to the delimitation of international frontiers and boundaries and to the name of any territory, city or area.

Source: UN FAO Aquastat database.

Box 17.1 ▷ Energy for water

The other half of the energy-water nexus concerns the energy requirements for supplying and treating water. Electricity is needed to power pumps that abstract (from ground and surface sources), transport, distribute and collect water. The amount needed depends on the distance to (or depth of) the water source. Water treatment processes, which convert water of various types – fresh, brackish, saline and waste – into water fit for a specific use, require electricity and, sometimes, heat. Desalination, a process that removes salt from water, is the most energy-intensive and expensive option for treating water and is used where alternatives are very limited, such as in the Middle East and Australia. Other energy needs associated with water occur at the point of end-use, often in households, primarily for water heating and clothes washing.

Looking ahead, several trends point to rising demands on energy from the water sector:

- Increasing water demand, as a result of population growth and improved standards of living.
- Scarcer freshwater supplies in the proximity of population centres, due to climate change. This means that water will have to be transported longer distances, pumped from greater depths or undergo additional treatment.
- More stringent standards for water treatment.
- A general shift in irrigation practices from surface or flood (relying on gravity) to pumped methods, which are more water-efficient but require energy for operation.

Water for energy linkages

Water is required to produce nearly all forms of energy. For primary fuels, water is used in resource extraction, irrigation of biofuels feedstock crops, fuel refining and processing, and transport. In power generation, water provides cooling and other process-related needs at thermal power plants; hydropower facilities harness its movement for electricity production. These uses can, in some cases, entail significant volumes of water. Additionally, they can have adverse effects on water quality via contamination by fluids that contain pollutants or physical alteration of the natural environment (Table 17.1).

Table 17.1 ▷ Key uses of water for energy and potential water quality impacts

	Uses	Potential water quality impacts
Primary energy production		
Oil and gas	Drilling, well completion and hydraulic fracturing. Injection into the reservoir in secondary and enhanced oil recovery. Oil sands mining and in-situ recovery. Upgrading and refining into products.	Contamination by tailings seepage, fracturing fluids, flowback or produced water (surface and groundwater).
Coal	Cutting and dust suppression in mining and hauling. Washing to improve coal quality. Re-vegetation of surface mines. Long-distance transport via coal slurry.	Contamination by tailings seepage, mine drainage or produced water (surface and groundwater).
Biofuels	Irrigation for feedstock crop growth. Wet milling, washing and cooling in the fuel conversion process.	Contamination by runoff containing fertilisers, pesticides and sediments (surface and groundwater). Wastewater produced by refining.
Power generation		
Thermal (fossil fuel, nuclear and bioenergy)	Boiler feed, *i.e.* the water used to generate steam or hot water. Cooling for steam-condensing. Pollutant scrubbing using emissions-control equipment.	Thermal pollution by cooling water discharge (surface water). Impact on aquatic ecosystems. Air emissions that pollute water downwind (surface water). Discharge of boiler blowdown, *i.e.* boiler feed that contains suspended solids.
Concentrating solar power and geothermal	System fluids or boiler feed, *i.e.* the water used to generate steam or hot water. Cooling for steam-condensing.	Thermal pollution by cooling water discharge (surface water). Impact on aquatic ecosystems.
Hydropower	Electricity generation. Storage in a reservoir (for operating hydro-electric dams or energy storage).	Alteration of water temperatures, flow volume/timing and aquatic ecosystems. Evaporative losses from the reservoir.

Water use per unit of energy produced is commonly discussed using two distinct measures: *withdrawal* and *consumption*. Withdrawal is the volume of water removed from a source; consumption is the volume of water withdrawn that is not returned to the source, *i.e.* it is evaporated or transported to another location. Discharge is the volume of water withdrawn that is returned to the source, often degraded by use (altered physically or chemically) and impacting water quality. Water withdrawals are, by definition, always greater than or equal to consumption and are, therefore, the first limit approached at an energy production facility when water availability is constrained. Consumptive use reduces the amount of water available to satisfy demands downstream and is an important consideration where water resources are strained to meet the needs of all users.

Primary energy production

Water needs for fossil fuels production – including the extraction, processing and transport phases of the fuel cycle – vary widely. Conventional natural gas entails minimal water use for drilling and processing and is generally much less water-intensive than producing other fossil fuels or biofuels (Figure 17.3). Shale gas developments use additional water for hydraulic fracturing, a well-stimulation technique that pumps fluids (water and sand, with chemical additives that aid the process) into shale formations at high pressure to crack the rock and release gas. Water requirements for shale gas depend on gas recovery rates, the number of hydraulic fracturing treatments performed and the use of water-recycling technologies. These factors vary from well to well, but can imply water needs many times greater than those for conventional gas. Additionally, public concern exists over potential water contamination risks associated with shale gas development, specifically the leakage of fracturing fluids, hydrocarbons or saline water into groundwater supplies and the handling and disposal of waste water. These hazards, which are also faced in conventional oil and gas development, can be responsibly addressed at a small additional cost using existing technologies and best practices (IEA, 2012).

Coal production uses water mainly for mining activities such as coal cutting and dust suppression. The amount of water needed depends on the characteristics of the mine, *i.e.* whether it is at the surface or underground, and processing and transport requirements. Washing coal increases its quality, but involves additional water. Washing is presently undertaken mostly only for export-quality grades of coal, but there is large scope for the practice to become more widespread given its potential to raise power plant efficiency, such as in India. Key water quality concerns for coal production are runoff from mine operations and tailings that can pollute surface and groundwater.

The amount of water needed for oil extraction is determined by the recovery technology applied, the geology of the oil field and its production history. Water requirements for conventional oil extraction are relatively minor, similar to those of conventional gas. Secondary recovery techniques that use water flooding to support reservoir pressure can have water needs about ten times those associated with primary recovery, which relies on natural support mechanisms. Producing synthetic crude oil from oil sands is comparatively more water-intensive, though in-situ recovery uses on average less than one-quarter of the amount used in surface mining (see Canada section). Refining crude oil into end-use products requires further water for cooling and chemical processes, the amount varying widely according to technologies employed (the cooling system, for example) and process configuration.

Figure 17.3 ▷ **Water use for primary energy production**

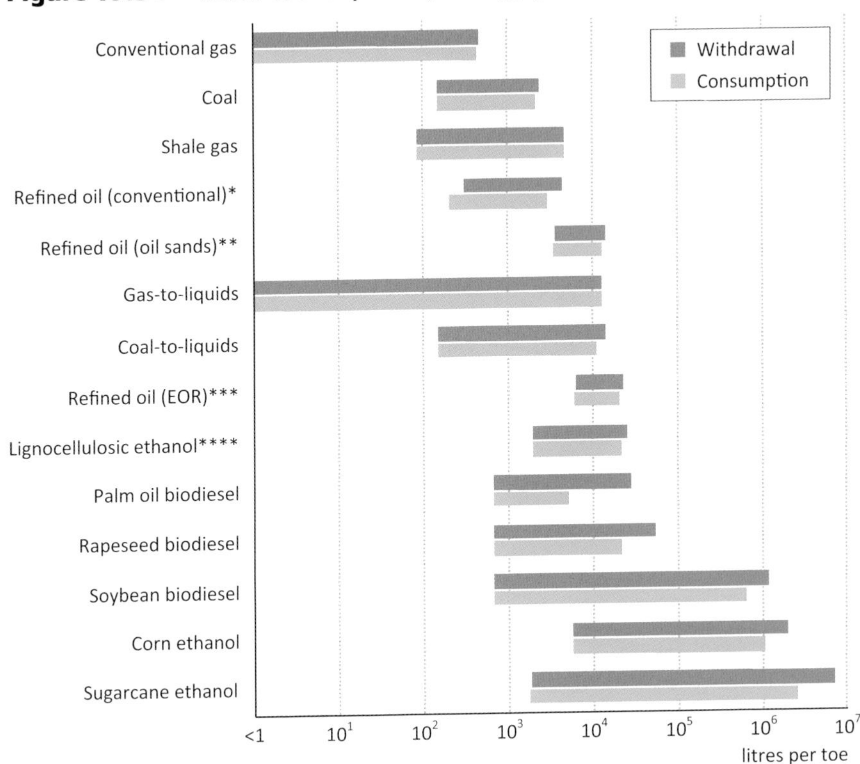

* The minimum is for primary recovery; the maximum is for secondary recovery. ** The minimum is for in-situ production; the maximum is for surface mining. *** Includes CO_2 injection, steam injection and alkaline injection and in-situ combustion. **** Excludes water use for crop residues allocated to food production.

Notes: Ranges shown are for "source-to-carrier" primary energy production, which includes withdrawals and consumption for extraction, processing and transport. Water use for biofuels production varies considerably because of differences in irrigation needs among regions and crops; the minimum for each crop represents non-irrigated crops whose only water requirements are for processing into fuels. EOR = enhanced oil recovery. For numeric ranges, see *www.worldenergyoutlook.org*.

Sources: Schornagel (2012); US DOE (2006); Gleick (1994); IEA analysis.

Biofuels require water for irrigating feedstock crops and for fuel conversion. Irrigation needs can range widely depending on the crop, the region in which it is grown and the water efficiency of irrigation technologies used. Growing feedstock crops that require minimal water or cultivating them in an area that receives ample rainfall can greatly reduce or eliminate water needs for irrigation. Rain-fed crops grown in Brazil and southeast Asia, for example, generally make lower demands on water resources than irrigated crops grown in parts of the United States. Advanced biofuels derived from waste products require little or no water for their growth as a fuel feedstock, as water used by these crops is allocated to the activity of primary value (food production, for example); water use figures would be higher for advanced biofuels if dedicated crops are grown.

Electricity generation

Thermal power plants – fossil fuel-based and nuclear – require water primarily for cooling. Per unit of energy produced, they are the energy sector's most intensive users of water. Water needs for thermal plants are determined by plant efficiency, access to alternative heat sinks and, in particular, the cooling system employed. Cooling needs that must be met by water can be reduced with higher plant efficiencies, which minimise the amount of waste heat produced per unit of electricity generated, and access to alternative heat sinks, such as the atmosphere (for example, a cooling tower). The choice of cooling system has the greatest impact on water requirements for a given type of thermal power generation. Two broad categories of cooling system are available: once-through and re-circulating, which is further divided into wet, dry and hybrid systems. Each involves trade-offs in terms of water use, impacts on water quality, plant efficiency and cost (Table 17.2):

- *Once-through* (or open-loop) systems withdraw freshwater (or non-freshwater), pass it through a steam condenser and return it at higher temperature to a nearby water body. A small fraction of the withdrawals are consumed through evaporation. The capital costs of once-through systems are lowest, compared with other cooling systems, but they require considerably higher water withdrawals and the large intake and subsequent discharge downstream at higher temperatures can be detrimental to aquatic life and ecosystems. These environmental impacts have led to increasingly stringent permitting requirements for once-through systems, for example in the United States, where existing systems are gradually being phased out.
- *Wet re-circulating* (or wet closed-loop) systems withdraw freshwater, and pass it through a steam condenser but, instead of being discharged downstream, the heated water is cooled in a *wet tower* or *pond*. Water not consumed by evaporation is returned to the steam condenser for reuse. Water withdrawals are much lower than in once-through systems, reducing exposure to risks posed by constrained water resources as well as environmental impacts (though periodic discharges are needed to prevent the accumulation of minerals and dissolved solids). Trade-offs relative to once-through systems include higher water consumption (as opposed to withdrawal) and greater land area requirements. Additionally, the cost of installing these systems is around 40% higher than for once-through systems (US DOE/NETL, 2008).
- *Dry* cooling systems use air flow through a cooling tower to condense steam. Their water requirements are minimal, compared with other systems, and they therefore are therefore better suited to dry climates. Their cost is about 3-4 times higher than for wet tower or pond systems, although the impact on the overall cost of the plant depends on its size and type. Because air is a less effective medium than water for cooling, dry cooling can affect power plant performance, reducing average generation by about 2-7% (more on hot days) depending on the type of plant (US EPA, 2009). Because dry cooling is less effective at high temperatures, it is sometimes installed in tandem with wet tower cooling to have a *hybrid* system that offers flexibility to operate during warm and cool periods.

Table 17.2 ▷ Power plant cooling system trade-offs

Cooling system	Advantages	Disadvantages
Once-through	Low water consumption. Mature technology. Lower capital cost.	High water withdrawals. Impact on ecosystem. Exposure to thermal discharge limits.
Wet tower	Significantly lower water withdrawal than once-through. Mature technology.	Higher water consumption than once-through. Lower power plant efficiency. Higher capital cost than once-through
Dry	Zero or minimal water withdrawal and consumption.	Higher capital cost relative to once-through and wet tower. Lower plant efficiency, particularly when ambient temperatures are high. Large land area requirements.
Hybrid	Lower capital cost than dry cooling. Reduced water consumption compared with wet tower. No efficiency penalty on hot days. Operational flexibility.	Higher capital cost than wet tower. Limited technology experience.

Source: Mielke, *et al.* (2010).

Water withdrawals per unit of electricity generated are highest for fossil-steam (coal-, gas- and oil-fired plants operating on a steam-cycle) and nuclear power plants with once-through cooling, at 75 000 - 450 000 litres per megawatt-hour (l/MWh) (Figure 17.4).[4] This is between 20-80 times higher than if wet tower cooling were used, although such systems increase water consumption. Combined-cycle gas turbines (CCGTs) generate less waste heat per unit of electricity produced thanks to higher thermal efficiency, and therefore require less cooling. Their water withdrawal and consumption are the lowest among thermal power plants, at 570 - 1 100 l/MWh using a wet cooling tower. Water use by CCGTs can be cut further when dry cooling is employed, though this comes at increased cost and a reduction in plant efficiency.

Water requirements for renewable electricity generating technologies range from negligible to comparable with thermal generation using wet tower cooling. Non-thermal renewables, such as wind and solar photovoltaic (PV) may use very small amounts of water, such as for cleaning or panel washing. This makes them well-suited for a future that may be both more carbon- and water-constrained. In addition to lower water use at the site of electricity generation, these renewable technologies have little or no water use associated with the production of fuel inputs and minimal impact on water quality compared to alternatives that discharge large volumes of heated cooling water or contaminants into the environment. Geothermal and concentrating solar power (CSP) technologies (Box 17.2) have water needs that range widely, depending on the particular generating technology and cooling system employed.

4. This range varies widely depending on plant efficiency. It accounts for values reported by operating coal-fired power plants in the United States (NETL, 2007), which can be considerably higher than estimates published in literature.

Figure 17.4 ▷ **Water use for electricity generation by cooling technology**

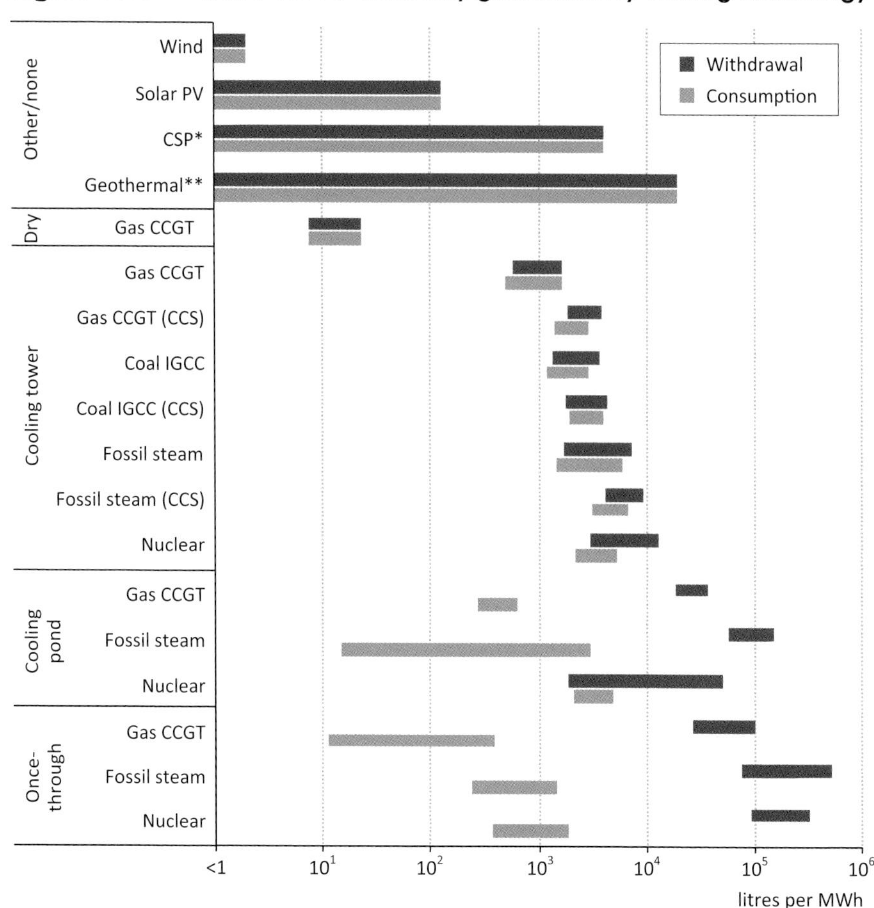

* Includes trough, tower and Fresnel technologies using tower, dry and hybrid cooling, and Stirling technology. ** Includes binary, flash and enhanced geothermal system technologies using tower, dry and hybrid cooling.

Notes: Ranges shown are for the operational phase of electricity generation, which includes cleaning, cooling and other process related needs; water used for the production of input fuels is excluded. Fossil steam includes coal-, gas- and oil-fired power plants operating on a steam cycle. Reported data from power plant operations are used for fossil-steam once-through cooling; other ranges are based on estimates summarised in the sources cited below. Solar PV = solar photovoltaic; CSP = concentrating solar power; CCGT = combined-cycle gas turbine; IGCC = integrated gasification combined-cycle; CCS = carbon capture and storage. For numeric ranges, see *www.worldenergyoutlook.org.*

Sources: Macknick (2011); US DOE/NETL (2007 and 2011); IEA analysis.

The high water requirements associated with other sources of low-carbon electricity mean that water availability must be strongly considered in plant siting. This applies, in particular, to nuclear plants and to fossil fuel-based plants fitted with carbon capture and storage (CCS) equipment. CCS-fitted power plants require additional water for the carbon dioxide (CO_2) capture process and to meet the higher cooling needs associated with reduced power plant efficiencies and consequently greater heat generation. Adding CCS equipment to power plants with a wet cooling tower is estimated to raise water withdrawals by between 60% for integrated gasification combined-cycle (IGCC) and 95% for CCGTs; consumption rises by similar amounts (US DOE/NETL, 2010).

Box 17.2 ▷ Water requirements for concentrating solar power

Concentrating solar power (CSP) is the most effective in areas with long hours of strong sunlight, but these areas tend to be drier and are more likely to face water scarcity challenges. CSP presently generates a near negligible share of global electricity output, but in the New Policies Scenario it grows quickly (23% per year on average), reaching 1% of global electricity generation in 2035. Most capacity additions are in the United States, China, India and South Africa, all of which contain regions of water scarcity.

The choice of technology and cooling system determines the level of water requirements for CSP. Using wet tower cooling, CSP based on parabolic trough, solar tower and Fresnel technologies can have water needs comparable to fossil fuel-based and nuclear power plants using the same cooling system. Some projects being built, in the Mojave Desert in California and in semi-arid areas of Shaanxi province in China, for example, are employing dry cooling to mitigate water constraints. However, dry cooling can appreciably lower plant efficiency and raise costs: in hot climates, CSP trough plants using dry cooling can see annual electricity production fall by 7% and electricity generating costs rise by about 10%. Solar tower technology has a higher conversion efficiency and therefore incurs a lower penalty when employing dry cooling (IEA, 2010).

CSP based on parabolic dish (Stirling engine) technology is cooled by air and requires no water for operation. Its land requirements are lower and conversion efficiency higher than other CSP technologies. However, its small capacity (less than 1 MW per dish) means that many of dishes are needed for utility-scale generation.

Hydropower is a major water user, relying on water passing through turbines to generate electricity. Water is consumed via seepage and evaporation from the reservoir created for hydropower facilities. Factors determining the amount consumed – climate, reservoir design and allocations to other uses – are highly site-specific and variable.[5] By one estimate, hydropower facilities in the United States consume 68 000 l/MWh on average, with a wide range that depends on the facility (Torcellini, *et al.*, 2003). This figure suggests that for certain facilities, hydropower plants with large reservoir storage can have some of the highest water consumption levels of any capacity type per unit of electricity generated. Run-of-river hydropower plants, however, store little water, leading to evaporation losses that are near zero.

5. Measurement approaches are not agreed upon and we therefore do not present ranges for water withdrawals and consumption for hydropower in Figure 17.4.

The vulnerability of energy to water constraints

Physical constraints on the availability of water for energy sector use span both quantity and quality issues: there may not be enough of it or that which is available may be of insufficient quality. These constraints may be natural or may arise from regulation of water use. Either way, they can pose challenges both to the reliability of existing operations and to proposed projects and, even where surmounted, impose additional costs for necessary adaptive measures. For example:

- River flows or reservoir levels can drop near or below water intake structures at thermal (nuclear and fossil fuel-based) power plants and hydropower facilities, curtailing or halting operations. This reduces available generating capacity and can threaten the reliability of the wider electricity system. Utilities may be forced to obtain electricity from other generators at higher cost.
- An oil and gas reservoir that requires water flooding to support production yet cannot secure ample water will experience declining field pressure and a fall in output. If water needs go unmet to the point that reservoir damage occurs, ultimate recovery from the reservoir may be lowered (such as in Iraq, see Chapter 14).
- Increased temperatures can diminish the effectiveness of water as a medium for cooling in thermal power plants, potentially lowering their thermal efficiency, and thereby lowering electricity output or, forcing the plants to shut down.

Regulations can impose limits on or increase the cost of water use by the energy sector. Depending on water rights, or the system of allocation in a given location, water may become difficult to access, potentially because of its prioritisation for competing uses, such as households or industry.[6] Environmental regulations may prohibit certain uses of water, do so under specific circumstances, or impose requirements on use – in containment, handling, treatment, discharge, disposal, etc. – that are more expensive. Shale gas development, for example, may become subject to more stringent regulation at added cost as a result of public concern over the potential of the process to contaminate water resources. Many power plants are required to limit the discharge of water heated in the cooling process when surface water temperatures approach a set threshold, causing the operators to scale-back output or even shut down.

The vulnerability of the energy sector to water constraints is widely spread geographically and across types of energy production (Table 17.3). Regions where water is scarce face obvious risks; but, even regions with ample water resources can face constraints related to droughts, heat waves, seasonal variation, climate change, regulations or some combination of these factors. Countries with a high proportion of their generating capacity in thermal plants with once-through cooling (using freshwater) and hydropower can be particularly exposed to fluctuations in water availability (Figure 17.5).

6. Water rights are typically categorised as: *riparian*, in which owners of land adjacent to a water source have equal rights of use; *prior appropriation*, in which water rights are acquired on the basis of beneficial use, rather than land ownership; or a hybrid of the two approaches. More progressive systems are being implemented that account for water's economic value.

Table 17.3 ▷ **Examples of water impacts on energy production**

Location (Year)	Description
	Power generation
India (2012)	A delayed monsoon raised electricity demand (for pumping groundwater for irrigation) and reduced hydro generation, contributing to blackouts lasting two days and affecting over 600 million people.
China (2011)	Drought limited hydro generation along the Yangtze river, contributing to higher coal demand (and prices) and forcing some provinces to implement strict energy efficiency measures and electricity rationing.
Vietnam, Philippines (2010)	The El Niño weather phenomenon caused a drought that lasted several months, reducing hydro generation and causing electricity shortages.
Southeast United States (2007)	During a drought, the Tennessee Valley Authority curtailed hydro generation to conserve water and reduced output from nuclear and fossil fuel-based plants.
Midwest United States (2006)	A heat wave forced nuclear plants to reduce their output because of the high water temperature of the Mississippi River.
France (2003)	An extended heat wave forced EdF to curtail nuclear power output equivalent to the loss of 4-5 reactors, costing an estimated €300 million to import electricity.
	Primary energy production
China (2008)	Dozens of planned coal-to-liquids (CTL) projects were abandoned, due in part to concerns they would place heavy burdens on scare water resources.
Australia, Bulgaria, Canada, France, United States	Public concern about the potential environmental impacts of unconventional gas production (including on water) has prompted additional regulation and, in some jurisdictions, temporary moratoria or bans on hydraulic fracturing.

Figure 17.5 ▷ **Share of power generation capacity with freshwater once-through cooling and hydro in selected countries, 2010**

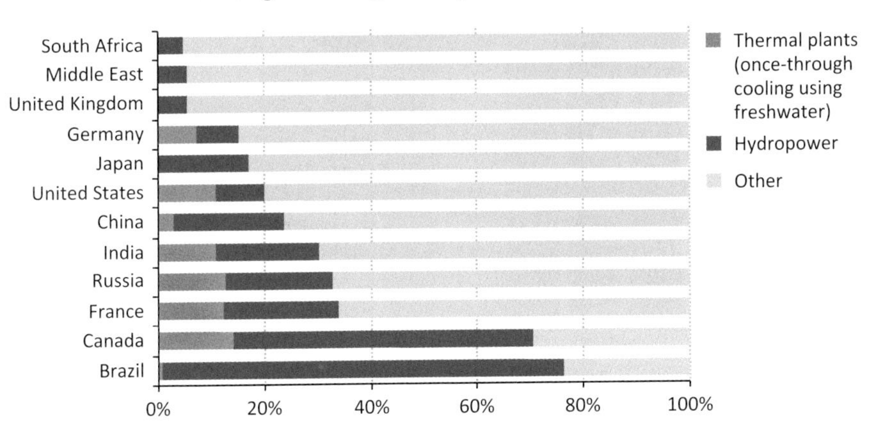

Sources: Platts (2012); IEA analysis.

Outlook for water requirements for energy production

We estimate that global freshwater withdrawals for energy production in 2010 were 583 billion cubic metres (bcm), some 15% of the world's total water withdrawals or exceeding the average annual discharge of the Ganges River in India. Water consumption by the energy sector was 66 bcm, or about 11% of energy-related water withdrawals. This estimate and the projections of future water requirements for energy production are based on the application of published water withdrawal and consumption factors for energy production to the *WEO-2012* energy supply projections in different scenarios (see Part A).[7]

Water requirements to support future energy production vary by scenario. The differences are largely a consequence of divergent trends related to energy demand (which necessitates more or less energy supply), the profile of the power generation mix and the cooling technologies used, and rates of production growth for biofuels. There is a general trend across the scenarios toward higher water consumption by the energy sector over 2010-2035, while the trend of withdrawals is more variable.

In the New Policies Scenario, global water withdrawals for energy production reach 690 bcm in 2035, an increase of about 20% over 2010, with growth slowing noticeably after 2020 (Figure 17.6 and Table 17.4). Withdrawals in the Current Policies Scenario (representing a pathway that assumes no change in existing energy-related policies) continue to rise throughout the projection period, climbing to 790 bcm in 2035, or 35% higher than in 2010. Energy-related water consumption grows over the *Outlook* period by about 85% in the New Policies Scenario and doubles in the Current Policies Scenario.

Figure 17.6 ▷ Global water use for energy production by scenario

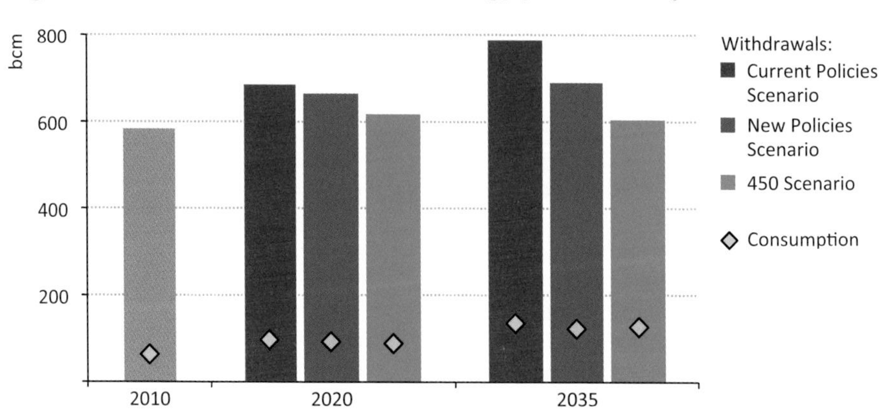

7. Freshwater requirements are quantified for the production of primary fuels (oil, gas, coal and biomass) consumed in all end-use sectors and for all forms of electricity generation, excluding hydropower (see footnote 5). Water factors are applied to energy production in each *WEO* region by fuel type and electricity generating (and cooling) technology. More information on the water factors used and key assumptions can be obtained at *www.worldenergyoutlook.org*.

Table 17.4 ▷ Global water use for energy production in the New Policies Scenario by region (bcm)

	Withdrawal				Consumption			
	2010	2020	2035	2010-35*	2010	2020	2035	2010-35*
OECD	307	316	302	-0.1%	30	39	46	1.7%
Americas	241	253	249	0.1%	21	29	38	2.5%
United States	206	214	212	0.1%	19	26	35	2.5%
Europe	61	57	49	-0.9%	8	8	6	-1.4%
Asia Oceania	5	6	5	0.1%	1	2	2	1.9%
Non-OECD	276	346	388	1.4%	35	56	76	3.1%
E. Europe/Eurasia	95	93	95	0.0%	4	5	6	1.5%
Asia	157	211	230	1.5%	21	34	43	2.9%
China	106	134	145	1.3%	16	26	30	2.5%
India	40	55	58	1.6%	4	6	9	3.5%
Middle East	3	4	5	1.3%	2	3	4	1.5%
Africa	5	7	8	1.8%	3	4	4	1.6%
Latin America	16	31	52	4.9%	5	11	19	5.6%
World	583	662	691	0.7%	66	95	122	2.5%
European Union	66	61	56	-0.7%	8	8	6	-1.2%

* Compound average annual growth rate.

Slower energy demand growth in the New Policies Scenario (averaging 1.2% per year, versus 1.5% in the Current Policies Scenario) plays a significant role in its comparatively lower water requirements. Differences in the global fleet of coal-fired power plants are notable in the two scenarios. In the New Policies Scenario, coal-fired generation is some 30% lower at the end of the *Outlook* period and inefficient plants (that withdraw large quantities of water) are retired more quickly. The power sector sees a continued trend toward wet cooling towers in both scenarios, but because older plants are more often based on traditional once-through systems, the move away from subcritical coal plants in the New Policies Scenario has a greater impact on water withdrawals. The expanded role of renewables, such as wind and solar PV, also reduces water withdrawals in the New Policies Scenario – their generation in 2035 is 25% and 60% higher, respectively, than in the Current Policies Scenario.

In addition to representing a low-carbon pathway, the 450 Scenario could also be a less water-intensive one. Withdrawals by the energy sector in this scenario reach about 600 bcm in 2035, only 4% above 2010 levels; consumption almost doubles. Compared with the New Policies Scenario, the 450 Scenario sees much more modest energy demand growth (averaging 0.6% per year) and a marked shift in the power sector away from coal-fired power plants and towards renewables. The 450 Scenario, nonetheless, widely deploys several technologies – including nuclear power, CCS-fitted power plants and conventional biofuels – whose high water use requirements must be taken into account when siting energy production facilities. Water withdrawals and consumption for biofuels expand the most in the 450 Scenario, even though the increase after 2020 is stemmed somewhat by penetration of non-irrigated advanced biofuels.

In the New Policies Scenario, water use for power generation – principally for cooling at thermal power plants – accounts for the bulk of water requirements for energy production worldwide, although the needs for biofuels also become much more significant as their production accelerates (Figure 17.7). Withdrawals for power generation in 2010 were some 540 bcm, over 90% of the total for energy production. These slowly rising requirements level off around 2015, before falling to 560 bcm at the end of the *Outlook* period. There are two counteracting forces at work: a reduction of generation by subcritical coal plants that use once-through cooling, particularly in the United States, China and European Union, cutting global withdrawals by coal-fired plants by almost 10%; and growth in generation from newly built nuclear power plants that use once-through cooling (for instance, some that are constructed inland in China), which expands water withdrawals for nuclear generators by a third. Consumption of water in the world's power sector rises by almost 40%, boosted by increased use of wet tower cooling in thermal capacity. Increasing shares of gas-fired and renewable generation play a significant role in constraining additional water use in many regions, as global electricity generation grows by some 70% over 2010-2035, much more than water withdrawal or consumption by the sector.

Figure 17.7 ▷ Global water use for energy production in the New Policies Scenario by fuel and power generation type

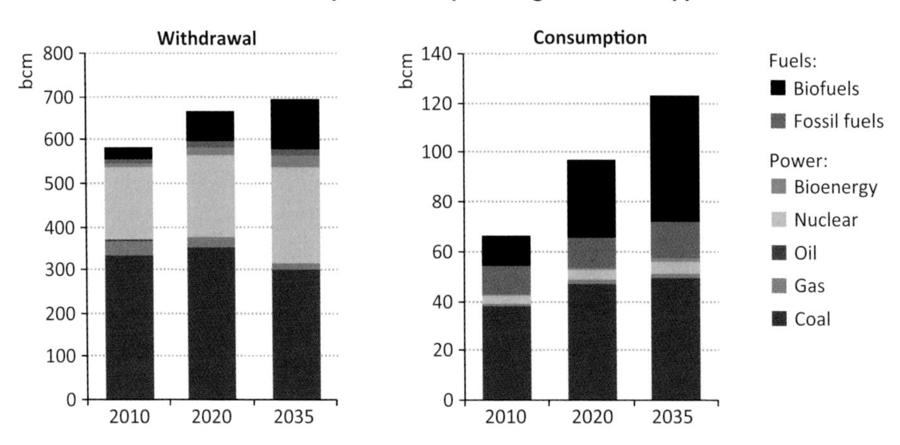

Energy-related water use rises as a direct consequence of steeply increasing global biofuels supply, which triples in the New Policies Scenario on government policies that mandate the use of biofuels. Water withdrawals for biofuels increase in line with global supply, from 25 bcm to 110 bcm over 2010-2035. However, consumption increases from 12 bcm to almost 50 bcm during that time, equalling the water consumption for power generation by the end of the *Outlook* period. These higher water requirements for biofuels production stem from the irrigation needs for feedstock crops for ethanol and biodiesel – primarily sugarcane, corn and soybean – in major producing regions, such as Brazil, the United States and China. Non-irrigated advanced biofuels from waste crops make inroads into the market after 2020, thereby serving to temper the growth in overall water needs for biofuels production in the New Policies Scenario.

The water-intensity of global withdrawals and consumption for energy production – that is, water withdrawals and consumption per unit of energy produced – head in opposite directions during the *Outlook* period (Figure 17.8). The withdrawal-intensity of global energy production falls by 23%, whereas consumption-intensity increases by almost 18%. As noted, this is primarily the result of an expected shift in the power sector away from traditional once-through cooling systems towards wet towers (that reduce withdrawals but raise consumption).

Figure 17.8 ▷ Water intensity of energy production for selected regions in the New Policies Scenario

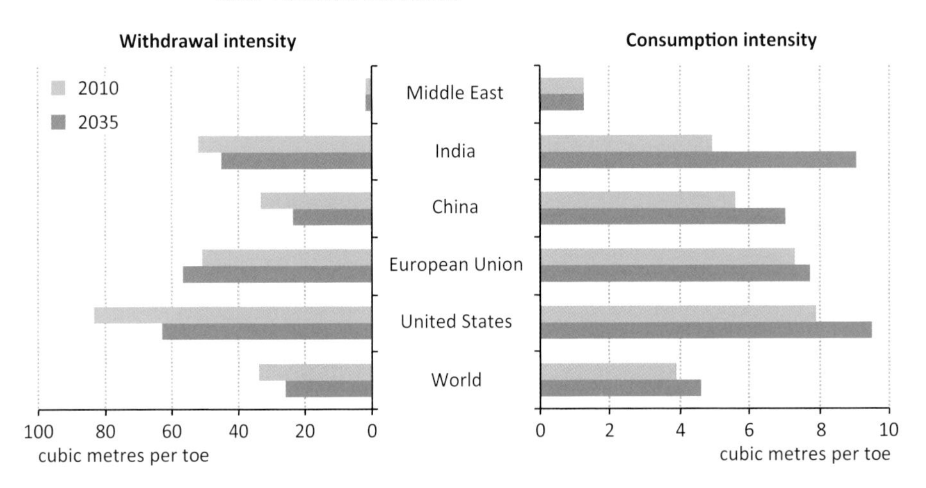

The largest users of water for energy production on a country and regional basis are the world's largest electricity generators: the United States, the European Union, China and India (Table 17.4). All have significant inland generating capacity to meet demand away from the coasts. On the other hand, countries such as Japan, Korea and Australia have minimal water requirements for energy because they can site virtually all of their power plants on the coasts and use seawater for cooling. Water scarcity is a major constraint on water use for energy production in the Middle East, where the energy sector's absolute water use and water intensity are strikingly low and remains so through the *Outlook* period. The region's fleet of power plants is adapted to scarce water conditions, employing significant capacity that depends little on freshwater availability, including combined-cycle gas turbines and combined water and power facilities, situated on the coast, to meet demand needs for both energy and freshwater.

Regional stress points

China

China's water resources are set to become more strained with the country's ongoing urbanisation and economic development. On a national basis China's renewable water resources per capita were 2 070 cubic metres in 2010, just above the level regarded in this analysis as indicating "water stress"; but resources vary widely across regions (Figure 17.9). Water withdrawals per capita amounted to about 460 cubic metres. Around 65% of China's water withdrawals are for irrigation, 23% for industry and 12% for municipal use (UN FAO, 2012). China's water challenges are exacerbated by geographical disparity between supply and demand: water is much more abundant in the south than in the north and west, where the country's water-intensive agriculture and industry sectors are concentrated. Limited water supplies and widespread pollution of river systems in parts of China have put increasing pressure on groundwater resources (IBRD, 2009).

Figure 17.9 ▷ Renewable water resources per capita and distribution of water-intensive energy production by type in China

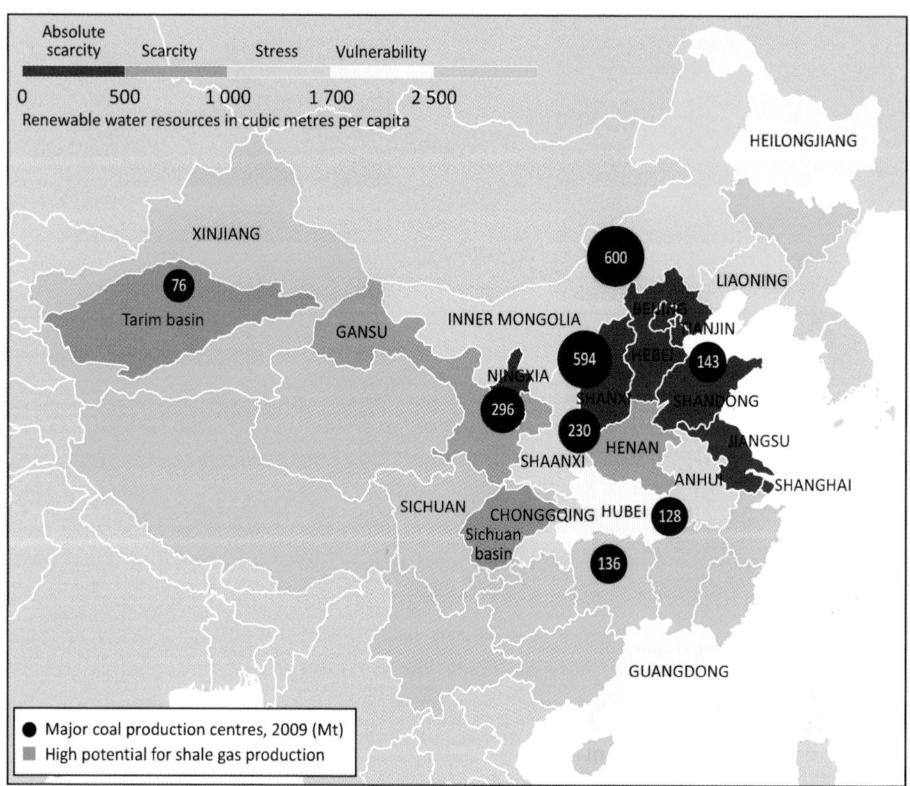

This map is without prejudice to the status of or sovereignty over any territory, to the delimitation of international frontiers and boundaries and to the name of any territory, city or area.

Notes: Although water resources in the Xinjiang Uygher Autonomous Region as a whole are above the national average, they are unevenly distributed. The Tarim Basin, which has high potential for shale gas production, is particularly arid. Sources: Water data from China National Bureau of Statistics; IEA analysis.

The Chinese government has identified water scarcity as a potential bottleneck to economic and social development. In response, it is pushing ahead with water pricing reforms and increasingly stringent regulations and enforcement procedures aimed at improving water conservation. China's 12th Five-Year Plan, ending in 2015, includes a target to cut water consumption per unit of value-added industrial output by 30%. Water withdrawal caps have been set (635 bcm in 2015 rising to 700 bcm in 2030) and targets have been introduced to raise water-use efficiencies to the level of developed nations by 2030. Alongside these efforts, China is building desalination capacity and developing major water infrastructure projects, most notably the massive south-to-north water transfer project, scheduled for completion in 2050, which is designed to divert 45 bcm per year of water (the annual combined flow of the Tigris and Euphrates Rivers) from southern rivers to the dry north. Notwithstanding these measures, the threat remains that the future of certain water-intensive industries – including some types of energy production – could be constrained in certain water-stressed regions by intense competition for water.

Figure 17.10 ▷ Water use for energy production in China in the New Policies Scenario

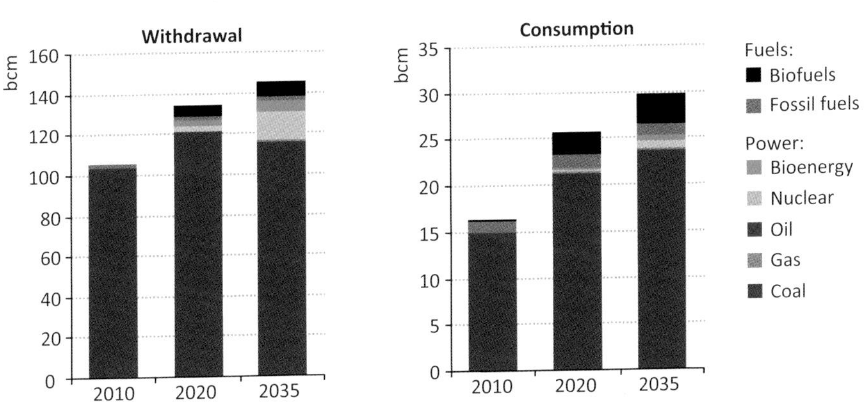

In the New Policies Scenario, water withdrawals for energy production in China rise by 38% between 2010 and 2035, or 40 bcm (Figure 17.10). Water consumed by the energy sector rises far more proportionately, 83% (14 bcm), increasing pressure on water resources. Power generation dominates water use in the energy sector. Water withdrawals by coal-fired power plants remain relatively steady, rising by 12% despite their electricity output increasing by about 65%, due to the deployment of increasingly efficient plants and the use of wet cooling towers. However, the use of wet cooling towers is the prime factor pushing up water consumption in the sector. A growing role for natural gas in the power mix tempers water requirements, but the inland expansion of nuclear power raises withdrawals towards the end of the *Outlook* period. The coal mining sector is another important source of incremental water use: withdrawals are projected to rise by 18% for an equivalent percentage increase in coal production, much of which is expected to come from the country's rich coal reserves in the dry north and west of the country. Water requirements per tonne of coal produced are expected to rise as coal mining operations move deeper underground and washing becomes more widespread.

Power generation

Water scarcity is already having an impact on power generation in parts of China. In coal-rich arid regions in the north, some power plants have turned to dry cooling rather than wet cooling tower systems. These systems are more expensive and reduce plant efficiency, but (as described earlier) they can sharply reduce the extent of water withdrawals and consumption. Coal is set to remain the cornerstone of China's electricity mix through the *Outlook* period, but government targets to diversify the structure of power generation will have implications for water consumption. Plans to install 150 gigawatts (GW) of wind by 2020 should alleviate some water scarcity concerns, particularly if this capacity is built in the north. But there are also targets to significantly expand nuclear capacity, a technology that is extremely water-intensive. While all existing nuclear plants in China use seawater for cooling, future plans include the development of inland nuclear power facilities – three are due to start construction during the course of the 12th Five-Year Plan – that will add to competition for scarce water resources where the plants are sited. China also has ambitions to increase production of CSP, which can be as water-intensive as fossil fuel-based and nuclear plants. As is often the nature of areas well-suited to CSP, China's sun belt is in a dry region: selection of particular sites will have to take into account the availability of water resources and the case for installing more expensive dry or hybrid dry/wet cooling systems.

Coal production and processing and shale gas

Water use in the coal sector – for mining, processing and cooling power plants – currently accounts for the largest share of industrial water use in China. The country's coal resources, which are being rapidly tapped to meet the increasing demand for electricity, are concentrated in the western and northern regions, where water is scarce. For example, Shanxi, Shaanxi, Inner Mongolia and Xinjiang hold 74% of China's total coal resources, but only 7% of the nation's water resources (IEA, 2008). Water shortages in these regions limit the scope to develop coal-fired power plants there to deliver electricity to demand centres in the east, thereby contributing to capacity bottlenecks in the transport of coal by truck or rail. Water scarcity also helps to explain the low share of China's coal production that is washed to improve combustion efficiency and reduce sulphur and particulate emissions from coal-fired combustion.

Water scarcity in coal-bearing regions accounts, in part, for the scaling back of China's plans to expand coal-to-liquids (CTL) production as a means of slowing its rising petroleum imports. Production of CTL is water intensive, the fuel processing phase withdrawing about 1.5 cubic metres of water per barrel of fuel produced (without recycling). This concern prompted the suspension of dozens of CTL projects in 2008. Nonetheless, several plants have gone ahead and other projects are in the start-up phase. In early 2011, China brought on-stream its first commercial CTL facility – the 24 thousand barrels per day (kb/d) Shenhua Group plant in Inner Mongolia – and there are plans to expand production, encouraged by the plant's performance and profitability.

At 36 trillion cubic metres (tcm), China is estimated to have the largest technically recoverable shale gas resources in the world.[8] Exploration and development is as yet in its infancy, but the government is actively seeking to develop these resources. In March 2012, China's National Development Reform Commission (NDRC) issued the National Shale Gas Development Plan for 2011-15, which envisions shale gas production reaching 60-100 bcm by 2020. However, some of the most promising basins are located in water scarce regions, which could constrain production growth (or raise developments costs). The Tarim Basin in the Xinjiang Uyghur Autonomous Region, for example, holds some of the country's largest shale gas deposits but suffers from severe water scarcity. Several other shale gas areas are likely to face strong competition for water, including the Ordos, Qinshui and North China basins. By contrast, most shale gas exploration and development in China has been concentrated in the Sichuan Basin, where water is more abundant.

India

India's water resources are set to come under further pressure as economic and population growth continue and people move into cities. The country's renewable water resources per capita were about 1 560 cubic metres in 2010, below the threshold for "water stress". Water withdrawals per capita totalled about 620 cubic metres. Around 90% of India's water withdrawal is for agriculture and livestock, with the remainder to municipalities (7%) and industries (2%) (UN FAO, 2012). Although the energy sector (primarily power generation under industry) accounts for only a minor share now, this is expected to increase as India's energy demand more than doubles over the *Outlook* period in the New Policies Scenario.

India's main sources of water – glacier melt and rainfall – are unevenly distributed. Most of the rainfall comes during the summer monsoon from June to September. A delayed or weak monsoon often results in water shortages, with severe drought having occurred in the past few years, particularly in 2009 and in 2012. The northwest and southern regions, which generally receive less rainfall, are especially susceptible, being host to significant power generating capacity that is highly dependent on water (Figure 17.11). Groundwater is pumped to supplement other water resources, a process incentivised by subsidised electricity prices. Withdrawal rates are highest in northern and southern regions. Extensive water pollution from sewage is one factor limiting water availability (for example, the lower stretch of Yamuna River from New Delhi is heavily polluted). River basins spread across states have witnessed interstate disputes over water rights. Following the last amendment, a decade ago, the government is currently drafting a new national water policy.

In the New Policies Scenario, water withdrawals for energy production in India grow by almost 50% between 2010 and 2035, or 19 bcm (Figure 17.12). Water consumption increases at a much faster rate, more than doubling over the period. The power sector, which accounts for the vast bulk of all water use by India's energy sector at present, remains

8. The estimate cited is by Advanced Resources International (ARI); China's Ministry of Land and Resources estimates 25 tcm shale gas, but this estimate does not include all provinces. The ARI estimate is used for consistency with our resource methodology for other countries.

the major source of incremental water use: it accounts for 98% of additional withdrawals and 95% of additional consumption during the *Outlook* period. The increase in withdrawals by coal-fired power plants slows between 2010 and 2035, as India's fleet of power plants becomes more efficient. Growth in the share of natural gas in power generation tempers water use in the power sector. Some additional water use results from expanding nuclear power output. Biofuels are a small source of incremental water use during the *Outlook* period.

Figure 17.11 ▷ Renewable water resources per capita and distribution of water-intensive energy production by type in India

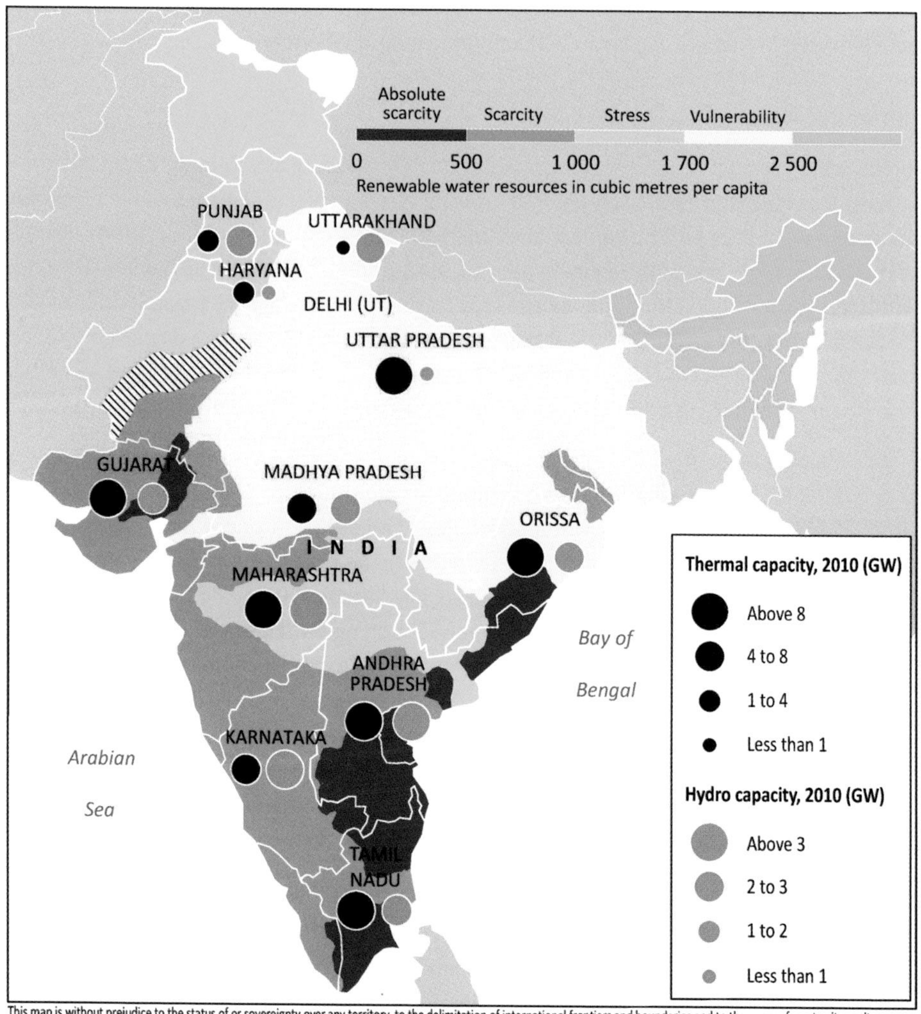

This map is without prejudice to the status of or sovereignty over any territory, to the delimitation of international frontiers and boundaries and to the name of any territory, city or area.

Notes: Per-capita water availability is shown by river basin. The striped area represents the inland drainage area of Rajasthan. Due to data limitations, some areas are not shaded.

Sources: Government of India (2010); Platts (2012); IEA analysis.

Figure 17.12 ▷ **Water use for energy production in India in the New Policies Scenario**

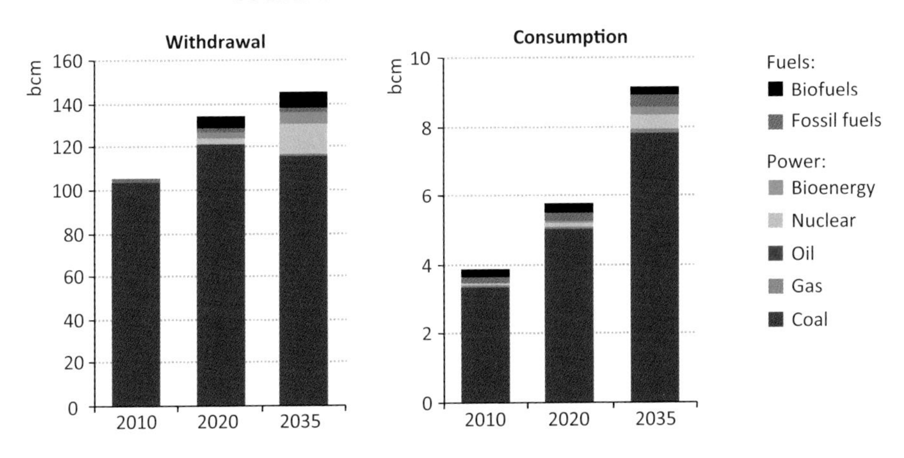

India relies heavily on coal-fired power plants, many of which employ subcritical technologies and run at low efficiencies of around 30%. India's warm climate and poor quality domestic coal add to cooling requirements for these plants, though power plant efficiencies could be boosted with greater coal washing. India's power sector has already faced constraints linked to water availability. Summer water shortages in 2010 caused the 2.3-GW Chandrapur coal-fired power station in Maharashtra to shut down, leading to power outages across the populous state. The plant again faced water shortages, due to the delayed monsoon, in mid-2012. Water shortages in northern India have at times reduced hydropower generation, exacerbating power shortages due to insufficient coal supply to power plants. Moreover, increased groundwater withdrawals (which use electric pumps) to support agriculture during periods of drought put an additional burden on strained electricity supply. To address water shortages, several power plants are now using treated wastewater and seawater for cooling. Under the government's 12th Five Year Plan over 2012-2017, coal allocations have been linked to water availability and priority is given to new power plants that use seawater, rather than freshwater, for cooling (Government of India, 2009). India's national solar mission and solar cities development programme foresee higher CSP output during the *Outlook* period, but some plants are being built in water-stressed areas, such as Rajasthan, where water could be a constraint.

United States

A temperate climate and moderate rainfall underpin renewable water resources in the United States that are well in excess of water use. Per capita annual renewable water resources were close to 10 000 cubic metres in 2010, while total per capita freshwater withdrawals were just over 1 500 cubic metres. Total water withdrawals in the United States more than doubled from 1950 to 1975, but have remained flat since then, despite ongoing population and economic growth. Total water withdrawals are split between power plant use (49%), irrigation (31%), municipal use (11%) and other (8%) (USGS, 2009).

In the New Policies Scenario, water withdrawals for energy production in the United States increase only slightly from just over 206 bcm in 2010 to just over 210 bcm in 2035 (Figure 17.13). Withdrawals for coal-fired power plants fall substantially with the retirement of old and less efficient stock and the continued phase out of once-through cooling systems. Water consumption for energy production rises by over 80% during the *Outlook* period, the increase occurring mostly in the next decade due to a substantial increase in the use of energy crops for biofuels production such as corn and soybean crops that generally require irrigation, leading to high water use.

Figure 17.13 ▷ Water use for energy production in the United States in the New Policies Scenario

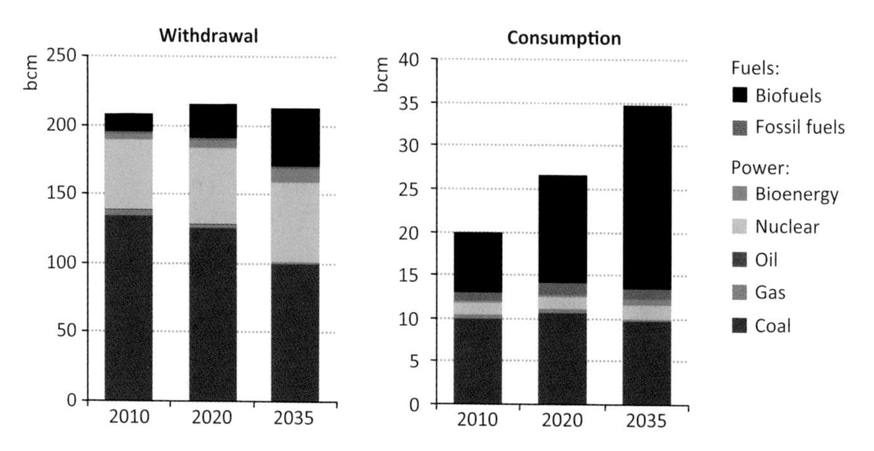

Power generation

Withdrawals for power plants have not increased since 1975, despite electricity generation more than doubling over the same period. This is partially due to a shift towards cooling systems which withdraw less water but consume more, *i.e.* the replacement of once-through systems with wet cooling towers, a trend that is expected to continue. Growing output from higher-efficiency gas-fired power generation in the past decades has also tempered water withdrawals.

Unlike the general water situation in the United States, water is scarce in the dry southwest and is a key concern for existing power plants as well as new builds. Water use already exceeds sustainable levels in the region, which faces growing water and electricity demand as the population increases. Low water availability has occasioned the reduction of output from coal-fired power plants and hydropower facilities in some cases, and put it seriously at risk in others, threatening blackouts and higher cost electricity. According to the US Energy Information Administration, electricity demand in the region is set to increase more than one-third by 2035 (US DOE/EIA, 2011). The water scarcity means that new generating capacity deployed to meet this demand will need to have limited dependence on water, favouring an expansion of gas-fired, wind and solar PV installations and the use of dry and hybrid cooling systems. The deployment of such technologies, backed by energy efficiency efforts, could dramatically cut water withdrawals and consumption by power generation and offer operational flexibility over a range of climatic conditions (Cooley, *et al.*, 2011).

Power generation facilities in other US regions – even where water is typically in ample supply – can be vulnerable to abnormal changes in water availability, such as during a drought or heat wave. For example, although renewable water resources in an average year are close to forty times total water consumption in the US southeast, droughts there in 2007 and 2010 forced operations to be scaled back at hydropower facilities and several coal-fired and nuclear power plants.

Light tight oil and unconventional gas

Water scarcity threatens to constrain burgeoning domestic oil and gas production from shale formations in some parts of the country. US light tight oil production has increased from 11 kb/d in 2005 to 840 kb/d in 2011 and is expected to continue this rapid expansion, reaching over 3.2 mb/d by 2025 in the New Policies Scenario (see Chapter 3, Box 3.3). These bullish prospects are underpinned by significant resources in the Bakken formation, spanning the western states of North Dakota and Montana, and in the Eagle Ford formation, within the state of Texas. In the Bakken, further expansion of production is contingent on water availability. Recent drilling activity has started to approach the limits of available water resources and developers believe that securing ample water will be a key challenge for continued development (MacPherson, 2011). Texas is one of the driest regions of the country and recent extended periods of drought have heightened concerns about water availability.

Concerns about water availability and the effect of production on water quality could also significantly slow the development of shale gas production in the United States. Shale gas production has increased by 45% per year between 2005 and 2011 and in the New Policies Scenario it is projected to grow to about 370 bcm in 2035, up from 193 bcm from in 2011. Hydraulic fracturing is a source of considerable public concern in the United States and some states, such as Vermont, New York, New Jersey and Maryland, have either banned its use or placed a moratorium on the practice while further study is undertaken. A major investigation is underway by the US Environmental Protection Agency to understand the potential impacts of hydraulic fracturing on drinking water sources, with a final report expected in 2014. This could greatly impact the rate of future development of shale gas production in the United States and elsewhere.

Canada

Although Canada is a water-rich country, with annual per capita renewable water resources in excess of 85 000 cubic metres, extensive use of water in the extraction and upgrading of oil sands (or bitumen) in parts of Alberta and Saskatchewan provinces could have significant implications for the oil production outlook and water resources in the surrounding area. Oil sands production has grown from 0.6 million barrels per day (mb/d) in 2000 to 1.6 mb/d in 2011 and is projected to increase to 4.3 mb/d in 2035 in the New Policies Scenario, making an important contribution to global oil supply and energy security (see Chapter 3).

There are two main techniques for producing oil sands, both of which use water. In mining operations, shallow deposits are extracted and the bitumen separated from the sand using hot water. Deeper deposits are typically produced using in-situ methods, in which steam is injected into the reservoir. Mining accounts for about 51% of oil sands production at present, although in-situ techniques are expected to account for the bulk of the projected increase. In both cases, the bitumen is converted into higher-value synthetic crude oil.

Net of recycling, we estimate that mining (plus upgrading) requires 0.9 cubic metres of water per barrel of synthetic crude oil produced while in-situ recovery requires 0.2 cubic metres of water per barrel produced. The majority of water required for mining operations is withdrawn from the Athabasca River, although the share fluctuates: the percentage was 85% in 2010, up from two-thirds the previous year. In-situ operations sourced just over 80% of their water needs from groundwater (such as deep saline aquifers) in 2010, making no withdrawals from the Athabasca River (CAPP, 2011). Presently, about half of the water withdrawn by in-situ operations is freshwater, but increasingly projects are sourcing water from saline aquifers. Based on expected production trends, we estimate that total water withdrawals for oil sands – including fresh and saline sources – will grow from about 220 million cubic metres (mcm) in 2010 to about 520 mcm in 2035 (Figure 17.14).

Realising large increases in oil sands production depends on reducing freshwater use as well as future water availability. Freshwater use can be cut by increasing reliance on saline aquifers or wastewater for water supply, greater use of water recycling and additional use of steam-less processes. Water availability does not present an immediate risk to operations: withdrawals from the Athabasca River for oil sands mining represent only 0.5% of its average annual flow and are about one-quarter of the amount allocated under existing regulations; and saline aquifers have provided increasing amounts of water as oil sands output has grown. Projected total (saline and fresh) water needs for oil sands production in 2035 would be greater than the existing threshold for withdrawals from the Athabasca River. While much of the water needs could be expected to come from saline aquifers, the implications of increasing withdrawals from these sources have not yet been fully assessed (in terms of aquifer size and ecosystem impact), emphasising the importance of achieving reductions in freshwater use with continued technology improvement. The seasonality of the Athabasca's flow could also pose operational challenges without careful management of withdrawals. During winter months, its flow can fall to around one-quarter of the summer peak, and mining companies are presently required to adhere to withdrawal limits that are adjusted weekly according to river conditions.

Figure 17.14 ▷ Canadian oil sands production and estimated water withdrawal by type in the New Policies Scenario

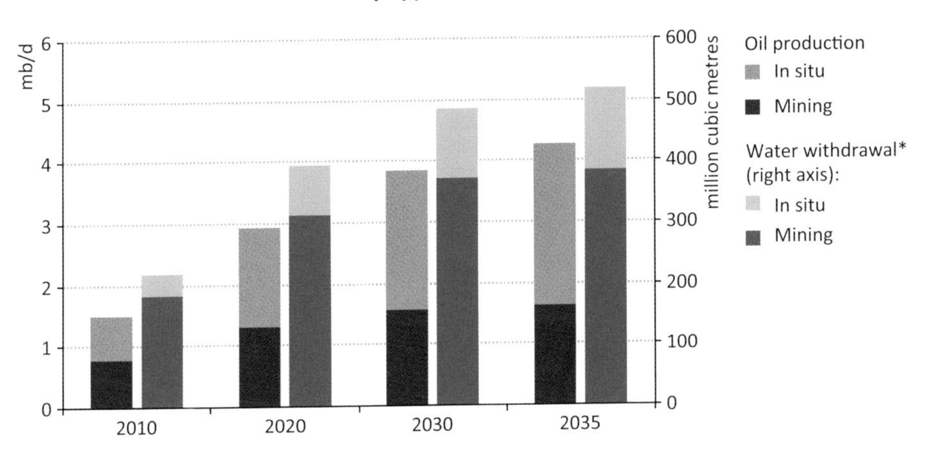

* Includes fresh and saline water.

Note: Water withdrawals are estimated by applying technology-specific water indicators to projections for mining and in-situ oil sands production.

Source: IEA analysis.

The impact of oil sands' production on water quality is also a critical issue. The discharge of untreated wastewater into rivers is prohibited, but there is concern that seepage from the vast tailing ponds already used to store degraded water could cause surface and groundwater pollution. For the protection of ecosystems, regulations mandate that rigorous monitoring is performed and prevention systems are in place to guard against seepage.

How serious is the water constraint?

The projections set out in *WEO-2012* assume that the associated water constraints can be overcome. However, there is no doubt that water is growing in importance as a criterion for assessing the physical, economic and environmental viability of energy projects. In an increasingly water-constrained world, the vulnerability of the energy sector to constraints in water availability can be expected to increase, as can issues around how the quality of water is affected by energy operations. These vulnerabilities and impacts are manageable in most cases, but better technology will need to be deployed and energy and water policies better integrated.

Several opportunities are available to improve the situation. In the power sector, these include greater reliance on renewable energy technologies that have minimal water requirements, such as solar PV and wind; improving the efficiency of power plants, for instance by shifting from subcritical coal to supercritical coal or IGCC plants; and deployment of more advanced cooling systems, including wet cooling towers, and dry and hybrid cooling. In biofuels production, biomass crops and locations that have the greatest water efficiency will be advantaged. More generally, the energy sector can look to exploit non-freshwater sources – saline water, treated wastewater, storm water and produced water from oil and gas operations – and adopt water re-use technologies. Importantly, assigning precious water resources a more appropriate economic value in regions where it is underpriced or even free would encourage more efficient use, not only in the energy sector but across the economy.

Measuring progress towards energy for all

Power to the people?

Highlights

- Since our last *Outlook*, the number of people without access to electricity globally has decreased by 50 million and the number without clean cooking facilities has declined by nearly 40 million. This has been realised despite the growth in world population and has been spurred by reported improvements in many countries, including India, Indonesia, Brazil, Thailand, South Africa and Ethiopia.

- Yet nearly 1.3 billion people remain without access to electricity and 2.6 billion still do not have access to clean cooking facilities. These people are mainly in either developing Asia or sub-Saharan Africa, and in rural areas. Just ten countries account for two-thirds of those without electricity and just three countries – India, China and Bangladesh – account for more than half of those without clean cooking facilities.

- In the New Policies Scenario, we project that close to 1 billion people will still be without electricity and 2.6 billion people will still be without clean cooking facilities in 2030. In the case of electricity, the number of people in developing Asia without access almost halves compared to 2010 (led by progress in India) and Latin America achieves universal access before 2030 but, in sub-Saharan Africa, a worsening trend persists until around 2025. For cooking, developing Asia sees a significant improvement (led by China), but the number of people without clean cooking facilities in India alone in 2030 is still twice the population of the United States today. In sub-Saharan Africa the picture worsens by around one-quarter by 2030.

- The UN Secretary-General's Sustainable Energy for All initiative has been vital in raising awareness of the urgent need to increase modern energy access. But the energy access funding commitments it had received by the time of the Rio+20 Summit were only equivalent to around 3% of the nearly $1 trillion in cumulative investment we estimate is needed to achieve universal access by 2030 in our Energy for All Case. Any concerns that achieving modern energy access for all would unduly magnify the challenges of energy security or climate change are unfounded, as it would only increase global energy demand by 1% in 2030 and CO_2 emissions by 0.6%.

- We present an Energy Development Index (EDI) for 80 countries, to aid policy makers in tracking progress towards providing modern energy access. It is a composite index that measures a country's energy development at household and community level. Our EDI results reveal a broad-based improvement in recent years. Countries showing some of the greatest improvements include China, Thailand, El Salvador, Argentina, Uruguay, Vietnam and Algeria. There are also a number of countries whose EDI scores are stubbornly low, such as Ethiopia, Liberia, Rwanda, Guinea, Uganda and Burkina Faso. As a region, sub-Saharan Africa scores least well.

Introduction

Has 2012 been a breakthrough year for modern energy access? A review of the last year reveals new focus, new commitments and new actions towards a goal of achieving universal energy access by 2030. The United Nations designation of 2012 as the Year of Sustainable Energy for All, coupled with the decision by the UN Secretary-General to include universal access to modern energy within his Sustainable Energy for All initiative (SE4All), has set the tone. At the Rio+20 Summit,[1] countries recognised the critical role of energy in the development process, committed themselves to measures to improve energy access and emphasised the need for further action. They noted the SE4All initiative, and stated their determination to act to make sustainable energy for all a reality, though they did not make a binding commitment to achieve universal modern energy access by 2030. The last year has raised the level of attention given to improving modern energy access and also the level of expectation about the ultimate results (Spotlight).

Even in a year intended to shine a light on energy access, challenges have continued to emerge. Higher oil prices (over $110/barrel in the first half of 2012) have helped push oil-import bills up in net-importing less-developed countries to an estimated 5.7% of gross domestic product, impacting on growth prospects. Energy expenditure in households is creeping higher in many countries and having a disproportionate impact on the poorest. Finance for energy access improvements often remains hard to secure and the necessary five-fold increase in investment, highlighted in *WEO-2011*, is far from being realised.

In this chapter, we report where we stand on universal modern energy access, based on a comprehensive update of our electricity and traditional biomass databases.[2] We then present projections for modern energy access in the New Policies Scenario, the central scenario in *WEO-2012*, together with an Energy for All Case that is designed to highlight what more needs to be done to put us on course to achieve universal access by 2030 and what some of the implications might be.[3] For our projections, we define energy access as a household having reliable and affordable access to clean cooking facilities and a first electricity supply connection, with a minimum level of consumption (250 kilowatt-hours [kWh] per year for a rural household and 500 kWh for an urban household) that increases over time to reach the regional average.[4] Our analysis takes into account the need for different technological solutions, such as grid, mini-grid and off-grid solutions for electricity, and advanced biomass cookstoves, liquefied petroleum gas (LPG) stoves and biogas systems for cooking. Additionally, this chapter covers the critical issue of tracking a country's energy development over time. We present an enhanced and expanded Energy Development Index (EDI) for 80 countries, a composite index that includes relevant indicators relating to household access to electricity and clean cooking facilities and

1. The United Nations Conference on Sustainable Development held in Rio de Janeiro, Brazil in June 2012.

2. We use 2010 data where available or an estimate based on latest available data.

3. While the *Outlook* period for *WEO-2012* is 2010 to 2035, analysis in this chapter is based exceptionally on the period 2010 to 2030, so as to be consistent with the goal of the Sustainable Energy for All initiative.

4. For more detail on our definition of energy access visit *www.worldenergyoutlook.org*.

to the use of modern energy for productive purposes (such as mechanical power) and public services (such as schools and hospitals). The objective is to provide an improved overall picture of a country's energy development. The EDI can support decision makers in ensuring that policy and financing commitments achieve maximum development impact.

SPOTLIGHT

What is the potential impact of new energy access commitments?

The UN Sustainable Energy for All initiative (SE4All) has had a big impact in raising global awareness of energy poverty and the urgent need to increase modern energy access. Over 150 commitments were submitted to the SE4All initiative across its three focus areas – energy access, energy efficiency and renewables – by the time of the Rio+20 Summit and more than 50 countries across Africa, Asia, Latin America and the Small Island Developing States confirmed their engagement (United Nations, 2012a).

The energy access commitments submitted vary significantly in terms of their size, scope and definition. For example, some are appliance driven goals (Solar Electric Light Fund, Nuru Energy LED lights programme, Toyola Energy cookstoves programme, Global LPG Partnership), some are capacity-driven commitments (ESMAP capacity building programme, Powering Agriculture Energy Grand Challenge and Schneider's BipBop programme), some focus more on additional financing (Energy+, OFID Energy for the Poor Initiative, African Development Bank investment programme and the GDF Suez Rassembleurs d'Energies programme) and some cut across different areas (the Rockefeller Foundation's "SPEED" initiative to demonstrate the potential for decentralised renewable energy provision in rural India) (United Nations, 2012b).

Across the three SE4All goals, the commitments equate to over $320 billion in direct investment. Of this total, around 10% is earmarked specifically for modern energy access, drawing the least investment of the three goals. Analysis of the commitments shows that, in line with our *WEO-2011* analysis, much of the energy access investment is sourced from multilateral development banks. Direct government sources were the second-largest source of energy access funding commitments, followed, some way behind, by the private sector. When compared to our projection that nearly $1 trillion of investment is required to achieve universal modern energy access by 2030 (see later section), it is clear that there is still a long way to go to achieve the financing required.

While the much-needed financing is not yet in place, there are encouraging signs with respect to achieving the necessary political commitment. Eight of the ten countries with the largest populations lacking electricity access today, and seven of the ten largest populations without clean cooking facilities, have signed up to the SE4All initiative. It is early days in the life of the SE4All initiative and, while significant additional funding and policy action is necessary, it has certainly had a positive impact in mobilising awareness and a greater unity of purpose to tackle this issue. Of paramount importance now is to ensure that it acts as a catalyst for even greater action in the future.

Global status of modern energy access

Hundreds of millions of people have attained modern energy access over the last two decades, especially in China and India. Rapid economic development in several developing countries, increasing urbanisation and ongoing energy access programmes have been important factors in this achievement. Despite this, in a world where the total population grows persistently, in 2010, nearly 1.3 billion people did not have access to electricity; though this is a reduction of 50 million, compared to our last *Outlook*, it is still close to one-fifth of the global population. Twice as many, around 2.6 billion people, relied on the traditional use of biomass for cooking (Table 18.1).[5]

Table 18.1 ▷ People without access to modern energy services by region, 2010 (million)

	Without access to electricity		Traditional use of biomass for cooking*	
	Population	Share of population	Population	Share of population
Developing countries	**1 265**	**24%**	**2 588**	**49%**
Africa	590	57%	698	68%
DR of Congo	58	85%	63	93%
Ethiopia	65	77%	82	96%
Kenya	33	82%	33	80%
Nigeria	79	50%	117	74%
Tanzania	38	85%	42	94%
Uganda	29	92%	31	96%
Other sub-Saharan Africa	286	66%	328	75%
North Africa	1	1%	2	1%
Developing Asia	628	18%	1 814	51%
Bangladesh	88	54%	149	91%
China	4	0%	387	29%
India	293	25%	772	66%
Indonesia	63	27%	128	55%
Pakistan	56	33%	111	64%
Philippines	16	17%	47	50%
Vietnam	2	2%	49	56%
Rest of developing Asia	106	34%	171	54%
Latin America	29	6%	65	14%
Middle East	18	9%	10	5%
World**	**1 267**	**19%**	**2 588**	**38%**

* IEA and World Health Organization databases. ** Includes OECD countries and Eastern Europe/Eurasia.

5. This chapter focuses on the traditional use of biomass for cooking, but there are also around 400 million people (not included in Table 18.1) that rely on coal for cooking and heating purposes, which causes air pollution and has serious potential health implications when used in traditional stoves. These people are mainly in China, but there are also significant numbers in South Africa and India.

Developing Asia and sub-Saharan Africa continue to account, together, for more than 95% of those without modern energy access. Across developing countries, the average electrification rate is 76%, increasing to around 92% in urban areas but only around 64% in rural areas. More than eight out of ten people without modern energy access live in rural areas, an important factor when seeking to identify the most appropriate solutions.

There are nearly 630 million people in developing Asia and nearly 590 million people in sub-Saharan Africa who lack access to electricity. Just ten countries – four in Asia and six in Africa – collectively account for nearly two-thirds of those deprived of electricity (Figure 18.1). While India has the largest population without electricity access, it has actually been a driving force in improving the trend in South Asia over the last decade, reducing the number of people without access to electricity by around 285 million. Large variations across the country persist, however: Goa and Himachal Pradesh, for example, report electricity use by around 97% of households, compared to only 16% in Bihar (Government of India, 2012). Other countries in developing Asia that report an improvement in the latest data include Indonesia, Myanmar, Nepal, Bangladesh and Pakistan. In sub-Saharan Africa, improvements in electricity access are reported in Ethiopia, Angola, Ivory Coast and Senegal, among others. Those countries with the lowest rate of electrification tend to be in sub-Saharan Africa.

Figure 18.1 ▷ Countries with the largest population without access to electricity, 2010

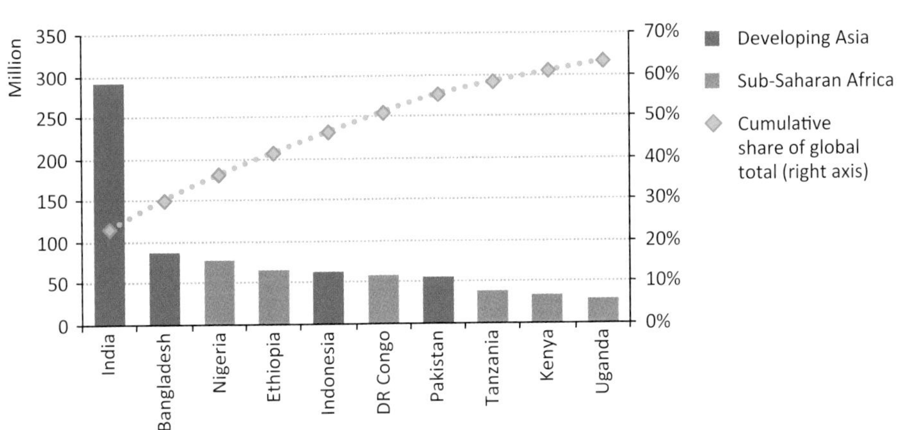

A number of new initiatives to increase access to electricity or lighting across various regions have been announced over the last year. These include, for example: the Global Lighting and Energy Access Partnership (Global LEAP), which is intended to catalyse markets for off-grid energy products and services; D.Light Design, which is committed to providing solar lamps to 30 million people in more than 40 countries by 2015; the Energising Development programme, which aims to provide modern energy access to eleven million people by 2014; and Lighting India, which plans to bring clean lighting services to two million people by the end of 2015.

More than half of the population of developing Asia – over 1.8 billion people – and around 80% of people in sub-Saharan Africa – nearly 700 million people – live without clean cooking facilities. The global population lacking clean cooking facilities is heavily skewed towards a small number of countries – India, China and Bangladesh alone account for more than half of the global total – and towards developing Asia, in which seven of the ten largest populations without access are to be found (Figure 18.2). In developing Asia, the largest single change to our data relates to India, where the latest census results have prompted a significant revision, decreasing our estimate by more than 60 million people (Government of India, 2012). Nonetheless, nearly two-thirds of India's population remains without clean cooking facilities. Large differences can be seen at state level in India, with 85% of households in Odisha relying mainly on traditional biomass for cooking, compared to around 40% in Punjab. In developing Asia, China, Pakistan, Thailand and Vietnam show notable improvements. In sub-Saharan Africa, improvements are reported in South Africa, Senegal, Uganda and Ivory Coast, among others. Data for Latin America suggest a broad-based improvement, with the number of people without clean cooking facilities falling in many countries. New data for the Middle East permit a more accurate estimate, with its largest population without clean cooking facilities being in Yemen.

Figure 18.2 ▷ Countries with the largest population relying on traditional use of biomass for cooking, 2010

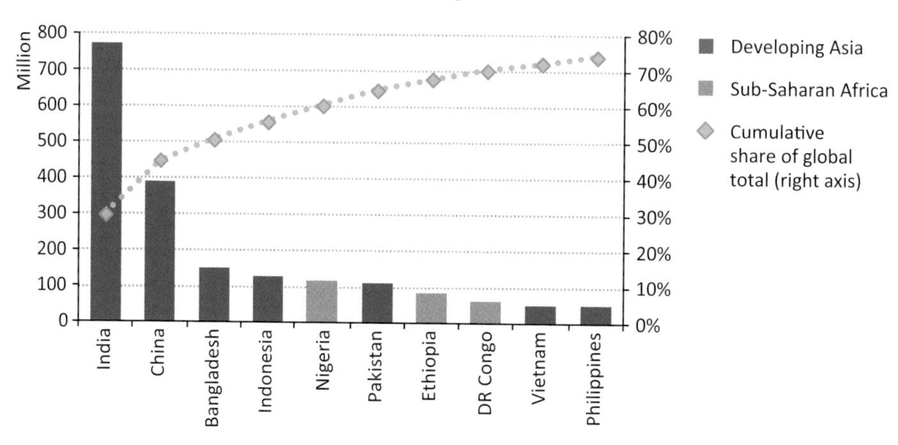

The inclusion of access to clean cooking facilities within the SE4All initiative was a welcome development, as was the launch in 2010 of the Global Alliance for Clean Cookstoves. A Global LPG Partnership has also been announced, which seeks to move at least 50 million people to LPG for cooking by 2018, with between $750 million and $1 billion of related investment. Another important development in the last year has been a new International Workshop Agreement (IWA), promulgated by the International Organization for Standardization (ISO), which provides guidance for rating cookstoves on four key performance indicators: fuel use/efficiency, total emissions, indoor emissions and safety (PCIA, 2012). It is the first international standard of its kind and it is hoped that it will be accepted as a benchmark against which to rate cookstove performance.

Outlook for energy access in the New Policies Scenario

Access to electricity

In our New Policies Scenario, the number of people without access to electricity is projected to decline to just over 990 million people in 2030, around 12% of the global population at that time (Table 18.2). Numbers larger than the population of China and the United States combined today – about 1.7 billion people – gain access to electricity over the projection period, but this achievement is counteracted, to a large extent, by global population growth. Our projection for the number of people without electricity access in 2030 is below one billion for the first time. The notable improvement, compared to *WEO-2011*, reflects a number of factors, including an improved economic outlook in many countries, stronger progress observed in some countries and reflected in our updated baseline, and a significant number of new commitments and policies aimed at improving electricity access. In the New Policies Scenario, total cumulative investment in electricity access is estimated to be $288 billion, or $14 billion per year on average.

Table 18.2 ▷ Number of people without access to electricity by region in the New Policies Scenario (million)

	2010				2030			
	Rural	Urban	Total	Share of population	Rural	Urban	Total	Share of population
Developing countries	1 081	184	1 265	24%	879	112	991	15%
Africa	475	114	590	57%	572	83	655	42%
Sub-Saharan Africa	474	114	589	68%	572	83	655	48%
Developing Asia	566	62	628	18%	305	29	334	8%
China	4	0	4	0%	0	0	0	0%
India	271	21	293	25%	144	8	153	10%
Rest of developing Asia	291	40	331	31%	161	20	181	14%
Latin America	23	6	29	6%	0	0	0	0%
Middle East	16	2	18	9%	0	0	0	0%
World	1 083	184	1 267	19%	879	112	991	12%

The number of people without electricity access in developing Asia is projected to nearly halve, going from around 630 million in 2010 to below 335 million in 2030. This continues an already positive trend, with China (reporting more than 99% access today) expected to reach universal access by the middle of this decade, and the remainder of East Asia having much reduced numbers without access in 2030. South Asia is also expected to see significant improvement, but India in 2030 continues to have the single largest population without electricity access, at around 150 million.

In sub-Saharan Africa, we project that the number of people without access to electricity will increase by around 11% to 655 million in 2030. Improved economic prospects and new commitments to action now suggest that the worsening trend will not extend beyond about 2025; but the prospect of improvement is fragile – it can still be upset by a change in economic fortunes, higher energy prices or a failure to implement policy action. Due to significant improvements elsewhere, sub-Saharan Africa accounts for an increasing share of the global population without electricity access, going from 46% in 2010 to 66% in 2030. North Africa is projected to achieve universal access by 2020.

We project universal access to electricity in Latin America to be achieved by around the mid-2020s. This change from our last *Outlook* reflects the progress that continues to be made, both in terms of general economic development and improving modern energy access. Brazil is a particularly strong example, as it pushes ahead with its commitment to achieve universal access to electricity by 2014 (Box 18.1). Other examples of programmes active in Latin America include Enabling Electricity, which focuses on commercially viable solutions for isolated communities, and the "Luz en Casa" (Light at Home) programme, which focuses on solar home systems in northern Peru.

Box 18.1 ▷ Brazil's Luz Para Todos ("Light for All") programme

Launched in 2003, the Luz Para Todos programme aims to achieve universal access to electricity in Brazil by 2014. It had provided access to an estimated 14.5 million people by late 2011 and Brazil can now boast an electrification rate of almost 99% (Ministry of Mines and Energy, 2010). The programme is directed by the Ministry of Mines and Energy, co-ordinated by Electrobrás (the holding company of the Brazilian electricity sector) and executed by the utilities and rural electrification co-operatives. It provides an electricity connection free of charge, together with three lamps and the installation of two outlets in each home. Tariffs are regulated at a "social" rate, with a 65% discount for monthly consumption below 30 kWh, a 40% discount from 31-100 kWh, 10% discount from 101-220 kWh and no discount above this level.

Those people who remain without electricity in Brazil represent a particular challenge, as they mostly live in the Amazon, where the population is thinly spread (about four inhabitants per square-kilometre) and where extension of the power grid is difficult. Recognising this, the Luz Para Todos programme has created a handbook including ideas for setting up decentralised renewable energy systems, such as collective action by citizens to install solar and biogas power systems. The Ministry of Mines and Energy estimates that the Luz Para Todos programme has generated nearly 300 000 new jobs and a survey reported an increase in income in more than one-third of households after receiving electricity access (Ministry of Mines and Energy, 2009). Gómez and Silveira (2010) also found that the arrival of electricity stimulated social programmes providing health services, education, water supply and sanitation in Brazil.

Many of the trends observed in the New Policies Scenario in terms of access to electricity are mirrored when looking at electricity consumption per capita (Figure 18.3). Strong economic growth helps China's electricity consumption per capita to more than double between 2010 and 2030, reaching the level of the European Union today. India sees a similar proportional increase, but from a much lower base; as a result, India's per capita electricity consumption in 2030 is still less than three-quarters the level in the United States in 1950. The increase in Latin America reflects successful action to reach relatively small, remote populations. Sub-Saharan Africa sees by far the smallest increase in electricity consumption per capita (in absolute terms) and a widening gap with the rest of the world.

Figure 18.3 ▷ Electricity consumption per capita in selected regions in the New Policies Scenario

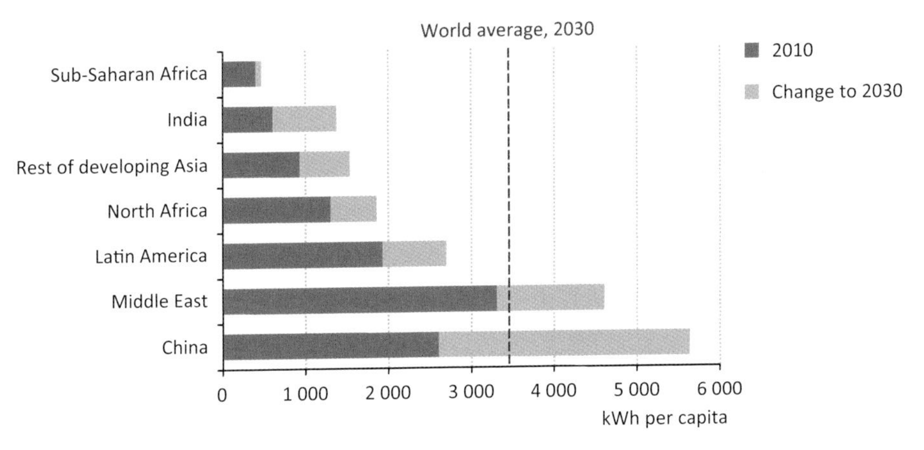

Note: Includes electricity consumption across all sectors of the economy.

Access to clean cooking facilities

The number of people without clean cooking facilities is projected to remain almost unchanged in our New Policies Scenario, continuing at around 2.6 billion in 2030 – more than 30% of the global population at that time (Table 18.3). China achieves the single biggest improvement, with almost 150 million fewer people lacking access to clean cooking facilities by 2030, mainly as a result of economic growth, urbanisation and deliberate policy intervention, such as action to expand natural gas networks. Over the *Outlook* period, we project that, on average, around $635 million per year will be invested in clean cooking facilities. Despite this effort, population growth limits the global achievement only to ensuring that there is no significant worsening of the situation between now and 2030.

The regional picture shows that developing Asia is projected to see a large reduction in the number of people without clean cooking facilities by 2030 – around 175 million. China and, to a lesser extent, India account for most of the net improvement (Figure 18.4), but India still has nearly 30% of the global population without clean cooking facilities in 2030. The story is grim in sub-Saharan Africa, where our projections reveal a worsening situation, with the number of people without clean cooking facilities increasing by more than one-quarter, reaching around 880 million in 2030.

Table 18.3 ▷ Number of people without clean cooking facilities by region in the New Policies Scenario (million)

	2010				2030			
	Rural	Urban	Total	Share of population	Rural	Urban	Total	Share of population
Developing countries	2 155	433	2 588	49%	2 139	456	2 595	39%
Africa	518	180	698	68%	629	257	886	56%
Sub-Saharan Africa	516	179	696	81%	627	256	883	65%
Developing Asia	1 580	234	1 814	51%	1 458	182	1 640	39%
China	345	42	387	29%	220	20	240	17%
India	698	75	772	66%	680	55	735	50%
Rest of developing Asia	538	117	655	61%	558	106	664	50%
Latin America	47	18	65	14%	45	18	62	11%
Middle East	9	1	10	5%	8	0	8	3%
World	2 155	433	2 588	38%	2 139	456	2 595	31%

Figure 18.4 ▷ Number of people without clean cooking facilities by region in the New Policies Scenario

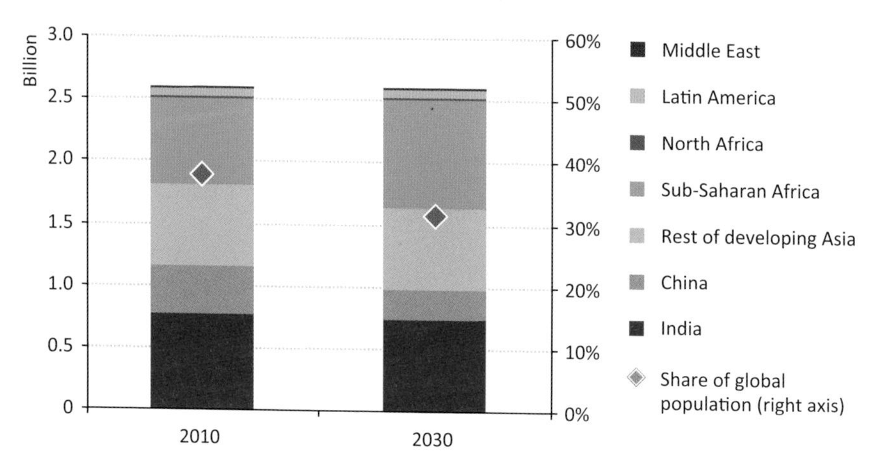

Energy for All Case

In our Energy for All Case, we examine the trajectory that would be required to achieve the goal of universal access to electricity and clean cooking facilities by 2030 and what the implications would be of doing so. We estimate that total investment of nearly $1 trillion ($979 billion) would be required to achieve universal energy access by 2030, an average of $49 billion per year (from 2011 to 2030). This requirement is small when compared to global energy-related infrastructure investment, equivalent to around 3% of the total.[6]

6. The additional investment in the Energy for All Case, compared to the New Policies Scenario, is equivalent to just over 2% of global energy-related infrastructure investment.

Our estimate includes both the $301 billion of investment we project to be forthcoming in the New Policies Scenario and the additional $678 billion that we estimate is required in the Energy for All Case. The additional investment required is derived from our analysis that seeks to match the most likely technical solutions within each region, given resource availability and government policies and measures.[7]

In the Energy for All Case, we find that around an additional $602 billion in investment is required to provide universal access to electricity by 2030, an average of $30 billion per year.[8] Sub-Saharan Africa accounts for 64% of the additional investment required, while developing Asia accounts for 36% (Figure 18.5). The additional investment provides electricity connections for almost 50 million people per year on average.

Figure 18.5 ▷ Average annual investment in modern energy access in selected regions, 2011-2030

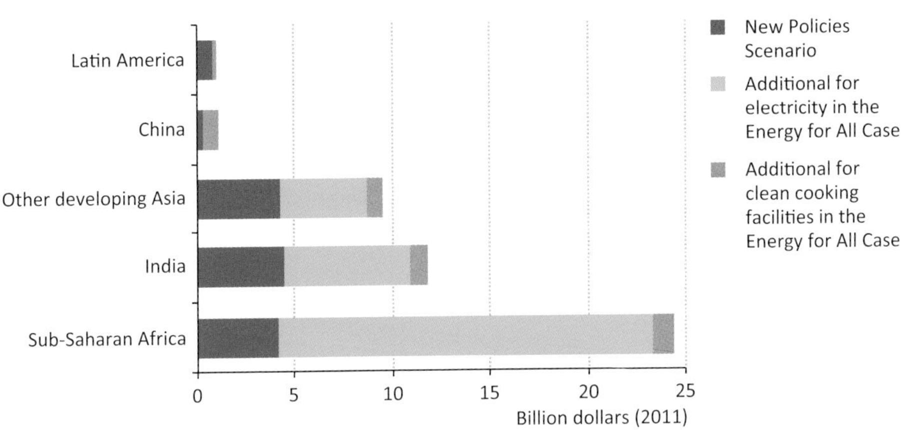

In our Energy for All Case, additional investment of just under $76 billion is required in order to achieve universal access to clean cooking facilities by 2030, an average of $3.8 billion per year. This investment provides clean cooking facilities to an additional 135 million people per year on average, through a combination of advanced biomass cookstoves, LPG stoves and biogas systems. Advanced biomass cookstoves and biogas systems are relatively more common solutions in rural areas whereas LPG stoves play a more significant role in urban

7. For more on financing and investment for modern energy access, see the *WEO-2011* special early excerpt "Energy for All: Financing Access for the Poor" (October, 2011), *www.worldenergyoutlook.org*.

8. To arrive at our estimate, we assess the required combination of on-grid, mini-grid and isolated off-grid solutions in each region. We take account of regional costs and consumer density in determining a regional cost per megawatt-hour (MWh). When delivered through an established grid, the cost per MWh is cheaper than other solutions, but extending the grid to sparsely populated, remote or mountainous areas can be very expensive and long distance transmission systems can have high technical losses. This results in grid extension being the most suitable option for all urban zones and around 30% of rural areas, but not in more remote rural areas. The remaining rural areas are connected either with mini-grids (65% of this share) or small, stand-alone off-grid solutions (the remaining 35%), which have no transmission and distribution costs.

areas. While the target population is much larger, and the operational challenge no less significant, it is striking how much lower the investment need is to provide universal access to clean cooking facilities, compared with electricity.

The Energy for All Case will require an increase in financing from all sources, including development banks, country governments, bilateral official development assistance and, perhaps most importantly, the private sector. Various forms of financing are required, from the large project level down to the micro level. However, money alone will not do the job. Adequate government policies and planning, regional and sectoral target setting, monitoring and evaluation, training and capacity building for engineers and local workforces (for implementation, maintenance and repair) are needed also. Where possible, plans need to provide for the supply of energy efficient lighting systems and electric appliances, such as telephone chargers, batteries, fridges and information technology equipment. If those appliances are not highly efficient, the volume of electricity initially available may not be sufficient to meet even basic needs.

Global primary energy demand is 167 Mtoe higher in 2030 in the Energy for All Case (Figure 18.6). Less than half of the additional energy demand for electricity generation comes from burning fossil fuels. While fossil fuels play a major role in on-grid electricity solutions, renewables dominate for mini-grid and off-grid solutions. By 2030, an additional 0.85 million barrels per day (mb/d) of LPG is estimated to be required for cooking in the Energy for All Case. The significant role of renewables in the Energy for All Case means that the overall impact on global CO_2 emissions is relatively small, increasing by around 0.6% in 2030.

Figure 18.6 ▷ Additional impact of the Energy for All Case compared with the New Policies Scenario

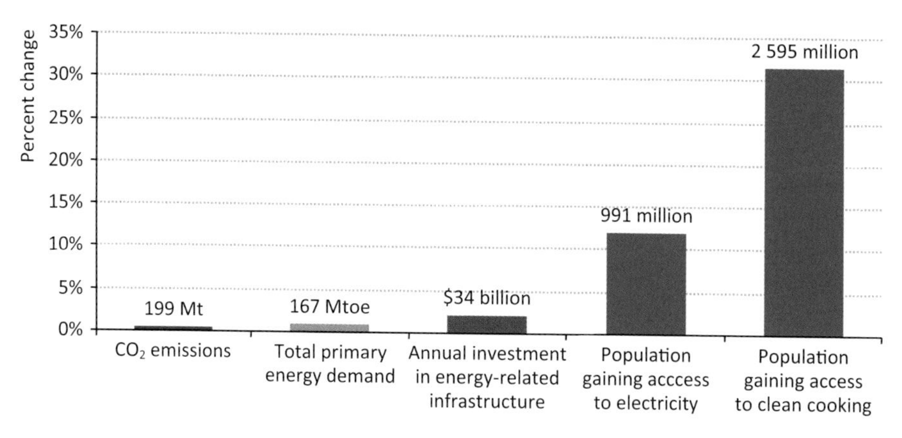

Notes: Percentages are as a share of global energy-related CO_2 emissions (2030); global primary energy demand (2030); global energy-related infrastructure investment (annual average, based on the New Policies Scenario) and global population (2030). Mt = million tonnes; Mtoe = million tonnes of oil equivalent.

Energy Development Index (EDI)[9]

An essential part of any successful initiative to achieve universal modern energy access will be to have the means to track progress, so as to be able to inform governments and other stakeholders of what is being achieved and what more needs to be done. Since 2004, the IEA has published an Energy Development Index (EDI), which is designed as a composite measure of a country's progress in making the transition to modern fuels and modern energy services. It is intended to help understanding of the role that energy can play in human development. This year, we have sought to improve the methodology of the EDI and present here updated and enhanced results for 80 countries.

Energy development framework

The perspective on modern energy access varies widely, from the individual user or supplier, through regional, national and supra-national levels. Our ambition for the EDI is to develop a multi-dimensional indicator that tracks energy development country-by-country, distinguishing between developments at the household level and at the community level (Figure 18.7). In the former, we focus on two key dimensions (as reflected earlier in this chapter), access to electricity and access to clean cooking facilities.[10] When looking at community level access (not to be confused with the term community services, which is sometimes used to describe health, education and other services), the categories are necessarily broader. In the case of public services, our focus is on the use of modern energy in schools, hospitals and clinics, water and sanitation, street lighting and other communal institutions or services. In the case of productive use, the focus is on modern energy use as part of economic activity, for example, agriculture (ploughing, irrigation and food processing), textiles and other manufacturing, etc. An additional aspect of modern energy use, captured to an extent within productive use, is transport. This is important because, particularly in the early stages of economic development, a significant share of energy consumed in the transport sector is used for productive economic purposes.

Within these broad categories, access to modern fuel and the appliances to utilise it are considered together *i.e.* a person has adequate access only if they have access to both. However, it is recognised that, in respect to both access to energy and access to appliances, there is also a progression. For instance, in the case of electricity, the first move might be from candles and batteries to solar lanterns, solar home systems or, possibly, a mini-grid. Similarly, first access is likely to involve only a small number of basic appliances, with greater diversity coming later. In addition, there are a number of issues that are sometimes referred to generically as "quality of supply". For any energy supply to provide a genuine

9. This analysis benefited from a roundtable meeting held by the IEA in Paris on 25 May 2012.

10. Access to heating is another important variable sometimes mentioned in this context. However, it is often excluded either due to the lack of data or because it is strongly related to cooking (the same means are often used for both). In some cases, clean cooking solutions can pose an additional challenge for policy makers because they are less effective at providing space heating than more traditional methods.

opportunity to use modern energy services there needs to be a technical possibility to use it (availability), a price that is not prohibitive (affordability), sufficient supply (adequacy) and a supply that is easy to use (and pay for), including being located nearby, available at desired hours of the day and safe to use (convenience). Importantly, the supply must be of the right quality (*e.g.* voltage level) and be usable for most of the time (reliability). At a more sophisticated level, it is also recognised that it may be desirable to track the quality of policies, regulations and institutions involved and, certainly, whether there is sufficient funding to support realisation of the objectives.

Figure 18.7 ▷ Energy development framework

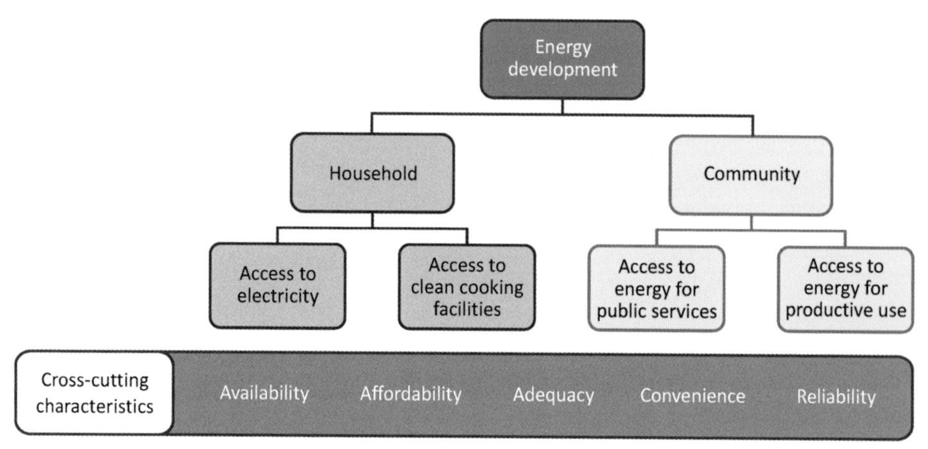

Notes: Household does not distinguish energy use as part of a micro-enterprise conducted within the home (which existing energy access data is often unable to identify). While very different in nature, energy for public services and for economic/productive purposes are grouped here under the community heading.

Focusing on the dimensions of modern energy access set out above makes it possible to identify a number of variables that can and should be monitored as a means of measuring energy development. However, adequate, regular, reliable and robust data are frequently not available. This is because data on many possible variables is typically collected (if at all) only as part of household or business surveys, which are often conducted on an infrequent basis in many countries and with a different prime focus. While this situation persists, some compromises have to be made: variables with some degree of explanatory value have to be used, despite imperfections. The main source is energy balance data from countries. Where possible, we use multiple data sources to cross-check figures. Our assessment of the strengths and weaknesses of the available possible indicators has led us to select those shown in Figure 18.8.[11] Our methodology leaves neutral any judgement on whether individual indicators are more or less important than others, ascribing an equal weighting to each in the calculation of new EDI (Box 18.2).

11. For more information, see our EDI methodology note at *www.worldenergyoutlook.org*.

Figure 18.8 ▷ **Composition of the new Energy Development Index**

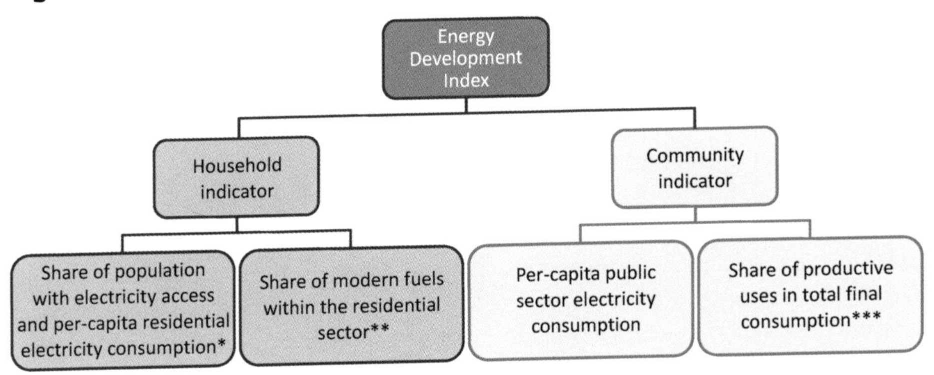

* The geometric mean of the two variables is taken. ** Excludes electricity to avoid double counting. *** Includes industry, agriculture, services, transport and other non-specified energy use.

Notes: All variables are normalised on a fixed scale before calculating the EDI. As the indicators are aggregated to reach the EDI score, they are averaged.

Box 18.2 ▷ **Areas for potential further development of the EDI**

The standardised definition and measurement of country data would help further improve the EDI. Another desirable improvement would be the inclusion of a stronger indicator for quality of supply or efficiency of energy use. Other factors include the affordability of modern energy (Winkler, *et al.* 2011), how efficiently it is consumed and the level of consumption by small and medium-size enterprises (Kooijman-Van Dijk and Clancy, 2010), given their crucial role in employment and economic growth.

An important issue for some countries or institutions might be the development of a "low carbon" EDI, in which the variables were specifically related to measurement of the role of renewables and low-carbon technologies in the energy system. This might be driven by a concern that energy development that relies on fossil fuels, even to a relatively limited extent, could result in the "lock-in" of these technologies, and their associated emissions, for decades to come. Examining either the level of electricity generated from renewables (or fossil fuels) as a proportion of overall electricity would be one obvious indicator. If this were unavailable, the proportion of renewable energy (or fossil fuels), in a country's energy mix could be a fall-back measure. The number of recorded sales/installations of solar home systems or other types of renewable technologies (or of various energy-efficient technologies) could also be measured. For cooking, it might be appropriate to examine not only the type of fuel used but also the efficiency of stoves that are commonly sold, as this will be an important determinant of fuel demand. In terms of environmental or impact indicators, the level of local pollution linked to burning hydrocarbons could and should be monitored, as well as the number of reported illnesses related to local pollution. In terms of public services, one could measure the number of public health centres and the percentage of schools or training centres that provide their services on the basis of renewable energy technologies.

Results from the Energy Development Index

Figure 18.9 ranks 80 countries according to their overall EDI score. It also shows the relative contribution of each of the constituent indicators discussed above and, where available, shows a country's EDI score in 2002 for comparison. Many of the countries with the highest EDI score are in the Middle East, North Africa or Latin America. Countries in sub-Saharan Africa represent a significant share of those in the lower half of our EDI country scores. Rankings for countries in developing Asia are more varied, with Malaysia and Thailand scoring particularly well, while Nepal and Myanmar score relatively poorly. Oil exporters typically score well, although Nigeria is a notable exception. Those countries with a low overall EDI ranking tend to have a low result on the clean cooking and public services indicators. The countries with a higher ranking generally have a more balanced contribution from all the indicators, although there are exceptions, such as the small contribution made by the clean cooking indicator to the overall score for South Africa and Thailand.

For the countries for which we have both 2002 and 2010 data (56 in total) a general improvement over time is observed (only two countries do not improve). The average score increases from 0.39 to 0.43 (on the overall index) and the median score from 0.36 to 0.42. Of the ten countries reporting the largest improvement in their EDI score, four are in developing Asia (China, Thailand, Vietnam and Malaysia), three are in Latin America (El Salvador, Argentina, Uruguay), one is in the Middle East (Jordan) and two are in North Africa (Algeria, Morocco). Looking across regions as a whole does suggest that, on average, the biggest improvements have taken place in East Asia and North Africa.

China shows one of the largest increases in EDI score over time, driven by improvements in electricity access, public services and productive uses. In the area of clean cooking, China also sees a moderate improvement reflecting, in part, the successful installation of an estimated 40 million biogas plants by the end of 2010 (SNV, 2011). In the case of Thailand, much of the improvement is attributable to a much higher score in the public services indicator. Vietnam also reports a strong increase in its EDI score over time, with the largest share being attributable to the improvement in the household electricity access indicator. While its efforts go back decades, Vietnam's rural electrification programme has been central to it achieving a national electrification rate of 98% in 2010. Important factors in realising its impressive gains, included harnessing its natural abundance of hydropower, recognising the key role of infrastructure and the importance of multiple funding sources, and sustaining strong public and political support for efforts to improve electrification (Asian Development Bank, 2011). South Africa's Integrated National Electrification Programme can clearly be seen to have helped its improved electricity indicator score. Several countries in Latin America have registered strong improvements in their EDI, including El Salvador, Argentina, Uruguay, Brazil and Ecuador.

Despite the general improvement, many of the countries with the lowest EDI ranking, based on 2010 data, are the same as those with the lowest ranking based on 2002 data. Ethiopia continues to have the lowest EDI in our ranking, although it has improved by nearly 30% compared to 2002. Ethiopia has signed up as a partner in the Norwegian-led

Figure 18.9 ▷ **Energy Development Index country results, 2010 (and 2002)**

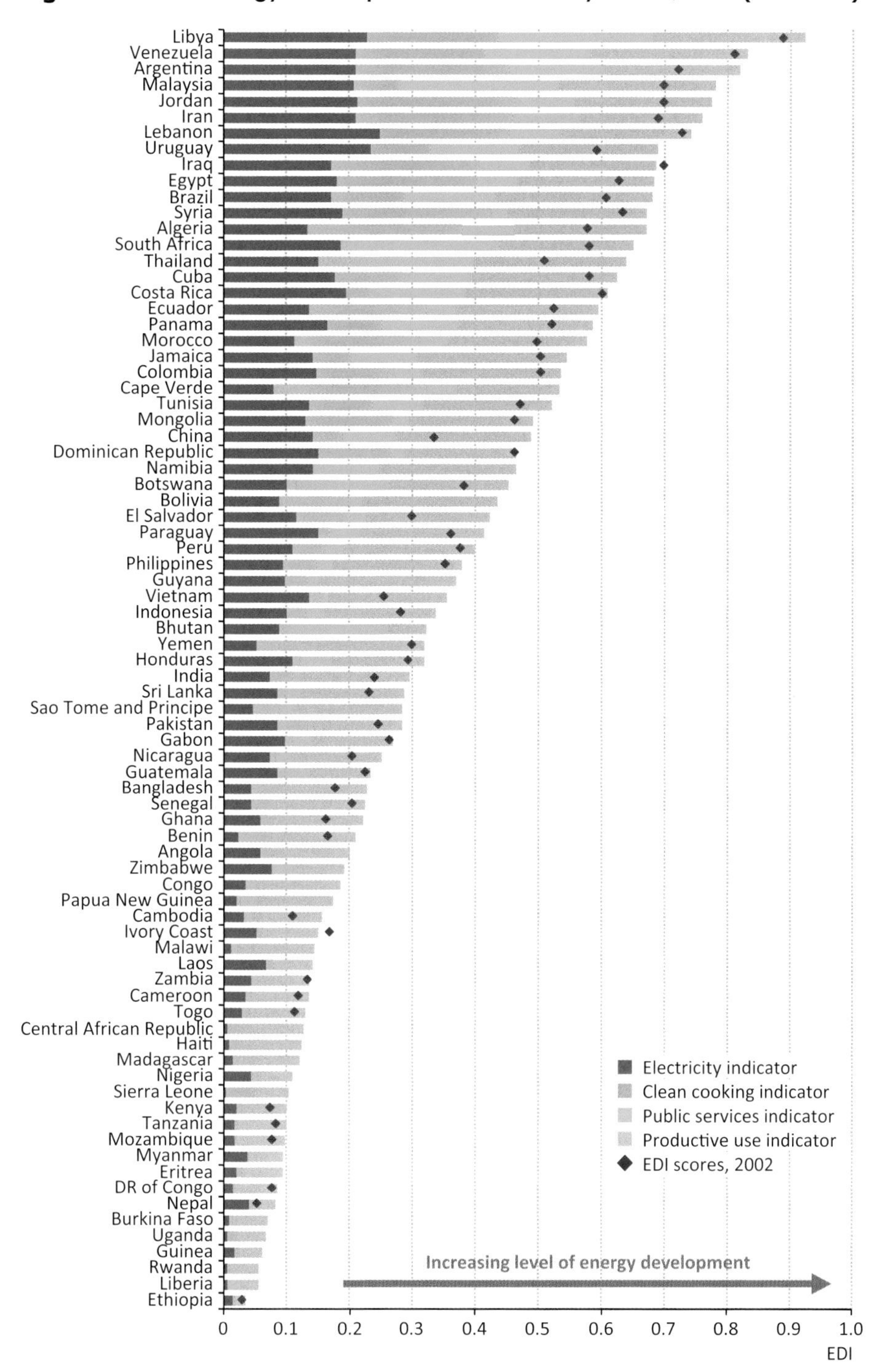

18

Energy+ Initiative, under which Ethiopia is to receive around $85 million[12] performance-based financing to support energy development, including increased distribution of clean cookstoves in rural areas. Kenya, which also has a relatively low EDI ranking, is to receive around $43 million under the same initiative, with the particular objectives of replacing paraffin lamps by lighting from solar power and increased adoption of clean cookstoves.

Nigeria, a country rich in hydrocarbon resources, continues to receive a low score in our EDI ranking. Despite large oil and gas export revenues, its EDI score reflects low modern fuel use for cooking and little electricity use for public services (Box 18.3). In the case of Ghana, whose electricity indicator has improved over time, the country has set itself the target of achieving universal access to electricity by 2020, in line with its National Energy Strategy of 2010, and has seen the electrification rate increase steadily. It fares less well on clean cooking, where a 2010 energy use survey, conducted by the Energy Commission, estimated that, despite improvement, around 40% of households still use firewood for cooking (Ghana Energy Commission, 2012), suggesting a need for increased focus.

Box 18.3 ▷ Africa: resource rich but modern energy poor

Africa is a continent full of energy resources, but it harvests only a little of these for its domestic use. North and West Africa have substantial oil and gas resources, while new exploration efforts have found significant resources also in East Africa (see Chapter 4); and South Africa is one of the world's largest suppliers of coal. Renewable energy resources are also abundant, with large hydropower potential in Central and East Africa, large geothermal energy potential in East Africa and favourable conditions for wind energy in North Africa, the Horn of Africa and South Africa. Solar energy potential is large across the continent and modern forms of biomass could also play a greater role in some areas. Despite this wealth of resources, Africa consumed less than one-quarter of the global average in modern energy per capita in 2011 while, at the same time, exporting more than half of the fossil fuels that it produced.

Africa's revenues from net energy exports are projected to increase from almost $280 billion per year to $415 billion in 2030. We estimate that achieving modern energy for all in Africa by 2030 would require investments of around $20 billion per year, or 5.5% of energy export revenues over the period. Over the projection period, Nigeria is projected to generate $105 billion per year in oil and gas revenues on average, while universal access to electricity and clean cooking facilities there would require investment of around $1 billion per year. In the case of Angola, the country would need to invest, on average, only 0.5% of its projected energy-export revenues in modern energy access in order to achieve universal access by 2030. For Mozambique, the story is of future potential, with the exploitation of new natural gas discoveries offering the opportunity to boost significantly efforts to provide modern energy access.

12. Based on NOK 0.17 = $1.

Iraq's EDI ranking serves to highlight the importance of including in a more sophisticated index an indicator capturing quality of supply (see Part C for more on Iraq's energy sector). While its ranking has worsened, when compared with 2002, it remains relatively high, reflecting a high rate of electrification (around 98%), relatively high consumption of electricity in the residential sector, significant use of either LPG or other modern fuels in cooking and relatively high levels of modern energy use in public services and productive sectors. However, what is not captured adequately is the unreliability of the electricity supply, entailing frequent power cuts, and unsatisfied electricity demand, compensated partly by reliance on expensive diesel generators. Were quality of supply factors reflected more directly within the EDI, we would expect Iraq to rank reasonably high, compared to many other countries, though not as high as at present.

The country and regional stories reflected in the EDI are generally confirmed when they are then compared to the scores of the UN Human Development Index (HDI) (Figure 18.10). Countries in sub-Saharan Africa once again tend towards the lower end of the spectrum, while those in the Middle East and North Africa tend to have both stronger EDI and HDI scores. There tends to be a more mixed picture in developing Asia and Latin America.

Figure 18.10 ▷ Comparison between the new Energy Development Index and the Human Development Index in 2010

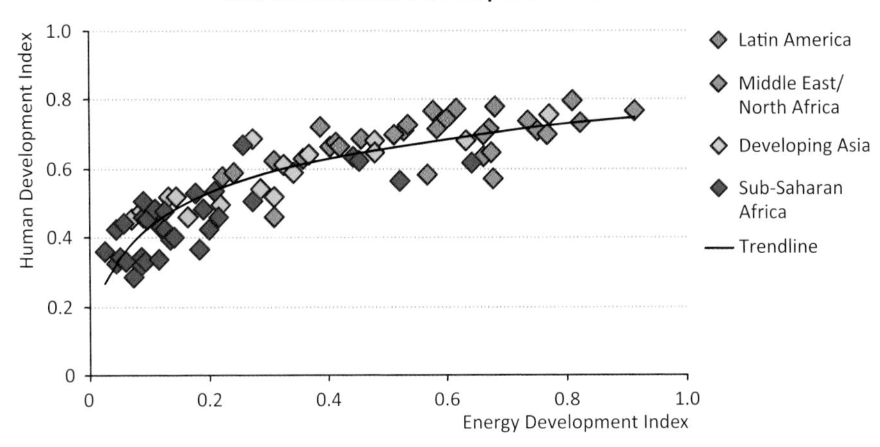

This EDI, presented by country, but also split by indicator and shown over time, should be a valuable aid to a range of decision makers in tracking progress in important elements of a country's energy development. The EDI will also become more valuable as data quality improves, reinforcing the need for initiatives, such as the UN Sustainable Energy for All initiative, to emphasise and support efforts to strengthen capacity in this area. However, the EDI should still be seen as part of a broader suite of indicators that might also provide coverage at project/programme, sub-national and regional levels. We will continue to update the EDI on a regular basis, reflecting the latest available data, and, whenever possible, seeking to expand our country coverage. We will also continue to review the range of data sets available and, in light of developments, consider if our EDI methodology can be strengthened further, to provide an even better measure of a country's energy development.

ANNEXES

Tables for Scenario Projections

General note to the tables

The tables detail projections for *energy demand*, gross *electricity generation* and *electrical capacity*, and *carbon-dioxide (CO_2) emissions* from fuel combustion. The following regions/countries are covered: World, OECD, OECD Americas, the United States, OECD Europe, the European Union, OECD Asia Oceania, Japan, non-OECD, Eastern Europe/Eurasia, Russia, non-OECD Asia, China, India, Africa, Latin America, Brazil, the Middle East and Iraq. The definitions for regions, fuels and sectors can be found in Annex C.

For all regions except Iraq, the tables present historical and projected data for the Current Policies, New Policies and 450 Scenarios. By convention, in the table headings CPS and 450 refer to Current Policies and 450 Scenarios respectively. The Iraq tables show historical and projected data for the Central Scenario and the High and Delayed Cases. With no existing plans for the use of coal and nuclear in Iraq these fuels have been omitted from the Iraq tables.

Data for *energy demand*, gross *electricity generation* and *CO_2 emissions* from fuel combustion up to 2010 are based on IEA statistics, published in *Energy Balances of OECD Countries, Energy Balances of non-OECD Countries* and *CO_2 Emissions from Fuel Combustion*. Historical data for *electrical capacity* is supplemented from the Platts World Electric Power Plants Database (December 2011 version) and the International Atomic Energy Agency PRIS database. Additional Iraq data has been sourced from direct communications with Iraqi authorities.

Both in the text of this book and in the tables, rounding may lead to minor differences between totals and the sum of their individual components. Growth rates are calculated on a compound average annual basis and are marked "n.a." when the base year is zero or the value exceeds 200%. Nil values are marked "-".

Definitional note to the tables

Total primary energy demand (TPED) is equivalent to power generation plus other energy sector excluding electricity and heat, plus total final consumption (TFC) excluding electricity and heat. TPED does not include ambient heat from heat pumps or electricity trade. Sectors comprising TFC include industry, transport, buildings (residential, services and non-specified other) and other (agriculture and non-energy use). Projected electrical capacity is the net result of existing capacity plus additions less retirements. Total CO_2 includes emissions from other energy sector in addition to the power generation and TFC sectors shown in the tables. CO_2 emissions and energy demand from international marine and aviation bunkers are included only at the world transport level. CO_2 emissions do not include emissions from industrial waste and non-renewable municipal waste.

World: New Policies Scenario

	Energy demand (Mtoe)							Shares (%)		CAAGR (%)
	1990	2010	2015	2020	2025	2030	2035	2010	2035	2010-35
TPED	**8 779**	**12 730**	**13 989**	**14 922**	**15 675**	**16 417**	**17 197**	**100**	**100**	**1.2**
Coal	2 231	3 474	3 945	4 082	4 131	4 180	4 218	27	25	0.8
Oil	3 230	4 113	4 352	4 457	4 521	4 578	4 656	32	27	0.5
Gas	1 668	2 740	2 993	3 266	3 536	3 820	4 106	22	24	1.6
Nuclear	526	719	751	898	1 003	1 073	1 138	6	7	1.9
Hydro	184	295	340	388	423	458	488	2	3	2.0
Bioenergy	903	1 277	1 408	1 532	1 642	1 755	1 881	10	11	1.6
Other renewables	36	112	200	299	418	554	710	1	4	7.7
Power generation	**2 986**	**4 839**	**5 409**	**5 912**	**6 336**	**6 768**	**7 226**	**100**	**100**	**1.6**
Coal	1 226	2 249	2 565	2 669	2 699	2 733	2 758	46	38	0.8
Oil	377	275	262	217	185	159	145	6	2	-2.5
Gas	581	1 102	1 179	1 281	1 397	1 527	1 657	23	23	1.6
Nuclear	526	719	751	898	1 003	1 073	1 138	15	16	1.9
Hydro	184	295	340	388	423	458	488	6	7	2.0
Bioenergy	59	109	146	205	268	337	423	2	6	5.6
Other renewables	32	90	166	254	361	481	618	2	9	8.0
Other energy sector	**911**	**1 396**	**1 519**	**1 578**	**1 626**	**1 676**	**1 730**	**100**	**100**	**0.9**
Electricity	*183*	*306*	*348*	*381*	*412*	*443*	*475*	*22*	*27*	*1.8*
TFC	**6 275**	**8 678**	**9 565**	**10 223**	**10 742**	**11 241**	**11 750**	**100**	**100**	**1.2**
Coal	773	853	970	982	984	983	976	10	8	0.5
Oil	2 593	3 557	3 813	3 984	4 108	4 219	4 336	41	37	0.8
Gas	942	1 329	1 464	1 612	1 740	1 864	1 993	15	17	1.6
Electricity	833	1 537	1 802	2 047	2 255	2 463	2 676	18	23	2.2
Heat	333	278	293	303	305	305	305	3	3	0.4
Bioenergy	795	1 103	1 188	1 250	1 294	1 335	1 373	13	12	0.9
Other renewables	4	22	33	45	57	72	91	0	1	5.9
Industry	**1 809**	**2 421**	**2 790**	**3 035**	**3 203**	**3 355**	**3 497**	**100**	**100**	**1.5**
Coal	476	676	784	799	809	818	822	28	23	0.8
Oil	324	321	345	356	358	357	354	13	10	0.4
Gas	358	463	521	600	655	702	748	19	21	1.9
Electricity	380	638	779	890	972	1 052	1 133	26	32	2.3
Heat	151	126	135	138	136	134	131	5	4	0.1
Bioenergy	119	196	224	251	273	292	309	8	9	1.8
Other renewables	0	0	1	1	1	1	1	0	0	2.7
Transport	**1 568**	**2 377**	**2 596**	**2 778**	**2 935**	**3 093**	**3 272**	**100**	**100**	**1.3**
Oil	1 472	2 201	2 379	2 517	2 626	2 732	2 850	93	87	1.0
Of which: Bunkers	*199*	*354*	*366*	*392*	*414*	*440*	*470*	*15*	*14*	*1.1*
Electricity	21	24	31	36	41	48	57	1	2	3.5
Biofuels	6	59	84	111	139	171	206	2	6	5.2
Other fuels	69	93	102	115	129	142	159	4	5	2.2
Buildings	**2 243**	**2 910**	**3 121**	**3 302**	**3 452**	**3 599**	**3 748**	**100**	**100**	**1.0**
Coal	240	124	130	124	115	105	96	4	3	-1.0
Oil	330	329	333	327	318	308	300	11	8	-0.4
Gas	431	616	661	711	757	807	856	21	23	1.3
Electricity	401	831	939	1 062	1 174	1 287	1 402	29	37	2.1
Heat	172	148	154	160	165	168	170	5	5	0.6
Bioenergy	665	841	872	877	870	857	840	29	22	-0.0
Other renewables	4	21	32	42	53	67	85	1	2	5.8
Other	**655**	**970**	**1 057**	**1 107**	**1 152**	**1 194**	**1 232**	**100**	**100**	**1.0**

World: Current Policies and 450 Scenarios

	Energy demand (Mtoe)						Shares (%)		CAAGR (%)	
	2020	2030	2035	2020	2030	2035	2035		2010-35	
	Current Policies Scenario			450 Scenario			CPS	450	CPS	450
TPED	**15 332**	**17 499**	**18 676**	**14 176**	**14 453**	**14 793**	**100**	**100**	**1.5**	**0.6**
Coal	4 417	5 115	5 523	3 569	2 580	2 337	30	16	1.9	-1.6
Oil	4 542	4 855	5 053	4 282	3 908	3 682	27	25	0.8	-0.4
Gas	3 341	3 999	4 380	3 078	3 278	3 293	23	22	1.9	0.7
Nuclear	886	1 013	1 019	939	1 360	1 556	5	11	1.4	3.1
Hydro	377	435	460	401	500	539	2	4	1.8	2.4
Bioenergy	1 504	1 664	1 741	1 568	2 003	2 235	9	15	1.2	2.3
Other renewables	265	419	501	340	823	1 151	3	8	6.2	9.8
Power generation	**6 173**	**7 432**	**8 136**	**5 455**	**5 640**	**5 982**	**100**	**100**	**2.1**	**0.9**
Coal	2 951	3 539	3 896	2 232	1 294	1 069	48	18	2.2	-2.9
Oil	229	180	173	194	111	91	2	2	-1.8	-4.3
Gas	1 306	1 609	1 808	1 178	1 204	1 126	22	19	2.0	0.1
Nuclear	886	1 013	1 019	939	1 360	1 556	13	26	1.4	3.1
Hydro	377	435	460	401	500	539	6	9	1.8	2.4
Bioenergy	198	293	346	221	438	577	4	10	4.7	6.9
Other renewables	226	362	434	291	733	1 025	5	17	6.5	10.2
Other energy sector	**1 616**	**1 786**	**1 885**	**1 504**	**1 471**	**1 459**	**100**	**100**	**1.2**	**0.2**
Electricity	*396*	*488*	*538*	*354*	*372*	*384*	*29*	*26*	*2.3*	*0.9*
TFC	**10 427**	**11 805**	**12 518**	**9 843**	**10 227**	**10 390**	**100**	**100**	**1.5**	**0.7**
Coal	1 025	1 075	1 090	925	870	846	9	8	1.0	-0.0
Oil	4 055	4 472	4 705	3 842	3 626	3 457	38	33	1.1	-0.1
Gas	1 654	1 937	2 080	1 542	1 695	1 779	17	17	1.8	1.2
Electricity	2 115	2 650	2 933	1 926	2 195	2 347	23	23	2.6	1.7
Heat	312	328	333	288	270	262	3	3	0.7	-0.2
Bioenergy	1 228	1 287	1 310	1 270	1 482	1 573	10	15	0.7	1.4
Other renewables	39	56	66	49	90	126	1	1	4.5	7.3
Industry	**3 137**	**3 594**	**3 807**	**2 904**	**3 108**	**3 194**	**100**	**100**	**1.8**	**1.1**
Coal	830	888	913	751	724	713	24	22	1.2	0.2
Oil	368	379	381	340	338	332	10	10	0.7	0.1
Gas	626	753	812	590	682	721	21	23	2.3	1.8
Electricity	924	1 147	1 260	838	945	994	33	31	2.8	1.8
Heat	142	144	145	134	118	113	4	4	0.5	-0.4
Bioenergy	247	282	295	250	298	316	8	10	1.6	1.9
Other renewables	1	1	1	1	3	5	0	0	2.7	9.5
Transport	**2 799**	**3 247**	**3 515**	**2 700**	**2 720**	**2 699**	**100**	**100**	**1.6**	**0.5**
Oil	2 553	2 924	3 145	2 424	2 234	2 086	89	77	1.4	-0.2
Of which: Bunkers	*394*	*451*	*486*	*382*	*399*	*406*	*14*	*15*	*1.3*	*0.6*
Electricity	35	45	52	37	68	107	1	4	3.1	6.2
Biofuels	98	142	170	128	287	358	5	13	4.4	7.5
Other fuels	113	136	148	111	131	148	4	5	1.9	1.9
Buildings	**3 367**	**3 745**	**3 931**	**3 140**	**3 220**	**3 283**	**100**	**100**	**1.2**	**0.5**
Coal	135	125	115	116	89	76	3	2	-0.3	-2.0
Oil	337	329	324	307	258	239	8	7	-0.1	-1.3
Gas	726	831	886	658	676	692	23	21	1.5	0.5
Electricity	1 093	1 378	1 530	991	1 109	1 166	39	36	2.5	1.4
Heat	166	179	184	151	148	144	5	4	0.9	-0.1
Bioenergy	873	851	831	872	857	850	21	26	-0.0	0.0
Other renewables	36	52	61	46	82	115	2	4	4.4	7.1
Other	**1 125**	**1 221**	**1 266**	**1 099**	**1 179**	**1 215**	**100**	**100**	**1.1**	**0.9**

World: New Policies Scenario

	Electricity generation (TWh)							Shares (%)		CAAGR (%)
	1990	2010	2015	2020	2025	2030	2035	2010	2035	2010-35
Total generation	**11 819**	**21 408**	**24 996**	**28 235**	**31 007**	**33 789**	**36 637**	**100**	**100**	**2.2**
Coal	4 426	8 687	10 242	10 897	11 212	11 565	11 908	41	33	1.3
Oil	1 336	1 000	967	787	679	601	555	5	2	-2.3
Gas	1 727	4 760	5 374	6 108	6 920	7 723	8 466	22	23	2.3
Nuclear	2 013	2 756	2 881	3 443	3 847	4 114	4 366	13	12	1.9
Hydro	2 144	3 431	3 950	4 513	4 924	5 323	5 677	16	15	2.0
Bioenergy	131	331	474	696	926	1 179	1 487	2	4	6.2
Wind	4	342	808	1 272	1 719	2 187	2 681	2	7	8.6
Geothermal	36	68	93	131	190	253	315	0	1	6.3
Solar PV	0	32	183	332	490	664	846	0	2	14.0
CSP	1	2	21	50	86	152	278	0	1	23.0
Marine	1	1	3	5	12	27	57	0	0	20.4

	Electrical capacity (GW)						Shares (%)		CAAGR (%)
	2010	2015	2020	2025	2030	2035	2010	2035	2010-35
Total capacity	**5 183**	**6 347**	**7 162**	**7 861**	**8 588**	**9 345**	**100**	**100**	**2.4**
Coal	1 649	2 012	2 119	2 171	2 250	2 327	32	25	1.4
Oil	435	428	354	301	261	245	8	3	-2.3
Gas	1 351	1 639	1 845	2 039	2 234	2 419	26	26	2.4
Nuclear	394	422	474	519	551	583	8	6	1.6
Hydro	1 033	1 184	1 348	1 467	1 583	1 684	20	18	2.0
Bioenergy	72	98	135	170	208	252	1	3	5.1
Wind	198	390	586	760	924	1 098	4	12	7.1
Geothermal	11	14	20	29	38	46	0	0	5.8
Solar PV	38	153	266	378	491	602	1	6	11.7
CSP	1	6	14	24	40	72	0	1	17.6
Marine	0	1	1	3	7	15	0	0	17.3

	CO_2 emissions (Mt)							Shares (%)		CAAGR (%)
	1990	2010	2015	2020	2025	2030	2035	2010	2035	2010-35
Total CO_2	**20 980**	**30 190**	**33 185**	**34 560**	**35 403**	**36 197**	**37 037**	**100**	**100**	**0.8**
Coal	8 335	13 105	14 901	15 350	15 391	15 360	15 287	43	41	0.6
Oil	8 836	10 893	11 546	11 863	12 080	12 292	12 573	36	34	0.6
Gas	3 808	6 192	6 738	7 347	7 932	8 545	9 176	21	25	1.6
Power generation	**7 481**	**12 495**	**13 849**	**14 338**	**14 545**	**14 738**	**14 951**	**100**	**100**	**0.7**
Coal	4 918	9 040	10 253	10 643	10 684	10 664	10 623	72	71	0.6
Oil	1 204	870	831	692	589	506	461	7	3	-2.5
Gas	1 359	2 585	2 765	3 003	3 272	3 569	3 868	21	26	1.6
TFC	**12 486**	**16 127**	**17 642**	**18 501**	**19 116**	**19 687**	**20 272**	**100**	**100**	**0.9**
Coal	3 278	3 769	4 308	4 362	4 361	4 351	4 321	23	21	0.5
Oil	7 075	9 367	10 044	10 507	10 835	11 135	11 462	58	57	0.8
Transport	*4 388*	*6 565*	*7 095*	*7 512*	*7 841*	*8 162*	*8 519*	*41*	*42*	*1.0*
Of which: Bunkers	*614*	*1 092*	*1 131*	*1 209*	*1 277*	*1 356*	*1 446*	*7*	*7*	*1.1*
Gas	2 133	2 992	3 290	3 632	3 920	4 201	4 490	19	22	1.6

World: Current Policies and 450 Scenarios

	Electricity generation (TWh)						Shares (%)		CAAGR (%)	
	2020	2030	2035	2020	2030	2035	2035		2010-35	
	Current Policies Scenario			450 Scenario			CPS	450	CPS	450
Total generation	**29 194**	**36 492**	**40 364**	**26 497**	**29 841**	**31 748**	**100**	**100**	**2.6**	**1.6**
Coal	12 048	15 015	16 814	9 105	5 483	4 364	42	14	2.7	-2.7
Oil	827	687	673	695	405	332	2	1	-1.6	-4.3
Gas	6 273	8 247	9 342	5 652	6 306	5 791	23	18	2.7	0.8
Nuclear	3 397	3 885	3 908	3 601	5 218	5 968	10	19	1.4	3.1
Hydro	4 390	5 055	5 350	4 658	5 816	6 263	13	20	1.8	2.4
Bioenergy	668	1 021	1 212	750	1 529	2 033	3	6	5.3	7.5
Wind	1 148	1 841	2 151	1 442	3 316	4 281	5	13	7.6	10.6
Geothermal	118	183	217	150	345	449	1	1	4.7	7.8
Solar PV	282	451	524	376	985	1 371	1	4	11.8	16.2
CSP	39	94	141	61	398	815	0	3	19.7	28.4
Marine	3	13	32	6	38	82	0	0	17.6	22.1

	Electrical capacity (GW)						Shares (%)		CAAGR (%)	
	2020	2030	2035	2020	2030	2035	2035		2010-35	
	Current Policies Scenario			450 Scenario			CPS	450	CPS	450
Total capacity	**7 184**	**8 717**	**9 481**	**7 048**	**8 589**	**9 512**	**100**	**100**	**2.4**	**2.5**
Coal	2 265	2 747	3 005	1 905	1 394	1 220	32	13	2.4	-1.2
Oil	359	276	265	345	230	205	3	2	-2.0	-3.0
Gas	1 861	2 308	2 544	1 765	2 036	2 148	27	23	2.6	1.9
Nuclear	476	526	524	495	699	796	6	8	1.1	2.9
Hydro	1 311	1 498	1 580	1 395	1 742	1 875	17	20	1.7	2.4
Bioenergy	128	183	211	143	264	338	2	4	4.4	6.4
Wind	527	781	890	655	1 337	1 658	9	17	6.2	8.9
Geothermal	18	27	32	23	51	65	0	1	4.2	7.3
Solar PV	227	341	384	303	720	966	4	10	9.7	13.8
CSP	11	25	38	17	107	219	0	2	14.6	22.9
Marine	1	3	8	2	10	22	0	0	14.7	19.1

	CO_2 emissions (Mt)						Shares (%)		CAAGR (%)	
	2020	2030	2035	2020	2030	2035	2035		2010-35	
	Current Policies Scenario			450 Scenario			CPS	450	CPS	450
Total CO_2	**36 281**	**41 177**	**44 090**	**31 449**	**24 861**	**22 055**	**100**	**100**	**1.5**	**-1.2**
Coal	16 663	19 104	20 515	13 205	7 556	5 620	47	25	1.8	-3.3
Oil	12 100	13 129	13 788	11 355	10 291	9 645	31	44	0.9	-0.5
Gas	7 518	8 943	9 786	6 889	7 014	6 790	22	31	1.8	0.4
Power generation	**15 556**	**18 329**	**20 112**	**12 183**	**6 696**	**4 704**	**100**	**100**	**1.9**	**-3.8**
Coal	11 767	13 990	15 334	8 810	3 722	2 144	76	46	2.1	-5.6
Oil	728	575	551	620	356	292	3	6	-1.8	-4.3
Gas	3 061	3 764	4 227	2 754	2 618	2 268	21	48	2.0	-0.5
TFC	**18 963**	**20 969**	**22 020**	**17 622**	**16 625**	**15 854**	**100**	**100**	**1.3**	**-0.1**
Coal	4 536	4 733	4 792	4 074	3 557	3 220	22	20	1.0	-0.6
Oil	10 698	11 868	12 536	10 094	9 363	8 815	57	56	1.2	-0.2
Transport	*7 617*	*8 732*	*9 396*	*7 234*	*6 684*	*6 251*	*43*	*39*	*1.4*	*-0.2*
Of which: Bunkers	*1 215*	*1 389*	*1 497*	*1 181*	*1 234*	*1 258*	*7*	*8*	*1.3*	*0.6*
Gas	3 728	4 369	4 692	3 454	3 705	3 819	21	24	1.8	1.0

OECD: New Policies Scenario

	Energy demand (Mtoe)							Shares (%)		CAAGR (%)
	1990	2010	2015	2020	2025	2030	2035	2010	2035	2010-35
TPED	**4 521**	**5 404**	**5 465**	**5 530**	**5 544**	**5 553**	**5 579**	**100**	**100**	**0.1**
Coal	1 081	1 086	1 076	1 037	985	911	827	20	15	-1.1
Oil	1 870	1 961	1 913	1 814	1 712	1 603	1 509	36	27	-1.0
Gas	843	1 317	1 362	1 427	1 481	1 537	1 598	24	29	0.8
Nuclear	451	596	566	604	607	622	641	11	11	0.3
Hydro	102	116	123	128	132	136	140	2	3	0.7
Bioenergy	147	264	314	362	411	466	526	5	9	2.8
Other renewables	29	63	111	158	216	278	339	1	6	7.0
Power generation	**1 719**	**2 267**	**2 286**	**2 353**	**2 402**	**2 450**	**2 491**	**100**	**100**	**0.4**
Coal	760	875	852	815	763	691	603	39	24	-1.5
Oil	154	71	63	39	30	24	20	3	1	-4.9
Gas	176	471	485	510	540	571	607	21	24	1.0
Nuclear	451	596	566	604	607	622	641	26	26	0.3
Hydro	102	116	123	128	132	136	140	5	6	0.7
Bioenergy	53	82	99	117	138	160	182	4	7	3.2
Other renewables	25	54	98	141	193	246	297	2	12	7.0
Other energy sector	**399**	**458**	**459**	**461**	**461**	**460**	**462**	**100**	**100**	**0.0**
Electricity	*105*	*127*	*129*	*131*	*133*	*135*	*136*	*28*	*29*	*0.3*
TFC	**3 109**	**3 691**	**3 776**	**3 822**	**3 832**	**3 833**	**3 851**	**100**	**100**	**0.2**
Coal	236	128	132	128	123	117	113	3	3	-0.5
Oil	1 592	1 767	1 741	1 680	1 600	1 509	1 433	48	37	-0.8
Gas	589	737	763	791	808	826	843	20	22	0.5
Electricity	552	807	848	894	936	974	1 008	22	26	0.9
Heat	43	63	66	68	69	70	71	2	2	0.5
Bioenergy	94	181	214	244	272	305	342	5	9	2.6
Other renewables	4	9	12	17	23	31	42	0	1	6.5
Industry	**830**	**829**	**870**	**889**	**897**	**901**	**903**	**100**	**100**	**0.3**
Coal	161	102	107	105	102	99	96	12	11	-0.3
Oil	169	116	113	108	102	96	89	14	10	-1.0
Gas	226	255	267	273	274	273	272	31	30	0.3
Electricity	222	259	277	288	294	300	304	31	34	0.6
Heat	15	24	24	24	23	23	23	3	2	-0.3
Bioenergy	37	72	82	92	101	110	119	9	13	2.0
Other renewables	0	0	0	0	0	1	1	0	0	2.4
Transport	**940**	**1 180**	**1 181**	**1 151**	**1 106**	**1 057**	**1 028**	**100**	**100**	**-0.5**
Oil	914	1 107	1 093	1 046	983	911	852	94	83	-1.0
Electricity	8	9	11	12	14	16	20	1	2	3.0
Biofuels	0	40	53	66	79	97	117	3	11	4.3
Other fuels	19	23	25	27	30	34	40	2	4	2.3
Buildings	**986**	**1 253**	**1 291**	**1 343**	**1 389**	**1 434**	**1 482**	**100**	**100**	**0.7**
Coal	71	22	21	19	17	15	13	2	1	-2.0
Oil	209	163	152	141	130	120	113	13	8	-1.5
Gas	304	425	435	455	467	480	490	34	33	0.6
Electricity	316	530	552	586	620	650	676	42	46	1.0
Heat	27	39	42	44	45	47	48	3	3	0.9
Bioenergy	56	66	77	83	89	95	103	5	7	1.8
Other renewables	4	8	11	15	21	28	38	1	3	6.4
Other	**353**	**430**	**434**	**438**	**441**	**441**	**438**	**100**	**100**	**0.1**

OECD: Current Policies and 450 Scenarios

	Energy demand (Mtoe)						Shares (%)		CAAGR (%)	
	2020	2030	2035	2020	2030	2035	2035		2010-35	
	Current Policies Scenario			450 Scenario			CPS	450	CPS	450
TPED	**5 629**	**5 829**	**5 945**	**5 328**	**5 061**	**5 033**	**100**	**100**	**0.4**	**-0.3**
Coal	1 107	1 104	1 104	918	503	454	19	9	0.1	-3.4
Oil	1 851	1 742	1 706	1 748	1 356	1 166	29	23	-0.6	-2.1
Gas	1 450	1 600	1 689	1 336	1 326	1 243	28	25	1.0	-0.2
Nuclear	599	604	584	623	726	777	10	15	-0.1	1.1
Hydro	127	133	136	131	144	149	2	3	0.6	1.0
Bioenergy	345	422	466	393	607	711	8	14	2.3	4.1
Other renewables	150	224	259	178	399	532	4	11	5.8	8.9
Power generation	**2 407**	**2 574**	**2 652**	**2 234**	**2 192**	**2 254**	**100**	**100**	**0.6**	**-0.0**
Coal	877	868	859	706	307	258	32	11	-0.1	-4.8
Oil	42	27	25	33	17	15	1	1	-4.1	-6.1
Gas	516	598	656	463	466	382	25	17	1.3	-0.8
Nuclear	599	604	584	623	726	777	22	34	-0.1	1.1
Hydro	127	133	136	131	144	149	5	7	0.6	1.0
Bioenergy	112	145	164	122	179	212	6	9	2.8	3.9
Other renewables	134	199	229	156	353	461	9	20	5.9	8.9
Other energy sector	**470**	**484**	**497**	**443**	**409**	**394**	**100**	**100**	**0.3**	**-0.6**
Electricity	*134*	*144*	*148*	*124*	*118*	*116*	*30*	*29*	*0.6*	*-0.4*
TFC	**3 881**	**4 018**	**4 097**	**3 714**	**3 554**	**3 498**	**100**	**100**	**0.4**	**-0.2**
Coal	134	128	125	121	101	95	3	3	-0.1	-1.2
Oil	1 712	1 642	1 621	1 624	1 282	1 112	40	32	-0.3	-1.8
Gas	806	856	877	752	734	734	21	21	0.7	-0.0
Electricity	912	1 016	1 066	862	906	929	26	27	1.1	0.6
Heat	70	74	76	63	60	59	2	2	0.8	-0.3
Bioenergy	232	276	302	270	427	499	7	14	2.1	4.1
Other renewables	15	25	30	21	45	71	1	2	5.1	8.8
Industry	**912**	**949**	**960**	**867**	**861**	**858**	**100**	**100**	**0.6**	**0.1**
Coal	109	106	104	99	86	82	11	10	0.1	-0.9
Oil	112	103	98	106	94	87	10	10	-0.7	-1.1
Gas	280	292	295	267	260	258	31	30	0.6	0.1
Electricity	296	318	326	280	285	286	34	33	0.9	0.4
Heat	24	24	24	23	21	20	3	2	0.0	-0.7
Bioenergy	90	105	113	92	114	124	12	14	1.8	2.2
Other renewables	0	1	1	0	1	1	0	0	2.4	2.8
Transport	**1 163**	**1 147**	**1 161**	**1 129**	**961**	**884**	**100**	**100**	**-0.1**	**-1.1**
Oil	1 063	1 020	1 015	1 015	733	592	87	67	-0.3	-2.5
Electricity	12	14	15	13	31	52	1	6	2.0	7.1
Biofuels	62	83	96	76	165	200	8	23	3.5	6.6
Other fuels	26	31	35	25	31	40	3	4	1.7	2.2
Buildings	**1 362**	**1 474**	**1 530**	**1 281**	**1 291**	**1 316**	**100**	**100**	**0.8**	**0.2**
Coal	21	18	17	18	12	10	1	1	-1.0	-3.1
Oil	146	129	122	129	91	78	8	6	-1.1	-2.9
Gas	464	495	507	425	406	399	33	30	0.7	-0.2
Electricity	595	675	715	561	582	584	47	44	1.2	0.4
Heat	45	50	52	40	39	38	3	3	1.2	-0.1
Bioenergy	77	85	89	89	119	140	6	11	1.2	3.1
Other renewables	14	22	28	20	42	66	2	5	5.0	8.8
Other	**445**	**447**	**445**	**437**	**442**	**441**	**100**	**100**	**0.1**	**0.1**

OECD: New Policies Scenario

	Electricity generation (TWh)							Shares (%)		CAAGR (%)
	1990	2010	2015	2020	2025	2030	2035	2010	2035	2010-35
Total generation	**7 629**	**10 848**	**11 349**	**11 910**	**12 430**	**12 888**	**13 297**	**100**	**100**	**0.8**
Coal	3 093	3 746	3 717	3 592	3 412	3 137	2 794	35	21	-1.2
Oil	697	309	278	166	129	106	90	3	1	-4.8
Gas	770	2 544	2 690	2 872	3 116	3 322	3 517	23	26	1.3
Nuclear	1 729	2 288	2 171	2 318	2 329	2 388	2 460	21	19	0.3
Hydro	1 182	1 351	1 430	1 486	1 538	1 582	1 622	12	12	0.7
Bioenergy	124	264	334	407	489	578	671	2	5	3.8
Wind	4	269	510	735	962	1 196	1 423	2	11	6.9
Geothermal	29	43	58	78	109	141	166	0	1	5.5
Solar PV	0	31	140	218	282	339	396	0	3	10.8
CSP	1	2	18	34	51	73	104	0	1	18.2
Marine	1	1	3	5	12	25	55	0	0	20.1

	Electrical capacity (GW)						Shares (%)		CAAGR (%)
	2010	2015	2020	2025	2030	2035	2010	2035	2010-35
Total capacity	**2 718**	**3 018**	**3 171**	**3 333**	**3 498**	**3 656**	**100**	**100**	**1.2**
Coal	669	687	640	589	548	499	25	14	-1.2
Oil	218	192	127	93	74	67	8	2	-4.6
Gas	819	918	998	1 080	1 144	1 196	30	33	1.5
Nuclear	327	324	321	315	320	328	12	9	0.0
Hydro	455	475	493	509	523	535	17	15	0.6
Bioenergy	49	60	71	84	97	111	2	3	3.3
Wind	136	226	315	400	480	555	5	15	5.8
Geothermal	7	9	11	16	21	24	0	1	4.9
Solar PV	37	121	182	229	266	301	1	8	8.8
CSP	1	5	9	14	19	27	0	1	13.2
Marine	0	1	1	3	6	14	0	0	17.2

	CO_2 emissions (Mt)							Shares (%)		CAAGR (%)
	1990	2010	2015	2020	2025	2030	2035	2010	2035	2010-35
Total CO_2	**11 116**	**12 340**	**12 239**	**11 920**	**11 481**	**10 907**	**10 362**	**100**	**100**	**-0.7**
Coal	4 155	4 182	4 109	3 930	3 659	3 270	2 840	34	27	-1.5
Oil	5 034	5 108	4 978	4 699	4 413	4 108	3 860	41	37	-1.1
Gas	1 928	3 050	3 153	3 291	3 408	3 528	3 662	25	35	0.7
Power generation	**3 965**	**4 872**	**4 775**	**4 593**	**4 380**	**4 064**	**3 723**	**100**	**100**	**-1.1**
Coal	3 067	3 541	3 438	3 274	3 023	2 657	2 248	73	60	-1.8
Oil	487	226	197	123	94	76	65	5	2	-4.9
Gas	411	1 105	1 140	1 197	1 264	1 331	1 411	23	38	1.0
TFC	**6 558**	**6 781**	**6 776**	**6 640**	**6 422**	**6 171**	**5 967**	**100**	**100**	**-0.5**
Coal	1 025	556	577	561	542	519	499	8	8	-0.4
Oil	4 184	4 525	4 440	4 253	4 015	3 747	3 525	67	59	-1.0
Transport	*2 681*	*3 266*	*3 222*	*3 084*	*2 899*	*2 686*	*2 513*	*48*	*42*	*-1.0*
Gas	1 349	1 700	1 759	1 825	1 865	1 905	1 944	25	33	0.5

OECD: Current Policies and 450 Scenarios

	Electricity generation (TWh)						Shares (%)		CAAGR (%)	
	2020	2030	2035	2020	2030	2035	2035		2010-35	
	Current Policies Scenario			450 Scenario			CPS	450	CPS	450
Total generation	**12 153**	**13 487**	**14 110**	**11 470**	**11 899**	**12 153**	**100**	**100**	**1.1**	**0.5**
Coal	3 882	4 000	4 056	3 136	1 387	1 118	29	9	0.3	-4.7
Oil	180	117	111	139	70	63	1	1	-4.0	-6.2
Gas	2 919	3 469	3 781	2 656	2 757	2 147	27	18	1.6	-0.7
Nuclear	2 299	2 316	2 240	2 392	2 787	2 982	16	25	-0.1	1.1
Hydro	1 474	1 549	1 578	1 521	1 671	1 730	11	14	0.6	1.0
Bioenergy	385	523	602	425	653	782	4	6	3.4	4.4
Wind	701	1 049	1 202	841	1 714	2 139	9	18	6.2	8.6
Geothermal	75	108	120	82	174	218	1	2	4.1	6.7
Solar PV	206	285	314	233	440	560	2	5	9.8	12.3
CSP	31	59	76	40	211	339	1	3	16.8	24.0
Marine	3	12	29	6	36	75	0	1	17.2	21.7

	Electrical capacity (GW)						Shares (%)		CAAGR (%)	
	2020	2030	2035	2020	2030	2035	2035		2010-35	
	Current Policies Scenario			450 Scenario			CPS	450	CPS	450
Total capacity	**3 176**	**3 509**	**3 654**	**3 187**	**3 598**	**3 877**	**100**	**100**	**1.2**	**1.4**
Coal	666	646	627	613	344	281	17	7	-0.3	-3.4
Oil	129	76	69	120	61	51	2	1	-4.5	-5.7
Gas	1 002	1 178	1 263	968	1 067	1 099	35	28	1.7	1.2
Nuclear	327	315	301	331	373	397	8	10	-0.3	0.8
Hydro	490	512	520	504	550	569	14	15	0.5	0.9
Bioenergy	68	89	100	74	109	128	3	3	2.9	3.9
Wind	303	433	486	357	663	798	13	21	5.2	7.4
Geothermal	11	15	17	12	25	31	0	1	3.5	6.1
Solar PV	172	227	243	195	340	414	7	11	7.8	10.2
CSP	9	16	21	11	55	88	1	2	12.0	18.7
Marine	1	3	8	2	9	20	0	1	14.4	18.7

	CO_2 emissions (Mt)						Shares (%)		CAAGR (%)	
	2020	2030	2035	2020	2030	2035	2035		2010-35	
	Current Policies Scenario			450 Scenario			CPS	450	CPS	450
Total CO_2	**12 354**	**12 288**	**12 347**	**11 006**	**7 416**	**6 087**	**100**	**100**	**0.0**	**-2.8**
Coal	4 209	4 092	4 017	3 414	1 123	659	33	11	-0.2	-7.1
Oil	4 801	4 520	4 450	4 521	3 401	2 869	36	47	-0.6	-2.3
Gas	3 345	3 676	3 880	3 071	2 892	2 558	31	42	1.0	-0.7
Power generation	**4 874**	**4 919**	**4 986**	**3 981**	**1 637**	**926**	**100**	**100**	**0.1**	**-6.4**
Coal	3 528	3 434	3 373	2 795	624	225	68	24	-0.2	-10.4
Oil	133	85	80	105	53	48	2	5	-4.1	-6.0
Gas	1 212	1 400	1 534	1 082	960	652	31	70	1.3	-2.1
TFC	**6 784**	**6 666**	**6 646**	**6 362**	**5 177**	**4 590**	**100**	**100**	**-0.1**	**-1.5**
Coal	584	561	548	527	412	351	8	8	-0.1	-1.8
Oil	4 339	4 130	4 073	4 104	3 100	2 603	61	57	-0.4	-2.2
Transport	*3 134*	*3 007*	*2 992*	*2 991*	*2 162*	*1 747*	*45*	*38*	*-0.3*	*-2.5*
Gas	1 861	1 975	2 025	1 731	1 665	1 637	30	36	0.7	-0.2

OECD Americas: New Policies Scenario

	Energy demand (Mtoe)							Shares (%)		CAAGR (%)
	1990	2010	2015	2020	2025	2030	2035	2010	2035	2010-35
TPED	**2 260**	**2 677**	**2 751**	**2 792**	**2 798**	**2 795**	**2 806**	**100**	**100**	**0.2**
Coal	490	538	523	518	506	476	447	20	16	-0.7
Oil	920	1 004	999	955	896	828	767	37	27	-1.1
Gas	517	693	736	770	788	815	846	26	30	0.8
Nuclear	180	244	253	266	273	280	286	9	10	0.6
Hydro	52	58	63	65	67	69	71	2	3	0.8
Bioenergy	82	116	135	157	184	218	256	4	9	3.2
Other renewables	19	25	42	62	85	108	134	1	5	6.9
Power generation	**853**	**1 092**	**1 116**	**1 161**	**1 194**	**1 222**	**1 245**	**100**	**100**	**0.5**
Coal	421	495	474	468	455	424	387	45	31	-1.0
Oil	47	24	20	13	10	8	7	2	1	-4.9
Gas	95	224	237	254	264	280	299	20	24	1.2
Nuclear	180	244	253	266	273	280	286	22	23	0.6
Hydro	52	58	63	65	67	69	71	5	6	0.8
Bioenergy	41	24	29	37	48	61	75	2	6	4.7
Other renewables	18	23	40	58	79	99	120	2	10	6.8
Other energy sector	**191**	**220**	**222**	**227**	**230**	**233**	**240**	**100**	**100**	**0.3**
Electricity	*56*	*65*	*67*	*70*	*72*	*73*	*74*	*30*	*31*	*0.5*
TFC	**1 548**	**1 833**	**1 903**	**1 920**	**1 914**	**1 901**	**1 902**	**100**	**100**	**0.1**
Coal	61	32	34	33	31	30	28	2	1	-0.5
Oil	809	925	933	902	853	791	739	50	39	-0.9
Gas	361	386	410	419	423	429	436	21	23	0.5
Electricity	272	390	411	437	460	480	500	21	26	1.0
Heat	3	7	7	6	5	5	4	0	0	-2.2
Bioenergy	41	92	105	119	136	157	181	5	10	2.8
Other renewables	0	2	2	4	6	9	14	0	1	8.8
Industry	**361**	**370**	**390**	**396**	**396**	**396**	**395**	**100**	**100**	**0.3**
Coal	51	30	33	32	30	29	27	8	7	-0.5
Oil	60	46	45	43	41	38	36	13	9	-1.0
Gas	138	144	151	152	150	148	146	39	37	0.1
Electricity	94	103	110	113	115	117	119	28	30	0.6
Heat	1	6	5	5	4	4	4	2	1	-1.8
Bioenergy	17	41	45	51	55	59	64	11	16	1.8
Other renewables	0	0	0	0	0	0	0	0	0	0.7
Transport	**562**	**702**	**721**	**701**	**669**	**634**	**613**	**100**	**100**	**-0.5**
Oil	543	656	666	638	594	541	497	93	81	-1.1
Electricity	1	1	1	2	2	3	5	0	1	6.3
Biofuels	-	26	34	40	50	64	80	4	13	4.5
Other fuels	18	19	20	21	23	26	31	3	5	2.0
Buildings	**461**	**571**	**596**	**622**	**644**	**666**	**690**	**100**	**100**	**0.8**
Coal	10	2	1	1	1	1	1	0	0	-3.1
Oil	64	51	46	42	37	32	29	9	4	-2.2
Gas	184	207	220	226	230	234	237	36	34	0.5
Electricity	176	284	299	320	340	357	373	50	54	1.1
Heat	2	1	1	1	1	1	0	0	0	-4.5
Bioenergy	24	24	26	28	30	33	37	4	5	1.7
Other renewables	0	2	2	3	5	8	13	0	2	8.6
Other	**164**	**190**	**197**	**202**	**205**	**205**	**205**	**100**	**100**	**0.3**

OECD Americas: Current Policies and 450 Scenarios

	Energy demand (Mtoe)						Shares (%)		CAAGR (%)	
	2020	2030	2035	2020	2030	2035	2035		2010-35	
	Current Policies Scenario			450 Scenario			CPS	450	CPS	450
TPED	**2 830**	**2 933**	**3 005**	**2 700**	**2 541**	**2 529**	**100**	**100**	**0.5**	**-0.2**
Coal	545	566	585	469	229	228	19	9	0.3	-3.4
Oil	970	913	895	928	707	594	30	23	-0.5	-2.1
Gas	770	826	860	718	739	687	29	27	0.9	-0.0
Nuclear	266	268	264	271	315	333	9	13	0.3	1.3
Hydro	65	68	70	66	70	71	2	3	0.7	0.8
Bioenergy	154	199	227	178	306	365	8	14	2.7	4.7
Other renewables	61	93	105	70	176	250	3	10	5.9	9.6
Power generation	**1 181**	**1 269**	**1 315**	**1 099**	**1 066**	**1 108**	**100**	**100**	**0.7**	**0.1**
Coal	493	508	516	423	185	179	39	16	0.2	-4.0
Oil	16	10	8	11	6	6	1	1	-4.2	-5.7
Gas	248	275	293	225	264	216	22	19	1.1	-0.1
Nuclear	266	268	264	271	315	333	20	30	0.3	1.3
Hydro	65	68	70	66	70	71	5	6	0.7	0.8
Bioenergy	37	55	67	39	67	83	5	8	4.2	5.1
Other renewables	57	85	95	64	160	221	7	20	5.8	9.4
Other energy sector	**231**	**247**	**260**	**218**	**208**	**203**	**100**	**100**	**0.7**	**-0.3**
Electricity	*71*	*77*	*79*	*67*	*64*	*63*	*31*	*31*	*0.8*	*-0.1*
TFC	**1 941**	**1 995**	**2 034**	**1 881**	**1 782**	**1 745**	**100**	**100**	**0.4**	**-0.2**
Coal	35	33	32	30	23	20	2	1	0.0	-1.8
Oil	914	874	864	879	678	575	42	33	-0.3	-1.9
Gas	424	440	448	399	378	377	22	22	0.6	-0.1
Electricity	441	492	516	422	443	459	25	26	1.1	0.7
Heat	6	5	5	6	4	4	0	0	-1.6	-2.6
Bioenergy	117	144	160	139	239	282	8	16	2.2	4.6
Other renewables	4	8	10	6	16	29	0	2	7.2	11.9
Industry	**406**	**416**	**419**	**383**	**373**	**369**	**100**	**100**	**0.5**	**-0.0**
Coal	33	32	31	29	22	20	7	5	0.1	-1.6
Oil	45	42	40	42	37	34	10	9	-0.6	-1.2
Gas	156	158	158	149	140	138	38	37	0.4	-0.2
Electricity	116	124	127	109	109	109	30	29	0.8	0.2
Heat	5	4	4	5	4	3	1	1	-1.2	-2.1
Bioenergy	49	56	59	50	62	66	14	18	1.5	1.9
Other renewables	0	0	0	0	0	0	0	0	0.7	0.7
Transport	**708**	**697**	**708**	**696**	**596**	**546**	**100**	**100**	**0.0**	**-1.0**
Oil	646	617	614	624	447	356	87	65	-0.3	-2.4
Electricity	1	2	2	2	12	26	0	5	2.5	13.3
Biofuels	40	56	66	50	114	137	9	25	3.7	6.8
Other fuels	21	23	26	20	23	28	4	5	1.3	1.7
Buildings	**626**	**677**	**703**	**600**	**607**	**623**	**100**	**100**	**0.8**	**0.3**
Coal	1	1	1	1	0	0	0	0	-1.3	-12.2
Oil	44	36	33	37	22	17	5	3	-1.7	-4.3
Gas	228	237	242	212	196	191	34	31	0.6	-0.3
Electricity	321	364	384	308	320	322	55	52	1.2	0.5
Heat	1	1	0	1	1	0	0	0	-4.4	-4.9
Bioenergy	27	31	34	34	52	66	5	11	1.3	4.1
Other renewables	3	7	9	6	15	27	1	4	7.1	11.9
Other	**202**	**205**	**204**	**202**	**206**	**207**	**100**	**100**	**0.3**	**0.3**

OECD Americas: New Policies Scenario

	Electricity generation (TWh)							Shares (%)		CAAGR (%)
	1990	2010	2015	2020	2025	2030	2035	2010	2035	2010-35
Total generation	**3 819**	**5 293**	**5 565**	**5 891**	**6 180**	**6 435**	**6 679**	**100**	**100**	**0.9**
Coal	1 796	2 132	2 055	2 051	2 025	1 919	1 791	40	27	-0.7
Oil	211	108	86	59	45	38	33	2	0	-4.6
Gas	406	1 221	1 346	1 464	1 560	1 677	1 804	23	27	1.6
Nuclear	687	935	973	1 020	1 046	1 076	1 098	18	16	0.6
Hydro	602	673	733	756	780	801	820	13	12	0.8
Bioenergy	91	89	113	149	195	251	311	2	5	5.1
Wind	3	106	198	283	369	458	553	2	8	6.8
Geothermal	21	24	32	43	56	68	80	0	1	4.9
Solar PV	0	3	21	47	74	102	129	0	2	15.9
CSP	1	1	8	19	28	39	51	0	1	17.6
Marine	0	0	0	0	2	5	8	0	0	25.6

	Electrical capacity (GW)						Shares (%)		CAAGR (%)
	2010	2015	2020	2025	2030	2035	2010	2035	2010-35
Total capacity	**1 308**	**1 396**	**1 461**	**1 530**	**1 594**	**1 660**	**100**	**100**	**1.0**
Coal	360	360	341	318	292	272	27	16	-1.1
Oil	94	82	52	35	29	25	7	1	-5.2
Gas	472	509	552	592	619	640	36	39	1.2
Nuclear	122	125	131	134	138	140	9	8	0.6
Hydro	193	200	207	213	219	223	15	13	0.6
Bioenergy	16	20	25	32	41	50	1	3	4.7
Wind	45	80	112	142	172	202	3	12	6.2
Geothermal	4	5	6	8	10	11	0	1	4.1
Solar PV	2	15	31	47	64	81	0	5	15.3
CSP	0	2	5	7	10	13	0	1	14.1
Marine	0	0	0	0	1	2	0	0	20.0

	CO_2 emissions (Mt)							Shares (%)		CAAGR (%)
	1990	2010	2015	2020	2025	2030	2035	2010	2035	2010-35
Total CO_2	**5 579**	**6 362**	**6 376**	**6 285**	**6 078**	**5 774**	**5 495**	**100**	**100**	**-0.6**
Coal	1 921	2 092	2 021	1 990	1 915	1 747	1 564	33	28	-1.2
Oil	2 469	2 675	2 666	2 537	2 367	2 172	2 008	42	37	-1.1
Gas	1 189	1 594	1 689	1 758	1 796	1 855	1 923	25	35	0.8
Power generation	**2 019**	**2 552**	**2 483**	**2 473**	**2 414**	**2 283**	**2 144**	**100**	**100**	**-0.7**
Coal	1 647	1 947	1 862	1 836	1 767	1 606	1 430	76	67	-1.2
Oil	150	82	65	44	32	27	23	3	1	-4.9
Gas	222	523	555	593	614	649	690	20	32	1.1
TFC	**3 213**	**3 427**	**3 507**	**3 424**	**3 279**	**3 108**	**2 966**	**100**	**100**	**-0.6**
Coal	270	136	147	142	136	129	122	4	4	-0.4
Oil	2 115	2 401	2 414	2 316	2 169	1 990	1 841	70	62	-1.1
Transport	*1 585*	*1 920*	*1 949*	*1 867*	*1 741*	*1 586*	*1 456*	*56*	*49*	*-1.1*
Gas	829	891	946	965	975	988	1 003	26	34	0.5

OECD Americas: Current Policies and 450 Scenarios

	Electricity generation (TWh)						Shares (%)		CAAGR (%)	
	2020	2030	2035	2020	2030	2035	2035		2010-35	
	Current Policies Scenario			450 Scenario			CPS	450	CPS	450
Total generation	**5 957**	**6 613**	**6 927**	**5 678**	**5 898**	**6 070**	**100**	**100**	**1.1**	**0.5**
Coal	2 174	2 330	2 409	1 892	877	817	35	13	0.5	-3.8
Oil	70	45	40	50	29	27	1	0	-3.9	-5.4
Gas	1 416	1 609	1 709	1 330	1 597	1 262	25	21	1.4	0.1
Nuclear	1 020	1 028	1 015	1 041	1 208	1 278	15	21	0.3	1.3
Hydro	756	794	810	763	810	829	12	14	0.7	0.8
Bioenergy	144	229	280	157	280	351	4	6	4.7	5.6
Wind	273	410	469	329	705	918	7	15	6.1	9.0
Geothermal	43	59	63	45	87	111	1	2	3.9	6.3
Solar PV	44	78	91	48	154	226	1	4	14.3	18.5
CSP	18	32	38	20	145	243	1	4	16.2	25.2
Marine	0	2	5	1	7	10	0	0	23.2	26.4

	Electrical capacity (GW)						Shares (%)		CAAGR (%)	
	2020	2030	2035	2020	2030	2035	2035		2010-35	
	Current Policies Scenario			450 Scenario			CPS	450	CPS	450
Total capacity	**1 466**	**1 598**	**1 655**	**1 452**	**1 617**	**1 781**	**100**	**100**	**0.9**	**1.2**
Coal	363	368	368	327	172	157	22	9	0.1	-3.3
Oil	51	29	25	48	27	23	1	1	-5.2	-5.5
Gas	542	592	613	535	592	609	37	34	1.1	1.0
Nuclear	131	132	130	133	154	163	8	9	0.3	1.2
Hydro	206	216	220	209	222	227	13	13	0.5	0.7
Bioenergy	24	37	45	27	46	57	3	3	4.3	5.2
Wind	108	156	176	129	257	326	11	18	5.6	8.2
Geothermal	6	9	9	7	12	16	1	1	3.2	5.5
Solar PV	29	50	58	32	97	141	3	8	13.7	17.8
CSP	5	9	11	6	37	62	1	3	13.2	21.5
Marine	0	0	1	0	2	2	0	0	17.8	20.9

	CO_2 emissions (Mt)						Shares (%)		CAAGR (%)	
	2020	2030	2035	2020	2030	2035	2035		2010-35	
	Current Policies Scenario			450 Scenario			CPS	450	CPS	450
Total CO_2	**6 438**	**6 453**	**6 515**	**5 874**	**3 764**	**3 021**	**100**	**100**	**0.1**	**-2.9**
Coal	2 097	2 139	2 157	1 781	361	170	33	6	0.1	-9.5
Oil	2 584	2 433	2 400	2 463	1 824	1 506	37	50	-0.4	-2.3
Gas	1 757	1 881	1 958	1 630	1 578	1 344	30	45	0.8	-0.7
Power generation	**2 567**	**2 657**	**2 718**	**2 198**	**802**	**418**	**100**	**100**	**0.3**	**-7.0**
Coal	1 935	1 984	2 006	1 640	259	86	74	21	0.1	-11.7
Oil	52	32	28	37	21	19	1	5	-4.2	-5.7
Gas	579	641	683	521	523	312	25	75	1.1	-2.0
TFC	**3 479**	**3 390**	**3 378**	**3 302**	**2 617**	**2 279**	**100**	**100**	**-0.1**	**-1.6**
Coal	150	143	139	129	91	74	4	3	0.1	-2.4
Oil	2 352	2 234	2 209	2 255	1 669	1 371	65	60	-0.3	-2.2
Transport	*1 891*	*1 806*	*1 797*	*1 828*	*1 310*	*1 043*	*53*	*46*	*-0.3*	*-2.4*
Gas	977	1 013	1 031	918	857	835	31	37	0.6	-0.3

United States: New Policies Scenario

	Energy demand (Mtoe)							Shares (%)		CAAGR (%)
	1990	2010	2015	2020	2025	2030	2035	2010	2035	2010-35
TPED	**1 915**	**2 214**	**2 246**	**2 260**	**2 240**	**2 206**	**2 187**	**100**	**100**	**-0.0**
Coal	460	503	484	478	466	441	417	23	19	-0.7
Oil	757	805	795	753	693	621	558	36	26	-1.5
Gas	438	556	583	596	603	614	628	25	29	0.5
Nuclear	159	219	222	234	238	243	247	10	11	0.5
Hydro	23	23	25	26	26	27	28	1	1	0.8
Bioenergy	62	90	106	125	149	178	209	4	10	3.4
Other renewables	14	18	31	47	64	82	101	1	5	7.0
Power generation	**750**	**937**	**943**	**976**	**997**	**1 010**	**1 021**	**100**	**100**	**0.3**
Coal	396	463	440	435	422	397	367	49	36	-0.9
Oil	27	11	8	5	4	4	3	1	0	-4.5
Gas	90	185	193	201	207	216	227	20	22	0.8
Nuclear	159	219	222	234	238	243	247	23	24	0.5
Hydro	23	23	25	26	26	27	28	2	3	0.8
Bioenergy	40	20	25	31	40	50	60	2	6	4.5
Other renewables	14	17	29	44	59	74	89	2	9	6.9
Other energy sector	**150**	**166**	**160**	**159**	**155**	**149**	**147**	**100**	**100**	**-0.5**
Electricity	*49*	*50*	*50*	*52*	*53*	*53*	*53*	*30*	*36*	*0.3*
TFC	**1 294**	**1 500**	**1 543**	**1 546**	**1 526**	**1 497**	**1 482**	**100**	**100**	**-0.0**
Coal	56	27	28	27	25	23	22	2	1	-0.9
Oil	683	749	749	716	664	598	543	50	37	-1.3
Gas	303	319	337	342	343	346	349	21	24	0.4
Electricity	226	327	340	359	376	390	403	22	27	0.8
Heat	2	7	6	6	5	4	4	0	0	-2.4
Bioenergy	23	70	81	94	108	128	149	5	10	3.1
Other renewables	0	2	2	3	5	8	13	0	1	8.5
Industry	**284**	**280**	**290**	**290**	**286**	**281**	**275**	**100**	**100**	**-0.1**
Coal	46	25	27	26	24	22	21	9	8	-0.8
Oil	44	30	28	26	23	21	19	11	7	-1.9
Gas	110	111	117	115	112	109	105	40	38	-0.2
Electricity	75	76	78	79	79	78	77	27	28	0.1
Heat	-	5	5	4	4	4	3	2	1	-1.9
Bioenergy	9	32	35	40	44	47	50	11	18	1.8
Other renewables	-	0	0	0	0	0	0	0	0	0.5
Transport	**488**	**583**	**598**	**576**	**544**	**504**	**480**	**100**	**100**	**-0.8**
Oil	472	541	547	518	474	418	371	93	77	-1.5
Electricity	0	1	1	1	2	3	4	0	1	7.9
Biofuels	-	25	32	39	48	61	77	4	16	4.5
Other fuels	15	16	17	18	20	23	28	3	6	2.2
Buildings	**389**	**486**	**503**	**523**	**539**	**554**	**571**	**100**	**100**	**0.6**
Coal	10	2	1	1	1	1	1	0	0	-3.0
Oil	48	37	31	26	21	16	12	8	2	-4.4
Gas	164	182	192	197	198	201	203	37	36	0.4
Electricity	152	251	261	279	296	309	322	52	56	1.0
Heat	2	1	1	1	1	1	0	0	0	-4.9
Bioenergy	14	12	13	15	17	19	21	3	4	2.2
Other renewables	0	2	2	3	5	7	12	0	2	8.5
Other	**133**	**151**	**153**	**157**	**158**	**157**	**156**	**100**	**100**	**0.1**

United States: Current Policies and 450 Scenarios

	Energy demand (Mtoe)						Shares (%)		CAAGR (%)	
	2020	2030	2035	2020	2030	2035	2035		2010-35	
	Current Policies Scenario			450 Scenario			CPS	450	CPS	450
TPED	**2 291**	**2 327**	**2 360**	**2 188**	**2 004**	**1 981**	**100**	**100**	**0.3**	**-0.4**
Coal	504	519	538	434	209	216	23	11	0.3	-3.3
Oil	764	700	677	738	539	441	29	22	-0.7	-2.4
Gas	593	610	618	556	566	513	26	26	0.4	-0.3
Nuclear	234	234	229	240	273	289	10	15	0.2	1.1
Hydro	25	27	27	26	28	29	1	1	0.8	0.9
Bioenergy	124	165	189	140	247	291	8	15	3.0	4.8
Other renewables	47	72	81	54	142	203	3	10	6.1	10.1
Power generation	**991**	**1 048**	**1 075**	**925**	**875**	**909**	**100**	**100**	**0.6**	**-0.1**
Coal	458	469	478	393	171	173	45	19	0.1	-3.8
Oil	6	4	4	5	3	3	0	0	-4.0	-4.9
Gas	192	201	205	179	216	171	19	19	0.4	-0.3
Nuclear	234	234	229	240	273	289	21	32	0.2	1.1
Hydro	25	27	27	26	28	29	3	3	0.8	0.9
Bioenergy	31	48	58	33	55	67	5	7	4.3	4.9
Other renewables	43	65	73	49	128	177	7	19	6.0	9.9
Other energy sector	**161**	**158**	**161**	**153**	**134**	**127**	**100**	**100**	**-0.1**	**-1.1**
Electricity	*52*	*55*	*56*	*49*	*46*	*45*	*35*	*36*	*0.5*	*-0.4*
TFC	**1 564**	**1 582**	**1 602**	**1 516**	**1 407**	**1 365**	**100**	**100**	**0.3**	**-0.4**
Coal	28	27	25	24	17	15	2	1	-0.2	-2.3
Oil	726	675	660	703	519	430	41	32	-0.5	-2.2
Gas	346	355	359	325	302	297	22	22	0.5	-0.3
Electricity	362	398	413	347	360	370	26	27	0.9	0.5
Heat	6	5	4	5	4	3	0	0	-1.8	-2.9
Bioenergy	92	117	131	107	192	224	8	16	2.5	4.7
Other renewables	3	7	9	5	14	26	1	2	7.0	11.7
Industry	**299**	**297**	**294**	**281**	**264**	**256**	**100**	**100**	**0.2**	**-0.4**
Coal	27	25	24	23	17	15	8	6	-0.2	-2.1
Oil	28	24	22	25	20	17	7	7	-1.2	-2.2
Gas	120	118	116	113	102	98	39	38	0.2	-0.5
Electricity	81	82	81	76	72	70	28	27	0.3	-0.3
Heat	5	4	4	4	3	3	1	1	-1.3	-2.4
Bioenergy	39	45	47	39	50	53	16	21	1.5	2.0
Other renewables	0	0	0	0	0	0	0	0	0.5	0.5
Transport	**583**	**566**	**572**	**574**	**481**	**437**	**100**	**100**	**-0.1**	**-1.2**
Oil	526	490	483	512	353	275	85	63	-0.5	-2.7
Electricity	1	1	1	2	11	23	0	5	2.7	15.2
Biofuels	39	54	64	43	97	113	11	26	3.8	6.1
Other fuels	18	20	23	17	20	26	4	6	1.4	1.8
Buildings	**526**	**562**	**580**	**505**	**504**	**514**	**100**	**100**	**0.7**	**0.2**
Coal	1	1	1	1	0	0	0	0	-1.3	-12.4
Oil	28	18	14	23	9	4	2	1	-3.8	-8.6
Gas	198	204	206	184	168	161	36	31	0.5	-0.5
Electricity	280	315	331	269	277	277	57	54	1.1	0.4
Heat	1	1	0	1	1	0	0	0	-4.8	-5.2
Bioenergy	14	18	20	21	36	47	3	9	1.9	5.5
Other renewables	3	6	8	5	13	25	1	5	7.1	11.9
Other	**157**	**157**	**155**	**157**	**159**	**158**	**100**	**100**	**0.1**	**0.2**

United States: New Policies Scenario

	Electricity generation (TWh)							Shares (%)		CAAGR (%)
	1990	2010	2015	2020	2025	2030	2035	2010	2035	2010-35
Total generation	**3 203**	**4 353**	**4 506**	**4 749**	**4 954**	**5 123**	**5 281**	**100**	**100**	**0.8**
Coal	1 700	1 994	1 910	1 907	1 883	1 802	1 698	46	32	-0.6
Oil	131	48	36	23	18	17	16	1	0	-4.3
Gas	382	1 018	1 107	1 170	1 230	1 297	1 382	23	26	1.2
Nuclear	612	839	853	900	915	933	947	19	18	0.5
Hydro	273	262	287	297	306	314	321	6	6	0.8
Bioenergy	86	75	97	128	167	212	259	2	5	5.1
Wind	3	95	166	233	301	371	440	2	8	6.3
Geothermal	16	18	24	32	42	50	58	0	1	4.9
Solar PV	0	3	20	44	67	91	113	0	2	15.5
CSP	1	1	8	17	25	33	43	0	1	16.8
Marine	-	-	-	-	1	3	4	-	0	n.a.

	Electrical capacity (GW)						Shares (%)		CAAGR (%)
	2010	2015	2020	2025	2030	2035	2010	2035	2010-35
Total capacity	**1 098**	**1 161**	**1 206**	**1 255**	**1 298**	**1 335**	**100**	**100**	**0.8**
Coal	338	336	318	295	274	257	31	19	-1.1
Oil	70	58	30	20	17	15	6	1	-6.0
Gas	425	453	487	518	532	540	39	40	1.0
Nuclear	106	108	114	116	118	119	10	9	0.5
Hydro	101	104	107	110	112	113	9	8	0.5
Bioenergy	12	15	20	26	33	40	1	3	5.0
Wind	40	68	93	117	140	161	4	12	5.7
Geothermal	3	4	5	6	7	8	0	1	4.0
Solar PV	2	13	28	42	55	68	0	5	14.9
CSP	0	2	5	7	9	11	0	1	13.3
Marine	-	-	-	0	1	1	-	0	n.a.

	CO_2 emissions (Mt)							Shares (%)		CAAGR (%)
	1990	2010	2015	2020	2025	2030	2035	2010	2035	2010-35
Total CO_2	**4 850**	**5 340**	**5 301**	**5 178**	**4 952**	**4 625**	**4 328**	**100**	**100**	**-0.8**
Coal	1 797	1 941	1 859	1 830	1 758	1 613	1 453	36	34	-1.2
Oil	2 042	2 117	2 097	1 972	1 806	1 603	1 436	40	33	-1.5
Gas	1 011	1 282	1 346	1 376	1 388	1 409	1 439	24	33	0.5
Power generation	**1 848**	**2 290**	**2 210**	**2 192**	**2 134**	**2 014**	**1 884**	**100**	**100**	**-0.8**
Coal	1 550	1 820	1 729	1 706	1 640	1 503	1 350	79	72	-1.2
Oil	88	37	29	17	13	12	12	2	1	-4.5
Gas	210	432	452	469	481	499	523	19	28	0.8
TFC	**2 730**	**2 788**	**2 830**	**2 732**	**2 576**	**2 385**	**2 230**	**100**	**100**	**-0.9**
Coal	245	112	119	114	108	101	94	4	4	-0.7
Oil	1 788	1 936	1 932	1 827	1 676	1 486	1 329	69	60	-1.5
Transport	*1 376*	*1 583*	*1 601*	*1 516*	*1 389*	*1 223*	*1 088*	*57*	*49*	*-1.5*
Gas	697	739	779	791	793	799	807	27	36	0.4

United States: Current Policies and 450 Scenarios

	Electricity generation (TWh)						Shares (%)		CAAGR (%)	
	2020	2030	2035	2020	2030	2035	2035		2010-35	
	Current Policies Scenario			450 Scenario			CPS	450	CPS	450
Total generation	**4 794**	**5 238**	**5 435**	**4 579**	**4 691**	**4 803**	**100**	**100**	**0.9**	**0.4**
Coal	2 019	2 146	2 224	1 762	815	792	41	16	0.4	-3.6
Oil	25	19	18	21	15	15	0	0	-3.8	-4.7
Gas	1 107	1 177	1 202	1 060	1 306	990	22	21	0.7	-0.1
Nuclear	900	900	880	921	1 049	1 110	16	23	0.2	1.1
Hydro	295	311	317	302	323	332	6	7	0.8	0.9
Bioenergy	125	201	246	136	238	293	5	6	4.8	5.6
Wind	231	337	378	282	600	752	7	16	5.7	8.6
Geothermal	32	45	47	33	65	83	1	2	4.0	6.4
Solar PV	42	73	86	44	139	202	2	4	14.3	18.2
CSP	17	30	35	18	136	228	1	5	15.9	24.9
Marine	-	1	3	1	5	6	0	0	n.a.	n.a.

	Electrical capacity (GW)						Shares (%)		CAAGR (%)	
	2020	2030	2035	2020	2030	2035	2035		2010-35	
	Current Policies Scenario			450 Scenario			CPS	450	CPS	450
Total capacity	**1 212**	**1 302**	**1 334**	**1 197**	**1 314**	**1 434**	**100**	**100**	**0.8**	**1.1**
Coal	337	338	339	303	158	146	25	10	0.0	-3.3
Oil	30	17	14	28	16	14	1	1	-6.1	-6.2
Gas	477	501	506	470	507	512	38	36	0.7	0.7
Nuclear	114	114	111	117	132	140	8	10	0.2	1.1
Hydro	106	111	112	108	115	117	8	8	0.4	0.6
Bioenergy	19	31	38	21	37	46	3	3	4.8	5.6
Wind	92	129	143	111	219	266	11	19	5.2	7.8
Geothermal	5	7	7	5	9	12	1	1	3.2	5.5
Solar PV	27	46	53	29	86	123	4	9	13.8	17.7
CSP	5	8	10	5	35	58	1	4	12.8	21.2
Marine	-	0	1	0	1	1	0	0	n.a.	n.a.

	CO_2 emissions (Mt)						Shares (%)		CAAGR (%)	
	2020	2030	2035	2020	2030	2035	2035		2010-35	
	Current Policies Scenario			450 Scenario			CPS	450	CPS	450
Total CO_2	**5 304**	**5 201**	**5 196**	**4 850**	**2 876**	**2 222**	**100**	**100**	**-0.1**	**-3.4**
Coal	1 930	1 955	1 976	1 639	295	142	38	6	0.1	-9.9
Oil	2 008	1 843	1 798	1 934	1 368	1 098	35	49	-0.7	-2.6
Gas	1 367	1 403	1 422	1 277	1 212	982	27	44	0.4	-1.1
Power generation	**2 267**	**2 313**	**2 349**	**1 957**	**649**	**309**	**100**	**100**	**0.1**	**-7.7**
Coal	1 799	1 831	1 858	1 526	217	79	79	26	0.1	-11.8
Oil	19	14	13	15	11	10	1	3	-4.0	-4.9
Gas	449	468	478	415	421	220	20	71	0.4	-2.7
TFC	**2 780**	**2 646**	**2 610**	**2 648**	**2 022**	**1 728**	**100**	**100**	**-0.3**	**-1.9**
Coal	120	113	108	104	69	54	4	3	-0.1	-2.9
Oil	1 858	1 713	1 672	1 794	1 265	1 010	64	58	-0.6	-2.6
Transport	*1 538*	*1 435*	*1 414*	*1 498*	*1 033*	*806*	*54*	*47*	*-0.5*	*-2.7*
Gas	802	820	829	750	687	664	32	38	0.5	-0.4

OECD Europe: New Policies Scenario

	Energy demand (Mtoe)							Shares (%)		CAAGR (%)
	1990	2010	2015	2020	2025	2030	2035	2010	2035	2010-35
TPED	**1 630**	**1 837**	**1 817**	**1 829**	**1 830**	**1 835**	**1 847**	**100**	**100**	**0.0**
Coal	452	307	300	277	246	217	186	17	10	-2.0
Oil	615	614	575	546	520	494	475	33	26	-1.0
Gas	260	469	452	481	509	528	550	26	30	0.6
Nuclear	205	239	230	222	214	215	213	13	12	-0.5
Hydro	38	48	48	51	53	54	55	3	3	0.6
Bioenergy	54	131	159	181	198	215	232	7	13	2.3
Other renewables	5	30	53	72	91	111	135	2	7	6.2
Power generation	**626**	**770**	**769**	**777**	**781**	**791**	**800**	**100**	**100**	**0.2**
Coal	278	221	214	193	165	140	111	29	14	-2.7
Oil	51	23	17	13	10	7	7	3	1	-4.9
Gas	41	165	154	168	187	198	212	21	27	1.0
Nuclear	205	239	230	222	214	215	213	31	27	-0.5
Hydro	38	48	48	51	53	54	55	6	7	0.6
Bioenergy	9	51	61	69	76	83	89	7	11	2.2
Other renewables	3	24	45	61	77	94	113	3	14	6.5
Other energy sector	**151**	**155**	**148**	**144**	**140**	**137**	**133**	**100**	**100**	**-0.6**
Electricity	*39*	*45*	*44*	*42*	*43*	*43*	*43*	*29*	*32*	*-0.2*
TFC	**1 130**	**1 288**	**1 290**	**1 313**	**1 329**	**1 341**	**1 357**	**100**	**100**	**0.2**
Coal	125	54	54	52	49	46	44	4	3	-0.8
Oil	523	545	515	493	473	453	437	42	32	-0.9
Gas	201	281	277	292	302	311	318	22	23	0.5
Electricity	193	271	284	298	311	323	333	21	25	0.8
Heat	40	51	55	57	58	60	61	4	5	0.7
Bioenergy	45	79	97	111	121	131	142	6	10	2.4
Other renewables	2	6	8	11	14	17	22	0	2	5.4
Industry	**324**	**298**	**309**	**318**	**321**	**324**	**325**	**100**	**100**	**0.3**
Coal	71	32	33	33	32	31	30	11	9	-0.3
Oil	59	37	35	33	31	29	27	12	8	-1.2
Gas	78	88	89	92	93	94	94	30	29	0.2
Electricity	88	100	108	112	114	115	116	34	36	0.6
Heat	14	16	16	17	17	17	17	5	5	0.1
Bioenergy	14	24	28	31	35	38	41	8	13	2.2
Other renewables	0	0	0	0	0	0	0	0	0	6.5
Transport	**268**	**338**	**323**	**320**	**313**	**306**	**302**	**100**	**100**	**-0.4**
Oil	262	316	296	284	271	259	250	94	83	-0.9
Electricity	5	6	7	8	8	9	11	2	3	2.3
Biofuels	0	13	18	25	29	32	36	4	12	4.1
Other fuels	1	2	3	3	4	5	5	1	2	3.2
Buildings	**406**	**506**	**515**	**536**	**554**	**572**	**591**	**100**	**100**	**0.6**
Coal	51	19	18	17	15	13	12	4	2	-1.9
Oil	97	72	67	62	56	51	47	14	8	-1.7
Gas	105	175	170	182	190	197	203	35	34	0.6
Electricity	96	160	165	173	184	193	202	32	34	0.9
Heat	24	35	38	40	41	43	44	7	8	1.0
Bioenergy	30	40	48	52	55	58	62	8	10	1.8
Other renewables	2	6	8	10	13	16	21	1	4	5.4
Other	**133**	**146**	**143**	**140**	**141**	**140**	**139**	**100**	**100**	**-0.2**

OECD Europe: Current Policies and 450 Scenarios

	Energy demand (Mtoe)						Shares (%)		CAAGR (%)	
	2020	2030	2035	2020	2030	2035	2035		2010-35	
	Current Policies Scenario			450 Scenario			CPS	450	CPS	450
TPED	**1 869**	**1 932**	**1 961**	**1 746**	**1 664**	**1 652**	**100**	**100**	**0.3**	**-0.4**
Coal	304	285	279	228	135	111	14	7	-0.4	-4.0
Oil	565	538	530	520	408	360	27	22	-0.6	-2.1
Gas	501	570	608	454	409	377	31	23	1.0	-0.9
Nuclear	214	196	171	225	252	271	9	16	-1.3	0.5
Hydro	50	53	54	52	58	60	3	4	0.5	0.9
Bioenergy	168	194	208	186	253	286	11	17	1.9	3.2
Other renewables	67	96	112	80	148	187	6	11	5.5	7.6
Power generation	**797**	**841**	**859**	**737**	**723**	**734**	**100**	**100**	**0.4**	**-0.2**
Coal	217	202	198	151	71	52	23	7	-0.4	-5.7
Oil	14	8	8	12	7	6	1	1	-4.3	-5.2
Gas	180	224	253	156	113	80	29	11	1.7	-2.8
Nuclear	214	196	171	225	252	271	20	37	-1.3	0.5
Hydro	50	53	54	52	58	60	6	8	0.5	0.9
Bioenergy	65	76	81	72	95	107	9	15	1.8	3.0
Other renewables	57	81	95	68	127	158	11	22	5.7	7.9
Other energy sector	**147**	**144**	**143**	**137**	**118**	**110**	**100**	**100**	**-0.3**	**-1.4**
Electricity	*44*	*46*	*47*	*40*	*38*	*37*	*33*	*34*	*0.3*	*-0.7*
TFC	**1 342**	**1 408**	**1 440**	**1 261**	**1 222**	**1 211**	**100**	**100**	**0.4**	**-0.2**
Coal	54	51	49	48	39	36	3	3	-0.4	-1.6
Oil	510	494	488	470	373	331	34	27	-0.4	-2.0
Gas	301	326	335	277	278	278	23	23	0.7	-0.0
Electricity	306	342	359	288	304	311	25	26	1.1	0.5
Heat	58	64	66	52	51	50	5	4	1.0	-0.1
Bioenergy	103	118	126	114	157	178	9	15	1.9	3.3
Other renewables	10	14	16	12	21	28	1	2	4.3	6.5
Industry	**327**	**343**	**349**	**309**	**309**	**308**	**100**	**100**	**0.6**	**0.1**
Coal	34	33	32	31	26	25	9	8	-0.0	-1.1
Oil	34	31	29	32	28	26	8	8	-0.9	-1.4
Gas	95	101	102	90	90	89	29	29	0.6	0.0
Electricity	116	124	126	109	111	111	36	36	0.9	0.4
Heat	17	17	18	16	15	15	5	5	0.4	-0.4
Bioenergy	32	38	42	32	39	43	12	14	2.2	2.3
Other renewables	0	0	0	0	0	0	0	0	6.5	7.4
Transport	**323**	**327**	**333**	**307**	**262**	**246**	**100**	**100**	**-0.1**	**-1.3**
Oil	291	288	289	270	198	167	87	68	-0.4	-2.5
Electricity	7	9	10	8	13	18	3	7	2.0	4.5
Biofuels	21	26	29	25	46	54	9	22	3.2	5.8
Other fuels	3	4	5	3	5	7	1	3	2.8	4.2
Buildings	**545**	**590**	**612**	**506**	**511**	**519**	**100**	**100**	**0.8**	**0.1**
Coal	18	15	15	15	10	9	2	2	-1.1	-3.0
Oil	64	55	51	56	39	33	8	6	-1.3	-3.1
Gas	188	206	212	170	169	169	35	33	0.8	-0.1
Electricity	178	204	218	166	176	177	36	34	1.2	0.4
Heat	41	46	48	36	35	35	8	7	1.3	-0.0
Bioenergy	47	50	52	51	62	68	8	13	1.1	2.2
Other renewables	9	13	16	11	20	27	3	5	4.3	6.5
Other	**148**	**147**	**146**	**140**	**139**	**139**	**100**	**100**	**-0.0**	**-0.2**

OECD Europe: New Policies Scenario

	Electricity generation (TWh)							Shares (%)		CAAGR (%)
	1990	2010	2015	2020	2025	2030	2035	2010	2035	2010-35
Total generation	**2 683**	**3 662**	**3 803**	**3 949**	**4 106**	**4 243**	**4 357**	**100**	**100**	**0.7**
Coal	1 040	908	922	829	711	597	461	25	11	-2.7
Oil	216	82	57	45	34	26	24	2	1	-4.8
Gas	168	869	805	874	997	1 062	1 119	24	26	1.0
Nuclear	787	916	881	850	820	825	816	25	19	-0.5
Hydro	446	556	564	590	611	628	643	15	15	0.6
Bioenergy	21	145	183	211	238	262	285	4	7	2.7
Wind	1	152	281	398	509	620	726	4	17	6.5
Geothermal	4	11	14	17	21	25	29	0	1	4.1
Solar PV	0	23	86	120	143	159	177	1	4	8.6
CSP	-	1	9	13	19	28	42	0	1	17.9
Marine	1	1	1	2	5	12	34	0	1	18.1

	Electrical capacity (GW)						Shares (%)		CAAGR (%)
	2010	2015	2020	2025	2030	2035	2010	2035	2010-35
Total capacity	**981**	**1 137**	**1 188**	**1 249**	**1 319**	**1 386**	**100**	**100**	**1.4**
Coal	205	216	183	157	141	120	21	9	-2.1
Oil	67	59	43	30	20	17	7	1	-5.3
Gas	231	273	284	315	346	374	24	27	1.9
Nuclear	138	130	125	119	119	117	14	8	-0.6
Hydro	194	205	213	221	226	232	20	17	0.7
Bioenergy	28	34	38	42	45	48	3	3	2.2
Wind	85	135	184	228	267	304	9	22	5.2
Geothermal	2	2	3	3	4	4	0	0	3.2
Solar PV	30	80	111	129	140	149	3	11	6.7
CSP	1	3	4	5	8	12	0	1	11.7
Marine	0	0	1	1	3	9	0	1	15.6

	CO_2 emissions (Mt)							Shares (%)		CAAGR (%)
	1990	2010	2015	2020	2025	2030	2035	2010	2035	2010-35
Total CO_2	**3 963**	**3 873**	**3 695**	**3 585**	**3 430**	**3 256**	**3 111**	**100**	**100**	**-0.9**
Coal	1 713	1 183	1 155	1 057	915	772	634	31	20	-2.5
Oil	1 673	1 614	1 502	1 423	1 347	1 273	1 219	42	39	-1.1
Gas	578	1 076	1 038	1 105	1 168	1 210	1 258	28	40	0.6
Power generation	**1 398**	**1 373**	**1 299**	**1 229**	**1 128**	**1 014**	**914**	**100**	**100**	**-1.6**
Coal	1 140	916	885	794	663	532	402	67	44	-3.2
Oil	164	72	54	42	31	23	21	5	2	-4.8
Gas	95	385	360	393	435	458	491	28	54	1.0
TFC	**2 387**	**2 309**	**2 215**	**2 182**	**2 134**	**2 081**	**2 041**	**100**	**100**	**-0.5**
Coal	534	234	236	228	217	205	197	10	10	-0.7
Oil	1 393	1 427	1 342	1 280	1 220	1 158	1 110	62	54	-1.0
Transport	*774*	*947*	*885*	*849*	*812*	*776*	*749*	*41*	*37*	*-0.9*
Gas	460	648	637	674	697	717	733	28	36	0.5

OECD Europe: Current Policies and 450 Scenarios

	Electricity generation (TWh)						Shares (%)		CAAGR (%)	
	2020	2030	2035	2020	2030	2035	2035		2010-35	
	Current Policies Scenario			450 Scenario			CPS	450	CPS	450
Total generation	**4 062**	**4 511**	**4 721**	**3 813**	**3 972**	**4 040**	**100**	**100**	**1.0**	**0.4**
Coal	938	904	915	637	272	180	19	4	0.0	-6.3
Oil	48	30	28	41	25	21	1	1	-4.2	-5.3
Gas	956	1 230	1 382	837	598	353	29	9	1.9	-3.5
Nuclear	819	752	654	863	968	1 038	14	26	-1.3	0.5
Hydro	583	614	625	607	676	703	13	17	0.5	0.9
Bioenergy	199	239	259	220	301	344	5	9	2.3	3.5
Wind	376	549	628	442	832	1 009	13	25	5.8	7.9
Geothermal	15	21	24	19	30	37	1	1	3.3	5.1
Solar PV	113	142	154	128	193	224	3	6	8.0	9.6
CSP	12	23	32	16	56	80	1	2	16.6	20.9
Marine	2	7	20	2	20	50	0	1	15.5	20.0

	Electrical capacity (GW)						Shares (%)		CAAGR (%)	
	2020	2030	2035	2020	2030	2035	2035		2010-35	
	Current Policies Scenario			450 Scenario			CPS	450	CPS	450
Total capacity	**1 182**	**1 319**	**1 384**	**1 223**	**1 416**	**1 505**	**100**	**100**	**1.4**	**1.7**
Coal	187	158	142	180	112	76	10	5	-1.5	-3.9
Oil	45	22	19	43	20	17	1	1	-4.8	-5.4
Gas	295	387	439	284	311	323	32	21	2.6	1.4
Nuclear	120	108	92	127	139	147	7	10	-1.6	0.3
Hydro	212	222	225	219	243	252	16	17	0.6	1.1
Bioenergy	36	42	45	39	51	57	3	4	1.8	2.9
Wind	177	246	275	204	347	404	20	27	4.8	6.4
Geothermal	2	3	3	3	4	5	0	0	2.5	4.2
Solar PV	104	125	130	119	169	187	9	12	6.1	7.6
CSP	3	6	9	5	16	23	1	1	10.4	14.6
Marine	0	2	5	1	6	14	0	1	13.2	17.5

	CO_2 emissions (Mt)						Shares (%)		CAAGR (%)	
	2020	2030	2035	2020	2030	2035	2035		2010-35	
	Current Policies Scenario			450 Scenario			CPS	450	CPS	450
Total CO_2	**3 785**	**3 756**	**3 796**	**3 244**	**2 332**	**1 975**	**100**	**100**	**-0.1**	**-2.7**
Coal	1 164	1 054	1 028	852	399	271	27	14	-0.6	-5.7
Oil	1 469	1 393	1 373	1 352	1 024	883	36	45	-0.6	-2.4
Gas	1 152	1 309	1 395	1 040	909	821	37	42	1.0	-1.1
Power generation	**1 356**	**1 344**	**1 388**	**1 014**	**477**	**290**	**100**	**100**	**0.0**	**-6.0**
Coal	891	796	776	610	211	108	56	37	-0.7	-8.2
Oil	45	27	24	39	22	19	2	7	-4.3	-5.2
Gas	420	521	587	365	244	162	42	56	1.7	-3.4
TFC	**2 253**	**2 244**	**2 244**	**2 064**	**1 715**	**1 556**	**100**	**100**	**-0.1**	**-1.6**
Coal	237	222	217	210	159	135	10	9	-0.3	-2.2
Oil	1 321	1 269	1 253	1 216	924	795	56	51	-0.5	-2.3
Transport	*873*	*860*	*865*	*808*	*593*	*498*	*39*	*32*	*-0.4*	*-2.5*
Gas	695	753	774	638	632	625	34	40	0.7	-0.1

European Union: New Policies Scenario

	Energy demand (Mtoe)							Shares (%)		CAAGR (%)
	1990	2010	2015	2020	2025	2030	2035	2010	2035	2010-35
TPED	**1 633**	**1 713**	**1 681**	**1 678**	**1 671**	**1 667**	**1 670**	**100**	**100**	**-0.1**
Coal	455	281	276	249	214	181	146	16	9	-2.6
Oil	601	569	524	492	465	438	417	33	25	-1.2
Gas	295	441	417	443	468	486	506	26	30	0.6
Nuclear	207	239	229	220	214	216	216	14	13	-0.4
Hydro	25	31	31	32	33	34	35	2	2	0.4
Bioenergy	46	130	158	180	197	214	231	8	14	2.3
Other renewables	3	22	45	62	80	98	119	1	7	6.9
Power generation	**644**	**726**	**717**	**715**	**713**	**716**	**721**	**100**	**100**	**-0.0**
Coal	286	212	204	179	148	119	86	29	12	-3.5
Oil	61	24	17	13	10	7	6	3	1	-5.2
Gas	54	148	134	147	164	174	189	20	26	1.0
Nuclear	207	239	229	220	214	216	216	33	30	-0.4
Hydro	25	31	31	32	33	34	35	4	5	0.4
Bioenergy	8	50	60	67	74	79	85	7	12	2.1
Other renewables	3	20	41	56	71	87	103	3	14	6.8
Other energy sector	**150**	**142**	**135**	**130**	**126**	**122**	**118**	**100**	**100**	**-0.7**
Electricity	*39*	*41*	*39*	*38*	*38*	*37*	*37*	*29*	*31*	*-0.5*
TFC	**1 124**	**1 194**	**1 187**	**1 201**	**1 211**	**1 218**	**1 227**	**100**	**100**	**0.1**
Coal	123	41	42	40	37	33	31	3	3	-1.1
Oil	499	501	467	442	421	399	380	42	31	-1.1
Gas	224	274	267	281	290	298	304	23	25	0.4
Electricity	185	244	252	262	272	281	288	20	23	0.7
Heat	54	53	56	58	60	61	62	4	5	0.6
Bioenergy	38	79	98	113	123	134	145	7	12	2.5
Other renewables	1	2	4	6	8	11	15	0	1	7.7
Industry	**341**	**273**	**281**	**287**	**289**	**290**	**290**	**100**	**100**	**0.2**
Coal	68	25	27	27	25	24	23	9	8	-0.4
Oil	57	34	32	30	29	27	24	13	8	-1.3
Gas	97	85	84	86	87	87	87	31	30	0.1
Electricity	85	89	94	97	98	98	99	33	34	0.4
Heat	19	16	16	16	16	16	16	6	5	0.0
Bioenergy	14	24	28	31	34	38	41	9	14	2.2
Other renewables	-	0	0	0	0	0	0	0	0	1.2
Transport	**257**	**319**	**303**	**296**	**287**	**277**	**270**	**100**	**100**	**-0.7**
Oil	251	297	275	260	246	231	218	93	81	-1.2
Electricity	5	6	7	7	8	9	10	2	4	2.2
Biofuels	0	13	19	26	30	34	38	4	14	4.3
Other fuels	1	2	3	3	4	4	4	1	2	2.4
Buildings	**394**	**470**	**477**	**495**	**512**	**527**	**544**	**100**	**100**	**0.6**
Coal	50	13	12	11	9	7	6	3	1	-3.1
Oil	89	64	59	54	50	45	42	14	8	-1.7
Gas	107	170	165	176	183	190	196	36	36	0.6
Electricity	90	145	148	154	163	170	176	31	32	0.8
Heat	33	37	40	42	44	45	47	8	9	0.9
Bioenergy	23	39	49	53	56	59	63	8	12	1.9
Other renewables	1	2	4	6	8	11	15	0	3	7.8
Other	**132**	**132**	**127**	**123**	**123**	**123**	**122**	**100**	**100**	**-0.3**

European Union: Current Policies and 450 Scenarios

	Energy demand (Mtoe)						Shares (%)		CAAGR (%)	
	2020	2030	2035	2020	2030	2035	2035		2010-35	
	Current Policies Scenario			450 Scenario			CPS	450	CPS	450
TPED	**1 716**	**1 756**	**1 775**	**1 606**	**1 530**	**1 518**	**100**	**100**	**0.1**	**-0.5**
Coal	269	239	227	204	120	98	13	6	-0.9	-4.1
Oil	512	483	473	470	362	317	27	21	-0.7	-2.3
Gas	464	528	565	419	380	349	32	23	1.0	-0.9
Nuclear	214	196	173	224	253	273	10	18	-1.3	0.5
Hydro	32	33	34	33	36	37	2	2	0.3	0.7
Bioenergy	167	193	206	186	250	282	12	19	1.9	3.2
Other renewables	58	83	97	70	129	162	5	11	6.0	8.2
Power generation	**732**	**759**	**772**	**681**	**671**	**681**	**100**	**100**	**0.2**	**-0.3**
Coal	198	173	163	140	68	51	21	7	-1.0	-5.5
Oil	14	8	7	12	7	6	1	1	-4.7	-5.4
Gas	158	201	230	139	102	72	30	11	1.8	-2.8
Nuclear	214	196	173	224	253	273	22	40	-1.3	0.5
Hydro	32	33	34	33	36	37	4	5	0.3	0.7
Bioenergy	63	73	77	70	90	102	10	15	1.7	2.9
Other renewables	52	75	87	63	114	140	11	21	6.0	8.1
Other energy sector	**133**	**129**	**127**	**123**	**106**	**98**	**100**	**100**	**-0.4**	**-1.4**
Electricity	*39*	*41*	*41*	*36*	*34*	*33*	*32*	*33*	*-0.0*	*-0.9*
TFC	**1 230**	**1 281**	**1 305**	**1 155**	**1 114**	**1 102**	**100**	**100**	**0.4**	**-0.3**
Coal	41	36	34	37	28	25	3	2	-0.7	-1.9
Oil	461	441	433	422	329	288	33	26	-0.6	-2.2
Gas	290	313	321	266	265	265	25	24	0.6	-0.1
Electricity	269	299	312	254	267	272	24	25	1.0	0.4
Heat	60	65	67	54	52	51	5	5	0.9	-0.2
Bioenergy	104	120	129	115	159	180	10	16	2.0	3.3
Other renewables	5	9	11	7	15	22	1	2	6.2	9.2
Industry	**296**	**308**	**312**	**279**	**278**	**276**	**100**	**100**	**0.5**	**0.0**
Coal	27	25	24	25	21	19	8	7	-0.2	-1.2
Oil	31	28	27	30	25	23	9	8	-1.0	-1.6
Gas	89	94	95	84	83	82	31	30	0.5	-0.1
Electricity	100	106	107	95	95	95	34	34	0.8	0.3
Heat	16	16	16	15	14	14	5	5	0.2	-0.5
Bioenergy	32	38	42	32	39	44	13	16	2.2	2.4
Other renewables	0	0	0	0	0	0	0	0	1.2	2.3
Transport	**300**	**299**	**302**	**285**	**240**	**224**	**100**	**100**	**-0.2**	**-1.4**
Oil	269	261	259	248	177	147	86	65	-0.6	-2.8
Electricity	7	8	9	8	12	17	3	7	1.9	4.3
Biofuels	21	26	29	26	47	55	10	24	3.2	5.8
Other fuels	3	4	4	3	5	6	1	3	2.1	3.6
Buildings	**504**	**544**	**564**	**468**	**474**	**481**	**100**	**100**	**0.7**	**0.1**
Coal	11	8	7	10	5	4	1	1	-2.4	-4.5
Oil	56	49	46	50	35	29	8	6	-1.3	-3.0
Gas	182	198	205	164	162	162	36	34	0.7	-0.2
Electricity	158	180	192	148	157	158	34	33	1.1	0.3
Heat	44	48	50	38	38	37	9	8	1.2	-0.0
Bioenergy	48	52	54	52	63	69	10	14	1.3	2.3
Other renewables	5	9	11	7	14	21	2	4	6.3	9.3
Other	**130**	**129**	**128**	**122**	**122**	**122**	**100**	**100**	**-0.1**	**-0.3**

European Union: New Policies Scenario

	Electricity generation (TWh)							Shares (%)		CAAGR (%)
	1990	2010	2015	2020	2025	2030	2035	2010	2035	2010-35
Total generation	**2 568**	**3 310**	**3 388**	**3 484**	**3 603**	**3 702**	**3 778**	**100**	**100**	**0.5**
Coal	1 050	862	870	760	628	498	341	26	9	-3.6
Oil	221	86	58	43	31	24	21	3	1	-5.4
Gas	191	758	660	723	838	903	960	23	25	0.9
Nuclear	795	917	878	845	821	828	830	28	22	-0.4
Hydro	286	366	362	374	385	394	404	11	11	0.4
Bioenergy	20	142	180	205	230	251	272	4	7	2.6
Wind	1	149	277	388	492	590	681	5	18	6.3
Geothermal	3	6	8	11	14	18	21	0	1	5.3
Solar PV	0	22	85	119	141	156	173	1	5	8.5
CSP	-	1	9	13	19	28	42	0	1	17.9
Marine	1	1	1	2	5	12	34	0	1	18.1

	Electrical capacity (GW)						Shares (%)		CAAGR (%)
	2010	2015	2020	2025	2030	2035	2010	2035	2010-35
Total capacity	**910**	**1 047**	**1 085**	**1 138**	**1 196**	**1 251**	**100**	**100**	**1.3**
Coal	202	210	174	147	129	104	22	8	-2.6
Oil	65	57	40	27	17	15	7	1	-5.8
Gas	216	249	257	285	314	340	24	27	1.8
Nuclear	138	130	125	120	120	120	15	10	-0.6
Hydro	145	151	156	160	164	169	16	13	0.6
Bioenergy	28	33	37	41	43	46	3	4	2.0
Wind	85	134	182	222	257	288	9	23	5.0
Geothermal	1	1	2	2	2	3	0	0	4.5
Solar PV	30	80	110	128	138	146	3	12	6.6
CSP	1	3	4	5	8	12	0	1	11.7
Marine	0	0	1	1	3	9	0	1	15.6

	CO_2 emissions (Mt)							Shares (%)		CAAGR (%)
	1990	2010	2015	2020	2025	2030	2035	2010	2035	2010-35
Total CO_2	**4 033**	**3 609**	**3 405**	**3 259**	**3 079**	**2 878**	**2 717**	**100**	**100**	**-1.1**
Coal	1 734	1 089	1 064	947	791	629	486	30	18	-3.2
Oil	1 642	1 508	1 384	1 297	1 217	1 137	1 074	42	40	-1.3
Gas	658	1 012	957	1 016	1 072	1 111	1 157	28	43	0.5
Power generation	**1 492**	**1 304**	**1 214**	**1 122**	**1 003**	**871**	**767**	**100**	**100**	**-2.1**
Coal	1 170	880	845	736	593	445	310	68	40	-4.1
Oil	195	76	55	42	30	23	20	6	3	-5.2
Gas	127	347	314	343	381	403	436	27	57	0.9
TFC	**2 370**	**2 132**	**2 031**	**1 986**	**1 931**	**1 870**	**1 819**	**100**	**100**	**-0.6**
Coal	526	181	190	181	168	155	146	8	8	-0.8
Oil	1 331	1 321	1 227	1 159	1 096	1 029	973	62	53	-1.2
Transport	*743*	*892*	*824*	*779*	*736*	*692*	*654*	*42*	*36*	*-1.2*
Gas	513	630	614	646	667	686	700	30	38	0.4

European Union: Current Policies and 450 Scenarios

	Electricity generation (TWh)						Shares (%)		CAAGR (%)	
	2020	2030	2035	2020	2030	2035	2035		2010-35	
	Current Policies Scenario			450 Scenario			CPS	450	CPS	450
Total generation	**3 588**	**3 944**	**4 106**	**3 373**	**3 500**	**3 545**	**100**	**100**	**0.9**	**0.3**
Coal	847	764	741	584	254	170	18	5	-0.6	-6.3
Oil	46	27	25	39	23	20	1	1	-4.8	-5.7
Gas	805	1 076	1 228	708	513	288	30	8	1.9	-3.8
Nuclear	819	752	664	861	972	1 045	16	29	-1.3	0.5
Hydro	375	389	394	383	418	433	10	12	0.3	0.7
Bioenergy	194	229	246	215	285	323	6	9	2.2	3.3
Wind	367	522	587	426	752	896	14	25	5.6	7.4
Geothermal	10	14	16	13	23	27	0	1	4.4	6.6
Solar PV	112	140	152	127	187	217	4	6	8.0	9.5
CSP	12	23	32	16	53	75	1	2	16.6	20.6
Marine	2	7	20	2	20	50	0	1	15.6	20.0

	Electrical capacity (GW)						Shares (%)		CAAGR (%)	
	2020	2030	2035	2020	2030	2035	2035		2010-35	
	Current Policies Scenario			450 Scenario			CPS	450	CPS	450
Total capacity	**1 083**	**1 197**	**1 250**	**1 116**	**1 276**	**1 349**	**100**	**100**	**1.3**	**1.6**
Coal	177	141	120	171	111	75	10	6	-2.0	-3.9
Oil	42	19	17	40	17	14	1	1	-5.3	-5.9
Gas	269	355	406	257	279	291	32	22	2.6	1.2
Nuclear	121	109	94	127	140	149	8	11	-1.5	0.3
Hydro	156	162	164	160	174	181	13	13	0.5	0.9
Bioenergy	35	40	42	38	49	54	3	4	1.7	2.7
Wind	174	237	261	198	318	364	21	27	4.6	6.0
Geothermal	1	2	2	2	3	4	0	0	3.6	5.7
Solar PV	104	124	129	117	165	181	10	13	6.1	7.5
CSP	3	6	9	4	15	21	1	2	10.4	14.4
Marine	0	2	6	1	6	14	0	1	13.2	17.5

	CO_2 emissions (Mt)						Shares (%)		CAAGR (%)	
	2020	2030	2035	2020	2030	2035	2035		2010-35	
	Current Policies Scenario			450 Scenario			CPS	450	CPS	450
Total CO_2	**3 438**	**3 340**	**3 341**	**2 955**	**2 105**	**1 781**	**100**	**100**	**-0.3**	**-2.8**
Coal	1 030	868	814	761	343	235	24	13	-1.2	-5.9
Oil	1 344	1 259	1 230	1 234	918	785	37	44	-0.8	-2.6
Gas	1 064	1 213	1 297	960	843	761	39	43	1.0	-1.1
Power generation	**1 228**	**1 168**	**1 186**	**931**	**447**	**285**	**100**	**100**	**-0.4**	**-5.9**
Coal	813	674	628	568	203	118	53	42	-1.3	-7.7
Oil	45	26	23	39	22	19	2	7	-4.6	-5.3
Gas	371	469	535	324	221	148	45	52	1.7	-3.4
TFC	**2 055**	**2 027**	**2 014**	**1 879**	**1 540**	**1 388**	**100**	**100**	**-0.2**	**-1.7**
Coal	187	165	157	166	116	95	8	7	-0.6	-2.5
Oil	1 201	1 142	1 118	1 103	823	701	56	51	-0.7	-2.5
Transport	*806*	*782*	*776*	*743*	*531*	*440*	*39*	*32*	*-0.6*	*-2.8*
Gas	667	720	739	610	601	592	37	43	0.6	-0.3

OECD Asia Oceania: New Policies Scenario

	Energy demand (Mtoe)							Shares (%)		CAAGR (%)
	1990	2010	2015	2020	2025	2030	2035	2010	2035	2010-35
TPED	**631**	**890**	**897**	**909**	**916**	**923**	**927**	**100**	**100**	**0.2**
Coal	138	241	253	243	233	218	194	27	21	-0.9
Oil	335	344	339	313	297	281	268	39	29	-1.0
Gas	66	155	174	176	184	193	202	17	22	1.1
Nuclear	66	114	83	117	120	127	142	13	15	0.9
Hydro	11	11	11	12	13	13	14	1	1	1.0
Bioenergy	10	17	21	25	29	33	37	2	4	3.2
Other renewables	4	8	16	24	40	58	70	1	8	8.9
Power generation	**241**	**405**	**400**	**414**	**426**	**438**	**446**	**100**	**100**	**0.4**
Coal	60	159	164	153	143	128	105	39	24	-1.7
Oil	56	24	26	12	10	8	7	6	2	-5.0
Gas	40	82	94	88	89	93	96	20	22	0.6
Nuclear	66	114	83	117	120	127	142	28	32	0.9
Hydro	11	11	11	12	13	13	14	3	3	1.0
Bioenergy	3	7	9	11	14	16	18	2	4	3.8
Other renewables	3	7	14	21	37	53	64	2	14	9.2
Other energy sector	**57**	**83**	**89**	**90**	**90**	**90**	**89**	**100**	**100**	**0.3**
Electricity	*11*	*17*	*18*	*19*	*19*	*19*	*19*	*21*	*21*	*0.4*
TFC	**431**	**570**	**583**	**588**	**589**	**590**	**592**	**100**	**100**	**0.2**
Coal	49	42	44	43	42	41	40	7	7	-0.2
Oil	261	297	293	285	275	265	256	52	43	-0.6
Gas	26	70	76	80	83	86	89	12	15	1.0
Electricity	86	145	152	159	165	171	175	26	30	0.7
Heat	0	5	5	5	5	5	6	1	1	0.5
Bioenergy	7	10	12	13	15	17	19	2	3	2.7
Other renewables	2	1	2	2	3	5	6	0	1	7.0
Industry	**145**	**161**	**172**	**176**	**179**	**182**	**183**	**100**	**100**	**0.5**
Coal	39	39	41	40	40	39	38	25	21	-0.1
Oil	51	33	33	32	30	28	27	20	15	-0.8
Gas	11	23	27	29	30	32	33	14	18	1.4
Electricity	40	56	60	63	65	67	69	35	38	0.8
Heat	-	2	2	2	2	2	2	1	1	-0.0
Bioenergy	5	7	8	10	11	12	14	4	8	2.7
Other renewables	0	0	0	0	0	0	0	0	0	1.3
Transport	**110**	**140**	**137**	**131**	**124**	**118**	**113**	**100**	**100**	**-0.9**
Oil	109	136	132	125	117	110	105	97	92	-1.0
Electricity	2	2	2	3	3	3	4	2	3	2.2
Biofuels	-	1	1	1	1	1	1	0	1	0.7
Other fuels	0	2	2	2	3	3	4	1	4	3.9
Buildings	**120**	**175**	**180**	**186**	**191**	**196**	**201**	**100**	**100**	**0.6**
Coal	10	1	2	1	1	1	1	1	0	-1.4
Oil	47	39	38	38	37	37	36	22	18	-0.3
Gas	15	43	45	47	48	49	50	25	25	0.6
Electricity	44	86	89	92	96	99	101	49	50	0.7
Heat	0	3	3	3	3	3	3	1	2	1.0
Bioenergy	2	2	3	3	4	4	4	1	2	2.9
Other renewables	1	1	1	2	2	3	5	1	2	7.0
Other	**56**	**94**	**94**	**95**	**95**	**95**	**95**	**100**	**100**	**0.0**

OECD Asia Oceania: Current Policies and 450 Scenarios

	Energy demand (Mtoe)						Shares (%)		CAAGR (%)	
	2020	2030	2035	2020	2030	2035	2035		2010-35	
	Current Policies Scenario			450 Scenario			CPS	450	CPS	450
TPED	**929**	**964**	**978**	**882**	**856**	**852**	**100**	**100**	**0.4**	**-0.2**
Coal	258	253	240	221	138	115	25	13	-0.0	-2.9
Oil	316	291	281	300	241	212	29	25	-0.8	-1.9
Gas	179	203	222	165	178	179	23	21	1.4	0.6
Nuclear	120	140	149	127	159	174	15	20	1.1	1.7
Hydro	12	12	12	13	16	17	1	2	0.6	1.9
Bioenergy	23	29	32	29	48	60	3	7	2.5	5.2
Other renewables	22	36	42	27	75	96	4	11	6.7	10.3
Power generation	**429**	**464**	**478**	**398**	**402**	**411**	**100**	**100**	**0.7**	**0.1**
Coal	167	158	144	132	50	28	30	7	-0.4	-6.8
Oil	12	9	9	9	3	3	2	1	-3.9	-7.8
Gas	89	99	110	81	89	86	23	21	1.2	0.2
Nuclear	120	140	149	127	159	174	31	42	1.1	1.7
Hydro	12	12	12	13	16	17	3	4	0.6	1.9
Bioenergy	10	14	16	12	18	21	3	5	3.2	4.5
Other renewables	20	33	38	24	67	82	8	20	7.0	10.3
Other energy sector	**92**	**94**	**94**	**88**	**83**	**81**	**100**	**100**	**0.5**	**-0.1**
Electricity	*19*	*21*	*21*	*17*	*16*	*16*	*23*	*19*	*0.8*	*-0.4*
TFC	**597**	**615**	**623**	**571**	**550**	**542**	**100**	**100**	**0.4**	**-0.2**
Coal	44	44	43	42	40	39	7	7	0.1	-0.3
Oil	288	275	269	275	231	206	43	38	-0.4	-1.4
Gas	82	90	95	76	78	79	15	15	1.2	0.5
Electricity	164	182	190	153	158	160	31	30	1.1	0.4
Heat	5	6	6	5	5	5	1	1	0.6	0.3
Bioenergy	12	15	16	17	30	39	3	7	1.9	5.6
Other renewables	2	3	4	3	8	14	1	3	4.9	10.4
Industry	**180**	**189**	**192**	**174**	**179**	**181**	**100**	**100**	**0.7**	**0.5**
Coal	42	42	41	40	38	37	21	20	0.2	-0.3
Oil	33	31	29	32	29	27	15	15	-0.5	-0.7
Gas	29	33	35	28	31	32	18	18	1.7	1.3
Electricity	64	70	73	62	65	66	38	37	1.0	0.7
Heat	2	2	2	2	2	2	1	1	-0.1	-0.1
Bioenergy	9	11	11	10	13	15	6	8	1.9	3.1
Other renewables	0	0	0	0	0	0	0	0	1.3	1.3
Transport	**132**	**123**	**120**	**127**	**103**	**92**	**100**	**100**	**-0.6**	**-1.7**
Oil	126	116	112	121	88	70	93	76	-0.8	-2.6
Electricity	3	3	3	3	6	8	3	9	1.9	5.5
Biofuels	1	1	1	1	6	9	1	10	0.3	11.7
Other fuels	2	3	4	2	4	4	3	5	3.9	4.2
Buildings	**191**	**207**	**215**	**175**	**172**	**174**	**100**	**100**	**0.8**	**-0.0**
Coal	2	1	1	1	1	1	1	1	-0.6	-1.8
Oil	39	38	38	35	30	28	18	16	-0.1	-1.4
Gas	48	52	54	43	41	40	25	23	0.9	-0.3
Electricity	96	107	112	87	86	84	52	48	1.1	-0.1
Heat	3	3	3	3	3	3	2	2	1.1	0.7
Bioenergy	3	4	4	4	5	6	2	3	2.3	4.0
Other renewables	1	2	3	3	7	12	1	7	4.5	11.0
Other	**95**	**96**	**95**	**95**	**96**	**95**	**100**	**100**	**0.1**	**0.1**

OECD Asia Oceania: New Policies Scenario

	Electricity generation (TWh)							Shares (%)		CAAGR (%)
	1990	2010	2015	2020	2025	2030	2035	2010	2035	2010-35
Total generation	**1 127**	**1 893**	**1 981**	**2 070**	**2 145**	**2 210**	**2 261**	**100**	**100**	**0.7**
Coal	257	707	739	712	677	621	541	37	24	-1.1
Oil	270	120	135	61	50	41	33	6	1	-5.0
Gas	197	454	539	534	560	583	594	24	26	1.1
Nuclear	255	437	317	448	462	487	546	23	24	0.9
Hydro	133	123	133	140	147	153	160	7	7	1.0
Bioenergy	12	29	38	47	56	65	74	2	3	3.8
Wind	-	11	32	55	85	118	144	1	6	10.7
Geothermal	4	9	12	18	33	48	57	0	3	7.9
Solar PV	0	5	33	51	66	78	89	0	4	12.3
CSP	-	0	1	3	4	7	10	0	0	36.9
Marine	-	-	1	3	5	8	13	-	1	n.a.

	Electrical capacity (GW)						Shares (%)		CAAGR (%)
	2010	2015	2020	2025	2030	2035	2010	2035	2010-35
Total capacity	**430**	**485**	**521**	**553**	**586**	**610**	**100**	**100**	**1.4**
Coal	105	112	115	115	114	108	24	18	0.1
Oil	57	51	33	28	25	25	13	4	-3.2
Gas	117	136	163	173	180	182	27	30	1.8
Nuclear	68	69	66	62	63	71	16	12	0.2
Hydro	68	71	73	76	78	80	16	13	0.6
Bioenergy	5	7	8	9	11	12	1	2	3.8
Wind	5	11	19	30	41	49	1	8	9.5
Geothermal	1	2	2	5	7	9	0	1	7.9
Solar PV	5	26	40	53	62	70	1	12	11.3
CSP	0	0	1	1	2	2	0	0	29.2
Marine	-	0	1	1	2	3	-	0	n.a.

	CO_2 emissions (Mt)							Shares (%)		CAAGR (%)
	1990	2010	2015	2020	2025	2030	2035	2010	2035	2010-35
Total CO_2	**1 574**	**2 105**	**2 169**	**2 050**	**1 973**	**1 877**	**1 756**	**100**	**100**	**-0.7**
Coal	521	906	933	884	830	751	642	43	37	-1.4
Oil	892	819	809	738	699	663	632	39	36	-1.0
Gas	161	380	426	428	444	463	482	18	27	1.0
Power generation	**548**	**947**	**994**	**891**	**838**	**767**	**666**	**100**	**100**	**-1.4**
Coal	280	678	691	644	592	518	416	72	62	-1.9
Oil	174	72	78	36	30	25	20	8	3	-4.9
Gas	94	198	225	210	215	223	230	21	35	0.6
TFC	**958**	**1 044**	**1 054**	**1 034**	**1 008**	**983**	**960**	**100**	**100**	**-0.3**
Coal	221	186	194	191	189	185	180	18	19	-0.1
Oil	676	697	684	657	627	598	574	67	60	-0.8
Transport	*321*	*399*	*388*	*368*	*345*	*324*	*308*	*38*	*32*	*-1.0*
Gas	61	161	176	186	193	200	207	15	22	1.0

OECD Asia Oceania: Current Policies and 450 Scenarios

	Electricity generation (TWh)						Shares (%)		CAAGR (%)	
	2020	2030	2035	2020	2030	2035	2035		2010-35	
	Current Policies Scenario			450 Scenario			CPS	450	CPS	450
Total generation	**2 134**	**2 363**	**2 462**	**1 979**	**2 029**	**2 043**	**100**	**100**	**1.1**	**0.3**
Coal	770	766	732	606	238	121	30	6	0.1	-6.8
Oil	61	43	43	48	16	15	2	1	-4.0	-8.0
Gas	548	631	690	490	562	533	28	26	1.7	0.6
Nuclear	460	536	571	487	611	666	23	33	1.1	1.7
Hydro	135	140	143	151	184	199	6	10	0.6	1.9
Bioenergy	41	56	63	48	73	87	3	4	3.2	4.5
Wind	51	90	106	69	176	212	4	10	9.4	12.5
Geothermal	16	28	33	19	57	70	1	3	5.6	8.8
Solar PV	48	66	70	56	94	109	3	5	11.3	13.3
CSP	2	4	6	3	9	16	0	1	33.9	39.4
Marine	2	3	5	3	9	15	0	1	n.a.	n.a.

	Electrical capacity (GW)						Shares (%)		CAAGR (%)	
	2020	2030	2035	2020	2030	2035	2035		2010-35	
	Current Policies Scenario			450 Scenario			CPS	450	CPS	450
Total capacity	**528**	**591**	**615**	**512**	**564**	**591**	**100**	**100**	**1.4**	**1.3**
Coal	117	121	117	106	61	48	19	8	0.5	-3.0
Oil	33	25	25	29	14	11	4	2	-3.2	-6.3
Gas	165	199	211	149	165	167	34	28	2.4	1.5
Nuclear	76	75	79	71	81	87	13	15	0.7	1.0
Hydro	71	73	74	76	86	90	12	15	0.3	1.1
Bioenergy	7	9	10	8	12	14	2	2	3.1	4.5
Wind	18	30	35	25	58	69	6	12	8.1	11.0
Geothermal	2	4	5	3	9	11	1	2	5.3	8.8
Solar PV	38	52	55	44	74	86	9	15	10.3	12.2
CSP	0	1	1	1	2	4	0	1	26.4	31.6
Marine	0	1	1	1	2	4	0	1	n.a.	n.a.

	CO_2 emissions (Mt)						Shares (%)		CAAGR (%)	
	2020	2030	2035	2020	2030	2035	2035		2010-35	
	Current Policies Scenario			450 Scenario			CPS	450	CPS	450
Total CO_2	**2 131**	**2 079**	**2 036**	**1 887**	**1 320**	**1 091**	**100**	**100**	**-0.1**	**-2.6**
Coal	948	899	832	781	363	218	41	20	-0.3	-5.5
Oil	747	694	677	705	553	480	33	44	-0.8	-2.1
Gas	435	487	527	401	404	393	26	36	1.3	0.1
Power generation	**951**	**917**	**881**	**769**	**357**	**218**	**100**	**100**	**-0.3**	**-5.7**
Coal	702	654	591	545	154	31	67	14	-0.5	-11.6
Oil	36	26	27	29	11	10	3	5	-3.8	-7.6
Gas	213	237	263	195	193	178	30	81	1.2	-0.4
TFC	**1 052**	**1 032**	**1 024**	**996**	**845**	**755**	**100**	**100**	**-0.1**	**-1.3**
Coal	197	196	193	187	162	141	19	19	0.1	-1.1
Oil	666	627	611	633	506	437	60	58	-0.5	-1.8
Transport	*370*	*341*	*330*	*355*	*259*	*205*	*32*	*27*	*-0.8*	*-2.6*
Gas	189	210	220	175	176	176	22	23	1.3	0.4

Japan: New Policies Scenario

	Energy demand (Mtoe)							Shares (%)		CAAGR (%)
	1990	2010	2015	2020	2025	2030	2035	2010	2035	2010-35
TPED	**439**	**497**	**472**	**465**	**456**	**450**	**447**	**100**	**100**	**-0.4**
Coal	77	115	114	103	99	96	92	23	21	-0.9
Oil	250	203	201	177	165	154	145	41	32	-1.3
Gas	44	86	99	96	98	101	102	17	23	0.7
Nuclear	53	75	34	59	52	45	45	15	10	-2.0
Hydro	8	7	7	8	8	9	9	1	2	1.1
Bioenergy	5	7	9	11	12	14	15	1	3	3.1
Other renewables	3	4	7	11	22	32	38	1	8	9.9
Power generation	**174**	**224**	**205**	**206**	**207**	**211**	**214**	**100**	**100**	**-0.2**
Coal	25	62	62	53	51	49	47	28	22	-1.1
Oil	51	18	22	9	7	6	4	8	2	-5.4
Gas	33	54	66	60	60	62	62	24	29	0.5
Nuclear	53	75	34	59	52	45	45	34	21	-2.0
Hydro	8	7	7	8	8	9	9	3	4	1.1
Bioenergy	2	4	6	8	9	10	11	2	5	3.8
Other renewables	1	3	7	10	20	30	35	1	16	10.4
Other energy sector	**38**	**45**	**41**	**38**	**35**	**33**	**30**	**100**	**100**	**-1.6**
Electricity	*7*	*9*	*9*	*9*	*9*	*9*	*9*	*21*	*30*	*-0.1*
TFC	**300**	**325**	**323**	**318**	**313**	**308**	**306**	**100**	**100**	**-0.2**
Coal	32	29	29	28	28	28	27	9	9	-0.2
Oil	184	171	167	158	149	141	135	53	44	-1.0
Gas	15	34	36	39	40	42	43	11	14	0.9
Electricity	64	86	86	88	90	91	93	27	30	0.3
Heat	0	1	1	1	1	1	1	0	0	1.6
Bioenergy	3	3	3	3	4	4	4	1	1	1.7
Other renewables	1	1	1	1	1	2	3	0	1	6.5
Industry	**103**	**90**	**92**	**92**	**92**	**91**	**91**	**100**	**100**	**0.0**
Coal	31	28	28	27	27	27	27	31	29	-0.2
Oil	37	23	23	22	20	19	18	26	20	-1.0
Gas	4	8	9	11	11	12	13	9	14	2.0
Electricity	29	29	29	29	29	30	30	32	33	0.1
Heat	-	-	-	-	-	-	-	-	-	n.a.
Bioenergy	3	3	3	3	4	4	4	3	5	1.7
Other renewables	-	-	-	-	-	-	-	-	-	n.a.
Transport	**72**	**77**	**73**	**67**	**61**	**55**	**51**	**100**	**100**	**-1.6**
Oil	70	75	71	65	59	52	49	98	95	-1.7
Electricity	1	2	2	2	2	2	3	2	5	1.9
Biofuels	-	-	-	-	-	-	-	-	-	n.a.
Other fuels	-	-	0	0	0	0	0	-	0	n.a.
Buildings	**84**	**114**	**114**	**116**	**119**	**121**	**124**	**100**	**100**	**0.3**
Coal	1	1	1	1	1	1	1	0	0	0.6
Oil	36	31	30	30	29	29	30	27	24	-0.1
Gas	11	26	27	28	28	29	30	23	24	0.5
Electricity	34	56	55	57	58	59	60	49	49	0.3
Heat	0	1	1	1	1	1	1	1	1	1.6
Bioenergy	0	0	0	0	0	0	0	0	0	6.5
Other renewables	1	1	1	1	1	2	3	0	2	6.8
Other	**41**	**43**	**43**	**42**	**42**	**41**	**40**	**100**	**100**	**-0.3**

Japan: Current Policies and 450 Scenarios

	Energy demand (Mtoe)						Shares (%)		CAAGR (%)	
	2020	2030	2035	2020	2030	2035	2035		2010-35	
	Current Policies Scenario			450 Scenario			CPS	450	CPS	450
TPED	**478**	**474**	**476**	**450**	**418**	**410**	**100**	**100**	**-0.2**	**-0.8**
Coal	112	109	107	92	56	48	22	12	-0.3	-3.5
Oil	179	160	154	169	132	118	32	29	-1.1	-2.1
Gas	96	104	110	89	88	85	23	21	1.0	-0.0
Nuclear	62	66	66	67	70	70	14	17	-0.5	-0.3
Hydro	8	8	8	8	10	11	2	3	0.6	1.8
Bioenergy	10	12	14	12	18	21	3	5	2.6	4.4
Other renewables	10	15	17	13	43	57	4	14	6.6	11.8
Power generation	**216**	**226**	**234**	**198**	**194**	**195**	**100**	**100**	**0.2**	**-0.5**
Coal	61	60	59	43	10	4	25	2	-0.2	-10.7
Oil	9	6	7	6	1	1	3	1	-3.9	-10.9
Gas	60	63	68	56	55	51	29	26	0.9	-0.3
Nuclear	62	66	66	67	70	70	28	36	-0.5	-0.3
Hydro	8	8	8	8	10	11	4	6	0.6	1.8
Bioenergy	6	9	10	7	10	11	4	6	3.3	3.8
Other renewables	10	14	16	11	38	48	7	24	7.1	11.8
Other energy sector	**39**	**34**	**32**	**37**	**30**	**27**	**100**	**100**	**-1.3**	**-2.0**
Electricity	*10*	*10*	*10*	*8*	*7*	*7*	*32*	*25*	*0.4*	*-1.2*
TFC	**324**	**323**	**324**	**308**	**285**	**278**	**100**	**100**	**-0.0**	**-0.6**
Coal	29	30	29	28	26	26	9	9	0.1	-0.4
Oil	160	147	142	153	125	113	44	40	-0.8	-1.7
Gas	39	43	45	36	36	37	14	13	1.1	0.2
Electricity	91	98	102	84	84	83	31	30	0.7	-0.2
Heat	1	1	1	1	1	1	0	0	1.8	1.2
Bioenergy	3	3	4	5	8	10	1	4	1.1	5.2
Other renewables	1	1	1	2	5	10	0	3	3.2	11.7
Industry	**94**	**95**	**96**	**91**	**90**	**90**	**100**	**100**	**0.2**	**-0.0**
Coal	28	29	29	27	26	25	30	28	0.1	-0.4
Oil	22	20	19	22	20	19	20	21	-0.7	-0.8
Gas	10	12	13	10	12	13	14	14	2.1	2.0
Electricity	30	31	32	29	28	28	33	32	0.4	-0.0
Heat	-	-	-	-	-	-	-	-	n.a.	n.a.
Bioenergy	3	3	4	3	4	5	4	5	1.1	2.2
Other renewables	-	-	-	-	-	-	-	-	n.a.	n.a.
Transport	**68**	**59**	**56**	**65**	**48**	**42**	**100**	**100**	**-1.3**	**-2.4**
Oil	66	56	53	63	44	35	95	84	-1.4	-3.0
Electricity	2	2	2	2	4	6	4	13	1.5	4.9
Biofuels	-	-	-	-	1	1	-	2	n.a.	n.a.
Other fuels	0	0	0	0	0	0	0	1	n.a.	n.a.
Buildings	**120**	**128**	**133**	**109**	**105**	**106**	**100**	**100**	**0.6**	**-0.3**
Coal	1	1	1	0	1	1	0	0	0.7	0.0
Oil	30	31	31	28	24	23	23	21	0.0	-1.2
Gas	28	30	31	25	24	23	24	22	0.7	-0.5
Electricity	59	65	68	54	51	49	51	46	0.8	-0.5
Heat	1	1	1	1	1	1	1	1	1.8	1.2
Bioenergy	0	0	0	0	0	1	0	1	-0.8	14.4
Other renewables	1	1	1	2	5	9	1	9	3.4	12.3
Other	**42**	**40**	**39**	**42**	**41**	**40**	**100**	**100**	**-0.4**	**-0.3**

Japan: New Policies Scenario

	Electricity generation (TWh)							Shares (%)		CAAGR (%)
	1990	2010	2015	2020	2025	2030	2035	2010	2035	2010-35
Total generation	**836**	**1 111**	**1 111**	**1 131**	**1 151**	**1 171**	**1 186**	**100**	**100**	**0.3**
Coal	117	304	310	279	271	262	255	27	22	-0.7
Oil	248	97	119	48	39	31	24	9	2	-5.5
Gas	167	305	389	379	395	411	408	27	34	1.2
Nuclear	202	288	132	227	198	174	174	26	15	-2.0
Hydro	89	82	87	91	97	102	108	7	9	1.1
Bioenergy	11	23	30	36	42	47	52	2	4	3.2
Wind	-	4	11	22	38	56	67	0	6	12.0
Geothermal	2	3	4	5	14	23	27	0	2	9.7
Solar PV	0	4	29	44	56	63	68	0	6	12.2
CSP	-	-	-	-	-	-	-	-	-	n.a.
Marine	-	-	-	-	0	1	4	-	0	n.a.

	Electrical capacity (GW)						Shares (%)		CAAGR (%)
	2010	2015	2020	2025	2030	2035	2010	2035	2010-35
Total capacity	**275**	**300**	**320**	**337**	**351**	**358**	**100**	**100**	**1.1**
Coal	46	48	49	50	50	48	17	13	0.1
Oil	50	44	27	24	21	21	18	6	-3.4
Gas	72	81	104	113	118	116	26	32	1.9
Nuclear	49	45	38	29	24	24	18	7	-2.8
Hydro	48	48	50	52	54	55	17	15	0.6
Bioenergy	4	5	6	7	8	8	1	2	3.2
Wind	2	5	9	16	22	25	1	7	10.1
Geothermal	1	1	1	2	4	5	0	1	9.0
Solar PV	4	23	35	45	51	54	1	15	11.5
CSP	-	-	-	-	-	-	-	-	n.a.
Marine	-	-	-	0	0	1	-	0	n.a.

	CO_2 emissions (Mt)							Shares (%)		CAAGR (%)
	1990	2010	2015	2020	2025	2030	2035	2010	2035	2010-35
Total CO_2	**1 063**	**1 138**	**1 166**	**1 049**	**1 006**	**972**	**943**	**100**	**100**	**-0.8**
Coal	293	425	424	383	368	355	344	37	37	-0.8
Oil	655	497	493	428	396	367	346	44	37	-1.4
Gas	115	215	248	238	242	250	253	19	27	0.6
Power generation	**363**	**460**	**499**	**405**	**388**	**377**	**363**	**100**	**100**	**-0.9**
Coal	128	275	273	234	220	209	200	60	55	-1.3
Oil	157	53	65	26	22	17	13	12	4	-5.4
Gas	78	132	161	146	146	150	150	29	41	0.5
TFC	**655**	**634**	**627**	**606**	**583**	**561**	**547**	**100**	**100**	**-0.6**
Coal	150	133	135	133	132	130	129	21	24	-0.1
Oil	470	421	407	383	357	334	318	66	58	-1.1
Transport	*208*	*222*	*210*	*192*	*172*	*154*	*143*	*35*	*26*	*-1.7*
Gas	35	80	85	90	94	97	101	13	18	0.9

Japan: Current Policies and 450 Scenarios

	Electricity generation (TWh)						Shares (%)		CAAGR (%)	
	2020	2030	2035	2020	2030	2035	2035		2010-35	
	Current Policies Scenario			450 Scenario			CPS	450	CPS	450
Total generation	**1 173**	**1 263**	**1 306**	**1 078**	**1 058**	**1 044**	**100**	**100**	**0.7**	**-0.2**
Coal	314	322	324	220	54	16	25	2	0.2	-11.1
Oil	48	33	34	34	6	6	3	1	-4.1	-10.5
Gas	384	422	444	349	361	336	34	32	1.5	0.4
Nuclear	239	252	253	255	268	268	19	26	-0.5	-0.3
Hydro	88	93	95	98	120	129	7	12	0.6	1.8
Bioenergy	31	41	46	36	47	52	4	5	2.7	3.3
Wind	20	36	42	33	100	116	3	11	9.9	14.5
Geothermal	5	7	9	5	27	35	1	3	4.9	10.9
Solar PV	43	56	59	48	73	79	4	8	11.6	12.9
CSP	-	-	-	-	-	-	-	-	n.a.	n.a.
Marine	-	0	1	-	2	6	0	1	n.a.	n.a.

	Electrical capacity (GW)						Shares (%)		CAAGR (%)	
	2020	2030	2035	2020	2030	2035	2035		2010-35	
	Current Policies Scenario			450 Scenario			CPS	450	CPS	450
Total capacity	**326**	**355**	**365**	**311**	**331**	**337**	**100**	**100**	**1.1**	**0.8**
Coal	49	52	53	42	19	16	15	5	0.6	-4.3
Oil	27	21	22	23	10	7	6	2	-3.3	-7.4
Gas	104	125	127	94	101	98	35	29	2.3	1.3
Nuclear	48	39	39	42	37	37	11	11	-0.9	-1.1
Hydro	49	50	51	51	57	59	14	18	0.3	0.8
Bioenergy	5	7	7	6	8	8	2	2	2.7	3.3
Wind	8	14	16	14	35	40	4	12	8.0	12.0
Geothermal	1	1	1	1	5	6	0	2	4.2	10.1
Solar PV	34	45	47	38	59	64	13	19	10.8	12.2
CSP	-	-	-	-	-	-	-	-	n.a.	n.a.
Marine	-	0	0	-	0	2	0	0	n.a.	n.a.

	CO_2 emissions (Mt)						Shares (%)		CAAGR (%)	
	2020	2030	2035	2020	2030	2035	2035		2010-35	
	Current Policies Scenario			450 Scenario			CPS	450	CPS	450
Total CO_2	**1 097**	**1 061**	**1 059**	**960**	**673**	**577**	**100**	**100**	**-0.3**	**-2.7**
Coal	424	418	413	333	164	119	39	21	-0.1	-5.0
Oil	434	387	374	406	307	270	35	47	-1.1	-2.4
Gas	239	256	272	221	202	187	26	32	0.9	-0.5
Power generation	**442**	**436**	**444**	**340**	**154**	**109**	**100**	**100**	**-0.1**	**-5.6**
Coal	270	264	260	187	35	4	58	3	-0.2	-16.0
Oil	26	18	20	18	3	3	4	3	-3.9	-10.9
Gas	147	154	165	136	117	103	37	94	0.9	-1.0
TFC	**617**	**590**	**581**	**583**	**486**	**437**	**100**	**100**	**-0.4**	**-1.5**
Coal	137	138	138	130	113	100	24	23	0.1	-1.2
Oil	389	352	339	370	290	256	58	58	-0.9	-2.0
Transport	*194*	*165*	*156*	*186*	*129*	*104*	*27*	*24*	*-1.4*	*-3.0*
Gas	90	100	104	83	83	82	18	19	1.1	0.1

Non-OECD: New Policies Scenario

	Energy demand (Mtoe)							Shares (%)		CAAGR (%)
	1990	2010	2015	2020	2025	2030	2035	2010	2035	2010-35
TPED	**4 058**	**6 972**	**8 158**	**9 001**	**9 716**	**10 424**	**11 147**	**100**	**100**	**1.9**
Coal	1 151	2 388	2 869	3 044	3 146	3 269	3 392	34	30	1.4
Oil	1 161	1 798	2 072	2 252	2 395	2 535	2 677	26	24	1.6
Gas	825	1 423	1 631	1 839	2 055	2 283	2 509	20	23	2.3
Nuclear	74	122	186	294	396	450	497	2	4	5.8
Hydro	83	179	217	260	291	322	349	3	3	2.7
Bioenergy	756	1 013	1 094	1 170	1 231	1 289	1 354	15	12	1.2
Other renewables	8	49	89	141	202	276	370	1	3	8.4
Power generation	**1 266**	**2 572**	**3 123**	**3 560**	**3 934**	**4 317**	**4 735**	**100**	**100**	**2.5**
Coal	466	1 373	1 713	1 855	1 936	2 042	2 155	53	46	1.8
Oil	223	204	199	179	155	135	124	8	3	-2.0
Gas	406	632	694	771	858	956	1 049	25	22	2.1
Nuclear	74	122	186	294	396	450	497	5	10	5.8
Hydro	83	179	217	260	291	322	349	7	7	2.7
Bioenergy	7	27	47	88	130	178	240	1	5	9.2
Other renewables	8	36	68	114	168	235	321	1	7	9.2
Other energy sector	**512**	**937**	**1 060**	**1 117**	**1 165**	**1 217**	**1 268**	**100**	**100**	**1.2**
Electricity	*78*	*178*	*219*	*250*	*279*	*309*	*340*	*19*	*27*	*2.6*
TFC	**2 966**	**4 634**	**5 422**	**6 009**	**6 495**	**6 967**	**7 428**	**100**	**100**	**1.9**
Coal	537	725	838	854	861	865	864	16	12	0.7
Oil	802	1 437	1 706	1 913	2 093	2 270	2 434	31	33	2.1
Gas	353	592	701	821	931	1 038	1 150	13	15	2.7
Electricity	281	730	955	1 154	1 319	1 489	1 668	16	22	3.4
Heat	291	215	227	235	236	235	234	5	3	0.3
Bioenergy	702	923	974	1 006	1 022	1 029	1 029	20	14	0.4
Other renewables	0	13	21	27	34	41	50	0	1	5.4
Industry	**979**	**1 592**	**1 920**	**2 146**	**2 306**	**2 454**	**2 595**	**100**	**100**	**2.0**
Coal	315	574	677	694	707	719	726	36	28	0.9
Oil	154	205	232	248	256	261	264	13	10	1.0
Gas	132	208	254	327	381	428	476	13	18	3.4
Electricity	158	379	502	602	678	752	829	24	32	3.2
Heat	136	102	111	115	113	111	109	6	4	0.2
Bioenergy	82	124	142	159	172	182	191	8	7	1.7
Other renewables	-	0	0	0	0	0	0	0	0	3.3
Transport	**429**	**844**	**1 049**	**1 235**	**1 414**	**1 595**	**1 773**	**100**	**100**	**3.0**
Oil	359	740	920	1 079	1 229	1 381	1 528	88	86	2.9
Electricity	13	15	20	24	27	32	38	2	2	3.8
Biofuels	6	18	31	45	59	74	89	2	5	6.5
Other fuels	50	70	77	88	98	108	119	8	7	2.1
Buildings	**1 256**	**1 658**	**1 831**	**1 959**	**2 063**	**2 165**	**2 266**	**100**	**100**	**1.3**
Coal	170	102	109	105	98	91	82	6	4	-0.9
Oil	121	167	181	186	188	188	187	10	8	0.5
Gas	126	191	226	256	290	327	365	12	16	2.6
Electricity	86	301	388	476	554	637	725	18	32	3.6
Heat	145	109	112	116	119	121	122	7	5	0.4
Bioenergy	608	775	795	794	782	762	737	47	33	-0.2
Other renewables	0	13	20	26	32	39	47	1	2	5.4
Other	**302**	**540**	**623**	**670**	**711**	**753**	**794**	**100**	**100**	**1.6**

Non-OECD: Current Policies and 450 Scenarios

	Energy demand (Mtoe)						Shares (%)		CAAGR (%)	
	2020	2030	2035	2020	2030	2035	2035		2010-35	
	Current Policies Scenario			450 Scenario			CPS	450	CPS	450
TPED	**9 310**	**11 219**	**12 245**	**8 466**	**8 973**	**9 321**	**100**	**100**	**2.3**	**1.2**
Coal	3 310	4 011	4 418	2 651	2 077	1 883	36	20	2.5	-0.9
Oil	2 298	2 662	2 861	2 151	2 153	2 109	23	23	1.9	0.6
Gas	1 891	2 400	2 691	1 742	1 951	2 050	22	22	2.6	1.5
Nuclear	287	409	435	316	634	778	4	8	5.2	7.7
Hydro	251	302	324	270	357	390	3	4	2.4	3.2
Bioenergy	1 159	1 241	1 274	1 175	1 377	1 491	10	16	0.9	1.6
Other renewables	115	194	242	162	425	619	2	7	6.6	10.7
Power generation	**3 766**	**4 859**	**5 484**	**3 221**	**3 449**	**3 728**	**100**	**100**	**3.1**	**1.5**
Coal	2 074	2 671	3 037	1 526	987	811	55	22	3.2	-2.1
Oil	187	154	148	161	95	77	3	2	-1.3	-3.8
Gas	790	1 011	1 152	715	738	745	21	20	2.4	0.7
Nuclear	287	409	435	316	634	778	8	21	5.2	7.7
Hydro	251	302	324	270	357	390	6	10	2.4	3.2
Bioenergy	86	149	182	99	259	365	3	10	8.0	11.0
Other renewables	92	163	206	135	380	563	4	15	7.3	11.7
Other energy sector	**1 146**	**1 302**	**1 387**	**1 061**	**1 062**	**1 064**	**100**	**100**	**1.6**	**0.5**
Electricity	*262*	*344*	*390*	*229*	*254*	*268*	*28*	*25*	*3.2*	*1.6*
TFC	**6 152**	**7 337**	**7 934**	**5 747**	**6 254**	**6 453**	**100**	**100**	**2.2**	**1.3**
Coal	891	947	965	804	769	751	12	12	1.1	0.1
Oil	1 949	2 380	2 598	1 835	1 945	1 939	33	30	2.4	1.2
Gas	848	1 081	1 203	790	961	1 046	15	16	2.9	2.3
Electricity	1 203	1 634	1 868	1 063	1 289	1 417	24	22	3.8	2.7
Heat	242	253	257	225	210	203	3	3	0.7	-0.2
Bioenergy	996	1 010	1 008	1 000	1 036	1 042	13	16	0.4	0.5
Other renewables	23	31	36	28	45	56	0	1	4.1	5.9
Industry	**2 225**	**2 645**	**2 847**	**2 037**	**2 247**	**2 335**	**100**	**100**	**2.4**	**1.5**
Coal	721	782	809	651	638	631	28	27	1.4	0.4
Oil	256	276	282	235	244	245	10	11	1.3	0.7
Gas	346	461	517	323	422	462	18	20	3.7	3.3
Electricity	628	829	935	558	660	708	33	30	3.7	2.5
Heat	117	120	121	111	97	93	4	4	0.7	-0.4
Bioenergy	157	177	183	158	184	192	6	8	1.6	1.8
Other renewables	0	0	0	0	2	4	0	0	3.3	13.7
Transport	**1 242**	**1 648**	**1 867**	**1 188**	**1 341**	**1 376**	**100**	**100**	**3.2**	**2.0**
Oil	1 096	1 453	1 644	1 027	1 102	1 087	88	79	3.2	1.5
Electricity	23	31	36	24	37	55	2	4	3.7	5.4
Biofuels	36	59	73	52	103	126	4	9	5.7	8.0
Other fuels	87	105	113	86	100	108	6	8	1.9	1.7
Buildings	**2 005**	**2 271**	**2 400**	**1 860**	**1 929**	**1 967**	**100**	**100**	**1.5**	**0.7**
Coal	114	107	98	99	77	66	4	3	-0.2	-1.7
Oil	191	200	202	178	167	161	8	8	0.8	-0.1
Gas	262	336	378	233	270	293	16	15	2.8	1.7
Electricity	499	703	816	430	527	583	34	30	4.1	2.7
Heat	120	129	132	111	109	106	5	5	0.8	-0.1
Bioenergy	796	766	741	783	738	710	31	36	-0.2	-0.4
Other renewables	22	29	33	26	40	49	1	2	4.0	5.6
Other	**680**	**773**	**820**	**662**	**737**	**774**	**100**	**100**	**1.7**	**1.4**

Non-OECD: New Policies Scenario

	Electricity generation (TWh)							Shares (%)		CAAGR (%)
	1990	2010	2015	2020	2025	2030	2035	2010	2035	2010-35
Total generation	**4 190**	**10 560**	**13 647**	**16 325**	**18 576**	**20 901**	**23 340**	**100**	**100**	**3.2**
Coal	1 333	4 940	6 525	7 306	7 800	8 428	9 114	47	39	2.5
Oil	639	691	689	621	550	495	465	7	2	-1.6
Gas	957	2 216	2 684	3 236	3 804	4 401	4 949	21	21	3.3
Nuclear	283	468	710	1 125	1 518	1 726	1 906	4	8	5.8
Hydro	962	2 079	2 520	3 027	3 386	3 741	4 054	20	17	2.7
Bioenergy	8	68	140	289	437	601	817	1	4	10.5
Wind	0	73	297	537	757	990	1 258	1	5	12.1
Geothermal	8	25	35	54	81	112	149	0	1	7.5
Solar PV	0	1	42	114	208	326	450	0	2	26.4
CSP	-	0	3	16	35	79	175	0	1	62.1
Marine	-	-	-	0	0	1	3	-	0	n.a.

	Electrical capacity (GW)						Shares (%)		CAAGR (%)
	2010	2015	2020	2025	2030	2035	2010	2035	2010-35
Total capacity	**2 465**	**3 329**	**3 992**	**4 528**	**5 090**	**5 689**	**100**	**100**	**3.4**
Coal	980	1 324	1 479	1 582	1 702	1 827	40	32	2.5
Oil	217	235	227	208	187	178	9	3	-0.8
Gas	532	721	847	959	1 089	1 223	22	22	3.4
Nuclear	68	98	153	204	231	255	3	4	5.5
Hydro	577	709	855	958	1 060	1 150	23	20	2.8
Bioenergy	23	38	64	86	111	142	1	2	7.5
Wind	62	164	271	360	444	544	3	10	9.1
Geothermal	4	6	9	13	17	22	0	0	7.0
Solar PV	1	32	84	149	225	302	0	5	24.2
CSP	0	1	5	10	21	45	0	1	32.1
Marine	-	-	0	0	0	1	-	0	n.a.

	CO_2 emissions (Mt)							Shares (%)		CAAGR (%)
	1990	2010	2015	2020	2025	2030	2035	2010	2035	2010-35
Total CO_2	**9 250**	**16 757**	**19 814**	**21 431**	**22 645**	**23 934**	**25 229**	**100**	**100**	**1.7**
Coal	4 181	8 923	10 792	11 420	11 731	12 090	12 448	53	49	1.3
Oil	3 189	4 693	5 438	5 955	6 389	6 828	7 268	28	29	1.8
Gas	1 880	3 141	3 585	4 057	4 524	5 017	5 514	19	22	2.3
Power generation	**3 516**	**7 623**	**9 074**	**9 745**	**10 165**	**10 674**	**11 228**	**100**	**100**	**1.6**
Coal	1 852	5 499	6 815	7 369	7 662	8 007	8 375	72	75	1.7
Oil	717	644	634	569	495	430	396	8	4	-1.9
Gas	948	1 481	1 625	1 806	2 008	2 238	2 457	19	22	2.0
TFC	**5 314**	**8 254**	**9 735**	**10 652**	**11 417**	**12 160**	**12 859**	**100**	**100**	**1.8**
Coal	2 253	3 213	3 731	3 800	3 819	3 832	3 822	39	30	0.7
Oil	2 277	3 750	4 473	5 045	5 543	6 033	6 491	45	50	2.2
Transport	*1 093*	*2 207*	*2 742*	*3 218*	*3 665*	*4 120*	*4 560*	*27*	*35*	*2.9*
Gas	784	1 292	1 531	1 807	2 055	2 296	2 546	16	20	2.8

Non-OECD: Current Policies and 450 Scenarios

	Electricity generation (TWh)						Shares (%)		CAAGR (%)	
	2020	2030	2035	2020	2030	2035	2035		2010-35	
	Current Policies Scenario			450 Scenario			CPS	450	CPS	450
Total generation	**17 040**	**23 005**	**26 255**	**15 026**	**17 941**	**19 595**	**100**	**100**	**3.7**	**2.5**
Coal	8 166	11 016	12 758	5 970	4 096	3 246	49	17	3.9	-1.7
Oil	647	569	562	556	336	269	2	1	-0.8	-3.7
Gas	3 354	4 777	5 562	2 996	3 549	3 643	21	19	3.7	2.0
Nuclear	1 099	1 569	1 668	1 209	2 431	2 986	6	15	5.2	7.7
Hydro	2 916	3 507	3 771	3 137	4 146	4 532	14	23	2.4	3.2
Bioenergy	283	498	610	325	876	1 251	2	6	9.2	12.4
Wind	447	791	949	602	1 602	2 142	4	11	10.8	14.5
Geothermal	43	76	98	67	171	231	0	1	5.7	9.4
Solar PV	76	166	209	144	545	811	1	4	22.6	29.4
CSP	8	35	65	21	188	476	0	2	55.8	68.7
Marine	-	1	3	0	3	7	0	0	n.a.	n.a.

	Electrical capacity (GW)						Shares (%)		CAAGR (%)	
	2020	2030	2035	2020	2030	2035	2035		2010-35	
	Current Policies Scenario			450 Scenario			CPS	450	CPS	450
Total capacity	**4 008**	**5 208**	**5 827**	**3 860**	**4 992**	**5 635**	**100**	**100**	**3.5**	**3.4**
Coal	1 599	2 101	2 378	1 292	1 049	939	41	17	3.6	-0.2
Oil	231	201	196	225	168	154	3	3	-0.4	-1.4
Gas	859	1 131	1 281	797	969	1 049	22	19	3.6	2.8
Nuclear	149	210	223	164	326	399	4	7	4.9	7.4
Hydro	821	986	1 061	891	1 192	1 306	18	23	2.5	3.3
Bioenergy	61	94	111	69	154	210	2	4	6.4	9.2
Wind	224	349	404	298	674	860	7	15	7.8	11.1
Geothermal	7	12	15	11	26	34	0	1	5.3	8.9
Solar PV	56	115	141	107	380	552	2	10	20.5	27.3
CSP	2	9	17	6	52	130	0	2	27.1	37.8
Marine	-	0	1	0	1	2	0	0	n.a.	n.a.

	CO_2 emissions (Mt)						Shares (%)		CAAGR (%)	
	2020	2030	2035	2020	2030	2035	2035		2010-35	
	Current Policies Scenario			450 Scenario			CPS	450	CPS	450
Total CO_2	**22 712**	**27 499**	**30 246**	**19 262**	**16 212**	**14 710**	**100**	**100**	**2.4**	**-0.5**
Coal	12 454	15 012	16 499	9 792	6 433	4 961	55	34	2.5	-2.3
Oil	6 085	7 220	7 841	5 653	5 656	5 518	26	38	2.1	0.6
Gas	4 174	5 267	5 906	3 817	4 122	4 232	20	29	2.6	1.2
Power generation	**10 682**	**13 410**	**15 125**	**8 202**	**5 059**	**3 779**	**100**	**100**	**2.8**	**-2.8**
Coal	8 238	10 556	11 961	6 015	3 098	1 919	79	51	3.2	-4.1
Oil	595	490	472	515	303	244	3	6	-1.2	-3.8
Gas	1 849	2 365	2 693	1 672	1 658	1 615	18	43	2.4	0.3
TFC	**10 964**	**12 914**	**13 877**	**10 079**	**10 214**	**10 006**	**100**	**100**	**2.1**	**0.8**
Coal	3 952	4 172	4 244	3 548	3 144	2 869	31	29	1.1	-0.5
Oil	5 144	6 349	6 966	4 809	5 029	4 954	50	50	2.5	1.1
Transport	*3 268*	*4 336*	*4 907*	*3 062*	*3 289*	*3 246*	*35*	*32*	*3.2*	*1.6*
Gas	1 868	2 393	2 667	1 722	2 041	2 182	19	22	2.9	2.1

E. Europe/Eurasia: New Policies Scenario

	Energy demand (Mtoe)							Shares (%)		CAAGR (%)
	1990	2010	2015	2020	2025	2030	2035	2010	2035	2010-35
TPED	**1 540**	**1 137**	**1 209**	**1 250**	**1 296**	**1 349**	**1 407**	**100**	**100**	**0.9**
Coal	367	216	223	221	219	220	227	19	16	0.2
Oil	472	223	247	255	260	262	265	20	19	0.7
Gas	602	574	604	617	642	669	695	51	49	0.8
Nuclear	59	76	83	95	103	112	116	7	8	1.7
Hydro	23	26	26	27	29	31	33	2	2	1.0
Bioenergy	17	20	26	29	34	39	48	2	3	3.5
Other renewables	0	1	2	5	10	15	22	0	2	14.2
Power generation	**743**	**552**	**565**	**579**	**597**	**620**	**648**	**100**	**100**	**0.6**
Coal	197	137	140	135	132	130	133	25	21	-0.1
Oil	126	17	15	13	10	9	8	3	1	-2.9
Gas	333	290	294	296	304	310	317	53	49	0.4
Nuclear	59	76	83	95	103	112	116	14	18	1.7
Hydro	23	26	26	27	29	31	33	5	5	1.0
Bioenergy	4	6	6	8	10	14	20	1	3	4.8
Other renewables	0	1	2	5	9	15	21	0	3	15.5
Other energy sector	**213**	**198**	**206**	**208**	**209**	**212**	**215**	**100**	**100**	**0.3**
Electricity	*35*	*38*	*40*	*42*	*43*	*45*	*48*	*19*	*22*	*1.0*
TFC	**1 058**	**715**	**776**	**816**	**855**	**895**	**935**	**100**	**100**	**1.1**
Coal	114	39	41	43	44	45	46	6	5	0.6
Oil	267	168	191	202	212	221	228	23	24	1.2
Gas	261	238	262	273	288	305	323	33	35	1.2
Electricity	127	105	115	125	134	144	154	15	17	1.6
Heat	277	150	150	154	155	157	158	21	17	0.2
Bioenergy	13	15	16	19	20	23	25	2	3	2.1
Other renewables	-	0	0	0	0	1	1	0	0	5.2
Industry	**391**	**207**	**219**	**230**	**240**	**251**	**264**	**100**	**100**	**1.0**
Coal	56	31	32	34	35	36	37	15	14	0.7
Oil	48	17	19	20	21	23	25	8	9	1.5
Gas	86	54	58	60	63	68	73	26	27	1.2
Electricity	75	47	51	56	60	65	69	23	26	1.6
Heat	125	57	57	58	58	58	58	28	22	0.1
Bioenergy	0	1	2	2	2	3	3	1	1	3.3
Other renewables	-	0	0	0	0	0	0	0	0	1.5
Transport	**162**	**143**	**165**	**176**	**186**	**195**	**201**	**100**	**100**	**1.4**
Oil	113	94	110	119	125	130	133	65	66	1.4
Electricity	12	9	10	11	12	13	14	6	7	1.8
Biofuels	0	1	1	2	2	2	3	1	1	4.8
Other fuels	37	40	44	45	47	49	51	28	25	1.0
Buildings	**386**	**269**	**286**	**300**	**312**	**324**	**335**	**100**	**100**	**0.9**
Coal	56	7	8	8	8	8	8	3	2	0.1
Oil	39	16	17	16	16	15	15	6	4	-0.2
Gas	111	100	110	116	123	129	135	37	40	1.2
Electricity	26	45	48	52	56	59	62	17	19	1.3
Heat	142	89	90	92	94	96	97	33	29	0.3
Bioenergy	12	12	13	14	15	17	18	4	5	1.6
Other renewables	-	0	0	0	0	1	1	0	0	4.8
Other	**118**	**96**	**106**	**111**	**117**	**125**	**134**	**100**	**100**	**1.3**

E. Europe/Eurasia: Current Policies and 450 Scenarios

	Energy demand (Mtoe)						Shares (%)		CAAGR (%)	
	2020	2030	2035	2020	2030	2035	2035		2010-35	
	Current Policies Scenario			450 Scenario			CPS	450	CPS	450
TPED	**1 284**	**1 424**	**1 508**	**1 189**	**1 189**	**1 212**	**100**	**100**	**1.1**	**0.3**
Coal	233	257	277	200	162	150	18	12	1.0	-1.5
Oil	259	271	277	248	241	234	18	19	0.9	0.2
Gas	638	715	762	569	521	506	51	42	1.1	-0.5
Nuclear	95	107	108	104	140	156	7	13	1.4	2.9
Hydro	27	30	32	29	38	40	2	3	0.8	1.8
Bioenergy	28	34	38	31	59	82	3	7	2.6	5.8
Other renewables	5	10	13	7	27	43	1	4	12.0	17.3
Power generation	**597**	**660**	**702**	**544**	**526**	**543**	**100**	**100**	**1.0**	**-0.1**
Coal	145	159	173	118	82	70	25	13	0.9	-2.6
Oil	14	9	8	12	8	8	1	1	-2.8	-3.2
Gas	305	333	355	265	200	177	51	33	0.8	-1.9
Nuclear	95	107	108	104	140	156	15	29	1.4	2.9
Hydro	27	30	32	29	38	40	5	7	0.8	1.8
Bioenergy	7	11	14	8	30	50	2	9	3.4	8.8
Other renewables	4	10	13	7	26	41	2	8	13.4	18.7
Other energy sector	**212**	**225**	**233**	**199**	**189**	**185**	**100**	**100**	**0.7**	**-0.3**
Electricity	*43*	*50*	*54*	*40*	*40*	*41*	*23*	*22*	*1.5*	*0.3*
TFC	**838**	**945**	**1 002**	**783**	**816**	**830**	**100**	**100**	**1.4**	**0.6**
Coal	45	49	52	41	39	39	5	5	1.1	-0.0
Oil	206	229	240	197	201	198	24	24	1.4	0.7
Gas	284	324	346	259	276	285	35	34	1.5	0.7
Electricity	130	158	174	118	130	137	17	17	2.0	1.1
Heat	156	165	169	148	143	140	17	17	0.5	-0.3
Bioenergy	17	20	21	20	26	29	2	4	1.5	2.8
Other renewables	0	0	0	0	1	1	0	0	2.2	7.4
Industry	**238**	**270**	**288**	**221**	**234**	**242**	**100**	**100**	**1.3**	**0.6**
Coal	35	38	40	32	31	31	14	13	1.1	0.0
Oil	20	24	26	19	22	24	9	10	1.7	1.3
Gas	66	77	82	57	64	68	28	28	1.7	0.9
Electricity	59	72	78	53	58	61	27	25	2.1	1.1
Heat	57	58	59	57	55	54	21	22	0.2	-0.2
Bioenergy	2	2	3	2	4	5	1	2	2.6	4.5
Other renewables	0	0	0	0	0	0	0	0	1.5	1.5
Transport	**176**	**198**	**207**	**171**	**172**	**167**	**100**	**100**	**1.5**	**0.6**
Oil	120	135	141	115	113	107	68	64	1.7	0.5
Electricity	10	13	14	11	14	16	7	10	1.7	2.4
Biofuels	1	1	1	2	3	4	1	2	1.4	6.0
Other fuels	45	49	51	43	41	40	24	24	1.0	0.1
Buildings	**310**	**348**	**367**	**283**	**289**	**289**	**100**	**100**	**1.3**	**0.3**
Coal	9	9	10	7	7	6	3	2	1.0	-0.6
Oil	17	17	17	16	14	13	4	5	0.2	-0.6
Gas	120	137	147	109	113	114	40	39	1.6	0.5
Electricity	54	66	72	49	51	52	20	18	1.9	0.6
Heat	96	103	105	87	85	82	29	28	0.7	-0.3
Bioenergy	14	16	17	15	18	20	4	7	1.3	2.1
Other renewables	0	0	0	0	1	1	0	0	1.4	7.1
Other	**113**	**129**	**139**	**108**	**122**	**131**	**100**	**100**	**1.5**	**1.3**

E. Europe/Eurasia: New Policies Scenario

	Electricity generation (TWh)							Shares (%)		CAAGR (%)
	1990	2010	2015	2020	2025	2030	2035	2010	2035	2010-35
Total generation	**1 894**	**1 681**	**1 818**	**1 961**	**2 087**	**2 228**	**2 374**	**100**	**100**	**1.4**
Coal	429	385	409	407	397	403	429	23	18	0.4
Oil	271	27	21	14	9	5	4	2	0	-7.1
Gas	702	671	759	828	895	945	980	40	41	1.5
Nuclear	226	289	315	365	395	428	445	17	19	1.7
Hydro	266	303	297	315	336	360	386	18	16	1.0
Bioenergy	0	4	7	13	21	36	59	0	2	11.7
Wind	-	1	8	13	22	32	45	0	2	14.6
Geothermal	0	1	1	4	9	13	19	0	1	15.7
Solar PV	-	0	1	2	3	5	7	0	0	25.8
CSP	-	-	-	-	-	-	-	-	-	n.a.
Marine	-	-	-	0	0	0	0	-	0	n.a.

	Electrical capacity (GW)						Shares (%)		CAAGR (%)
	2010	2015	2020	2025	2030	2035	2010	2035	2010-35
Total capacity	**422**	**443**	**465**	**486**	**513**	**545**	**100**	**100**	**1.0**
Coal	109	104	95	91	85	84	26	15	-1.0
Oil	25	24	18	11	5	4	6	1	-7.5
Gas	153	167	187	204	223	238	36	44	1.8
Nuclear	42	45	51	55	59	61	10	11	1.5
Hydro	91	96	101	107	114	120	22	22	1.1
Bioenergy	1	2	3	4	6	10	0	2	9.3
Wind	1	4	6	10	14	19	0	3	11.2
Geothermal	0	0	1	1	2	3	0	0	14.6
Solar PV	0	1	2	3	4	6	0	1	21.6
CSP	-	-	-	-	-	-	-	-	n.a.
Marine	-	-	0	0	0	0	-	0	n.a.

	CO_2 emissions (Mt)							Shares (%)		CAAGR (%)
	1990	2010	2015	2020	2025	2030	2035	2010	2035	2010-35
Total CO_2	**3 997**	**2 615**	**2 759**	**2 805**	**2 865**	**2 929**	**3 014**	**100**	**100**	**0.6**
Coal	1 336	818	838	831	820	816	835	31	28	0.1
Oil	1 257	535	593	613	628	642	651	20	22	0.8
Gas	1 404	1 262	1 329	1 362	1 416	1 471	1 528	48	51	0.8
Power generation	**1 982**	**1 306**	**1 323**	**1 301**	**1 296**	**1 297**	**1 322**	**100**	**100**	**0.0**
Coal	799	571	584	565	549	541	553	44	42	-0.1
Oil	405	55	49	41	34	29	26	4	2	-2.9
Gas	778	681	690	695	713	728	743	52	56	0.4
TFC	**1 900**	**1 176**	**1 294**	**1 358**	**1 417**	**1 477**	**1 533**	**100**	**100**	**1.1**
Coal	526	227	235	246	251	256	262	19	17	0.6
Oil	784	422	482	510	532	550	561	36	37	1.1
Transport	*361*	*276*	*324*	*349*	*369*	*383*	*392*	*23*	*26*	*1.4*
Gas	591	526	577	602	634	671	710	45	46	1.2

E. Europe/Eurasia: Current Policies and 450 Scenarios

	Electricity generation (TWh)						Shares (%)		CAAGR (%)	
	2020	2030	2035	2020	2030	2035	2035		2010-35	
	Current Policies Scenario			450 Scenario			CPS	450	CPS	450
Total generation	**2 036**	**2 442**	**2 674**	**1 857**	**2 008**	**2 094**	**100**	**100**	**1.9**	**0.9**
Coal	441	505	566	349	230	176	21	8	1.6	-3.1
Oil	15	6	4	14	5	3	0	0	-7.2	-8.6
Gas	876	1 105	1 229	716	591	501	46	24	2.5	-1.2
Nuclear	361	409	413	398	537	598	15	29	1.4	3.0
Hydro	312	349	370	340	443	470	14	22	0.8	1.8
Bioenergy	12	28	39	16	92	160	1	8	9.9	16.3
Wind	13	28	38	15	80	138	1	7	13.9	19.9
Geothermal	4	8	11	6	21	33	0	2	13.1	18.2
Solar PV	1	3	4	3	9	13	0	1	22.8	29.1
CSP	-	-	-	-	-	-	-	-	n.a.	n.a.
Marine	-	0	0	0	0	1	0	0	n.a.	n.a.

	Electrical capacity (GW)						Shares (%)		CAAGR (%)	
	2020	2030	2035	2020	2030	2035	2035		2010-35	
	Current Policies Scenario			450 Scenario			CPS	450	CPS	450
Total capacity	**470**	**512**	**538**	**433**	**480**	**526**	**100**	**100**	**1.0**	**0.9**
Coal	102	100	101	85	56	51	19	10	-0.3	-3.0
Oil	18	5	4	18	5	4	1	1	-7.5	-7.6
Gas	188	219	233	151	147	145	43	28	1.7	-0.2
Nuclear	51	57	57	56	75	82	11	16	1.2	2.7
Hydro	100	110	116	108	136	144	22	27	1.0	1.8
Bioenergy	3	5	7	3	16	27	1	5	7.6	13.5
Wind	6	12	16	7	33	56	3	11	10.6	16.2
Geothermal	0	1	2	1	3	4	0	1	12.1	17.1
Solar PV	2	3	3	3	8	12	1	2	18.9	24.9
CSP	-	-	-	-	-	-	-	-	n.a.	n.a.
Marine	-	0	0	0	0	0	0	0	n.a.	n.a.

	CO_2 emissions (Mt)						Shares (%)		CAAGR (%)	
	2020	2030	2035	2020	2030	2035	2035		2010-35	
	Current Policies Scenario			450 Scenario			CPS	450	CPS	450
Total CO_2	**2 913**	**3 203**	**3 395**	**2 587**	**2 172**	**2 009**	**100**	**100**	**1.0**	**-1.0**
Coal	877	962	1 030	741	505	436	30	22	0.9	-2.5
Oil	626	670	690	595	573	547	20	27	1.0	0.1
Gas	1 411	1 571	1 674	1 251	1 093	1 026	49	51	1.1	-0.8
Power generation	**1 365**	**1 476**	**1 576**	**1 154**	**756**	**627**	**100**	**100**	**0.8**	**-2.9**
Coal	605	663	716	493	288	235	45	38	0.9	-3.5
Oil	44	30	27	41	28	25	2	4	-2.8	-3.2
Gas	717	783	833	620	440	367	53	59	0.8	-2.4
TFC	**1 399**	**1 565**	**1 651**	**1 293**	**1 277**	**1 247**	**100**	**100**	**1.4**	**0.2**
Coal	253	277	292	229	199	183	18	15	1.0	-0.9
Oil	519	574	597	494	487	466	36	37	1.4	0.4
Transport	*353*	*398*	*416*	*339*	*334*	*314*	*25*	*25*	*1.7*	*0.5*
Gas	628	714	763	570	591	598	46	48	1.5	0.5

Russia: New Policies Scenario

	Energy demand (Mtoe)							Shares (%)		CAAGR (%)
	1990	2010	2015	2020	2025	2030	2035	2010	2035	2010-35
TPED	**880**	**710**	**750**	**774**	**802**	**837**	**875**	**100**	**100**	**0.8**
Coal	191	115	120	120	122	123	128	16	15	0.4
Oil	264	139	154	160	163	165	168	20	19	0.7
Gas	367	389	403	407	420	438	454	55	52	0.6
Nuclear	31	45	51	60	64	69	74	6	8	2.0
Hydro	14	14	14	15	17	18	20	2	2	1.4
Bioenergy	12	7	7	8	10	13	18	1	2	3.9
Other renewables	0	0	1	3	7	10	15	0	2	15.2
Power generation	**444**	**372**	**380**	**391**	**406**	**423**	**443**	**100**	**100**	**0.7**
Coal	105	71	74	75	77	80	85	19	19	0.7
Oil	62	11	10	10	8	7	7	3	2	-2.0
Gas	228	226	225	223	225	228	229	61	52	0.1
Nuclear	31	45	51	60	64	69	74	12	17	2.0
Hydro	14	14	14	15	17	18	20	4	5	1.4
Bioenergy	4	4	4	5	7	9	14	1	3	4.6
Other renewables	0	0	1	3	7	10	15	0	3	15.2
Other energy sector	**127**	**120**	**124**	**125**	**125**	**128**	**131**	**100**	**100**	**0.3**
Electricity	*21*	*25*	*27*	*28*	*29*	*31*	*34*	*21*	*26*	*1.2*
TFC	**625**	**448**	**482**	**502**	**523**	**546**	**570**	**100**	**100**	**1.0**
Coal	55	19	20	21	20	20	20	4	3	0.1
Oil	145	104	117	123	129	134	139	23	24	1.2
Gas	143	146	160	165	174	185	197	32	35	1.2
Electricity	71	62	68	74	80	86	92	14	16	1.6
Heat	203	115	114	116	117	118	119	26	21	0.1
Bioenergy	8	2	3	3	3	3	4	1	1	2.0
Other renewables	-	-	0	0	0	0	0	-	0	n.a.
Industry	**209**	**132**	**138**	**142**	**147**	**154**	**163**	**100**	**100**	**0.8**
Coal	15	15	16	16	16	16	16	11	10	0.3
Oil	25	11	12	13	14	15	17	8	10	1.6
Gas	30	33	35	35	37	40	43	25	27	1.1
Electricity	41	28	31	33	36	38	41	21	25	1.6
Heat	98	45	44	44	44	44	44	34	27	-0.0
Bioenergy	-	0	0	0	1	1	1	0	0	3.0
Other renewables	-	-	-	-	-	-	-	-	-	n.a.
Transport	**116**	**97**	**110**	**116**	**122**	**127**	**131**	**100**	**100**	**1.2**
Oil	73	56	65	70	74	76	77	58	59	1.2
Electricity	9	7	8	9	9	11	12	8	9	1.9
Biofuels	-	-	-	-	-	-	-	-	-	n.a.
Other fuels	34	33	37	38	39	41	42	34	32	0.9
Buildings	**228**	**149**	**157**	**163**	**169**	**173**	**178**	**100**	**100**	**0.7**
Coal	40	4	4	4	4	3	3	3	2	-1.1
Oil	12	6	7	6	5	5	5	4	3	-1.3
Gas	57	44	49	52	55	58	61	30	34	1.3
Electricity	15	26	28	30	32	34	35	17	20	1.3
Heat	98	67	67	68	70	71	71	45	40	0.2
Bioenergy	7	2	2	2	2	2	3	1	1	1.5
Other renewables	-	-	0	0	0	0	0	-	0	n.a.
Other	**72**	**70**	**78**	**81**	**86**	**92**	**99**	**100**	**100**	**1.4**

Russia: Current Policies and 450 Scenarios

	Energy demand (Mtoe)						Shares (%)		CAAGR (%)	
	2020	2030	2035	2020	2030	2035	2035		2010-35	
	Current Policies Scenario			450 Scenario			CPS	450	CPS	450
TPED	**796**	**887**	**945**	**736**	**726**	**742**	**100**	**100**	**1.2**	**0.2**
Coal	128	147	157	106	80	72	17	10	1.3	-1.9
Oil	161	167	171	156	155	153	18	21	0.8	0.4
Gas	421	470	504	375	334	319	53	43	1.0	-0.8
Nuclear	60	68	71	68	85	97	7	13	1.8	3.2
Hydro	15	18	19	17	23	24	2	3	1.2	2.1
Bioenergy	8	11	13	9	29	45	1	6	2.5	7.8
Other renewables	3	7	9	5	20	32	1	4	13.0	18.7
Power generation	**404**	**452**	**484**	**368**	**348**	**360**	**100**	**100**	**1.1**	**-0.1**
Coal	82	100	110	63	43	37	23	10	1.8	-2.6
Oil	10	7	7	10	7	7	1	2	-1.9	-2.1
Gas	228	244	259	200	146	124	54	35	0.5	-2.4
Nuclear	60	68	71	68	85	97	15	27	1.8	3.2
Hydro	15	18	19	17	23	24	4	7	1.2	2.1
Bioenergy	5	7	9	6	24	39	2	11	3.0	9.1
Other renewables	3	7	9	5	20	32	2	9	13.0	18.7
Other energy sector	**127**	**136**	**143**	**120**	**113**	**111**	**100**	**100**	**0.7**	**-0.3**
Electricity	*29*	*34*	*38*	*27*	*28*	*29*	*27*	*26*	*1.7*	*0.6*
TFC	**516**	**580**	**616**	**482**	**499**	**507**	**100**	**100**	**1.3**	**0.5**
Coal	21	22	23	19	16	16	4	3	0.6	-0.8
Oil	124	137	143	121	125	124	23	25	1.3	0.7
Gas	173	198	213	158	169	176	35	35	1.5	0.8
Electricity	77	96	106	70	77	82	17	16	2.1	1.1
Heat	118	124	127	111	106	104	21	20	0.4	-0.4
Bioenergy	3	3	4	3	5	6	1	1	1.7	3.7
Other renewables	0	0	0	0	0	0	0	0	n.a.	n.a.
Industry	**147**	**166**	**178**	**138**	**145**	**152**	**100**	**100**	**1.2**	**0.6**
Coal	16	17	18	15	13	13	10	8	0.7	-0.6
Oil	12	15	17	12	15	16	9	11	1.7	1.5
Gas	39	46	49	34	40	43	28	28	1.6	1.1
Electricity	35	43	48	32	35	37	27	24	2.2	1.1
Heat	43	44	46	44	42	41	26	27	0.1	-0.3
Bioenergy	0	1	1	1	1	1	0	1	2.8	5.8
Other renewables	-	-	-	-	-	-	-	-	n.a.	n.a.
Transport	**117**	**128**	**133**	**114**	**113**	**109**	**100**	**100**	**1.3**	**0.5**
Oil	70	77	79	69	68	64	60	58	1.4	0.5
Electricity	8	10	12	9	12	14	9	12	1.9	2.5
Biofuels	-	-	-	-	-	-	-	-	n.a.	n.a.
Other fuels	38	40	42	36	34	32	32	29	0.9	-0.2
Buildings	**170**	**191**	**202**	**152**	**151**	**148**	**100**	**100**	**1.2**	**-0.0**
Coal	4	4	4	4	3	2	2	2	0.3	-2.0
Oil	7	6	6	6	5	4	3	3	-0.5	-1.6
Gas	54	63	69	48	50	51	34	34	1.8	0.6
Electricity	32	39	43	28	28	28	21	19	2.1	0.3
Heat	71	76	78	64	61	59	39	40	0.6	-0.5
Bioenergy	2	2	2	2	3	4	1	3	1.2	3.0
Other renewables	0	0	0	0	0	0	0	0	n.a.	n.a.
Other	**82**	**95**	**102**	**79**	**90**	**97**	**100**	**100**	**1.5**	**1.3**

Russia: New Policies Scenario

	Electricity generation (TWh)							Shares (%)		CAAGR (%)
	1990	2010	2015	2020	2025	2030	2035	2010	2035	2010-35
Total generation	**1 082**	**1 036**	**1 121**	**1 209**	**1 287**	**1 380**	**1 477**	**100**	**100**	**1.4**
Coal	157	166	183	186	197	215	243	16	16	1.5
Oil	129	9	8	7	4	2	2	1	0	-6.7
Gas	512	521	562	591	617	639	645	50	44	0.9
Nuclear	118	170	193	230	245	264	281	16	19	2.0
Hydro	166	166	168	180	196	214	233	16	16	1.4
Bioenergy	0	3	5	8	13	23	41	0	3	11.3
Wind	-	0	1	3	7	11	15	0	1	38.8
Geothermal	0	1	1	3	7	11	16	0	1	14.7
Solar PV	-	-	0	0	1	1	1	-	0	n.a.
CSP	-	-	-	-	-	-	-	-	-	n.a.
Marine	-	-	-	-	-	-	-	-	-	n.a.

	Electrical capacity (GW)						Shares (%)		CAAGR (%)
	2010	2015	2020	2025	2030	2035	2010	2035	2010-35
Total capacity	**236**	**247**	**266**	**284**	**306**	**329**	**100**	**100**	**1.3**
Coal	52	48	44	42	40	42	22	13	-0.8
Oil	6	6	5	3	2	1	3	0	-6.5
Gas	105	115	129	142	156	165	44	50	1.8
Nuclear	24	27	32	34	36	38	10	12	1.8
Hydro	48	49	52	56	61	66	20	20	1.3
Bioenergy	1	1	2	2	4	7	0	2	8.3
Wind	0	1	2	3	4	6	0	2	26.1
Geothermal	0	0	0	1	1	2	0	1	13.8
Solar PV	-	0	0	1	1	1	-	0	n.a.
CSP	-	-	-	-	-	-	-	-	n.a.
Marine	-	-	-	-	-	-	-	-	n.a.

	CO_2 emissions (Mt)							Shares (%)		CAAGR (%)
	1990	2010	2015	2020	2025	2030	2035	2010	2035	2010-35
Total CO_2	**2 179**	**1 624**	**1 701**	**1 726**	**1 769**	**1 816**	**1 871**	**100**	**100**	**0.6**
Coal	687	443	459	464	474	485	506	27	27	0.5
Oil	625	315	346	360	366	371	376	19	20	0.7
Gas	866	867	896	902	929	960	989	53	53	0.5
Power generation	**1 162**	**872**	**880**	**875**	**889**	**903**	**926**	**100**	**100**	**0.2**
Coal	432	304	317	320	332	344	365	35	39	0.7
Oil	198	36	34	32	27	23	22	4	2	-1.9
Gas	532	532	528	522	529	536	538	61	58	0.0
TFC	**960**	**689**	**754**	**782**	**810**	**841**	**871**	**100**	**100**	**0.9**
Coal	253	132	135	137	135	134	134	19	15	0.1
Oil	389	243	274	289	300	310	315	35	36	1.0
Transport	*217*	*165*	*191*	*206*	*216*	*223*	*226*	*24*	*26*	*1.3*
Gas	318	314	345	356	375	397	422	46	48	1.2

Russia: Current Policies and 450 Scenarios

	Electricity generation (TWh)						Shares (%)		CAAGR (%)	
	2020	2030	2035	2020	2030	2035	2035		2010-35	
	Current Policies Scenario			450 Scenario			CPS	450	CPS	450
Total generation	**1 252**	**1 529**	**1 693**	**1 144**	**1 243**	**1 303**	**100**	**100**	**2.0**	**0.9**
Coal	207	270	310	153	103	82	18	6	2.5	-2.8
Oil	7	2	2	7	2	2	0	0	-6.7	-6.8
Gas	615	758	843	511	404	320	50	25	1.9	-1.9
Nuclear	230	260	270	261	326	372	16	29	1.9	3.2
Hydro	178	206	222	193	267	280	13	21	1.2	2.1
Bioenergy	7	17	25	9	72	126	1	10	9.1	16.5
Wind	4	9	12	4	50	90	1	7	37.8	49.3
Geothermal	3	7	9	5	18	27	1	2	12.4	17.3
Solar PV	0	0	0	1	2	3	0	0	n.a.	n.a.
CSP	-	-	-	-	-	-	-	-	n.a.	n.a.
Marine	-	-	-	-	0	0	-	0	n.a.	n.a.

	Electrical capacity (GW)						Shares (%)		CAAGR (%)	
	2020	2030	2035	2020	2030	2035	2035		2010-35	
	Current Policies Scenario			450 Scenario			CPS	450	CPS	450
Total capacity	**264**	**295**	**312**	**237**	**266**	**294**	**100**	**100**	**1.1**	**0.9**
Coal	46	47	51	37	22	20	16	7	-0.1	-3.8
Oil	5	2	1	5	2	1	0	0	-6.6	-6.5
Gas	125	144	150	97	86	80	48	27	1.4	-1.1
Nuclear	32	35	36	36	45	50	12	17	1.6	3.0
Hydro	51	59	63	55	75	78	20	27	1.1	2.0
Bioenergy	2	3	4	2	12	21	1	7	6.4	13.2
Wind	2	4	5	2	20	36	2	12	25.5	35.9
Geothermal	0	1	1	1	2	4	0	1	11.5	16.4
Solar PV	0	0	0	1	2	3	0	1	n.a.	n.a.
CSP	-	-	-	-	-	-	-	-	n.a.	n.a.
Marine	-	-	-	-	0	0	-	0	n.a.	n.a.

	CO_2 emissions (Mt)						Shares (%)		CAAGR (%)	
	2020	2030	2035	2020	2030	2035	2035		2010-35	
	Current Policies Scenario			450 Scenario			CPS	450	CPS	450
Total CO_2	**1 795**	**1 992**	**2 119**	**1 583**	**1 273**	**1 143**	**100**	**100**	**1.1**	**-1.4**
Coal	497	583	630	404	246	201	30	18	1.4	-3.1
Oil	364	381	391	352	340	324	18	28	0.9	0.1
Gas	934	1 028	1 098	827	687	618	52	54	1.0	-1.3
Power generation	**920**	**1 027**	**1 101**	**769**	**476**	**371**	**100**	**100**	**0.9**	**-3.4**
Coal	351	429	471	270	140	107	43	29	1.8	-4.1
Oil	33	24	23	32	23	21	2	6	-1.8	-2.0
Gas	536	574	608	467	313	243	55	65	0.5	-3.1
TFC	**804**	**889**	**938**	**748**	**734**	**713**	**100**	**100**	**1.2**	**0.1**
Coal	140	146	150	127	101	89	16	13	0.5	-1.5
Oil	292	318	329	283	281	268	35	38	1.2	0.4
Transport	*207*	*227*	*234*	*202*	*200*	*187*	*25*	*26*	*1.4*	*0.5*
Gas	372	425	458	338	352	356	49	50	1.5	0.5

Non-OECD Asia: New Policies Scenario

	Energy demand (Mtoe)							Shares (%)		CAAGR (%)
	1990	2010	2015	2020	2025	2030	2035	2010	2035	2010-35
TPED	**1 589**	**3 936**	**4 808**	**5 400**	**5 875**	**6 351**	**6 839**	**100**	**100**	**2.2**
Coal	694	2 034	2 496	2 658	2 754	2 874	2 989	52	44	1.6
Oil	319	865	1 037	1 155	1 258	1 368	1 469	22	21	2.1
Gas	69	323	424	545	662	784	918	8	13	4.3
Nuclear	10	38	90	179	256	294	330	1	5	9.1
Hydro	24	84	110	140	156	173	187	2	3	3.2
Bioenergy	466	548	574	607	627	645	674	14	10	0.8
Other renewables	7	43	77	117	162	212	272	1	4	7.7
Power generation	**328**	**1 519**	**1 985**	**2 338**	**2 626**	**2 919**	**3 237**	**100**	**100**	**3.1**
Coal	226	1 167	1 490	1 627	1 711	1 819	1 928	77	60	2.0
Oil	45	50	41	31	24	21	20	3	1	-3.6
Gas	16	139	170	209	256	309	371	9	11	4.0
Nuclear	10	38	90	179	256	294	330	2	10	9.1
Hydro	24	84	110	140	156	173	187	6	6	3.2
Bioenergy	0	11	26	59	91	125	170	1	5	11.7
Other renewables	7	31	58	93	133	177	231	2	7	8.4
Other energy sector	**165**	**470**	**551**	**582**	**606**	**635**	**666**	**100**	**100**	**1.4**
Electricity	*26*	*98*	*129*	*152*	*173*	*195*	*218*	*21*	*33*	*3.3*
TFC	**1 221**	**2 569**	**3 118**	**3 506**	**3 809**	**4 107**	**4 399**	**100**	**100**	**2.2**
Coal	397	653	761	774	778	782	779	25	18	0.7
Oil	241	728	902	1 035	1 152	1 272	1 381	28	31	2.6
Gas	31	136	194	276	344	407	473	5	11	5.1
Electricity	83	447	627	779	899	1 024	1 156	17	26	3.9
Heat	14	65	77	81	80	78	76	3	2	0.6
Bioenergy	455	528	539	538	525	509	493	21	11	-0.3
Other renewables	0	12	19	24	29	35	41	0	1	4.9
Industry	**401**	**1 040**	**1 307**	**1 481**	**1 597**	**1 702**	**1 802**	**100**	**100**	**2.2**
Coal	239	519	619	633	644	654	660	50	37	1.0
Oil	53	100	116	124	126	127	128	10	7	1.0
Gas	8	54	81	137	174	203	232	5	13	6.0
Electricity	51	269	376	461	524	585	649	26	36	3.6
Heat	11	45	54	57	55	53	51	4	3	0.4
Bioenergy	39	53	61	70	75	79	83	5	5	1.8
Other renewables	-	0	0	0	0	0	0	0	0	3.3
Transport	**111**	**365**	**488**	**607**	**725**	**851**	**976**	**100**	**100**	**4.0**
Oil	97	340	453	559	662	772	875	93	90	3.9
Electricity	1	5	9	12	15	18	22	1	2	6.2
Biofuels	-	2	7	12	19	27	38	1	4	12.0
Other fuels	12	18	19	24	30	35	40	5	4	3.3
Buildings	**589**	**867**	**971**	**1 039**	**1 088**	**1 134**	**1 182**	**100**	**100**	**1.2**
Coal	111	87	94	89	83	75	67	10	6	-1.0
Oil	33	93	103	105	105	103	100	11	8	0.3
Gas	5	34	53	72	95	119	144	4	12	6.0
Electricity	22	148	209	270	321	376	435	17	37	4.4
Heat	3	20	22	24	25	25	25	2	2	1.0
Bioenergy	415	473	472	456	432	402	371	55	31	-1.0
Other renewables	0	12	18	23	28	34	40	1	3	5.0
Other	**120**	**297**	**352**	**379**	**399**	**420**	**439**	**100**	**100**	**1.6**

Non-OECD Asia: Current Policies and 450 Scenarios

	Energy demand (Mtoe)						Shares (%)		CAAGR (%)	
	2020	2030	2035	2020	2030	2035	2035		2010-35	
	Current Policies Scenario			450 Scenario			CPS	450	CPS	450
TPED	**5 628**	**6 946**	**7 655**	**5 038**	**5 375**	**5 620**	**100**	**100**	**2.7**	**1.4**
Coal	2 903	3 555	3 929	2 298	1 773	1 598	51	28	2.7	-1.0
Oil	1 176	1 428	1 564	1 114	1 188	1 187	20	21	2.4	1.3
Gas	549	780	916	543	760	872	12	16	4.3	4.0
Nuclear	173	268	291	189	436	552	4	10	8.5	11.3
Hydro	131	156	167	147	199	218	2	4	2.8	3.9
Bioenergy	601	611	613	611	704	759	8	14	0.4	1.3
Other renewables	95	148	175	135	316	435	2	8	5.8	9.7
Power generation	**2 502**	**3 364**	**3 847**	**2 078**	**2 258**	**2 471**	**100**	**100**	**3.8**	**2.0**
Coal	1 831	2 400	2 742	1 326	838	682	71	28	3.5	-2.1
Oil	35	26	25	28	17	15	1	1	-2.8	-4.6
Gas	201	291	353	212	307	360	9	15	3.8	3.9
Nuclear	173	268	291	189	436	552	8	22	8.5	11.3
Hydro	131	156	167	147	199	218	4	9	2.8	3.9
Bioenergy	58	102	124	67	183	254	3	10	10.3	13.5
Other renewables	74	121	145	110	279	390	4	16	6.4	10.7
Other energy sector	**599**	**684**	**734**	**550**	**548**	**551**	**100**	**100**	**1.8**	**0.6**
Electricity	*160*	*219*	*253*	*137*	*155*	*166*	*34*	*30*	*3.9*	*2.1*
TFC	**3 601**	**4 351**	**4 732**	**3 346**	**3 672**	**3 800**	**100**	**100**	**2.5**	**1.6**
Coal	808	857	871	728	696	678	18	18	1.2	0.2
Oil	1 052	1 328	1 473	1 001	1 107	1 120	31	29	2.9	1.7
Gas	288	422	489	271	386	439	10	12	5.2	4.8
Electricity	814	1 131	1 304	709	871	963	28	25	4.4	3.1
Heat	86	89	88	77	67	63	2	2	1.2	-0.1
Bioenergy	533	498	477	535	510	493	10	13	-0.4	-0.3
Other renewables	21	27	30	25	36	45	1	1	3.6	5.3
Industry	**1 539**	**1 844**	**1 990**	**1 396**	**1 540**	**1 597**	**100**	**100**	**2.6**	**1.7**
Coal	657	713	736	594	582	575	37	36	1.4	0.4
Oil	127	133	134	116	119	119	7	7	1.2	0.7
Gas	147	219	253	141	209	234	13	15	6.4	6.1
Electricity	480	643	730	423	506	544	37	34	4.1	2.9
Heat	61	62	61	54	43	39	3	2	1.2	-0.6
Bioenergy	66	73	75	69	79	82	4	5	1.4	1.8
Other renewables	0	0	0	0	2	4	0	0	3.3	13.8
Transport	**613**	**883**	**1 032**	**592**	**724**	**764**	**100**	**100**	**4.2**	**3.0**
Oil	567	810	943	540	631	642	91	84	4.2	2.6
Electricity	12	17	21	12	21	34	2	4	6.0	8.1
Biofuels	10	21	29	16	41	53	3	7	10.8	13.4
Other fuels	24	34	38	24	31	35	4	5	3.0	2.7
Buildings	**1 066**	**1 195**	**1 257**	**981**	**992**	**1 004**	**100**	**100**	**1.5**	**0.6**
Coal	98	90	81	84	63	53	6	5	-0.3	-2.0
Oil	108	110	109	99	88	83	9	8	0.6	-0.5
Gas	73	116	140	63	95	113	11	11	5.8	4.9
Electricity	286	424	500	238	300	336	40	34	5.0	3.3
Heat	25	26	27	24	24	24	2	2	1.2	0.8
Bioenergy	456	403	372	449	389	357	30	36	-1.0	-1.1
Other renewables	20	25	28	23	33	39	2	4	3.6	4.9
Other	**383**	**430**	**454**	**377**	**416**	**435**	**100**	**100**	**1.7**	**1.5**

Non-OECD Asia: New Policies Scenario

	Electricity generation (TWh)							Shares (%)		CAAGR (%)
	1990	2010	2015	2020	2025	2030	2035	2010	2035	2010-35
Total generation	**1 271**	**6 325**	**8 780**	**10 816**	**12 459**	**14 165**	**15 972**	**100**	**100**	**3.8**
Coal	730	4 273	5 769	6 505	6 994	7 613	8 268	68	52	2.7
Oil	162	158	135	100	76	69	66	3	0	-3.4
Gas	59	658	842	1 081	1 368	1 693	2 050	10	13	4.7
Nuclear	39	145	345	685	982	1 127	1 268	2	8	9.1
Hydro	274	981	1 278	1 624	1 819	2 013	2 171	16	14	3.2
Bioenergy	1	23	75	195	308	426	577	0	4	13.7
Wind	0	66	270	483	673	864	1 063	1	7	11.8
Geothermal	7	19	26	40	57	78	100	0	1	6.8
Solar PV	0	1	37	95	167	249	330	0	2	25.1
CSP	-	-	3	8	15	32	76	-	0	n.a.
Marine	-	-	-	0	0	1	3	-	0	n.a.

	Electrical capacity (GW)						Shares (%)		CAAGR (%)
	2010	2015	2020	2025	2030	2035	2010	2035	2010-35
Total capacity	**1 446**	**2 105**	**2 630**	**3 039**	**3 456**	**3 880**	**100**	**100**	**4.0**
Coal	826	1 161	1 313	1 413	1 534	1 654	57	43	2.8
Oil	66	65	62	58	54	51	5	1	-1.1
Gas	156	225	286	346	414	493	11	13	4.7
Nuclear	21	46	91	130	149	168	1	4	8.7
Hydro	300	398	505	569	633	684	21	18	3.3
Bioenergy	14	25	46	63	81	101	1	3	8.3
Wind	58	152	248	325	393	464	4	12	8.6
Geothermal	3	4	6	9	12	15	0	0	6.4
Solar PV	1	28	71	122	177	229	0	6	23.4
CSP	-	1	2	4	8	19	-	0	n.a.
Marine	-	-	0	0	0	1	-	0	n.a.

	CO_2 emissions (Mt)							Shares (%)		CAAGR (%)
	1990	2010	2015	2020	2025	2030	2035	2010	2035	2010-35
Total CO_2	**3 568**	**10 540**	**12 990**	**14 204**	**15 106**	**16 121**	**17 120**	**100**	**100**	**2.0**
Coal	2 563	7 690	9 465	10 058	10 376	10 766	11 136	73	65	1.5
Oil	871	2 147	2 594	2 934	3 246	3 589	3 910	20	23	2.4
Gas	134	704	931	1 213	1 484	1 766	2 074	7	12	4.4
Power generation	**1 066**	**5 135**	**6 426**	**7 021**	**7 415**	**7 913**	**8 441**	**100**	**100**	**2.0**
Coal	886	4 652	5 895	6 431	6 738	7 120	7 508	91	89	1.9
Oil	143	160	133	101	78	69	64	3	1	-3.6
Gas	37	324	398	489	598	724	869	6	10	4.0
TFC	**2 344**	**4 983**	**6 063**	**6 673**	**7 169**	**7 659**	**8 104**	**100**	**100**	**2.0**
Coal	1 617	2 851	3 349	3 402	3 413	3 419	3 403	57	42	0.7
Oil	668	1 838	2 293	2 659	2 986	3 326	3 639	37	45	2.8
Transport	*290*	*1 013*	*1 351*	*1 666*	*1 972*	*2 300*	*2 611*	*20*	*32*	*3.9*
Gas	59	294	421	612	770	913	1 063	6	13	5.3

Non-OECD Asia: Current Policies and 450 Scenarios

	Electricity generation (TWh)						Shares (%)		CAAGR (%)	
	2020	2030	2035	2020	2030	2035	2035		2010-35	
	Current Policies Scenario			450 Scenario			CPS	450	CPS	450
Total generation	**11 316**	**15 691**	**18 095**	**9 827**	**11 916**	**13 123**	**100**	**100**	**4.3**	**3.0**
Coal	7 303	10 011	11 638	5 263	3 570	2 831	64	22	4.1	-1.6
Oil	108	81	78	90	56	54	0	0	-2.8	-4.2
Gas	1 032	1 548	1 876	1 100	1 712	2 025	10	15	4.3	4.6
Nuclear	662	1 027	1 117	727	1 672	2 119	6	16	8.5	11.3
Hydro	1 527	1 820	1 945	1 706	2 319	2 534	11	19	2.8	3.9
Bioenergy	194	344	416	222	622	874	2	7	12.2	15.6
Wind	395	687	809	542	1 337	1 670	4	13	10.6	13.8
Geothermal	30	49	62	51	119	155	0	1	4.7	8.6
Solar PV	61	118	141	117	428	613	1	5	20.9	28.2
CSP	2	6	11	9	79	244	0	2	n.a.	n.a.
Marine	-	1	2	0	2	4	0	0	n.a.	n.a.

	Electrical capacity (GW)						Shares (%)		CAAGR (%)	
	2020	2030	2035	2020	2030	2035	2035		2010-35	
	Current Policies Scenario			450 Scenario			CPS	450	CPS	450
Total capacity	**2 628**	**3 539**	**3 993**	**2 544**	**3 371**	**3 767**	**100**	**100**	**4.1**	**3.9**
Coal	1 424	1 908	2 172	1 142	927	821	54	22	3.9	-0.0
Oil	61	55	51	61	51	46	1	1	-1.0	-1.4
Gas	283	404	477	291	411	476	12	13	4.6	4.6
Nuclear	88	136	148	96	221	281	4	7	8.1	10.9
Hydro	475	569	610	532	734	806	15	21	2.9	4.0
Bioenergy	43	67	77	49	111	147	2	4	7.1	9.9
Wind	202	306	347	272	567	667	9	18	7.4	10.2
Geothermal	5	8	10	8	18	24	0	1	4.4	8.3
Solar PV	45	84	98	89	309	435	2	12	19.3	26.6
CSP	0	2	3	3	21	63	0	2	n.a.	n.a.
Marine	-	0	1	0	0	1	0	0	n.a.	n.a.

	CO_2 emissions (Mt)						Shares (%)		CAAGR (%)	
	2020	2030	2035	2020	2030	2035	2035		2010-35	
	Current Policies Scenario			450 Scenario			CPS	450	CPS	450
Total CO_2	**15 231**	**18 951**	**21 068**	**12 593**	**10 348**	**9 265**	**100**	**100**	**2.8**	**-0.5**
Coal	11 015	13 424	14 801	8 574	5 645	4 339	70	47	2.7	-2.3
Oil	2 994	3 772	4 199	2 814	3 044	3 054	20	33	2.7	1.4
Gas	1 222	1 755	2 068	1 205	1 659	1 872	10	20	4.4	4.0
Power generation	**7 816**	**10 206**	**11 661**	**5 779**	**3 400**	**2 444**	**100**	**100**	**3.3**	**-2.9**
Coal	7 235	9 442	10 756	5 193	2 658	1 615	92	66	3.4	-4.1
Oil	111	85	79	91	54	49	1	2	-2.8	-4.6
Gas	470	680	826	495	687	780	7	32	3.8	3.6
TFC	**6 888**	**8 158**	**8 779**	**6 328**	**6 477**	**6 351**	**100**	**100**	**2.3**	**1.0**
Coal	3 542	3 726	3 780	3 175	2 819	2 573	43	41	1.1	-0.4
Oil	2 707	3 485	3 901	2 556	2 815	2 830	44	45	3.1	1.7
Transport	*1 691*	*2 415*	*2 814*	*1 611*	*1 882*	*1 916*	*32*	*30*	*4.2*	*2.6*
Gas	639	947	1 099	597	843	948	13	15	5.4	4.8

China: New Policies Scenario

	Energy demand (Mtoe)							Shares (%)		CAAGR (%)
	1990	2010	2015	2020	2025	2030	2035	2010	2035	2010-35
TPED	**881**	**2 416**	**3 020**	**3 359**	**3 574**	**3 742**	**3 872**	**100**	**100**	**1.9**
Coal	534	1 602	1 931	1 969	1 966	1 976	1 967	66	51	0.8
Oil	123	420	536	614	668	702	712	17	18	2.1
Gas	13	92	163	254	326	392	455	4	12	6.6
Nuclear	-	19	65	140	201	229	256	1	7	10.9
Hydro	11	62	80	101	108	114	118	3	3	2.6
Bioenergy	200	206	205	217	219	219	227	9	6	0.4
Other renewables	0	16	40	65	87	110	138	1	4	9.0
Power generation	**180**	**1 004**	**1 356**	**1 576**	**1 729**	**1 865**	**1 993**	**100**	**100**	**2.8**
Coal	153	884	1 122	1 175	1 199	1 239	1 269	88	64	1.5
Oil	16	9	8	7	7	6	6	1	0	-1.9
Gas	1	20	45	69	94	122	148	2	7	8.2
Nuclear	-	19	65	140	201	229	256	2	13	10.9
Hydro	11	62	80	101	108	114	118	6	6	2.6
Bioenergy	-	5	15	41	59	74	93	1	5	12.4
Other renewables	0	4	22	43	61	81	103	0	5	13.8
Other energy sector	**100**	**349**	**402**	**417**	**420**	**420**	**419**	**100**	**100**	**0.7**
Electricity	*15*	*64*	*85*	*96*	*104*	*112*	*119*	*18*	*28*	*2.5*
TFC	**672**	**1 506**	**1 876**	**2 099**	**2 235**	**2 335**	**2 402**	**100**	**100**	**1.9**
Coal	320	514	582	564	541	514	478	34	20	-0.3
Oil	88	357	472	554	614	656	677	24	28	2.6
Gas	9	57	98	161	204	237	269	4	11	6.4
Electricity	41	300	439	544	612	676	736	20	31	3.6
Heat	13	64	76	80	79	77	74	4	3	0.6
Bioenergy	200	201	191	176	160	145	134	13	6	-1.6
Other renewables	0	12	18	22	26	30	34	1	1	4.3
Industry	**245**	**714**	**904**	**1 001**	**1 046**	**1 074**	**1 090**	**100**	**100**	**1.7**
Coal	181	401	463	447	430	411	385	56	35	-0.2
Oil	21	48	59	63	63	63	63	7	6	1.1
Gas	2	16	35	79	100	111	120	2	11	8.3
Electricity	30	203	293	356	398	435	472	29	43	3.4
Heat	11	45	54	56	55	53	50	6	5	0.4
Bioenergy	-	-	-	-	-	-	-	-	-	n.a.
Other renewables	-	0	0	0	0	0	0	0	0	3.3
Transport	**35**	**184**	**268**	**351**	**420**	**477**	**517**	**100**	**100**	**4.2**
Oil	25	169	249	323	384	431	460	92	89	4.1
Electricity	1	3	8	10	12	15	19	2	4	7.0
Biofuels	-	1	2	6	10	16	22	1	4	12.4
Other fuels	10	10	10	12	14	15	16	6	3	1.6
Buildings	**314**	**452**	**517**	**552**	**568**	**579**	**586**	**100**	**100**	**1.0**
Coal	95	68	72	67	61	53	44	15	7	-1.8
Oil	7	49	54	53	50	45	39	11	7	-0.9
Gas	2	24	41	56	74	92	110	5	19	6.2
Electricity	6	81	123	162	185	208	226	18	39	4.2
Heat	2	19	22	23	24	24	24	4	4	0.9
Bioenergy	200	200	188	170	149	129	111	44	19	-2.3
Other renewables	0	11	17	21	25	29	33	3	6	4.4
Other	**78**	**156**	**187**	**196**	**201**	**205**	**208**	**100**	**100**	**1.2**

China: Current Policies and 450 Scenarios

	Energy demand (Mtoe)						Shares (%)		CAAGR (%)	
	2020	2030	2035	2020	2030	2035	2035		2010-35	
	Current Policies Scenario			450 Scenario			CPS	450	CPS	450
TPED	**3 519**	**4 144**	**4 406**	**3 106**	**3 077**	**3 070**	**100**	**100**	**2.4**	**1.0**
Coal	2 148	2 429	2 561	1 719	1 248	1 054	58	34	1.9	-1.7
Oil	621	734	766	589	577	531	17	17	2.4	0.9
Gas	253	379	446	254	391	449	10	15	6.5	6.6
Nuclear	134	211	232	150	326	408	5	13	10.5	13.0
Hydro	96	108	113	104	118	122	3	4	2.4	2.7
Bioenergy	216	205	197	219	256	281	4	9	-0.2	1.2
Other renewables	51	79	91	71	162	226	2	7	7.2	11.2
Power generation	**1 697**	**2 180**	**2 411**	**1 393**	**1 418**	**1 486**	**100**	**100**	**3.6**	**1.6**
Coal	1 322	1 627	1 785	969	599	459	74	31	2.9	-2.6
Oil	8	7	6	7	5	4	0	0	-1.4	-3.5
Gas	63	105	134	71	133	162	6	11	7.8	8.6
Nuclear	134	211	232	150	326	408	10	27	10.5	13.0
Hydro	96	108	113	104	118	122	5	8	2.4	2.7
Bioenergy	42	67	74	44	106	142	3	10	11.3	14.3
Other renewables	32	56	66	48	131	189	3	13	11.8	16.6
Other energy sector	**430**	**456**	**467**	**392**	**355**	**336**	**100**	**100**	**1.2**	**-0.1**
Electricity	*101*	*128*	*140*	*86*	*88*	*88*	*30*	*26*	*3.2*	*1.3*
TFC	**2 162**	**2 495**	**2 618**	**1 987**	**2 030**	**2 004**	**100**	**100**	**2.2**	**1.2**
Coal	589	559	528	532	457	413	20	21	0.1	-0.9
Oil	561	688	731	532	540	508	28	25	2.9	1.4
Gas	166	241	274	159	225	250	10	12	6.5	6.1
Electricity	570	759	850	490	562	597	32	30	4.2	2.8
Heat	85	87	86	76	65	61	3	3	1.2	-0.2
Bioenergy	174	138	123	176	150	139	5	7	-1.9	-1.5
Other renewables	19	23	25	22	31	36	1	2	2.9	4.5
Industry	**1 043**	**1 168**	**1 212**	**941**	**966**	**959**	**100**	**100**	**2.1**	**1.2**
Coal	462	441	421	420	366	334	35	35	0.2	-0.7
Oil	65	64	64	59	61	60	5	6	1.1	0.9
Gas	84	118	129	86	125	136	11	14	8.6	8.8
Electricity	372	482	536	323	369	386	44	40	4.0	2.6
Heat	60	62	61	53	42	39	5	4	1.2	-0.6
Bioenergy	-	-	-	-	-	-	-	-	n.a.	n.a.
Other renewables	0	0	0	0	2	4	0	0	3.3	13.8
Transport	**352**	**497**	**555**	**339**	**381**	**374**	**100**	**100**	**4.5**	**2.9**
Oil	326	457	506	310	328	306	91	82	4.5	2.4
Electricity	10	14	18	10	17	27	3	7	6.8	8.6
Biofuels	4	11	15	8	26	32	3	9	10.5	14.0
Other fuels	12	15	16	11	10	9	3	3	1.8	-0.4
Buildings	**568**	**618**	**633**	**512**	**482**	**467**	**100**	**100**	**1.4**	**0.1**
Coal	75	66	55	63	43	32	9	7	-0.9	-3.0
Oil	53	47	42	50	37	30	7	6	-0.6	-2.0
Gas	56	88	104	48	70	81	16	17	6.0	4.9
Electricity	171	243	276	140	158	164	44	35	5.0	2.9
Heat	24	25	25	23	23	23	4	5	1.1	0.7
Bioenergy	170	127	108	167	124	107	17	23	-2.4	-2.5
Other renewables	18	22	24	21	28	32	4	7	3.0	4.2
Other	**199**	**212**	**218**	**194**	**201**	**203**	**100**	**100**	**1.4**	**1.1**

China: New Policies Scenario

	Electricity generation (TWh)							Shares (%)		CAAGR (%)
	1990	2010	2015	2020	2025	2030	2035	2010	2035	2010-35
Total generation	**650**	**4 247**	**6 107**	**7 445**	**8 337**	**9 170**	**9 945**	**100**	**100**	**3.5**
Coal	471	3 297	4 410	4 759	4 948	5 219	5 453	78	55	2.0
Oil	49	13	14	13	12	12	11	0	0	-1.0
Gas	3	83	211	346	493	660	809	2	8	9.5
Nuclear	-	74	248	539	772	879	983	2	10	10.9
Hydro	127	722	930	1 179	1 254	1 321	1 372	17	14	2.6
Bioenergy	-	11	46	145	211	261	325	0	3	14.3
Wind	0	45	218	393	529	653	762	1	8	12.0
Geothermal	-	0	1	3	6	9	13	0	0	19.3
Solar PV	0	1	27	65	103	131	153	0	2	22.6
CSP	-	-	2	5	10	23	61	-	1	n.a.
Marine	-	-	-	-	0	1	2	-	0	n.a.

	Electrical capacity (GW)						Shares (%)		CAAGR (%)
	2010	2015	2020	2025	2030	2035	2010	2035	2010-35
Total capacity	**997**	**1 444**	**1 811**	**2 038**	**2 228**	**2 398**	**100**	**100**	**3.6**
Coal	671	885	982	1 031	1 079	1 122	67	47	2.1
Oil	15	11	11	10	10	10	2	0	-1.8
Gas	35	72	106	137	172	204	3	9	7.3
Nuclear	11	33	70	101	115	128	1	5	10.4
Hydro	213	285	360	383	404	420	21	18	2.7
Bioenergy	6	13	30	42	50	58	1	2	9.9
Wind	45	123	200	254	293	326	4	14	8.3
Geothermal	0	0	0	1	1	2	0	0	18.7
Solar PV	1	20	50	78	98	113	0	5	21.4
CSP	-	1	1	3	6	14	-	1	n.a.
Marine	-	-	-	0	0	0	-	0	n.a.

	CO_2 emissions (Mt)							Shares (%)		CAAGR (%)
	1990	2010	2015	2020	2025	2030	2035	2010	2035	2010-35
Total CO_2	**2 289**	**7 214**	**8 952**	**9 532**	**9 839**	**10 108**	**10 224**	**100**	**100**	**1.4**
Coal	1 945	6 009	7 275	7 407	7 374	7 369	7 281	83	71	0.8
Oil	317	1 004	1 314	1 551	1 725	1 849	1 909	14	19	2.6
Gas	27	201	364	574	740	890	1 034	3	10	6.8
Power generation	**650**	**3 625**	**4 590**	**4 853**	**4 984**	**5 168**	**5 302**	**100**	**100**	**1.5**
Coal	597	3 546	4 459	4 667	4 741	4 861	4 936	98	93	1.3
Oil	51	31	27	25	23	22	19	1	0	-1.9
Gas	2	48	105	161	220	285	347	1	7	8.2
TFC	**1 552**	**3 316**	**4 038**	**4 341**	**4 511**	**4 591**	**4 570**	**100**	**100**	**1.3**
Coal	1 297	2 284	2 607	2 528	2 422	2 299	2 139	69	47	-0.3
Oil	237	905	1 212	1 448	1 625	1 754	1 821	27	40	2.8
Transport	*73*	*505*	*742*	*963*	*1 144*	*1 285*	*1 371*	*15*	*30*	*4.1*
Gas	17	127	219	364	463	539	610	4	13	6.5

China: Current Policies and 450 Scenarios

	Electricity generation (TWh)						Shares (%)		CAAGR (%)	
	2020	2030	2035	2020	2030	2035	2035		2010-35	
	Current Policies Scenario			450 Scenario			CPS	450	CPS	450
Total generation	**7 804**	**10 314**	**11 523**	**6 701**	**7 565**	**7 968**	**100**	**100**	**4.1**	**2.5**
Coal	5 342	6 837	7 629	3 861	2 501	1 824	66	23	3.4	-2.3
Oil	14	12	11	13	10	8	0	0	-0.8	-1.9
Gas	311	552	713	359	747	918	6	12	9.0	10.1
Nuclear	515	809	892	575	1 252	1 564	8	20	10.5	13.0
Hydro	1 112	1 260	1 320	1 212	1 373	1 420	11	18	2.4	2.7
Bioenergy	148	236	258	153	367	490	2	6	13.3	16.2
Wind	325	545	620	442	994	1 201	5	15	11.1	14.1
Geothermal	1	3	6	3	14	21	0	0	15.3	21.6
Solar PV	34	55	64	76	249	342	1	4	18.4	26.6
CSP	1	5	10	6	57	176	0	2	n.a.	n.a.
Marine	-	1	2	-	1	2	0	0	n.a.	n.a.

	Electrical capacity (GW)						Shares (%)		CAAGR (%)	
	2020	2030	2035	2020	2030	2035	2035		2010-35	
	Current Policies Scenario			450 Scenario			CPS	450	CPS	450
Total capacity	**1 823**	**2 354**	**2 587**	**1 732**	**2 120**	**2 277**	**100**	**100**	**3.9**	**3.4**
Coal	1 080	1 358	1 491	859	666	563	58	25	3.2	-0.7
Oil	11	9	9	11	10	9	0	0	-2.2	-2.0
Gas	104	169	208	104	177	215	8	9	7.4	7.6
Nuclear	67	105	116	75	163	203	4	9	10.0	12.5
Hydro	340	386	404	370	420	435	16	19	2.6	2.9
Bioenergy	30	45	49	31	67	84	2	4	9.1	11.5
Wind	165	239	261	220	413	468	10	21	7.3	9.8
Geothermal	0	1	1	0	2	3	0	0	14.9	20.8
Solar PV	26	40	46	60	188	256	2	11	17.1	25.4
CSP	0	1	2	2	14	41	0	2	n.a.	n.a.
Marine	-	0	0	-	0	1	0	0	n.a.	n.a.

	CO_2 emissions (Mt)						Shares (%)		CAAGR (%)	
	2020	2030	2035	2020	2030	2035	2035		2010-35	
	Current Policies Scenario			450 Scenario			CPS	450	CPS	450
Total CO_2	**10 251**	**11 968**	**12 727**	**8 419**	**6 205**	**4 948**	**100**	**100**	**2.3**	**-1.5**
Coal	8 108	9 166	9 648	6 368	3 877	2 632	76	53	1.9	-3.2
Oil	1 571	1 942	2 067	1 479	1 479	1 373	16	28	2.9	1.3
Gas	572	860	1 013	573	849	943	8	19	6.7	6.4
Power generation	**5 426**	**6 709**	**7 388**	**3 998**	**2 113**	**1 208**	**100**	**100**	**2.9**	**-4.3**
Coal	5 253	6 440	7 052	3 807	1 810	869	95	72	2.8	-5.5
Oil	27	24	22	24	17	13	0	1	-1.4	-3.4
Gas	147	245	314	166	287	326	4	27	7.8	8.0
TFC	**4 473**	**4 876**	**4 943**	**4 106**	**3 809**	**3 475**	**100**	**100**	**1.6**	**0.2**
Coal	2 630	2 486	2 349	2 367	1 914	1 629	48	47	0.1	-1.3
Oil	1 466	1 843	1 973	1 381	1 398	1 304	40	38	3.2	1.5
Transport	*972*	*1 360*	*1 505*	*923*	*978*	*912*	*30*	*26*	*4.5*	*2.4*
Gas	377	548	621	358	497	542	13	16	6.6	6.0

India: New Policies Scenario

	Energy demand (Mtoe)							Shares (%)		CAAGR (%)
	1990	2010	2015	2020	2025	2030	2035	2010	2035	2010-35
TPED	**317**	**691**	**837**	**974**	**1 120**	**1 300**	**1 516**	**100**	**100**	**3.2**
Coal	103	283	371	442	502	572	657	41	43	3.4
Oil	61	166	192	216	249	304	372	24	25	3.3
Gas	11	53	63	76	96	120	148	8	10	4.2
Nuclear	2	7	11	20	32	42	52	1	3	8.5
Hydro	6	10	13	16	21	26	30	1	2	4.6
Bioenergy	133	170	182	192	201	209	219	25	14	1.0
Other renewables	0	2	6	11	18	27	37	0	2	12.4
Power generation	**70**	**257**	**327**	**402**	**478**	**565**	**671**	**100**	**100**	**3.9**
Coal	56	201	254	299	333	373	425	78	63	3.1
Oil	4	11	10	8	6	5	4	4	1	-3.6
Gas	3	26	31	42	55	70	89	10	13	5.0
Nuclear	2	7	11	20	32	42	52	3	8	8.5
Hydro	6	10	13	16	21	26	30	4	5	4.6
Bioenergy	-	1	3	6	14	24	36	0	5	14.6
Other renewables	0	2	5	10	16	25	34	1	5	12.7
Other energy sector	**20**	**55**	**74**	**88**	**105**	**126**	**151**	**100**	**100**	**4.1**
Electricity	*7*	*22*	*30*	*39*	*49*	*60*	*72*	*40*	*48*	*4.9*
TFC	**252**	**462**	**550**	**634**	**724**	**841**	**978**	**100**	**100**	**3.0**
Coal	42	76	102	122	142	164	188	16	19	3.7
Oil	53	132	157	185	220	275	340	29	35	3.8
Gas	6	23	27	29	35	42	51	5	5	3.2
Electricity	18	61	85	111	139	172	212	13	22	5.1
Heat	-	-	-	-	-	-	-	-	-	n.a.
Bioenergy	133	169	179	186	187	185	183	37	19	0.3
Other renewables	0	0	1	1	2	2	3	0	0	10.1
Industry	**70**	**152**	**193**	**236**	**277**	**324**	**375**	**100**	**100**	**3.7**
Coal	29	62	86	106	126	149	175	41	47	4.2
Oil	10	26	28	32	35	37	40	17	11	1.8
Gas	0	7	8	9	11	14	16	5	4	3.1
Electricity	9	28	38	51	65	80	98	18	26	5.2
Heat	-	-	-	-	-	-	-	-	-	n.a.
Bioenergy	23	29	33	37	40	43	47	19	12	1.9
Other renewables	-	-	-	-	-	-	-	-	-	n.a.
Transport	**27**	**55**	**68**	**85**	**113**	**162**	**225**	**100**	**100**	**5.8**
Oil	24	52	62	76	101	145	200	93	89	5.6
Electricity	0	1	2	2	2	2	2	2	1	2.7
Biofuels	-	0	2	2	3	6	10	0	4	17.2
Other fuels	2	2	3	4	7	10	13	4	6	7.2
Buildings	**137**	**198**	**219**	**234**	**246**	**258**	**272**	**100**	**100**	**1.3**
Coal	11	14	16	16	15	15	14	7	5	-0.0
Oil	11	23	27	30	33	36	40	12	15	2.2
Gas	0	0	0	1	2	3	5	0	2	23.7
Electricity	4	21	30	40	52	66	83	11	31	5.6
Heat	-	-	-	-	-	-	-	-	-	n.a.
Bioenergy	111	140	145	146	143	136	127	71	47	-0.4
Other renewables	0	0	1	1	2	2	3	0	1	9.8
Other	**17**	**56**	**70**	**80**	**88**	**97**	**106**	**100**	**100**	**2.6**

India: Current Policies and 450 Scenarios

	Energy demand (Mtoe)						Shares (%)		CAAGR (%)	
	2020	2030	2035	2020	2030	2035	2035		2010-35	
	Current Policies Scenario			450 Scenario			CPS	450	CPS	450
TPED	**1 013**	**1 407**	**1 680**	**904**	**1 089**	**1 233**	**100**	**100**	**3.6**	**2.3**
Coal	478	701	862	371	330	340	51	28	4.6	0.7
Oil	225	322	399	210	279	317	24	26	3.6	2.6
Gas	77	113	130	81	123	149	8	12	3.7	4.2
Nuclear	20	35	43	21	67	92	3	8	7.6	11.0
Hydro	15	21	23	17	39	48	1	4	3.5	6.6
Bioenergy	190	201	205	191	215	228	12	18	0.7	1.2
Other renewables	9	15	18	12	36	59	1	5	9.3	14.5
Power generation	**431**	**646**	**798**	**348**	**402**	**475**	**100**	**100**	**4.6**	**2.5**
Coal	332	489	611	238	153	137	77	29	4.6	-1.5
Oil	11	9	8	7	3	3	1	1	-1.6	-5.1
Gas	39	62	73	46	71	87	9	18	4.2	5.0
Nuclear	20	35	43	21	67	92	5	19	7.6	11.0
Hydro	15	21	23	17	39	48	3	10	3.5	6.6
Bioenergy	6	18	25	8	35	53	3	11	12.9	16.4
Other renewables	8	13	16	11	34	55	2	11	9.3	14.8
Other energy sector	**91**	**135**	**164**	**83**	**110**	**126**	**100**	**100**	**4.5**	**3.4**
Electricity	*41*	*66*	*82*	*35*	*48*	*56*	*50*	*45*	*5.4*	*3.8*
TFC	**647**	**873**	**1 025**	**611**	**774**	**868**	**100**	**100**	**3.2**	**2.6**
Coal	125	174	205	114	144	164	20	19	4.0	3.1
Oil	190	288	363	181	254	291	35	34	4.1	3.2
Gas	32	44	49	30	45	54	5	6	3.1	3.5
Electricity	114	181	225	102	149	180	22	21	5.4	4.4
Heat	-	-	-	-	-	-	-	-	n.a.	n.a.
Bioenergy	184	183	180	183	179	175	18	20	0.3	0.1
Other renewables	1	2	3	1	3	4	0	0	9.1	11.2
Industry	**240**	**336**	**393**	**223**	**293**	**332**	**100**	**100**	**3.9**	**3.2**
Coal	109	159	190	99	132	153	48	46	4.6	3.7
Oil	33	40	43	30	35	37	11	11	2.0	1.4
Gas	11	16	17	8	13	15	4	4	3.4	2.7
Electricity	52	83	103	49	71	85	26	26	5.4	4.6
Heat	-	-	-	-	-	-	-	-	n.a.	n.a.
Bioenergy	35	39	40	36	42	43	10	13	1.3	1.6
Other renewables	-	-	-	-	-	-	-	-	n.a.	n.a.
Transport	**88**	**169**	**236**	**86**	**152**	**191**	**100**	**100**	**6.0**	**5.1**
Oil	79	153	216	76	131	160	91	84	5.9	4.6
Electricity	2	2	2	2	3	5	1	2	2.7	5.8
Biofuels	2	5	9	3	7	10	4	5	16.7	17.5
Other fuels	4	9	10	5	12	15	4	8	6.0	7.9
Buildings	**239**	**269**	**287**	**223**	**233**	**239**	**100**	**100**	**1.5**	**0.8**
Coal	16	15	15	15	12	11	5	5	0.2	-1.0
Oil	31	39	44	28	32	34	15	14	2.7	1.6
Gas	1	3	4	1	4	6	1	3	22.5	24.9
Electricity	43	71	91	34	51	63	32	26	6.0	4.5
Heat	-	-	-	-	-	-	-	-	n.a.	n.a.
Bioenergy	147	139	131	144	131	121	46	51	-0.2	-0.6
Other renewables	1	2	2	1	2	3	1	1	8.8	10.3
Other	**80**	**99**	**108**	**79**	**97**	**105**	**100**	**100**	**2.7**	**2.6**

India: New Policies Scenario

	Electricity generation (TWh)							Shares (%)		CAAGR (%)
	1990	2010	2015	2020	2025	2030	2035	2010	2035	2010-35
Total generation	**289**	**960**	**1 326**	**1 734**	**2 175**	**2 691**	**3 298**	**100**	**100**	**5.1**
Coal	192	653	888	1 095	1 264	1 472	1 741	68	53	4.0
Oil	10	26	25	20	17	14	13	3	0	-2.7
Gas	10	118	158	224	306	400	518	12	16	6.1
Nuclear	6	26	41	77	123	160	201	3	6	8.5
Hydro	72	114	148	191	245	302	352	12	11	4.6
Bioenergy	-	2	8	18	43	76	113	0	3	17.4
Wind	0	20	49	83	124	166	211	2	6	9.9
Geothermal	-	-	0	0	1	1	2	-	0	n.a.
Solar PV	-	0	7	23	49	90	134	0	4	41.5
CSP	-	-	1	2	4	8	13	-	0	n.a.
Marine	-	-	-	0	0	0	1	-	0	n.a.

	Electrical capacity (GW)						Shares (%)		CAAGR (%)
	2010	2015	2020	2025	2030	2035	2010	2035	2010-35
Total capacity	**189**	**328**	**422**	**527**	**663**	**808**	**100**	**100**	**6.0**
Coal	101	190	221	248	293	341	53	42	5.0
Oil	7	8	8	9	8	8	4	1	0.3
Gas	21	37	52	69	86	109	11	14	6.9
Nuclear	5	6	12	18	23	29	2	4	7.7
Hydro	40	48	62	80	99	115	21	14	4.3
Bioenergy	3	4	6	9	13	19	1	2	8.5
Wind	13	27	44	62	80	97	7	12	8.4
Geothermal	-	0	0	0	0	0	-	0	n.a.
Solar PV	0	5	16	32	58	85	0	11	30.0
CSP	-	0	1	1	2	4	-	0	n.a.
Marine	-	-	0	0	0	0	-	0	n.a.

	CO_2 emissions (Mt)							Shares (%)		CAAGR (%)
	1990	2010	2015	2020	2025	2030	2035	2010	2035	2010-35
Total CO_2	**582**	**1 635**	**2 054**	**2 415**	**2 775**	**3 247**	**3 830**	**100**	**100**	**3.5**
Coal	396	1 093	1 424	1 688	1 903	2 156	2 471	67	65	3.3
Oil	166	429	495	561	659	824	1 026	26	27	3.6
Gas	21	113	134	167	213	267	333	7	9	4.4
Power generation	**235**	**872**	**1 088**	**1 281**	**1 438**	**1 624**	**1 869**	**100**	**100**	**3.1**
Coal	215	776	984	1 158	1 289	1 444	1 647	89	88	3.1
Oil	11	35	31	25	19	16	14	4	1	-3.6
Gas	8	61	73	98	129	165	208	7	11	5.0
TFC	**330**	**702**	**891**	**1 055**	**1 247**	**1 516**	**1 834**	**100**	**100**	**3.9**
Coal	175	315	437	525	609	706	817	45	45	3.9
Oil	146	343	403	473	568	723	910	49	50	4.0
Transport	*74*	*156*	*187*	*230*	*303*	*436*	*603*	*22*	*33*	*5.6*
Gas	9	44	51	56	70	87	106	6	6	3.6

India: Current Policies and 450 Scenarios

	Electricity generation (TWh)						Shares (%)		CAAGR (%)	
	2020	2030	2035	2020	2030	2035	2035		2010-35	
	Current Policies Scenario			450 Scenario			CPS	450	CPS	450
Total generation	**1 799**	**2 864**	**3 565**	**1 590**	**2 280**	**2 743**	**100**	**100**	**5.4**	**4.3**
Coal	1 207	1 916	2 446	895	692	627	69	23	5.4	-0.2
Oil	28	23	20	17	10	9	1	0	-1.1	-4.3
Gas	212	349	409	253	415	516	11	19	5.1	6.1
Nuclear	77	134	165	82	257	355	5	13	7.6	11.0
Hydro	173	239	272	196	454	564	8	21	3.5	6.6
Bioenergy	17	54	76	25	113	175	2	6	15.5	19.5
Wind	63	100	118	88	196	257	3	9	7.4	10.8
Geothermal	0	1	1	1	2	4	0	0	n.a.	n.a.
Solar PV	22	48	57	30	119	170	2	6	36.7	42.8
CSP	0	0	1	3	22	66	0	2	n.a.	n.a.
Marine	-	0	1	0	1	1	0	0	n.a.	n.a.

	Electrical capacity (GW)						Shares (%)		CAAGR (%)	
	2020	2030	2035	2020	2030	2035	2035		2010-35	
	Current Policies Scenario			450 Scenario			CPS	450	CPS	450
Total capacity	**402**	**624**	**751**	**407**	**653**	**791**	**100**	**100**	**5.7**	**5.9**
Coal	221	350	435	186	177	176	58	22	6.0	2.3
Oil	8	8	8	8	8	7	1	1	0.1	-0.3
Gas	52	79	90	63	94	109	12	14	6.0	6.8
Nuclear	12	20	24	12	37	51	3	6	6.9	10.2
Hydro	56	78	88	64	148	184	12	23	3.2	6.3
Bioenergy	5	10	14	7	19	29	2	4	7.0	10.2
Wind	33	48	55	46	87	106	7	13	5.9	8.7
Geothermal	0	0	0	0	0	1	0	0	n.a.	n.a.
Solar PV	15	32	38	20	76	107	5	13	25.9	31.2
CSP	0	0	0	1	6	22	0	3	n.a.	n.a.
Marine	-	0	0	0	0	0	0	0	n.a.	n.a.

	CO_2 emissions (Mt)						Shares (%)		CAAGR (%)	
	2020	2030	2035	2020	2030	2035	2035		2010-35	
	Current Policies Scenario			450 Scenario			CPS	450	CPS	450
Total CO_2	**2 579**	**3 779**	**4 654**	**2 125**	**2 113**	**2 238**	**100**	**100**	**4.3**	**1.3**
Coal	1 824	2 648	3 254	1 404	1 100	1 059	70	47	4.5	-0.1
Oil	588	880	1 110	544	743	851	24	38	3.9	2.8
Gas	167	251	289	177	270	328	6	15	3.8	4.4
Power generation	**1 411**	**2 067**	**2 559**	**1 045**	**720**	**684**	**100**	**100**	**4.4**	**-1.0**
Coal	1 285	1 895	2 366	917	547	475	92	69	4.6	-1.9
Oil	34	28	23	21	11	9	1	1	-1.6	-5.1
Gas	92	145	169	107	163	200	7	29	4.2	4.9
TFC	**1 087**	**1 602**	**1 962**	**1 003**	**1 294**	**1 443**	**100**	**100**	**4.2**	**2.9**
Coal	535	748	882	483	548	578	45	40	4.2	2.5
Oil	489	763	979	461	655	756	50	52	4.3	3.2
Transport	*239*	*460*	*649*	*230*	*393*	*483*	*33*	*33*	*5.9*	*4.6*
Gas	63	90	101	58	91	109	5	8	3.4	3.7

Africa: New Policies Scenario

	Energy demand (Mtoe)							Shares (%)		CAAGR (%)
	1990	2010	2015	2020	2025	2030	2035	2010	2035	2010-35
TPED	**388**	**690**	**750**	**819**	**877**	**932**	**984**	**100**	**100**	**1.4**
Coal	74	112	116	128	133	135	137	16	14	0.8
Oil	86	149	162	174	182	190	201	22	20	1.2
Gas	30	86	99	117	128	139	148	12	15	2.2
Nuclear	2	3	3	3	8	14	16	0	2	6.7
Hydro	5	9	11	14	18	22	26	1	3	4.4
Bioenergy	191	328	356	377	396	413	427	48	43	1.1
Other renewables	0	2	3	6	11	19	29	0	3	12.5
Power generation	**68**	**141**	**160**	**186**	**209**	**233**	**259**	**100**	**100**	**2.5**
Coal	39	63	70	78	81	82	83	45	32	1.1
Oil	11	20	19	18	16	13	12	14	5	-1.9
Gas	11	44	52	63	69	74	78	31	30	2.3
Nuclear	2	3	3	3	8	14	16	2	6	6.7
Hydro	5	9	11	14	18	22	26	6	10	4.4
Bioenergy	0	0	2	4	7	11	16	0	6	17.2
Other renewables	0	1	3	5	10	18	28	1	11	12.5
Other energy sector	**57**	**94**	**97**	**105**	**111**	**115**	**118**	**100**	**100**	**0.9**
Electricity	*5*	*10*	*12*	*14*	*16*	*17*	*19*	*11*	*16*	*2.5*
TFC	**291**	**513**	**562**	**610**	**651**	**691**	**727**	**100**	**100**	**1.4**
Coal	20	21	21	22	22	22	22	4	3	0.1
Oil	71	125	139	155	168	181	194	24	27	1.8
Gas	9	27	29	32	35	38	41	5	6	1.8
Electricity	22	48	57	68	78	89	102	9	14	3.0
Heat	-	-	-	-	-	-	-	-	-	n.a.
Bioenergy	170	292	316	333	347	359	367	57	50	0.9
Other renewables	-	0	0	0	1	1	1	0	0	13.3
Industry	**60**	**86**	**95**	**107**	**115**	**122**	**128**	**100**	**100**	**1.6**
Coal	14	12	13	13	13	13	13	14	10	0.5
Oil	14	13	14	16	16	17	17	15	13	1.1
Gas	5	12	13	15	17	19	20	14	16	2.2
Electricity	12	20	23	27	29	32	34	24	26	2.0
Heat	-	-	-	-	-	-	-	-	-	n.a.
Bioenergy	16	29	32	36	39	41	43	33	34	1.7
Other renewables	-	-	-	-	-	-	-	-	-	n.a.
Transport	**36**	**78**	**88**	**99**	**109**	**119**	**129**	**100**	**100**	**2.0**
Oil	36	76	86	97	106	116	126	98	97	2.0
Electricity	0	0	1	1	1	1	1	1	1	2.4
Biofuels	-	-	0	0	0	0	0	-	0	n.a.
Other fuels	0	1	1	2	2	2	2	2	2	1.8
Buildings	**179**	**321**	**347**	**370**	**390**	**410**	**429**	**100**	**100**	**1.2**
Coal	3	8	8	8	7	7	7	2	2	-0.4
Oil	13	20	21	22	24	27	29	6	7	1.5
Gas	1	6	6	6	7	8	9	2	2	2.0
Electricity	9	25	31	38	45	53	62	8	15	3.7
Heat	-	-	-	-	-	-	-	-	-	n.a.
Bioenergy	153	262	282	296	306	315	321	82	75	0.8
Other renewables	-	0	0	0	0	0	1	0	0	9.3
Other	**15**	**29**	**31**	**35**	**37**	**39**	**41**	**100**	**100**	**1.4**

Africa: Current Policies and 450 Scenarios

	Energy demand (Mtoe)						Shares (%)		CAAGR (%)	
	2020	2030	2035	2020	2030	2035	2035		2010-35	
	Current Policies Scenario			450 Scenario			CPS	450	CPS	450
TPED	**833**	**962**	**1 026**	**785**	**849**	**878**	**100**	**100**	**1.6**	**1.0**
Coal	134	152	163	119	111	105	16	12	1.5	-0.3
Oil	177	198	212	159	150	146	21	17	1.4	-0.1
Gas	119	151	164	109	110	98	16	11	2.6	0.5
Nuclear	3	8	9	3	20	28	1	3	4.2	9.2
Hydro	13	21	25	15	22	27	2	3	4.1	4.5
Bioenergy	380	418	432	373	403	414	42	47	1.1	0.9
Other renewables	5	13	21	7	33	58	2	7	11.0	15.6
Power generation	**192**	**247**	**280**	**172**	**207**	**230**	**100**	**100**	**2.8**	**2.0**
Coal	84	96	106	70	60	53	38	23	2.1	-0.7
Oil	19	16	15	17	10	9	5	4	-1.2	-3.0
Gas	64	83	91	56	49	36	32	15	2.9	-0.9
Nuclear	3	8	9	3	20	28	3	12	4.2	9.2
Hydro	13	21	25	15	22	27	9	12	4.1	4.5
Bioenergy	3	10	15	5	14	20	5	9	16.7	18.2
Other renewables	5	12	20	6	32	57	7	25	10.9	15.7
Other energy sector	**106**	**119**	**124**	**103**	**108**	**109**	**100**	**100**	**1.1**	**0.6**
Electricity	*15*	*19*	*21*	*13*	*15*	*16*	*17*	*15*	*3.0*	*1.8*
TFC	**619**	**710**	**753**	**587**	**630**	**646**	**100**	**100**	**1.5**	**0.9**
Coal	22	23	23	21	20	20	3	3	0.3	-0.2
Oil	156	187	204	142	144	143	27	22	2.0	0.5
Gas	33	40	43	31	35	38	6	6	1.9	1.4
Electricity	70	95	110	64	81	91	15	14	3.4	2.6
Heat	-	-	-	-	-	-	-	-	n.a.	n.a.
Bioenergy	337	364	372	328	347	352	49	54	1.0	0.7
Other renewables	0	1	1	0	1	2	0	0	12.6	14.2
Industry	**110**	**130**	**139**	**102**	**113**	**117**	**100**	**100**	**1.9**	**1.2**
Coal	13	14	14	12	12	12	10	10	0.7	0.1
Oil	16	18	19	15	14	14	14	12	1.5	0.3
Gas	16	20	22	15	18	19	16	16	2.5	1.9
Electricity	28	35	38	26	29	31	27	26	2.5	1.6
Heat	-	-	-	-	-	-	-	-	n.a.	n.a.
Bioenergy	37	43	46	35	39	41	33	35	1.9	1.4
Other renewables	-	-	-	-	-	-	-	-	n.a.	n.a.
Transport	**99**	**122**	**134**	**90**	**89**	**87**	**100**	**100**	**2.2**	**0.5**
Oil	97	119	131	86	85	82	98	94	2.2	0.3
Electricity	1	1	1	1	1	1	0	1	1.8	4.1
Biofuels	-	-	-	1	2	3	-	3	n.a.	n.a.
Other fuels	2	2	2	1	1	2	2	2	1.6	0.6
Buildings	**374**	**417**	**437**	**361**	**389**	**401**	**100**	**100**	**1.2**	**0.9**
Coal	8	7	7	7	7	6	2	2	-0.3	-0.7
Oil	23	27	30	21	24	26	7	6	1.7	1.1
Gas	7	8	9	6	7	7	2	2	2.1	1.0
Electricity	39	55	65	35	48	55	15	14	3.8	3.2
Heat	-	-	-	-	-	-	-	-	n.a.	n.a.
Bioenergy	298	319	324	291	303	306	74	76	0.9	0.6
Other renewables	0	0	1	0	1	1	0	0	9.4	10.8
Other	**35**	**41**	**43**	**34**	**39**	**40**	**100**	**100**	**1.6**	**1.3**

Africa: New Policies Scenario

	Electricity generation (TWh)							Shares (%)		CAAGR (%)
	1990	2010	2015	2020	2025	2030	2035	2010	2035	2010-35
Total generation	**316**	**662**	**783**	**940**	**1 080**	**1 224**	**1 386**	**100**	**100**	**3.0**
Coal	165	260	298	337	352	358	367	39	26	1.4
Oil	41	80	79	77	67	57	54	12	4	-1.6
Gas	45	199	251	315	354	382	409	30	29	2.9
Nuclear	8	12	13	13	32	53	61	2	4	6.7
Hydro	56	105	127	163	206	254	306	16	22	4.4
Bioenergy	0	1	5	12	24	38	55	0	4	18.3
Wind	-	2	4	8	14	22	37	0	3	11.5
Geothermal	0	1	2	4	7	10	14	0	1	9.4
Solar PV	-	0	2	7	15	27	41	0	3	34.3
CSP	-	-	0	4	10	22	42	-	3	n.a.
Marine	-	-	-	-	-	-	-	-	-	n.a.

	Electrical capacity (GW)						Shares (%)		CAAGR (%)
	2010	2015	2020	2025	2030	2035	2010	2035	2010-35
Total capacity	**145**	**190**	**228**	**268**	**310**	**360**	**100**	**100**	**3.7**
Coal	41	50	59	66	72	79	28	22	2.7
Oil	24	28	29	26	24	23	17	7	-0.1
Gas	49	70	82	92	101	107	34	30	3.2
Nuclear	2	2	2	4	7	8	1	2	6.1
Hydro	27	33	43	54	66	79	18	22	4.5
Bioenergy	1	2	3	5	7	10	1	3	9.3
Wind	1	2	4	6	10	16	1	4	11.3
Geothermal	0	0	1	1	2	2	0	1	9.8
Solar PV	0	1	4	9	16	24	0	7	28.5
CSP	-	0	1	3	6	11	-	3	n.a.
Marine	-	-	-	-	-	-	-	-	n.a.

	CO_2 emissions (Mt)							Shares (%)		CAAGR (%)
	1990	2010	2015	2020	2025	2030	2035	2010	2035	2010-35
Total CO_2	**544**	**952**	**1 041**	**1 153**	**1 212**	**1 240**	**1 265**	**100**	**100**	**1.1**
Coal	234	329	358	391	395	374	349	35	28	0.2
Oil	248	437	474	516	547	577	611	46	48	1.4
Gas	62	186	209	246	269	289	306	20	24	2.0
Power generation	**212**	**410**	**453**	**508**	**519**	**500**	**482**	**100**	**100**	**0.6**
Coal	152	246	273	304	308	287	262	60	54	0.3
Oil	35	61	59	57	49	41	38	15	8	-1.9
Gas	25	103	121	147	162	173	182	25	38	2.3
TFC	**301**	**502**	**547**	**600**	**645**	**690**	**733**	**100**	**100**	**1.5**
Coal	82	83	85	87	87	88	87	17	12	0.2
Oil	201	362	402	445	484	521	558	72	76	1.8
Transport	*105*	*227*	*257*	*288*	*317*	*346*	*375*	*45*	*51*	*2.0*
Gas	18	57	60	67	74	81	88	11	12	1.7

Africa: Current Policies and 450 Scenarios

	Electricity generation (TWh)						Shares (%)		CAAGR (%)	
	2020	2030	2035	2020	2030	2035	2035		2010-35	
	Current Policies Scenario			450 Scenario			CPS	450	CPS	450
Total generation	**972**	**1 311**	**1 509**	**882**	**1 106**	**1 237**	**100**	**100**	**3.4**	**2.5**
Coal	362	430	480	310	266	216	32	17	2.5	-0.7
Oil	81	68	64	70	46	41	4	3	-0.9	-2.6
Gas	329	449	502	274	254	196	33	16	3.8	-0.1
Nuclear	13	32	34	13	77	108	2	9	4.2	9.2
Hydro	155	240	288	169	261	316	19	26	4.1	4.5
Bioenergy	11	35	50	18	48	70	3	6	17.8	19.4
Wind	8	19	27	9	45	92	2	7	10.0	15.6
Geothermal	4	8	12	4	15	22	1	2	8.6	11.5
Solar PV	6	19	28	10	40	64	2	5	32.2	36.6
CSP	3	13	24	5	54	112	2	9	n.a.	n.a.
Marine	-	-	-	-	0	0	-	0	n.a.	n.a.

	Electrical capacity (GW)						Shares (%)		CAAGR (%)	
	2020	2030	2035	2020	2030	2035	2035		2010-35	
	Current Policies Scenario			450 Scenario			CPS	450	CPS	450
Total capacity	**233**	**322**	**375**	**221**	**312**	**382**	**100**	**100**	**3.9**	**3.9**
Coal	62	81	93	55	58	60	25	16	3.3	1.5
Oil	29	25	24	28	23	23	6	6	-0.1	-0.2
Gas	88	119	134	74	82	81	36	21	4.1	2.0
Nuclear	2	4	5	2	10	15	1	4	3.6	8.5
Hydro	41	62	74	45	68	82	20	22	4.2	4.6
Bioenergy	3	7	9	4	9	12	2	3	9.0	10.2
Wind	4	8	12	4	20	40	3	10	9.8	15.4
Geothermal	1	1	2	1	2	3	0	1	9.1	11.7
Solar PV	4	11	17	6	24	37	5	10	26.7	30.6
CSP	1	3	6	2	15	30	2	8	n.a.	n.a.
Marine	-	-	-	-	0	0	-	0	n.a.	n.a.

	CO_2 emissions (Mt)						Shares (%)		CAAGR (%)	
	2020	2030	2035	2020	2030	2035	2035		2010-35	
	Current Policies Scenario			450 Scenario			CPS	450	CPS	450
Total CO_2	**1 191**	**1 383**	**1 492**	**1 051**	**866**	**740**	**100**	**100**	**1.8**	**-1.0**
Coal	414	464	501	352	192	108	34	15	1.7	-4.4
Oil	526	604	649	473	454	443	44	60	1.6	0.1
Gas	251	315	341	227	220	189	23	26	2.5	0.1
Power generation	**537**	**616**	**668**	**453**	**265**	**152**	**100**	**100**	**2.0**	**-3.9**
Coal	325	372	410	269	119	43	61	28	2.1	-6.7
Oil	61	49	46	52	32	29	7	19	-1.2	-3.0
Gas	151	195	213	131	114	80	32	53	2.9	-1.0
TFC	**608**	**716**	**772**	**554**	**556**	**544**	**100**	**100**	**1.7**	**0.3**
Coal	88	91	91	83	73	64	12	12	0.4	-1.0
Oil	451	541	588	407	409	402	76	74	2.0	0.4
Transport	*289*	*355*	*390*	*257*	*253*	*244*	*51*	*45*	*2.2*	*0.3*
Gas	69	84	92	65	73	77	12	14	1.9	1.2

Latin America: New Policies Scenario

	Energy demand (Mtoe)							Shares (%)		CAAGR (%)
	1990	2010	2015	2020	2025	2030	2035	2010	2035	2010-35
TPED	**331**	**586**	**675**	**740**	**801**	**856**	**905**	**100**	**100**	**1.8**
Coal	15	22	31	34	36	36	35	4	4	1.8
Oil	150	256	285	299	306	311	318	44	35	0.9
Gas	52	125	140	156	174	195	213	21	23	2.1
Nuclear	2	6	8	10	15	18	18	1	2	4.8
Hydro	30	58	68	77	85	92	99	10	11	2.2
Bioenergy	81	116	136	155	172	186	197	20	22	2.2
Other renewables	1	3	6	9	13	19	26	1	3	8.4
Power generation	**66**	**148**	**167**	**187**	**205**	**224**	**244**	**100**	**100**	**2.0**
Coal	3	6	12	13	13	11	10	4	4	1.9
Oil	14	30	26	23	20	16	15	20	6	-2.9
Gas	14	36	34	38	40	46	51	24	21	1.4
Nuclear	2	6	8	10	15	18	18	4	8	4.8
Hydro	30	58	68	77	85	92	99	39	41	2.2
Bioenergy	2	10	13	17	20	24	28	6	12	4.4
Other renewables	1	3	5	8	12	17	23	2	9	8.5
Other energy sector	**56**	**95**	**110**	**117**	**124**	**131**	**133**	**100**	**100**	**1.4**
Electricity	*8*	*18*	*20*	*23*	*25*	*27*	*28*	*19*	*21*	*1.7*
TFC	**251**	**434**	**506**	**559**	**607**	**650**	**689**	**100**	**100**	**1.9**
Coal	6	11	13	14	14	14	14	3	2	1.0
Oil	122	202	233	250	262	272	282	47	41	1.3
Gas	24	59	70	80	91	102	114	14	16	2.6
Electricity	35	74	87	99	111	122	133	17	19	2.4
Heat	-	-	-	-	-	-	-	-	-	n.a.
Bioenergy	64	88	102	115	128	138	144	20	21	2.0
Other renewables	-	0	1	1	2	2	3	0	0	8.2
Industry	**85**	**146**	**170**	**188**	**205**	**219**	**232**	**100**	**100**	**1.9**
Coal	6	11	13	13	14	14	14	7	6	0.9
Oil	20	35	40	42	43	44	44	24	19	0.9
Gas	15	28	33	39	46	53	60	19	26	3.2
Electricity	16	31	37	42	46	50	53	21	23	2.2
Heat	-	-	-	-	-	-	-	-	-	n.a.
Bioenergy	27	41	47	52	56	59	61	28	26	1.5
Other renewables	-	-	-	-	-	-	-	-	-	n.a.
Transport	**71**	**138**	**166**	**184**	**200**	**214**	**227**	**100**	**100**	**2.0**
Oil	65	117	135	146	153	160	168	84	74	1.5
Electricity	0	0	0	0	1	1	1	0	0	4.4
Biofuels	6	15	23	31	38	44	47	11	21	4.7
Other fuels	0	6	7	7	8	9	11	4	5	2.3
Buildings	**68**	**97**	**107**	**117**	**126**	**135**	**143**	**100**	**100**	**1.6**
Coal	0	0	0	0	0	0	0	0	0	5.7
Oil	17	17	19	19	20	20	21	17	15	0.9
Gas	6	12	14	16	17	19	20	13	14	2.0
Electricity	17	40	46	53	59	66	72	41	50	2.4
Heat	-	-	-	-	-	-	-	-	-	n.a.
Bioenergy	28	28	27	28	28	28	27	29	19	-0.1
Other renewables	-	0	1	1	2	2	3	0	2	8.2
Other	**27**	**53**	**63**	**70**	**76**	**82**	**87**	**100**	**100**	**2.0**

Latin America: Current Policies and 450 Scenarios

	Energy demand (Mtoe)						Shares (%)		CAAGR (%)	
	2020	2030	2035	2020	2030	2035	2035		2010-35	
	Current Policies Scenario			450 Scenario			CPS	450	CPS	450
TPED	**755**	**892**	**957**	**709**	**768**	**787**	**100**	**100**	**2.0**	**1.2**
Coal	36	43	44	31	28	26	5	3	2.8	0.6
Oil	307	330	340	279	253	235	36	30	1.1	-0.3
Gas	168	224	254	144	151	152	27	19	2.9	0.8
Nuclear	10	13	14	12	18	19	1	2	3.6	4.9
Hydro	77	91	97	77	92	99	10	13	2.1	2.2
Bioenergy	148	174	186	155	199	218	19	28	1.9	2.6
Other renewables	9	16	21	11	26	37	2	5	7.6	10.1
Power generation	**195**	**246**	**274**	**177**	**197**	**208**	**100**	**100**	**2.5**	**1.4**
Coal	14	15	15	12	7	5	5	2	3.7	-0.8
Oil	24	21	21	16	8	6	8	3	-1.5	-6.0
Gas	45	69	82	34	23	13	30	6	3.4	-4.1
Nuclear	10	13	14	12	18	19	5	9	3.6	4.9
Hydro	77	91	97	77	92	99	35	48	2.1	2.2
Bioenergy	17	22	25	17	25	30	9	14	4.0	4.6
Other renewables	8	15	19	9	24	35	7	17	7.7	10.3
Other energy sector	**121**	**138**	**144**	**109**	**111**	**112**	**100**	**100**	**1.7**	**0.7**
Electricity	*24*	*29*	*32*	*21*	*23*	*23*	*22*	*21*	*2.2*	*1.0*
TFC	**565**	**669**	**716**	**539**	**594**	**609**	**100**	**100**	**2.0**	**1.4**
Coal	14	16	17	13	12	11	2	2	1.7	0.2
Oil	256	286	297	239	226	212	42	35	1.5	0.2
Gas	81	105	118	77	94	103	17	17	2.8	2.2
Electricity	103	132	146	95	112	119	20	20	2.8	1.9
Heat	-	-	-	-	-	-	-	-	n.a.	n.a.
Bioenergy	109	128	136	115	149	162	19	27	1.8	2.5
Other renewables	1	2	2	1	2	2	0	0	7.5	7.4
Industry	**193**	**232**	**248**	**182**	**208**	**219**	**100**	**100**	**2.1**	**1.6**
Coal	14	16	16	12	11	11	6	5	1.6	0.1
Oil	43	48	48	40	42	41	20	19	1.3	0.7
Gas	40	55	64	37	48	53	26	24	3.4	2.6
Electricity	45	56	61	41	47	49	24	23	2.7	1.9
Heat	-	-	-	-	-	-	-	-	n.a.	n.a.
Bioenergy	52	58	59	52	61	64	24	29	1.4	1.8
Other renewables	-	-	-	-	-	-	-	-	n.a.	n.a.
Transport	**183**	**215**	**231**	**176**	**182**	**178**	**100**	**100**	**2.1**	**1.0**
Oil	150	169	177	137	119	105	77	59	1.7	-0.4
Electricity	0	1	1	0	1	1	0	1	4.2	6.3
Biofuels	25	36	43	31	52	61	18	34	4.2	5.8
Other fuels	7	9	11	7	9	11	5	6	2.2	2.5
Buildings	**119**	**140**	**150**	**112**	**124**	**128**	**100**	**100**	**1.7**	**1.1**
Coal	0	0	0	0	0	0	0	0	5.8	5.1
Oil	20	21	22	19	19	19	15	15	1.1	0.6
Gas	16	20	21	15	16	17	14	14	2.3	1.4
Electricity	54	70	77	50	59	63	51	49	2.6	1.8
Heat	-	-	-	-	-	-	-	-	n.a.	n.a.
Bioenergy	28	27	27	27	27	26	18	21	-0.1	-0.2
Other renewables	1	2	2	1	2	2	2	2	7.5	7.4
Other	**70**	**82**	**87**	**69**	**80**	**84**	**100**	**100**	**2.0**	**1.9**

Latin America: New Policies Scenario

	Electricity generation (TWh)							Shares (%)		CAAGR (%)
	1990	2010	2015	2020	2025	2030	2035	2010	2035	2010-35
Total generation	**489**	**1 069**	**1 245**	**1 416**	**1 576**	**1 726**	**1 868**	**100**	**100**	**2.3**
Coal	9	22	49	55	55	51	46	2	2	3.0
Oil	64	134	118	106	90	76	68	13	4	-2.7
Gas	45	173	183	215	246	284	316	16	17	2.4
Nuclear	10	22	29	39	58	67	70	2	4	4.8
Hydro	354	672	796	896	990	1 074	1 148	63	61	2.2
Bioenergy	7	40	51	64	76	88	103	4	5	3.8
Wind	-	3	13	29	39	48	60	0	3	12.9
Geothermal	1	3	5	6	8	11	15	0	1	6.3
Solar PV	-	0	2	5	12	20	30	0	2	35.6
CSP	-	-	-	-	1	6	11	-	1	n.a.
Marine	-	-	-	-	-	-	-	-	-	n.a.

	Electrical capacity (GW)						Shares (%)		CAAGR (%)
	2010	2015	2020	2025	2030	2035	2010	2035	2010-35
Total capacity	**237**	**289**	**337**	**382**	**423**	**465**	**100**	**100**	**2.7**
Coal	4	9	10	10	10	9	2	2	2.9
Oil	34	36	36	34	31	31	14	7	-0.3
Gas	41	57	72	85	100	112	17	24	4.1
Nuclear	3	4	5	8	9	9	1	2	4.8
Hydro	146	166	187	207	225	240	62	52	2.0
Bioenergy	7	9	11	13	15	17	3	4	3.4
Wind	1	5	11	15	18	22	1	5	12.2
Geothermal	1	1	1	1	2	2	0	0	5.6
Solar PV	0	1	4	8	14	20	0	4	29.6
CSP	-	-	-	0	1	3	-	1	n.a.
Marine	-	-	-	-	-	-	-	-	n.a.

	CO_2 emissions (Mt)							Shares (%)		CAAGR (%)
	1990	2010	2015	2020	2025	2030	2035	2010	2035	2010-35
Total CO_2	**579**	**1 067**	**1 218**	**1 296**	**1 350**	**1 402**	**1 453**	**100**	**100**	**1.2**
Coal	46	80	122	128	126	118	110	7	8	1.3
Oil	416	698	776	812	833	850	871	65	60	0.9
Gas	117	289	320	355	392	434	472	27	32	2.0
Power generation	**90**	**208**	**223**	**226**	**216**	**211**	**210**	**100**	**100**	**0.0**
Coal	15	30	61	65	61	52	45	14	22	1.7
Oil	44	94	83	73	61	51	46	45	22	-2.9
Gas	32	84	79	88	94	107	119	40	57	1.4
TFC	**425**	**744**	**860**	**931**	**989**	**1 042**	**1 094**	**100**	**100**	**1.6**
Coal	29	47	57	60	61	61	61	6	6	1.0
Oil	342	569	653	698	731	758	784	76	72	1.3
Transport	*194*	*347*	*404*	*434*	*456*	*477*	*501*	*47*	*46*	*1.5*
Gas	54	128	150	173	197	223	249	17	23	2.7

Latin America: Current Policies and 450 Scenarios

	Electricity generation (TWh)						Shares (%)		CAAGR (%)	
	2020	2030	2035	2020	2030	2035	2035		2010-35	
	Current Policies Scenario			450 Scenario			CPS	450	CPS	450
Total generation	**1 476**	**1 874**	**2 060**	**1 350**	**1 568**	**1 657**	**100**	**100**	**2.7**	**1.8**
Coal	59	66	69	47	28	22	3	1	4.6	-0.1
Oil	111	98	98	74	38	30	5	2	-1.2	-5.9
Gas	274	446	529	178	141	73	26	4	4.6	-3.4
Nuclear	39	50	53	48	69	72	3	4	3.6	4.9
Hydro	893	1 060	1 126	890	1 075	1 157	55	70	2.1	2.2
Bioenergy	63	82	92	65	91	108	4	7	3.4	4.1
Wind	28	45	52	31	70	105	3	6	12.3	15.4
Geothermal	6	10	13	7	16	21	1	1	5.8	7.8
Solar PV	4	14	20	8	30	52	1	3	33.3	38.6
CSP	-	3	7	2	9	15	0	1	n.a.	n.a.
Marine	-	-	-	-	0	2	-	0	n.a.	n.a.

	Electrical capacity (GW)						Shares (%)		CAAGR (%)	
	2020	2030	2035	2020	2030	2035	2035		2010-35	
	Current Policies Scenario			450 Scenario			CPS	450	CPS	450
Total capacity	**345**	**433**	**474**	**328**	**405**	**454**	**100**	**100**	**2.8**	**2.6**
Coal	10	11	11	9	7	6	2	1	3.8	1.2
Oil	37	35	36	35	29	29	8	6	0.2	-0.6
Gas	80	115	133	60	69	72	28	16	4.8	2.3
Nuclear	5	7	7	6	9	10	1	2	3.7	4.9
Hydro	187	222	236	186	225	242	50	53	1.9	2.0
Bioenergy	11	14	15	12	15	17	3	4	3.0	3.6
Wind	11	17	19	12	26	38	4	8	11.6	14.7
Geothermal	1	2	2	1	2	3	0	1	5.2	7.0
Solar PV	3	9	13	6	19	32	3	7	27.4	32.1
CSP	-	1	2	1	2	4	0	1	n.a.	n.a.
Marine	-	-	-	-	0	0	-	0	n.a.	n.a.

	CO_2 emissions (Mt)						Shares (%)		CAAGR (%)	
	2020	2030	2035	2020	2030	2035	2035		2010-35	
	Current Policies Scenario			450 Scenario			CPS	450	CPS	450
Total CO_2	**1 355**	**1 558**	**1 654**	**1 194**	**1 077**	**997**	**100**	**100**	**1.8**	**-0.3**
Coal	136	145	146	114	78	65	9	7	2.5	-0.8
Oil	836	910	941	754	675	618	57	62	1.2	-0.5
Gas	382	503	566	326	324	314	34	32	2.7	0.3
Power generation	**252**	**297**	**328**	**185**	**108**	**71**	**100**	**100**	**1.8**	**-4.2**
Coal	69	71	70	55	29	21	21	29	3.5	-1.4
Oil	77	66	66	51	26	20	20	29	-1.4	-6.0
Gas	106	161	193	79	54	30	59	42	3.4	-4.1
TFC	**957**	**1 102**	**1 164**	**883**	**854**	**814**	**100**	**100**	**1.8**	**0.4**
Coal	63	70	72	54	46	41	6	5	1.7	-0.5
Oil	718	801	832	664	614	565	71	69	1.5	-0.0
Transport	*447*	*503*	*527*	*408*	*355*	*311*	*45*	*38*	*1.7*	*-0.4*
Gas	176	231	260	165	194	208	22	26	2.9	1.9

Brazil: New Policies Scenario

	Energy demand (Mtoe)							Shares (%)		CAAGR (%)
	1990	2010	2015	2020	2025	2030	2035	2010	2035	2010-35
TPED	**138**	**262**	**309**	**346**	**381**	**413**	**444**	**100**	**100**	**2.1**
Coal	10	14	22	22	21	18	17	6	4	0.7
Oil	59	104	117	125	129	133	139	40	31	1.2
Gas	3	23	25	32	42	53	66	9	15	4.3
Nuclear	1	4	4	7	9	12	12	1	3	4.9
Hydro	18	35	40	44	47	51	54	13	12	1.8
Bioenergy	48	81	99	114	129	140	149	31	33	2.4
Other renewables	- 0	1	2	3	4	6	7	0	2	10.9
Power generation	**22**	**58**	**68**	**80**	**92**	**103**	**113**	**100**	**100**	**2.7**
Coal	2	3	8	8	7	5	4	6	3	0.8
Oil	1	4	2	2	2	2	2	6	2	-3.0
Gas	0	7	5	6	10	14	19	11	17	4.2
Nuclear	1	4	4	7	9	12	12	7	11	4.9
Hydro	18	35	40	44	47	51	54	60	47	1.8
Bioenergy	1	6	8	12	14	15	17	10	15	4.3
Other renewables	- 0	0	1	2	3	4	6	0	5	14.8
Other energy sector	**26**	**40**	**47**	**51**	**54**	**57**	**58**	**100**	**100**	**1.5**
Electricity	*3*	*10*	*11*	*12*	*13*	*15*	*16*	*24*	*27*	*1.9*
TFC	**112**	**211**	**249**	**277**	**305**	**332**	**358**	**100**	**100**	**2.1**
Coal	4	7	9	9	9	9	9	3	2	0.9
Oil	53	94	108	116	121	125	131	45	37	1.4
Gas	2	13	16	20	25	31	38	6	11	4.5
Electricity	18	38	44	50	57	63	70	18	19	2.5
Heat	-	-	-	-	-	-	-	-	-	n.a.
Bioenergy	34	59	71	81	93	102	108	28	30	2.5
Other renewables	-	0	1	1	1	1	1	0	0	5.7
Industry	**40**	**79**	**94**	**105**	**116**	**127**	**137**	**100**	**100**	**2.2**
Coal	4	7	8	9	9	9	9	9	6	0.8
Oil	8	12	13	14	15	15	16	15	11	1.1
Gas	1	9	11	15	19	23	29	11	21	4.9
Electricity	10	17	21	23	26	29	31	22	23	2.3
Heat	-	-	-	-	-	-	-	-	-	n.a.
Bioenergy	17	34	40	44	48	51	53	43	39	1.7
Other renewables	-	-	-	-	-	-	-	-	-	n.a.
Transport	**33**	**70**	**85**	**95**	**105**	**113**	**121**	**100**	**100**	**2.2**
Oil	27	54	62	66	69	71	76	77	63	1.4
Electricity	0	0	0	0	0	0	0	0	0	4.7
Biofuels	6	14	20	26	33	38	41	20	34	4.4
Other fuels	0	2	2	2	3	3	4	3	3	2.6
Buildings	**23**	**34**	**38**	**42**	**46**	**50**	**53**	**100**	**100**	**1.8**
Coal	-	-	-	-	-	-	-	-	-	n.a.
Oil	6	7	8	8	8	9	9	21	17	1.0
Gas	0	0	1	1	1	1	1	1	3	4.2
Electricity	8	19	21	25	28	31	34	54	64	2.5
Heat	-	-	-	-	-	-	-	-	-	n.a.
Bioenergy	9	8	8	7	7	7	7	23	13	-0.4
Other renewables	-	0	1	1	1	1	1	1	3	5.6
Other	**16**	**27**	**32**	**36**	**39**	**43**	**46**	**100**	**100**	**2.2**

Brazil: Current Policies and 450 Scenarios

	Energy demand (Mtoe)						Shares (%)		CAAGR (%)	
	2020	2030	2035	2020	2030	2035	2035		2010-35	
	Current Policies Scenario			450 Scenario			CPS	450	CPS	450
TPED	**352**	**430**	**467**	**334**	**377**	**393**	**100**	**100**	**2.3**	**1.6**
Coal	23	23	22	19	12	11	5	3	1.8	-1.3
Oil	128	142	147	120	105	95	31	24	1.4	-0.3
Gas	38	70	88	27	39	44	19	11	5.5	2.7
Nuclear	7	9	10	7	12	12	2	3	3.8	4.9
Hydro	44	51	53	43	50	53	11	14	1.7	1.7
Bioenergy	109	131	140	114	153	168	30	43	2.2	3.0
Other renewables	3	5	7	3	6	8	1	2	10.3	11.2
Power generation	**85**	**114**	**128**	**76**	**90**	**97**	**100**	**100**	**3.2**	**2.1**
Coal	9	7	6	7	2	1	5	1	2.8	-5.9
Oil	2	2	2	2	1	1	2	1	-2.3	-4.9
Gas	11	27	36	3	4	5	28	5	7.0	-1.3
Nuclear	7	9	10	7	12	12	7	13	3.8	4.9
Hydro	44	51	53	43	50	53	42	55	1.7	1.7
Bioenergy	11	14	16	12	16	18	12	18	4.0	4.6
Other renewables	2	4	5	2	5	7	4	7	14.2	15.3
Other energy sector	**52**	**60**	**63**	**49**	**52**	**52**	**100**	**100**	**1.8**	**1.0**
Electricity	*13*	*16*	*17*	*11*	*13*	*13*	*28*	*25*	*2.4*	*1.2*
TFC	**280**	**341**	**369**	**270**	**306**	**320**	**100**	**100**	**2.3**	**1.7**
Coal	9	11	11	8	7	7	3	2	1.7	-0.4
Oil	119	133	138	111	99	90	37	28	1.6	-0.2
Gas	20	33	41	19	28	34	11	11	4.7	3.9
Electricity	53	69	76	49	58	63	21	20	2.9	2.1
Heat	-	-	-	-	-	-	-	-	n.a.	n.a.
Bioenergy	78	94	102	82	113	126	28	39	2.2	3.1
Other renewables	1	1	1	1	1	1	0	0	5.2	5.2
Industry	**107**	**133**	**146**	**101**	**120**	**129**	**100**	**100**	**2.5**	**2.0**
Coal	9	10	11	8	7	6	7	5	1.7	-0.5
Oil	15	17	18	13	14	14	12	11	1.6	0.8
Gas	15	25	31	14	21	25	21	19	5.2	4.2
Electricity	24	32	35	23	27	29	24	22	2.8	2.0
Heat	-	-	-	-	-	-	-	-	n.a.	n.a.
Bioenergy	44	50	52	44	52	55	35	43	1.7	1.9
Other renewables	-	-	-	-	-	-	-	-	n.a.	n.a.
Transport	**94**	**112**	**120**	**93**	**98**	**98**	**100**	**100**	**2.2**	**1.3**
Oil	69	76	79	63	47	38	66	39	1.6	-1.4
Electricity	0	0	0	0	0	1	0	1	4.5	6.7
Biofuels	23	32	36	26	47	55	30	56	3.9	5.6
Other fuels	2	3	4	3	4	4	3	4	2.7	2.8
Buildings	**43**	**52**	**57**	**40**	**45**	**47**	**100**	**100**	**2.0**	**1.3**
Coal	-	-	-	-	-	-	-	-	n.a.	n.a.
Oil	8	9	10	8	8	8	17	17	1.2	0.6
Gas	1	1	2	1	1	1	3	2	5.2	3.1
Electricity	26	34	37	24	28	30	65	63	2.8	1.9
Heat	-	-	-	-	-	-	-	-	n.a.	n.a.
Bioenergy	7	7	7	7	7	7	12	15	-0.5	-0.5
Other renewables	1	1	1	1	1	1	2	3	5.2	5.2
Other	**36**	**43**	**46**	**36**	**43**	**46**	**100**	**100**	**2.2**	**2.2**

Brazil: New Policies Scenario

	Electricity generation (TWh)							Shares (%)		CAAGR (%)
	1990	2010	2015	2020	2025	2030	2035	2010	2035	2010-35
Total generation	**223**	**515**	**603**	**693**	**784**	**872**	**956**	**100**	**100**	**2.5**
Coal	5	11	33	34	30	23	19	2	2	2.0
Oil	5	16	9	8	7	8	8	3	1	-2.6
Gas	0	36	31	41	65	96	127	7	13	5.1
Nuclear	2	15	16	26	36	45	48	3	5	4.9
Hydro	207	403	462	508	551	590	624	78	65	1.8
Bioenergy	4	31	40	51	59	64	70	6	7	3.2
Wind	-	2	10	23	30	35	40	0	4	12.3
Geothermal	-	-	-	-	-	-	-	-	-	n.a.
Solar PV	-	-	1	2	5	10	15	-	2	n.a.
CSP	-	-	-	-	-	3	5	-	1	n.a.
Marine	-	-	-	-	-	-	-	-	-	n.a.

	Electrical capacity (GW)						Shares (%)		CAAGR (%)
	2010	2015	2020	2025	2030	2035	2010	2035	2010-35
Total capacity	**114**	**138**	**166**	**191**	**216**	**242**	**100**	**100**	**3.1**
Coal	2	5	6	5	4	4	2	1	2.0
Oil	6	7	8	8	8	9	5	4	1.3
Gas	9	15	23	33	43	55	8	23	7.5
Nuclear	2	2	3	5	6	6	2	3	4.9
Hydro	88	96	106	116	124	131	78	54	1.6
Bioenergy	5	7	9	10	11	11	5	5	3.0
Wind	1	4	9	12	13	15	1	6	11.6
Geothermal	-	-	-	-	-	-	-	-	n.a.
Solar PV	-	1	2	4	7	10	-	4	n.a.
CSP	-	-	-	-	1	1	-	1	n.a.
Marine	-	-	-	-	-	-	-	-	n.a.

	CO_2 emissions (Mt)							Shares (%)		CAAGR (%)
	1990	2010	2015	2020	2025	2030	2035	2010	2035	2010-35
Total CO_2	**194**	**388**	**463**	**498**	**525**	**552**	**589**	**100**	**100**	**1.7**
Coal	29	52	87	87	80	70	63	13	11	0.8
Oil	159	284	319	338	351	363	379	73	64	1.2
Gas	7	52	57	72	94	119	146	13	25	4.2
Power generation	**12**	**45**	**62**	**64**	**65**	**65**	**70**	**100**	**100**	**1.8**
Coal	8	18	45	44	37	27	21	40	30	0.7
Oil	4	12	6	5	5	5	5	26	8	-3.0
Gas	0	15	11	15	23	33	44	35	62	4.2
TFC	**166**	**318**	**370**	**400**	**424**	**449**	**481**	**100**	**100**	**1.7**
Coal	18	31	38	39	39	39	38	10	8	0.9
Oil	144	258	296	316	329	340	357	81	74	1.3
Transport	*81*	*161*	*187*	*200*	*206*	*214*	*228*	*51*	*47*	*1.4*
Gas	5	29	36	45	56	70	86	9	18	4.5

Brazil: Current Policies and 450 Scenarios

	Electricity generation (TWh)						Shares (%)		CAAGR (%)	
	2020	2030	2035	2020	2030	2035	2035		2010-35	
	Current Policies Scenario			450 Scenario			CPS	450	CPS	450
Total generation	**724**	**949**	**1 056**	**663**	**791**	**848**	**100**	**100**	**2.9**	**2.0**
Coal	35	31	28	26	7	3	3	0	3.7	-4.9
Oil	8	9	10	7	5	5	1	1	-2.1	-4.6
Gas	74	183	242	16	29	29	23	3	7.9	-0.9
Nuclear	26	34	36	29	45	48	3	6	3.8	4.9
Hydro	507	588	621	505	586	621	59	73	1.7	1.7
Bioenergy	50	60	64	51	66	72	6	9	2.9	3.4
Wind	23	34	37	25	40	46	3	5	12.0	13.0
Geothermal	-	-	-	-	-	-	-	-	n.a.	n.a.
Solar PV	2	9	13	4	12	17	1	2	n.a.	n.a.
CSP	-	2	4	-	3	5	0	1	n.a.	n.a.
Marine	-	-	-	-	0	1	-	0	n.a.	n.a.

	Electrical capacity (GW)						Shares (%)		CAAGR (%)	
	2020	2030	2035	2020	2030	2035	2035		2010-35	
	Current Policies Scenario			450 Scenario			CPS	450	CPS	450
Total capacity	**169**	**219**	**243**	**158**	**192**	**208**	**100**	**100**	**3.1**	**2.4**
Coal	6	5	5	5	3	3	2	1	3.1	1.0
Oil	8	8	9	7	6	6	4	3	1.3	-0.1
Gas	27	48	60	15	19	21	25	10	7.9	3.6
Nuclear	3	4	5	4	6	6	2	3	3.9	4.9
Hydro	106	123	130	106	123	131	54	63	1.6	1.6
Bioenergy	9	10	10	9	11	12	4	6	2.7	3.1
Wind	9	13	14	10	15	17	6	8	11.4	12.2
Geothermal	-	-	-	-	-	-	-	-	n.a.	n.a.
Solar PV	2	6	9	3	8	11	4	5	n.a.	n.a.
CSP	-	0	1	-	1	1	0	1	n.a.	n.a.
Marine	-	-	-	-	0	0	-	0	n.a.	n.a.

	CO_2 emissions (Mt)						Shares (%)		CAAGR (%)	
	2020	2030	2035	2020	2030	2035	2035		2010-35	
	Current Policies Scenario			450 Scenario			CPS	450	CPS	450
Total CO_2	**529**	**635**	**687**	**458**	**398**	**365**	**100**	**100**	**2.3**	**-0.2**
Coal	92	89	86	74	39	31	13	8	2.1	-2.0
Oil	350	389	404	324	275	242	59	66	1.4	-0.6
Gas	86	157	197	61	83	93	29	25	5.5	2.4
Power generation	**78**	**108**	**125**	**46**	**23**	**18**	**100**	**100**	**4.2**	**-3.5**
Coal	46	38	34	36	9	4	27	21	2.7	-5.9
Oil	5	6	6	5	3	3	5	18	-2.3	-4.9
Gas	26	64	85	6	10	11	68	61	7.0	-1.3
TFC	**415**	**485**	**519**	**380**	**346**	**321**	**100**	**100**	**2.0**	**0.0**
Coal	42	46	47	34	28	24	9	8	1.7	-1.0
Oil	327	365	379	302	257	225	73	70	1.5	-0.6
Transport	*207*	*229*	*238*	*190*	*142*	*113*	*46*	*35*	*1.6*	*-1.4*
Gas	46	74	93	43	61	72	18	22	4.8	3.7

Middle East: New Policies Scenario

	Energy demand (Mtoe)							Shares (%)		CAAGR (%)
	1990	2010	2015	2020	2025	2030	2035	2010	2035	2010-35
TPED	**210**	**624**	**715**	**792**	**867**	**935**	**1 012**	**100**	**100**	**1.9**
Coal	1	2	3	3	3	4	4	0	0	2.4
Oil	135	306	341	370	389	403	424	49	42	1.3
Gas	72	314	365	405	449	496	535	50	53	2.1
Nuclear	-	-	2	6	13	13	16	-	2	n.a.
Hydro	1	2	2	3	3	3	4	0	0	3.7
Bioenergy	0	0	1	2	3	5	8	0	1	11.8
Other renewables	0	0	1	3	6	11	22	0	2	21.9
Power generation	**61**	**212**	**246**	**271**	**297**	**322**	**347**	**100**	**100**	**2.0**
Coal	0	0	0	0	1	1	1	0	0	7.5
Oil	27	87	98	93	85	75	70	41	20	-0.9
Gas	32	123	144	166	189	217	232	58	67	2.6
Nuclear	-	-	2	6	13	13	16	-	5	n.a.
Hydro	1	2	2	3	3	3	4	1	1	3.7
Bioenergy	-	0	1	1	2	4	7	0	2	34.9
Other renewables	0	0	0	2	4	8	18	0	5	32.9
Other energy sector	**22**	**81**	**96**	**106**	**115**	**123**	**136**	**100**	**100**	**2.1**
Electricity	*4*	*14*	*17*	*20*	*22*	*25*	*27*	*17*	*20*	*2.5*
TFC	**146**	**402**	**460**	**518**	**572**	**624**	**678**	**100**	**100**	**2.1**
Coal	0	1	1	1	2	2	2	0	0	2.9
Oil	102	212	240	271	299	323	349	53	51	2.0
Gas	28	131	147	161	173	186	199	33	29	1.7
Electricity	15	56	70	82	96	109	123	14	18	3.2
Heat	-	-	-	-	-	-	-	-	-	n.a.
Bioenergy	0	0	0	1	1	1	1	0	0	4.2
Other renewables	0	0	1	1	2	3	3	0	1	13.7
Industry	**42**	**112**	**128**	**140**	**150**	**159**	**168**	**100**	**100**	**1.6**
Coal	0	1	1	1	2	2	2	1	1	2.9
Oil	19	40	44	47	49	50	51	35	30	1.0
Gas	19	60	69	75	81	86	91	54	54	1.7
Electricity	3	11	14	16	18	21	24	10	14	3.1
Heat	-	-	-	-	-	-	-	-	-	n.a.
Bioenergy	0	0	0	0	0	0	0	0	0	1.4
Other renewables	-	0	0	0	0	0	0	0	0	3.5
Transport	**48**	**119**	**142**	**169**	**194**	**216**	**240**	**100**	**100**	**2.8**
Oil	48	114	135	159	182	203	226	96	94	2.8
Electricity	-	0	0	0	0	0	0	0	0	0.5
Biofuels	-	-	-	-	-	-	-	-	-	n.a.
Other fuels	-	5	7	10	12	13	15	4	6	4.3
Buildings	**33**	**104**	**119**	**133**	**146**	**161**	**177**	**100**	**100**	**2.1**
Coal	-	0	0	0	0	0	0	0	0	-1.6
Oil	18	21	22	23	23	23	22	20	13	0.2
Gas	3	40	43	45	48	52	57	38	32	1.4
Electricity	11	43	53	63	73	83	94	41	53	3.2
Heat	-	-	-	-	-	-	-	-	-	n.a.
Bioenergy	0	0	0	0	1	1	1	0	0	5.5
Other renewables	0	0	1	1	2	3	3	0	2	13.5
Other	**23**	**65**	**71**	**76**	**82**	**87**	**93**	**100**	**100**	**1.4**

Middle East: Current Policies and 450 Scenarios

	Energy demand (Mtoe)						Shares (%)		CAAGR (%)	
	2020	2030	2035	2020	2030	2035	2035		2010-35	
	Current Policies Scenario			450 Scenario			CPS	450	CPS	450
TPED	**811**	**994**	**1 099**	**745**	**792**	**825**	**100**	**100**	**2.3**	**1.1**
Coal	3	4	4	3	3	4	0	0	2.6	2.0
Oil	380	434	468	350	321	308	43	37	1.7	0.0
Gas	416	529	594	376	410	422	54	51	2.6	1.2
Nuclear	6	13	13	6	20	23	1	3	n.a.	n.a.
Hydro	2	3	4	3	4	5	0	1	3.5	4.7
Bioenergy	2	4	5	4	11	17	0	2	9.9	15.5
Other renewables	2	6	10	3	23	46	1	6	18.3	25.6
Power generation	**280**	**342**	**381**	**249**	**260**	**277**	**100**	**100**	**2.4**	**1.1**
Coal	0	1	1	0	0	0	0	0	8.3	5.4
Oil	95	82	80	88	51	38	21	14	-0.3	-3.2
Gas	174	235	271	149	160	160	71	58	3.2	1.0
Nuclear	6	13	13	6	20	23	4	8	n.a.	n.a.
Hydro	2	3	4	3	4	5	1	2	3.5	4.7
Bioenergy	1	3	4	1	6	11	1	4	32.0	37.6
Other renewables	1	5	8	2	19	40	2	15	28.8	37.2
Other energy sector	**108**	**136**	**153**	**100**	**104**	**107**	**100**	**100**	**2.6**	**1.1**
Electricity	*21*	*27*	*30*	*19*	*21*	*21*	*20*	*20*	*3.0*	*1.7*
TFC	**529**	**661**	**730**	**491**	**543**	**568**	**100**	**100**	**2.4**	**1.4**
Coal	1	2	2	1	2	2	0	0	3.1	2.8
Oil	278	349	384	257	268	267	53	47	2.4	0.9
Gas	162	190	206	153	170	181	28	32	1.8	1.3
Electricity	86	118	135	77	95	106	18	19	3.5	2.5
Heat	-	-	-	-	-	-	-	-	n.a.	n.a.
Bioenergy	1	1	1	2	5	6	0	1	4.2	11.9
Other renewables	1	1	2	1	4	6	0	1	11.3	15.9
Industry	**144**	**169**	**181**	**136**	**153**	**160**	**100**	**100**	**1.9**	**1.4**
Coal	1	2	2	1	2	2	1	1	3.1	2.9
Oil	49	53	55	45	47	47	30	29	1.3	0.7
Gas	77	90	97	73	84	88	53	55	1.9	1.5
Electricity	17	23	28	15	20	23	15	14	3.7	2.9
Heat	-	-	-	-	-	-	-	-	n.a.	n.a.
Bioenergy	0	0	0	0	0	0	0	0	1.6	1.7
Other renewables	0	0	0	0	0	0	0	0	3.5	5.0
Transport	**171**	**231**	**264**	**159**	**174**	**180**	**100**	**100**	**3.2**	**1.7**
Oil	162	220	252	148	154	152	95	85	3.2	1.2
Electricity	0	0	0	0	1	2	0	1	0.5	19.0
Biofuels	-	-	-	2	4	5	-	3	n.a.	n.a.
Other fuels	9	11	12	10	16	20	5	11	3.6	5.7
Buildings	**136**	**170**	**188**	**122**	**135**	**144**	**100**	**100**	**2.4**	**1.3**
Coal	0	0	0	0	0	0	0	0	-1.6	-1.6
Oil	24	25	24	23	21	20	13	14	0.5	0.3
Gas	46	54	61	40	40	42	32	29	1.7	0.2
Electricity	65	89	101	58	69	76	54	53	3.5	2.3
Heat	-	-	-	-	-	-	-	-	n.a.	n.a.
Bioenergy	0	1	1	0	1	1	0	1	5.5	5.7
Other renewables	1	1	2	1	4	5	1	4	11.0	15.8
Other	**78**	**91**	**97**	**74**	**81**	**84**	**100**	**100**	**1.6**	**1.0**

Middle East: New Policies Scenario

	Electricity generation (TWh)							Shares (%)		CAAGR (%)
	1990	2010	2015	2020	2025	2030	2035	2010	2035	2010-35
Total generation	**219**	**824**	**1 020**	**1 192**	**1 374**	**1 558**	**1 740**	**100**	**100**	**3.0**
Coal	0	0	1	2	3	3	3	0	0	9.1
Oil	101	291	336	325	308	288	273	35	16	-0.3
Gas	106	514	648	796	940	1 097	1 194	62	69	3.4
Nuclear	-	-	7	23	52	52	62	-	4	n.a.
Hydro	12	18	23	29	35	39	44	2	3	3.7
Bioenergy	-	0	2	4	7	13	23	0	1	34.9
Wind	0	0	2	4	9	23	53	0	3	26.0
Geothermal	-	-	-	-	-	-	-	-	-	n.a.
Solar PV	-	-	1	5	12	24	42	-	2	n.a.
CSP	-	-	-	4	9	19	46	-	3	n.a.
Marine	-	-	-	-	-	-	-	-	-	n.a.

	Electrical capacity (GW)						Shares (%)		CAAGR (%)
	2010	2015	2020	2025	2030	2035	2010	2035	2010-35
Total capacity	**214**	**302**	**331**	**353**	**387**	**439**	**100**	**100**	**2.9**
Coal	0	0	1	1	1	1	0	0	6.3
Oil	68	82	82	79	73	69	32	16	0.1
Gas	133	201	220	232	252	273	62	62	2.9
Nuclear	-	1	3	7	7	8	-	2	n.a.
Hydro	13	15	19	21	24	25	6	6	2.8
Bioenergy	0	0	1	1	2	4	0	1	17.4
Wind	0	1	2	4	10	23	0	5	24.7
Geothermal	-	-	-	-	-	-	-	-	n.a.
Solar PV	-	1	3	6	13	23	-	5	n.a.
CSP	-	-	1	3	6	14	-	3	n.a.
Marine	-	-	-	-	-	-	-	-	n.a.

	CO_2 emissions (Mt)							Shares (%)		CAAGR (%)
	1990	2010	2015	2020	2025	2030	2035	2010	2035	2010-35
Total CO_2	**562**	**1 583**	**1 807**	**1 972**	**2 112**	**2 243**	**2 376**	**100**	**100**	**1.6**
Coal	1	7	8	12	14	16	18	0	1	4.0
Oil	397	876	1 002	1 079	1 134	1 170	1 224	55	52	1.3
Gas	163	700	797	882	963	1 057	1 135	44	48	2.0
Power generation	**167**	**563**	**649**	**688**	**720**	**753**	**772**	**100**	**100**	**1.3**
Coal	0	1	2	4	6	7	7	0	1	7.4
Oil	91	273	311	297	272	240	222	49	29	-0.8
Gas	76	288	336	387	441	507	543	51	70	2.6
TFC	**344**	**850**	**970**	**1 090**	**1 197**	**1 292**	**1 394**	**100**	**100**	**2.0**
Coal	1	4	5	6	7	8	9	1	1	2.9
Oil	281	559	643	732	811	877	948	66	68	2.1
Transport	*143*	*344*	*407*	*481*	*551*	*613*	*682*	*40*	*49*	*2.8*
Gas	62	286	322	352	380	407	437	34	31	1.7

Middle East: Current Policies and 450 Scenarios

	Electricity generation (TWh)						Shares (%)		CAAGR (%)	
	2020	2030	2035	2020	2030	2035	2035		2010-35	
	Current Policies Scenario			450 Scenario			CPS	450	CPS	450
Total generation	**1 241**	**1 685**	**1 918**	**1 112**	**1 343**	**1 484**	**100**	**100**	**3.4**	**2.4**
Coal	2	3	4	2	2	2	0	0	9.9	7.0
Oil	332	316	318	308	191	142	17	10	0.4	-2.8
Gas	842	1 228	1 426	728	852	848	74	57	4.2	2.0
Nuclear	23	52	52	23	75	88	3	6	n.a.	n.a.
Hydro	28	38	42	31	48	56	2	4	3.5	4.7
Bioenergy	4	9	13	5	22	38	1	3	32.0	37.6
Wind	3	13	22	5	69	137	1	9	21.7	30.8
Geothermal	-	-	-	-	-	-	-	-	n.a.	n.a.
Solar PV	4	13	17	6	38	69	1	5	n.a.	n.a.
CSP	3	13	23	4	46	105	1	7	n.a.	n.a.
Marine	-	-	-	-	-	0	-	0	n.a.	n.a.

	Electrical capacity (GW)						Shares (%)		CAAGR (%)	
	2020	2030	2035	2020	2030	2035	2035		2010-35	
	Current Policies Scenario			450 Scenario			CPS	450	CPS	450
Total capacity	**332**	**403**	**446**	**334**	**424**	**506**	**100**	**100**	**3.0**	**3.5**
Coal	1	1	1	1	1	1	0	0	7.0	4.8
Oil	84	81	81	82	60	52	18	10	0.7	-1.0
Gas	220	273	304	221	259	274	68	54	3.4	2.9
Nuclear	3	7	7	3	10	12	2	2	n.a.	n.a.
Hydro	18	23	25	19	28	32	6	6	2.7	3.7
Bioenergy	1	2	2	1	4	6	0	1	14.9	19.6
Wind	2	6	10	2	29	59	2	12	20.4	29.5
Geothermal	-	-	-	-	-	-	-	-	n.a.	n.a.
Solar PV	2	7	10	3	20	37	2	7	n.a.	n.a.
CSP	1	4	7	2	14	34	2	7	n.a.	n.a.
Marine	-	-	-	-	-	0	-	0	n.a.	n.a.

	CO_2 emissions (Mt)						Shares (%)		CAAGR (%)	
	2020	2030	2035	2020	2030	2035	2035		2010-35	
	Current Policies Scenario			450 Scenario			CPS	450	CPS	450
Total CO_2	**2 022**	**2 404**	**2 638**	**1 837**	**1 750**	**1 699**	**100**	**100**	**2.1**	**0.3**
Coal	12	17	20	11	13	14	1	1	4.4	2.8
Oil	1 102	1 263	1 361	1 017	910	855	52	50	1.8	-0.1
Gas	908	1 124	1 257	809	826	831	48	49	2.4	0.7
Power generation	**712**	**814**	**891**	**632**	**531**	**485**	**100**	**100**	**1.9**	**-0.6**
Coal	4	7	9	4	4	4	1	1	8.2	5.4
Oil	302	260	254	280	162	121	29	25	-0.3	-3.2
Gas	406	547	628	347	364	359	71	74	3.2	0.9
TFC	**1 111**	**1 374**	**1 511**	**1 020**	**1 050**	**1 050**	**100**	**100**	**2.3**	**0.9**
Coal	6	8	9	6	7	7	1	1	3.0	2.1
Oil	749	948	1 049	688	703	691	69	66	2.5	0.9
Transport	*489*	*665*	*760*	*446*	*465*	*461*	*50*	*44*	*3.2*	*1.2*
Gas	356	417	453	326	340	351	30	33	1.9	0.8

Iraq: Central Scenario

	Energy demand (Mtoe)							Shares (%)		CAAGR (%)
	1990	2010	2015	2020	2025	2030	2035	2010	2035	2010-35
TPED	**21**	**38**	**77**	**113**	**131**	**145**	**160**	**100**	**100**	**5.9**
Oil	19	32	64	75	76	84	92	82	58	4.4
Gas	2	6	12	37	55	60	66	17	41	9.8
Hydro	0	0	0	0	1	1	1	1	1	4.4
Other renewables	0	0	0	0	0	0	0	0	0	n.a.
Power generation	**6**	**14**	**37**	**50**	**51**	**53**	**56**	**100**	**100**	**5.9**
Oil	5	8	28	20	8	6	6	62	11	-1.4
Gas	1	5	9	30	43	45	49	35	87	9.8
Hydro	0	0	0	0	1	1	1	3	2	4.4
Other renewables	-	-	-	-	0	0	0	-	0	n.a.
Other energy sector	**2**	**4**	**11**	**19**	**23**	**25**	**28**	**100**	**100**	**8.1**
Electricity	*0*	*2*	*6*	*8*	*8*	*9*	*10*	*50*	*34*	*6.4*
TFC	**15**	**26**	**41**	**61**	**77**	**89**	**99**	**100**	**100**	**5.6**
Oil	12	22	31	47	58	67	74	85	75	5.0
Gas	1	1	2	4	7	9	11	4	11	9.8
Electricity	2	3	7	10	12	13	14	11	15	6.7
Other fuels	0	0	0	0	0	0	0	0	0	n.a.
Industry	**3**	**3**	**6**	**9**	**12**	**14**	**15**	**100**	**100**	**6.4**
Oil	2	2	2	3	3	2	2	61	12	-0.3
Gas	1	1	2	4	7	9	10	24	66	10.7
Electricity	1	0	2	2	3	3	3	14	21	8.1
Other fuels	-	-	-	-	-	-	-	-	-	n.a.
Transport	**7**	**15**	**23**	**37**	**46**	**52**	**59**	**100**	**100**	**5.6**
Oil	7	15	23	37	46	52	59	100	100	5.6
Buildings	**3**	**7**	**11**	**14**	**16**	**18**	**19**	**100**	**100**	**4.2**
Oil	2	4	5	6	7	7	8	64	40	2.2
Gas	-	0	0	0	0	0	0	1	1	3.1
Electricity	1	2	5	8	9	10	11	35	58	6.4
Other fuels	0	0	0	0	0	0	0	0	2	n.a.
Other	**1**	**0**	**1**	**1**	**3**	**6**	**6**	**100**	**100**	**10.8**

Iraq: High and Delayed Cases

	Energy demand (Mtoe)						Shares (%)		CAAGR (%)	
	2020	2030	2035	2020	2030	2035	2035		2010-35	
	High Case			Delayed Case			High	Delayed	High	Delayed
TPED	**144**	**173**	**187**	**74**	**102**	**110**	**100**	**100**	**6.5**	**4.3**
Oil	91	104	114	56	64	69	61	63	5.3	3.2
Gas	52	67	71	17	37	39	38	36	10.1	7.5
Hydro	1	1	1	1	1	1	1	1	3.7	4.6
Other renewables	0	0	0	0	0	0	0	0	n.a.	n.a.
Power generation	**55**	**56**	**61**	**32**	**41**	**43**	**100**	**100**	**6.2**	**4.8**
Oil	12	7	9	18	13	14	15	33	0.3	2.2
Gas	42	48	51	13	26	28	83	64	10.0	7.3
Hydro	1	1	1	1	1	1	2	3	3.7	4.6
Other renewables	-	0	0	-	0	0	0	0	n.a.	n.a.
Other energy sector	**26**	**31**	**33**	**12**	**16**	**18**	**100**	**100**	**8.8**	**6.2**
Electricity	*8*	*9*	*10*	*6*	*7*	*7*	*30*	*40*	*6.6*	*5.3*
TFC	**84**	**109**	**118**	**43**	**62**	**67**	**100**	**100**	**6.3**	**3.9**
Oil	66	83	90	32	44	47	76	71	5.8	3.2
Gas	6	11	12	3	7	8	10	12	10.5	8.6
Electricity	12	15	16	8	10	11	14	17	7.1	5.5
Other fuels	0	0	0	0	0	0	0	0	n.a.	n.a.
Industry	**13**	**18**	**19**	**6**	**11**	**11**	**100**	**100**	**7.2**	**5.0**
Oil	4	3	3	2	1	1	14	11	0.9	-2.1
Gas	6	11	12	3	7	8	65	70	11.4	9.5
Electricity	3	4	4	2	2	2	22	20	8.9	6.3
Other fuels	-	-	-	-	-	-	-	-	n.a.	n.a.
Transport	**54**	**67**	**73**	**24**	**31**	**34**	**100**	**100**	**6.5**	**3.3**
Oil	54	67	73	24	31	34	100	100	6.5	3.3
Buildings	**15**	**19**	**21**	**12**	**15**	**16**	**100**	**100**	**4.5**	**3.4**
Oil	7	8	8	6	6	7	39	43	2.4	1.7
Gas	0	0	0	0	0	0	1	1	3.3	2.6
Electricity	9	11	12	6	8	9	59	55	6.7	5.4
Other fuels	0	0	0	0	0	0	2	2	n.a.	n.a.
Other	**1**	**6**	**6**	**1**	**5**	**5**	**100**	**100**	**10.9**	**10.5**

Iraq: Central Scenario

	Electricity generation (TWh)							Shares (%)		CAAGR (%)
	1990	2010	2015	2020	2025	2030	2035	2010	2035	2010-35
Total generation	**24**	**51**	**139**	**202**	**230**	**249**	**277**	**100**	**100**	**7.0**
Oil	21	29	100	72	30	23	22	57	8	-1.2
Gas	-	17	33	124	193	215	242	33	87	11.2
Hydro	3	5	5	5	7	11	14	9	5	4.5
Other renewables	-	-	-	-	0	0	0	-	0	n.a.

	Electrical capacity (GW)						Shares (%)		CAAGR (%)
	2010	2015	2020	2025	2030	2035	2010	2035	2010-35
Total capacity	**16**	**36**	**60**	**70**	**76**	**83**	**100**	**100**	**6.8**
Oil	10	28	36	36	35	34	59	41	5.2
Gas	4	6	22	31	37	45	26	54	9.9
Hydro	2	2	2	2	3	4	14	5	2.5
Other renewables	-	-	-	0	0	0	-	0	n.a.

	CO_2 emissions (Mt)							Shares (%)		CAAGR (%)
	1990	2010	2015	2020	2025	2030	2035	2010	2035	2010-35
Total CO_2	**59**	**104**	**219**	**304**	**336**	**365**	**402**	**100**	**100**	**5.5**
Oil	55	93	193	224	219	235	260	89	65	4.2
Gas	4	12	27	80	117	130	142	11	35	10.5
Power generation	**19**	**35**	**107**	**126**	**115**	**116**	**123**	**100**	**100**	**5.1**
Oil	17	26	88	62	25	20	19	74	15	-1.4
Gas	2	9	19	63	90	96	104	26	85	10.1
TFC	**37**	**65**	**96**	**146**	**180**	**204**	**228**	**100**	**100**	**5.2**
Oil	35	62	91	137	164	182	204	96	89	4.9
Transport	*21*	*44*	*69*	*111*	*136*	*153*	*175*	*68*	*77*	*5.7*
Gas	2	2	6	10	16	21	24	4	11	9.8

Iraq: High and Delayed Cases

	Electricity generation (TWh)						Shares (%)		CAAGR (%)	
	2020	2030	2035	2020	2030	2035	2035		2010-35	
	High Case			Delayed Case			High	Delayed	High	Delayed
Total generation	**233**	**275**	**301**	**142**	**188**	**208**	**100**	**100**	**7.4**	**5.8**
Oil	45	24	32	67	49	53	11	26	0.4	2.4
Gas	180	240	258	63	126	140	85	67	11.5	8.8
Hydro	8	11	12	12	12	15	4	7	3.9	4.7
Other renewables	-	0	0	-	0	0	0	0	n.a.	n.a.

	Electrical capacity (GW)						Shares (%)		CAAGR (%)	
	2020	2030	2035	2020	2030	2035	2035		2010-35	
	High Case			Delayed Case			High	Delayed	High	Delayed
Total capacity	**69**	**80**	**86**	**48**	**58**	**63**	**100**	**100**	**6.9**	**5.6**
Oil	39	30	31	37	38	41	36	64	4.8	5.9
Gas	27	46	52	9	17	19	60	30	10.5	6.1
Hydro	2	3	3	2	3	4	4	6	1.6	2.0
Other renewables	-	0	0	-	0	0	0	0	n.a.	n.a.

	CO_2 emissions (Mt)						Shares (%)		CAAGR (%)	
	2020	2030	2035	2020	2030	2035	2035		2010-35	
	High Case			Delayed Case			High	Delayed	High	Delayed
Total CO_2	**380**	**439**	**477**	**202**	**256**	**277**	**100**	**100**	**6.3**	**4.0**
Oil	270	297	326	166	178	194	68	70	5.2	3.0
Gas	110	142	151	36	78	83	32	30	10.8	8.2
Power generation	**125**	**121**	**133**	**84**	**95**	**102**	**100**	**100**	**5.4**	**4.3**
Oil	39	21	28	58	42	45	21	44	0.3	2.2
Gas	86	100	105	26	54	57	79	56	10.2	7.5
TFC	**206**	**257**	**278**	**100**	**133**	**144**	**100**	**100**	**6.0**	**3.3**
Oil	193	231	250	93	116	126	90	87	5.7	2.8
Transport	*161*	*198*	*217*	*72*	*91*	*101*	*78*	*70*	*6.6*	*3.4*
Gas	13	26	28	7	17	19	10	13	10.5	8.6

Policies and measures by scenario

World Energy Outlook 2012 (*WEO-2012*) presents projections for four scenarios, which are primarily differentiated by the underlying assumptions about government policies.

The ***Current Policies Scenario*** is based on the perpetuation, without change, of the government policies and measures that had been enacted by mid-2012.

The ***New Policies Scenario*** – our central scenario – takes into account broad policy commitments and plans that have already been implemented to address energy-related challenges as well as those that have been announced, even where the specific measures to implement these commitments have yet to be introduced. It assumes only cautious implementation of current commitments and plans.

The ***450 Scenario*** sets out an energy pathway that is consistent with a 50% chance of meeting the goal of limiting the increase in average global temperature to 2 °C compared with pre-industrial levels. For the period to 2020, the 450 Scenario assumes more vigorous policy action to implement fully the Cancun Agreements than is assumed in the New Policies Scenario. After 2020, OECD countries and other major economies are assumed to set economy-wide emissions targets for 2035 and beyond to collectively ensure an emissions trajectory consistent with stabilisation of the greenhouse-gas concentration at 450 parts per million.

The ***Efficient World Scenario***, which has been developed especially for *WEO-2012*, is based on the core assumption that policies are put in place to allow the market to realise the potential of all known energy efficiency measures which are economically viable. The detailed methodology and assumptions for this scenario are presented and discussed in Chapters 9-11.

A number of the policy commitments and plans that were included in the New Policies Scenario in *WEO-2011* have since been enacted, so are now included in the Current Policies Scenario in this *Outlook*. These include, for example:

- Heavy-duty vehicle fuel-efficiency standards in the United States.
- Feed-in tariffs for renewable energy technologies in Japan.
- Programmes that put a price on CO_2 emissions (either through cap-and-trade programmes or carbon taxes) in New Zealand from 2010, in Australia as of mid-2012 and in Korea as of 2015.
- The gradual phase-out of fossil-fuel subsidies in a number of non-OECD countries.

The key policies that are assumed to be adopted in each of the main scenarios of *WEO-2012* are presented below, by sector and region. The policies are cumulative. That is, measures listed under the New Policies Scenario supplement those under the Current Policies Scenario, and measures listed under the 450 Scenario supplement those under the New Policies Scenario. The following tables start with broad cross-cutting policy frameworks and are followed by more detailed policy assumptions by sector as they have been adopted in this year's *Outlook*.

Table B.1 ▷ Cross-cutting policy assumptions by scenario for selected regions

	Current Policies Scenario	New Policies Scenario	450 Scenario
OECD			• Staggered introduction of CO_2 prices in all countries. • $100 billion annual financing provided to non-OECD countries by 2020.
United States	• State-level renewable portfolio standards (RPS) that include the option of using energy efficiency as a means of compliance. • Regional Greenhouse Gas Initiative (RGGI): mandatory cap-and-trade scheme covering fossil-fuel power plants in nine north-eastern states including recycling of revenues for energy efficiency and renewable energy investments.	• State-wide cap-and-trade scheme in California with binding commitments from 2013.	• 17% reduction in greenhouse-gas (GHG) emissions compared with 2005 by 2020. • CO_2 pricing implemented from 2020.
Japan[1]			• 25% reduction in GHG emissions compared with 1990 by 2020. • CO_2 pricing implemented from 2020.
European Union	• EU-level target to reduce GHG emissions by 20% in 2020, relative to 1990. • EU Emissions Trading System. • Renewables to reach a share of 20% in energy demand in 2020.	• Partial implementation of the EU-level target to reduce primary energy consumption by 20% in 2020. ○ Partial implementation of the EU Energy Efficiency Directive. ○ National Energy Efficiency Action Plans.	• 30% reduction in GHG emissions compared with 1990 by 2020. • EU ETS strengthened in line with the 2050 roadmap. • Full implementation of the EU Energy Efficiency Directive.
Australia and New Zealand	• Australia: Clean Energy Future Package - carbon prices through taxes/ETS as of mid-2015. • New Zealand: ETS from 2010.	• Australia: 5% reduction in GHG emissions compared with 2000 by 2020. • New Zealand: 10% cut in GHG emissions compared with 1990 by 2020.	• Australia: 25% reduction in GHG emissions compared with 2000 by 2020. • New Zealand: 20% reduction in GHG emissions compared with 1990 by 2020.
Korea	• Cap-and-trade scheme from 2015 (CO_2 emissions reductions of 4% compared with 2005 by 2020).	• 30% reduction in GHG emissions compared with business-as-usual by 2020.	• 30% reduction in GHG emissions compared with business-as-usual by 2020. • Higher CO_2 prices.

1. Japan released the Innovative Strategy for Energy and the Environment in September 2012, which aims to increase energy efficiency and the use of renewables, thereby reducing reliance on nuclear power and fossil fuels. Although not all of the details of the new strategy were available as this analysis was completed, it includes a target of electricity savings of 10% by 2030 compared with 2010.

Table B.1 ▷ Cross-cutting policy assumptions by scenario for selected regions (continued)

	Current Policies Scenario	New Policies Scenario	450 Scenario
Non-OECD	• Fossil-fuel subsidies are phased out in countries that already have policies in place to do so.	• Fossil-fuel subsidies are phased out in all net-importing regions by 2020 (at the latest) and in net-exporting regions where specific policies have already been announced.	• Finance for domestic mitigation. • Fossil-fuel subsidies are phased out by net-importers by 2020 and by exporters by 2035.*
Russia	• Gradual real increases in residential gas and electricity prices (1% per year) and in gas prices in industry (1.5% per year). • Implementation of 2009 energy efficiency legislation.	• 15% reduction in GHG emissions by 2020, compared with 1990. • 2% per year real rise in residential gas and electricity prices. • Industrial gas prices reach export prices (minus taxes and transport) in 2020. • Partial implementation of the 2010 energy efficiency state programme.	• 25% reduction in GHG emissions by 2020, compared with 1990. • Quicker rise in residential gas and electricity prices. • CO_2 pricing from 2020. • More support for nuclear and renewables. • Full implementation of the 2010 energy efficiency state programme.
China	• Implementation of measures in the 12th Five-Year Plan, including a 17% cut in CO_2 intensity by 2015 and a 16% reduction in energy intensity by 2015 compared with 2010.	• 40% reduction in CO_2 intensity compared with 2005 by 2020. • CO_2 pricing from 2020. • Share of 15% of non-fossil fuel in total supply by 2020.	• 45% reduction in CO_2 intensity compared with 2005 by 2020; higher CO_2 pricing. • Reduction of local air pollutants from 2010 to 2015 (8% for sulphur dioxide, 10% for nitrogen oxides).
India	• Trading of renewable energy certificates. • National solar mission and national mission on enhanced energy efficiency. • 11th Five-Year Plan (2007-2012).	• 20% reduction in CO_2 intensity compared with 2005 by 2020.	• 25% reduction in CO_2 intensity compared with 2005 by 2020.
Brazil	• 2011 National Energy Plan.	• 36% reduction in GHG emissions compared with business-as-usual by 2020.	• 39% reduction in GHG emissions compared with business-as-usual by 2020. • CO_2 pricing from 2020.

*Except in the Middle East where subsidisation rates are assumed to decline to a maximum of 20% by 2035.

Note: Pricing of CO_2 emissions is either by an emissions trading scheme (ETS) or taxes.

Table B.2 ▷ Power sector policies and measures as modelled by scenario in selected regions

	Current Policies Scenario	New Policies Scenario	450 Scenario
OECD			
United States	• State-level renewable portfolio standards (RPS) and support for renewables prolonged over the projection period. • Lifetimes of most US nuclear plants extended beyond 60 years. • Funding for CCS (demonstration-scale).	• Shadow price of carbon adopted from 2015, affecting investment decisions in power generation. • Mercury and Air Toxics Standard (MATS). • Cross-State Air Pollution Rule (CSAPR) regulating sulphur oxides and nitrogen oxides emissions. • Extension and strengthening of support for renewables and nuclear, including loan guarantees.	• CO_2 pricing implemented from 2020. • Extended support to renewables, nuclear and CCS. • Efficiency and emission standards preventing refurbishment of old inefficient plants.
Japan	• Support for renewables generation. • Decommissioning of units 1-4 of Fukushima Daiichi nuclear power plant.	• Shadow price of carbon adopted from 2015, affecting investment decisions in power generation. • Lifetime of nuclear power plants limited to 40 years for plants built until 1990 and 50 years for all others. • Increased support for renewables generation.	• CO_2 pricing implemented from 2020. • Share of low-carbon electricity generation to increase by 2020 and expand further by 2030. • Expansion of renewables support. • Introduction of CCS to coal-fired power generation.

European Union	• Climate and Energy Package:
	  ○ Emissions Trading System.
	  ○ Support for renewables sufficient to reach 20% share of energy demand in 2020.
	  ○ Financial support for CCS, including use of credits from the Emissions Trading System New Entrants' Reserve.
	• Early retirement of all nuclear plants in Germany by the end of 2022.
	• Removal of some barriers to combined heat and power (CHP) plants resulting from the Cogeneration Directive 2004.
		• Extended and strengthened support to renewables-based electricity generation technologies.
		• Further removal of barriers to CHP due to partial implementation of the Energy Efficiency Directive.
			• Emissions Trading System strengthened in line with the 2050 roadmap.
			• Reinforcement of government support in favour of renewables.
			• Expanded support measures for CCS.

Note: CCS = carbon capture and storage.

Table B.2 ▷ Power sector policies and measures as modelled by scenario in selected regions (continued)

	Current Policies Scenario	New Policies Scenario	450 Scenario
Non-OECD			
Russia	• Competitive wholesale electricity market.	• State support to the nuclear and hydropower sectors; a support mechanism for non-hydro renewables introduced from 2014.	• CO_2 pricing implemented from 2020. • Stronger support for nuclear power and renewables.
China	• Implementation of measures in 12th Five-Year Plan. • Start construction of 40 GW of new nuclear plants by 2015. • Start construction of 120 GW of hydropower by 2015. • Wind capacity additions of 70 GW by 2015. • Solar additions of 5 GW by 2015. • Priority given to gas use to 2015.	• 12th Five-Year Plan renewables targets for 2015 are exceeded. • 70 to 80 GW of nuclear capacity by 2020. • 200 GW of wind capacity by 2020. • Solar capacity of 50 GW by 2020; subsidies for building-integrated PV projects. • 30 GW of biomass capacity by 2020. • CO_2 pricing implemented from 2020.	• Higher CO_2 pricing. • Enhanced support for renewables. • Continued support to nuclear capacity additions post 2020. • Deployment of CCS from around 2020.
India	• Renewable Energy Certificate trade for all eligible grid-connected renewable-based electricity generation technologies. • National solar mission target of 20 GW of solar PV capacity by 2022. • Increased use of supercritical coal technology.	• Renewable energy support policies and targets, including small hydro. • Coal-fired power stations energy efficiency mandates.	• Renewables (excluding large hydro) to reach 15% of installed capacity by 2020. • Expanded support to renewables, nuclear and efficient coal. • Deployment of CCS from around 2020.
Brazil	• Increase of wind, biomass and hydro (small and large) capacity. • Auctions for renewables-based power generation.	• Enhanced deployment of renewables technologies. • Implementation of measures in the Ten-Year Plan for Energy Expansion (PDE2020).	• CO_2 pricing implemented from 2020. • Further increases of generation from renewable sources.

Note: CCS = carbon capture and storage.

Table B.3 ▷ Transport sector policies and measures as modelled by scenario in selected regions

OECD	Current Policies Scenario	New Policies Scenario
United States	• CAFE standards: 34.5 miles-per-gallon for PLDVs by 2016, and further strengthening thereafter. • Renewables Fuel Standard. • Truck standards for each model year from 2014 to 2018 reduce average on-road fuel consumption by up to 18% in 2018.	• CAFE standards: 54.5 miles-per-gallon for PLDVs by 2025. • Renewables Fuel Standard. • Truck standards for each model year from 2014 to 2018 reduce average on-road fuel consumption by up to 21% in 2018. • Support to natural gas in road freight. • Increase of ethanol blending mandates.
Japan	• Fuel economy target for PLDVs: 16.8 kilometres per litre (km/l) by 2015. • Average fuel economy target for road freight vehicles: 7.09 km/l by 2015. • Fiscal incentives for hybrid and electric vehicles; subsidies for electric vehicles.	• Fuel economy target for PLDVs: 20.3 km/l by 2020. • Target share of next generation vehicles 50% by 2020.
European Union	• Climate and Energy Package: renewables to reach 10% of transport energy demand in 2020; CO_2 emission standards for PLDVs by 2015 (130 g CO_2/km through efficiency measures, additional 10 g CO_2/km by alternative fuels). • Support to biofuels. • Emissions Trading System to include aviation from 2012.	• More stringent emission target for PLDVs (95 g CO_2/km by 2020), and further strengthening post 2020. • Emission target for LCVs (147 g CO_2/km by 2020), and further strengthening post 2020. • Enhanced support to alternative fuels.

450 Scenario	
All OECD	
On-road emission targets for PLDVs in 2035	65 g CO_2/km
Light-commercial vehicles	Full technology spill-over from PLDVs.
Medium- and heavy-freight traffic	45% more efficient by 2035 than in New Policies Scenario.
Aviation	50% efficiency improvements by 2035 (compared with 2010) and support for the use of biofuels.
Other sectors such as maritime and rail	National policies and measures.
Fuels	Retail fuel prices kept at a level similar to New Policies Scenario.
Alternative clean fuels	Enhanced support to alternative fuels.

Notes: CAFE = Corporate Average Fuel Economy; PLDVs = passenger light-duty vehicles; LCV = light-commercial vehicles.

Table B.3 ▷ Transport sector policies and measures as modelled by scenario in selected regions (continued)

	Current Policies Scenario	New Policies Scenario
Non-OECD		
China	• Subsidies for hybrid and electric vehicles. • Promotion of fuel-efficient cars. • Ethanol blending mandates 10% in selected provinces.	• Fuel economy target PLDVs: 6.9 l/100 km by 2015, 5.0 l/100 km by 2020. • Extended subsidies for the purchase of alternative vehicles. • Increased biofuels blending.
India	• Support for alternative fuel vehicles.	• Extended support for alternative fuel vehicles. • Proposed auto fuel efficiency standards to reduce average test-cycle fuel consumption by 1.3% per year between 2010 and 2020. • Increased utilisation of natural gas in road transport.
Brazil	• Ethanol targets in road transport 20% to 25%. • Biodiesel blending mandate of 5%. • Voluntary fuel efficiency labelling scheme for PLDVs.	• Increase of ethanol targets and biodiesel mandates. • Local renewable fuel targets for urban transport.

450 Scenario	
All non-OECD	
On-road emission targets for PLDVs in 2035	85 g CO_2/km
Light-commercial vehicles	Full technology spill-over from PLDVs.
Medium- and heavy-freight traffic	45% more efficient by 2035 than in New Policies Scenario.
Aviation	50% efficiency improvements by 2035 (compared with 2010) and support for the use of biofuels.
Other sectors such as maritime and rail	National policies and measures.
Fuels	Retail fuel prices kept at a level similar to New Policies Scenario.
Alternative clean fuels	Enhanced support to alternative fuels.

Note: PLDVs = passenger light-duty vehicles.

Table B.4 ▷ Industry sector policies and measures as modelled by scenario in selected regions

OECD	Current Policies Scenario	New Policies Scenario	450 Scenario (All OECD)
United States	• Support for high-energy efficiency technologies. • Energy Star Program for Industry.	• Tax reduction and funding for efficient technologies. • R&D in low-carbon technologies. • Energy certification programme.	• CO_2 pricing introduced from 2025 at the latest in all countries. • International sectoral agreements with energy intensity targets for iron and steel and cement industries. • Enhanced energy efficiency standards. • Policies to support the introduction of CCS in industry.
Japan	• Energy efficiency benchmarking. • Tax credits for investments in energy efficiency. • Mandatory energy management for large business operators. • Top-runner programme setting minimum energy standards, including for lighting, space heating, and transformers.	• Maintenance and strengthening of top-end/low carbon efficiency standards by: ○ Higher efficiency CHP systems. ○ Promotion of state-of-the-art technology and faster replacement of aging equipment.	
European Union	• Emissions Trading System. • Effort sharing decision: ○ Energy Performance Directive for Buildings. ○ Eco-Design Directive (including minimum standards for electric motors). • Voluntary energy efficiency agreements in several countries.	• Partial implementation of Energy Efficiency Directive. • Directive on energy end-use efficiency and energy efficiency, including the development of: ○ Inverters for electric motors. ○ High-efficiency co-generation. ○ Mechanical vapour compression. ○ Innovations in industrial processes.	

Notes: ETS = emissions trading system; R&D = research and development; CHP = combined heat and power; CCS = carbon capture and storage.

Table B.4 ▷ Industry sector policies and measures as modelled by scenario in selected regions (continued)

<table>
<tr><th></th><th>Current Policies Scenario</th><th>New Policies Scenario</th><th>450 Scenario</th></tr>
<tr><td>Non-OECD</td><td></td><td></td><td>All non-OECD</td></tr>
<tr><td>Russia</td><td>• Competitive wholesale electricity market price.
• Mandatory energy audits for energy-intensive industries.
• Complete phase-out of open hearth furnaces in the iron and steel industry.</td><td>• Industrial gas prices reach the equivalent of export prices (minus taxes and transportation) in 2020.
• Elaboration of comprehensive federal and regional legislation on energy savings, including implementation of energy management systems.</td><td rowspan="4">• CO_2 pricing introduced as of 2020 in Russia, China, Brazil and South Africa.
• Wider hosting of international offset projects.
• International sectoral agreements with targets for iron and steel and cement industries.
• Enhanced energy-efficiency standards.
• Policies to support the introduction of CCS in industry.</td></tr>
<tr><td>China</td><td>• Priority given to gas use to 2015 (12th Five-Year Plan).
• Partial implementation of the Top-10 000 energy-consuming enterprises programme.
• Small plant closures and phasing out of outdated production capacity.</td><td>• Full implementation of the Top-10 000 energy-consuming enterprises programme.
• CO_2 pricing implemented from 2020.
• Contain the expansion of energy-intensive industries.
• Partial implementation of reduction in energy intensity of large-scale companies by 21% during the 12th Five-Year Plan period (2011-2015).
• Enhanced use of energy service companies and energy performance contracting.
• Industrial Energy Performance Standard.
• Mandatory adoption of coke dry quenching and top-pressure turbines in new iron and steel plants.
• Support non-blast furnace iron making.</td></tr>
<tr><td>India</td><td>• Perform Achieve and Trade (PAT) mechanism, targeting a 5% reduction in energy use by 2015 compared with 2010.</td><td>• Further implementation of National Mission for Enhanced Energy Efficiency recommendations including:
○ Enhancement of cost-effective improvements in energy efficiency in energy-intensive large industries and facilities through tradable certificates.
○ Financing mechanism for demand-side management programmes.
○ Development of fiscal instruments to promote energy efficiency.
• Mandatory adoption of coke dry quenching and top-pressure turbines in new iron and steel plants.</td></tr>
<tr><td>Brazil</td><td>• Encourage investment and R&D in energy efficiency.
• National Climate Change Plan including improvements in energy efficiency.</td><td>• Higher use of charcoal in blast furnaces as a substitute for coal.
• Implementation of measures included in the 2010 energy efficiency state programme.</td></tr>
</table>

Table B.5 ▷ Buildings sector policies and measures as modelled by scenario in selected regions

	Current Policies Scenario	New Policies Scenario	450 Scenario
OECD			
United States	• AHAM-ACEEE Multi-Product Standards Agreement. • American Recovery and Reinvestment Act (2009): funding for energy efficiency and renewables. • Energy Star: federal tax credits for consumer energy efficiency; new appliance efficiency standards. • Energy Improvement and Extension Act of 2008.	• Extensions to 2025 of tax credits for energy-efficient equipment (including furnaces, boilers, air conditioners, air and ground source heat pumps, water heaters and windows), and for solar PV and solar thermal water heaters. • Budget proposals 2011 - institute programmes to make commercial buildings 20% more efficient by 2020; tax credit for renewable energy deployment. • Mandatory energy requirements in building codes in some states.	• Mandatory energy requirements in building codes in all states by 2020. • Extension of energy efficiency grants to end of projection period. • Zero-energy buildings initiative.
Japan	• Top Runner Programme. • Long-Term Outlook on Energy Supply and Demand (2009): energy savings using demand-side management.	• Extension of the Top Runner Programme to cover more products. • High-efficiency lighting: 100% in public facilities by 2020; 100% of lighting stock by 2030. • Voluntary buildings labelling; national voluntary equipment labelling programmes. • Net zero-energy buildings by 2030 for all new construction. • Increased introduction of gas and renewable energy.	• Rigorous and mandatory building energy codes for all new and existing buildings. • Net zero-energy buildings by 2025 for all new construction.
European Union	• Energy Performance of Buildings Directive. • Eco-Design and Energy Labelling Directive. • EU-US Energy Star Agreement: energy labelling of appliances.	• Energy Efficiency Directive. • Building energy performance requirements for new buildings (zero-energy buildings by 2021) and for existing buildings when extensively renovated. 3% renovation rate of central government buildings. • Mandatory energy labelling for sale or rental of all buildings and some appliances, lighting and equipment. • Further product groups in EcoDesign Directive. • Phase-out of incandescent light bulbs.	• Zero-carbon footprint for all new buildings as of 2015; enhanced energy efficiency in all existing buildings.

Notes: ACEEE = American Council for an Energy-Efficient Economy; AHAM = Association of Home Appliance Manufacturers.

Table B.5 ▷ Buildings sector policies and measures as modelled by scenario in selected regions (continued)

	Current Policies Scenario	New Policies Scenario	450 Scenario
Non-OECD			
Russia	• Implementation of 2009 energy efficiency legislation.	• Gradual above-inflation increase in residential electricity and gas prices. • New building codes, meter installations and refurbishment programmes, leading to efficiency gains in space heating (relative to Current Policies Scenario). • Information and awareness on energy efficiency classes for appliances. • Phase-out of incandescent >100 Watt light bulbs.	• Faster liberalisation of gas and electricity prices. • Extension and reinforcement of all measures included in the 2010 energy efficiency state programme; mandatory building codes by 2030 and phase-out of inefficient equipment and appliances by 2030.
China	• Civil Construction Energy Conservation Design Standard. • Appliance standards and labelling programme.	• Energy-efficient buildings to account for 30% of all new construction projects by 2020. • Civil Construction Energy Conservation Design Standard: heating energy consumption per unit area of existing buildings to be reduced by 65% in cold cold regions; 50% in hot-in-summer and cold-in-winter regions compared to 1980-1981 levels. New buildings: 65% improvement in all regions. • Energy Price Policy (reform heating price to be based on actual consumption, rather than on living area supplied). • Mandatory energy efficiency labels for appliances and equipment. • Labelling mandatory for new, large commercial and governmental buildings in big cities. • Mandatory codes for all new large residential buildings in big cities.	• More stringent implementation of Civil Construction Energy Conservation Design Standard. • Mandatory energy efficiency labels for all appliances and also for building shell.
India	• Measures under national solar mission. • Energy Conservation Building Code 2007, with voluntary requirements for commercial and residential buildings.	• Mandatory standards and labels for room air conditioners and refrigerators, voluntary for 5 other products. • Phase out of incandescent light bulbs by 2020. • Voluntary Star Ratings for the services sector.	• Mandatory energy conservation standards and labelling requirements for all equipment and appliances by 2025. • Increased penetration of energy efficient lighting.

Definitions

This annex provides general information on terminology used throughout *WEO-2012* including: units and general conversion factors; definitions of fuels, processes and sectors; regional and country groupings; abbreviations and acronyms.

Units

Area	Ha	hectare
	GHa	giga-hectare (1 hectare x 10^9)
	km^2	square kilometre
Emissions	ppm	parts per million (by volume)
	Gt CO_2-eq	gigatonnes of carbon-dioxide equivalent (using 100-year global warming potentials for different greenhouse gases)
	kg CO_2-eq	kilogrammes of carbon-dioxide equivalent
	g CO_2/km	grammes of carbon dioxide per kilometre
	g CO_2/kWh	grammes of carbon dioxide per kilowatt-hour
Energy	Mtce	million tonnes of coal equivalent (equals 0.7 Mtoe)
	boe	barrels of oil equivalent
	toe	tonne of oil equivalent
	ktoe	kilotonne of oil equivalent
	Mtoe	million tonnes of oil equivalent
	MBtu	million British thermal units
	kcal	kilocalorie (1 calorie x 10^3)
	Gcal	gigacalorie (1 calorie x 10^9)
	MJ	megajoule (1 joule x 10^6)
	GJ	gigajoule (1 joule x 10^9)
	TJ	terajoule (1 joule x 10^{12})
	PJ	petajoule (1 joule x 10^{15})
	EJ	exajoule (1 joule x 10^{18})
	kWh	kilowatt-hour
	MWh	megawatt-hour
	GWh	gigawatt-hour
	TWh	terawatt-hour
Gas	mcm	million cubic metres

	bcm	billion cubic metres
	tcm	trillion cubic metres
	scf	standard cubic foot
Mass	kg	kilogramme (1 000 kg = 1 tonne)
	kt	kilotonnes (1 tonne x 10^3)
	Mt	million tonnes (1 tonne x 10^6)
	Gt	gigatonnes (1 tonne x 10^9)
Monetary	$ million	1 US dollar x 10^6
	$ billion	1 US dollar x 10^9
	$ trillion	1 US dollar x 10^{12}
Oil	b/d	barrels per day
	kb/d	thousand barrels per day
	mb/d	million barrels per day
	mpg	miles per gallon
Power	W	watt (1 joule per second)
	kW	kilowatt (1 Watt x 10^3)
	MW	megawatt (1 Watt x 10^6)
	GW	gigawatt (1 Watt x 10^9)
	GW_{th}	gigawatt thermal (1 Watt x 10^9)
	TW	terawatt (1 Watt x 10^{12})

General conversion factors for energy

Convert to:	**TJ**	**Gcal**	**Mtoe**	**MBtu**	**GWh**
From:	multiply by:				
TJ	1	238.8	2.388×10^{-5}	947.8	0.2778
Gcal	4.1868×10^{-3}	1	10^{-7}	3.968	1.163×10^{-3}
Mtoe	4.1868×10^{4}	10^{7}	1	3.968×10^{7}	11 630
MBtu	1.0551×10^{-3}	0.252	2.52×10^{-8}	1	2.931×10^{-4}
GWh	3.6	860	8.6×10^{-5}	3 412	1

Currency conversions

Exchange rates (2011)	1 US Dollar equals:
Australian Dollar	0.97
British Pound	0.62
Canadian Dollar	0.99
Chinese Yuan	6.47
Euro	0.72
Indian Rupee	46.26
Japanese Yen	79.84
Korean Won	1 107.81
Russian Ruble	29.42

Definitions

Advanced biofuels

Advanced biofuels comprise different emerging and novel conversion technologies that are currently in the research and development, pilot or demonstration phase. This definition differs from the one used for "Advanced Biofuels" in the US legislation, which is based on a minimum 50% lifecycle greenhouse-gas reduction and which, therefore, includes sugarcane ethanol.

Advanced biomass cookstoves

Advanced biomass cookstoves are biomass gasifier-operated cooking stoves that run on solid biomass, such as wood chips and briquettes. These cooking devices have significantly lower emissions and higher efficiencies than the traditional biomass cookstoves (three-stone fires) currently used largely in developing countries

Agriculture

Includes all energy used on farms, in forestry and for fishing.

Biodiesel

Biodiesel is a diesel-equivalent, processed fuel made from the transesterification (a chemical process which removes the glycerine from the oil) of both vegetable oils and animal fats.

Bioenergy

Refers to the energy content in solid, liquid and gaseous products derived from biomass feedstocks and biogas. This includes biofuels for transport and products (e.g. wood chips, pellets, black liquor) to produce electricity and heat. Municipal solid waste and industrial waste are also included.

Biofuels

Biofuels are fuels derived from biomass or waste feedstocks and include ethanol and biodiesel. They can be classified as conventional and advanced biofuels according to the technologies used to produce them and their respective maturity.

Biogas

A mixture of methane and carbon dioxide produced by bacterial degradation of organic matter and used as a fuel.

Brown coal

Includes lignite and sub-bituminous coal where lignite is defined as non-agglomerating coal with a gross calorific value less than 4 165 kilocalories per kilogramme (kcal/kg) and sub-bituminous coal is defined as non-agglomerating coal with a gross calorific value between 4 165 kcal/kg and 5 700 kcal/kg.

Buildings

The buildings sector includes energy used in residential, commercial and institutional buildings, and non-specified other. Building energy use includes space heating and cooling, water heating, lighting, appliances and cooking equipment.

Bunkers

Includes both international marine bunkers and international aviation bunkers.

Capacity credit

Capacity credit refers to the proportion of capacity that can be reliably expected to generate electricity during times of peak demand in the grid to which it is connected.

Clean coal technologies

Clean coal technologies (CCTs) are designed to enhance the efficiency and the environmental acceptability of coal extraction, preparation and use.

Coal

Coal includes both primary coal (including hard coal and brown coal) and derived fuels (including patent fuel, brown-coal briquettes, coke-oven coke, gas coke, gas-works gas, coke-oven gas, blast-furnace gas and oxygen steel furnace gas). Peat is also included.

Coalbed methane

Methane found in coal seams. Coalbed methane (CBM) is a source of unconventional natural gas.

Coal-to-liquids

Coal-to-liquids (CTL) refers to the transformation of coal into liquid hydrocarbons. It can be achieved through either coal gasification into syngas (a mixture of hydrogen and carbon monoxide), combined using the Fischer-Tropsch or methanol-to-gasoline synthesis process to produce liquid fuels, or through the less developed direct-coal liquefaction technologies in which coal is directly reacted with hydrogen.

Coking coal

Coking coal is a type of hard coal that can be used in the production of coke, which is capable of supporting a blast furnace charge.

Condensates

Condensates are liquid hydrocarbon mixtures recovered from associated or nonassociated gas reservoirs. They are composed of C5 and higher carbon number hydrocarbons and normally have an API gravity between 50° and 85°.

Conventional biofuels

Conventional biofuels include well-established technologies that are producing biofuels on a commercial scale today. These biofuels are commonly referred to as firstgeneration and include sugarcane ethanol, starchbased ethanol, biodiesel, Fatty Acid Methyl Esther (FAME) and Straight Vegetable Oil (SVO). Typical feedstocks used in these mature processes include sugarcane and sugar beet, starch bearing grains, like corn and wheat, and oil crops, like canola and palm, and in some cases, animal fats.

Electricity generation

Defined as the total amount of electricity generated by power only or combined heat and power plants including generation required for own use. This is also referred to as gross generation.

Ethanol

Although ethanol can be produced from a variety of fuels, in this publication, ethanol refers to bio-ethanol only. Ethanol is produced from fermenting any biomass high in carbohydrates. Today, ethanol is made from starches and sugars, but second generation technologies will allow it to be made from cellulose and hemicellulose, the fibrous material that makes up the bulk of most plant matter.

Gas

Gas includes natural gas, both associated and non-associated with petroleum deposits, but excludes natural gas liquids.

Gas-to-liquids

Gas-to-liquids refers to a process featuring reaction of methane with oxygen or steam to produce syngas (a mixture of hydrogen and carbon monoxide) followed by synthesis of liquid products (such as diesel and naphtha) from the syngas using Fischer-Tropsch catalytic synthesis. The process is similar to those used in coal-to-liquids.

Hard coal

Coal of gross calorific value greater than 5 700 kilocalories per kilogramme on an ash-free but moist basis. Hard coal can be further disaggregated into anthracite, coking coal and other bituminous coal.

Heat energy

Heat is obtained from the combustion of fuels, nuclear reactors, geothermal reservoirs, capture of sunlight, exothermic chemical processes and heat pumps which can extract it from ambient air and liquids. It may be used for heating or cooling, or converted into mechanical energy for transport vehicles or electricity generation. Commercial heat sold is reported under total final consumption with the fuel inputs allocated under power generation.

Heavy petroleum products

Heavy petroleum products include heavy fuel oil.

Hydropower

Hydropower refers to the energy content of the electricity produced in hydropower plants, assuming 100% efficiency. It excludes output from pumped storage and marine (tide and wave) plants.

Industry

The industry sector includes fuel used within the manufacturing and construction industries. Key industry sectors include iron and steel, chemical and petrochemical, non-metallic minerals, and pulp and paper. Use by industries for the transformation of energy into another form or for the production of fuels is excluded and reported separately under other energy sector. Consumption of fuels for the transport of goods is reported as part of the transport sector.

International aviation bunkers

Includes the deliveries of aviation fuels to aircraft for international aviation. Fuels used by airlines for their road vehicles are excluded. The domestic/international split is determined on the basis of departure and landing locations and not by the nationality of the airline. For many countries this incorrectly excludes fuels used by domestically owned carriers for their international departures.

International marine bunkers

Covers those quantities delivered to ships of all flags that are engaged in international navigation. The international navigation may take place at sea, on inland lakes and waterways, and in coastal waters. Consumption by ships engaged in domestic navigation is excluded. The domestic/international split is determined on the basis of port of departure and port of arrival, and not by the flag or nationality of the ship. Consumption by fishing vessels and by military forces is also excluded and included in residential, services and agriculture.

Light petroleum products

Light petroleum products include liquefied petroleum gas (LPG), naphtha and gasoline.

Lignocellusosic feedstock

Lignocellulosic crops refers to those crops cultivated to produce biofuels from their cellulosic or hemicellulosic components, which include switchgrass, poplar and miscanthus.

Lower heating value

Lower heating value is the heat liberated by the complete combustion of a unit of fuel when the water produced is assumed to remain as a vapour and the heat is not recovered.

Middle distillates

Middle distillates include jet fuel, diesel and heating oil.

Modern biomass

Includes all biomass with the exception of traditional biomass.

Modern renewables

Includes all types of renewables with the exception of traditional biomass.

Natural decline rate

The base production decline rate of an oil or gas field without intervention to enhance production.

Natural gas liquids

Natural gas liquids (NGLs) are the liquid or liquefied hydrocarbons produced in the manufacture, purification and stabilisation of natural gas. These are those portions of natural gas which are recovered as liquids in separators, field facilities, or gas processing plants. NGLs include but are not limited to ethane, propane, butane, pentane, natural gasoline and condensates.

Non-energy use

Fuels used for chemical feedstocks and non-energy products. Examples of non-energy products include lubricants, paraffin waxes, coal tars and oils as timber preservatives.

Nuclear

Nuclear refers to the primary heat equivalent of the electricity produced by a nuclear plant with an average thermal efficiency of 33%.

Observed decline rate

The production decline rate of an oil or gas field after all measures have been taken to maximise production. It is the aggregation of all the production increases and declines of new and mature oil or gas fields in a particular region.

Oil

Oil includes crude oil, condensates, natural gas liquids, refinery feedstocks and additives, other hydrocarbons (including emulsified oils, synthetic crude oil, mineral oils extracted from bituminous minerals such as oil shale, bituminous sand and oils from coal liquefaction) and petroleum products (refinery gas, ethane, LPG, aviation gasoline, motor gasoline, jet fuels, kerosene, gas/diesel oil, heavy fuel oil, naphtha, white spirit, lubricants, bitumen, paraffin waxes and petroleum coke).

Other energy sector

Covers the use of energy by transformation industries and the energy losses in converting primary energy into a form that can be used in the final consuming sectors. It includes losses by gas works, petroleum refineries, coal and gas transformation and liquefaction. It also includes energy used in coal mines, in oil and gas extraction and in electricity and heat production. Transfers and statistical differences are also included in this category.

Power generation

Power generation refers to fuel use in electricity plants, heat plants and combined heat and power (CHP) plants. Both main activity producer plants and small plants that produce fuel for their own use (autoproducers) are included.

Renewables

Includes bioenergy, geothermal, hydropower, solar photovoltaic (PV), concentrating solar power (CSP), wind and marine (tide and wave) energy for electricity and heat generation.

R/P ratio

Reserves-to-production (R/P) ratio is based on the sum of probable and proven reserves and the last year of available data for production.

Self-sufficiency

Self-sufficiency is indigenous production divided by total primary energy demand (TPED).

Total final consumption

Total final consumption (TFC) is the sum of consumption by the different end-use sectors. TFC is broken down into energy demand in the following sectors: industry (including manufacturing and mining), transport, buildings (including residential and services) and other (including agriculture and non-energy use). It excludes international marine and aviation bunkers, except at world level where it is included in the transport sector.

Total primary energy demand

Total primary energy demand (TPED) represents domestic demand only and is broken down into power generation, other energy sector and total final consumption.

Traditional biomass

Traditional biomass refers to the use of fuelwood, charcoal, animal dung and agricultural residues in stoves with very low efficiencies.

Transport

Fuels and electricity used in the transport of goods or persons within the national territory irrespective of the economic sector within which the activity occurs. This includes fuel and electricity delivered to vehicles using public roads or for use in rail vehicles; fuel delivered to vessels for domestic navigation; fuel delivered to aircraft for domestic aviation; and energy consumed in the delivery of fuels through pipelines. Fuel delivered to international marine and aviation bunkers is presented only at the world level and is excluded from the transport sector at the domestic level.

Regional and country groupings

Africa

Algeria, Angola, Benin, Botswana, Cameroon, Congo, Democratic Republic of Congo, Côte d'Ivoire, Egypt, Eritrea, Ethiopia, Gabon, Ghana, Kenya, Libya, Morocco, Mozambique, Namibia, Nigeria, Senegal, South Africa, Sudan, United Republic of Tanzania, Togo, Tunisia, Zambia, Zimbabwe and other African countries (Burkina Faso, Burundi, Cape Verde, Central African Republic, Chad, Comoros, Djibouti, Equatorial Guinea, Gambia, Guinea, Guinea-Bissau, Lesotho, Liberia, Madagascar, Malawi, Mali, Mauritania, Mauritius, Niger, Reunion, Rwanda, Sao Tome and Principe, Seychelles, Sierra Leone, Somalia, Swaziland and Uganda).

Annex I Parties to the United Nations Framework Convention on Climate Change

Australia, Austria, Belarus, Belgium, Bulgaria, Canada, Croatia, Czech Republic, Denmark, Estonia, Finland, France, Germany, Greece, Hungary, Iceland, Ireland, Italy, Japan, Latvia, Liechtenstein, Lithuania, Luxembourg, Monaco, Netherlands, New Zealand, Norway, Poland, Portugal, Romania, Russian Federation, Slovak Republic, Slovenia, Spain, Sweden, Switzerland, Turkey, Ukraine, United Kingdom and United States.

APEC (Asia-Pacific Economic Cooperation)

Australia, Brunei Darussalam, Canada, Chile, China, Chinese Taipei, Indonesia, Hong Kong (China), Japan, Korea, Malaysia, Mexico, New Zealand, Papua New Guinea, Peru, Philippines, Russia, Singapore, Thailand, United States and Vietnam.

ASEAN

Brunei Darussalam, Cambodia, Indonesia, Laos, Malaysia, Myanmar, Philippines, Singapore, Thailand and Vietnam.

Caspian

Armenia, Azerbaijan, Georgia, Kazakhstan, Kyrgyz Republic, Tajikistan, Turkmenistan and Uzbekistan.

China

Refers to the People's Republic of China, including Hong Kong.

Developing countries

Non-OECD Asia, Middle East, Africa and Latin America regional groupings.

Eastern Europe/Eurasia

Albania, Armenia, Azerbaijan, Belarus, Bosnia and Herzegovina, Bulgaria, Croatia, Georgia, Kazakhstan, Kyrgyz Republic, Latvia, Lithuania, the former Yugoslav Republic of Macedonia, the Republic of Moldova, Romania, Russian Federation, Serbia , Tajikistan, Turkmenistan, Ukraine and Uzbekistan. For statistical reasons, this region also includes Cyprus, Gibraltar and Malta.

European Union

Austria, Belgium, Bulgaria, Cyprus, Czech Republic, Denmark, Estonia, Finland, France, Germany, Greece, Hungary, Ireland, Italy, Latvia, Lithuania, Luxembourg, Malta, Netherlands, Poland, Portugal, Romania, Slovak Republic, Slovenia, Spain, Sweden and United Kingdom.

G-8

Canada, France, Germany, Italy, Japan, Russian Federation, United Kingdom and United States.

G-20

G-8 countries and Argentina, Australia, Brazil, China, India, Indonesia, Mexico, Saudi Arabia, South Africa, Korea, Turkey and the European Union.

Latin America

Argentina, Bolivia, Brazil, Chile, Colombia, Costa Rica, Cuba, the Dominican Republic, Ecuador, El Salvador, Guatemala, Haiti, Honduras, Jamaica, Netherlands Antilles, Nicaragua, Panama, Paraguay, Peru, Trinidad and Tobago, Uruguay, Venezuela and other Latin American countries (Antigua and Barbuda, Aruba, Bahamas, Barbados, Belize, Bermuda, British Virgin Islands, Cayman Islands,. Dominica, Falkland Islands, French Guyana, Grenada, Guadeloupe, Guyana, Martinique, Montserrat, St. Kitts and Nevis, Saint Lucia, Saint Pierre et Miquelon, St. Vincent and the Grenadines, Suriname and Turks and Caicos Islands).

Middle East

Bahrain, Islamic Republic of Iran, Iraq, Jordan, Kuwait, Lebanon, Oman, Qatar, Saudi Arabia, Syrian Arab Republic, United Arab Emirates and Yemen. It includes the neutral zone between Saudi Arabia and Iraq.

Non-OECD Asia

Bangladesh, Brunei Darussalam, Cambodia, China, Chinese Taipei, India, Indonesia, the Democratic People's Republic of Korea, Malaysia, Mongolia, Myanmar, Nepal, Pakistan, the Philippines, Singapore, Sri Lanka, Thailand, Vietnam and other non-OECD Asian countries (Afghanistan, Bhutan, Cook Islands, East Timor, Fiji, French Polynesia, Kiribati, Laos, Macau, Maldives, New Caledonia, Papua New Guinea, Samoa, Solomon Islands, Tonga and Vanuatu).

North Africa

Algeria, Egypt, Libya, Morocco and Tunisia.

OECD

Includes OECD Europe, OECD Americas and OECD Asia Oceania regional groupings.

OECD Americas

Canada, Chile, Mexico and United States.

OECD Asia Oceania

Australia, Japan, Korea and New Zealand.

OECD Europe

Austria, Belgium, Czech Republic, Denmark, Estonia, Finland, France, Germany, Greece, Hungary, Iceland, Ireland, Italy, Luxembourg, Netherlands, Norway, Poland, Portugal, Slovak Republic, Slovenia, Spain, Sweden, Switzerland, Turkey and United Kingdom. For statistical reasons, this region also includes Israel.

OPEC

Algeria, Angola, Ecuador, Islamic Republic of Iran, Iraq, Kuwait, Libya, Nigeria, Qatar, Saudi Arabia, United Arab Emirates and Venezuela.

Other Asia

Non-OECD Asia regional grouping excluding China and India.

Sub-Saharan Africa

Africa regional grouping excluding the North African regional grouping.

Abbreviations and Acronyms

APEC	Asia-Pacific Economic Cooperation
API	American Petroleum Institute
ASEAN	Association of Southeast Asian Nations
BTL	biomass-to-liquids
BGC	Basrah Gas Company
BGR	German Federal Institute for Geosciences and Natural Resources
CAAGR	compound average annual growth rate
CAFE	corporate average fuel economy (standards in the United States)
CBM	coalbed methane
CER	Certified emission reduction
CCGT	combined-cycle gas turbine
CCS	carbon capture and storage
CDM	Clean Development Mechanism (under the Kyoto Protocol)
CFL	compact fluorescent lamp
CH_4	methane
CHP	combined heat and power; the term co-generation is sometimes used
CMM	coal mine methane
CNG	compressed natural gas
CO	carbon monoxide
CO_2	carbon dioxide
CO_2-eq	carbon-dioxide equivalent

COP	Conference of Parties (UNFCCC)
CPC	Caspian Pipeline Consortium
CPS	Current Policies Scenario
CSP	concentrating solar power
CSS	cyclic steam stimulation
CSSF	common seawater supply facility
CTL	coal-to-liquids
CV	calorific value
E&P	exploration and production
EDI	Energy Development Index
EOR	enhanced oil recovery
EPA	Environmental Protection Agency (United States)
EPC	engineering, procurement and construction
ESCO	energy service company
EU	European Union
EUA	European Union allowances
EU ETS	European Union Emissions Trading System
EV	electric vehicle
EWS	Efficient World Scenario
FAO	Food and Agriculture Organization of the United Nations
FDI	foreign direct investment
FFV	flex-fuel vehicle
FOB	free on board
GCV	gross calorific value
GDP	gross domestic product
GHG	greenhouse gases
GT	gas turbine
GTL	gas-to-liquids
HDI	Human Development Index
HDV	heavy-duty vehicles
HFO	heavy fuel oil
IAEA	International Atomic Energy Agency
ICE	internal combustion engine
ICT	information and communication technologies
IGCC	integrated gasification combined-cycle
IIASA	International Institute for Applied Systems Analysis
IMF	International Monetary Fund
INOC	Iraq National Oil Company
IOC	international oil company

IPC	Iraq Petroleum Company
IPCC	Intergovernmental Panel on Climate Change
IPP	independent power producer
KRG	Kurdistan Regional Government
LCV	light commercial vehicle
LDV	light-duty vehicle
LED	light-emitting diode
LHV	lower heating value
LNG	liquefied natural gas
LPG	liquefied petroleum gas
LRMC	long-run marginal cost
LTO	light tight oil
LULUCF	land use, land-use change and forestry
MER	market exchange rate
MDGs	Millennium Development Goals
MEPS	minimum energy performance standards
N_2O	nitrous oxide
NCV	net calorific value
NEA	Nuclear Energy Agency (an agency within the OECD)
NGL	natural gas liquids
NGV	natural gas vehicle
NOC	national oil company
NOx	nitrogen oxides
NPS	New Policies Scenario
OCGT	open-cycle gas turbine
ODI	outward foreign direct investment
OECD	Organisation for Economic Co-operation and Development
OPEC	Organization of the Petroleum Exporting Countries
PHEV	plug-in hybrid vehicle
PLDV	passenger light-duty vehicle
PM	particulate matter
$PM_{2.5}$	particulate matter with a diameter of 2.5 micrometres or less
PPP	purchasing power parity
PSA	production-sharing agreement
PV	photovoltaic
RD&D	research, development and demonstration
RDD&D	research, development, demonstration and deployment
RRR	remaining recoverable resource
SAGD	steam-assisted gravity drainage

SCO	synthetic crude oil
SO_2	sulphur dioxide
SOMO	State Oil Marketing Organization of Iraq
SPM	single-point mooring
SRMC	short-run marginal cost
T&D	transmission and distribution
TFC	total final consumption
TPED	total primary energy demand
UAE	United Arab Emirates
UCG	underground coal gasification
UN	United Nations
UNDP	United Nations Development Programme
UNEP	United Nations Environment Programme
UNFCCC	United Nations Framework Convention on Climate Change
UNIDO	United Nations Industrial Development Organization
UNPD	United Nations Population Division
URR	ultimate recoverable resource
US	United States
USC	ultra-supercritical
USGS	United States Geological Survey
WEO	World Energy Outlook
WEM	World Energy Model
WHO	World Health Organization
WTI	West Texas Intermediate
WTO	World Trade Organization
WTW	well-to-wheel

References

Part A: Global Trends

Chapter 1: Understanding the scenarios

Baron, R., *et al.* (2012), "Policy Options for Low Carbon Power Generation in China - Designing an Emissions Trading System for China's Electricity Sector", *International Energy Agency Insight Paper*, OECD/IEA, Paris.

IMF (International Monetary Fund) (2012a), *World Economic Outlook Update: Coping with High Debt and Sluggish Growth*, IMF, Washington, DC, October.

– (2012b), *World Economic Outlook Update: New Setbacks, Further Policy Action Needed*, IMF, Washington, DC, July.

OECD (Organisation for Economic Co-operation and Development) (2012), *OECD Economic Outlook*, No. 90, OECD, Paris, May.

UNPD (United Nations Population Division) (2011), *World Population Prospects: The 2010 Revision*, United Nations, New York.

Chapter 2: Energy projections to 2035

BGR (*Bundesanstalt für Geowissenschaften und Rohstoffe* – German Federal Institute for Geosciences and Natural Resources) (2011), *Energierohstoffe 2011, Reserven, Ressourcen, Verfügbarkeit, Tabellen* (Energy Resources 2011, Reserves, Resources, Availability, Tables), BGR, Hannover, Germany.

IEA (International Energy Agency) (2011), *World Energy Outlook 2011*, OECD/IEA, Paris.

– (2012), *Golden Rules for a Golden Age of Gas: World Energy Outlook Special Report on Unconventional Gas*, OECD/IEA, Paris.

NEA (Nuclear Energy Agency, OECD) and IAEA (International Atomic Energy Agency) (2010), *Uranium 2009: Resources, Production and Demand*, OECD, Paris.

OECD (Organisation for Economic Co-operation and Development) (2011), *Inventory of Estimated Budgetary Support and Tax Expenditures Relatng to Fossil Fuels in Selected OECD Countries*, OECD, Paris.

O&GJ (*Oil and Gas Journal*) (2011), "Worldwide Look at Reserves and Production", *Oil and Gas Journal*, Vol. 109, Issue 49, Pennwell Corporation, Oklahoma City, United States.

USGS (United States Geological Survey) (2000), *World Petroleum Assessment*, USGS, Boulder, United States.

– (2012a), *Variability of Distributions of Well-Scale Estimated Ultimate Recovery for Continuous (Unconventional) Oil and Gas Resources in the United States,* USGS Oil and Gas Assessment Team, USGS Open-File Report 2012–1118, Boulder, United States.

– (2012b), *An Estimate of Undiscovered Conventional Oil and Gas Resources of the World*, USGS, Boulder, United States.

Chapter 3: Oil market outlook

BGR (*Bundesanstalt für Geowissenschaften und Rohstoffe* – German Federal Institute for Geosciences and Natural Resources), *Energierohstoffe 2011, Reserven, Ressourcen, Verfügbarkeit, Tabellen* (Energy Resources 2011, Reserves, Resources, Availability, Tables), BGR, Hannover, Germany.

BP (2012), *BP Statistical Review of World Energy 2012*, BP, London.

Hill, N., *et al.* (2011), *Reduction and Testing of Greenhouse-Gas Emissions from Heavy Duty Vehicles – Lot 1: Strategy*, European Commission, DG Climate Action, Brussels.

IEA (International Energy Agency) (2008), *World Energy Outlook 2008*, OECD/IEA, Paris.

– (2009), *Transport, Energy and CO_2: Moving Toward Sustainability*, OECD/IEA, Paris.

– (2010), *World Energy Outlook 2010*, OECD/IEA, Paris.

– (2011), *World Energy Outlook 2011*, OECD/IEA, Paris.

– (2012a), *Energy Technology Perspectives 2012 – Pathways to a Clean Energy System*, OECD/IEA, Paris.

– (2012b), *Technology Roadmap: Fuel Economy of Road Vehicles*, OECD/IEA, Paris.

– (2012c), *Policy Pathway: Improving the Fuel Economy of Road Vehicles*, OECD/IEA, Paris.

– (2012d), *Medium-Term Oil Market Report*, OECD/IEA, Paris.

– (2012e), *Monthly Oil Market Report*, OECD/IEA, Paris, September.

O&GJ (*Oil and Gas Journal*) (2011), "Worldwide Look at Reserves and Production", *Oil and Gas Journal*, Vol. 109, Issue 49, Pennwell Corporation, Oklahoma City, United States.

Secretaria de Energia (Mexico) (2012), *Estrategia Nacional de Energia 2012-2026* (National Energy Strategy), Secretaria de Energia, Mexico City.

SPE (Society of Petroleum Engineers) (2007), *Petroleum Resources Management System*, Society of Petroleum Engineers, Houston, United States.

USGS (United States Geological Survey) (2000), *World Petroleum Assessment*, USGS, Boulder, United States.

– (2008), "Circum-Arctic Resources Appraisal: Estimates of Undiscovered Oil and Gas North of the Arctic Circle", *Fact Sheet 2008-3049*, USGS, Boulder, United States.

– (2012a), "An Estimate of Undiscovered Oil and Gas Resources of the World", *Fact Sheet FS2012-3042*, USGS, Boulder, United States.

– (2012b), "Assessment of Potential Additions to Conventional Oil and Gas Resources of the World (Outside the United States) from Reserve Growth", *Fact Sheet FS2012-3052*, USGS, Boulder, United States.

Chapter 4: Gas market outlook

BGR (*Bundesanstalt für Geowissenschaften und Rohstoffe* – German Federal Institute for Geosciences and Natural Resources) (2011), *Energierohstoffe 2011, Reserven, Ressourcen, Verfügbarkeit, Tabellen* (Energy Resources 2011, Reserves, Resources, Availability, Tables), BGR, Hannover, Germany.

BP (2012), *BP Statistical Review of World Energy 2012*, BP, London.

IEA (International Energy Agency) (2011), *World Energy Outlook 2011*, OECD/IEA, Paris.

– (2012a), *Golden Rules for a Golden Age of Gas: World Energy Outlook Special Report on Unconventional Gas,* OECD/IEA, Paris.

– (2012b), *Medium-Term Natural Gas Market Report*, OECD/IEA, Paris.

GIIGNL (International Group of LNG Importers) (2012), *The LNG Industry in 2011*, GIIGNL, Paris.

O&GJ (*Oil and Gas Journal*) (2011), "Worldwide Look at Reserves and Production", *Oil and Gas Journal*, Vol. 109, Issue 49, Pennwell Corporation, Oklahoma City, United States.

US DOE/EIA (US Department of Energy/Energy Information Administration) (2011), *World Shale Gas Resources: An Initial Assessment of 14 Regions Outside the United States*, USDOE/EIA-0383, US DOE, Washington, DC.

– (2012), *Annual Energy Outlook 2012*, USDOE/EIA-0383, US DOE, Washington, DC.

USGS (United States Geological Survey) (2000), *World Petroleum Assessment*, USGS, Boulder, United States.

– (2012a), *Variability of Distributions of Well-Scale Estimated Ultimate Recovery for Continuous (Unconventional) Oil and Gas Resources in the United States,* USGS Oil and Gas Assessment Team, Open-File Report 2012–1118, USGS, Boulder, United States.

– (2012b), *An Estimate of Undiscovered Conventional Oil and Gas Resources of the World*, USGS, Boulder, United States.

– (2012c), "Assessment of Potential Additions to Conventional Oil and Gas Resources of the World (Outside the United States) from Reserve Growth", *Fact Sheet FS2012-3052*, USGS, Boulder, United States.

Chapter 5: Coal market outlook

BGR (*Bundesanstalt für Geowissenschaften und Rohstoffe* – German Federal Institute for Geosciences and Natural Resources) (2011), *Energierohstoffe 2011, Reserven, Ressourcen, Verfügbarkeit, Tabellen* (Energy Resources 2011, Reserves, Resources, Availability, Tables), BGR, Hannover, Germany.

BREE (Australian Bureau of Resources and Energy Economics) (2012), *Australian Bulk Commodity Exports and Infrastructure – Outlook to 2025*, BREE, Canberra.

IEA (International Energy Agency) (2011a), *World Energy Outlook 2011,* OECD/IEA, Paris.

– (2011b), *Medium-Term Coal Market Report,* OECD/IEA, Paris.

McCloskey (2008-2012), *McCloskey Coal Reports*, McCloskey Group, London.

US DOE/EIA (US Department of Energy/Energy Information Administration) (2012), *Monthly Energy Review*, US DOE, Washington, DC, May.

Chapter 6: Power sector outlook

Autorita'per l'Energia Elettrica e il Gas (Authority for Electricity and Gas) (2012), *Annual Report on the State of Services and Activity Performed*, Autorita'per l'Energia Elettrica e il Gas, Rome.

BDEW (*Bundesverband der Energie- und Wasserwirtschaft e.V* – German Association of Energy and Water Industries) (2011), *Erneuerbare Energien und das EEG: Zahlen, Fakten, Grafiken 2011* (Renewable Energy and the EEG: Facts, Figures, Graphics 2011), BDEW, Berlin.

CRE (Commission de Régulation de l'Énergie – French Energy Regulatory Commission) (2011), *Observatoire des Marchés de l'Electricité et du Gaz, 4e trimestre 2011* – Electricity and Gas Market Observations, Q4 2011, CRE, Paris.

IEA (International Energy Agency) (2009), *Projected Costs of Electricity Generation*, OECD/IEA, Paris.

– (2011), *Empowering Customer Choice in Electricity Markets*, OECD/IEA, Paris.

– (2012), *Electricity Prices and Taxes – Quarterly Statistics, Second Quarter 2012*, OECD/IEA, Paris.

Mount, T., *et al.* (2012), "The Hidden System Costs of Wind Generation in a Deregulated Electricity Market", *Energy Journal*, Volume 33, Issue 1, International Association for Energy Economics, Cleveland, United States.

Ofgem (UK Office of the Gas and Electricity Markets) (2012), "Household Energy Bills Explained (Updated)", *Factsheet 97*, Ofgem, London.

Chapter 7: Renewable energy outlook

EC (European Commission) (2011), *Energy Roadmap 2050*, Communication from the Commission to the European Parliament, the Council, the European Economic and Social Committee and the Committee of the Regions, EC, Brussels.

EU (European Union) (2011), *Renewables Make the Difference,* EU, Brussels.

Frankfurt School UNEP Collaborating Centre and Bloomberg New Energy Finance (2012), *Global Trends in Renewable Energy Investment*, Frankfurt School of Finance and Management, Frankfurt, Germany.

F.O. Licht (2012), *World Ethanol & Biofuels Report*, F.O. Licht, London.

IEA (International Energy Agency) (2011a), *Technology Roadmap, Biofuels for Transport*, OECD/IEA, Paris.

– (2011b), *Solar Energy Perspectives*, OECD/IEA, Paris.

– (2012), *Renewable Energy Medium-Term Market Report 2012*, OECD/IEA, Paris.

IFPRI (International Food Policy Research Institute) and CEPII (The French Center for Research and Studies on the World Economy) (2010), *Global Trade and Environmental Impact Study of the EU Biofuels Mandate*, European Commission, Brussels.

IIASA (International Institute for Applied Systems Analysis) (2012), *Global Energy Assessment*, IIASA, Laxenburg, Austria.

IPCC (Intergovernmental Panel on Climate Change) (2011), *Special Report on Renewable Energy Sources and Climate Change Mitigation*, IPCC, Geneva.

Lamers, P. (2012), "Developments in International Solid Biofuel Trade - An Analysis of Volumes, Policies, and Market Factors", *Renewable and Sustainable Energy Reviews*, Vol. 16, Elsevier, Amsterdam, pp. 3176-3199.

REN21 (Renewable Energy Policy Network for the 21st Century) (2012), *Renewables 2012 Global Status Report*, REN21 Secretariat, Paris.

UNEP (United Nations Environment Programme) (2010), *Africa Water Atlas*, UNEP, Nairobi.

USDA (United States Department of Agriculture) (2011), *People's Republic of China, Biofuels Annual,* GAIN Report No. 12044, USDA, Washington, DC.

WEC (World Energy Council) (2010), *2010 Survey of Energy Resources,* World Energy Council, London.

Chapter 8: Climate change mitigation and the 450 Scenario

Höhne, N., *et al.* (2012), *Reality Gap: Some Countries Progress in National Policies, but Many Risk Failing to Meet Pledges*, Ecofys, Climate Analytics and Potsdam Institute for Climate Impact Research, Berlin.

IEA (International Energy Agency) (2011), *World Energy Outlook 2011*, OECD/IEA, Paris.

– (2012), *Golden Rules for a Golden Age of Gas: World Energy Outlook Special Report on Unconventional Gas*, OECD/IEA, Paris.

IIASA (International Institute for Applied Systems Analysis) (2012), *Emissions of Air Pollutants for the World Energy Outlook 2012 Energy Scenarios*, IIASA, Laxenburg, Austria.

IPCC (Intergovernmental Panel on Climate Change) (2012), *Managing the Risks of Extreme Events and Disasters to Advance Climate Change Adaptation*, Field, C., *et al.*(eds.), Cambridge University Press, Cambridge, United Kingdom and New York.

Meinshausen, M., *et al.* (2009), "Greenhouse-gas Emission Targets for Limiting Global Warming to 2 °C", *Nature*, Vol. 458, Nature Publishing Group, London, pp. 1158-1163.

OECD (Organisation for Economic Co-operation and Development) (2012), *OECD Environmental Outlook to 2050*, OECD, Paris.

Oliver, J., G. Janssens-Maenhout and J. Peters (2012), *Trends in Global CO_2 Emissions*, PBL Netherlands Environmental Assessment Agency and Joint Research Centre, The Hague, Netherlands and Ispra, Italy.

Rogelj, J., M. Meinshausen and R. Knutti (2012), "Global Warming Under Old and New Scenarios using IPCC Climate Sensitivity Range Estimates", *Nature Climate Change*, Vol. 2, No. 6, Nature Publishing Group, London, pp. 248-253.

UNEP (United Nations Environment Programme) (2011), *The Emissions Gap Report*, UNEP, Paris.

Vuuren, D. van, *et al.* (2011), "RCP2.6: Exploring the Possibility to Keep Global Mean Temperature Increase Below 2 °C", *Climatic Change*, Vol. 109, Nos. 1-2, Springer, Heidelberg, Germany, pp. 95-116.

US DOE (US Department of Energy) (2010), *Critical Materials Strategy*, US DOE, Washington, DC.

WMO (World Meteorological Organization) (2012), "WMO's Annual Statement Confirms 2011 as 11th Warmest on Record - Climate Change Accelerated in 2001-2010", Press Release No. 943, WMO, Geneva, 23 March.

Part B: Energy Efficiency Focus

Chapter 9: Energy efficiency: the current state of play

AGF (Advisory Group of Climate Change Financing) (2010), *Report of the Secretary-General's High-Level Advisory Group on Climate Change Financing*, United Nations, New York.

ACEEE (American Council for an Energy-Efficient Economy) (2011), *The 2011 State Energy Efficiency Scorecard*, ACEEE, Washington, DC.

Ang, B., F. Zhang and K. Choi (1998), "Factorizing Changes in Energy and Environmental Indicators through Decomposition", *Energy*, Vol. 23, No. 6, Elsevier, Amsterdam, pp.489-495.

BPIE (Buildings Performance Institute Europe) (2011), *Principles for Nearly Zero-Energy Buildings*, Buildings Performance Institute Europe, Ixelles, Belgium.

Capital E (2012), *Energy Efficiency Financing: Models and Strategies*, *www.cap-e.com/Capital-E/Energy_Efficiency_Financing_files/Energy_Efficiency_Financing-Models_and%20Strategies.pdf*, accessed 2 October 2012.

ClimateWorks (2012), *Industrial Energy Efficiency Data Analysis Project: Barriers and Policy Analysis,* ClimateWorks, Clayton, Australia.

CPI (Climate Policy Initiative) (2012), *Annual Review of Low Carbon Development in China 2011-2012*, Climate Policy Initiative, Beijing.

EBRD (European Bank for Reconstruction and Development) (2011), *The Low Carbon Transition*, EBRD, London.

GEA (Global Energy Assessment) (2012), *Toward a Sustainable Future*, GEA, Laxenburg, Austria.

HSBC (2011), *Sizing India's Climate Economy*, HSBC Securities and Capital Markets (India) Private Limited, Mumbai.

IEA (International Energy Agency) (2009), *Implementing Energy Efficiency Policies: Are IEA Member Countries on Track?*, OECD/IEA, Paris.

– (2010), *Monitoring, Verification and Enforcement Policy Pathway*, OECD/IEA, Paris.

– (2011a), *IEA 2011 Scoreboard: Implementing Energy Efficiency Policy: Progress and Challenges in IEA Member Countries*, OECD/IEA, Paris.

– (2011b), *25 Energy Efficiency Policy Recommendations - 2011 Update*, OECD/IEA, Paris.

– (2011c), *World Energy Outlook 2011*, OECD/IEA, Paris.

– (2011d), *Energy Efficiency and Carbon Pricing*, OECD/IEA, Paris.

– (2012a), *Spreading the Net: The Multiple Benefits of Energy Efficiency Improvements*, OECD/IEA, Paris.

– (2012b), *Policy Pathways: Improving the Fuel Economy of Road Vehicles – A Policy Package*, OECD/IEA, Paris.

– (2012c), *Progress Implementing the IEA 25 Energy Efficiency Policy Recommendations: 2011 Evaluation*, OECD/IEA, Paris.

Joskow, P. and D. Marron (1992), "What Does a Negawatt Really Cost? Evidence from Utility Conservation Programs", *Energy Journal*, Vol. 13, No. 4, International Association for Energy Economics, Cleveland, United States, pp. 41-75.

Kesicki, F. (2012), "Marginal Abatement Cost Curves: Combining Energy System Modelling and Decomposition Analysis", *Environmental Modelling and Assessment*, Springer, Heidelberg, Germany, forthcoming.

METI (Ministry of Economy, Trade and Industry of Japan) (2011a), Presentation to the IEA on the Status of post-Fukushima Japan, 12-13 September 2011, Paris.

– (2011b), *Measures for Electric Power Supply and Demand – Summer 2011* (in Japanese), METI, *www.meti.go.jp/earthquake/electricity_supply/0513_electricity_supply_03_01.pdf*, accessed 2 October 2012.

– (2011c), *Progress Report on Measures for Electric Power Supply and Demand - Summer 2011* (in Japanese), METI, *www.meti.go.jp/press/2011/10/20111014009/20111014009-2.pdf*, accessed 2 October 2012.

Mundaca, L. and L. Neij (2006), *Transaction Costs of Energy Efficiency Projects: A Review of Quantitative Estimations*, Euro White Cert Project, Intelligent Energy Programme European Commission, Brussels.

Sathaye, J. and S. Murtishaw (2004), *Market Failures, Consumer Preferences, and Transaction Costs in Energy Efficiency Purchase Decisions*, Public Interest Energy Research, Berkeley, United States.

UNIDO (United Nations Industrial Development Organization) (2011), *Industrial Development Report*, UNIDO, Vienna.

Chapter 10: A blueprint for a more energy-efficient world

Bergh, J. van den (2011), "Energy Conservation More Effective with Rebound Policy", *Environmental and Resource Economics*, Vol. 48, No. 1, Springer, Heidelberg, Germany, pp. 43-58.

IIASA (International Institute for Applied Systems Analysis) (2012), *Emissions of Air Pollutants for the World Energy Outlook 2012 Energy Scenarios*, IIASA, Laxenburg, Austria.

Nadel, S. (2012), *The Rebound Effect: Large or Small?*, White Paper, American Council for an Energy-Efficient Economy, Washington, DC.

OECD (Organisation for Economic Co-operation and Development) (2008), "An Overview of the OECD ENV-Linkages Model", OECD Economics Department Working Paper, No. 653, OECD, Paris.

– (2012), *Economic Implications of the IEA Efficient World Scenario*, OECD, Paris, forthcoming.

Sorrell, S. (2007), *The Rebound Effect: An Assessment of the Evidence for Economy-wide Energy Savings from Improved Energy Efficiency*, UK Energy Research Centre, London.

Turner, K. (2012), *Rebound Effects from Increased Energy Efficiency: A Time to Pause and Reflect*, University of Stirling, Scotland, United Kingdom.

US DOE (US Department of Energy) (2012), "Appliance and Equipment Standards Translate to Huge Energy Savings", US DOE, *www1.eere.energy.gov/buildings/appliance_standards/*, accessed September 2012.

WHO (World Health Organization) (2011), *Tackling the Global Clean Air Challenge*, World Health Organization, Geneva.

Chapter 11: Unlocking energy efficiency at the sectoral level

China Daily (2008), "Chinese Per Capita Housing Space Triples in 20 Years", *China Daily*, Beijing, 17 March.

Crist, P. (2012), "Mitigating Greenhouse-Gas Emissions from Shipping: Potential, Cost and Strategies", in Asariotis, R. and H. Benamara, *Maritime Transport and the Climate Change Challenge*, Earthscan, Abingdon, United Kingdom, p.327.

CSI (Cement Sustainability Initiative) and ECRA (European Cement Research Academy) (2009), *Development of State of the Art-Techniques in Cement Manufacturing: Trying to Look Ahead*, CSI/ECRA Technology Paper, CSI, Geneva.

ECEEE (Eurpoean Council for an Energy-Efficient Economy) (2011), *Energy Efficiency First: The Foundation of a Low-Carbon Society,* ECEEE Summer Study, Stockholm.

Econoler, *et al.* (2011), *Cooling Benchmarking Study Report*, Collaborative Labeling and Appliance Standards Program (CLASP), Brussels.

GBPN (Global Buildings Performance Network) and CEU (Central European University) (2012), *Best Practice Policies for Low Carbon & Energy Buildings - Based on Scenario Analysis*, GBPN and CEU, Paris and Budapest.

IATA (International Air Transport Association) (2009), *The IATA Technology Roadmap Report Third Edition*, IATA, Geneva.

IEA (International Energy Agency) (2009a), *Gadgets and Gigawatts - Policies for Energy Efficient Electronics*, OECD/IEA, Paris.

– (2009b), *World Energy Outlook 2009*, OECD/IEA, Paris.

– (2010), *Energy Technology Perspectives 2010*, OECD/IEA, Paris.

– (2011a), *World Energy Outlook 2011*, OECD/IEA, Paris.

– (2011b), *Energy Efficiency Policy Opportunities for Electric Motor-Driven Systems*, OECD/IEA, Paris.

– (2011c), *Smart Grids Technology Roadmap*, OECD/IEA, Paris.

– (2012a), *Energy Technology Perspectives 2012 – Pathways to a Clean Energy System*, OECD/IEA, Paris.

– (2012b), *Technology Roadmap: Fuel Economy of Road Vehicles*, OECD/IEA, Paris.

– (2012c), *Policy Pathway: Improving the Fuel Economy of Road Vehicles*, OECD/IEA, Paris.

IIP (Insitute for Industrial Productivity) (2012), *Insights into Industrial Energy Efficiency Policy Packages - Sharing Best Practices from Six Countries*, IIP, Washington, DC.

LBNL (Lawrence Berkeley National Laboratory) (2010), *Assessment of China's Energy-Saving and Emission-Reduction Accomplishments and Opportunities During the 11th Five-Year Plan*, LBNL, Berkeley, United States.

– (2012), *Bottom-Up Energy Analysis System - Methodology and Results*, LBNL and The Collaborative Labeling and Appliance Standards Program, Berkeley, United States.

NREL (National Renewable Energy Laboratory) (2007), *Assessment of the Technical Potential for Achieving Net Zero-Energy Buildings in the Commercial Sector*, NREL, Boulder, United States.

Sivak, M. (2009), "Potential Energy Demand for Cooling in the 50 Largest Metropolitan Areas of the World: Implications for Developing Countries", *Energy Policy*, Vol. 37, Elsevier, Amsterdam, pp.1382-1384.

UNIDO (United Nations Industrial Development Organization) (2011), *Industrial Development Report 2011*, UNIDO, Vienna.

US EPA (US Environmental Protection Agency) (2012), Combined Heat and Power Partnership - Funding Resources, *www.epa.gov/chp/funding/funding.html*, accessed September 2012.

US White House (2012), Executive Order: Accelerating Investment in Industrial Energy Efficiency, *www.whitehouse.gov/the-press-office/2012/08/30/executive-order-accelerating-investment-industrial-energy-efficiency*, accesssed September 2012.

Waide, P. (2011), *Opportunities for Success and CO_2 Savings from Appliance Energy Efficiency Harmonization*, Navigant Consulting and CLASP, London and Brussels.

World Steel Association (2012), *World Steel in Figures 2011,* World Steel Association, Brussels.

Chapter 12: Pathways to energy efficiency

IIASA (International Institute for Applied Systems Analysis) (2012), *Emissions of Air Pollutants for the World Energy Outlook 2012 Energy Scenarios*, IIASA, Laxenburg, Austria.

OECD (Organisation for Economic Development and Co-operation) (2012), *Economic Implications of the IEA Efficient World Scenario*, OECD, Paris, forthcoming.

Part C: Iraq Energy Outlook

Chapter 13: Iraq today: energy and the economy

Arab Union of Electricity (2010), *Statistical Bulletin 2010*, Issue 19, Arab Union of Electricity, Amman, Jordan.

Genel Energy (2011), *Creation of a Regional Champion*, Investor Presentation, Genel Energy, *www.genelenergy.com/admin/resimler/detay_resim/basin_dokumanlar/genel-vallares_analyst_presentation_sep2011.pdf*, accessed September 2012.

IKN (Iraq Knowledge Network) (2012), *Iraq Knowledge Network Survey 2011*, Iraq Central Statistics Oraganization/Kurdistan Region Statistics Organization/United Nations, Baghdad.

Ministry of Planning of Iraq (2010), *National Development Plan, 2010-2014*, Ministry of Planning of Iraq, Baghdad.

Parsons Brinckerhoff (2009a), *Masterplan for the KRG Electricity Sector*, Parsons Brinckerhoff, Erbil, Iraq.

– (2009b), *Final Report on the Survey of Private Generation in the Baghdad Governorate*, Parsons Brinckerhoff, Baghdad.

– (2010), *Iraq Electricity Masterplan*, Parsons Brinckerhoff, Baghdad.

United Nations Inter-Agency Information and Analysis Unit (2012), *Landmines and Unexploded Ordnances Factsheet in Iraq*, United Nations, New York.

World Bank (2012), *Doing Business 2012: Doing Business in a More Transparent World*, World Bank, Washington, DC.

Chapter 14: Iraq oil and gas resources and supply potential

Al-Naqib, F., *et al.* (1971), *Water Drive Performance of the Fractured Kirkuk Field of Northern Iraq*, Society of Petroleum Engineers Conference Paper, SPE 3437-MS, New Orleans, United States.

BP (2012), *BP Statistical Review of World Energy 2012*, BP, London.

Mohammed, W., M. Al-Jawad and D. Al-Shamma (2010), *Reservoir Flow Simulation Study for a Sector in Main Pay South Rumaila Oil Field*, Society of Petroleum Engineers Conference Paper, SPE 126427-MS, Mumbai.

O&GJ (*Oil and Gas Journal*) (2011a), "Worldwide Look at Reserves and Production", *Oil and Gas Journal*, Vol. 109, Issue 49, Pennwell Corporation, Oklahoma City, United States.

– (2011b), "A Veteran Revisits Iraq's Oil Resource and Lists Implications of the Magnitude", *Oil and Gas Journal*, Vol. 109, Issue 23, Pennwell Corporation, Oklahoma City, United States.

OPEC (Organization of Petroleum Exporting Countries) (2012), *Annual Statistical Bulletin 2012*, OPEC, Vienna.

Chapter 15: Iraq: fuelling future reconstruction and growth

IKN (Iraq Knowledge Network) (2012), *Iraq Knowledge Network Survey 2011*, Iraq Central Statistics Oraganization/Kurdistan Region Statistics Organization/United Nations, Baghdad.

Iraq Central Statistics Organization (2012), *Annual Abstract of Statistics 2010-2011*, Central Statistics Organization, Baghdad.

Iraq Ministry of Planning (2010), *Iraq National Development Plan, 2010-2014*, Iraq Ministry of Planning, Baghdad.

Parsons Brinckerhoff (2009a), *Masterplan for the KRG Electricity Sector*, Parsons Brinckerhoff, Erbil, Iraq.

– (2009b), *Final Report on the Survey of Private Generation in the Baghdad Governorate*, Parsons Brinckerhoff, Baghdad.

– (2010), *Iraq Electricity Masterplan*, Parsons Brinckerhoff, Baghdad.

Chapter 16: Implications of Iraq's energy development

Humphreys, M., J. Sachs and J. Stiglitz (eds.) (2007), *Escaping the Resource Curse*, Columbia University Press, New York.

Mott MacDonald (2010), *Supplying the EU Natural Gas Market*, Mott MacDonald, Croydon, United Kingdom.

Part D: Special Topics

Chapter 17: Water for energy

CAPP (Canadian Association of Petroleum Producers) (2011), *Responsible Canadian Energy 2010 Report*, CAPP, Calgary, Canada.

Cooley, H., J. Fulton and P. Gleick (2011), *Water for Energy: Future Water Needs for Electricity in the Intermountain West*, Pacific Institute, Oakland, United States.

Falkenmark, M., J. Lundquist and C. Widstrand (1989), "Macro-scale Water Scarcity Requires Micro-scale Approaches: Aspects of Vulnerability in Semi-Arid Development", *Natural Resources Forum*, Vol. 13, No. 4, pp. 258-267.

Gleick, P. (1994), "Water and Energy", *Annual Review of Energy and Environment*, Vol. 36, No. 3, Annual Reviews, Palo Alto, United States, pp. 267-299.

Government of India (2009), Coal Linkage Policy for 12th Plan Projects, Ministry of Power, *www.powermin.nic.in/whats_new/pdf/Coal_linkage_policy_for_12th_plan_projects.pdf*, accessed 17 May 2012.

– (2010), *Water and Related Statistics*, Central Water Commission, *www.cwc.nic.in/ ISO_DATA_Bank/ISO_Home_Page.htm*, accessed 4 April 2012.

IBRD (International Bank for Reconstruction and Development) (2009), *Addressing China's Water Scarcity*, World Bank, Washington, DC.

IEA (International Energy Agency) (2008), *Cleaner Coal in China*, OECD/IEA, Paris.

– (2010), *Concentrating Solar Power Technology Roadmap,* OECD/IEA, Paris.

– (2012), *Golden Rules for a Golden Age of Gas: World Energy Outlook Special Report on Unconventional Gas,* OECD/IEA, Paris.

IPCC (Intergovernmental Panel on Climate Change) (2008), *Climate Change and Water*, Technical Paper, IPCC, Geneva.

IWMI (International Water Management Institute) (2007), *Water for Food, Water for Life: A Comprehensive Assessment of Water Management in Agriculture*, Earthscan and IWMI, London and Colombo, Sri Lanka.

Macknick, J., *et al.* (2011), *A Review of Operational Water Consumption and Withdrawal Factors for Electricity Generating Technologies*, National Renewable Energy Laboratory, Golden, United States.

MacPherson, J. (2011), "Water Key to North Dakota Oil Patch Growth", *Businessweek*, Bloomberg, 25 March.

Mielke, E., *et al.* (2010), *Water Consumption of Energy Resource Extraction, Processing and Conversion*, Belfer Center for International Affairs, Harvard Kennedy School of government, Cambridge, United States.

Platts (2012), *World Electric Power Plants Database*, McGraw-Hill Companies, accessed July 2012.

Torcellini, P., N. Long and R. Judkoff (2003), *Consumptive Water Use for US Power Production*, National Renewable Energy Laboratory, Golden, United States.

Schornagel, J., *et al.* (2012), "Water Accounting for (Agro) Industrial Operations and its Application to Energy Pathways", *Resources, Conservation and Recycling*, Vol. 61, Elsevier, Amsterdam, pp. 1-15.

Shiklomanov, I. (1993), "World Fresh Water Resources", *Water in Crisis: A Guide to the World's Fresh Water Resources*, Gleick, P. (ed.), Oxford University Press, New York.

UNDP (United Nations Development Programme) (2006), *Human Development Report 2006*, Palgrave Macmillan, New York.

UN FAO (United Nations Food and Agriculture Organization) (2010), *The Wealth of Waste: The Economics of Wastewater Use in Agriculture*, UN FAO, Rome.

– (2012), *Aquastat Database*, UN FAO, *www.fao.org/nr/water/aquastat/main/index.stm*, accessed 10 September 2012.

US DOE (US Department of Energy) (2006), *Energy Demands on Water Resources*, US DOE, Washington, DC.

US DOE/EIA (US Department of Energy/Energy Information Administration) (2011), "WECC – Southwest Region Data", *Annual Energy Outlook 2011*, US DOE, Washington, DC.

US DOE/NETL (US Department of Energy/National Energy Technology Laboratory) (2007), *2007 Coal Power Plant Database*, NETL, *www.netl.doe.gov/energy-analyses/hold/technology.html*, accessed 12 April 2012.

– (2008), *Water Requirements for Existing and Future Thermoelectric Plant Technologies*, NETL, Pittsburgh, United States.

– (2010), *Cost and Performance Baseline for Fossil Energy Plants*, NETL, Pittsburgh, United States.

– (2011), *Estimating Freshwater Needs to Meet Future Thermal Generation Requirements*, NETL, Pittsburgh, United States.

USGS (US Geological Survey) (2009), *Estimated Use of Freshwater in the United States in 2005*, USGS, Washington, DC.

US EPA (US Environmental Protection Agency) (2009), *Clean Water Act – Section 316b Chapter 3*, US EPA, Washington, DC.

WWAP (World Water Assessment Programmme) (2012), *The United Nations World Water Development Report 4: Managing Water under Uncertainty and Risk*, United Nations Educational Scientific and Cultural Organization, Paris.

Chapter 18: Energy for all

Asian Development Bank (2011), *Energy for All: Viet Nam's Success in Increasing Access to Energy through Rural Electrification*, Asian Development Bank, Manila, Philippines.

Ghana Energy Commission (2012), *Ghana Action Plan for Sustainable Energy for All by 2030*, Ghana Energy Commission, Accra, Ghana.

Gómez, M. and S. Silveira (2010), "Rural Electrification of the Brazilian Amazon – Achievements and Lessons", *Energy Policy*, Vol. 38, No. 10, Elsevier, Amsterdam, pp. 6251-6260.

Government of India (2012), *Census of India 2011*, Ministry of Home Affairs, Government of India, New Delhi.

Kooijman-van Dijk, A. and J. Clancy (2010), "Impacts of Electricity Access to Rural Enterprises in Bolivia, Tanzania and Vietnam", *Energy for Sustainable Development*, Vol. 14, No. 1, Elsevier, Amsterdam, pp. 14-21.

Ministry of Mines and Energy (2009), *Pesquisa Quantitativa Domiciliar de Avaliação – Principais Resultados*, Government of Brazil, Brasilia.

– (2010), *Light for All: A Historic Landmark*, Government of Brazil, Brasilia.

PCIA (Partnership for Clean Indoor Air) (2012), "International Workshop Agreement on Cookstoves Unanimously Approved", March 2012, PCIA online, *www.pciaonline.org/news/cookstoves-iwa-unanimously-approved*, accessed 14 September 2012.

SNV (2011), SNV Newsletter, September 2011, SNV World online, *www.snvworld.org/sites/www.snvworld.org/files/publications/snv_domestic_biogas_newsletter_issue_5_september_2011.pdf*, accessed 1 October 2012.

United Nations (2012a), Commitments: Sustainable Energy for All, *www.sustainableenergyforall.org/actions-commitments/commitments*, accessed 24 September 2012.

– (2012b), "Rio+20: Ban Announces More than 100 Commitments on Sustainable Energy", 21 June, UN News Centre, *www.un.org/apps/news/story.asp?NewsID=42297# .UGqt4ZH_mnB*, accessed 19 July 2012.

Winkler, H., *et al.* (2011), "Access and Affordability of Electricity in Developing Countries", *World Development*, Vol. 39, No. 6, Elsevier, Amsterdam, pp. 1037–1050.

Online bookshop

Buy IEA publications online:

www.iea.org/books

PDF versions available at 20% discount

Books published before January 2011 - except statistics publications - are freely available in pdf

International Energy Agency • 9 rue de la Fédération • 75739 Paris Cedex 15, France

iea

Tel: +33 (0)1 40 57 66 90

E-mail:
books@iea.org

The paper used for this document has received certification from the Programme for the Endorsement of Forest Certification (PEFC) for being produced respecting PEFC's ecological, social and ethical standards. PEFC is an international non-profit, non-governmental organization dedicated to promoting Sustainable Forest Management (SFM) through independent third-party certification.

IEA PUBLICATIONS, 9 rue de la Fédération, 75739 PARIS CEDEX 15
Layout in France by Easy Catalogue - Printed in France by Corlet, November 2012
(61 2012 25 1P1) ISBN: 978 92 64 18084 0
Photo credits: GraphicObsession